OXFORD MATHEMATICAL MONOGRAPHS

A. Belleni-Morante: *Applied semigroups and evolution equations*
I. G. Macdonald: *Symmetric functions and Hall polynomials*
J. W. P. Hirschfeld: *Projective geometries over finite fields*
N. Woodhouse: *Geometric quantization*
A. M. Arthurs: *Complementary variational principles* Second edition
P. L. Bhatnagar: *Nonlinear waves in one-dimensional dispersive systems*
N. Aronszajn, T. M. Creese, and L. J. Lipkin: *Polyharmonic functions*
J. A. Goldstein: *Semigroups of linear operators*
M. Rosenblum and J. Rovnyak: *Hardy classes and operator theory*
J. W. P. Hirschfeld: *Finite projective spaces of three dimensions*
K. Iwasawa: *Local class field theory*
A. Pressley and G. Segal: *Loop groups*
J. C. Lennox and S. E. Stonehewer: *Subnormal subgroups of groups*
D. E. Edmunds and W. D. Evans: *Spectral theory and differential operators*
Wang Jianhua: *The theory of games*
S. Omatu and J. H. Seinfeld: *Distributed parameter systems: theory and applications*
D. Holt and W. Plesken: *Perfect groups*
J. Hilgert, K. H. Hofmann, and J. D. Lawson: *Lie groups, convex cones, and semigroups*
S. Dineen: *The Schwarz lemma*
B. Dwork: *Generalized hypergeometric functions*
R. J. Baston and M. G. Eastwood: *The Penrose transform: its interaction with representation theory*
S. K. Donaldson and P. B. Kronheimer: *The geometry of four-manifolds*
T. Petrie and J. Randall: *Connections, definite forms, and four-manifolds*
R. Henstock: *The general theory of integration*
D. W. Robinson: *Elliptic operators and Lie groups*
A. G. Werschulz: *The computational complexity of differential and integral equations*
J. B. Griffiths: *Colliding plane waves in general relativity*
P. N. Hoffman and J. F. Humphreys: *Projective representations of symmetric groups*
I. Györi and G. Ladas: *Oscillation theory of delay differential equations: with applications*
J. W. P. Hirschfeld and J. A. Thas: *General Galois geometries*

General Galois Geometries

by

J. W. P. HIRSCHFELD
University of Sussex

and

J. A. THAS
University of Ghent

CLARENDON PRESS · OXFORD
1991

Oxford University Press, Walton Street, Oxford OX2 6DP
Oxford New York Toronto
Delhi Bombay Calcutta Madras Karachi
Petaling Jaya Singapore Hong Kong Tokyo
Nairobi Dar es Salaam Cape Town
Melbourne Auckland
and associated companies in
Berlin Ibadan

Oxford is a trade mark of Oxford University Press

Published in the United States
by Oxford University Press, New York

A catalogue record for this book is available from the British Library

Library of Congress Cataloging in Publication Data
Hirschfeld, J. W. P. (James William Peter), 1940–
General Galois geometries / by J. W. P. Hirschfeld and J. A. Thas.
(Oxford mathematical monographs)
Continues: Projective geometries over finite fields (1979) and
Finite projective spaces of three dimensions (1985).
Includes bibliographical references and indexes.
1. Geometry, Projective. 2. Modular fields. I. Thas, J. A.
(Joseph Adolphe) II. Title. III. Series.
QA471.H58 1991 516′.5—dc20 91-16464
ISBN 0–19–853537–6

Typeset by Integral Typesetting, Gorleston, Norfolk
Printed in Great Britain by
Bookcraft (Bath) Ltd
Midsomer Norton, Avon

TO OUR FAMILIES

Adrienne	Laurette
Rachel	Inge
Benjamin	Koen

PREFACE

This book is the third and last volume of a treatise on projective spaces over a finite field, also known as Galois geometries. The first volume, *Projective geometries over finite fields* (Hirschfeld 1979), consists of Parts I to III and contains Chapters 1 to 14 and Appendices I and II. The second volume, *Finite projective spaces of three dimensions* (Hirschfeld 1985), consists of Part IV and contains Chapters 15 to 21 and Appendices III to V. The present volume comprises Part V and contains Chapters 22 to 27 and Appendices VI and VII. The scheme of the treatise is indicated by the titles of the Parts: Part I Introduction; Part II Elementary properties of $PG(n, q)$; Part III $PG(1, q)$ and $PG(2, q)$; Part IV $PG(3, q)$; Part V $PG(n, q)$.

Outline of the book

There are three themes within the book: (a) properties of algebraic varieties over a finite field; (b) the determination of various constants arising from the combinatorics of Galois spaces such as the maximum number of points of a subset under certain linear independence conditions; (c) the identification in Galois spaces of various incidence structures.

Many of the results on theme (a) could be equally well stated over an arbitrary field. However, over a finite field, counting arguments come more into play. A significant number of theorems count certain sets and establish the existence of combinatorial structures. Most of Chapters 22 to 25 is on theme (a), whereas Chapter 26 is on theme (c) and Chapter 27 is for the most part on (b).

Chapter 22 on quadrics develops their properties and gives one way of characterizing them. Chapter 23 on Hermitian varieties similarly develops their properties and characterizes them in the course of describing all sets of type $(1, r, q + 1)$. This chapter is the one on algebraic varieties most different from the classical case, as Hermitian manifolds over the complex numbers are not algebraic varieties. Chapter 24 on Grassmann varieties and Chapter 25 on Veronese and Segre varieties most closely follow a classical model in the description of their properties. Although most of the characterizations of the Veronesean of quadrics resemble classical theorems over the complex numbers, the characterization of Grassmannians is quite different. This is because the Grassmannian characterization is in terms of an incidence structure, a topic which was studied over the real and complex numbers only for the entire projective space rather than any substructure, whereas the Veronesean is studied as a subset of $PG(n, q)$ in terms of sections by subspaces and, for $n = 2$, also in terms of tangent planes. Chapter 26 begins with polar spaces, thereby unifying the subjects of Chapters 22 and 23, and it goes on to consider the special case of generalized quadrangles and structures which are natural developments. This is the only chapter in which there is no attempt to prove every theorem given, although proofs are given of results used in other chapters. Chapter 27 generalizes to an arbitrary dimension results of Chapters 18 and 21 from the previous volume: an upper bound is found for the size of a k-cap and the maximum size of a k-arc is found under some restrictions on n and q; the corresponding arcs are generally normal rational curves.

The book is conceived as a work of reference and does not have any exercises. However, each individual chapter is self-contained and is suitable for a course of lectures. Apart from Chapter 26, complete proofs are given for nearly all results. The last section of each chapter contains all references as well as remarks both on the chapter itself and on related aspects that are not covered.

This volume may be considered as developing over finite fields aspects of the three volumes of Hodge and Pedoe (1947, 1953, 1954), particularly regarding quadrics and Grassmannians. Burau (1961) is also an appropriate analogy for quadrics, Grassmannians, Veroneseans, and Segre varieties.

Status of the subject

Apart from being an interesting and exciting area in combinatorics with beautiful results, Galois geometries have many applications to coding theory, algebraic geometry, design theory, graph theory, and group theory. As an example, the theory of linear maximum distance separable codes (MDS codes) is equivalent to the theory of arcs in $PG(n, q)$; so all results of §§27.5–27.7 can be expressed in terms of linear MDS codes. Finite projective geometry is essential for finite algebraic geometry, and finite algebraic curves are used to construct interesting classes of codes, the Goppa codes, now also known as algebraic geometry codes. Many interesting designs and graphs are constructed from finite Hermitian varieties, finite quadrics, finite Grassmannians, and finite normal rational curves. Further, most of the objects studied in this book have an interesting group; the classical groups and other finite simple groups appear in this way.

Currently there are several international journals on combinatorics and geometry publishing a large number of papers on Galois geometries: for example, *Journal of Combinatorial Theory Series A*, *Geometriae Dedicata*, *Journal of Geometry*, *Ars Combinatoria*, *Combinatorica*, *European Journal of Combinatorics*, as well as the recently founded *Designs, Codes and Cryptography*, *Journal of Algebraic Combinatorics* and the conference series *Annals of Discrete Mathematics*. Every year there are many conferences on combinatorics, and a considerable part of some of them is devoted to Galois geometries: one may cite the *Finite Geometries* sessions at Oberwolfach, the series of conferences from *Combinatorics '81* to *Combinatorics '90* in Italy, the Pingree Park conference in Colorado on *Finite Buildings, Related Geometries and Applications*, the conference on *Finite Geometry and Combinatorics* at Deinze in Belgium, and the three Isle of Thorns conferences in England on *Finite Geometries and Designs*.

Finite vector spaces and hence also finite projective spaces are of great importance for theoretical computer science. So, in most syllabuses of a computer science degree, there is a course on discrete mathematics with a section on combinatorial structures.

Related topics

There are some interesting topics either not covered or only touched upon in the three volumes. In the *Handbook of incidence geometry* (North-Holland, to appear), edited by F. Buekenhout, surveys of several of these topics will be given.

Spreads and partial spreads in $PG(n, q)$ are considered in Chapters 4 and 17, in §25.6, and in Appendix VI. They can be used to construct nets, projective planes, designs, strongly regular graphs, codes, partial geometries, semi-partial geometries and other structures. Related to spreads and partial spreads are blocking sets, for which only the plane case is considered in Chapter 13. For the theory of spreads, partial spreads, and blocking sets in n dimensions, see §§8–10 of the chapter 'Projective geometry over a finite field' by J. A. Thas in the Handbook.

Finite non-Desarguesian planes are not discussed in the treatise. For references see the chapters in the Handbook on 'Projective planes' by A. Beutelspacher and 'Translation planes' by M. J. Kallaher.

Flocks of quadrics in $PG(3, q)$ are not considered at all. These appear to be the key objects for the construction of some new classes of translation planes and generalized quadrangles. Their classification in the hyperbolic case led to the classification of all maximal exterior sets of hyperbolic quadrics and to the solution of a classical problem on finite inversive planes of odd order. As a by-product, the first computer-free proof of the uniqueness of the inversive plane of order seven was obtained. For further details, see the Handbook chapter 'Projective geometry over a finite field' by Thas.

In Chapter 26, the embedded finite generalized quadrangles, that is the finite classical generalized quadrangles, are considered. Generalized quadrangles are the rank 2 polar spaces, the point of view of Chapter 26, but are also the generalized n-gons with $n = 4$. Generalized n-gons for $n > 4$ are not mentioned in the treatise. For these, see the Handbook chapter 'Generalized polygons' by Thas. Generalized quadrangles can be generalized to partial and semi-partial geometries; but only the embedded cases are considered in Chapter 26. In the Handbook, the chapter 'Some classes of rank 2 geometries' by F. De Clerck contains a survey.

Although null (symplectic) polarities are mentioned in Appendix VI, they are not discussed in detail, nor are pseudo polarities: references are given there.

The treatise contains only a few group-theoretical results; also, theorems on graphs and designs are rare. Apart from the Handbook, the books by Dembowski (1968), Beth, Jungnickel, and Lenz (1984), Hughes and Piper (1985), Cameron and van Lint (1980), and Brouwer, Cohen, and Neumaier (1989) may be consulted.

Codes are considered only in §27.2. For an introduction, see van Lint (1982) or Hill (1986); for further results and geometrical connections, see Cameron and van Lint (1980), MacWilliams and Sloane (1977), and Peterson and Weldon (1972). For an introduction to Goppa's algebraic geometry codes, see van Lint and van der Geer (1988), Goppa (1988), Hirschfeld (1984, 1990a), Moreno (1991), and the Bibliography.

The language of diagram geometries, introduced by Buekenhout, is not used; only in §24.5 is a geometry belonging to some diagram explicitly mentioned. For a survey, see the Handbook chapter 'Foundations of incidence geometry' by Buekenhout.

Acknowledgements

The typing of the manuscript was done in Sussex by Pauline Steele, Georgette Beech, Christine Coles, and Sue Bullock and in Ghent by Zita Oost. We are extremely grateful to them for their diligence and patience. The project was greatly assisted by research grants from the National Fund for Scientific Research of Belgium and the Royal Society of London. The second author is also very grateful to Professor J. Bilo who initiated him into the mysteries of classical geometry thereby providing him with the necessary background to appreciate the beauty of Galois geometry.

Brighton and Ghent J.W.P.H.
January 1991 J.A.T.

CONTENTS

PART V *PG(n, q)*

TERMINOLOGY

$V(n+1, K)$ is $(n+1)$-*dimensional vector space* over the field K and is taken to be the set of vectors $X = (x_0, \ldots, x_n)$, $x_i \in K$. Correspondingly, $PG(n, K)$ is *n-dimensional projective space* over K and is the set of elements, called *points*, $\mathbf{P}(X)$ with $X \in V(n+1, K)\backslash\{0\}$. When $K = GF(q)$, the finite field of q elements, also called the Galois field of q elements, then $V(n+1, K)$ is written $V(n+1, q)$ and $PG(n, K)$ is written $PG(n, q)$. Often $GF(q)$ is written γ.

A *projectivity* (or *projective transformation*) from S_1 to S_2, with S_1 and S_2 n-dimensional projective spaces over $GF(q)$, is a mapping $\mathfrak{T}: S_1 \to S_2$ such that $\mathbf{P}(X)\mathfrak{T} = \mathbf{P}(XT)$ for all $X \neq 0$ and some non-singular $(n+1) \times (n+1)$ matrix T. The group of projectivities from $PG(n, q)$ to itself is denoted $PGL(n+1, q)$. A *collineation* from S_1 to S_2 is a mapping $\mathfrak{T}: S_1 \to S_2$ such that $\mathbf{P}(X)\mathfrak{T} = \mathbf{P}(X^{\sigma}T)$ with σ an automorphism of $GF(q)$. Mostly the properties considered are invariant under $PGL(n+1, q)$.

A *reciprocity* of $PG(n, q)$ is a collineation $\mathfrak{T}$ from $PG(n, q)$ to its dual space; if $\mathfrak{T}$ is a projectivity, then the reciprocity is a *correlation* of $PG(n, q)$.

A *subspace of dimension r* in $PG(n, q)$ is a $PG(r, q)$ and is written Π_r: this notation is used both specifically and generically. Then Π_0 is a point of $PG(n, q)$, Π_1 is a *line*, Π_2 is a *plane*, Π_3 is a *solid*, Π_{n-1} is a *prime* or *hyperplane*. A subspace written π_r can have any dimension. In $PG(n, q)$, the vertices of the simplex of reference are denoted $\mathbf{U}_0, \mathbf{U}_1, \ldots, \mathbf{U}_n$, where $\mathbf{U}_i$ has a one in the $(i+1)$th coordinate place and zeros elsewhere, and $\mathbf{U}$ is the unit point. Dually, $\mathbf{u}_0, \mathbf{u}_1, \ldots, \mathbf{u}_n$ are prime faces of the simplex of reference and $\mathbf{u}$ is the unit prime.

The ring $\Gamma = \gamma[x_0, \ldots, x_n]$ is the ring of polynomials in the indeterminates $x_0, \ldots, x_n$ over γ. For $F_1, \ldots, F_r$ non-zero forms (homogeneous polynomials) in Γ, the *variety*

$$\mathbf{V}(F_1, \ldots, F_r) = \{\mathbf{P}(X) \in PG(n, q) \mid F_1(X) = \cdots = F_r(X) = 0\}.$$

A variety $\mathbf{V}(F)$ is called a *primal* or *hypersurface*. A primal in $PG(2, q)$ is a (*plane*)(*algebraic*) *curve*; *a primal in* $PG(3, q)$ is a *surface*. If the primals $\mathscr{F}_1$ and $\mathscr{F}_2$ are projectively equivalent, then write $\mathscr{F}_1 \sim \mathscr{F}_2$. In §27.7 a more sophisticated notion of algebraic variety is required.

In keeping with the terminology of Chapter 8, in $PG(2, q)$ an *oval* is a $(q+1)$-arc for q odd and a $(q+2)$-arc for q even. Other authors use *hyperoval* or *complete oval* in the latter case.

For more detailed explanation of the foregoing, see Chapter 2. An index of notation is provided at the end of this volume.

PART V

PG(n, q)

22

QUADRICS

22.1 Canonical forms

Quadrics were introduced in Chapter 5. The properties of plane quadrics were developed in Chapters 6 and 7. The properties of quadrics in three dimensions were developed in Chapters 15 and 16. Quadrics in five dimensions were also considered in Chapters 15, 17, and 20.

First the essential definitions are recalled. Let F in $\gamma[x_0, \ldots, x_n]$, where

$$F = \sum_{i=0}^{n} a_i x_i^2 + \sum_{i<j} a_{ij} x_i x_j,$$

be a quadratic form which is *non-degenerate*; that is, F is not reducible to a form in fewer than $n + 1$ variables by a linear transformation. The variety $\mathbf{V}(F)$ is a *non-singular quadric*. Under projectivities of $PG(n, q)$ there are one or two distinct non-singular quadrics $\mathcal{Q}_n = \mathbf{V}(F)$ according as n is even or odd. Equivalently, the projective linear group $PGL(n + 1, q)$ acting on all non-singular quadrics in $PG(n, q)$ has one or two orbits as n is even or odd. Throughout the chapter, the notation $\mathcal{Q}_n$ is used for non-singular quadrics and $\mathcal{W}_n$ for general quadrics.

For n even, $\mathcal{Q}_n \sim \mathcal{P}_n$, where

$$\mathcal{P}_n = \mathbf{V}(x_0^2 + x_1 x_2 + \cdots + x_{n-1} x_n), \text{ parabolic.}$$

For n odd, $\mathcal{Q}_n \sim \mathcal{H}_n$ or $\mathcal{E}_n$, where

$$\mathcal{H}_n = \mathbf{V}(x_0 x_1 + x_2 x_3 + \cdots + x_{n-1} x_n), \text{ hyperbolic;}$$

$$\mathcal{E}_n = \mathbf{V}(f(x_0, x_1) + x_2 x_3 + \cdots + x_{n-1} x_n), \text{ elliptic;}$$

here f is irreducible over γ.

In each of the three cases, we write $\mathcal{Q}_n = \mathbf{V}(F_n)$, where F_n is the corresponding quadratic form.

For the method to reduce F to canonical form, see §5.1.

Suppose now that the form F may be degenerate. Then the quadric $\mathcal{W}_n = \mathbf{V}(F)$ may be singular and is a *cone* $\Pi_k \mathcal{Q}_s$, the join of the *vertex* Π_k to a non-singular quadric $\mathcal{Q}_s$ in Π_s with $\Pi_k \cap \Pi_s = \Pi_{-1}$ and $k + s = n - 1$.

If F is reduced to canonical form F_s, then

$$\mathcal{W}_n = \Pi_{n-s-1} \mathcal{Q}_s = \mathbf{V}(F_s),$$

where the vertex $\Pi_k = \Pi_{n-s-1} = \mathbf{V}(x_0, \ldots, x_s)$ is the space of singular points and $\mathcal{Q}_s$ is non-singular in $\Pi_s = \mathbf{V}(x_{s+1}, \ldots, x_n)$.

When $k = -1$, then $\mathscr{W}_n = \Pi_{-1}\mathscr{Q}_n = \mathscr{Q}_n$ and is non-singular.

Lemma 22.1.1: *The number of projectively distinct quadrics in $PG(n, q)$ is*

$$\tfrac{1}{2}[3n + 1 + (n + 1, 2)],$$

of which $(n + 1, 2)$ are non-singular and

$$\tfrac{1}{2}[3n + 1 - (n + 1, 2)],$$

are singular.

Proof: Each quadric may be written as $\Pi_{n-r-1}\mathscr{Q}_r$ for r in $\bar{\mathbf{N}}_n$. For n even, there is one non-singular quadric for each of $r = 0, 2, \ldots, n$ and two non-singular quadrics for each of $r = 1, 3, 5, \ldots, n - 1$. Hence the total number of quadrics is $\frac{1}{2}(n + 2) + 2 . \frac{1}{2}n = \frac{1}{2}(3n + 2)$. For n odd, there are two quadrics for each of $r = 1, 3, 5, \ldots, n$ and one for each of $r = 0, 2, \ldots, n - 1$. Hence the total number is $2 . \frac{1}{2}(n + 1) + \frac{1}{2}(n + 1) = \frac{1}{2}(3n + 3)$. □

For $n \leq 5$ the quadrics in $PG(n, q)$ are now described; for $n = 5$, only the non-singular quadrics are listed.

$PG(1, q)$	$\mathscr{H}_1$	$= \mathbf{V}(x_0x_1)$ is two points $\mathbf{U}_0, \mathbf{U}_1$
	$\mathscr{E}_1$	$= \mathbf{V}(f(x_0, x_1))$ is empty
	$\Pi_0\mathscr{P}_0$	$= \mathbf{V}(x_0^2)$ is a single point, the join of $\mathbf{\Pi}_0 = \mathbf{U}_1$ to the empty quadric $\mathscr{P}_0$ in $\mathbf{u}_1$
$PG(2, q)$	$\mathscr{P}_2$	$= \mathbf{V}(x_0^2 + x_1x_2)$ is a *conic*, comprising $q+1$ points no three of which are collinear
	$\Pi_0\mathscr{H}_1$	$= \mathbf{V}(x_0x_1)$ is a line pair
	$\Pi_0\mathscr{E}_1$	$= \mathbf{V}(f(x_0, x_1))$ is a single point $\mathbf{U}_2$
	$\Pi_1\mathscr{P}_0$	$= \mathbf{V}(x_0^2)$ is a line
$PG(3, q)$	$\mathscr{H}_3$	$= \mathbf{V}(x_0x_1 + x_2x_3)$ consists of $(q + 1)^2$ points on $2(q + 1)$ lines, two through each point
	$\mathscr{E}_3$	$= \mathbf{V}(f(x_0, x_1) + x_2x_3)$ consists of $q^2 + 1$ points no three of which are collinear
	$\Pi_0\mathscr{P}_2$	$= \mathbf{V}(x_0^2 + x_1x_2)$ is a cone comprising the join of a point to a conic
	$\Pi_1\mathscr{H}_1$	$= \mathbf{V}(x_0x_1)$ is a plane pair
	$\Pi_1\mathscr{E}_1$	$= \mathbf{V}(f(x_0, x_1))$ is a line
	$\Pi_2\mathscr{P}_0$	$= \mathbf{V}(x_0^2)$ is a plane
$PG(4, q)$	$\mathscr{P}_4$	$= \mathbf{V}(x_0^2 + x_1x_2 + x_3x_4)$ consists of $(q + 1)(q^2 + 1)$ points on $(q + 1)(q^2 + 1)$ lines with $q + 1$ lines through each point

	$\Pi_0\mathscr{H}_3$	$= \mathbf{V}(x_0x_1 + x_2x_3)$ is a cone comprising the join of a point to the hyperbolic surface $\mathscr{H}_3$, that is $q(q+1)^2+1$ points in $2(q+1)$ concurrent planes
	$\Pi_0\mathscr{E}_3$	$= \mathbf{V}(f(x_0, x_1) + x_2x_3)$ is a cone comprising the join of a point to the elliptic surface $\mathscr{E}_3$, that is $q(q^2+1)+1$ points on q^2+1 concurrent lines
	$\Pi_1\mathscr{P}_2$	$= \mathbf{V}(x_0^2 + x_1x_2)$ is the join of a line to a conic, and so consists of $q+1$ planes through a line, no three in a solid
	$\Pi_2\mathscr{H}_1$	$= \mathbf{V}(x_0x_1)$ is a pair of solids
	$\Pi_2\mathscr{E}_1$	$= \mathbf{V}(f(x_0, x_1))$ is a plane
	$\Pi_3\mathscr{P}_0$	$= \mathbf{V}(x_0^2)$ is a solid
$PG(5, q)$	$\mathscr{H}_5$	$= \mathbf{V}(x_0x_1 + x_2x_3 + x_4x_5)$ consists of $(q^2+1)(q^2+q+1)$ points on $2(q+1)(q^2+1)$ planes with $2(q+1)$ planes through a point
	$\mathscr{E}_5$	$= \mathbf{V}(f(x_0, x_1) + x_2x_3 + x_4x_5)$ consists of $(q+1)(q^3+1)$ points on $(q^2+1)(q^3+1)$ lines with q^2+1 lines through a point

The properties of the singular quadrics follow inductively from the properties of non-singular quadrics in lower dimension. So, for the most part, it is reasonable to concentrate on the properties of non-singular quadrics.

22.2 Invariants

In the next theorem, two invariants are given. One, Δ, decides whether the quadric $\mathscr{W}_n$ is singular or not; the other, α, decides whether $\mathscr{W}_n$ is hyperbolic or elliptic in the odd-dimensional non-singular case. In these invariants, factors 1/2 and 1/4 appear. This means that, even in the characteristic 2 case, when the rest of the formula is evaluated in general, the factors 2 and 4 that appear must be cancelled. The invariant Δ is usually called the *discriminant* and α the *Arf invariant.*

First we summarize some results on quadratic equations over γ. Here we also write $\gamma_0 = \gamma\backslash\{0\}$; see §§1.4 and 1.8 for more details.

Define

$$C(t) = \tfrac{1}{2}(1 - t^{(q-1)/2}),\ t \in \gamma_0,\ q \text{ odd};$$

$$C(t) = t + t^2 + t^{2^2} + \cdots + t^{2^{h-1}},\ t \in \gamma,\ q = 2^h.$$

Then, for q odd,

$$\begin{aligned}\mathscr{C}_0 &= \{c \in \gamma \mid x^2 - c = 0 \text{ has two solutions}\}\\ &= \{t \in \gamma_0 \mid C(t) = 0\},\end{aligned}$$

$$\mathscr{C}_1 = \{c \in \gamma \,|\, x^2 - c = 0 \text{ has no solutions}\}$$
$$= \{t \in \gamma_0 \,|\, C(t) = 1\}.$$

For $q = 2^h$,

$$\mathscr{C}_0 = \{c \in \gamma \,|\, x^2 + x + c = 0 \text{ has two solutions}\}$$
$$= \{t \in \gamma \,|\, C(t) = 0\},$$
$$\mathscr{C}_1 = \{c \in \gamma \,|\, x^2 + x + c = 0 \text{ has no solutions}\}$$
$$= \{t \in \gamma \,|\, C(t) = 1\}.$$

Also, for q odd, $|\mathscr{C}_0| = |\mathscr{C}_1| = \frac{1}{2}(q-1)$; for q even, $|\mathscr{C}_0| = |\mathscr{C}_1| = \frac{1}{2}q$.

Another way of phrasing the above is to consider the group homomorphisms

$$\gamma_0 \xrightarrow{\mu} \gamma_0 \xrightarrow{\rho} \mathbf{Z}_2 \quad \text{for } q \text{ odd},$$

$$\gamma \xrightarrow{\sigma} \gamma \xrightarrow{\rho} \mathbf{Z}_2 \quad \text{for } q \text{ even},$$

where γ is regarded as the additive group and γ_0 the multiplicative group of the field, with

$$t\mu = t^2, \qquad t\sigma = t + t^2, \qquad t\rho = C(t).$$

Then $\mu\rho = 0$, $\sigma\rho = 0$, $\ker \rho = \mathscr{C}_0$.

As before, let $\mathscr{W}_n = \mathbf{V}(F)$ with

$$F = \sum a_i x_i^2 + \sum_{i<j} a_{ij} x_i x_j.$$

Define $A = [a_{ij}]$, where $a_{ii} = 2a_i$, $a_{ji} = a_{ij}$ for $i < j$.

Let $B = [b_{ij}]$, where $b_{ii} = 0$, $b_{ji} = -b_{ij} = -a_{ij}$ for $i < j$. Then, with $X = (x_0, x_1, \ldots, x_n)$,

$$F = \tfrac{1}{2} XAX^* = \tfrac{1}{2} X(A + B)X^*.$$

When q is even the formulae for Δ and α in the next theorem should be interpreted as follows. If, in A and B, the terms a_i and a_{ij} are replaced by indeterminates z_i and z_{ij}, and Δ and α are evaluated as rational functions over $\mathbf{Z}$, then z_i and z_{ij} can be specialized to a_i and a_{ij} to give the result. In the lemma following the theorem, α is obtained for small dimensions.

Theorem 22.2.1: (i) *$\mathscr{W}_n$ is singular or not according as Δ is zero or not, where*

$$\Delta = \begin{cases} \frac{1}{2}|A|, & n \text{ even} \\ |A|, & n \text{ odd}. \end{cases}$$

(ii) *For n odd, the non-singular quadric $\mathscr{W}_n$ is hyperbolic or elliptic according* *as* $\alpha \in \mathscr{C}_0$ *or* $\mathscr{C}_1$, *where*

$$\alpha = \begin{cases} (-1)^{(n+1)/2}|A|, & q \text{ odd} \\ \{|B| - (-1)^{(n+1)/2}|A|\}/\{4|B|\}, & q \text{ even}. \end{cases}$$

Proof: Under a projectivity $\mathbf{P}(X) \to \mathbf{P}(XT^{-1})$, we have

$$\mathbf{V}(\tfrac{1}{2}XAX^*) = \mathbf{V}(\tfrac{1}{2}X(A+B)X^*)$$

$$\to \mathbf{V}(\tfrac{1}{2}XTAT^*X^*) = \mathbf{V}(\tfrac{1}{2}XT(A+B)T^*X^*).$$

So in (i), $\Delta \to \Delta|T|^2$; thus, both Δ and $\Delta|T|^2$ are zero or neither is. In (ii), for q odd, $\alpha \to \alpha|T|^2$; thus α and $\alpha|T|^2$ are both squares or both non-squares. For q even, α is fixed under the projectivity. Hence the invariance of the conditions has been established.

It now suffices to examine the invariants for the canonical forms:

$$\Pi_{n-s-1}\mathscr{P}_s = \mathbf{V}(x_0^2 + x_1x_2 + \cdots + x_{s-1}x_s)$$

$$\Pi_{n-s-1}\mathscr{H}_s = \mathbf{V}(x_0x_1 + x_2x_3 + \cdots + x_{s-1}x_s)$$

$$\Pi_{n-s-1}\mathscr{E}_s = \mathbf{V}(x_0^2 + x_0x_1 + dx_1^2 + x_2x_3 + \cdots + x_{s-1}x_s)$$

with $x^2 + x + d$ irreducible.

$$\Pi_{n-s-1}\mathscr{P}_s\colon \tfrac{1}{2}|\mathrm{A}| = \begin{cases} (-1)^{n/2}, & s = n \\ 0, & s < n. \end{cases}$$

$$\Pi_{n-s-1}\mathscr{H}_s\colon |\mathrm{A}| = \begin{cases} (-1)^{(n+1)/2}, & s = n \\ 0, & s < n; \end{cases}$$

$$|B| = 1, \qquad s = n.$$

$$\Pi_{n-s-1}\mathscr{E}_s\colon |\mathrm{A}| = \begin{cases} (1-4d)(-1)^{(n+1)/2}, & s = n \\ 0, & s < n; \end{cases}$$

$$|B| = 1, \qquad s = n.$$

So $\Delta \neq 0$ for $s = n$ and $\Delta = 0$ for $s < n$. With $s = n$, the invariant α is given by the following table:

$\mathscr{W}_n$	q odd	q even
$\mathscr{H}_n$	1	0
$\mathscr{E}_n$	$1-4d$	d

Since $x^2 + x + d$ is irreducible, it follows from the table and the formulae for C that, when $\mathscr{W}_n = \mathscr{H}_n$, we have $C(\alpha) = 0$ and, when $\mathscr{W}_n = \mathscr{E}_n$, we have $C(\alpha) = 1$. □

Lemma 22.2.2: *For q even the invariant α is given, modulo $\gamma\sigma$, as follows:*

(i) $n = 1$, $\alpha = a_0 a_1 / a_{01}^2$;
(ii) $n = 3$,

$$\alpha = \frac{\sum a_i a_{jk} a_{jl} a_{kl} + \sum a_i a_j a_{kl}^2 + (\prod a_{ij}) \sum (a_{kl} a_{mn})^{-1}}{(\sum a_{ij} a_{kl})^2},$$

where the summands in the numerator have four, six, three terms respectively and that of the denominator also has three terms.

Proof:
(i)

$$A = \begin{bmatrix} 2a_0 & a_{01} \\ a_{01} & 2a_1 \end{bmatrix}, \qquad B = \begin{bmatrix} 0 & a_{01} \\ -a_{01} & 0 \end{bmatrix}.$$

So $|A| = 4a_0 a_1 - a_{01}^2$, $|B| = a_{01}^2$,

$$\alpha = \frac{|B| + |A|}{4|B|} = \frac{a_0 a_1}{a_{01}^2}.$$

(ii)

$$A = \begin{bmatrix} 2a_0 & a_{01} & a_{02} & a_{03} \\ a_{01} & 2a_1 & a_{12} & a_{13} \\ a_{02} & a_{12} & 2a_2 & a_{23} \\ a_{03} & a_{13} & a_{23} & 2a_3 \end{bmatrix}, \qquad B = \begin{bmatrix} 0 & a_{01} & a_{02} & a_{03} \\ -a_{01} & 0 & a_{12} & a_{13} \\ -a_{02} & -a_{12} & 0 & a_{23} \\ -a_{03} & -a_{13} & -a_{23} & 0 \end{bmatrix},$$

$\varphi = a_{01}a_{23} - a_{02}a_{13} + a_{03}a_{12}$ (φ is the Pfaffian of the matrix B). Then

$$|B| = \varphi^2,$$

$$\begin{aligned} |A| = \sum a_{ij}^2 a_{kl}^2 &- 2a_{01}a_{23}a_{02}a_{13} - 2a_{01}a_{23}a_{03}a_{12} \\ &- 2a_{02}a_{13}a_{03}a_{12} + 4\sum a_i a_{jk} a_{jl} a_{kl} \\ &- 4\sum a_i a_j a_{kl}^2 + 16 a_0 a_1 a_2 a_3. \end{aligned}$$

So, in $\mathbf{Z}(\{a_i, a_{ij}\})$,

$$\alpha = \frac{|B| - |A|}{4|B|} = \frac{\sum a_i a_j a_{kl}^2 - \sum a_i a_{jk} a_{jl} a_{kl} - 4a_0 a_1 a_2 a_3 + a_{01}a_{23}a_{03}a_{12}}{\varphi^2}.$$

Hence, over γ,

$$\alpha = \frac{\sum a_i a_j a_{kl}^2 + \sum a_i a_{jk} a_{jl} a_{kl} + a_{01}a_{23}a_{03}a_{12}}{\varphi^2}.$$

However,

$$\frac{a_{01}a_{23}a_{02}a_{13} + a_{02}a_{13}a_{03}a_{12}}{\varphi^2} = \frac{a_{02}a_{13}}{\varphi} + \left(\frac{a_{02}a_{13}}{\varphi}\right)^2.$$

So, modulo $\gamma\sigma = \{t + t^2 \mid t \in \gamma\}$,

$$\alpha = \frac{\sum a_i a_j a_{kl}^2 + \sum a_i a_{jk} a_{jl} a_{kl} + (\prod a_{ij})\{(a_{01}a_{23})^{-1} + (a_{02}a_{31})^{-1} + (a_{03}a_{12})^{-1}\}}{\varphi^2}. \qquad \square$$

22.3 Tangency and polarity

Consider the non-singular quadric $\mathcal{Q}_n = \mathbf{V}(F)$. Let $P = \mathbf{P}(A)$ and $Q = \mathbf{P}(B)$, where $A = (a_0, \ldots, a_n)$ and $B = (b_0, \ldots, b_n)$, $A \neq B$. Then

$$F(A + tB) = F(A) + tG(A, B) + t^2F(B), \qquad (22.1)$$

where

$$G(A, B) = F(A + B) - F(A) - F(B).$$

The line l is a *tangent* to $\mathcal{Q}_n$ if $|l \cap \mathcal{Q}_n| = 1$.

Lemma 22.3.1: *Let $P = \mathbf{P}(A) \in \mathcal{Q}_n$.*

(i) *If $Q \notin \mathcal{Q}_n$, then $G(A, B) = 0 \Leftrightarrow PQ$ is a tangent to $\mathcal{Q}_n$.*
(ii) *If $Q \in \mathcal{Q}_n$, then $G(A, B) = 0 \Leftrightarrow PQ \subset \mathcal{Q}_n$.*
(iii) *$G(A, B) \neq 0 \Leftrightarrow |PQ \cap \mathcal{Q}_n| = 2$.*

Proof: Since $P \in \mathcal{Q}_n$, equation (22.1) becomes

$$F(A + tB) = tG(A, B) + t^2F(B).$$

The point $\mathbf{P}(A + tB) \in \mathcal{Q}_n$ if and only if

$$0 = tG(A, B) + t^2F(B). \qquad (22.2)$$

The solution $t = 0$ of (22.2) corresponds to P. Parts (i), (ii), and (iii) now follow. $\square$

Corollary 1: *For q even, if one of P and Q is not on $\mathcal{Q}_n$, then PQ is a tangent $\Leftrightarrow G(A, B) = 0$.*

Proof: When $F(A + tB) = 0$, equation (22.1) becomes

$$0 = F(A) + tG(A, B) + t^2F(B). \qquad (22.3)$$

If $G(A, B) = 0$, then this becomes

$$0 = F(A) + t^2 F(B) \tag{22.4}$$

which has just one solution. Conversely, if (22.3) has just one solution, the coefficient of t must be zero. □

A point $\mathbf{P}(A)$ is a *nucleus* of $\mathscr{Q}_n$ if $G(A, B) = 0$ for all points $\mathbf{P}(B)$.

Corollary 2: (i) *The quadric $\mathscr{Q}_n$ has a nucleus if and only if q and n are both even.*

(ii) *For q even, $\mathscr{P}_n$ in canonical form has precisely one nucleus $N = \mathbf{U}_0$.*

Proof: This follows immediately from the forms for $G(A, B)$. □

It should be noted that (ii) applies in the case $n = 0$. The empty quadric $\mathscr{P}_0$ has the point $\mathbf{U}_0$ as nucleus.

If $G(A, B) = 0$, the points $P = \mathbf{P}(A)$ and $Q = \mathbf{P}(B)$ are *conjugate*. If P is not a nucleus, then, with

$$G(A, X) = F(A + X) - F(A) - F(X),$$

the prime $\mathbf{V}(G(A, X))$ is the *polar prime* of P. When $P \in \mathscr{Q}_n$, the prime $\mathbf{V}(G(A, X))$ is the *tangent prime to $\mathscr{Q}_n$ at P* and is denoted $T_P = T_P(\mathscr{Q}_n)$. If P is the nucleus of $\mathscr{Q}_n$, then $\mathbf{V}(G(A, X)) = \Pi_n$.

Theorem 22.3.2: (i) *$T_P(\mathscr{Q}_n)$ comprises the points on the tangents to $\mathscr{Q}_n$ at P and the lines on $\mathscr{Q}_n$ through P.*

(ii) *$T_P(\mathscr{Q}_n)$ contains any subspace Π_m such that $P \in \Pi_m \subset \mathscr{Q}_n$.*

Proof: (i) This follows from Lemma 22.3.1(i) and (ii).

(ii) Since every line through P of $\mathscr{Q}_n$ lies in the tangent space and every point of Π_m lies on such a line, so $\Pi_m \subset T_P(\mathscr{Q}_n)$. □

Lemma 22.3.1 and its Corollary 1 also hold for general quadrics.

When $\mathscr{W}_n = \Pi_k \mathscr{Q}_t = \mathbf{V}(F)$ is an arbitrary quadric and $P \in \mathscr{W}_n$, the *tangent space to $\mathscr{W}_n$ at P*, denoted $T_P(\mathscr{W}_n)$, is $\mathbf{V}(G(A, X))$.

Corollary: (i) *$T_P(\mathscr{W}_n)$ contains the vertex Π_k.*

(ii) *If $P \in \Pi_k$, then $T_P(\mathscr{W}_n) = \Pi_n$, the whole space.* □

It should be noted that if we write $F_{(i)} = \partial F/\partial x_i$ so that $F_{(i)}(A)$ is the partial derivative of F with respect to x_i evaluated at A, then

$$G(A, X) = \sum F_{(i)}(A) x_i.$$

Thus the tangent space to a quadric as defined here coincides with that for an arbitrary primal; see §2.6. One can also define a nucleus **P**(A) of $\mathcal{Q}_n$ as a point at which $F_{(i)}(A) = 0$, all i.

With the canonical forms of §22.1 and $f(x_0, x_1) = x_0^2 + x_0x_1 + dx_1^2$, the linear form $G(A, X)$ is as follows:

$$\mathcal{Q}_n = \mathcal{P}_n, \quad G(A, X) = 2a_0x_0 + (a_1x_2 + a_2x_1) + \cdots + (a_nx_{n-1} + a_{n-1}x_n);$$

$$\mathcal{Q}_n = \mathcal{H}_n, \quad G(A, X) = (a_1x_0 + a_0x_1) + (a_3x_2 + a_2x_3) + \cdots + (a_nx_{n-1} + a_{n-1}x_n);$$

$$\mathcal{Q}_n = \mathcal{E}_n, \quad G(A, X) = (2a_0 + a_1)x_0 + (a_0 + 2da_1)x_1 + (a_3x_2 + a_2x_3) + \cdots + (a_nx_{n-1} + a_{n-1}x_n).$$

Lemma 22.3.3: *Let $\mathcal{Q}_n$ be a non-singular quadric.*

(i) *If q and n are not both even, the correspondence*

$$\mathbf{P}(A) \leftrightarrow \mathbf{V}(G(A, X))$$

is a polarity. For q odd, the set of self-polar points is $\mathcal{Q}_n$. For q even, the polarity is null and every point in $PG(n, q)$ is self-polar.

(ii) *If q and n are both even, the tangent primes to $\mathcal{P}_n$ are concurrent at the nucleus $N = \mathbf{U}_0$.* □

Even though the points and tangent primes of $\mathcal{P}_n$ are not related by a polarity for q even, the following lemma plus Theorem 22.3.2(ii) are strong enough to prove facts about $\mathcal{P}_n$ for q even which follow from the polar theory for all other $\mathcal{Q}_n$.

Lemma 22.3.4: *The tangent primes at $r + 1$ independent points of a Π_r lying on $\mathcal{Q}_n$ are themselves independent.*

Proof: When $(q, n) \not\equiv 0 \pmod 2$, this follows from Lemma 22.3.3(i). This leaves the case that q is even and $\mathcal{Q}_n = \mathcal{P}_n$. With $A = (a_0, a_1, \ldots, a_n)$ and $P = \mathbf{P}(A)$, then $T_P(\mathcal{P}_n) = \mathbf{V}(G(A, X)) = \mathbf{V}(A^\tau X^*)$, where $A^\tau = (0, a_2, a_1, a_4, a_3, \ldots, a_n, a_{n-1})$.

Suppose that the points $P_i = \mathbf{P}(A_i)$, $i = 0, \ldots, r$, span Π_r on $\mathcal{P}_n$; that is, they are independent. If the corresponding tangent primes $\mathbf{V}(A_i^\tau X^*)$ are dependent, so are the $r + 1$ points $\mathbf{P}(A_i^\tau)$. Hence, under the projectivity fixing $\mathcal{P}_n$ given by

$$x_0 \to x_0, \qquad x_{2j-1} \leftrightarrow x_{2j}, \qquad j = 1, 2, \ldots, \tfrac{1}{2}n,$$

the points $\mathbf{P}(A_i^\sigma)$, where $A^\sigma = (0, a_1, a_2, \ldots, a_n)$, are dependent and hence lie in a Π_{r-1}. But $\mathbf{P}(A^\sigma)$ is the projection of $\mathbf{P}(A)$ from the nucleus $\mathbf{U}_0$ of $\mathcal{P}_n$

onto $\mathbf{u}_0$. Since the $r+1$ points $\mathbf{P}(A_i^\sigma)$ lie in Π_{r-1}, so the $r+1$ points $\mathbf{P}(A_i)$ lie in the r-space $\mathbf{U}_0\Pi_{r-1}$: this follows from the fact that $\mathbf{P}(A_i^\sigma)$ lies on $\mathbf{U}_0\mathbf{P}(A_i)$. However, by hypothesis, the $r+1$ points lie in a Π_r on $\mathscr{P}_n$ and are independent. As Π_r cannot be $\mathbf{U}_0\Pi_{r-1}$, we have a contradiction. □

If $\Pi_r \subset \mathscr{Q}_n$, the *tangent space at* (or *of*) Π_r is the intersection of the tangent primes at $r+1$ independent points $P_0, \ldots, P_r$ of Π_r: in symbols,

$$T_{\Pi_r}(\mathscr{Q}_n) = \bigcap T_{P_i}(\mathscr{Q}_n).$$

Corollary 1: (i) *The tangent space of* Π_r *on* $\mathscr{Q}_n$ *is a* Π_{n-r-1} *containing* Π_r.
(ii) $\mathscr{Q}_n \supset \Pi_r \supset \Pi_s \Rightarrow T_{\Pi_s}(\mathscr{Q}_n) \supset \Pi_r$.

Proof: (i) By the lemma, the tangent space is the intersection of $r+1$ independent primes and so a Π_{n-r-1}. By Theorem 22.3.2(ii), it contains Π_r.
(ii) This follows from Theorem 22.3.2(i). □

In the subsequent results, Π'_{n-r-1} is the tangent space of Π_r on $\mathscr{Q}_n$ and Π'_{n-m-1} is the polar space of Π_m for the cases in which a polarity occurs.

If $\Pi_r \subset \mathscr{W}_n$, the *tangent space*

$$T_{\Pi_r}(\mathscr{W}_n) = \bigcap_{P \in \Pi_r} T_P(\mathscr{W}_n).$$

Corollary 2: *If* $\mathscr{W}_n = \Pi_k \mathscr{Q}_t$ *and* $\Pi_r \subset \mathscr{W}_n$ *so that* $\Pi_r = \Pi_s \Pi_e$ *with* $\Pi_s \subset \mathscr{Q}_t$ *and* $\Pi_e \subset \Pi_k$, *then*

$$T_{\Pi_r}(\mathscr{W}_n) = T_{\Pi_s}(\mathscr{Q}_t)\Pi_k$$

and has dimension $t - s + k$. □

A tangent line meets $\mathscr{Q}_n$ precisely in a point. We now investigate what happens in general when a subspace Π_m meets $\mathscr{Q}_n$ in a subspace Π_r that is not the whole of Π_m.

Lemma 22.3.5: *Suppose* $\Pi_m \not\subset \mathscr{Q}_n$ *and* $\Pi_m \cap \mathscr{Q}_n = \Pi_r$. *Then the following hold*:

(i) $\Pi_m \subset T_P(\mathscr{Q}_n)$ *for all* P *in* Π_r, *whence* $\Pi_m \subset \Pi'_{n-r-1}$;
(ii) *either* (a) $m = r+1$ *and* $\Pi_m \cap \mathscr{Q}_n = \Pi_{m-1}\mathscr{P}_0$,
or (b) $m = r+2$ *and* $\Pi_m \cap \mathscr{Q}_n = \Pi_{m-2}\mathscr{E}_1$;
(iii) *when* q *is odd*, $\Pi_m \cap \Pi'_{n-m-1} = \Pi_r$ *and* $\Pi_m\Pi'_{n-m-1} = \Pi'_{n-r-1}$;
(iv) *when* q *is even with* n *odd and* $m = r+1$, *then* $\Pi_m \subset \Pi'_{n-m-1}$;
(v) *when* q *is even with* n *odd and* $m = r+2$, *then* $\Pi_m \cap \Pi'_{n-m-1} = \Pi_r$ *and* $\Pi_m\Pi'_{n-m-1} = \Pi'_{n-r-1}$.

Proof: (i) If $P \in \Pi_r$, then any line l through P in Π_m either lies in Π_r and so on $\mathcal{Q}_n$ or meets Π_r and so $\mathcal{Q}_n$ in the single point P. Thus l is a tangent through P or a line of $\mathcal{Q}_n$ through P; hence $l \subset T_P(\mathcal{Q}_n)$.

(ii) A quadric $\Pi_{n-s-1}\mathcal{Q}_s$ is a Π_r if and only if $\mathcal{Q}_s$ is empty; that is, $\mathcal{Q}_s = \mathcal{P}_0$ or $\mathcal{E}_1$. Hence, if $\Pi_m \cap \mathcal{Q}_n = \Pi_r$, then either $\Pi_r = \Pi_{m-1}\mathcal{P}_0$ or $\Pi_{m-2}\mathcal{E}_1$. We think of $\Pi_{m-1}\mathcal{P}_0$ as a repeated Π_{m-1} and of $\Pi_{m-2}\mathcal{E}_1$ as a Π_{m-2} which is the intersection of two Π_{m-1} lying over $GF(q^2)$.

(iii) Π'_{n-m-1} is the intersection of the polar primes of points of Π_m. So every point of $\Pi_m \cap \Pi'_{n-m-1}$ lies in its own polar prime. Hence $\Pi_m \cap \Pi'_{n-m-1} \subset \Pi_r$. Since $\Pi_m \subset \Pi'_{n-r-1}$ by (i), so $\Pi'_{n-m-1} \supset \Pi_r$. But, by hypothesis, $\Pi_m \supset \Pi_r$: so $\Pi_r \subset \Pi_m \cap \Pi'_{n-m-1}$. Thus $\Pi_m \cap \Pi'_{n-m-1} = \Pi_r$ and consequently $\Pi_m\Pi'_{n-m-1} = \Pi'_{n-r-1}$.

(iv) Since q is even, the polarity defined by $\mathcal{Q}_n$ is a null polarity. Thus, if two particular points on a line l are conjugate, then any two points on l are conjugate and l is self-polar; that is, l lies in its polar space. Hence, the self-polar lines are the lines of $\mathcal{Q}_n$ and the tangents to $\mathcal{Q}_n$: see Corollary 1 to Lemma 22.3.1.

Now, if $r = m - 1$, then every line in Π_m through a point P of $\Pi_m \backslash \Pi_r$ meets Π_r and is tangent to $\mathcal{Q}_n$. So P is conjugate to every point of Π_m, and hence $\Pi_m \subset \Pi'_{n-m-1}$.

(v) Now, with $r = m - 2$, let $P \in \Pi_m \backslash \Pi_r$ and let $Q \in \Pi_m \backslash \{P\}$. Then PQ is self-polar if and only if $PQ \cap \Pi_r = \Pi_0$. However, through P, there is a line of Π_m missing Π_r. But R in Π_m is in Π'_{n-m-1} if and only if R is conjugate to every point Q of Π_m. Hence $\Pi_m \cap \Pi'_{n-m-1} = \Pi_r$ and $\Pi_m\Pi'_{n-m-1} = \Pi'_{n-r-1}$. □

Lemma 22.3.6: *If $\Pi_m \not\subset \mathcal{Q}_n$, then the following are equivalent:*

(i) *Π_r is the largest subspace on $\mathcal{Q}_n$ such that $\Pi_m \subset T_P(\mathcal{Q}_n)$ for all P in Π_r;*
(ii) *Π_r is the largest subspace on $\mathcal{Q}_n$ such that $\Pi_m \subset \Pi'_{n-r-1}$;*
(iii) *Π_r is the singular space of $\Pi_m \cap \mathcal{Q}_n$.*

Proof: P is in the singular space of $\Pi_m \cap \mathcal{Q}_n$
$\Leftrightarrow$ every line through P in Π_m is a tangent or line of $\mathcal{Q}_n$
$\Leftrightarrow \Pi_m \subset T_P(\mathcal{Q}_n)$.

Hence Π_r is the singular space of $\Pi_m \cap \mathcal{Q}_n$

$$\Leftrightarrow \Pi_m \subset \bigcap_{P \in \Pi_r} T_P(\mathcal{Q}_n) = \Pi'_{n-r-1}. \quad \square$$

When the conditions of this lemma hold, then Π_m *touches* $\mathcal{Q}_n$ *along* Π_r.

Corollary: *In the case that n and q are not both even, let Π_m and its polar space Π'_{n-m-1} be not contained in $\mathcal{Q}_n$. If $\Pi_m \cap \mathcal{Q}_n = \Pi_k\mathcal{Q}_t$, then the polar space Π'_{n-m-1} satisfies the following:*

(i) $\Pi'_{n-m-1} \cap \mathscr{Q}_n$ *has singular space* Π_k;

(ii) $\Pi_m \cap \Pi'_{n-m-1} \cap \mathscr{Q}_n = \Pi_k$;

(iii) $$\Pi_m \cap \Pi'_{n-m-1} = \begin{cases} \Pi_k, & q \text{ odd, or } q \text{ even with } t \text{ odd} \\ \Pi_k N, & q \text{ even with } t \text{ even} \end{cases}$$

where N *is the nucleus of* $\mathscr{Q}_t$.

Proof: The hypothesis means that the set of points P in $\Pi_m \cap \mathscr{Q}_n$ such that $|l \cap \mathscr{Q}_n| = 1$ or $q + 1$ for every line l through P in Π_m is Π_k. Any such line l lies in $T_P(\mathscr{Q}_n)$ but lies in Π'_{n-m-1} only if it lies in the polar prime of every point Q in Π_m. This occurs only if l lies in Π_k or does not lie in Π_k and $\Pi_m l \cap \Pi_t$ lies in the polar prime of every point Q in Π_t, where $\mathscr{Q}_t \subset \Pi_t$. These two possibilities give the respective cases for q even in (iii). □

Lemma 22.3.7: *Suppose that* $\Pi_m \not\subset \mathscr{Q}_n$. *Let* $\Pi_m \cap \mathscr{Q}_n = \Pi_k \mathscr{Q}_t$ *and let* Π_d *be any subspace on* $\Pi_k \mathscr{Q}_t$ *containing* Π_k. *Then*

$$\Pi_m \Pi'_{n-d-1} = \Pi'_{n-k-1},$$

where Π'_{n-d-1} *is the tangent space to* $\mathscr{Q}_n$ *at* Π_d *and* Π'_{n-k-1} *the tangent space at* Π_k.

Proof: Since $\Pi_d \supset \Pi_k$, so $\Pi'_{n-d-1} \subset \Pi'_{n-k-1}$. By Lemma 22.3.6, $\Pi_m \subset \Pi'_{n-k-1}$. So $\Pi_m \Pi'_{n-d-1} \subset \Pi'_{n-k-1}$.

To prove the converse we consider two cases.

(*a*) *Suppose* $\Pi_k \mathscr{Q}_t$ *spans* Π_m

The space $\Pi_m \cap \Pi'_{n-d-1}$ is the tangent space of $\Pi_k \mathscr{Q}_t$ at Π_d. Since $\Pi_d \supset \Pi_k$, the dimension of $\Pi_m \cap \Pi'_{n-d-1}$ is $t - (d - k - 1) + k = m - d + k$ by Lemma 22.3.4, Corollary 2. Hence the dimension of $\Pi_m \Pi'_{n-d-1}$ is

$$m + (n - d - 1) - (m - d + k) = n - k - 1.$$

Hence $\Pi_m \Pi'_{n-d-1} = \Pi'_{n-k-1}$.

(*b*) $\Pi_k \mathscr{Q}_t$ *does not span* Π_m

Then, as in Lemma 22.3.5, $\Pi_k \mathscr{Q}_t = \Pi_k$. So $\Pi_k = \Pi_d$ and $\Pi_m \subset \Pi'_{n-d-1}$. The result follows. □

Two quadrics have the *same character* if they are both parabolic, both hyperbolic or both elliptic. An absolute definition of character is given in §22.4.

Lemma 22.3.8: *For $n \geq 2$, the tangent prime T_p at a point P of $\mathscr{Q}_n$ meets $\mathscr{Q}_n$ in a cone $P\mathscr{Q}_{n-2}$, where $\mathscr{Q}_n$ and $\mathscr{Q}_{n-2}$ have the same character.*

Proof: Choose $P = \mathbf{U}_n$. Also, let $\mathbf{U}_{n-1}$ be in $\mathscr{Q}_n$ and let $\mathbf{U}_0, \mathbf{U}_1, \ldots, \mathbf{U}_{n-2}$ be in T_P. So $T_P = \mathbf{u}_{n-1}$ and

$$\mathscr{Q}_n = \mathbf{V}(F(x_0, \ldots, x_{n-1}) + x_{n-1}x_n)$$

where F contains no term in x_{n-1}^2. If $F = \sum_{i<j} a_{ij}x_ix_j$, substitute x_n for

$$a_{0,n-1}x_0 + a_{1,n-1}x_1 + \cdots + a_{n-2,n-1}x_{n-2} + x_n.$$

So

$$\mathscr{Q}_n = \mathbf{V}(F(x_0, x_1, \ldots, x_{n-2}, 0) + x_{n-1}x_n) \tag{22.5}$$

and

$$\mathscr{Q}_n \cap T_P = \mathbf{V}(x_{n-1}, F(x_0, \ldots, x_{n-2}, 0)) = P\mathscr{Q}_{n-2}$$

where

$$\mathscr{Q}_{n-2} = \mathbf{V}(x_{n-1}, x_n, F(x_0, \ldots, x_{n-2}, 0)). \tag{22.6}$$

Reference to the canonical forms in §22.1 shows that $\mathscr{Q}_n$ and $\mathscr{Q}_{n-2}$ have the same character, since the quadratic forms in (22.5) and (22.6) which define them differ by $x_{n-1}x_n$. □

22.4 Generators

A subspace of maximum dimension on a quadric $\mathscr{W}_n$ is a *generator*; its dimension $g = g(\mathscr{W}_n)$ is the *projective index* of $\mathscr{W}_n$. The more classical *Witt index* is $g + 1$: we do not use this.

Lemma 22.4.1: (i) *For $\mathscr{Q}_n$ and $\mathscr{Q}_{n-2}$, non-singular quadrics of the same character,*

$$g(\mathscr{Q}_n) = g(\mathscr{Q}_{n-2}) + 1.$$

(ii)

$\mathscr{Q}_n$	$\mathscr{P}_n$	$\mathscr{H}_n$	$\mathscr{E}_n$
g	$\frac{1}{2}(n-2)$	$\frac{1}{2}(n-1)$	$\frac{1}{2}(n-3)$

(iii) *Any subspace on $\mathscr{Q}_n$ lies in a generator.*

Proof: (i) This follows directly from Lemma 22.3.8 and Theorem 22.3.2(ii).

(ii) This follows from (i) and the knowledge of $\mathscr{Q}_n$ for low n; namely, $g(\mathscr{P}_0) = -1 = g(\mathscr{E}_1)$, $g(\mathscr{H}_1) = 0$.

(iii) Induction on n and a similar argument to (i) gives the result. □

For $\mathscr{Q}_n$, we now define the *character* $w = w(\mathscr{Q}_n)$ by

$$w = 2g - n + 3. \tag{22.7}$$

Lemma 22.4.2:

$\mathscr{Q}_n$	$\mathscr{P}_n$	$\mathscr{H}_n$	$\mathscr{E}_n$
w	1	2	0

□

This lemma justifies the name parabolic, hyperbolic, and elliptic for the respective quadrics. Sometimes it is convenient to invert (22.7) to give

$$g = \tfrac{1}{2}(n - 3 + w). \tag{22.8}$$

Lemma 22.4.3: *A generator of $\mathscr{W}_n = \Pi_k \mathscr{Q}_t$ is the join of the vertex Π_k to a generator of $\mathscr{Q}_t$.* □

We now define the *character* $w = w(\mathscr{W}_n)$ of an arbitrary quadric $\mathscr{W}_n = \Pi_k \mathscr{Q}_t$. Recall that

k = dimension of singular space of $\mathscr{W}_n$,
n = dimension of space in which $\mathscr{W}_n$ is defined by a quadratic form,
g = projective index of $\mathscr{W}_n$.

Define

$$w = 2g - k - n + 2. \tag{22.9}$$

This agrees with (22.7) in the non-singular case, when $k = -1$.

Lemma 22.4.4: *For a quadric $\mathscr{W}_n$, the constants g and w are as follows:*

$\mathscr{W}_n$	$\Pi_{n-t-1}\mathscr{P}_t$	$\Pi_{n-t-1}\mathscr{H}_t$	$\Pi_{n-t-1}\mathscr{E}_t$
g	$n - \frac{1}{2}(t + 2)$	$n - \frac{1}{2}(t + 1)$	$n - \frac{1}{2}(t + 3)$
w	1	2	0

□

It should be noted that the character of $\mathscr{W}_n = \Pi_k \mathscr{Q}_t$ is the same as for the base $\mathscr{Q}_t$. Putting $g = k = n$, we get $w = 2$. One may consistently write $\Pi_n = \Pi_n \mathscr{H}_{-1}$, and include the whole space Π_n as the quadric $\mathbf{V}(0)$. This becomes relevant when sections of a quadric by a subspace are considered.

Corollary: *A quadric $\mathscr{W}_n = \Pi_{n-t-1}\mathscr{Q}_t$ of character w has projective index*

$$g = n - \tfrac{1}{2}(t + 3 - w). \quad \square \tag{22.10}$$

Lemma 22.4.5: *If* $\Pi_m \subset \mathcal{Q}_n$ *and* Π'_{n-m-1} *is the tangent space of* Π_m, *then* $\Pi'_{n-m-1} \cap \mathcal{Q}_n = \Pi_m \mathcal{Q}'_{n-2m-2}$, *where* $\mathcal{Q}'_{n-2m-2}$ *has the same character as* $\mathcal{Q}_n$.

Proof: For P in Π_m, every line through P in Π'_{n-m-1} is a tangent or a line of $\mathcal{Q}_n$. So Π_m lies in the singular space of $\Pi'_{n-m-1} \cap \mathcal{Q}_n$. It must be shown that the singular space is no bigger.

Suppose $\Pi_m = \mathbf{U}_0\mathbf{U}_1 \ldots \mathbf{U}_m$. Then $\mathcal{Q}_n = \mathbf{V}(F)$ with

$$F = x_0 f_0 + \cdots + x_m f_m + g,$$

where each f_i and g are forms in $x_{m+1}, \ldots, x_n$. Since $\mathcal{Q}_n$ is non-singular, the forms $f_0, \ldots, f_m$ are linearly independent. Hence by a change of coordinates

$$F = x_0 x_{m+1} + \cdots + x_m x_{2m+1} + g'(x_{m+1}, \ldots, x_n).$$

The non-singularity of F considered as a form in $x_0, \ldots, x_n$ is equivalent to the non-singularity of $G = g'(0, \ldots, 0, x_{2m+2}, \ldots, x_n)$ considered as a form in $x_{2m+2}, \ldots, x_n$. Thus in $\Pi'_{n-m-1} = \mathbf{V}(x_{m+1}, \ldots, x_{2m+1})$, the equation of $\Pi'_{n-m-1} \cap \mathcal{Q}_n$ is $G = 0$.

It follows that $\Pi'_{n-m-1} \cap \mathcal{Q}_n = \Pi_m \mathcal{Q}'_{n-2m-2}$. Any Π_r lying on $\mathcal{Q}_n$ and containing Π_m lies in Π'_{n-m-1}, by Lemma 22.3.4, Corollary 1(ii). Hence $g(\mathcal{Q}_n) = g(\Pi_m \mathcal{Q}'_{n-2m-2})$. So, if w and w' are the respective characters of $\mathcal{Q}_n$ and $\mathcal{Q}'_{n-2m-2}$, then

$$\tfrac{1}{2}(n - 3 + w) = n - m - 1 - \tfrac{1}{2}(n - 2m - 2 + 3 - w'),$$

whence $w = w'$. $\square$

For $m = 0$, the result was given by Lemma 22.3.8.

We now consider some numerical properties of the generators before describing the whole system. Let $\mathcal{G} = \mathcal{G}(\mathcal{Q}_n)$ be the set of generators of $\mathcal{Q}_n$. Define

$$\rho(d, n; w) = |\{\Pi_g \in \mathcal{G} \mid \Pi_g \supset \text{a fixed } \Pi_d\}|$$
$$\lambda(d, n; w) = |\{\Pi_g \in \mathcal{G} \mid \Pi_g \text{ meets a fixed generator in some } \Pi_d\}|$$
$$\mu(c) = \mu(c, n; w) = |\{\Pi_g \in \mathcal{G} \mid \Pi_g \text{ meets a fixed generator in a fixed } \Pi_{g-c}\}|$$
$$\kappa(n; w) = |\mathcal{G}|.$$

In the subsequent results, the following numerical notation is frequently used:

$$[r, s]_+ = \begin{cases} (q^r + 1)(q^{r+1} + 1) \ldots (q^s + 1) & \text{for } s \geq r \\ 1 & \text{for } s < r; \end{cases}$$

$$[r, s]_- = \begin{cases} (q^r - 1)(q^{r+1} - 1) \ldots (q^s - 1) & \text{for } s \geq r \\ 1 & \text{for } s < r. \end{cases}$$

Theorem 22.4.6:

$$\kappa(n; w) = [2 - w, \tfrac{1}{2}(n - w + 1)]_+ = [2 - w, g + 2 - w]_+$$

$$= \begin{cases} [2, s + 1]_+ & \textit{for} \quad \mathscr{E}_{2s+1} \\ [0, s]_+ & \textit{for} \quad \mathscr{H}_{2s+1} \\ [1, s]_+ & \textit{for} \quad \mathscr{P}_{2s}. \end{cases}$$

Proof: We count the set $\{(P, \Pi_g) | P \in \Pi_g \in \mathscr{G}\}$ in two ways. By Theorem 22.3.2 and Lemma 22.3.8, the set $\{(P_0, \Pi_g) | P_0 \in \Pi_g \in \mathscr{G}\}$ for a fixed point P_0 has size $\kappa(n - 2; w)$. Hence

$$\kappa(n - 2; w)|\mathscr{Q}_n| = \kappa(n; w)|\Pi_g|.$$

The size of $\mathscr{Q}_n$ is given in §5.2 and the result then follows by induction. □

Theorem 22.4.7:

$$\rho(d, n; w) = [2 - w, \tfrac{1}{2}(n - 1 - 2d - w)]_+.$$

Proof: By Lemma 22.3.4, Corollary 1, the tangent space Π'_{n-d-1} at Π_d to $\mathscr{Q}_n$ contains all generators through Π_d. By Lemma 22.4.5, $\Pi'_{n-d-1} \cap \mathscr{Q}_n = \Pi_d \mathscr{Q}_{n-2d-2}$ and has the same character w as $\mathscr{Q}_n$. Hence each generator of $\mathscr{Q}_n$ through Π_d is the join of Π_d to a generator of $\mathscr{Q}_{n-2d-2}$ and conversely. So

$$\rho(d, n; w) = \kappa(n - 2d - 2; w). \quad \square$$

It may be noted that, when $d = g$,

$$\rho(d, n; w) = [2 - w, 1 - w]_+ = 1,$$

confirming that the only generator containing a given Π_g is Π_g itself.

Lemma 22.4.8:

$$\mu(c, n; w) = q^{c(c+3-2w)/2}.$$

Proof: The only generator meeting Π_g in Π_g is Π_g itself; hence $\mu(0) = 1$. Now we proceed by induction on c and assume the formula true for all values less than c.

The number of generators meeting Π_g in at least the fixed space Π_{g-c} is

$$\rho(g - c, n; w) = [2 - w, c + 1 - w]_+.$$

So to find $\mu(c)$ we need to subtract from $\rho(g - c, n; w)$ the number of generators meeting Π_g in a $(g - i)$-space containing Π_{g-c} for all i such that $0 \leq i < c$. So

$$\mu(c) = \rho(g - c, n; w) - \sum_{i=0}^{c-1} \mu(i)\chi(g - c, g - i; g, q)$$

where

$$\chi(g-c, g-i; g, q) = \text{number of } \Pi_{g-i} \text{ through } \Pi_{g-c} \text{ in } \Pi_g$$
$$= [c-i+1, c]_-/[1, i]_-,$$

§3.1. Hence

$$\mu(c) = [2-w, c+1-w]_+ - \sum_{i=0}^{c-1} q^{i(i+3-2w)/2}[c-i+1, c]_-/[1, i]_-$$

which gives the result after some manipulation. □

Lemma 22.4.9: *For* $-1 \le d \le g$,

$$\lambda(d, n; w) = q^{c(c+3-2w)/2}[g-d+1, g+1]_-/[1, d+1]_-$$

where $c = g - d$, $g = \frac{1}{2}(n-3+w)$.

Proof: For $0 \le d \le g$,

$$\lambda(d, n; w) = (\text{number of generators meeting } \Pi_g \text{ in a given } \Pi_d)$$
$$\times (\text{number of } \Pi_d \text{ in } \Pi_g)$$
$$= \mu(c)\phi(d; g, q).$$

From §3.1,

$$\phi(d; g, q) = [g-d+1, g+1]_-/[1, d+1]_-.$$

For $d = -1$,

$$\lambda(-1, n; w) = \kappa(n; w) - \sum_{i=0}^{g} \lambda(i, n; w)$$
$$= q^{(g+1)(g+4-2w)/2}$$
$$= \mu(g+1). \quad \square$$

In §16.3, the theory of stereographic projection of a quadric and an ovaloid of $PG(3, q)$ was explained. Here we consider the *stereographic projection* of a non-singular quadric $\mathscr{Q}_n$ onto a prime from a point P_0 on the quadric. Precisely the same argument applies to a variety of degree $d > 2$ if P_0 is taken to be a point of multiplicity $d - 1$.

Let P_0 be any point of $\mathscr{Q}_n$ and Π_{n-1} a fixed prime not containing P_0. Let $\mathscr{V} = T_{P_0}(\mathscr{Q}_n)$ be the tangent prime at P_0, let $\mathscr{W} = \mathscr{V} \cap \mathscr{Q}_n$ be the tangent cone, let $\mathscr{V}' = \Pi_{n-1} \cap \mathscr{V}$, and $\mathscr{W}' = \Pi_{n-1} \cap \mathscr{W}$. For example, when $\mathscr{Q}_n = \mathscr{H}_3$, then Π_{n-1} is a plane, $\mathscr{V}$ is a plane meeting $\mathscr{H}_3$ in a line pair $\mathscr{W}$, and $\mathscr{V}'$ is a line meeting $\mathscr{H}_3$ in a point pair $\mathscr{W}'$.

For P in $\mathscr{Q}_n\backslash\{P_0\}$, define $P' = P_0P \cap \Pi_{n-1}$. This gives the correspondence

$$P \to P', \qquad P_0 \to \mathscr{V}'.$$

Analytically, let $P_0 = \mathbf{U}_0$, $\Pi_{n-1} = \mathbf{u}_0$,

$$\mathscr{Q}_n = \mathbf{V}(x_0F_1(x_1, \ldots, x_n) + F_2(x_1, \ldots, x_n))$$

where $\deg F_i = i$. Then $\mathscr{V} = \mathbf{V}(F_1)$ and $\mathscr{W} = \mathbf{V}(F_1, F_2)$. If $P = \mathbf{P}(a_0, \ldots, a_n)$, then $P' = \mathbf{P}(0, a_1, \ldots, a_n)$. Conversely, if $P' = \mathbf{P}(0, a_1, \ldots, a_n)$, then $P = \mathbf{P}(a_0, \ldots, a_n)$ with $a_0 = -F_2(a_1, \ldots, a_n)/F_1(a_1, \ldots, a_n)$.

Now, we describe the effect of stereographic projection on the generators of $\mathscr{Q}_n$.

Let Π_g be a generator of $\mathscr{Q}_n$ and let $P_0 \in \mathscr{Q}_n\backslash\Pi_g$. Let Π_{n-1} be a fixed prime not containing P_0; then Π_g projects from P_0 to $\Pi'_g = P_0\Pi_g \cap \Pi_{n-1}$. Now, Π_g does not lie in $\mathscr{V} = T_{P_0}(\mathscr{Q}_n)$, as otherwise $P_0\Pi_g$ would lie on $\mathscr{Q}_n$. So Π_g meets $\mathscr{V}$ in a Π_{g-1} lying in $\mathscr{V}$ and $\mathscr{Q}_n$, and so in $\mathscr{W} = \mathscr{V} \cap \mathscr{Q}_n$. Hence the projection of Π_g onto Π_{n-1} from P_0 is a Π'_g not lying in $\mathscr{V}' = \Pi_{n-1} \cap \mathscr{V}$ but meeting $\mathscr{W}' = \Pi_{n-1} \cap \mathscr{W}$ in a Π'_{g-1}. Conversely, a space Π'_g of Π_{n-1} not on $\mathscr{V}'$ but containing a space Π'_{g-1} of $\mathscr{W}'$ is joined to P_0 by a Π_{g+1}, which contains the generator $P_0\Pi'_{g-1}$ of $\mathscr{Q}_n$; the space Π_{g+1} meets $\mathscr{Q}_n$ residually in a Π_g not containing P_0 which projects from P_0 to the space Π'_g. However, $\mathscr{V}'$ is a Π_{n-2} in Π_{n-1} and $\mathscr{W}'$ a quadric in $\mathscr{V}'$. From Lemma 22.3.8, $\mathscr{W} = P_0\mathscr{Q}_{n-2}$ where $\mathscr{Q}_{n-2}$ has the same character as $\mathscr{Q}_n$; hence $\mathscr{W}'$ is a $\mathscr{Q}_{n-2}$.

Now we establish the existence of a partition of the generators of $\mathscr{H}_n$ into two sets.

Given two generators Π_g and $\bar{\Pi}_g$ of $\mathscr{H}_n$, define Π_g to be equivalent to $\bar{\Pi}_g$ if $\Pi_g \cap \bar{\Pi}_g = \Pi_t$ with g and t of the same parity. It will be shown that this relation is an equivalence relation. Trivially, the relation is reflexive and symmetric. Stereographic projection is used to show the transitivity. The key lemma follows.

Lemma 22.4.10: *If two generators $\Pi_g^{(1)}$ and $\Pi_g^{(2)}$ of $\mathscr{H}_{2g+1}$ intersect in Π_{g-1}, then a third generator $\Pi_g^{(3)}$ intersects $\Pi_g^{(1)}$ and $\Pi_g^{(2)}$ in spaces whose dimensions have different parity.*

Proof: Since $\Pi_g^{(1)} \cap \Pi_g^{(2)} = \Pi_{g-1}$, so $\Pi_g^{(1)}\Pi_g^{(2)} = \Pi_{g+1}$ and $\Pi_{g+1} \cap \mathscr{H}_{2g+1} = \Pi_{g-1}\mathscr{H}_1$, which consists of the pair $\Pi_g^{(1)}$ and $\Pi_g^{(2)}$. By Lemma 22.3.6, Π_{g+1} touches $\mathscr{H}_{2g+1}$ along Π_{g-1}. Let $\Pi_{g+1} \cap \Pi_g^{(3)} = \Pi_m$; then $m \geq 0$ since $2g+1$ is the dimension of the ambient space. Now either (a) Π_m lies in exactly one of $\Pi_g^{(1)}$, $\Pi_g^{(2)}$, or (b) Π_m lies in Π_{g-1}. In case (b), the polar space Π'_{2g-m} of Π_m contains both Π_{g+1} and $\Pi_g^{(3)}$. So, $\Pi_{g+1} \cap \Pi_g^{(3)} = \Pi_k$, where $k \geq g + (g+1) - (2g-m) = m+1 > m$, a contradiction. So (a) holds. Suppose therefore that $\Pi_m \subset \Pi_g^{(1)}$, whence $\Pi_g^{(1)} \cap \Pi_g^{(3)} = \Pi_m$. Since Π_{g-1} and Π_m are both contained in $\Pi_g^{(1)}$, so $\Pi_{g-1} \cap \Pi_m = \Pi_l$ with $l = (g-1) + m - g = m - 1$. Hence $\Pi_g^{(2)} \cap \Pi_g^{(3)} = \Pi_{m-1}$. □

Lemma 22.4.11: *If the generators Π_g and $\bar{\Pi}_g$ of $\mathscr{H}_n$ with intersection Π_t are projected from a point P_0 in $\mathscr{H}_n \backslash (\Pi_g \cup \bar{\Pi}_g)$ to spaces Π'_g and $\bar{\Pi}'_g$ containing the generators Π'_{g-1} and $\bar{\Pi}'_{g-1}$ of $\mathscr{W}' = \mathscr{H}_{n-2}$ with $\Pi'_{g-1} \cap \bar{\Pi}'_{g-1} = \Pi'_{s-1}$, then t and s have the same parity.*

Proof: Take a point P_0 in $\mathscr{H}_n \backslash (\Pi_g \cup \bar{\Pi}_g)$ and project $\mathscr{H}_n$ stereographically onto Π_{n-1}. Let Π'_g and $\bar{\Pi}'_g$ be the projections of Π_g and $\bar{\Pi}_g$, and let Π'_{g-1} and $\bar{\Pi}'_{g-1}$ be the spaces in which Π'_g and $\bar{\Pi}'_g$ meet $\mathscr{V}'$, where the notation is that described above; also let $\Pi'_{g-1} \cap \bar{\Pi}'_{g-1} = \Pi'_{s-1}$.

As in the above description of stereographic projection, $P_0\Pi'_g \cap \mathscr{H}_n = \Pi_g + P_0\Pi'_{g-1}$ and $P_0\bar{\Pi}'_g \cap \mathscr{H}_n = \bar{\Pi}_g + P_0\bar{\Pi}'_{g-1}$. Now $\Pi_g \cap \bar{\Pi}_g = \Pi_t$ and $P_0\Pi'_{g-1} \cap P_0\bar{\Pi}'_{g-1} = P_0\Pi'_{s-1}$, which are of respective dimensions t and s. Both these are of different parity to $\dim(\Pi_g \cap P_0\bar{\Pi}'_{g-1})$, by Lemma 22.4.10, when the two triples of generators $(\bar{\Pi}_g, P_0\bar{\Pi}'_{g-1}, \Pi_g)$ and $(\Pi_g, P_0\Pi'_{g-1}, P_0\bar{\Pi}'_{g-1})$ are considered. So s and t have the same parity. □

Theorem 22.4.12: *When $\mathscr{Q}_n = \mathscr{H}_n$, the relation on the generators is an equivalence relation with two equivalence classes.*

Proof: It remains to prove that the relation is transitive. If, on $\mathscr{H}_n$, we have $\Pi_g^{(1)}$ equivalent to $\Pi_g^{(2)}$ and $\Pi_g^{(2)}$ equivalent to $\Pi_g^{(3)}$, and if, by projection at the ith stage, $\Pi_{g-i}^{(j)} \subset \mathscr{H}_{n-2i}$ corresponds to $\Pi_g^{(j)}$, then the parity of $(g-i) - \dim(\Pi_{g-i}^{(1)} \cap \Pi_{g-i}^{(2)})$ is the same for all i, as is the parity of $(g-i) - \dim(\Pi_{g-i}^{(2)} \cap \Pi_{g-i}^{(3)})$, by Lemma 22.4.11. Successive projection gives three lines l_1, l_2, l_3 on $\mathscr{H}_3$. As $\Pi_g^{(1)}$ is equivalent to $\Pi_g^{(2)}$, so $g - \dim(\Pi_g^{(1)} \cap \Pi_g^{(2)})$ is even; therefore $1 - \dim(l_1 \cap l_2)$ is even. Thus l_1 and l_2 are the same line or are skew. Similarly, l_2 and l_3 are the same or skew. So l_1, l_2, l_3 belong to the same regulus of $\mathscr{H}_3$. Hence the dimension of $l_1 \cap l_3$ is 1 or -1. Thus $1 - \dim(l_1 \cap l_3)$ is even, and so is $g - \dim(\Pi_g^{(1)} \cap \Pi_g^{(3)})$. We conclude that $\Pi_g^{(1)}$ is equivalent to $\Pi_g^{(3)}$.

From Lemma 22.4.10 it follows that there are exactly two equivalence classes. □

Each equivalence class is called a *system of generators*.

Corollary: *Let Π_g and $\bar{\Pi}_g$ be distinct generators of $\mathscr{H}_n$, $n = 2g + 1$. Their possible intersections are as follows*:

$$n = 4s + 1,\ g = 2s,$$

$$\dim(\Pi_g \cap \bar{\Pi}_g) = \begin{cases} 0, 2, 4, \ldots, 2s-2 & \textit{same system} \\ -1, 1, 3, \ldots, 2s-1 & \textit{different systems}; \end{cases}$$

$$n = 4s + 3,\ g = 2s + 1,$$

$$\dim(\Pi_g \cap \bar{\Pi}_g) = \begin{cases} -1, 1, 3, \ldots, 2s-1 & \textit{same system} \\ 0, 2, 4, \ldots, 2s & \textit{different systems.} \end{cases} \qquad \square$$

For dimensions up to nine of hyperbolic quadrics, the following table gives all the occurring dimensions of intersections of distinct generators.

$\mathcal{Q}_n$	*Dimension of generator*	*Same system*	*Different systems*
$\mathcal{H}_1$	0	–	−1
$\mathcal{H}_3$	1	−1	0
$\mathcal{H}_5$	2	0	−1, 1
$\mathcal{H}_7$	3	−1, 1	0, 2
$\mathcal{H}_9$	4	0, 2	−1, 1, 3

22.5 Numbers of subspaces on a quadric

Let $N(m; n, w)$ be the number of subspaces Π_m on the quadric $\mathcal{Q}_n$ of character w. In §22.4, the number of generators of $\mathcal{Q}_n$ was determined; that is,

$$\kappa(n; w) = N(g; n, w)$$

where $g = \frac{1}{2}(n + w - 3)$.

We also write $N(\Pi_m, \mathcal{W}_n)$ for the number of m-spaces on the quadric $\mathcal{W}_n$; so

$$N(m; n, w) = N(\Pi_m, \mathcal{Q}_n).$$

Theorem 22.5.1:

$$\begin{aligned} N(m; n, w) &= [\tfrac{1}{2}(n + 1 - w) - m, \tfrac{1}{2}(n + 1 - w)]_+ \\ &\quad \times [\tfrac{1}{2}(n - 1 + w) - m, \tfrac{1}{2}(n - 1 + w)]_- \\ &\quad \div [1, m + 1]_- \qquad (22.11) \\ &= [g + 2 - w - m, g + 2 - w]_+ \\ &\quad \times [g + 1 - m, g + 1]_- \\ &\quad \div [1, m + 1]_-. \qquad (22.12) \end{aligned}$$

(a) *m-spaces on particular quadrics*

$$\begin{aligned} N(m; 2s - 1, 0) &= N(\Pi_m, \mathcal{E}_{2s-1}) \\ &= [s - m, s]_+[s - 1 - m, s - 1]_-/[1, m + 1]_-; \qquad (22.13) \end{aligned}$$

$$N(m; 2s-1, 2) = N(\Pi_m, \mathscr{H}_{2s-1})$$
$$= [s-1-m, s-1]_+[s-m, s]_-/[1, m+1]_-; \qquad (22.14)$$
$$N(m; 2s, 1) = N(\Pi_m, \mathscr{P}_{2s})$$
$$= [s-m, s]_+[s-m, s]_-/[1, m+1]_-. \qquad (22.15)$$

(b) *Points on general and particular quadrics*

$$N(0; n, w) = (q^{(n+1-w)/2} + 1)(q^{(n-1+w)/2} - 1)/(q-1); \qquad (22.16)$$
$$= (q^n - 1)/(q-1) + (w-1)q^{(n-1)/2}; \qquad (22.17)$$
$$N(\Pi_0, \mathscr{E}_{2s-1}) = (q^s + 1)(q^{s-1} - 1)/(q-1); \qquad (22.18)$$
$$N(\Pi_0, \mathscr{H}_{2s-1}) = (q^{s-1} + 1)(q^s - 1)/(q-1); \qquad (22.19)$$
$$N(\Pi_0, \mathscr{P}_{2s}) = (q^s + 1)(q^s - 1)/(q-1). \qquad (22.20)$$

(c) *Generators on a quadric*

$$\kappa(n; w) = [2-w, g+2-w]_+ \qquad (22.21)$$
$$= [2-w; \tfrac{1}{2}(n+1-w)]_+; \qquad (22.22)$$
$$\kappa(2s-1; 0) = [2, s]_+; \qquad (22.23)$$
$$\kappa(2s-1; 2) = [0, s-1]_+; \qquad (22.24)$$
$$\kappa(2s; 1) = [1, s]_+. \qquad (22.25)$$

Proof: First the number of points on $\mathscr{Q}_n$ is calculated. By Lemma 22.3.8,

$$N(0; n, w) = q^{n-1} + 1 + qN(0; n-2, w). \qquad (22.26)$$

However, we know that $N(0; 1, 0) = N(0; 0, 1) = 0$ and $N(0; 1, 2) = 2$, whence induction gives (22.16) and (22.17).

Now, by Theorem 22.3.2(ii), if $P \in \Pi_m \subset \mathscr{Q}_n$, then $\Pi_m \subset T_P(\mathscr{Q}_n) \cap \mathscr{Q}_n$, which by Lemma 22.3.8 is $P\mathscr{Q}_{n-2}$ of the same character. So Π_m meets $\mathscr{Q}_{n-2}$ in Π_{m-1} and, conversely, every Π_{m-1} on $\mathscr{Q}_{n-2}$ determines a Π_m on $\mathscr{Q}_n$ through P, Hence

$$N(m; n, w) = N(0; n, w)N(m-1; n-2, w)/\theta(m), \qquad (22.27)$$

where $\theta(m) = (q^{m+1} - 1)/(q-1)$. Induction and (22.16) give the result. □

Corollary:

$$N(\Pi_0, \Pi_{n-t-1}\mathscr{Q}_t) = (q^n - 1)/(q-1) + (w-1)q^{(2n-t-1)/2}. \quad \square$$

Proof: The joins of two points of the base to the vertex give $(n-t)$-spaces intersecting in the vertex Π_{n-t-1}. Hence

$$N(\Pi_0, \Pi_{n-t-1}\mathcal{Q}_t) = N(\Pi_0, \mathcal{Q}_t)(\theta(n-t) - \theta(n-t-1)) + \theta(n-t-1),$$

which gives the answer. □

22.6 The orthogonal groups

The group of projectivities of $PG(n, q)$ is $PGL(n+1, q)$. Let $G(\mathcal{Q}_n)$ be the subgroup of $PGL(n+1, q)$ fixing the form defining $\mathcal{Q}_n$ up to a scalar multiple. This is actually the same as the group fixing $\mathcal{Q}_n$, providing $\mathcal{Q}_n \neq \mathcal{E}_1$. The group $G(\mathcal{Q}_n)$ is called *orthogonal* and is also denoted by $PGO(n+1, q)$, $PGO_+(n+1, q)$, or $PGO_-(n+1, q)$ according as $\mathcal{Q}_n$ is $\mathcal{P}_n$, $\mathcal{H}_n$, or $\mathcal{E}_n$. Also let $\mathcal{N}(\mathcal{Q}_n)$ be the set of all quadrics in $PG(n, q)$ projectively equivalent to $\mathcal{Q}_n$; that is, $\mathcal{N}(\mathcal{Q}_n)$ is the orbit of $\mathcal{Q}_n$ under the action of $PGL(n+1, q)$. Again $\mathcal{E}_1$ is a special case and is considered here as a pair of conjugate points in $PG(1, q^2)$. First we calculate $|G(\mathcal{Q}_n)|$ and $|\mathcal{N}(\mathcal{Q}_n)|$.

Lemma 22.6.1: *The number of quadrics $\mathcal{Q}_n$ in $PG(n, q)$, $n \geq 2$, containing a given $\Pi_0\mathcal{Q}_{n-2}$ as a tangent cone is $q^n(q-1)$.*

Proof: Let $\Pi_0\mathcal{Q}_{n-2}$ have vertex $\Pi_0 = \mathbf{U}_{n-1}$ and base $\mathcal{Q}_{n-2} = \mathbf{V}(x_n, x_{n-1}, f_2(x_0, \ldots, x_{n-2}))$. So the prime containing $\Pi_0\mathcal{Q}_{n-2}$ is $\mathbf{u}_n$. Any quadric containing $\mathbf{U}_{n-1}$ and $\mathcal{Q}_{n-2}$ has the form

$$\mathcal{Q} = \mathbf{V}(x_n(a_0x_0 + \cdots + a_{n-1}x_{n-1} + a_nx_n) + x_{n-1}(b_0x_0 + \cdots + b_{n-2}x_{n-2}) + f_2).$$

The tangent prime to $\mathcal{Q}$ at $\mathbf{U}_{n-1}$ is

$$\mathbf{V}(a_{n-1}x_n + b_0x_0 + \cdots + b_{n-2}x_{n-2}).$$

Since this is $\mathbf{u}_n$, so

$$b_0 = \cdots = b_{n-2} = 0, \qquad a_{n-1} \neq 0.$$

Thus

$$\mathcal{Q} = \mathbf{V}(x_n(a_0x_0 + \cdots + a_{n-1}x_{n-1} + a_nx_n) + f_2)$$

which is non-singular since $a_{n-1} \neq 0$. Therefore, a_{n-1} may be chosen in $q-1$ ways and every other a_i in q ways, giving $q^n(q-1)$ possibilities for $\mathcal{Q}$. □

In this proof, when $\mathcal{Q}_n = \mathcal{E}_3$, then $\Pi_0\mathcal{Q}_{n-2} = \Pi_0\mathcal{E}_1$ is a pair of conjugate intersecting lines in the quadratic extension.

Theorem 22.6.2:

$\mathscr{Q}_n$	$\lvert G(\mathscr{Q}_n)\rvert$	$\lvert \mathscr{N}(\mathscr{Q}_n)\rvert$
$\mathscr{P}_n$	$q^{n^2/4} \prod_{i=1}^{n/2} (q^{2i}-1)$	$q^{n(n+2)/4} \prod_{i=1}^{n/2} (q^{2i+1}-1)$
$\mathscr{H}_n$	$2q^{(n^2-1)/4}(q^{(n+1)/2}-1) \prod_{i=1}^{(n-1)/2} (q^{2i}-1)$	$\frac{1}{2}q^{(n+1)^2/4}(q^{(n+1)/2}+1) \prod_{i=1}^{(n-1)/2} (q^{2i+1}-1)$
$\mathscr{E}_n$	$2q^{(n^2-1)/4}(q^{(n+1)/2}+1) \prod_{i=1}^{(n-1)/2} (q^{2i}-1)$	$\frac{1}{2}q^{(n+1)^2/4}(q^{(n+1)/2}-1) \prod_{i=1}^{(n-1)/2} (q^{2i+1}-1)$

Proof: First $|\mathscr{N}(\mathscr{Q}_n)|$ is calculated by counting the set $\{(\mathscr{Q}_n, S) \mid \mathscr{Q}_n$ a non-singular quadric, S a tangent cone of $\mathscr{Q}_n\}$ in two ways. Let M be the number of cones $\Pi_0\mathscr{Q}_{n-2}$ for a fixed character w in $PG(n, q)$. Then

$$|\mathscr{N}(\mathscr{Q}_n)|N(0; n, w) = Mq^n(q-1).$$

However,

$$\begin{aligned} M &= \text{number of } \Pi_{n-1} \text{ in } PG(n, q) \\ &\quad \times \text{number of } \Pi_0 \text{ in } \Pi_{n-1} \\ &\quad \times \text{number of } \mathscr{Q}_{n-2} \text{ in a fixed } \Pi_{n-2} \text{ of } \Pi_{n-1} \\ &= \theta(n)\theta(n-1)|\mathscr{N}(\mathscr{Q}_{n-2})|. \end{aligned}$$

Thus

$$\begin{aligned} |\mathscr{N}(\mathscr{Q}_n)| &= \theta(n)\theta(n-1)|\mathscr{N}(\mathscr{Q}_{n-2})|q^n(q-1)/N(0; n, w) \\ &= \frac{(q^{n+1}-1)(q^n-1)q^n|\mathscr{N}(\mathscr{Q}_{n-2})|}{(q^{(n+1-w)/2}+1)(q^{(n-1+w)/2}-1)}. \end{aligned}$$

Since $|\mathscr{N}(\mathscr{P}_0)| = 1$, $|\mathscr{N}(\mathscr{E}_1)| = \frac{1}{2}q(q-1)$, $|\mathscr{N}(\mathscr{H}_1)| = \frac{1}{2}q(q+1)$, induction now gives $|\mathscr{N}(\mathscr{Q}_n)|$.

Finally,

$$|PGL(n+1, q)| = |G(\mathscr{Q}_n)|\,|\mathscr{N}(\mathscr{Q}_n)|$$

in each case, where

$$|PGL(n+1, q)| = q^{n(n+1)/2} \prod_{i=2}^{n+1} (q^i - 1). \quad \square$$

For the orders of groups associated to these orthogonal groups, see Appendix III of Part IV in the previous volume.

Consider the following involutory transformations which fix $\mathscr{Q}_n$. Let Q be any point of $PG(n, q)\backslash\mathscr{Q}_n$ with the only restriction that, for q and n even, Q is not the nucleus of $\mathscr{Q}_n$. Let

$$\mu_Q: \mathscr{Q}_n \to \mathscr{Q}_n$$

be defined as follows. For P in $\mathscr{Q}_n$,

$$P\mu_Q = P \text{ if } PQ \text{ is a tangent to } \mathscr{Q}_n;$$

$$P\mu_Q = P' \text{ if } PQ \text{ meets } \mathscr{Q}_n \text{ again at } P'.$$

Lemma 22.6.3: *μ_Q can be extended to an element of $G(\mathscr{Q}_n)$.*

Proof: Let $\mathscr{Q}_n = \mathbf{V}(F)$, $Q = \mathbf{P}(B) \notin \mathscr{Q}_n$, $P = \mathbf{P}(A) \in \mathscr{Q}_n$. If $P' \in PQ$, then $P' = \mathbf{P}(A + tB)$. If $P' \in \mathscr{Q}_n$, then, as in (22.3),

$$F(A + tB) = F(A) + tG(A, B) + t^2F(B) = 0.$$

Since $P \in \mathscr{Q}_n$, so $F(A) = 0$. Hence

$$tG(A, B) + t^2F(B) = 0.$$

The solution $t = 0$ corresponds to P and the solution $t = -G(A, B)/F(B)$ corresponds to P'. Hence $P' = \mathbf{P}(A - G(A, B)B/F(B))$, which is the same point as P when $G(A, B) = 0$; that is, when P lies in the polar prime of Q. In any case, μ_Q is given by

$$\mathbf{P}(X) \to \mathbf{P}\left(X - \frac{G(X, B)}{F(B)}B\right).$$

Thus μ_Q can be extended to an element of $G(\mathscr{Q}_n)$. □

Since the identity is the only element of $G(\mathscr{Q}_n)$ which fixes Q and all points of $\mathscr{Q}_n$, the extension of μ_Q is necessarily unique. This extension is a perspectivity with centre Q; the axis contains all points P of $\mathscr{Q}_n$ for which PQ is tangent to $\mathscr{Q}_n$.

The extension of μ_Q is also denoted μ_Q.

Theorem 22.6.4: *$G(\mathscr{Q}_n)$ acts transitively on $\mathscr{Q}_n$.*

Proof: Let P, P' be any two points of $\mathscr{Q}_n$. If $PP' \not\subset \mathscr{Q}_n$, let Q be any point on $PP'\backslash\{P, P'\}$. Then μ_Q maps P to P'.

If $PP' \subset \mathscr{Q}_n$, choose P'' such that neither PP'' nor $P'P''$ lies on $\mathscr{Q}_n$. The point P'' exists, since otherwise $\mathscr{Q}_n$ would be singular. Now, choose Q on $PP''\backslash\{P, P''\}$ and R on $P'P''\backslash\{P', P''\}$. Then $P\mu_Q\mu_R = P''\mu_R = P'$. □

Consider the quadric $\mathscr{Q}_n$ of character w. Let $\mathscr{S} = \mathscr{S}(m, t, v; n, w)$ be the set of m-spaces Π_m in $PG(n, q)$ with $m \neq n$ such that $\Pi_m \cap \mathscr{Q}_n$ is of type $\Pi_{m-t-1}\mathscr{Q}_t$ where $\mathscr{Q}_t$ has character v. In §22.8, we determine the size $N(m, t, v; n, w) = N(\Pi_{m-t-1}\mathscr{Q}_t, \mathscr{Q}_n)$ of the set $\mathscr{S}(m, t, v; n, w)$, and in particular decide when the set is empty. Here we give the number of orbits of $\mathscr{S}(m, t, v; n, w)$ under the action of $G(\mathscr{Q}_n)$.

First we consider $PG(1, q)$. The quadric $\mathscr{H}_1$ consists of two points and $G(\mathscr{H}_1) = PGO_+(2, q)$ has order $2(q-1)$. As $PGL(2, q)$ acts triply transitively on $PG(1, q)$, §6.1, there is a projectivity fixing both points of $\mathscr{H}_1$ and moving P_1 to P_2, where P_1 and P_2 are any points off $\mathscr{H}_1$. So $PGO_+(2, q)$ acts transitively on the points off $\mathscr{H}_1$.

The quadric $\mathscr{E}_1$ is empty in $PG(1, q)$ but consists of two conjugate points on $PG(1, q^2)$. So, if $\mathscr{E}_1 = \mathbf{V}(F)$ with $F = x^2 - bx + c = (x - \alpha)(x - \alpha^q)$, then in non-homogeneous coordinates the projectivity $\mathfrak{T}\colon t \to t'$ of $PG(1, q^2)$ given by

$$tt'\{e + e' - (\alpha^q + \alpha)\} - (t + t')\{ee' - \alpha^{q+1}\} + \{(\alpha + \alpha^q)ee' - \alpha^{q+1}(e + e')\} = 0$$

is an involution with pairs (α, α^q) and (e, e'). It therefore fixes $\mathscr{E}_1$ and takes e to e'. Thus $PGO_-(2, q)$ acts transitively on the points of $PG(1, q)$ off $\mathscr{E}_1$. The group $G(\mathscr{E}_1) = PGO_-(2, q)$ has order $2(q + 1)$.

Next we examine the conic $\mathscr{P}_2$. Let

$$\mathcal{O}_1 = \mathscr{S}(0, -1, 2; 2, 1) = \{\text{points on } \mathscr{P}_2\},$$

$$\mathcal{O}_2 = \mathscr{S}(0, 0, 1; 2, 1) = \{\text{points off } \mathscr{P}_2\},$$

$$\mathcal{O}_3 = \mathscr{S}(1, 0, 1; 2, 1) = \{\text{tangents to } \mathscr{P}_2\},$$

$$\mathcal{O}_4 = \mathscr{S}(1, 1, 2; 2, 1) = \{\text{bisecants of } \mathscr{P}_2\},$$

$$\mathcal{O}_5 = \mathscr{S}(1, 1, 0; 2, 1) = \{\text{external lines of } \mathscr{P}_2\}.$$

Theorem 22.6.4 says that $G(\mathscr{P}_2)$ acts transitively on $\mathcal{O}_3$. In fact, $G(\mathscr{P}_2)$ acts triply transitively on $\mathcal{O}_1$ and $\mathcal{O}_3$, Lemma 7.2.3, Corollary 8. We recall that, for q odd,

$$\mathcal{O}_2 = \mathcal{O}_2^+ \cup \mathcal{O}_2^-,$$

where

$$\mathcal{O}_2^+ = \{\text{external points of } \mathscr{P}_2\}, \qquad \mathcal{O}_2^- = \{\text{internal points of } \mathscr{P}_2\};$$

here, a point Q off $\mathscr{P}_2$ is external or internal according as it lies on two or no tangents of $\mathscr{P}_2$, §8.2. For q even,

$$\mathcal{O}_2 = \{N\} \cup \mathcal{O}_2',$$

where N is the nucleus, the meet of all the tangents, and each point of $\mathcal{O}_2'$ lies on precisely one tangent.

Lemma 22.6.5: (i) *$G(\mathscr{P}_2)$ acts transitively on $\mathcal{O}_4$ and $\mathcal{O}_5$;*

(ii) *$G(\mathscr{P}_2)$ has two orbits on $\mathcal{O}_2$, namely $\mathcal{O}_2^+$ and $\mathcal{O}_2^-$ for q odd, and $\{N\}$ and $\mathcal{O}'_2$ for q even.*

Proof: (a) For q odd, consider the action of $G(\mathscr{P}_2)$ on $\mathcal{O}_2$, the points off $\mathscr{P}_2 = \mathbf{V}(x_0^2 + x_1x_2)$. Since each point of $\mathcal{O}_2^+$ is the intersection of two tangents, so $G(\mathscr{P}_2)$ is transitive on $\mathcal{O}_2^+$, the external points, and similarly or by the polarity on $\mathcal{O}_4$, the bisecants.

Any external line contains an external point. So, to show the transitivity of $G(\mathscr{P}_2)$ on $\mathcal{O}_5$ and, similarly or by the polarity, on $\mathcal{O}_2^-$, it suffices to show the transitivity on the external lines through a particular external point. Let $\mathbf{U}_0$ be this point. Then the line $l(t) = \mathbf{V}(x_1 + tx_2)$ is a bisecant or an external line as t is a non-zero square or a non-square. The projectivity $\mathfrak{T}_c$ given by $\mathbf{P}(x_0, x_1, x_2)\mathfrak{T}_c = \mathbf{P}(cx_0, x_1, c^2x_2)$ fixes $\mathscr{P}_2$ and transforms $l(t)$ to $l(t/c^2)$. So we can pass from any bisecant through $\mathbf{U}_0$ to any other bisecant through $\mathbf{U}_0$ and any external line through $\mathbf{U}_0$ to any other external line through $\mathbf{U}_0$.

(b) For q even, $G(\mathscr{P}_2)$ is similarly transitive on $\mathcal{O}_4$. Since any point of $\mathcal{O}'_2$ is the meet of a tangent and a bisecant, the triple transitivity of $G(\mathscr{P}_2)$ on $\mathcal{O}_1$ ensures the transitivity on $\mathcal{O}'_2$.

To show the transitivity of $G(\mathscr{P}_2)$ on $\mathcal{O}_5$, it suffices to consider the external lines through a particular point of $\mathcal{O}'_2$. Let $Q = \mathbf{P}(1, 0, 1)$ with $\mathscr{P}_2 = \mathbf{V}(x_0^2 + x_1x_2)$ as above. Then the line $l(t) = \mathbf{V}(x_0 + x_2 + tx_1)$ contains Q and meets $\mathscr{P}_2$ where $x_2^2 + x_1x_2 + t^2x_1 = 0$. So $l(t)$ is a bisecant or an external line according as t^2 and so t is in $\mathscr{C}_0$ or $\mathscr{C}_1$, §22.2. Now, the projectivity $\mathfrak{S}_b$ given by $\mathbf{P}(x_0, x_1, x_2)\mathfrak{S}_b = \mathbf{P}(x_0 + bx_1, x_1, b^2x_1 + x_2)$ fixes Q and $\mathscr{P}_2$ and transforms $l(t)$ to $l(t + b + b^2)$. As $t + b + b^2$ is in the same $\mathscr{C}_i$ as t, we can pass from any external line through Q to any other. □

For a section $\Pi_{m-t-1}\mathscr{Q}_t$ of $\mathscr{Q}_n$, let $T = n + t - 2m$.

Theorem 22.6.6: *For given m, t, v, n, w, the set $\mathscr{S}(m, t, v; n, w)$ acted on by $G(\mathscr{Q}_n)$ is either empty or has*

(a) *one orbit when* (i) *n is odd or* (ii) *n is even and t is odd or* (iii) *n is even, t is even, and $T = 0$;*
(b) *two orbits when n is even, t is even, and $T > 0$.*

Proof:

(1) *$t = m$ with $(v, w) \neq (1, 1)$*

First, assume that $m \geq 2$.

Let $\Pi_m^{(i)} \cap \mathscr{Q}_n = \mathscr{W}_m^{(i)}$ with $\Pi_m^{(i)}$ in $\mathscr{S}$, $i = 1, 2$, and let $P \in \mathscr{W}_m^{(1)} \cap \mathscr{W}_m^{(2)}$. Project $\mathscr{Q}_n$ from P onto a prime Π_{n-1} not containing P, as in §22.4. Then

$\mathscr{Q}_n$ determines a quadric $\mathscr{W}' = \mathscr{Q}_{n-2}$ in Π_{n-2}, and $\mathscr{W}_m^{(1)}$ and $\mathscr{W}_m^{(2)}$ give quadrics $\mathscr{R}_1$ and $\mathscr{R}_2$ of the same type but in dimension $m-2$. By induction there is a projectivity $\mathfrak{T}$ of Π_{n-2} fixing $\mathscr{Q}_{n-2}$ and mapping $\mathscr{R}_1$ to $\mathscr{R}_2$. Let $\Pi_m^{(i)} \cap \Pi_{n-1} = \Pi_{m-1}^{(i)}$, $i = 1, 2$. Extend $\mathfrak{T}$ to Π_{n-1} and let $\Pi_{m-1}^{(1)}\mathfrak{T} = \Pi_{m-1}^{(1)'}$. In Π_{n-1} there is an elation $\mathfrak{T}'$ with axis Π_{n-2} mapping $\Pi_{m-1}^{(1)'}$ to $\Pi_{m-1}^{(2)}$. Hence $\mathfrak{T}\mathfrak{T}'$ maps $\mathscr{R}_1$ to $\mathscr{R}_2$, $\Pi_{m-1}^{(1)}$ to $\Pi_{m-1}^{(2)}$, and $\mathscr{Q}_{n-2}$ to itself.

Taking $P = \mathbf{U}_n$ and the tangent prime $T_P = \mathbf{u}_{n-1}$, the quadric $\mathscr{Q}_n = \mathbf{V}(F)$ with

$$F = f(x_0, \ldots, x_{n-1}) + x_{n-1}x_n.$$

By a linear transformation

$$F = g(x_0, \ldots, x_{n-2}) + x_{n-1}x_n.$$

Then $\mathscr{Q}_{n-2} = \mathbf{V}(g, x_{n-1}, x_n)$. So $\mathfrak{T}\mathfrak{T}'$ is given by a linear transformation on $x_0, \ldots, x_{n-1}$ taking g to λg and x_{n-1} to $\lambda' x_{n-1}$. Extending $\mathfrak{T}\mathfrak{T}'$ by $x_n \to (\lambda/\lambda')x_n$ gives a projectivity $\mathfrak{S}$ fixing P and $\mathscr{Q}_n$ as well as mapping $\mathscr{W}_m^{(1)}$ to $\mathscr{W}_m^{(2)}$.

Next, let $\mathscr{W}_m^{(1)} \cap \mathscr{W}_m^{(2)} = \emptyset$. Since $G(\mathscr{Q}_n)$ acts transitively on $\mathscr{Q}_n$, there exists $\mathfrak{S}_1$ in $G(\mathscr{Q}_n)$ for which $\mathscr{W}_m^{(1)}\mathfrak{S}_1 = \mathscr{W}_m^{(3)}$ meets $\mathscr{W}_m^{(2)}$. Then application of the preceding argument gives an element $\mathfrak{S}_2$ of $G(\mathscr{Q}_n)$ with $\mathscr{W}_m^{(3)}\mathfrak{S}_2 = \mathscr{W}_m^{(2)}$. Hence $\mathfrak{S}_1\mathfrak{S}_2$ is the required element of $G(\mathscr{Q}_n)$ taking $\mathscr{W}_m^{(1)}$ to $\mathscr{W}_m^{(2)}$.

Since induction was used the small cases have still to be considered.

First assume $m = t = 0$. Then the section is a point off the quadric $\mathscr{Q}_n$, and n is odd. For $n = 1$ the group $G(\mathscr{Q}_1)$ acts transitively on the set of all points off $\mathscr{Q}_1$, as discussed after Theorem 22.6.4. So let $n \geq 3$. Assume that P_1 and P_2 are points off $\mathscr{Q}_n$ and let $\Pi_{n-1}^{(1)}$ and $\Pi_{n-1}^{(2)}$ be the polar primes of P_1 and P_2. Then $\Pi_{n-1}^{(1)} \cap \mathscr{Q}_n$ and $\Pi_{n-1}^{(2)} \cap \mathscr{Q}_n$ are non-singular quadrics $\mathscr{P}_{n-1}^{(1)}$ and $\mathscr{P}_{n-1}^{(2)}$ as $n-1$ is even. It is sufficient to show that there is an element $\mathfrak{T}$ in $G(\mathscr{Q}_n)$ with $\mathscr{P}_{n-1}^{(1)}\mathfrak{T} = \mathscr{P}_{n-1}^{(2)}$. By induction, as in a previous argument, we are now reduced to the case $n = 1$ and $m = t = 0$.

Now let $m = t = 1$; then n is even. For $n = 2$ there is nothing to prove by Theorem 22.6.5. So let $n \geq 4$. If $\mathscr{W}_1^{(1)}$ and $\mathscr{W}_1^{(2)}$ are hyperbolic and meet at P, then by projecting from P and applying a previous argument we see that $G(\mathscr{Q}_n)$ contains an element which maps $\mathscr{W}_1^{(1)}$ onto $\mathscr{W}_1^{(2)}$. If $\mathscr{W}_1^{(1)} \cap \mathscr{W}_1^{(2)} = \emptyset$, then proceed as in the case $\mathscr{W}_m^{(1)} \cap \mathscr{W}_m^{(2)} = \emptyset$. Finally, assume that $\mathscr{W}_1^{(1)}$ and $\mathscr{W}_1^{(2)}$ are elliptic. Let $\mathscr{W}_1^{(1)} = \{P_1, P_1'\}$ and $\mathscr{W}_1^{(2)} = \{P_2, P_2'\}$, with P_i, P_i' conjugate in a quadratic extension of the field γ. Let P_i'' be a point of P_iP_i' off $\mathscr{Q}_n$, $i = 1, 2$. By the preceding paragraph there is an element $\mathfrak{T}$ in $G(\mathscr{Q}_n)$ which maps P_1'' onto P_2''. Let $P_1\mathfrak{T} = R_1$, $P_1'\mathfrak{T} = R_1'$, $R_1P_2' \cap R_1'P_2 = \{Q\}$, and $R_1P_2 \cap R_1'P_2' = \{Q'\}$. If Q and Q' are on $\mathscr{Q}_n$, then the plane containing R_1, R_1', P_2, P_2' is on $\mathscr{Q}_n$; so P_2'' is on $\mathscr{Q}_n$, a contradiction. Assume therefore that Q is not on $\mathscr{Q}_n$. Then $R_1\mu_Q = P_2'$ and $R_1'\mu_Q = P_2$. Thus $\mathfrak{T}\mu_Q$ maps $\mathscr{W}_1^{(1)}$ to $\mathscr{W}_1^{(2)}$.

(2) $t = m$ *with* $w = v = 1$

First, assume $m \geq 2$.

Let $\mathscr{S}_P$ be the set of all elements of $\mathscr{S}$ containing the point P, and let G_P be the subgroup of $G(\mathscr{Q}_n)$ fixing P. By projection of $\mathscr{Q}_n$ from P onto a Π_{n-1} not containing P and by using induction on m as in (1) G_P has two orbits on $\mathscr{S}_P$.

Suppose that $\mathscr{W}_m^{(1)}$ and $\mathscr{W}_m^{(2)}$ are in $\mathscr{S}_P$ and also in one orbit $\mathscr{O}$ of $G(\mathscr{Q}_n)$; then $\mathscr{W}_m^{(1)}\mathfrak{T} = \mathscr{W}_m^{(2)}$ for some $\mathfrak{T}$ in $G(\mathscr{Q}_n)$. Let $P\mathfrak{T} = Q \in \mathscr{W}_m^{(2)}$. We show that there exists $\mathfrak{T}'$ in $G(\mathscr{Q}_n)$ fixing $\mathscr{W}_m^{(2)}$ and mapping Q to P. If PQ is not a line of $\mathscr{W}_m^{(2)}$ then μ_R, with R on PQ but not on $\mathscr{W}_m^{(2)}$, fixes $\mathscr{Q}_n$, fixes $\mathscr{W}_m^{(2)}$, and maps Q to P. Now, let PQ be a line of $\mathscr{W}_m^{(2)}$. Consider a point R on $\mathscr{W}_m^{(2)}$ such that neither PR nor QR is on $\mathscr{W}_m^{(2)}$. Further, let A be a point on PR but not on $\mathscr{W}_m^{(2)}$ and let B be a point on QR but not on $\mathscr{W}_m^{(2)}$. Then $\mu_A\mu_B$ fixes both $\mathscr{Q}_n$ and $\mathscr{W}_m^{(2)}$, and maps P to Q. Hence $\mathscr{W}_m^{(2)}$ is in the orbit $\mathscr{O}_P$ of $\mathscr{W}_m^{(1)}$ under G_P. Since $G(\mathscr{Q}_n)$ acts transitively on $\mathscr{Q}_n$, it follows immediately that the number of orbits $\mathscr{O}$ of $\mathscr{S}$ under $G(\mathscr{Q}_n)$ is the number of orbits $\mathscr{O}_P$ of $\mathscr{S}_P$ under G_P.

Since induction was used, the small cases still need to be considered. So, assume that $m = t = 0$; then the section is a point off $\mathscr{Q}_n$ and n is even.

First, let q be odd. Let P be a point off $\mathscr{Q}_n$, let Π_{n-1} be the polar prime of P, and let $\mathscr{Q}_n \cap \Pi_{n-1} = \mathscr{Q}_{n-1}$. By (1) and Theorem 22.6.5, $G(\mathscr{Q}_n)$ has two orbits on the set of all sections $\mathscr{Q}_{n-1}$. Hence $G(\mathscr{Q}_n)$ has two orbits on the set of all points off $\mathscr{Q}_n$.

Now suppose that q is even. By Theorem 22.6.5 we may assume that $n \geq 4$. One orbit consists of a single point, the nucleus N. So take distinct points P_1 and P_2 not in $\mathscr{Q}_n \cup \{N\}$. Let $\mathscr{C}_i$ be a conic on $\mathscr{Q}_n$ with nucleus P_i, $i = 1, 2$. It suffices to show that there exists $\mathfrak{T}$ in $G(\mathscr{Q}_n)$ with $\mathscr{C}_1\mathfrak{T} = \mathscr{C}_2$. We can choose $\mathscr{C}_1$ and $\mathscr{C}_2$ in such a way that $P \in \mathscr{C}_1 \cap \mathscr{C}_2$. Project $\mathscr{Q}_n$ from P onto a prime Π_{n-1} not containing P. In Π_{n-1} this gives a $\mathscr{Q}_{n-2}$ with nucleus N'. The tangents to $\mathscr{C}_1$ and $\mathscr{C}_2$ at P meet Π_{n-1} in points P_1' and P_2' distinct from N'. By induction on n, the group $G(\mathscr{Q}_{n-2})$ contains an element $\mathfrak{T}'$ with $P_1'\mathfrak{T}' = P_2'$. As in (1), $\mathfrak{T}'$ can be extended to an element of $G(\mathscr{Q}_n)$ which fixes both $\mathscr{Q}_n$ and P and maps $\mathscr{C}_1$ to $\mathscr{C}_2$. Hence this extension maps P_1 to P_2. The smallest case, where $n = 2$ and $m = t = 0$, is contained in Theorem 22.6.5.

(3) $-1 \leq t \leq m - 1$

First, let $t = -1$. Then $\Pi_m^{(1)} \subset \mathscr{Q}_n$ and $\Pi_m^{(2)} \subset \mathscr{Q}_n$. For $m = 0$ there is nothing to prove, by Theorem 22.6.4. So assume $m > 0$. Let $P \in \Pi_m^{(1)} \cap \Pi_m^{(2)}$. By projection of $\mathscr{Q}_n$ onto a prime not containing P and by using induction on m as in (1), we see that there is an element $\mathfrak{T}$ of $G(\mathscr{Q}_n)$ which fixes P and maps $\Pi_m^{(1)}$ to $\Pi_m^{(2)}$. Now assume that $\Pi_m^{(1)} \cap \Pi_m^{(2)} = \Pi_{-1}$. If $P_1 \in \Pi_m^{(1)}$, then there is a point P_2 in $T_{P_1}(\mathcal{Q}_n) \cap \Pi_m^{(2)}$ since $m > 0$. Take $\Pi_m^{(3)} \subset \mathcal{Q}_n$ with $P_1P_2 \subset \Pi_m^{(3)}$. Then there exist $\mathfrak{T}_1$ and $\mathfrak{T}_2$ in $G(\mathcal{Q}_n)$ such that $\mathfrak{T}_1$ maps $\Pi_m^{(1)}$ to $\Pi_m^{(3)}$ and that $\mathfrak{T}_2$ maps $\Pi_m^{(2)}$ to $\Pi_m^{(3)}$. Hence $\mathfrak{T}_1\mathfrak{T}_2^{-1}$ maps $\Pi_m^{(1)}$ to $\Pi_m^{(2)}$.

Next, let $t \geq 0$. Define $\mathcal{S}_{\Pi_{m-t-1}}$ to be the set of all elements of $\mathcal{S}$ containing the subspace Π_{m-t-1} as vertex of a section of $\mathcal{Q}_n$ by an m-space, and let $G_{\Pi_{m-t-1}}$ be the subgroup of $G(\mathcal{Q}_n)$ fixing Π_{m-t-1}. Suppose that $G_{\Pi_{m-t-1}}$ has M orbits on $\mathcal{S}_{\Pi_{m-t-1}}$. Project $\mathcal{Q}_n$ from Π_{m-t-1} onto Π_{n-m+t}, with $\Pi_{m-t-1} \cap \Pi_{n-m+t} = \emptyset$. Then $\mathcal{Q}_n$ determines a quadric $\mathcal{Q}_{n-2(m-t)}$ of Π_{n-m+t}, whose equations we now determine.

Let $P_0, \ldots, P_{m-t-1}$ be linearly independent points of Π_{m-t-1}. Take $P_i = \mathbf{U}_i$ and the tangent prime $T_{P_i}(\mathcal{Q}_n) = \mathbf{u}_{m-t+i}$, for $i = 0, 1, \ldots, m-t-1$; then $\mathcal{Q}_n = \mathbf{V}(F)$ with

$$F = f(x_{2m-2t}, \ldots, x_n) + a_0x_0x_{m-t} + \cdots + a_{m-t-1}x_{m-t-1}x_{2m-2t-1}.$$

This gives

$$\mathcal{Q}_{n-2(m-t)} = \mathbf{V}(f, x_0, \ldots, x_{2m-2t-1}).$$

Let $\mathcal{Q}_t^{(1)}$ and $\mathcal{Q}_t^{(2)}$ be sections of $\mathcal{Q}_{n-2(m-t)}$ by subspaces $\Pi_t^{(1)}$ and $\Pi_t^{(2)}$ of $\Pi_{n-2(m-t)}$, where $\mathcal{Q}_t^{(1)}$ and $\mathcal{Q}_t^{(2)}$ are of character v and belong to the same orbit of $G(\mathcal{Q}_{n-2(m-t)})$. Let $\mathfrak{T}$ be an element of $G(\mathcal{Q}_{n-2(m-t)})$ mapping $\mathcal{Q}_t^{(1)}$ to $\mathcal{Q}_t^{(2)}$; it is therefore given by a linear transformation on $x_{2m-2t}, \ldots, x_n$ taking f to λf. Extending $\mathfrak{T}$ by $x_0 \to x_0, \ldots, x_{m-t-1} \to x_{m-t-1}, x_{m-t} \to \lambda x_{m-t}, \ldots, x_{2m-2t-1} \to \lambda x_{2m-2t-1}$ gives a projectivity fixing Π_{m-t-1} and $\mathcal{Q}_n$ as well as mapping $\Pi_{m-t-1}\mathcal{Q}_t^{(1)}$ to $\Pi_{m-t-1}\mathcal{Q}_t^{(2)}$. If M' is the number of orbits of $G(\mathcal{Q}_{n-2(m-t)})$ on the set of all sections $\mathcal{Q}_t$ of character v of $\mathcal{Q}_{n-2(m-t)}$, then it follows that $M \leq M'$. But, clearly, $M \geq M'$, and so $M = M'$.

Suppose that $\mathcal{W}_m^{(1)}$ and $\mathcal{W}_m^{(2)}$ are in $\mathcal{S}_{\Pi_{m-t-1}}$ and also in one orbit $\mathcal{O}$ of $G(\mathcal{Q}_n)$. Since Π_{m-t-1} is the vertex of $\mathcal{W}_m^{(1)}$ and $\mathcal{W}_m^{(2)}$ every element $\mathfrak{T}$ of $G(\mathcal{Q}_n)$ mapping $\mathcal{W}_m^{(1)}$ to $\mathcal{W}_m^{(2)}$ fixes Π_{m-t-1} and so is in $G_{\Pi_{m-t-1}}$. Hence $\mathcal{W}_m^{(2)}$ is in the orbit $\mathcal{O}_{\Pi_{m-t-1}}$ of $\mathcal{W}_m^{(1)}$ under $G_{\Pi_{m-t-1}}$. Since $G(\mathcal{Q}_n)$ acts transitively on the set of all $(m-t-1)$-spaces on $\mathcal{Q}_n$, it follows immediately that the number of orbits $\mathcal{O}$ of $\mathcal{S}$ under $G(\mathcal{Q}_n)$ is the same as the number M of orbits $\mathcal{O}_{\Pi_{m-t-1}}$ of $\mathcal{S}_{\Pi_{m-t-1}}$ under $G_{\Pi_{m-t-1}}$. Hence the number of orbits of $\mathcal{S}$ under $G(\mathcal{Q}_n)$ is the number of orbits of $G(\mathcal{Q}_{n-2(m-t)})$ on the set of all sections $\mathcal{Q}_t$ of character v of $\mathcal{Q}_{n-2(m-t)}$. When $n = 2m - t$, then $\mathcal{Q}_{n-2(m-t)} = \mathcal{Q}_t$ and so $\mathcal{S}$ has just one orbit under $G(\mathcal{Q}_n)$. When $n \neq 2m - t$, and so $n - 2(m-t) > t$, then the number of orbits of $G(\mathcal{Q}_{n-2(m-t)})$ on the set of all sections of character v was calculated in (1) and (2): one orbit when (i) n is odd or (ii) n is even and t is odd; two orbits when n is even and t is even. □

22.7 The polarity reconsidered

We now give the complete generalization of Lemmas 22.3.8 and 22.4.5, and describe sections of $\mathcal{Q}_n$ by subspaces Π and Π' which are polar under the polarity of $\mathcal{Q}_n$.

Let $\mathcal{Q}_n$ have character w and projective index g, and let a section $\Pi_{m-t-1}\mathcal{Q}_t$ have character v and projective index f.

Lemma 22.7.1: (i) $T \geq v - w$;
(ii) $T + w - v$ *is even.*

Proof: From (22.8) and (22.10),

$$g = \tfrac{1}{2}(n - 3 + w), \qquad f = m - \tfrac{1}{2}(t + 3 - v).$$

So $g - f = \frac{1}{2}(T + w - v)$ and (ii) follows. However, $g \geq f$ since $\Pi_{m-t-1}\mathcal{Q}_t$ lies on $\mathcal{Q}_n$; hence we obtain (i). □

Theorem 22.7.2: *Let* $\Pi = \Pi_m$ *have polar space* $\Pi' = \Pi'_{n-m-1}$ *with respect to* $\mathcal{Q}_n$. *Then the possibilities for* $\Pi \cap \mathcal{Q}_n = \Pi_k\mathcal{V}$ *and* $\Pi' \cap \mathcal{Q}_n = \Pi_k\mathcal{V}'$ *are listed in the following table.*

q	$\mathcal{Q}_n$	$\Pi_k\mathcal{V}$	$\Pi_k\mathcal{V}'$
All	$\mathcal{H}_n$	$\Pi_{m-t-1}\mathcal{H}_t$	$\Pi_{m-t-1}\mathcal{H}_{T-1}$
		$\Pi_{m-t-1}\mathcal{E}_t$	$\Pi_{m-t-1}\mathcal{E}_{T-1}$
		$\Pi_{m-t-1}\mathcal{P}_t$	$\Pi_{m-t-1}\mathcal{P}_{T-1}$
All	$\mathcal{E}_n$	$\Pi_{m-t-1}\mathcal{H}_t$	$\Pi_{m-t-1}\mathcal{E}_{T-1}$
		$\Pi_{m-t-1}\mathcal{E}_t$	$\Pi_{m-t-1}\mathcal{H}_{T-1}$
		$\Pi_{m-t-1}\mathcal{P}_t$	$\Pi_{m-t-1}\mathcal{P}_{T-1}$
Odd	$\mathcal{P}_n$	$\Pi_{m-t-1}\mathcal{H}_t$	$\Pi_{m-t-1}\mathcal{P}_{T-1}$
		$\Pi_{m-t-1}\mathcal{E}_t$	$\Pi_{m-t-1}\mathcal{P}_{T-1}$
		$\Pi_{m-t-1}\mathcal{P}_t$	$\left\{\begin{array}{l}\Pi_{m-t-1}\mathcal{H}_{T-1}\\ \Pi_{m-t-1}\mathcal{E}_{T-1}\end{array}\right.$

Proof: By the corollary to Lemma 22.3.6 and Lemma 22.4.5, $\Pi' \cap \mathcal{Q}_n$ has the same singular space Π_k as $\Pi \cap \mathcal{Q}_n$. So, if $\Pi' \cap \mathcal{Q}_n = \Pi_{m-t-1}\mathcal{Q}_s$, then

$$s = (n - m - 1) - (m - t - 1) - 1 = n + t - 2m - 1 = T - 1.$$

Now, it suffices to look at particular cases of each type of subspace, using the standard equations, as for a given trio m, t, and v, there are at most two orbits, by Theorem 22.6.5. □

Corollary: *In the theorem, let* $\mathcal{Q}_n$, $\mathcal{V}$, $\mathcal{V}'$ *have respective characters* w, v, v' *and respective projective indices* g, f, f'. *Then*

(i) $$v' = |2 - w - v|$$

unless $w = v = 1$, *in which case* $v' = 0$ *or* 2;

(ii) $$f + f' - g = k - 1 + \tfrac{1}{2}(v + v' - w).$$

Proof: (i) This follows from the theorem.

(ii) From (22.7) and (22.9),

$$w = 2g - n + 3$$
$$v = 2f - k - m + 2$$
$$v' = 2f' - k - (n - m - 1) + 2 = 2f' - n + m - k + 3.$$

Elimination of m and n gives the formula. □

To add something comparable to Theorem 22.7.2 for $\mathscr{P}_n$ with q even, the following result is available; it only repeats a particular case of Lemma 22.4.5(ii).

Lemma 22.7.3: *In the notation of Theorem 22.7.2, with tangent space replacing polar space, the following holds*:

q even	$\mathscr{Q}_n$	$\Pi_k\mathscr{V}$	$\Pi_k\mathscr{V}'$	
	$\mathscr{P}_n$	$\Pi_m\mathscr{H}_{-1}$	$\Pi_m\mathscr{P}_{n-2m-2}$	□

The next result gives more information on the tangency properties when $\mathscr{Q}_n$ does not have a polarity.

Theorem 22.7.4: *With q even, let N be the nucleus of the parabolic quadric $\mathscr{P}_n$.*

(i) *Every section of $\mathscr{P}_n$ through N is parabolic.*

(ii) *There is a bijection between m-spaces Π_m through the nucleus N of $\mathscr{P}_n = \mathbf{V}(x_0^2 + x_1x_2 + \cdots + x_{n-1}x_n)$ and $(m-1)$-spaces Π_{m-1} in $\mathbf{u}_0$*:

$$\Pi_m \to \Pi_{m-1} = \Pi_m \cap \mathbf{u}_0,$$

$$\Pi_{m-1} \to \Pi_m = \mathbf{U}_0\Pi_{m-1}.$$

Here, $\Pi_m \cap \mathscr{P}_n$ is a $\Pi_{m-t-1}\mathscr{P}_t$ with Π_m containing N if and only if, with $\mathscr{H}_{n-1} = \mathbf{u}_0 \cap \mathscr{P}_n$, $\Pi_{m-1} \cap \mathscr{H}_{n-1}$ is a $\Pi_{m-t-1}\mathscr{H}_{t-1}$ or a $\Pi_{m-t-1}\mathscr{E}_{t-1}$ or a $\Pi_{m-t-2}\mathscr{P}_t$.

Proof: If $\Pi_m \cap \mathscr{P}_n = \Pi_{m-t-1}\mathscr{P}_t$ and $\Pi_{m-1} = \Pi_m \cap \mathbf{u}_0$, then Π_{m-1} meets $\mathscr{H}_{n-1}$ in a section $\Pi_{m-s-1}\mathscr{Q}_{s-1}$, which we reduce to canonical form by a projectivity μ of $PG(n, q)$ fixing $\mathbf{u}_0$ and $\mathbf{U}_0$. Hence μ has the matrix

$$\begin{bmatrix} 1 & Z \\ Z^* & M \end{bmatrix},$$

where $Z = (0, 0, \ldots, 0)$ and M is an $n \times n$ matrix with no further restriction. So the possible canonical forms for $\mathcal{Q}_{s-1}$ and correspondingly $\mathcal{P}_t$ are given by the following table.

Form for $\mathcal{Q}_{s-1}$	$\Pi_{m-s-1}\mathcal{Q}_{s-1}$	Form for $\mathcal{P}_t$	$\Pi_{m-t-1}\mathcal{P}_t$
$x_1x_2 + \cdots + x_{s-1}x_s$	$\Pi_{m-s-1}\mathcal{H}_{s-1}$	$x_0^2 + x_1x_2 + \cdots + x_{s-1}x_s$	$\Pi_{m-s-1}\mathcal{P}_s$
$f(x_1, x_2) + \cdots + x_{s-1}x_s$	$\Pi_{m-s-1}\mathcal{E}_{s-1}$	$x_0^2 + f(x_1, x_2) + \cdots + x_{s-1}x_s$	$\Pi_{m-s-1}\mathcal{P}_s$
$x_1^2 + x_2x_3 + \cdots + x_{s-1}x_s$	$\Pi_{m-s-1}\mathcal{P}_{s-1}$	$(x_0 + x_1)^2 + x_2x_3 + \cdots + x_{s-1}x_s$	$\Pi_{m-s}\mathcal{P}_{s-1}$

This proves (ii) and so, *a fortiori*, (i). □

22.8 Sections of non-singular quadrics

As in §22.7, let $\mathcal{S}(m, t, v; n, w)$ be the set of m-spaces Π_m such that $\Pi_m \cap \mathcal{Q}_n$ is of type $\Pi_{m-t-1}\mathcal{Q}_t$, where $\mathcal{Q}_n$ and $\mathcal{Q}_t$ have characters w and v respectively. The size of this set is denoted by

$$N(m, t, v; n, w) \quad \text{or} \quad N(\Pi_{m-t-1}\mathcal{Q}_t, \mathcal{Q}_n)$$

and this number will be calculated. As special cases, the formula gives the number of Π_m lying on $\mathcal{Q}_n$ (here $\Pi_m \cap \mathcal{Q}_n = \Pi_m\mathcal{H}_{-1}$) as well as the number of points Π_0 not on $\mathcal{Q}_n$ (here $\Pi_0 \cap \mathcal{Q}_n = \Pi_{-1}\mathcal{P}_0$). The formula gives the size of the orbits when $G(\mathcal{Q}_n)$ operates on the lattice of subspaces of $PG(n, q)$, apart from the case $w = v = 1$. This fact is contained in Theorem 22.6.6: when $(w, v) \neq (1, 1)$, the sections of $\mathcal{Q}_n$ for a triple (m, t, v) form a single orbit under $G(\mathcal{Q}_n)$; when $(w, v) = (1, 1)$, the sections for a given pair (m, t) form one or two orbits, and the size of these orbits will also be determined.

To obtain the general result, some special cases are first required. These are, however, subsumed in the general result. In §22.5 we determined

$$N(m; n, w) = N(m, -1, 2; n, w),$$

which is the number of m-spaces on $\mathcal{Q}_n$ of character w.

The next special cases required are the numbers of bisecants, tangent lines, and skew lines to $\mathcal{Q}_n$; the total number of lines on $\mathcal{Q}_n$ has already been determined in §22.5.

Lemma 22.8.1:

(i) $N(\Pi_1, \mathscr{Q}_n) = [\tfrac{1}{2}(n-1-w), \tfrac{1}{2}(n+1-w)]_+$
$$\times [\tfrac{1}{2}(n-3+w), \tfrac{1}{2}(n-1+w)]_-/[1,2]_-; \quad (22.28)$$

(ii) $N(\Pi_0\mathscr{P}_0, \mathscr{Q}_n) = \{(q^n-1)/(q-1) + (w-1)q^{(n-1)/2}\}$
$$\times \{q^{n-2} - (w-1)q^{(n-3)/2}\}; \quad (22.29)$$

(iii) $N(\mathscr{H}_1, \mathscr{Q}_n) = \tfrac{1}{2}q^{n-1}\{(q^n-1)/(q-1) + (w-1)q^{(n-1)/2}\}; \quad (22.30)$

(iv) $N(\mathscr{E}_1, \mathscr{Q}_n) = \tfrac{1}{2}q^{n-1}(q^{\{n+(w-1)^2\}/2} - w^2 + w + 1)$
$$\times (q^{\{n-(w-1)^2\}/2} + w^2 - 3w + 1)/(q+1). \quad (22.31)$$

Proof: The number of lines on $\mathscr{Q}_n$ through a point P is $N(0; n-2, w)$, Lemma 22.3.8. Hence the number of tangents through P is $\theta(n-2) - N(0; n-2, w)$ and the total number of tangents is

$$N(0; n, w)\{\theta(n-2) - N(0; n-2, w)\};$$

this gives (ii).

The number of bisecants through a point P of $\mathscr{Q}_n$ is $\theta(n-1) - \theta(n-2) = q^{n-1}$; hence

$$N(\mathscr{H}_1, \mathscr{Q}_n) = N(0; n, w)q^{n-1}/2.$$

The total number of lines in $PG(n, q)$ is, from Theorem 3.1.1,

$$\phi(1; n, q) = [n, n+1]_-/[1,2]_-. \quad (22.32)$$

Hence (iv) is obtained from the formula

$$N(\mathscr{E}_1, \mathscr{Q}_n) = \phi(1; n, q) - N(\Pi_1, \mathscr{Q}_n) - N(\Pi_0\mathscr{P}_0, \mathscr{Q}_n) - N(\mathscr{H}_1, \mathscr{Q}_n). \quad \square$$

Corollary:

(i) $N(1, 1, 0; n, 0) = N(\mathscr{E}_1, \mathscr{E}_n)$
$$= \tfrac{1}{2}q^{n-1}(q^{(n+1)/2} + 1)(q^{(n-1)/2} + 1)/(q+1); \quad (22.33)$$

(ii) $N(1, 1, 0; n, 2) = N(\mathscr{E}_1, \mathscr{H}_n)$
$$= \tfrac{1}{2}q^{n-1}(q^{(n+1)/2} - 1)(q^{(n-1)/2} - 1)/(q+1); \quad (22.34)$$

(iii) $N(1, 1, 0; n, 1) = N(\mathscr{E}_1, \mathscr{P}_n) = \tfrac{1}{2}q^{n-1}(q^n - 1)/(q+1). \quad \square \quad (22.35)$

Theorem 22.8.2: *The number of sections* $\Pi_{m-t-1}\mathcal{Q}_t$ *of character* v *on* $\mathcal{Q}_n$ *of character* w *is*

$$N(m,t,v;\,n,w)$$
$$= q^{1/2\{T[t+1+vw(2-v)(2-w)]-v(2-v)(w-1)^2\}}$$
$$\times[\tfrac{1}{2}\{T+v+(1+3v-2v^2)w-v(2-v)w^2\},\tfrac{1}{2}(n+1-w)]_+$$
$$\times[\tfrac{1}{2}\{T+2-v-(1-5v+2v^2)w-v(2-v)w^2\},\tfrac{1}{2}(n-1+w)]_-$$
$$\div\{[v(2-v),\tfrac{1}{2}(t+1-v)]_+[1,\tfrac{1}{2}(t-1+v)]_-[1,m-t]_-\}, \qquad (22.36)$$

where $T = n + t - 2m$.

Proof: We consider the spaces Π_d lying on $\mathcal{Q}_n$ and the spaces Π_m meeting $\mathcal{Q}_n$ in a quadric $\Pi_k\mathcal{Q}_t$ of projective index d and character v. Let N_0 be the number of such Π_m through a Π_d. Also, let $N'(\Pi_d, \Pi_k\mathcal{Q}_t)$ be the number of Π_d on $\Pi_k\mathcal{Q}_t$ containing Π_k. Then, counting pairs (Π_d, Π_m) gives

$$N(\Pi_k\mathcal{Q}_t, \mathcal{Q}_n) = N_0 N(\Pi_d, \mathcal{Q}_n)/N'(\Pi_d, \Pi_k\mathcal{Q}_t). \qquad (22.37)$$

Here

$$m = k + t + 1.$$

From (22.9),

$$v = 2d - (m - t - 1) - m + 2,$$

whence

$$d = m - \tfrac{1}{2}(t - v + 3). \qquad (22.38)$$

To find N_0, we consider all $\mathcal{W}_m = \Pi_k\mathcal{Q}_t$ on $\mathcal{Q}_n$ through a particular Π_d. Let Π'_{n-d-1} be the tangent space of Π_d with respect to $\mathcal{Q}_n$. Then, by Lemma 22.4.5,

$$\Pi'_{n-d-1} \cap \mathcal{Q}_n = \Pi_d\mathcal{Q}_{n-2d-2},$$

which has the same character w and projective index g as $\mathcal{Q}_n$.

The required spaces Π_m must satisfy the following properties:

(a) $\mathcal{W}_m$ has projective index at most d;
(b) Π_m touches $\mathcal{Q}_n$ along Π_k.

For (a), it is necessary and sufficient that Π_m contains none of the $(d+1)$-spaces through Π_d on $\Pi_d\mathcal{Q}_{n-2d-2}$. If (a) is assumed, then it is necessary and sufficient for (b) that $\Pi_m\Pi'_{n-d-1} = \Pi'_{n-k-1}$, using Lemma 22.3.7. This is equivalent to $\Pi_m \cap \Pi'_{n-d-1} = \Pi_r$, where

$$r = m + (n - d - 1) - (n - k - 1) = m - d + k. \qquad (22.39)$$

The space Π_r contains Π_d and, to satisfy (a), $\Pi_r \cap \Pi_d\mathcal{Q}_{n-2d-2} = \Pi_d$. If $r > d$, so $\Pi_r \cap \mathcal{Q}_{n-2d-2}$ is either $\mathcal{P}_0$ or $\mathcal{E}_1$; hence

$$d \le r \le d + 2. \qquad (22.40)$$

We distinguish the three cases that $\mathscr{W}_m$ is parabolic, hyperbolic, or elliptic; that is, $v = 1, 2$, or 0 where $v = 2d - k - m + 2$, as in (22.9).

(1) $\mathscr{W}_m$ *parabolic*

Since $2 + 2d - k - m = 1$ and $r = m - d + k$, so $r = d + 1$. Therefore, by (a), Π_r is any one of the Π_{d+1} which lie in Π'_{n-d-1} and contain Π_d without being on $\Pi_d\mathscr{Q}_{n-2d-2}$. So the number of Π_r is, in the notation of §3.1,

$$\chi(d, d+1; n-d-1, q) - N(\Pi_0, \mathscr{Q}_{n-2d-2})$$
$$= \theta(n-2d-2) - N(\Pi_0, \mathscr{Q}_{n-2d-2}) = N_1. \quad (22.41)$$

The spaces Π_m are those m-spaces containing such a $\Pi_r = \Pi_{d+1}$ so that $\Pi_m \cap \Pi'_{n-d-1} = \Pi_r$. If Π_r is fixed, the number of these Π_m is

$$\begin{aligned}\psi_{12}(d+1, n-d-1, m; n, q)\\ &= q^{(m-d-1)(n-2d-2)}[2d+3-m, d+1]_-/[1, m-d-1]_-\\ &= N_2 \quad (22.42)\end{aligned}$$

by Theorem 3.1.2. Thus

$$\begin{aligned}N_0 &= N_1 N_2\\ &= q^{1/2\{T(t+1)-(w-1)^2\}}(q^{1/2\{T+(w-1)^2\}} - w + 1)\\ &\quad \times [m-t+1, m-\tfrac{1}{2}t]_-/[1, \tfrac{1}{2}t]_-, \quad (22.43)\end{aligned}$$

where N_1 has been evaluated using (22.16) and d has been eliminated by (22.38). Here $n - 2d - 2 = n - 2m + t - v + 3 - 2 = T$.

(2) $\mathscr{W}_m$ *hyperbolic*

Since $2 + 2d - k - m = 2$ and $r = m - d + k$, so $r = d$ and $\Pi_r = \Pi_d$. Thus the spaces Π_m are those m-spaces such that $\Pi_d = \Pi_m \cap \Pi'_{n-d-1}$. By Theorem 3.1.2, their number is

$$\begin{aligned}\psi_{12}(d, n-d-1, m; n, q)\\ &= q^{(m-d)(n-2d-1)}[2d+2-m, d+1]_-/[1, m-d]_-\\ &= q^{T(t+1)/2}[m-t+1, m-\tfrac{1}{2}(t-1)]_-/[1, \tfrac{1}{2}(t+1)]_-\\ &= N_0 \quad (22.44)\end{aligned}$$

using (22.38) with $v = 2$.

(3) $\mathscr{W}_m$ *elliptic*

Since $2 + 2d - k - m = 0$ and $r = m - d + k$, so $r = d + 2$. Then, from (a), Π_r is one of the spaces Π_{d+2} in Π'_{n-d-1} through the vertex Π_d of $\Pi_d\mathscr{Q}_{n-2d-2}$

but not containing any point of the base $\mathcal{Q}_{n-2d-2}$. So Π_{d+2} is the join of Π_d and a line external to $\mathcal{Q}_{n-2d-2}$. Thus the number of Π_r is the number of lines of $PG(n-2d-2, q)$ meeting $\mathcal{Q}_{n-2d-2}$ in some $\mathcal{E}_1$, that is $N(\mathcal{E}_1, \mathcal{Q}_{n-2d-2})$. Now, (22.38) with $v = 0$ gives

$$d = m - \tfrac{1}{2}(t+3),$$

whence

$$n - 2d - 2 = n - 2m + (t+3) - 2 = T + 1.$$

So, from (22.31),

$$\begin{aligned} N(\mathcal{E}_1, \mathcal{Q}_{n-2d-2}) &= \tfrac{1}{2}q^T(q^{\{T+1+(w-1)^2\}/2} - w^2 + w + 1) \\ &\qquad \times (q^{\{T+1-(w-1)^2\}/2} + w^2 - 3w + 1)/(q+1). \\ &= N_1. \end{aligned} \tag{22.45}$$

The spaces Π_m are those m-spaces meeting Π'_{n-d-1} in such a $\Pi_r = \Pi_{d+2}$. If Π_r is fixed, the number of Π_m is, by Theorem 3.1.2,

$$\begin{aligned} &\psi_{12}(d+2, n-d-1, m; n, q) \\ &\quad = q^{(m-d-2)(n-2d-3)}[2d+4-m, d+1]_-/[1, m-d-2]_- \\ &\quad = q^{T(t-1)/2}[m-t+1, m-\tfrac{1}{2}(t+1)]_-/[1, \tfrac{1}{2}(t-1)]_- \\ &\quad = N_2. \end{aligned} \tag{22.46}$$

Thus

$$\begin{aligned} N_0 &= N_1 N_2 \\ &= \tfrac{1}{2}q^{T(t+1)/2}[m-t+1, m-\tfrac{1}{2}(t+1)]_- N_3 \\ &\quad \div \{(q+1)[1, \tfrac{1}{2}(t-1)]_-\}, \end{aligned} \tag{22.47}$$

where

$$N_3 = \begin{cases} [\tfrac{1}{2}T, \tfrac{1}{2}T+1]_+ & \text{for } w = 0 \\ [\tfrac{1}{2}T, \tfrac{1}{2}T+1]_- & \text{for } w = 2 \\ [T, T+1]_- & \text{for } w = 1. \end{cases}$$

This completes the calculation of N_0 in the three cases $v = 1, 2, 0$.

To apply the formula (22.37), we require $N(\Pi_d, \mathcal{Q}_n)$ and $N'(\Pi_d, \Pi_k\mathcal{Q}_t)$. By (22.38) and the definition,

$$N(\Pi_d, \mathcal{Q}_n) = N(m - \tfrac{1}{2}(t - v + 3); n, w)$$

and

$$N'(\Pi_d, \Pi_k\mathcal{Q}_t) = N(\Pi_{d-k-1}, \mathcal{Q}_t) = N(\tfrac{1}{2}(t + v - 3); t, v).$$

Thus, using (22.11) in Theorem 22.5.1,

$$N(\Pi_d, \mathcal{Q}_n)/N'(\Pi_d, \Pi_k\mathcal{Q}_t)$$

$$= \frac{[\tfrac{1}{2}(T+4-w-v), \tfrac{1}{2}(n+1-w)]_+[\tfrac{1}{2}(T+2+w-v), \tfrac{1}{2}(n-1+w)]_-[1, \tfrac{1}{2}(t-1+v)]_-}{[2-v, \tfrac{1}{2}(t+1-v)]_+[1, \tfrac{1}{2}(t-1+v)]_-[1, m-\tfrac{1}{2}(t-v+1)]_-}$$

$$= N_4. \tag{22.48}$$

So (22.37) becomes

$$N(\Pi_k\mathcal{Q}_t, \mathcal{Q}_n) = N(m, t, v; n, w) = N_0N_4.$$

Thus, from (22.43) and (22.48) with $v = 1$,

$$N(m, t, 1; n, w) = q^{\{T(t+1+2w-w^2)-(w-1)^2\}/2}$$
$$\times \frac{[\tfrac{1}{2}(T+1+2w-w^2), \tfrac{1}{2}(n+1-w)]_+[\tfrac{1}{2}(T+1+2w-w^2), \tfrac{1}{2}(n-1+w)]_-}{[1, t/2]_+[1, t/2]_-[1, m-t]_-}. \tag{22.49}$$

From (22.44) and (22.48) with $v = 2$,

$$N(m, t, 2; n, w)$$
$$= q^{T(t+1)/2}\frac{[\tfrac{1}{2}(T+2-w), \tfrac{1}{2}(n+1-w)]_+[\tfrac{1}{2}(T+w), \tfrac{1}{2}(n-1+w)]_-}{[0, \tfrac{1}{2}(t-1)]_+[1, \tfrac{1}{2}(t+1)]_-[1, m-t]_-}. \tag{22.50}$$

From (22.47) and (22.48) with $v = 0$,

$$N(m, t, 0; n, w)$$
$$= q^{T(t+1)/2}\frac{[\tfrac{1}{2}(T+w), \tfrac{1}{2}(n+1-w)]_+[\tfrac{1}{2}(T+2-w), \tfrac{1}{2}(n-1+w)]_-}{[0, \tfrac{1}{2}(t+1)]_+[1, \tfrac{1}{2}(t-1)]_-[1, m-t]_-}. \tag{22.51}$$

The substitution of $v = 1, 2, 0$ in the 'big formula' (22.36) gives (22.49), (22.50), (22.51) respectively. Thus (22.36) is established. □

It may be noted that all special cases of (22.36) required for its proof are immediately retrievable from the general formula.

Example: The number of conics on a quadric $\mathcal{P}_n$, n even:

$$v = w = 1, \qquad m = t = 2, \qquad T = n-2,$$

$$N(\mathcal{P}_2, \mathcal{P}_n) = q^{2(n-2)}[\tfrac{1}{2}n, \tfrac{1}{2}n]_+[\tfrac{1}{2}n, \tfrac{1}{2}n]_-$$
$$\div \{[1, 1]_+[1, 1]_-[1, 0]_-\}$$
$$= q^{2(n-2)}(q^n - 1)/(q^2 - 1).$$

We now consider under what conditions on the parameters m, t, v, n, w, and $T = n + t - 2m$ the quadric $\mathcal{Q}_n$ of character w has a section $\Pi_{m-t-1}\mathcal{Q}_t$ of character v. First, we list the properties of the parameters that are contained within their definition:

(a) $n - w$ is odd;
(b) $t - v$ is odd;
(c) $T + w - v$ is even;
(d) $n > m \geq 0$;
(e) $m \geq t \geq 1 - v$;
(f) $n \geq (w-1)^2$.

In Lemma 22.7.1 it was also shown that $T \geq v - w$.

Theorem 22.8.3: *Subject to (a)–(f), a quadric $\mathcal{Q}_n$ of character w has a section $\Pi_{m-t-1}\mathcal{Q}_t$ of character v if and only if*

$$T \geq |w - v|.$$

Proof: $[r, s]_- = 0$, $r, s \in \mathbf{N}$, if and only if $r = 0$. So, from (22.36), the number $N(m, t, v; n, w) > 0$ if and only if

$$T + 2 - v - (1 - 5v + 2v^2)w - v(2 - v)w^2 > 0;$$

that is, $T > f(v, w)$ where $f(v, w)$ is given by the following table.

v	w: 0	1	2
0	−2	−1	0
1	−1	−2	−1
2	0	−1	−2

Since $T + w - v$ is even, the minimum value $g(v, w)$ of T is given as follows:

$$g(v, w) = f(v, w) + 1 \quad \text{if } f(v, w) + 1 - (w - v) \text{ is even}$$
$$g(v, w) = f(v, w) + 2 \quad \text{if } f(v, w) + 1 - (w - v) \text{ is odd}.$$

Hence $g(v, w)$ is given by the following table.

v	w: 0	1	2
0	0	1	2
1	1	0	1
2	2	1	0

Thus $g(v, w) = |w - v|$. □

Corollary 1: *The quadric $\mathscr{Q}_n$ of character w has a section $\Pi_{m-t-1}\mathscr{Q}_t$ of character v if and only if*

(i) *$n - w$ is odd and $n \geq (w-1)^2$;*
(ii) *$t - v$ is odd;*
(iii) *$n > m \geq t \geq \max(2m - n + |w - v|, 1 - v)$.* □

Corollary 2: *The quadric $\mathscr{Q}_n$ of character w has a section $\Pi_k\mathscr{Q}_t$ of character v if and only if*

(i) *$n - w$ is odd and $n \geq (w-1)^2$;*
(ii) *$t - v$ is odd;*
(iii) *$1 - v \leq t < n - k - 1$;*
(iv) *$-2 \leq 2k \leq n - t - |w - v| - 2$.* □

Theorem 22.8.4: *For a given $\mathscr{Q}_n$ the number of projectively distinct pairs $(\Pi_m, \Pi_{m-t-1}\mathscr{Q}_t)$ where $\Pi_m \cap \mathscr{Q}_n = \Pi_{m-t-1}\mathscr{Q}_t$ is as follows:*

	$\mathscr{Q}_n$		
Type of section	$\mathscr{H}_n$	$\mathscr{E}_n$	$\mathscr{P}_n$
Hyperbolic	$\frac{1}{8}(n^2-1)+n$	$\frac{1}{8}(n-1)(n+5)$	$\frac{1}{8}n(n+6)$
Elliptic	$\frac{1}{8}(n^2-1)$	$\frac{1}{8}(n-1)(n+5)$	$\frac{1}{8}n(n+2)$
Parabolic	$\frac{1}{8}(n+1)(n+3)$	$\frac{1}{8}(n+1)(n+3)$	$\frac{1}{8}n(n+6)$
Total number	$\frac{1}{8}(3n+1)(n+1)+n$	$\frac{1}{8}(3n+7)(n-1)+n$	$\frac{1}{8}n(3n+14)$

Proof: For each m such that $0 \leq m \leq n-1$, Corollary 1 to Theorem 22.8.3 allows us to count the values of t for which a section $\Pi_{m-t-1}\mathscr{Q}_t$ of character v exists. □.

By way of example, we list the different sections for the three quadrics $\mathscr{E}_5, \mathscr{P}_6, \mathscr{H}_7$.

	Hyperbolic	Elliptic	Parabolic
$\mathscr{E}_5$			
$m = 4$		$\Pi_0\mathscr{E}_3$	$\mathscr{P}_4$
$m = 3$	$\mathscr{H}_3$	$\mathscr{E}_3, \Pi_1\mathscr{E}_1$	$\Pi_0\mathscr{P}_2$
$m = 2$	$\Pi_0\mathscr{H}_1$	$\Pi_0\mathscr{E}_1$	$\mathscr{P}_2, \Pi_1\mathscr{P}_0$
$m = 1$	$\mathscr{H}_1, \Pi_1$	$\mathscr{E}_1$	$\Pi_0\mathscr{P}_0$
$m = 0$	Π_0		$\mathscr{P}_0$

	Hyperbolic	Elliptic	Parabolic
$\mathscr{P}_6$			
$m = 5$	$\mathscr{H}_5$	$\mathscr{E}_5$	$\Pi_0\mathscr{P}_4$
$m = 4$	$\Pi_0\mathscr{H}_3$	$\Pi_0\mathscr{E}_3$	$\mathscr{P}_4, \Pi_1\mathscr{P}_2$
$m = 3$	$\mathscr{H}_3, \Pi_1\mathscr{H}_1$	$\mathscr{E}_3, \Pi_1\mathscr{E}_1$	$\Pi_0\mathscr{P}_2, \Pi_2\mathscr{P}_0$
$m = 2$	$\Pi_0\mathscr{H}_1, \Pi_2$	$\Pi_0\mathscr{E}_1$	$\mathscr{P}_2, \Pi_1\mathscr{P}_0$
$m = 1$	$\mathscr{H}_1, \Pi_1$	$\mathscr{E}_1$	$\Pi_0\mathscr{P}_0$
$m = 0$	Π_0		$\mathscr{P}_0$
$\mathscr{H}_7$			
$m = 6$	$\Pi_0\mathscr{H}_5$		$\mathscr{P}_6$
$m = 5$	$\mathscr{H}_5, \Pi_1\mathscr{H}_3$	$\mathscr{E}_5$	$\Pi_0\mathscr{P}_4$
$m = 4$	$\Pi_0\mathscr{H}_3, \Pi_2\mathscr{H}_1$	$\Pi_0\mathscr{E}_3$	$\mathscr{P}_4, \Pi_1\mathscr{P}_2$
$m = 3$	$\mathscr{H}_3, \Pi_1\mathscr{H}_1, \Pi_3$	$\mathscr{E}_3, \Pi_1\mathscr{E}_1$	$\Pi_0\mathscr{P}_2, \Pi_2\mathscr{P}_0$
$m = 2$	$\Pi_0\mathscr{H}_1, \Pi_2$	$\Pi_0\mathscr{E}_1$	$\mathscr{P}_2, \Pi_1\mathscr{P}_0$
$m = 1$	$\mathscr{H}_1, \Pi_1$	$\mathscr{E}_1$	$\Pi_0\mathscr{P}_0$
$m = 0$	Π_0		$\mathscr{P}_0$

Since the sections of non-singular quadrics have been determined, it is possible to say precisely what are the sections of a singular quadric.

Theorem 22.8.5: *If, for a fixed t, the non-singular quadric $\mathscr{Q}_n$ has a section $\Pi_i\mathscr{Q}_t$ of character v, then $\Pi_k\mathscr{Q}_n$ has a section $\Pi_j\mathscr{Q}_t$ of character v for all j with* $i \leq j \leq i + k + 1$.

Proof: A section of $\Pi_k\mathscr{Q}_n$ is the join of a section Π_s of Π_k to a section $\Pi_i\mathscr{Q}_t$ of $\mathscr{Q}_n$; this join is $\Pi_{s+1+i}\mathscr{Q}_t$, where s may vary from -1 to k. □

22.9 Parabolic sections of parabolic quadrics

Part of Theorem 22.6.6 is that, in the operation of $G(\mathscr{Q}_n)$ on the set $\mathscr{S}(m, t, v; n, w)$ of m-spaces Π_m meeting $\mathscr{Q}_n$ of character w in a section $\Pi_{m-t-1}\mathscr{Q}_t$ of character v, the only case that two orbits may occur is when $v = w = 1$. The geometrical explanation of this phenomenon is different for q odd and q even. From (22.36),

$$N(m, t, 1; n, 1) = q^{T(t+2)/2}[\tfrac{1}{2}(T + 2), \tfrac{1}{2}n]_+ \times [\tfrac{1}{2}(T + 2), \tfrac{1}{2}n]_- \div \{[1, \tfrac{1}{2}t]_+[1, \tfrac{1}{2}t]_-[1, m - t]_-\} \quad (22.52)$$

where, as before,

$$T = n + t - 2m.$$

This is the number of Π_m such that

$$\Pi_m \cap \mathscr{P}_n = \Pi_{m-t-1}\mathscr{P}_t. \quad (22.53)$$

In this section we only consider such Π_m.

For q odd, as in Theorem 22.7.2, the polar Π'_{n-m-1} of Π_m meets $\mathscr{P}_n$ in either an elliptic or a hyperbolic section. If $\Pi'_{n-m-1} \cap \mathscr{P}_n = \Pi_{m-t-1}\mathscr{E}_{T-1}$, then Π_m is *internal*; if $\Pi'_{n-m-1} \cap \mathscr{P}_n = \Pi_{m-t-1}\mathscr{H}_{T-1}$, then Π_m is *external*. This conforms with the notion of internal and external points of a conic as in §8.2. Accordingly, we write $N_-(m, t, 1; n, 1)$ for the number of internal Π_m and $N_+(m, t, 1; n, 1)$ for the number of external Π_m such that (22.53) holds. Hence

$$N_-(m, t, 1; n, 1) + N_+(m, t, 1; n, 1) = N(m, t, 1; n, 1). \qquad (22.54)$$

Theorem 22.9.1:

(i) $N_-(m, t, 1; n, 1) = q^{T(t+1)/2}[\tfrac{1}{2}(T+2), \tfrac{1}{2}n]_+[\tfrac{1}{2}T, \tfrac{1}{2}n]_-$

$$\div \{[0, \tfrac{1}{2}t]_+[1, \tfrac{1}{2}t]_-[1, m-t]_-\}; \qquad (22.55)$$

(ii) $N_+(m, t, 1; n, 1) = q^{T(t+1)/2}[\tfrac{1}{2}T, \tfrac{1}{2}n]_+[\tfrac{1}{2}(T+2), \tfrac{1}{2}n]_-$

$$\div \{[0, \tfrac{1}{2}t]_+[1, \tfrac{1}{2}t]_-[1, m-t]_-\}. \qquad (22.56)$$

Proof: By definition,

$$N_-(m, t, 1; n, 1) = N(n-m-1, T-1, 0; n, 1),$$

$$N_+(m, t, 1; n, 1) = N(n-m-1, T-1, 2; n, 1).$$

Application of (22.36) and some manipulation give the required answers, for which (22.52) and (22.54) provide a check. □

Corollary: *For q odd, the set $\mathscr{S}(m, t, 1; n, 1)$ has one or two orbits under $G(\mathscr{P}_n)$ according as $T = 0$ or $T > 0$. In the former case, the sections of the orbit are all external and have polar sections $\Pi'_{n-m-1} \cap \mathscr{P}_n = \Pi_r$, where $r = n - m - 1 = m - t - 1$.*

Proof: This follows from Theorem 26.6.6, (22.55), and (22.56). When $T = 0$, $\Pi'_{n-m-1} \cap \mathscr{P}_n = \Pi_{m-t-1}\mathscr{H}_{T-1} = \Pi_r$. □

For q even, a space Π_m such that (22.53) holds either does or does not contain the nucleus N of $\mathscr{P}_n$. If Π_m does contain N, it is called *nuclear*; if Π_m does not contain N, it is called *non-nuclear*. Accordingly we write $N_0(m, t, 1; n, 1)$ for the number of nuclear Π_m and $N_1(m, t, 1; n, 1)$ for the number of non-nuclear Π_m such that $\Pi_m \cap \mathscr{P}_n = \Pi_{m-t-1}\mathscr{P}_t$. So

$$N_0(m, t, 1; n, 1) + N_1(m, t, 1; n, 1) = N(m, t, 1; n, 1). \qquad (22.57)$$

Theorem 22.9.2:

(i) $N_0(m, t, 1; n, 1) = q^{tT/2}[\frac{1}{2}(T+2), \frac{1}{2}n]_+$

$$\times [\tfrac{1}{2}(T+2), \tfrac{1}{2}n]_- \div \{[1, \tfrac{1}{2}t]_+[1, \tfrac{1}{2}t]_-[1, m-t]_-\}. \quad (22.58)$$

(ii) $N_1(m, t, 1; n, 1) = q^{tT/2}[\frac{1}{2}T, \frac{1}{2}n]_+[\frac{1}{2}T, \frac{1}{2}n]_-$

$$\div \{[1, \tfrac{1}{2}t]_+[1, \tfrac{1}{2}t]_-[1, m-t]_-\}. \quad (22.59)$$

Proof: From Theorem 22.7.4(ii),

$$\begin{aligned} N_0(m, t, 1; n, 1) &= N(\Pi_{m-t-1}\mathscr{H}_{t-1}, \mathscr{H}_{n-1}) + N(\Pi_{m-t-1}\mathscr{E}_{t-1}, \mathscr{H}_{n-1}) \\ &\quad + N(\Pi_{m-t-2}\mathscr{P}_t, \mathscr{H}_{n-1}) \\ &= N(m-1, t-1, 2; n-1, 2) + N(m-1, t-1, 0; n-1, 2) \\ &\quad + N(m-1, t, 1; n-1, 2). \end{aligned}$$

Applying (22.36) gives (22.58); then (22.57) gives (22.59). □

Corollary: *For q even, the set $\mathscr{S}(m, t, 1; n, 1)$ has one or two orbits under $G(\mathscr{P}_n)$ according as $T = 0$ or $T > 0$. In the former case, the sections of the orbit are all nuclear and each is the section of $\mathscr{P}_n$ by the tangent space of a Π_{m-t-1} lying on $\mathscr{P}_n$.*

Proof: This follows from Theorem 22.6.6, (22.58), and (22.59). When $T = 0$, $n - 2(m - t - 1) - 2 = t$. So, by Lemma 22.7.3, the tangent space of a Π_{m-t-1} on $\mathscr{P}_n$ meets $\mathscr{P}_n$ in a section $\Pi_{m-t-1}\mathscr{P}_t$. □

The corollaries to Theorems 22.9.1 and 22.9.2 for q odd and even respectively can be combined, as follows.

Theorem 22.9.3: *The set $\mathscr{S}(m, t, 1; n, 1)$ has one or two orbits under $G(\mathscr{P}_n)$ according as $T = 0$ or $T > 0$. In the former case, each element of the orbit is the section of $\mathscr{P}_n$ by the tangent space at a Π_{m-t-1} lying on $\mathscr{P}_n$.* □

Example: *Orbits of parabolic sections of $\mathscr{P}_4$.* The five types of parabolic section of $\mathscr{P}_4$ are as follows:

(a) $\mathscr{P}_0$, $m = 0$, $T = 4$, a point off $\mathscr{P}_4$;
(b) $\Pi_0\mathscr{P}_0$, $m = 1$, $T = 2$, a (tangent) line meeting $\mathscr{P}_4$ in a point;
(c) $\Pi_1\mathscr{P}_0$, $m = 2$, $T = 0$, a plane meeting $\mathscr{P}_4$ in a line;
(d) $\mathscr{P}_2$, $m = 2$, $T = 2$, a plane meeting $\mathscr{P}_4$ in a conic;
(e) $\Pi_0\mathscr{P}_2$, $m = 3$, $T = 0$, a (tangent) solid meeting $\mathscr{P}_4$ in a cone.

In both cases (c) and (e), for q odd or even, there is a single orbit of size $(q+1)(q^2+1)$. In case (a), for q odd, there are $\frac{1}{2}q^2(q^2-1)$ internal points and $\frac{1}{2}q^2(q^2 + 1)$ external points; for q even, there is the nucleus and $q^4 - 1$ non-nuclear points. In case (b), for q odd, there are $\frac{1}{2}q(q^4 - 1)$ internal tangents and $\frac{1}{2}q(q + 1)^2(q^2 + 1)$ external tangents; for q even, there are $(q + 1)(q^2 + 1)$ nuclear tangents and $(q + 1)(q^4 - 1)$ non-nuclear tangents. In case (d), for q odd, there are $\frac{1}{2}q^3(q - 1)(q^2 + 1)$ internal conics and $\frac{1}{2}q^3(q + 1)(q^2 + 1)$ external conics; for q even, there are $q^2(q^2 + 1)$ nuclear conics and $q^2(q^4 - 1)$ non-nuclear conics.

22.10 The characterization of quadrics

In this section, non-singular quadrics in $\Sigma = PG(n, q)$ are characterized purely in terms of their intersections with lines of Σ. The characterization also applies to infinite fields with only slight rewording. First, we require a number of definitions.

(1) A set $\mathscr{K}$ in Σ is of *type* $(r_1, r_2, \ldots, r_s)$ if $|l \cap \mathscr{K}| \in \{r_1, r_2, \ldots, r_s\}$ for all lines l.

(2) Let $\mathscr{K}$ be a set of type $(0, 1, 2, q + 1)$. A line meeting $\mathscr{K}$ in i points is an *i-secant*. The alternative terms for a 0-secant, 1-secant, 2-secant, and $(q + 1)$-secant are *external line*, *unisecant*, *bisecant*, and *line on* (or *of*) $\mathscr{K}$; *tangent* is also used for *unisecant* in this treatise. Some authors use 'tangent' to mean 1-secant or $(q + 1)$-secant, but we shall avoid this usage. However, in the context of this section, it is convenient to have a single term for this idea. So a *B-line* is defined to be a 1-secant or a $(q + 1)$-secant.

(3) $\mathscr{K}$ is a *quadratic* set if

(a) $\mathscr{K}$ is of type $(0, 1, 2, q + 1)$;
(b) for each P in $\mathscr{K}$, the union of B-lines through P together with P form the *tangent space* $T_P = T_P(\mathscr{K})$ which is either a prime or Σ itself.

In (b), if P is not specifically included in $T_P(\mathscr{K})$, there is a difficulty when $n = 1$ and $\mathscr{K}$ consists of two points.

(4) The point P of the quadratic set $\mathscr{K}$ is *singular* if $T_P = \Sigma$; in other words, there is no bisecant through P. If $\mathscr{K}$ has a singular point it is *singular*.

(5) If a quadratic set does not contain a line, it is an *ovoid*.

(6) A *perspectivity* of Σ is a projectivity fixing all lines through a certain point P_0, the *centre*. A quadratic set is *perspective* if there is a non-identity perspectivity with centre Q fixing $\mathscr{K}$ for every Q in Σ not in $\mathscr{K} \cup \bigcap T_P$. Every non-singular quadric in Σ is perspective, by Lemma 22.6.3.

Lemma 22.10.1: *Let $\mathscr{K}$ be a quadratic set and Π_s a subspace of Σ. Then $\mathscr{K}' = \mathscr{K} \cap \Pi_s$ is a quadratic set in Π_s for which $T_P(\mathscr{K}') = T_P(\mathscr{K}) \cap \Pi_s$, where P is any point of $\mathscr{K}'$.*

Proof: If l is a line of Σ, then it meets Π_s in 0, 1, or $q+1$ points. Hence there are the following possibilities for $|(l \cap \Pi_s) \cap \mathscr{K}'|$:

$|l \cap \Pi_s|$	$|l \cap \mathscr{K}|$: 0	1	2	$q+1$
0	0	0	0	0
1	0	0, 1	0, 1	0, 1
$q+1$	0	1	2	$q+1$

So $\mathscr{K}'$ is a quadratic set. The other part follows similarly. □

Corollary: *If Π is a prime and $\mathscr{K}$ is non-singular, then $\Pi \cap \mathscr{K}$ has a singular point P if and only if $\Pi = T_P(\mathscr{K})$.* □

Theorem 22.10.2: *In $\Sigma = PG(n, q)$, $n \geq 2$, a set $\mathscr{K}$ is a quadratic set if and only if each plane section is a quadratic set.*

Proof: One implication is included in Lemma 22.10.1. Suppose therefore that every plane section of $\mathscr{K}$ is quadratic. Let l be a line and π a plane containing l. Since $\mathscr{K} \cap l = (\mathscr{K} \cap \pi) \cap l$, it follows that $\mathscr{K}$ is of type $(0, 1, 2, q+1)$. Further, l is a B-line of $\mathscr{K}$ if and only if l is a B-line of $\mathscr{K} \cap \pi$ for every plane π containing l.

For each P in $\mathscr{K}$, let the union of the B-lines through P be T_P. If l_1 and l_2 are two of these B-lines, let $\pi = l_1 l_2$; then $\pi \cap \mathscr{K}$ is a quadratic set whose tangent prime at P is π. So π is contained in T_P and T_P is a subspace. As each plane containing P has a line in T_P, it follows that T_P is a prime or the whole of Σ. □

Some properties of singular quadratic sets are now developed.

Theorem 22.10.3: *The set of singular points of a quadratic set is a subspace.*

Proof: Let P, Q be distinct singular points of the quadratic set $\mathscr{K}$ and let R be any point of PQ. As PQ is a B-line at P, so $PQ \subset \mathscr{K}$ and $R \in \mathscr{K}$. Let l be any line through R. It must be shown that l is a B-line. We may take $R \neq P$ and $l \neq PQ$. If l contains a point S in $\mathscr{K}$ with $S \neq R$, the tangent prime T_S contains P and Q since PS and QS are lines of $\mathscr{K}$. So T_S contains the line PQ and the point R. Hence $RS = l$ is a B-line and lies in $\mathscr{K}$. So R is singular. □

The next result shows that the theory of quadratic sets is entirely dependent on the theory of non-singular ones. The structure is similar to quadrics.

Theorem 22.10.4: *If $\mathcal{K}$ is a quadratic set, then $\mathcal{K}$ is a cone $\Pi_s\mathcal{K}'$, where Π_s is the subspace of singular points of $\mathcal{K}$ and $\mathcal{K}'$ is a non-singular quadratic set in a subspace Π_{n-s-1} disjoint to Π_s.*

Proof: Let Π_{n-s-1} be any subspace disjoint from Π_s and let $\mathcal{K}' = \mathcal{K} \cap \Pi_{n-s-1}$. If $\mathcal{K}'$ has a singular point P, then $\Pi_{n-s-1} \subset T_P(\mathcal{K})$, by Lemma 22.10.1. As Π_s also lies in T_P, so $T_P = \Sigma$ and $P \in \Pi_s$, a contradiction. If $Q \in \mathcal{K}\backslash(\Pi_s \cup \Pi_{n-s-1})$, then $Q\Pi_s \cap \Pi_{n-s-1} \neq \emptyset$. So there is a line P_0P_1 with P_0 in Π_s, P_1 in Π_{n-s-1}, and Q in P_0P_1. So $P_1 \in \mathcal{K}'$, $Q \in \Pi_s\mathcal{K}'$, and hence $\mathcal{K} \subset \Pi_s\mathcal{K}'$. However, every point of $\Pi_s\mathcal{K}'$ is also in $\mathcal{K}$. □

Theorem 22.10.5: *If $\mathcal{K}$ is a quadratic set in $\Sigma = PG(n, q)$, $n \geq 2$, then every plane section of $\mathcal{K}$ is singular or empty if and only if $\mathcal{K}$ is a subspace or the union of two primes.*

Proof: In Σ, if $\mathcal{K}$ is a subspace, then for any plane Π_2, the intersection with $\mathcal{K}$ is Π_2, Π_1, Π_0, or Π_{-1}; if $\mathcal{K} = \Pi_{n-1} \cup \Pi'_{n-1}$, then $\mathcal{K} \cap \Pi_2 = \Pi_2$ or $\Pi_1 \cup \Pi'_1$. So, in both cases, a plane section of $\mathcal{K}$ is singular or empty.

To prove the converse, suppose $\mathcal{K}$ is not a subspace. So there are points P and P' with PP' not contained in $\mathcal{K}$. Each plane π through PP' meets $\mathcal{K}$ in two lines l and l' with P on l and P' on l'. The line l is the only B-line of $\mathcal{K}$ through P in π. Hence the tangent space of $\mathcal{K}$ at P is the prime $\Pi_{n-1} = \bigcup_\pi l$, where the union is taken over all planes π. Similarly the tangent space of $\mathcal{K}$ at P' is the prime $\Pi'_{n-1} = \bigcup_\pi l'$. Since $\mathcal{K} = \bigcup (l \cup l')$, so $\mathcal{K} = \Pi_{n-1} \cup \Pi'_{n-1}$. □

Theorem 22.10.6: *If $\mathcal{K}$ is a quadratic set in Σ other than a subspace, then the smallest subspace containing $\mathcal{K}$ is Σ.*

Proof: $\mathcal{K}$ must have at least one non-singular point P, whence T_P is a prime. So every line through P not in T_P contains a second point of $\mathcal{K}$. Thus any subspace containing $\mathcal{K}$ contains every line not in T_P and so must be Σ. □

A *generator* of $\mathcal{K}$ is a subspace contained in $\mathcal{K}$ and maximal with respect to inclusion. Any subspace contained in $\mathcal{K}$ will be referred to as a *subgenerator* of $\mathcal{K}$.

Theorem 22.10.7: *Let $\mathcal{K}'$ be a subset of the quadratic set $\mathcal{K}$ such that $PQ \subset \mathcal{K}$ for all P and Q in $\mathcal{K}'$. Then the subspace spanned by $\mathcal{K}'$ is a subgenerator of $\mathcal{K}$.*

Proof: We proceed by induction on $m = |\mathcal{K}'|$. The result is immediate for $m = 0, 1, 2$. So, let $m > 2$, let $P \in \mathcal{K}'$, and let Π_s be the subspace spanned by

$\mathscr{K}'\backslash\{P\}$. By the induction hypothesis, Π_s is a subgenerator of $\mathscr{K}$. Every point Q of the subspace $P\Pi_s$ spanned by $\mathscr{K}'$ lies on a line PP' with P' in Π_s. The tangent prime T_P contains $\mathscr{K}'\backslash\{P\}$ and hence Π_s. So every line PP' is a B-line at P and, since it contains two points of $\mathscr{K}$, lies in $\mathscr{K}$; thus $Q \in \mathscr{K}$. □

Theorem 22.10.8: *Let $\mathscr{K}$ be a quadratic set with a subgenerator Π and a point P of $\mathscr{K}$ such that $P\Pi$ is not a subgenerator. Then*

(i) *the union Π_P of the lines of $\mathscr{K}$ through P and a point Q of Π is a subgenerator*;

(ii) $\Pi_P \cap \Pi = T_P \cap \Pi$ *and this subspace is a prime in Π and in Π_P*;

(iii) $\dim \Pi_P = \dim \Pi$;

(iv) *if Π is a generator of $\mathscr{K}$, so is Π_P.*

Proof: First, $\Pi_P = P\Pi \cap T_P$. So Π_P is a subspace and $\Pi_P \cap \Pi = T_P \cap \Pi$. Further, Π_P is a subgenerator by Theorem 22.10.7, where $\mathscr{K}' = (\Pi_P \cap \Pi) \cup \{P\}$ in the notation used there. As Π_P cannot contain Π, so $\Pi_P \cap \Pi$ is a prime of Π since T_P is a prime of Σ. Hence $P(\Pi_P \cap \Pi) = \Pi_P$ is a prime of $P\Pi$ and $\Pi_P \cap \Pi$ is a prime of Π_P. This proves (i) and (ii); part (iii) now follows.

Suppose Π_P is a generator and Π is not. Then Π is properly contained in a generator Π'. Hence, by (ii), $\Pi'_P \cap \Pi'$ is a prime in Π'. So there is a line of $\mathscr{K}$ through P not in Π_P, whence Π_P is not a generator. Thus, if Π_P is a generator, so is Π. The converse result (iv) follows by symmetry. For, let Q be a point of $\Pi\backslash\Pi_P$; then $Q\Pi_P$ is not a subgenerator, unless $P\Pi$ is. As $(\Pi_P)_Q = \Pi$, the result is proved. □

Theorem 22.10.9: *The generators of a quadratic set $\mathscr{K}$ all have the same dimension.*

Proof: Suppose Π and Π' are generators of respective dimensions r and r' with $r < r'$. Let $B = \{P_0, \ldots, P_r\}$ be a generating set for Π. For each P_i, let Π'_i be the union of lines of $\mathscr{K}$ joining P_i to a point of Π'. Then $\Pi'_i \cap \Pi'$ is either a prime of Π' or Π' itself, by Theorem 22.10.8. As $r < r'$, the intersection of the Π'_i is non-empty since its codimension in Π' is at most r. Let P be in this intersection. Then $P\Pi$ is a subgenerator by Theorem 22.10.7 and so Π is not a generator. □

As for quadrics, the dimension g of a generator of $\mathscr{K}$ is called the *projective index* of $\mathscr{K}$. It should be noted that this is one less than the *Witt index.*

Lemma 22.10.10: *If Π is a subgenerator of a non-singular quadratic set $\mathscr{K}$, then there exists a generator disjoint from Π.*

Proof: If Π' is a generator meeting Π and $\dim(\Pi' \cap \Pi) = j$, it suffices to show that there exists a generator Π'' such that $\dim(\Pi'' \cap \Pi) = j - 1$. There exists a point P in $\mathscr{K}$ such that $P(\Pi \cap \Pi')$ is not a subgenerator; otherwise every point of $\Pi \cap \Pi'$ would be singular for $\mathscr{K}$. Hence $(\Pi \cap \Pi')_P$ is a subgenerator whose intersection with $\Pi \cap \Pi'$ has dimension $j - 1$, by Theorem 22.10.8. By the same result Π'_P is a generator. Clearly, Π'_P contains $(\Pi \cap \Pi')_P$. Since $P(\Pi \cap \Pi')$ is not a subgenerator,

$$\Pi \cap \Pi' \not\subset \Pi'_P. \tag{22.60}$$

So

$$(\Pi \cap \Pi')_P \cap (\Pi \cap \Pi') = \Pi'_P \cap (\Pi \cap \Pi'). \tag{22.61}$$

It will now be shown that

$$\Pi'_P \cap \Pi \cap \Pi' = \Pi \cap \Pi'_P. \tag{22.62}$$

Assume on the contrary that there is a point Q of $\Pi \cap \Pi'_P$ not in Π'. Let Ω be the union of $\Pi \cap \Pi'$, $\Pi'_P \cap \Pi'$, and $\{Q\}$. The join of any two points of Ω is a subgenerator. By Theorem 22.10.7, Ω spans a subgenerator $\mathscr{K}'$. Since $\mathscr{K}'$ contains the prime $\Pi'_P \cap \Pi'$ of Π'_P and the point Q of Π'_P, so $\Pi'_P \subset \mathscr{K}'$. Since Π'_P is a generator, so $\Pi'_P = \mathscr{K}'$. Hence $\Pi'_P = \mathscr{K}' \supset \Pi \cap \Pi'$, contradicting (22.60). So (22.62) holds. Now, from (22.61),

$$(\Pi \cap \Pi')_P \cap (\Pi \cap \Pi') = \Pi \cap \Pi'_P.$$

Since $\dim((\Pi \cap \Pi')_P \cap (\Pi \cap \Pi')) = j - 1$, so $\dim(\Pi \cap \Pi'_P) = j - 1$. Thus there is a generator $\Pi'' = \Pi'_P$ with $\dim(\Pi \cap \Pi'') = j - 1$. □

Corollary: *If $\mathscr{K}$ is a non-singular quadratic set in $PG(n, q)$, its projective index g satisfies $2g \leq n - 1$.*

Proof: Let Π and Π' be disjoint generators of $\mathscr{K}$; then $\dim \Pi\Pi' \leq n$. So $2g = \dim \Pi + \dim \Pi' = \dim \Pi\Pi' + \dim(\Pi \cap \Pi') \leq n - 1$. □

Lemma 22.10.11: *If Π is a subgenerator of the non-singular quadratic set $\mathscr{K}$ and Π is contained in the generator Γ, then there exists a generator Γ' such that $\Gamma \cap \Gamma' = \Pi$.*

Proof: This is by induction on $\dim \Pi = m$. If $m = -1$, the result is that of the previous lemma. Suppose now that the property is satisfied for dimension $m - 1$, $m \geq 0$. Let Π' be a prime of Π and let P be a point of $\Pi \backslash \Pi'$. There exists a generator Γ_1 such that $\Gamma \cap \Gamma_1 = \Pi'$ since $\dim \Pi' = m - 1$. Then $\Gamma' = (\Gamma_1)_P$ is the required generator. □

Theorem 22.10.12: *Every subgenerator of a non-singular quadratic set $\mathscr{K}$ is the intersection of the generators containing it.*

Proof: This follows from the preceding lemma. □

Lemma 22.10.13: *If $\mathscr{K}$ is a quadratic set in $PG(n, q)$, $n \geq 2$, which is not a subspace, then any collineation σ fixing every point of $\mathscr{K}$ is the identity.*

Proof: Since $\mathscr{K}$ is not a subspace, it contains non-singular points P and Q. Hence σ fixes every line through P other than a tangent line. It follows that σ also fixes these tangent lines. So σ is a perspectivity with centre P. Similarly, it is a perspectivity with centre Q. If R is any point not on the line PQ, the lines RP and RQ are fixed. So R is fixed and σ is the identity. □

Lemma 22.10.14: *If $\mathscr{K}$ is a quadratic set in $\Sigma = PG(n, q)$, $n \geq 2$, and P in $\Sigma \backslash \mathscr{K}$ is a point not lying on all tangent primes, then there is at most one perspectivity other than the identity with centre P which fixes $\mathscr{K}$.*

Proof: Since P exists, $\mathscr{K}$ is not a subspace. Let σ and σ' be perspectivities with centre P fixing $\mathscr{K}$. Let P_1P_2 be a bisecant of $\mathscr{K}$ through P with P_1, P_2 in $\mathscr{K}$. First, let $P_i\sigma' = P_i$, $i = 1, 2$. Then σ' fixes both T_{P_1} and T_{P_2}, whence σ' is the identity, a contradiction. So $P_1\sigma' = P_2$ and $P_2\sigma' = P_1$. Analogously, $P_1\sigma = P_2$ and $P_2\sigma = P_1$. So $\sigma^{-1}\sigma'$ is a perspectivity with centre P fixing $\mathscr{K}$, P_1, and P_2. Again, $\sigma^{-1}\sigma'$ is the identity, implying that $\sigma = \sigma'$. □

Theorem 22.10.15: *If $\mathscr{K}$ is a quadratic set other than a subspace and every plane section of $\mathscr{K}$ is the empty set, a point, a line, or a conic, then $\mathscr{K}$ is a quadric.*

Proof: This is by induction on the dimension n. The result is in the hypotheses when $n = 1$ or 2. So let $n \geq 3$. Let $P'Q' \cap \mathscr{K} = \{P', Q'\}$. A prime Π through $P'Q'$ therefore does not meet $\mathscr{K}$ in a subspace. So, by induction, $\Pi \cap \mathscr{K}$ is a quadric.

If all points of $\mathscr{K}$ outside Π were singular, they would be contained in a subspace Π'. So $\mathscr{K} \subset \Pi \cup \Pi'$, whence $\mathscr{K}$ has bisecants through each point of $(\mathscr{K} \cap \Pi')\backslash\Pi$, a contradiction. So $\mathscr{K}$ has a non-singular point P not in Π. Choose coordinates such that (i) $P = \mathbf{U}_0$, (ii) $\Pi = \mathbf{u}_0$, (iii) $T_P = \mathbf{u}_n$. Then

$$\Pi \cap \mathscr{K} = \mathbf{V}\left(x_0, \sum_1^n{}' a_{ij}x_ix_j\right).$$

The summation sign $\sum'$ indicates summation over all i and j with $i \leq j$, whereas $\sum''$ indicates summation with $i < j$.

Consider the pencil of quadrics $\mathscr{F}_t$, $t \neq 0$, where

$$\mathscr{F}_t = \mathbf{V}\left(tx_0x_n + \sum_1^n{}' a_{ij}x_ix_j\right).$$

Each $\mathscr{F}_t$ contains $\Pi \cap \mathscr{K}$, passes through P, and has $T_P(\mathscr{K})$ as the tangent prime at P.

Now, let Q be a point of $\Pi \cap \mathscr{K}$ not lying in T_P; suppose $Q = \mathbf{P}(0, c_1, \ldots, c_n)$, $c_n \neq 0$. Not all points of $\Pi \cap \mathscr{K}$ outside T_P are singular for $\Pi \cap \mathscr{K}$. So, let Q be non-singular for $\Pi \cap \mathscr{K}$; then

$$T_Q(\mathscr{F}_t) = \mathbf{V}\left(tc_n x_0 + 2\sum_1^n a_{ii}c_i x_i + \sum_1^n{}'' a_{ij}(c_i x_j + c_j x_i) \right).$$

However,

$$T_Q(\mathscr{K}) = \mathbf{V}\left(\lambda x_0 + 2\sum_1^n a_{ii}c_i x_i + \sum_1^n{}'' a_{ij}(c_i x_j + c_j x_i) \right), \quad \text{with } \lambda \neq 0.$$

So there exists some t, say $t = b$, such that $T_Q(\mathscr{F}_b) = T_Q(\mathscr{K})$. It will be shown that $\mathscr{K} = \mathscr{F}_b$.

Let R be a point of $\mathscr{K}$ other than P and not in Π. The points P, Q, R are not collinear; so the plane $\alpha = PQR$ meets $\mathscr{K}$ in a conic $\mathscr{C}$ and $\mathscr{F}_b$ in a conic $\mathscr{C}'$. Let $l = \alpha \cap \Pi$; then P and R do not belong to the line l, although Q does. There are two cases.

(a) *l contains a point Q' in $\mathscr{K}$ with $Q' \neq Q$.* Then $\mathscr{C}$ and $\mathscr{C}'$ have the points P, Q, Q' in common as well as the tangents at P and Q, by Lemma 22.10.1; thus $\mathscr{C} = \mathscr{C}'$. Let A be any point outside the union of the planes α through PQ meeting Π in a tangent l to $\mathscr{K}$, and so in particular outside the prime $PT_Q(\Pi \cap \mathscr{K})$ which is the union of the planes β through PQ meeting Π in a B-line to $\mathscr{K}$. Then A belongs to $\mathscr{K}$ if and only if it belongs to $\mathscr{F}_b$.

(b) *l is a tangent to $\mathscr{K}$.* Then RP belongs neither to $\mathscr{K}$ nor to $\mathscr{F}_b$. Let S be a point of $\mathscr{K}$ not in the prime $PT_Q(\Pi \cap \mathscr{K})$. The plane PRS meets $\mathscr{K}$ and $\mathscr{F}_b$ in conics $\mathscr{D}$ and $\mathscr{D}'$ which coincide outside the line PR. Since RP is contained in neither $\mathscr{D}$ nor $\mathscr{D}'$, so $\mathscr{D} = \mathscr{D}'$; hence $R \in \mathscr{F}_b$.

Thus $\mathscr{K} \subset \mathscr{F}_b$. Similarly $\mathscr{F}_b \subset \mathscr{K}$. This concludes the proof. □

Theorem 22.10.16: *In $PG(n, q)$, a perspective quadratic set $\mathscr{K}$ which is not a subspace is a quadric.*

Proof: If the set is the union of two subspaces, neither of which is contained in the other, then it follows, for example from Theorem 22.10.5, that $\mathscr{K}$ is the union of two distinct primes and so is a quadric. From now on, assume that this is not the case.

(a) $n = 2$. Here $\mathscr{K}$ must be non-singular and non-empty. So it has no three points collinear and is therefore a $(q + 1)$-arc. Let $\mathbf{U}_0, \mathbf{U}_1, \mathbf{U}_2$ be points of $\mathscr{K}$ and let $\mathbf{U}_0\mathbf{U}_2$ and $\mathbf{U}_1\mathbf{U}_2$ be the respective tangents at $\mathbf{U}_0$ and $\mathbf{U}_1$. Every point of $\mathbf{u}_2$ not in $\mathscr{K}$ is of the form $\mathbf{P}(1, m, 0)$ with $m \neq 0$. There exists a non-identity perspectivity σ with centre $\mathbf{P}(1, m, 0)$ fixing $\mathscr{K}$ and $\mathbf{U}_2$ but interchanging $\mathbf{U}_0$ and $\mathbf{U}_1$. Hence the axis of σ is $\mathbf{V}(mx_0 + x_1)$; if $p = 2$, the axis contains $\mathbf{P}(1, m, 0)$ and, if $p \neq 2$, the axis passes through the harmonic conjugate $\mathbf{P}(1, -m, 0)$ of $\mathbf{P}(1, m, 0)$ with respect to $\mathbf{U}_0$ and $\mathbf{U}_1$. Thus $\mathbf{U}$ is mapped to $\mathbf{P}(1, m^2, -m)$, whence $\mathscr{K} = \mathbf{V}(x_0x_1 - x_2^2)$.

(b) $n > 2$. Let Π be a plane such that $\Pi \cap \mathscr{K}$ is not a subspace. If $\Pi \cap \mathscr{K}$ is non-singular, it is a conic by (a). If $\Pi \cap \mathscr{K}$ is singular, it is a line pair and therefore a quadric. Hence, by Theorem 22.10.15, $\mathscr{K}$ is a quadric. □

To reach the final characterization of quadrics, more properties of perspective quadratic sets are required. For the next theorem, let P, Q be distinct points of a quadratic set $\mathscr{K}$. Define $S_{PQ} = T_P \cap T_Q$. For the remainder of this section, $\mathscr{K}$ is always non-singular.

Lemma 22.10.17: (i) *S_{PQ} has codimension 2 in Σ.*

(ii) *If A, B, C are collinear points of $\mathscr{K}$, then $S_{AB} = S_{AC}$.*

Proof: (i) If $T_P = T_Q$, then $PQ \subset \mathscr{K}$. We first show that if R is any other point on PQ, then $T_P = T_R$. Let $A \in T_P \backslash PQ$; then either AP and AQ are both tangents or both lines of $\mathscr{K}$.

(a) Suppose AP, AQ are tangents. The plane APR lies in T_P. If AR is a bisecant, then it meets $\mathscr{K}$ in another point R'. So PR' and QR' are lines of $\mathscr{K}$. Hence any line l through R other than AR meets PR' and QR' in a point of $\mathscr{K}$; so $l \subset \mathscr{K}$. Thus the plane APR lies on $\mathscr{K}$. So AR is not a bisecant and is hence a tangent.

(b) Suppose AP, AQ are lines of $\mathscr{K}$. Again, any line l through R other than AR contains distinct points of $\mathscr{K}$ on AP and AQ and so lies on $\mathscr{K}$. Hence the plane APQ lies on $\mathscr{K}$. So AR is a line of $\mathscr{K}$.

From (a) and (b), we conclude that $T_P \subset T_R$. Since $\mathscr{K}$ is non-singular, $T_R = T_P$ for all R on PQ. Let $A \notin T_P \cup \mathscr{K}$. Then AR is a bisecant for every R on PQ; so AP, AQ, AR all meet $\mathscr{K}$ again at P_1, Q_1, R_1 respectively. If P_1, Q_1, R_1 are not collinear, then each side of the triangle $P_1Q_1R_1$ meets the line PQ in a point of $\mathscr{K}$. So these sides are lines of $\mathscr{K}$ and the plane of the triangle lies on $\mathscr{K}$; thus A lies on $\mathscr{K}$. So P_1, Q_1, R_1 are collinear and R_1 lies on P_1Q_1 for every R on PQ.

Let $PQ \cap P_1Q_1 = S$. Then AS is a tangent to $\mathscr{K}$. Thus $A \in T_S$ and $T_S = \Sigma$. So $\mathscr{K}$ is singular, a contradiction. Hence $T_P \neq T_Q$, and so S_{PQ} has codimension 2.

(ii) If $D \in S_{AB}$ and $D \notin \mathscr{K}$, then AD and BD are tangents. If DC is not a tangent, it meets $\mathscr{K}$ again at C'. The plane $\alpha = ABD$ lies in S_{AB}, hence $C' \in S_{AB}$, and AC' and BC' lie in $\mathscr{K}$. This gives enough points of α on $\mathscr{K}$ to mean that α lies in $\mathscr{K}$, a contradiction. If $D \in \mathscr{K} \backslash AB$, then $ABD \subset \mathscr{K}$. In both cases, $D \in T_C$. □

Lemma 22.10.18: *If PQ is a bisecant of $\mathscr{K}$, then S_{PQ} is spanned by $S_{PQ} \cap \mathscr{K}$ providing $S_{PQ} \cap \mathscr{K}$ is non-empty.*

Proof: $S_{PQ} \cap \mathcal{K}$ is non-singular, for a singular point R would be such that T_R contains S_{PQ} as well as P and Q; so R would be singular for $\mathcal{K}$, since S_{PQ} and P span S_P. Now, Theorem 22.10.8 gives the result. □

Lemma 22.10.19: *If σ is a perspectivity fixing $\mathcal{K}$ with centre R not in $\mathcal{K}$ and $P\sigma = Q$, where $P, Q \in \mathcal{K}$ and $P \neq Q$, then the axis of σ contains S_{PQ}.*

Proof: Since $P\sigma = Q$, so $T_P\sigma = T_Q$. Hence $T_P \cap T_Q = S_{PQ}$ is in the axis of σ. □

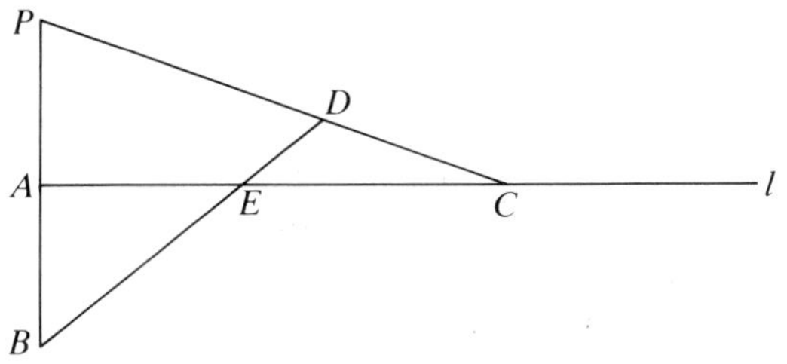

Figure 22.1

Let AB be a bisecant of $\mathcal{K}$, with A, B in $\mathcal{K}$, and let $P \in AB \backslash \mathcal{K}$. Let l be a line of $\mathcal{K}$ through A; then $Bl \not\subset \mathcal{K}$. By Theorem 22.10.8, there exists a unique E on l such that $BE \subset \mathcal{K}$. Let C be a point of $l \backslash \{A, E\}$ and let $D = PC \cap BE$. See fig. 22.1.

Lemma 22.10.20: (i) $S_{AB} \neq S_{BC}$;
(ii) $S_{AB}S_{CD}$ *is a prime.*

Proof: (i) If $S_{AB} = S_{BC}$, then $S_{AB} = S_{AC} = S_{CD}$ and T_B contains A.
(ii) $S_{AB} \cap S_{CD} = T_A \cap T_C \cap T_B \cap T_D$

$$= T_A \cap T_C \cap T_E \cap T_B \quad \text{since } T_B \cap T_D = T_E \cap T_B$$

$$= T_A \cap T_B \cap T_C \quad \text{since } T_C \cap T_E = T_A \cap T_C,$$

using Lemma 22.10.17(ii) twice. By (i), $\dim(T_A \cap T_B \cap T_C) = n - 3$. So $S_{AB}S_{CD}$ is a prime. □

Lemma 22.10.21: *Let σ be the perspectivity with centre P, axis $S_{AB}S_{CD}$, and such that $A\sigma = B$. Then σ fixes $\mathcal{K}$.*

Proof: First we show that $A, B \notin S_{AB}S_{CD}$ in order to guarantee the existence of σ. Suppose that $A \in S_{AB}S_{CD}$. As $A \notin S_{AB}$ we have $AS_{AB} = T_A = S_{AB}S_{CD}$. Hence $S_{CD} \subset T_A$ and so $S_{AC} = S_{CD}$. By Lemma 22.10.17(ii), $S_{AC} \subset T_E$; so

$S_{CD} \subset T_E$. Hence $S_{ED} = S_{CD}$. Again by Lemma 22.10.17(ii), $S_{ED} \subset T_B$ and so $S_{CD} \subset T_B$. Consequently S_{CD} is contained in T_C, T_D, T_A, T_B. Hence $S_{AB} = S_{BC} = S_{CD}$, contradicting Lemma 22.10.20(i). Analogously, $B \notin S_{AB}S_{CD}$.

Since $E \in S_{AB} \cap S_{CD}$ so E is on the axis. The perspectivity σ depends only on $\mathscr{K}$, P, A, B, C; so we may write

$$\sigma = \sigma(\mathscr{K}, P, A, B, C).$$

Now, σ maps every line of $\mathscr{K}$ through A to a line through B and every line of $\mathscr{K}$ through C to a line through D. Hence $T_A\sigma = (AS_{AB})\sigma = BS_{AB} = T_B$; analogously, $T_C\sigma = T_D$.

Choose Q in $\mathscr{K}$ not in the plane ABC.

(a) *QE is not a line of $\mathscr{K}$*

Let $l' = QF$, with F on AC, be on $\mathscr{K}$. Consider the solid $\Pi_3 = ABCQ$. One can show that $\Pi_3 \cap \mathscr{K}$ is non-singular. It follows that every point of $\Pi_3 \cap \mathscr{K}$ lies on at most two lines of $\Pi_3 \cap \mathscr{K}$. For any point R of $\Pi_3 \cap \mathscr{K}$ such that RE is not a line of $\mathscr{K}$, the tangent prime T_R intersects AE and BE in points A' and B'; then RA' and RB' are lines of $\mathscr{K}$. If R' in $\Pi_3 \cap \mathscr{K}$ is such that $R'E$ is a line of $\mathscr{K}$, then choose E' such that neither $E'E$ nor $E'R'$ are lines of $\mathscr{K}$. Then the previous argument can be applied with R' for R and E' for E. Thus each point of $\Pi_3 \cap \mathscr{K}$ lies on exactly two lines of $\Pi_3 \cap \mathscr{K}$. Hence $\Pi_3 \cap \mathscr{K}$ is a hyperbolic quadric, by Theorem 16.2.6.

In Π_3 there exists a perspectivity σ' with centre P and axis $\Pi_3 \cap S_{AB}S_{CD}$ fixing $\Pi_3 \cap \mathscr{K}$ such that $A\sigma' = B$. Hence $\sigma|_{\Pi_3} = \sigma'$ and $Q\sigma \in \mathscr{K}$. It also follows that $B\sigma = A$, and so $T_B\sigma = T_A$. Since E is non-singular, such a point Q always exists.

(b) *QE is a line of $\mathscr{K}$*

If QA and QB are lines of $\mathscr{K}$, then $Q \in S_{AB}$ and so $Q\sigma = Q$. Thus we may assume that QA is not on $\mathscr{K}$. We may also assume that $Q \notin S_{AB}S_{CD}$ since then $Q\sigma = Q$. There exists a point E' in $S_{AB} \cup S_{CD}$ such that $E' \in \mathscr{K}$ and QE' is a bisecant of $\mathscr{K}$; otherwise T_Q contains S_{AB} and S_{CD} by Lemma 22.10.18 ($S_{AB} \cap \mathscr{K}$ and $S_{CD} \cap \mathscr{K}$ are not empty since they both contain E), which is excluded by the fact that Q is not singular nor in $S_{AB}S_{CD}$.

Let $E' \in S_{AB}$. Through Q there is a line QA' in $\mathscr{K}$ meeting AE' in A'. Then $A' \neq E'$ and $A' \neq A$. Consider the perspectivity $\sigma_1 = \sigma(\mathscr{K}, P, A, B, A')$; so $A'\sigma_1 = B'$ where $B' = PA' \cap E'B$. Applying the result of the previous paragraph to σ_1 gives that $Q\sigma_1 \in \mathscr{K}$. As σ and σ_1 both transpose T_A and T_B, the map $\sigma\sigma_1^{-1}$ is the identity, whence $\sigma_1 = \sigma$. □

Theorem 22.10.22: *In $PG(n, q)$, a non-singular quadratic set containing a line is perspective.*

Proof: This merely restates the previous lemma in a different form. □

Theorem 22.10.23: *In $PG(n, q)$, a non-singular quadratic set $\mathscr{K}$ is a quadric or an ovoid. If $\mathscr{K}$ is an ovoid, then it is one of the following:*

(i) *a $(q + 1)$-arc in $PG(2, q)$;*
(ii) *an ovaloid of $PG(3, q)$, $q > 2$;*
(iii) *an elliptic quadric in $PG(3, 2)$.*

Proof: The first part follows from Theorems 22.10.16 and 22.10.22. The second part follows from Theorem 26.3.32. □

22.11 Notes and references

§§22.1–22.9. These sections continue the material in Chapter 5. They are based on Hirschfeld (1986), although many details come from Segre (1959a). In particular, the proof of Theorem 22.8.2 is based on Segre's treatment, although the formula (22.36) is an amalgamation of several formulae established by matrix-theoretic methods by Dai and Feng (1965) and by Feng and Dai (1965). The nature of the transitivity of the orthogonal group, as expounded in Theorem 22.6.6, seems more difficult in the projective space than in the vector space case. See Higman (1978), Artin (1957), Dieudonné (1971) for the vector space case, and Dye (1970a) for the connection between the two cases. The size of the stabilizer in the orthogonal group of an isotropic subspace in the vector space has been calculated by Derr (1980).

The principle of triality

On the hyperbolic quadric $\mathscr{H}_7$, known sometimes as the Study quadric after the discoverer of the principle, consider the two systems of generators $\mathscr{A}$ and $\mathscr{B}$. From Theorem 22.4.6,

$$|\mathscr{A}| = |\mathscr{B}| = \tfrac{1}{2}\kappa(7; 2) = \tfrac{1}{2}[0, 3]_+ = (q + 1)(q^2 + 1)(q^3 + 1).$$

From Theorem 22.5.1(b),

$$|\mathscr{H}_7| = N(0; 7, 2) = (q^3 + 1)(q^4 - 1)/(q - 1) = \tfrac{1}{2}\kappa(7; 2).$$

Then it can be shown that the solids in $\mathscr{A}$ correspond to the points of a quadric $\mathscr{H}'_7$ of $PG'(7, q)$ such that solids in $\mathscr{A}$ containing a given line are mapped to the points of a line of $\mathscr{H}'_7$; similarly, the solids in $\mathscr{B}$ correspond to the points of a quadric $\mathscr{H}''_7$ of $PG''(7, q)$. A *triality* is a permutation τ of order 3 of $\mathscr{H}_7 \cup \mathscr{A} \cup \mathscr{B}$ such that $\mathscr{H}_7\tau = \mathscr{A}$, $\mathscr{A}\tau = \mathscr{B}$, $\mathscr{B}\tau = \mathscr{H}_7$, and such that incidence is preserved, where *incidence* is defined as follows: (a) a point is incident with a solid if it lies in the solid; (b) two points are incident if their join lies on $\mathscr{H}_7$; (c) two solids of the same system are incident if they meet in a line; (d) two solids of different systems are incident if they meet in a plane. Then

$$\mathscr{H}_7 \overset{\tau}{\rightarrow} \mathscr{A} \overset{\tau}{\rightarrow} \mathscr{B} \overset{\tau}{\rightarrow} \mathscr{H}_7$$

is induced by collineations

$$PG(7, q) \overset{\sigma_1}{\rightarrow} PG'(7, q) \overset{\sigma_2}{\rightarrow} PG''(7, q) \overset{\sigma_3}{\rightarrow} PG(7, q)$$

such that

$$\mathscr{H}_7\sigma_1 = \mathscr{H}'_7, \qquad \mathscr{H}'_7\sigma_2 = \mathscr{H}''_7, \qquad \mathscr{H}''_7\sigma_3 = \mathscr{H}_7.$$

Trialities were classified by Tits (1959). For an introduction and further references, see Hirschfeld and Thas (1985).

§22.10. This section follows Buekenhout (1969a). Two restrictions are put on a subset $\mathscr{K}$ of $\Sigma = PG(n, q)$:

(a) $\mathscr{K}$ is of type $(0, 1, 2, q+1)$;

(b) $\mathscr{K}$ has a tangent space at each point P, which is either a prime or Σ itself.

Here we describe what happens if condition (b) is dropped. In what follows, condition (a) still holds. So the definitions of §22.10 still apply for a set $\mathscr{K}$ of type $(0, 1, 2, q+1)$. A point P of $\mathscr{K}$ is *singular* if there is no bisecant of $\mathscr{K}$ through it. The set of singular points of $\mathscr{K}$ is the *singular space* of $\mathscr{K}$, and $\mathscr{K}$ is *singular* or *non-singular* according as it has singular points or not.

Lemma 22.11.1: *The singular space $\mathscr{S}$ of $\mathscr{K}$ is a subspace of Σ.*

Proof: Let $P, Q \in \mathscr{S}$; then it must be shown that any other point R of PQ is also singular. Since P is singular, so $PQ \subset \mathscr{K}$. Let $A \in \mathscr{K} \backslash PQ$. Then the three lines AP, AQ, PQ lie on $\mathscr{K}$ and hence the plane APQ also lies on $\mathscr{K}$; in particular, $AR \subset \mathscr{K}$. So R is singular. □

Theorem 22.11.2: *If $\mathscr{K}$ is a set of type* $(0, 1, 2, q+1)$ *in Σ with singular space Π_r, then $\mathscr{K}$ is a cone $\Pi_r\mathscr{K}'$, where $\mathscr{K}'$ is a non-singular set of type* $(0, 1, 2, q+1)$ *in a subspace Π_{n-r-1} skew to Π_r.*

Proof: Let Π_{n-r-1} be any space of dimension $n-r-1$ skew to the singular space Π_r and let $\mathscr{K}' = \Pi_{n-r-1} \cap \mathscr{K}$. Now take A in Π_r and A' in $\mathscr{K}'$; then $AA' \subset \mathscr{K}$. For any point P of $\mathscr{K} \backslash (\Pi_r \cup \mathscr{K}')$, the line $AP \subset \mathscr{K}$. The space $P\Pi_r$ meets Π_{n-r-1} in a point B. Hence $B \in \mathscr{K}'$. So $\mathscr{K}$ is the cone $\Pi_r\mathscr{K}'$.

It must still be shown that $\mathscr{K}'$ is non-singular; we already know that it is of type $(0, 1, 2, q+1)$. Suppose $\mathscr{K}'$ has a singular point Q. Let $P \in \mathscr{K} \backslash \Pi_r$ and let $B = P\Pi_r \cap \Pi_{n-r-1}$. If $B = Q$, then $PQ \subset \mathscr{K}$. If $B \neq Q$, then, with

$A = BP \cap \Pi_r$, the planes APQ and ABQ coincide. However, $BQ \subset \mathscr{K}$ and A is singular; so $ABQ \subset \mathscr{K}$. Thus $PQ \subset \mathscr{K}$ and Q is a singular point of $\mathscr{K}$, which is a contradiction. □

The complication of further characterizations, even when the set $\mathscr{K}$ is non-singular, is illustrated by the work of Lefèvre(-Percsy) (1975).

The following two results are due to Tallini (1956a, 1957a) and characterize k-sets of type $(0, 1, 2, q + 1)$ with large k, but allowing singularities.

Theorem 22.11.3: *A k-set of type* $(0, 1, 2, q + 1)$ *in* $PG(n, q)$, $n \geq 3$, $q > 2$, *such that* $\theta(n) > k \geq \theta(n - 1)$ *is one of the following*:

(1) $\Pi_{n-1} \cup \Pi'_r$ *for some* $r = -1, 0, 1, \ldots, n - 1$;
(2) $\Pi_t\mathscr{P}_{n-t-1}$ *for some* $t = -1, 1, \ldots, n - 3$ *when n is even and some* $t = 0, 2, \ldots, n - 3$ *when n is odd*;
(3) $\Pi_t\mathscr{H}_{n-t-1}$ *for some* $t = -1, 1, \ldots, n - 4$ *when n is odd and some* $t = 0, 2, \ldots, n - 4$ *when n is even*;
(4) $\mathscr{W}_n \cup \Pi_r$, *where* $\mathscr{W}_n$ *is one of the quadrics* (2) *and* $\Pi_r \subset \Pi_t N$ *with N the nucleus of a base* $\mathscr{P}_{n-t-1}$, *and q is even*;
(5) $\Pi_{n-3}\mathscr{K}' \cup \Pi_r$, *where* $\mathscr{K}'$ *is a* $(q + 1)$*-arc in a* Π_2 *skew to* Π_{n-3} *and* $\Pi_r \subset \Pi_{n-3}N$ *with N the nucleus of* $\mathscr{K}'$, *and q is even.*

Theorem 22.11.4: *In* $PG(n, q)$, $q > 3$, $n \geq 4$, *a k-set of type* $(0, 1, 2, q + 1)$ *with* $k = \theta(n - 1) - q^{g+1}$, *where g is the largest dimension of a subspace in* $\mathscr{K}$, *is one of the following*:

(1) $\Pi_t\mathscr{E}_{n-t-1}$ *for some* $t = -1, 1, \ldots, n - 4$ *when n is odd and some* $t = 0, 2, \ldots, n - 4$ *when n is even*;
(2) $\Pi_{n-4}\mathscr{K}'$, *where* $\mathscr{K}'$ *is a* $(q^2 + 1)$*-cap in* $PG(3, q)$ *skew to* Π_{n-4} *and q is even*;
(3) Π_{n-2}. □

The last theorem was improved by Lefèvre-Percsy (1980) as follows.

Theorem 22.11.5: *In* $PG(n, q)$, $q > 3$, $n \geq 3$, *q odd, a k-set of type* $(0, 1, 2, q + 1)$ *with* $\theta(n) > k > \theta(n - 1) - q^{n-2} + q^{n-3}$ *is either a quadric or* $\Pi_{n-1} \cup \Pi'_r$ *for some* $r = -1, 0, 1, \ldots, n - 1$. □

From the previous three results, we can immediately give the following result for the non-singular case.

Theorem 22.11.6: *In* $PG(n, q)$ *with* $n \geq 4$ *and* $q > 2$, *let* $\mathscr{K}$ *be a non-singular k-set of type* $(0, 1, 2, q + 1)$.

(i) *If* $\theta(n) > k \geq \theta(n-1)$, *then one of the following holds*:

(a) $k = \theta(n-1)$, *n is even and* $\mathcal{K} = \mathcal{P}_n$;
(b) $k = \theta(n-1) + q^{(n-1)/2}$, *n is odd and* $\mathcal{K} = \mathcal{H}_n$;
(c) $k = \theta(n-1) + 1$ *and* $\mathcal{K} = \Pi_t \mathcal{P}_{n-t-1} \cup \{N\}$ *or* $\mathcal{K} = \Pi_{n-3}\mathcal{K}' \cup \{N\}$, *where* $\mathcal{K}'$ *is a* $(q+1)$*-arc in a* Π_2 *skew to* Π_{n-3}, *and N is the nucleus of* $\mathcal{P}_{n-t-1}$ *and* $\mathcal{K}'$ *in the two cases.*

(ii) *If* $k = \theta(n-1) - q^{g+1}$, *where g is the largest dimension of a subspace in* $\mathcal{K}$, *then n is odd*, $g + 1 = (n-1)/2$, *and* $\mathcal{K} = \mathcal{E}_n$.

(iii) *If q is odd*, $q > 3$ *and* $\theta(n) > k > \theta(n-1) - q^{n-2} + q^{n-3}$, *then*

(a) *for n even*, $\mathcal{K} = \mathcal{P}_n$;
(b) *for n odd*, $\mathcal{K} = \mathcal{H}_n$ or $\mathcal{E}_n$. □

23

HERMITIAN VARIETIES

23.1 Introduction

Over $\gamma = GF(q)$, q square, a Hermitian form F is an element of $\gamma[x_0, x_1, x_2, \ldots, x_n]$ such that $F = XH\bar{X}^*$, where $\bar{H}^* = H$ and $H \neq 0$; thus $h_{ii} \in GF(\sqrt{q})$ and $\bar{h}_{ji} = h_{ij}$ for $H = [h_{ij}]$, where $\bar{x} = x^{\sqrt{q}}$. As in §5.1, F can be reduced by a non-singular linear transformation to the canonical form

$$F_r = x_0\bar{x}_0 + \cdots + x_r\bar{x}_r.$$

The variety $\mathbf{V}(F_r)$ in $PG(n, q)$ is a Hermitian variety which is non-singular when $r = n$. The Hermitian variety $\mathbf{V}(F_n)$ is written $\mathscr{U}_n$ or $\mathscr{U}_{n,q}$; that is,

$$\mathscr{U}_n = \mathbf{V}(x_0\bar{x}_0 + x_1\bar{x}_1 + \cdots + x_n\bar{x}_n). \tag{23.1}$$

Similarly to quadrics, $\mathbf{V}(F_r) = \Pi_{n-r-1}\mathscr{U}_r$, where $\mathscr{U}_r$ is the non-singular Hermitian variety in the r-space $\mathbf{U}_0\mathbf{U}_1 \cdots \mathbf{U}_r = \mathbf{V}(x_{r+1}, \ldots, x_n)$ and $\Pi_{n-r-1} = \mathbf{V}(x_0, \ldots, x_r) = \mathbf{U}_{r+1}\mathbf{U}_{r+2} \cdots \mathbf{U}_n$; that is, the points of $\Pi_{n-r-1}\mathscr{U}_r$ comprise all the points of the lines joining any point of Π_{n-r-1} to any point of $\mathscr{U}_r$. As for quadrics, Π_{n-r-1} is the vertex of the cone and $\mathscr{U}_r$ a base.

Theorem 23.1.1: *There are $n + 1$ projectively distinct Hermitian varieties in $PG(n, q)$.*

Proof: From above, there is one variety $\Pi_{n-r-1}\mathscr{U}_r$ for each r in $\{0, 1, \ldots, n\}$. □

Lemma 23.1.2: *A section of a Hermitian variety is still a Hermitian variety.*

Proof: If $\mathscr{U} = \mathbf{V}(F)$ is Hermitian with F in $\gamma[x_0, \ldots, x_n]$, then a section by a prime $\mathbf{V}(L)$, where $L = x_0 - a_1x_1 - \cdots - a_nx_n$, is $\mathscr{U}' = \mathbf{V}(L, G)$, where

$$G(x_1, \ldots, x_n) = F(a_1x_1 + \cdots + a_nx_n, x_1, \ldots, x_n).$$

As G is a Hermitian form, so $\mathscr{U}'$ is a Hermitian variety in $\mathbf{V}(L)$. Thus the result follows by induction. □

The behaviour of low-dimensional varieties gives a feeling for higher-dimensional ones. In Table 23.1, all types up to five dimensions are described.

Table 23.1

$PG(0, q)$	$\mathcal{U}_0 = \mathbf{V}(x_0\bar{x}_0)$ is empty
$PG(1, q)$	$\mathcal{U}_1 = \mathbf{V}(x_0\bar{x}_0 + x_1\bar{x}_1)$ is $\sqrt{q} + 1$ points forming a subline $PG(1, \sqrt{q})$
	$\Pi_0\mathcal{U}_0 = \mathbf{V}(x_0\bar{x}_0)$ is the single point $\mathbf{U}_1$
$PG(2, q)$	$\mathcal{U}_2 = \mathbf{V}(x_0\bar{x}_0 + x_1\bar{x}_1 + x_2\bar{x}_2)$ is a unital (Hermitian arc) comprising $q\sqrt{q} + 1$ points; through each point of $\mathcal{U}_2$ there is a unique line meeting $\mathcal{U}_2$ in a $\Pi_0\mathcal{U}_0$, whereas all other lines meet $\mathcal{U}_2$ in a $\mathcal{U}_1$
	$\Pi_0\mathcal{U}_1 = \mathbf{V}(x_0\bar{x}_0 + x_1\bar{x}_1)$ comprises $\sqrt{q} + 1$ lines concurrent at $\mathbf{U}_2$
	$\Pi_1\mathcal{U}_0 = \mathbf{V}(x_0\bar{x}_0)$ is the single line $\mathbf{u}_0$
$PG(3, q)$	$\mathcal{U}_3 = \mathbf{V}(x_0\bar{x}_0 + x_1\bar{x}_1 + x_2\bar{x}_2 + x_3\bar{x}_3)$ comprises $(q + 1)(q\sqrt{q} + 1)$ points on $(\sqrt{q} + 1)(q\sqrt{q} + 1)$ lines; there are as many plane sections $\Pi_0\mathcal{U}_1$ as points and the remaining plane sections are of type $\mathcal{U}_2$
	$\Pi_0\mathcal{U}_2 = \mathbf{V}(x_0\bar{x}_0 + x_1\bar{x}_1 + x_2\bar{x}_2)$ is a cone comprising the join of the vertex $\mathbf{U}_3$ to a Hermitian curve
	$\Pi_1\mathcal{U}_1 = \mathbf{V}(x_0\bar{x}_0 + x_1\bar{x}_1)$ is $\sqrt{q} + 1$ collinear planes
	$\Pi_2\mathcal{U}_0 = \mathbf{V}(x_0\bar{x}_0)$ is the plane $\mathbf{u}_0$
$PG(4, q)$	$\mathcal{U}_4 = \mathbf{V}(x_0\bar{x}_0 + x_1\bar{x}_1 + x_2\bar{x}_2 + x_3\bar{x}_3 + x_4\bar{x}_4)$ comprises $(q + 1) \times (q^2\sqrt{q} + 1)$ points on $(q\sqrt{q} + 1)(q^2\sqrt{q} + 1)$ lines with $q\sqrt{q} + 1$ lines through each point
	$\Pi_0\mathcal{U}_3 = \mathbf{V}(x_0\bar{x}_0 + x_1\bar{x}_1 + x_2\bar{x}_2 + x_3\bar{x}_3)$ is a cone with vertex $\mathbf{U}_4$ and base a Hermitian surface $\mathcal{U}_3$; its generators are planes
	$\Pi_1\mathcal{U}_2 = \mathbf{V}(x_0\bar{x}_0 + x_1\bar{x}_1 + x_2\bar{x}_2)$ is a cone with vertex the line $\mathbf{U}_3\mathbf{U}_4$ and base a Hermitian curve; its generators are planes
	$\Pi_2\mathcal{U}_1 = \mathbf{V}(x_0\bar{x}_0 + x_1\bar{x}_1)$ comprises $\sqrt{q} + 1$ solids through the plane $\mathbf{U}_2\mathbf{U}_3\mathbf{U}_4$
	$\Pi_3\mathcal{U}_0 = \mathbf{V}(x_0\bar{x}_0)$ is the solid $\mathbf{u}_0$
$PG(5, q)$	$\mathcal{U}_5 = \mathbf{V}(\sum_{i=0}^{5} x\bar{x}_i)$ comprises $(q^2 + q + 1)(q^2\sqrt{q} + 1)$ points on $(\sqrt{q} + 1)(q\sqrt{q} + 1)(q^2\sqrt{q} + 1)$ planes
	$\Pi_0\mathcal{U}_4 = \mathbf{V}(\sum_{i=0}^{4} x_i\bar{x}_i)$ is a cone with vertex $\mathbf{U}_5$ and base $\mathcal{U}_4$; its generators are planes
	$\Pi_1\mathcal{U}_3 = \mathbf{V}(\sum_{i=0}^{3} x_i\bar{x}_i)$ is a cone with vertex the line $\mathbf{U}_4\mathbf{U}_5$ and base $\mathcal{U}_3$; its generators are solids
	$\Pi_2\mathcal{U}_2 = \mathbf{V}(x_0\bar{x}_0 + x_1\bar{x}_1 + x_2\bar{x}_2)$ comprises $q\sqrt{q} + 1$ solids through the plane $\mathbf{U}_3\mathbf{U}_4\mathbf{U}_5$
	$\Pi_3\mathcal{U}_1 = \mathbf{V}(x_0\bar{x}_0 + x_1\bar{x}_1)$ is $\sqrt{q} + 1$ primes through the solid $\mathbf{u}_0 \cap \mathbf{u}_1$
	$\Pi_4\mathcal{U}_0 = \mathbf{V}(x_0\bar{x}_0)$ is the solid $\mathbf{u}_0$

23.2 Tangency and polarity

The notation $\mathcal{U}_n$ or $\mathcal{U}_{n,q}$ will also be used for any non-singular Hermitian variety of $PG(n, q)$. Consider the non-singular Hermitian variety $\mathcal{U}_n = \mathbf{V}(F)$. Let $P = \mathbf{P}(A)$ and $Q = \mathbf{P}(B)$, where $A = (a_0, \ldots, a_n)$ and $B = (b_0, \ldots, b_n)$. If

$F(X) = XH\bar{X}^*$, then

$$F(A + tB) = (A + tB)H(\bar{A} + \bar{t}\bar{B})^*$$
$$= AH\bar{A}^* + tBH\bar{A}^* + \bar{t}AH\bar{B}^* + t\bar{t}BH\bar{B}^*.$$

So $\mathbf{P}(A + tB)$ lies on $\mathscr{U}_n$ if and only if

$$0 = F(A) + tG(A, B) + \overline{tG(A, B)} + t\bar{t}F(B) \tag{23.2}$$

where $G(A, B) = BH\bar{A}^*$.

The line l is a *tangent* to $\mathscr{U}_n$ if $|l \cap \mathscr{U}_n| = 1$.

Lemma 23.2.1: *Let* $P = \mathbf{P}(A)$ *and* $Q = \mathbf{P}(B)$.

(i) *If* $F(A)F(B) - G(A, B)\overline{G(A, B)} \neq 0$, *then* $|PQ \cap \mathscr{U}_n| = \sqrt{q} + 1$.

(ii) *If* $F(A)F(B) - G(A, B)\overline{G(A, B)} = 0$, *then*

(a) $F(A) = F(B) = G(A, B) = 0 \Rightarrow PQ \subset \mathscr{U}_n$;

(b) *otherwise* $|PQ \cap \mathscr{U}_n| = 1$.

(iii) *Suppose* $P \in \mathscr{U}_n$.

(a) *If* $Q \notin \mathscr{U}_n$, *then* PQ *is a tangent to* $\mathscr{U}_n \Leftrightarrow G(A, B) = 0$.

(b) *If* $Q \in \mathscr{U}_n$, *then* PQ *is a line of* $\mathscr{U}_n \Leftrightarrow G(A, B) = 0$.

Proof: (i), (ii) With $F(A) = a$, $F(B) = b$, and $G(A, B) = \delta$, (23.2) becomes

$$bt\bar{t} + \bar{\delta}\bar{t} + \delta t + a = 0; \tag{23.3}$$

here $\bar{b} = b$ and $\bar{a} = a$.

First, let $a = b = 0$. If $\delta = 0$, then $PQ \subset \mathscr{U}_n$. Now, let $\delta \neq 0$. Then (23.3) has $t = 0$ and $t = \infty$ as solutions corresponding to P and Q on $\mathscr{U}_n$. Every other t satisfies

$$t^{\sqrt{q} - 1} = -\delta^{-(\sqrt{q} - 1)}, \tag{23.4}$$

which has $\sqrt{q} - 1$ solutions by §1.5(v). Thus $|PQ \cap \mathscr{U}_n| = \sqrt{q} + 1$.

If $b = 0$ and $a \neq 0$, then the substitution $t \to t^{-1}$ gives an equation of the form (23.3) with $b \neq 0$.

Finally, if $b \neq 0$, the substitution $t \to t - \bar{\delta}/b$ transforms (23.3) to

$$t\bar{t} + (ab - \delta\bar{\delta})/b^2 = 0. \tag{23.5}$$

If $ab - \delta\bar{\delta} = 0$, then $|PQ \cap \mathscr{U}_n| = 1$; if $ab - \delta\bar{\delta} \neq 0$, then, again by §1.5(v), (23.5) has $\sqrt{q} + 1$ solutions, which means that $|PQ \cap \mathscr{U}_n| = \sqrt{q} + 1$.

(iii) This follows immediately from (i) and (ii) with $F(A) = 0$. □

When $G(A, B) = 0$, the points $P = \mathbf{P}(A)$ and $Q = \mathbf{P}(B)$ are *conjugate*. With $X = (x_0, x_1, \ldots, x_n)$ the prime $\mathbf{V}(G(X, A))$ is the *polar prime* of P, and is denoted $P\mathfrak{U}$ (the image of P under the Hermitian polarity $\mathfrak{U}$ with matrix H). If $\Pi_r = P_0P_1 \cdots P_r$, then the *polar space* $\Pi_{n-r-1} = \Pi_r\mathfrak{U}$ of Π_r is

$P_0\mathfrak{U} \cap P_1\mathfrak{U} \cap \cdots \cap P_r\mathfrak{U}$. This is independent of the choice of $P_0, \ldots, P_r$ in Π_r in the sense that if $\Pi_r = P_0 \cdots P_r = Q_0 \cdots Q_r$, then $P_0\mathfrak{U} \cap \cdots \cap P_r\mathfrak{U} = Q_0\mathfrak{U} \cap \cdots \cap Q_r\mathfrak{U}$. Two spaces are *conjugate* if they are contained in polar spaces.

When $P \in \mathscr{U}_n$, the polar prime of P is also called the *tangent prime at P* and written T_P or $T_P(\mathscr{U}_n)$. Similarly, if $\Pi_r = P_0 \cdots P_r \subset \mathscr{U}_n$, then the *tangent space at* Π_r is $T_{\Pi_r} = \Pi_{n-r-1} = \bigcap_i T_{P_i}$.

For $\mathscr{U}_n$ in canonical form, that is $\mathscr{U}_n = \mathbf{V}(\sum x_i\bar{x}_i)$,

$$G(X, A) = \sum \bar{a}_i x_i. \tag{23.6}$$

Lemma 23.2.2: *If $\Pi_r \subset \mathscr{U}_n$, then the tangent space at Π_r is a Π_{n-r-1} which contains Π_r and any Π_s on $\mathscr{U}_n$ through Π_r.*

Proof: This follows from Lemma 23.2.1 and the definition of tangent space. □

It should be observed that the polarity of $\mathscr{U}_n$ lies behind its reduction to the canonical form (23.1). For, let $\mathscr{U}$ be any non-singular Hermitian variety with polarity $\mathfrak{U}$ in $\Pi_n = PG(n, q)$. Choose P_0 in $\Pi_n\backslash\mathscr{U}$ and let $\pi_0 = P_0\mathfrak{U}$. Now, choose P_1 in $\pi_0\backslash\mathscr{U}$ and let $\pi_1 = P_1\mathfrak{U}$. Thus, choose $P_0, P_1, \ldots, P_n$ so that $P_0, \ldots, P_i$ span an i-space and so that P_i lies in $(\pi_0 \cap \pi_1 \cap \cdots \cap \pi_{i-1})\backslash\mathscr{U}$. Take $\mathbf{U}_i = P_i$, $i = 0, \ldots, n$, the vertices of the simplex of reference. Hence

$$\mathscr{U} = \mathbf{V}(c_0x_0\bar{x}_0 + \cdots + c_nx_n\bar{x}_n).$$

Now, a suitable choice of the unit point or, equivalently, the transformation $x_i' = d_ix_i$, $i = 0, \ldots, n$, with $d_i\bar{d}_i = c_i$, gives the required form

$$\mathscr{U} = \mathbf{V}(x_0\bar{x}_0 + x_1\bar{x}_1 + \cdots + x_n\bar{x}_n)$$

where the primes have been omitted.

When considering spaces lying on $\mathscr{U}$, another canonical form is useful. Since $\mathscr{U}_n$ is projectively unique, it can also be written as $\mathbf{V}(G_n)$, where

$$G_n(x_0, \ldots, x_n) = \begin{cases} x_0\bar{x}_0 + (x_1\bar{x}_2 + x_2\bar{x}_1) + \cdots + (x_{n-1}\bar{x}_n + x_n\bar{x}_{n-1}), & n \text{ even} \\ (x_0\bar{x}_1 + x_1\bar{x}_0) + \cdots + (x_{n-1}\bar{x}_n + x_n\bar{x}_{n-1}), & n \text{ odd}. \end{cases} \tag{23.7}$$

Lemma 23.2.3: *The tangent prime at a point P of $\mathscr{U}_n$ is a cone $\Pi_0\mathscr{U}_{n-2}$.*

Proof: Let $P = \mathbf{U}_0$ and the tangent prime $T_P = \mathbf{u}_1$. Also choose $\mathbf{U}_1 \in \mathscr{U}_n$. Then by a suitable choice of unit point $\mathscr{U}_n = \mathbf{V}(F)$, where

$$F = x_0\bar{x}_1 + x_1\bar{x}_0 + F'(x_2, \ldots, x_n).$$

Then F' can be reduced to canonical form so that

$$F = \bar{x}_0x_1 + x_0\bar{x}_1 + x_2\bar{x}_2 + \cdots + x_n\bar{x}_n.$$

So

$$\begin{aligned}\mathcal{U}_n \cap T_P &= \mathbf{V}(x_1, x_2\bar{x}_2 + \cdots + x_n\bar{x}_n)\\ &= \{\mathbf{P}(\lambda, 0, x_2, \ldots, x_n) | x_2\bar{x}_2 + \cdots + x_n\bar{x}_n = 0\}\\ &= \mathbf{U}_0\mathcal{U}_{n-2}. \quad \square\end{aligned}$$

Let $\mu_n = |\mathcal{U}_n|$.

Corollary: *The tangent prime at P meets $\mathcal{U}_n$ in $q\mu_{n-2} + 1$ points.* □

Theorem 23.2.4:

$$\mu_n = \{(\sqrt{q})^{n+1} + (-1)^n\}\{(\sqrt{q})^n - (-1)^n\}/(q-1) \tag{23.8}$$

$$= \theta(n-1) + \{q^n - (-\sqrt{q})^n\}/(\sqrt{q}+1). \tag{23.9}$$

Proof: Any line through a point P of $\mathcal{U}_n$ not in T_P meets $\mathcal{U}_n$ in $\sqrt{q}$ other points. Hence

$$\begin{aligned}\mu_n &= \sqrt{q}\{\theta(n-1) - \theta(n-2)\} + q\mu_{n-2} + 1\\ &= q^{n-1}\sqrt{q} + 1 + q\mu_{n-2}.\end{aligned}$$

Put $\alpha_n = \mu_n/(\sqrt{q})^n$. Then

$$\alpha_n = (\sqrt{q})^{n-1} + (1/\sqrt{q})^n + \alpha_{n-2}.$$

Separate calculations for n even and odd give the desired result. □

Now let $\mu_n^{(t)} = |\Pi_t\mathcal{U}_{n-t-1}|$. Thus $\mu_n = \mu_n^{(-1)}$.

Corollary:

$$\mu_n^{(t)} = \theta(n-1) + (q^n - (-\sqrt{q})^{n+t+1})/(\sqrt{q}+1). \tag{23.10}$$

Proof: Any two points of the base $\mathcal{U}_{n-t-1}$ are joined to the vertex Π_t by two Π_{t+1} having precisely the vertex in common. Therefore

$$\mu_n^{(t)} = \theta(t) + \{\theta(t+1) - \theta(t)\}\mu_{n-t-1}$$

and the result follows. □

Let $\mathcal{N}(\mathcal{U}_n)$ be the set of non-singular Hermitian varieties in $PG(n, q)$.

Lemma 23.2.5: *The number of Hermitian varieties $\mathcal{U}_n$ containing a given $\Pi_0\mathcal{U}_{n-2}$ as tangent cone is*

$$(q-1)q^{n-1}\sqrt{q}.$$

Proof: Let $\Pi_0 = \mathbf{U}_{n-1}$ and $\mathscr{U}_{n-2} = \mathbf{V}(x_n, x_{n-1}, F_{n-2})$ where $F_{n-2} = x_0\bar{x}_0 + \cdots + x_{n-2}\bar{x}_{n-2}$. The prime containing $\Pi_0\mathscr{U}_{n-2}$ is $\mathbf{u}_n$. Any Hermitian variety containing $\mathbf{U}_{n-1}$ and $\mathscr{U}_{n-2}$ has the form

$$\mathscr{U} = \mathbf{V}(x_n\bar{f} + \bar{x}_n f + x_{n-1}\bar{g} + \bar{x}_{n-1}g + F_{n-2}),$$

where f is linear in $x_0, \ldots, x_n$ and g is linear in $x_0, \ldots, x_{n-2}$. Since the tangent prime at $\mathbf{U}_{n-1}$ is $\mathbf{u}_n$, so the only term involving $\bar{x}_{n-1}$ is $cx_n\bar{x}_{n-1}$ and hence the only term involving x_{n-1} is $\bar{c}x_{n-1}\bar{x}_n$; thus $g = 0$.

The non-singularity of $\mathscr{U}$ means that $c \neq 0$. Thus the number of choices for the form f and thus for $\mathscr{U}$ is $(q-1)q^{n-1}\sqrt{q}$. □

Lemma 23.2.6: *The number of cones $\Pi_0\mathscr{U}_{n-2}$ in $PG(n, q)$ is $\theta(n)\theta(n-1) \times |\mathscr{N}(\mathscr{U}_{n-2})|$.*

Proof: The number of cones = number of Π_{n-1} in $PG(n, q)$ × number of Π_0 in Π_{n-1} × number of $\mathscr{U}_{n-2}$ in a fixed Π_{n-2} of Π_{n-1}. □

Define the notation:

$$\overline{[r, s]} = \prod_{i=r}^{i=s} \{(\sqrt{q})^i - (-1)^i\} \quad \text{for } r \leq s \cdot$$

$$\overline{[r, s]} = 1 \quad \text{for } r > s.$$

Theorem 23.2.7:

$$|\mathscr{N}(\mathscr{U}_n)| = q^{n(n+1)/4}[2, n+1]_-/\overline{[2, n+1]}.$$

Proof: Count $\{(\mathscr{U}_n, \mathscr{S}) | \mathscr{U}_n$ a Hermitian variety, $\mathscr{S}$ a tangent cone of $\mathscr{U}_n\}$ in two ways:

$$|\mathscr{N}(\mathscr{U}_n)|\mu_n = \theta(n)\theta(n-1)|\mathscr{N}(\mathscr{U}_{n-2})|(q-1)q^{n-1}\sqrt{q}.$$

Hence

$$|\mathscr{N}(\mathscr{U}_n)| = \frac{\theta(n)\theta(n-1)(q-1)q^{n-1}\sqrt{q}}{\mu_n}|\mathscr{N}(\mathscr{U}_{n-2})|.$$

Repetition of this recurrence relation gives the result, after a separate calculation for n even and odd; it is only required that $|\mathscr{N}(\mathscr{U}_0)| = 1$ and $|\mathscr{N}(\mathscr{U}_1)| = \sqrt{q}(q+1)$, §6.2. □

$G = G(\mathscr{U}_n)$, the *unitary* group, is the group of projectivities fixing $\mathscr{U}_n$.

Corollary: $|G| = |G(\mathscr{U}_n)| = |PGU(n+1, q)| = q^{n(n+1)/4}\overline{[2, n+1]}$.

Proof: All $\mathcal{U}_n$ are projectively equivalent; in other words $\mathcal{N}(\mathcal{U}_n)$ is a single orbit under $PGL(n+1, q)$. So

$$|G| = |PGL(n+1, q)|/|\mathcal{N}(\mathcal{U}_n)|.$$

As $|PGL(n+1, q)| = q^{n(n+1)/2}[2, n+1]_-$, the result follows. □

Lemma 23.2.8: *If $\Pi_m \cap \mathcal{U}_n = \Pi_v \mathcal{U}_s$, then the polar space Π'_{n-m-1} of Π_m satisfies the following:*

(i) *$\Pi'_{n-m-1} \cap \mathcal{U}_n$ also has singular space Π_v;*
(ii) *$\Pi_m \cap \Pi'_{n-m-1} = \Pi_v$.*

Proof: $P \in \Pi_v$ if and only if every line l through P in Π_m is either a tangent or lies in $\mathcal{U}_n$. This is precisely the condition for P to be conjugate to every point of Π_m. Hence $\Pi_v \subset \Pi'_{n-m-1}$. Since $P \in \Pi_v \subset \Pi_m$ is conjugate to every point of Π'_{n-m-1}, so Π_v belongs to the singular space Π'_w of $\Pi'_{n-m-1} \cap \mathcal{U}_n$. Similarly, $\Pi'_w \subset \Pi_v$ and consequently $\Pi'_w = \Pi_v$.

Since every point P of $\Pi_m \cap \Pi'_{n-m-1}$ is conjugate to all points of $\Pi_m \cup \Pi'_{n-m-1}$, it belongs to the singular space Π_v of $\Pi_m \cap \mathcal{U}_n$ and of $\Pi'_{n-m-1} \cap \mathcal{U}_n$. Hence $\Pi_v = \Pi_m \cap \Pi'_{n-m-1}$. □

Corollary: *If $\Pi_m \cap \mathcal{U}_n = \Pi_v \mathcal{U}_s$, then $\Pi'_{n-m-1} \cap \mathcal{U}_n = \Pi_v \mathcal{U}_{T-1}$, where Π_m and Π'_{n-m-1} are polar spaces, and $T = n - 2m + s$.* □

23.3 Generators and subgenerators

A *generator* of $\mathcal{U}_n$ is a subspace of maximum dimension lying on $\mathcal{U}_n$. A *subgenerator* of $\mathcal{U}_n$ is any subspace lying on $\mathcal{U}_n$. The dimension of a generator is the *projective index* of $\mathcal{U}_n$, as for quadrics, and is denoted $g = g(\mathcal{U}_n)$.

Lemma 23.3.1: $g(\mathcal{U}_n) = [\frac{1}{2}(n-1)]$.

Proof: If Π_g is a generator, then by Lemma 23.2.2 it lies in its own tangent space Π'_{n-g-1}. So $g \le n - g - 1$ and hence $g \le \frac{1}{2}(n-1)$.

When n is even, $\mathbf{V}(x_0, x_1, x_3, \ldots, x_{n-1})$ is a space of dimension $\frac{1}{2}n - 1$ on

$$\mathbf{V}(x_0\bar{x}_0 + (x_1\bar{x}_2 + x_2\bar{x}_1) + \cdots + (x_{n-1}\bar{x}_n + x_n\bar{x}_{n-1}));$$

when n is odd, $\mathbf{V}(x_0, x_2, x_4, \ldots, x_{n-1})$ is a space of dimension $\frac{1}{2}(n-1)$ on

$$\mathbf{V}((x_0\bar{x}_1 + x_1\bar{x}_0) + \cdots + (x_{n-1}\bar{x}_n + x_n\bar{x}_{n-1})).$$

So the upper bound is achieved. □

Now, the number of subgenerators of a given dimension is calculated.

Theorem 23.3.2: *The number of* Π_r *on* $\mathscr{U}_n$ *is*

$$\nu_{r,n} = N(\Pi_r, \mathscr{U}_n) = \overline{[n - 2r, n + 1]}/[1, r + 1]_-. \tag{23.11}$$

Proof: We count the following set in two ways:

$$\{(\Pi_r, \Pi_{r-1}) | \Pi_r \subset \mathscr{U}_n, \Pi_{r-1} \subset \Pi_r\}.$$

Hence

$$\nu_{r,n}\theta(r) = \nu_{r-1,n}M, \tag{23.12}$$

where M is the number of Π_r on $\mathscr{U}_n$ through a given Π_{r-1}.

To calculate M, consider the polar Π'_{n-r} of Π_{r-1}. It contains Π_{r-1} and meets $\mathscr{U}_n$ in $\Pi_{r-1}\mathscr{U}_{n-2r}$ with the same Π_{r-1} as vertex, by Lemma 23.2.8. So take a Π_{n-2r} skew to Π_{r-1} in Π_{n-r}. It meets $\mathscr{U}_n$ in a $\mathscr{U}_{n-2r}$ whose points joined to Π_{r-1} form the Π_r of $\mathscr{U}_n$ containing Π_{r-1}. Thus $M = \mu_{n-2r}$.

Now, (23.12) becomes

$$\nu_{r,n}\theta(r) = \nu_{r-1,n}\mu_{n-2r}. \tag{23.13}$$

Iteration gives that

$$\nu_{r,n} = \frac{\mu_{n-2r}\mu_{n-2(r-1)} \cdots \mu_n}{\theta(r)\theta(r-1) \cdots \theta(0)}.$$

Since $\theta(i) = (q^{i+1} - 1)/(q - 1)$ and $\mu(j)$ is given by (23.8), the result follows. □

23.4 Sections of $\mathscr{U}_n$

If $\Pi_m \subset PG(n, q)$, then from Theorem 23.1.1 there are $m + 1$ types of Hermitian varieties in Π_m, namely $\Pi_{m-s-1}\mathscr{U}_s$ for $s = 0, \ldots, m$. However, when the intersection of Π_m with $\mathscr{U}_m$ is considered, it may also occur that Π_m lies entirely on $\mathscr{U}_n$. Suppose therefore that

$$\Pi_m \cap \mathscr{U}_n = \Pi_v\mathscr{U}_s; \tag{23.14}$$

then

$$s + v = m - 1. \tag{23.15}$$

The section is non-singular when $v = -1$, and Π_m lies entirely on $\mathscr{U}_n$ when $s = -1$. The parameters n, m, v, s satisfy the following:

$$-1 \leq v \leq m, \tag{23.16}$$

$$-1 \leq s \leq m, \tag{23.17}$$

$$0 \leq m \leq n - 1. \tag{23.18}$$

The question now is to determine for what values of v and s there is a section as in (23.14).

Lemma 23.4.1: *There is a section* $\Pi_m \cap \mathscr{U}_n = \Pi_v \mathscr{U}_s$ *if and only if*

$$T \geq 0, \tag{23.19}$$

where

$$T = n - 2m + s. \tag{23.20}$$

Proof: When $m \leq [\frac{1}{2}(n-1)]$, the condition $T \geq 0$ means that $m \geq s \geq -1$, so that (23.19) is equivalent to (23.17); then it must be shown that $\Pi_v \mathscr{U}_s$ exists for each s in $I = \{-1, 0, 1, \ldots, m\}$ or equivalently for each v in I. When $m > [\frac{1}{2}(n-1)]$, condition (23.19) means that $m \geq s \geq 2m - n$ or equivalently $-1 \leq v \leq n - m - 1$. Thus, if the result is established for $m \leq [\frac{1}{2}(n-1)]$, the polarity gives the result for $m > [\frac{1}{2}(n-1)]$.

Now, let $\Pi_m = \mathbf{U}_0 \mathbf{U}_1 \cdots \mathbf{U}_m = \mathbf{V}(x_{m+1}, \ldots, x_n)$ with $m \leq [\frac{1}{2}(n-1)]$. For $-1 \leq s \leq m$, write

$$K_s = \sum_{i=0}^{s} x_i \bar{x}_i + \sum_{j=s+1}^{m} (\bar{x}_j x_{2m+1-j} + x_j \bar{x}_{2m+1-j}) + \sum_{i=m+1}^{n} x_i \bar{x}_i.$$

Then $\mathbf{V}(K_s)$ is non-singular and

$$\mathbf{V}(K_s) \cap \Pi_m = \mathbf{V}\left(\sum_{i=0}^{s} x_i \bar{x}_i, x_{m+1}, \ldots, x_n\right)$$
$$= \Pi_v \mathscr{U}_s. \quad \square$$

Corollary: *For fixed m and n, the number of projectively distinct sections* $\Pi_m \cap \mathscr{U}_n$ *of* $\mathscr{U}_n$ *is*

$m + 2$ *when* $m \leq [\frac{1}{2}(n-1)]$,

$n - m + 1$ *when* $m > [\frac{1}{2}(n-1)]$. $\square$

The next result looks at the orbits of subspaces Π_m under the action of the unitary group $G(\mathscr{U}_n) = PGU(n+1, q)$.

Theorem 23.4.2: (i) *Two subspaces are in the same orbit of* $G(\mathscr{U}_n)$ *if and only if they have the same parameters m and s.*

(ii) *If there is a projectivity* $\mathfrak{T}: \Pi_m \to \Pi'_m$ *such that* $(\Pi_m \cap \mathscr{U}_n)\mathfrak{T} = \Pi'_m \cap \mathscr{U}_n$, *then* $\mathfrak{T}$ *can be extended to an element of* $G(\mathscr{U}_n)$.

Proof: We reduce $\mathscr{U}_n$ and Π_m simultaneously to canonical form. Consider the section $\Pi_v \mathscr{U}_s$, where $\mathscr{U}_s \subset \Pi_s$; thus Π_v and Π_s are skew. The polar of Π_m is Π'_t with

$$t = n - m - 1.$$

By Lemma 23.2.8, Π'_t contains Π_v. Choose in Π'_t a space Π'_w skew to Π_v with $\Pi'_w\Pi_v = \Pi'_t$; so

$$w = t - v - 1 = n - m - v - 2.$$

Since $w \geq -1$, (23.19) is satisfied. By construction, Π'_w is conjugate to Π_m. The polar Π_d of Π'_w has dimension

$$d = n - w - 1 = m + v + 1.$$

Also Π_d contains Π_m. Choose in Π_d a space Π'_v skew to Π_m with $\Pi'_v\Pi_m = \Pi_d$ in a way that is specified below.

The set of points P of Π_m such that P is conjugate to every point of Π_m is Π_v, and Π_v is the same set with respect to Π'_t. Hence, when $w \neq -1$, Π'_w meets $\mathscr{U}_n$ in a non-singular Hermitian variety. So Π'_w and its polar space Π_d are skew, as any intersection would be singular for such a Hermitian variety, Lemma 23.2.8. Thus $\Pi'_w\Pi_d = PG(n, q)$ and $\Pi_d \cap \mathscr{U}_n$ is non-singular.

Take in Π_d the polar space Π_{2v+1} of Π_s with respect to $\mathscr{U}_n \cap \Pi_d$. In Π_{2v+1} take Π'_v skew to Π_v with $\Pi'_v \cap \mathscr{U}_n$ non-singular. Then, with $S_m = \Pi'_v\Pi_s$, we have $S_m \cap \Pi_v = \varnothing$ and $S_m \cap \mathscr{U}_n$ is non-singular. Also, in S_m, the polar of Π_s with respect to $S_m \cap \mathscr{U}_n$ is the v-space Π'_v, which is skew to both Π_v and Π_s. Thus $\Pi'_v \cap \mathscr{U}_n$ is non-singular and the polarity $\mathfrak{U}$ of $\mathscr{U}_n$ induces a reciprocity (§2.1) between Π'_v and Π_v.

Thus there are four mutually skew spaces

$$\Pi_s, \Pi_v, \Pi'_w, \Pi'_v.$$

By construction, the spaces Π_s, Π'_w, Π'_v are skew and mutually conjugate. Their join is a space Π_e with

$$e = s + w + v + 2 = m + w + 1,$$

and $\Pi_e \cap \mathscr{U}_n$ is non-singular. For, if P is singular on $\Pi_e \cap \mathscr{U}_n$, then P is conjugate to all points of Π'_w and to all points of S_m; so $P \in \Pi'_w \cap S_m$, a contradiction.

As in the derivation of the canonical form following Lemma 23.2.2, in the three spaces Π_s, Π'_w, Π'_v one can choose $s + 1$, $w + 1$, $v + 1$ points respectively so that they form $e + 1$ independent points in Π_e with any two of these points conjugate.

The reciprocity induced between Π'_v and Π_v by $\mathfrak{U}$ transforms the $v + 1$ points of Π'_v, which may be considered as vertices of a simplex, to $v + 1$ faces of a simplex in Π_v; hence the vertices of this simplex in Π_v form $v + 1$ independent points. These $v + 1$ and the above $e + 1$ give $n + 1$ independent points. Thus, in a suitable order, these points are the vertices of the simplex of reference of a coordinate system. Correspondingly $\Pi_m = \Pi_v\Pi_s$ has equations

$$x_{m+1} = x_{m+2} = \cdots = x_n = 0.$$

Table 23.2

	Π_s	Π_v	Π'_w	Π'_v
Π_s	H_1	0	0	0
Π_v	0	0	0	H_4
Π'_w	0	0	H_2	0
Π'_v	0	$\bar{H}_4$	0	H_3

The equation of $\mathscr{U}_n$ becomes $XH\bar{X}^* = 0$ with $\bar{H}^* = H$, where H denotes the matrix of Table 23.2; here H_1, H_2, H_3, H_4 are square diagonal submatrices, the three matrices H_1, H_2, and H_3 have all their elements in $GF(\sqrt{q})$, and the elements of $\bar{H}_4$ are the conjugates of the elements of H_4. The zero submatrices indicate the conjugacy of the spaces bordering the matrix. A suitable choice of the unit point allows H_i to be reduced to the identity matrix of the appropriate order. Thus the simultaneous reduction of $\mathscr{U}_n$ and $\Pi_m \cap \mathscr{U}_n$ to canonical form has been achieved.

The first part of the theorem now follows since the canonical forms are determined by the integers n, m, v.

To prove the second part, let us observe that in the reduction to canonical form, the $m + 1$ reference points chosen in Π_m consist of $v + 1$ in Π_v and $s + 1$ in Π_s every two of which are conjugate. Now, we show that for any choice of $v + 1$ independent points $P_0, \ldots, P_v$ in Π_v there exists Π'_v in Π_{2v+1} and independent points $P'_0, \ldots, P'_v$ in Π'_v such that $P'_i \notin \mathscr{U}_n$, $\Pi_v \cap \Pi'_v = \emptyset$, $\Pi'_v \cap \mathscr{U}_n$ is non-singular, $P'_0, \ldots, P'_v$ are mutually conjugate, and P_i is conjugate to P'_j for $i \neq j$. Therefore it is sufficient to show that for any $v + 1$ independent points $Q_0, \ldots, Q_v$ in Π_v there is a projectivity $\mathfrak{S}$ fixing $\mathscr{U}_n \cap \Pi_{2v+1}$ and Π_v and with $P_i\mathfrak{S} = Q_i$, $i = 0, 1, \ldots, v$. Choose a space Π''_v on $\mathscr{U}_n \cap \Pi_{2v+1}$ skew to Π_v and choose independent points $P''_0, \ldots, P''_v$ in Π''_v with P_i conjugate to P''_j, $i \neq j$. With respect to the reference points $P_0, \ldots, P_v$, $P''_0, \ldots, P''_v$, and a suitable unit point, $\mathscr{U}_n \cap \Pi_{2v+1}$ is represented by the canonical form G_{2v+1} as in (23.7). This proves the existence of the projectivity $\mathfrak{S}$.

Since the projectivity $\mathfrak{T}$ from Π_m to Π'_m transforms the chosen $(m + 1)$-tuple in Π_m to such an $(m + 1)$-tuple in Π'_m, by the preceding paragraph the reference points and the unit point may be chosen in a new system of coordinates $(x'_0, \ldots, x'_n)$ so that Π'_m has equations

$$x'_{m+1} = \cdots = x'_n = 0$$

and so that $\mathfrak{T}$ has equations

$$x'_0 : x'_1 : \cdots : x'_m = c_0 x_0 : c_1 x_1 : \cdots : c_m x_m$$

while $\mathscr{U}_n$ preserves its equation in these coordinates apart from changing x

to x'. Since $\Pi_m \cap \mathcal{U}_n$ and $\Pi'_m \cap \mathcal{U}_n$ have the same equation, apart from changing x to x', we have $c_0\bar{c}_0 = \cdots = c_s\bar{c}_s$. The projectivity $\mathfrak{T}$ has an extension to $PG(n, q)$ with equations

$$x'_i = \begin{cases} c_i x_i, & i = 0, \ldots, m \\ c_0 x_i, & i = m+1, \ldots, m+w+1 \\ c_{i+m-n} x_i, & i = m+w+2, \ldots, n; \end{cases}$$

this is an element of $G(\mathcal{U}_n)$. Thus the theorem is established. □

Now, the size of the orbits in the previous theorem must be determined. Let $N(\Pi_v \mathcal{U}_{m-v-1}, \mathcal{U}_n)$ be the number of m-spaces Π_m meeting $\mathcal{U}_n$ in a section of type $\Pi_v \mathcal{U}_{m-v-1}$. This number was determined in Theorem 23.3.2 in the case that $v = m$, that is when Π_m lies on $\mathcal{U}_n$.

Theorem 23.4.3:

$$N(\Pi_v \mathcal{U}_s, \mathcal{U}_n) = q^{(1/2)T(s+1)} \overline{[s+2, n+1]} / \{\overline{[1, T]}[1, v+1]_-\}$$

where

$$s + v = m - 1 \tag{23.15}$$

$$T = n - 2m + s. \tag{23.20}$$

Proof: $N(\Pi_v \mathcal{U}_s, \mathcal{U}_n)$ is equal to the product of $N(\Pi_v, \mathcal{U}_n)$ and the number of sections $\Pi_v \mathcal{U}_s$ with given Π_v.

So let Π_v be given and let Π_{n-v-1} be skew to Π_v with $\Pi_{n-v-1} \cap \mathcal{U}_n$ non-singular; put $\Pi_{n-v-1} \cap \mathcal{U}_n = \mathcal{U}_{n-v-1}$. If Π'_{n-v-1} is the tangent space of $\mathcal{U}_n$ at Π_v, then $\Pi'_{n-v-1} \cap \mathcal{U}_{n-v-1}$ is a non-singular $\mathcal{U}_{n-2v-2}$. Each section $\Pi_v \mathcal{U}_s$ is contained in Π'_{n-v-1}, by Lemma 23.2.2, and $\Pi_v \mathcal{U}_s \cap \mathcal{U}_{n-2v-2}$ is a non-singular $\mathcal{U}'_s$. Conversely, each non-singular $\mathcal{U}'_s$ on $\mathcal{U}_{n-2v-2}$ defines a section $\Pi_v \mathcal{U}'_s$ of the prescribed type. Hence the number of sections $\Pi_v \mathcal{U}_s$ with given Π_v is $N(\Pi_{-1} \mathcal{U}_s, \mathcal{U}_{n-2v-2})$.

Now, we calculate $N(\Pi_{-1} \mathcal{U}_s, \mathcal{U}_r) = \rho_{s,r}$. Count in two ways the set

$$\{(\Pi_s, \Pi_{s+1}) \mid \Pi_s \subset \Pi_{s+1}, \Pi_s \cap \mathcal{U}_r \text{ non-singular}, \Pi_{s+1} \cap \mathcal{U}_r \text{ non-singular}\}.$$

For a given Π_s, the number of such pairs (Π_s, Π_{s+1}) is the number of points of Π'_{r-s-1} not on $\mathcal{U}_r \cap \Pi'_{r-s-1}$, with Π'_{r-s-1} the polar space of Π_s. Hence

$$\rho_{s,r}\{\theta(r-s-1) - \mu_{r-s-1}\} = \rho_{s+1,r}\{\theta(s+1) - \mu_{s+1}\}.$$

Since $\rho_{r-1,r}$ is the number of non-tangent primes to $\mathcal{U}_r$ in $PG(r, q)$, so

$$\rho_{r-1,r} = \theta(r) - \mu_r.$$

Hence

$$\rho_{s,r} = \frac{\{\theta(s+1) - \mu_{s+1}\}\{\theta(s+2) - \mu_{s+2}\} \cdots \{\theta(r) - \mu_r\}}{\{\theta(1) - \mu_1\}\{\theta(2) - \mu_2\} \cdots \{\theta(r-s-1) - \mu_{r-s-1}\}}.$$

By (23.9),

$$\rho_{s,r} = \frac{\{q^{s+1}\sqrt{q} + (-\sqrt{q})^{s+1}\}\{q^{s+2}\sqrt{q} + (-\sqrt{q})^{s+2}\}\cdots\{q^{r}\sqrt{q} + (-\sqrt{q})^{r}\}}{\{\sqrt{q} + 1\}\{q\sqrt{q} + (-\sqrt{q})\}\{q^{2}\sqrt{q} + (-\sqrt{q})^{2}\}\cdots\{q^{r-s-1}\sqrt{q} + (-\sqrt{q})^{r-s-1}\}}$$

$$= q^{(1/2)(s+1)(r-s)} \frac{\{(\sqrt{q})^{s+2} - (-1)^{s+2}\}\cdots\{(\sqrt{q})^{r+1} - (-1)^{r+1}\}}{\{\sqrt{q} + 1\}\{(\sqrt{q})^{2} - 1\}\cdots\{(\sqrt{q})^{r-s} - (-1)^{r-s}\}}$$

$$= q^{(1/2)(s+1)(r-s)}\overline{[s+2, r+1]}/\overline{[1, r-s]}.$$

Hence, for $r = n - 2v - 2$,

$$\rho_{s,n-2v-2} = q^{(1/2)(s+1)T}\overline{[s+2, n-2v-1]}/\overline{[1, T]}.$$

Since $N(\Pi_v, \mathscr{U}_n) = \overline{[n-2v, n+1]}/[1, v+1]_-$ by (23.11), we finally have that

$$N(\Pi_v \mathscr{U}_s, \mathscr{U}_n) = q^{(1/2)T(s+1)}\overline{[s+2, n+1]}/\{\overline{[1, T]}[1, v+1]_-\}. \quad \square$$

23.5 The characterization of Hermitian varieties

This is a continuation of the treatment in §19.4, where Hermitian surfaces were characterized as subsets of $PG(3, q)$ meeting every line in 1, r, or $q + 1$ points with some further restrictions. Some definitions and results are recalled. A subset $\mathscr{K}$ of $PG(n, q)$ is a $k_{r,n,q}$ if r is a fixed integer with $1 \leq r \leq q$ such that

(i) $|\mathscr{K}| = k$;
(ii) $|l \cap \mathscr{K}| = 1, r$, or $q + 1$ for each line l;
(iii) $|l \cap \mathscr{K}| = r$ for some line l.

It is clear that a $k_{1,n,q}$ is a prime. A $k_{2,n,2}$ is the complement of a cap with at least one unisecant; the only cap of $PG(n, 2)$ with no unisecant is the complement of a prime.

From now on, assume that $q > 2$.

From Theorem 19.4.4 there are seven types of plane section $\mathscr{K}'$, $|\mathscr{K}'| = k'$, of such a $\mathscr{K}$:

I. a Hermitian arc (unital), that is a set of type $(1, \sqrt{q} + 1)$ with $r = \sqrt{q} + 1$, $k' = q\sqrt{q} + 1$;
II. a subplane $PG(2, \sqrt{q})$, that is a set of type $(1, \sqrt{q} + 1)$ with $r = \sqrt{q} + 1$, $k' = q + \sqrt{q} + 1$;
III. a set of type $(0, r - 1)$ plus an external line, whence $k' = (r - 1)q + r$ and $r - 1 | q$;
IV. the complement of a set of type $(0, q + 1 - r)$, whence $k' = r(q + 1)$ and $(q + 1 - r) | q$;

V. r concurrent lines, whence $k' = rq + 1$;
VI. a single line, $k' = q + 1$;
VII. a plane, $k' = q^2 + q + 1$.

A point P of $\mathscr{K}$ is *singular* if every line through P is either a unisecant or a line of $\mathscr{K}$. The set $\mathscr{K}$ is called *singular* or *non-singular* according as it has singular points or not. In $PG(3, q)$, the fundamental result is contained in Theorem 19.5.13 and §19.6.

Theorem 23.5.1: *Let $\mathscr{K}$ be a $k_{r,3,q}$ in $\Pi_3 = PG(3, q)$.*
(i) *When $r = 1$, $\mathscr{K}$ is a plane.*
(ii) *When $r = 2$, then $\mathscr{K}$ is one of*
(a) $\Pi_2 \cup \Pi_0$, (b) $\Pi_2 \cup \Pi_1$, (c) $\Pi_2 \cup \Pi_2'$.
(iii) *When $r = q$, then $\mathscr{K}$ is one of*
(a) $(\Pi_3 \backslash \Pi_2) \cup \Pi_1$ *with* $\Pi_1 \subset \Pi_2$, (b) $(\Pi_3 \backslash \Pi_1) \cup \Pi_0$ *with* $\Pi_0 \subset \Pi_1$, (c) $\Pi_3 \backslash \Pi_0$.
(iv) *When $3 \le r \le q - 1$, then one of the following occurs.*
(a) *If $\mathscr{K}$ is singular, then $\mathscr{K}$ is r planes through a line or a cone $\Pi_0\mathscr{K}'$ with base $\mathscr{K}'$ a plane of type* I, II, III, *or* IV *as above.*
(b) *If $\mathscr{K}$ is non-singular, then*
(1) *for q odd, $r = \sqrt{q} + 1$ and $\mathscr{K} = \mathscr{U}_{3,q}$;*
(2) *for q even, and $q > 4$, either $r = \sqrt{q} + 1$ and $\mathscr{K} = \mathscr{U}_{3,q}$ or $n = \frac{1}{2}q + 1$ and $\mathscr{K} = \mathscr{R}_3$, the projection of a quadric;*
(3) *for $q = 4$, $\mathscr{K} = \mathscr{U}_{3,4}$ or $\mathscr{K} = \mathscr{R}_3$ or $\mathscr{K}$ contains sections of type* II. □

A similar result is true in $PG(n, q)$ and we begin by dealing with singular sets. As for $n = 3$, the study of singular sets $k_{r,n,q}$ reduces to that of non-singular ones.

Lemma 23.5.2: *The singular points of a $k_{r,n,q}$ form a subspace.* □

The subspace of singular points of $\mathscr{K}$ is the *singular space* of $\mathscr{K}$.

Theorem 23.5.3: *If $\mathscr{K}$ is a singular $k_{r,n,q}$ with singular space Π_d, then one of the following holds:*
(i) *$d = n - 1$ and $\mathscr{K}$ is a prime;*
(ii) *$d = n - 2$ and $\mathscr{K}$ is r primes through Π_d with $r > 1$;*
(iii) *$d \le n - 3$ and $\mathscr{K} = \Pi_d\mathscr{K}'$ where $\mathscr{K}'$ is a non-singular $k_{r,n-d-1,q}$.* □

Now we consider the cases $r = 2$ and $r = q$.

Theorem 23.5.4: *If $\mathscr{K}$ is a $k_{2,n,q}$, then $\mathscr{K} = \Pi_{n-1} \cup \Pi_i'$ for some Π_i' not contained in Π_{n-1}.*

Proof: (a) $n = 2$. Suppose $\mathcal{K}$ contains no line. Then $\mathcal{K}$ is a k-arc meeting every line of the plane, a contradiction. So $\mathcal{K}$ contains a line Π_1. If $\mathcal{K}\backslash\Pi_1$ contains a 3-arc then $\mathcal{K}$ is the whole plane. Hence it follows that $\mathcal{K} = \Pi_1 \cup \Pi'_1$ or $\Pi_1 \cup \Pi_0$.

(b) $n > 2$

(i) $\mathcal{K}$ *is non-singular*

We proceed by induction on n. Since not every plane is contained in $\mathcal{K}$, some plane meets $\mathcal{K}$ in a line or a $k'_{2,2,q}$. Hence $\mathcal{K}$ has a unisecant l with point of contact Q.

Let Π_{n-1} be a prime of $PG(n, q)$ not containing Q and let $\mathcal{K}'$ be the projection of $\mathcal{K}\backslash\{Q\}$ from Q onto Π_{n-1}. If l' is any line of Π_{n-1}, then $Ql' \cap \mathcal{K} = \Pi_2, \Pi_1 \cup \Pi'_1, \Pi_1 \cup \Pi_0$, or Π_1, by (a). Hence $|l' \cap \mathcal{K}'| = 1, 2$, or $q + 1$. Let $l \cap \Pi_{n-1} = \{R\}$; then at least one line m through R in Π_{n-1} meets $\mathcal{K}'$ in two points. For, otherwise, every plane through l meets $\mathcal{K}$ in a line and so Q is singular, a contradiction. So $\mathcal{K}'$ is a $k'_{2,n-1,q}$. Hence, by the induction hypothesis, $\mathcal{K}' = \Pi_{n-2} \cup \Pi'_d$ for $0 \le d \le n - 2$. Consequently, $\mathcal{K} \subset Q\Pi_{n-2} \cup Q\Pi'_d$.

(1) $d < n - 2$. If $S \in Q\Pi_{n-2}\backslash\mathcal{K}$, then there is a line through S with no point in $\mathcal{K}$, a contradiction; so $Q\Pi_{n-2} \subset \mathcal{K}$. If P_1 and P_2 are points of $\mathcal{K}\backslash Q\Pi_{n-2}$, the line P_1P_2 meets $Q\Pi_{n-2}$ and so contains a third point of $\mathcal{K}$; therefore $P_1P_2 \subset \mathcal{K}$. Hence, if $|\mathcal{K}\backslash Q\Pi_{n-2}| \ge 2$, then $\mathcal{K}\backslash Q\Pi_{n-2}$ is an affine subspace of the affine space $PG(n, q)\backslash Q\Pi_{n-2}$ (since $q > 2$). In this case, $\mathcal{K} = Q\Pi_{n-2} \cup \Pi''_t$ with $t \ge 1$; since $\mathcal{K}$ is non-singular, $Q\Pi_{n-2} \cap \Pi''_t = \emptyset$, a contradiction. Therefore $\mathcal{K} = \Pi_{n-1} \cup \Pi_0$ with Π_0 not in Π_{n-1}.

(2) $d = n - 2$. If $Q\Pi_{n-2} \not\subset \mathcal{K}$ and $Q\Pi'_{n-2} \not\subset \mathcal{K}$, then there is a line with no point in $\mathcal{K}$, a contradiction. So, let $Q\Pi_{n-2} \subset \mathcal{K}$. Now, proceed exactly as in (1).

(ii) $\mathcal{K}$ *is singular*

The result follows from Theorem 23.5.3. □

Now we consider the case $r = q$.

Theorem 23.5.5: *If* $\mathcal{K}$ *is a* $k_{q,n,q}$ *in* $\Pi_n = PG(n, q)$, $q \ge 3$, *then* $\mathcal{K} = (\Pi_n\backslash\Pi_i) \cup \Pi_{i-1}$ $(0 \le i \le n - 1)$ *with* $\Pi_{i-1} \subset \Pi_i$.

Proof: Let Q be a point of $\Pi_n\backslash\mathcal{K}$. If l_1 and l_2 are two unisecants through Q with points of contact P_1 and P_2, the line P_1P_2 either belongs to $\mathcal{K}$ or is a q-secant. Since $q \ge 3$, there is another point P_3 of $\mathcal{K}$ on P_1P_2. Each line l of the plane $\pi = l_1l_2$ through P_3 other than P_1P_2 and QP_3 is a unisecant of $\mathcal{K}$ since $l \cap l_1$ and $l \cap l_2$ are not in $\mathcal{K}$. It follows that every line of π through P_1 other than P_1P_2 is also a unisecant and that all points of $\mathcal{K}$ in π lie on P_1P_2. As $\mathcal{K}$ has no external lines, so $\mathcal{K} \cap \pi = P_1P_2$. So it has been shown that, for any two unisecants through Q, all the lines of the pencil determined by these two are unisecants and that the plane of the pencil meets $\mathscr{K}$ in a line. Hence the unisecants through Q generate a Π_i meeting $\mathscr{K}$ in a Π_{i-1}. Since every line through Q not in this Π_i is a q-secant, it follows that $\mathscr{K}$ consists of the points of Π_{i-1} plus the points not in Π_i. □

The previous two theorems mean that the rest of the characterization can be restricted to

$$3 \leq r \leq q - 1. \tag{23.21}$$

Lemma 23.5.6: *If $\mathscr{K}$ is a $k_{r,n,q}$ with $3 \leq r \leq q-1$ such that $\mathscr{K}$ has a section $\mathscr{K}'$ by a plane π containing a triangle of lines of $\mathscr{K}$ with $\pi \not\subset \mathscr{K}$, then one of the following occurs:*

(i) $\mathscr{K} = \Pi_{n-3}\mathscr{K}'$, *where $\mathscr{K}'$ is a section of type* IV;

(ii) $q = 2^h$, $r = \frac{1}{2}q + 1$, *and the singular space of $\mathscr{K}$ has dimension at most* $n - 4$.

Proof: Since π is not contained in $\mathscr{K}$, the section $\mathscr{K}'$ must be of type IV; that is, $\mathscr{K}'$ is the complement in π of a maximal arc of type $(0, q+1-r)$. So, by Theorem 12.2.1, $q \equiv 0 \pmod{q - r + 1}$; hence, with $q = p^h$,

$$p^h - r + 1 = p^m \tag{23.22}$$

for some m with $0 < m < h$. Thus

$$r - 1 = p^m(p^{h-m} - 1). \tag{23.23}$$

This gives two possibilities:

(a) $r - 1 \neq p^m$ in which case every plane section of $\mathscr{K}$ is of type IV, V, VI, or VII;

(b) $r - 1 = p^m$ and $p^{h-m} - 1 = 1$, whence $p = 2$, $m = h - 1$, and $r = \frac{1}{2}q + 1$.

In case (a), let Π_3 be a solid through π. By the corollary to Lemma 19.4.7, there is a unisecant l_1 of $\Pi_3 \cap \mathscr{K}$ and so of $\mathscr{K}$ with point of contact P. Let l be any line through P other than l_1 and let $\pi_1 = ll_1$. The section $\pi_1 \cap \mathscr{K}$ cannot be of type IV or VII since they do not have unisecants. So $\pi_1 \cap \mathscr{K}$ is of type V or VI, that is r lines of a pencil or a single line; in both these cases, P is singular. So every solid through π meets the singular space S of $\mathscr{K}$, while S does not meet π since $\pi \cap \mathscr{K}$ is non-singular. Hence $S = \Pi_{n-3}$ and $\mathscr{K} = \Pi_{n-3}\mathscr{K}'$.

In case (b), (i) may also occur. If it does not, then the dimension of S is at most $n - 4$. □

This section continues the investigation under the hypothesis that a $k_{r,n,q}$ has no section of type IV. The case when such a section occurs is investigated in §23.6.

Lemma 23.5.7: *Let $\mathscr{K}$ be a $k_{r,n,q}$ with $3 \le r \le q-1$ such that* (a) *$\mathscr{K}$ contains a prime Π_{n-1} with $\mathscr{K} \ne \Pi_{n-1}$,* (b) *$\mathscr{K}$ has no plane section of type* IV. *Then $\mathscr{K}$ is the union of r primes of a pencil or $\mathscr{K} = \Pi_{n-3}\mathscr{K}'$ with $\mathscr{K}'$ a plane section of type* III.

Proof: Let π be a plane in Π_{n-1} and let P be a point of $\mathscr{K}\backslash\Pi_{n-1}$.

First we show that the solid $P\pi$ contains a line l of $\mathscr{K}$ not in π. Suppose otherwise and consider the set $\mathscr{K}' = (\mathscr{K} \cap P\pi)\backslash\pi$. It is a k'-set of type $(0, r-1)$ in $P\pi$ with $2 \le r-1 \le q-1$. By Lemma 19.4.7, such a set $\mathscr{K}'$ does not exist.

Let $P' = l \cap \pi$, let l' be any line through P' other than l and let $\pi' = ll'$. The plane π' meets Π_{n-1} in a line $m \ne l$. So π' has the two lines l and m in common with $\mathscr{K}$ in a section of type V or VII; in the former case, the r lines contain P'. Hence l' is either a unisecant or a line of $\mathscr{K}$; so P' is singular. Thus $\mathscr{K}$ is singular and its singular space S is necessarily in Π_{n-1}. Also every plane in Π_{n-1} meets S. Thus the dimension of S is at least $n-3$. Now Theorem 23.5.3 gives the result. □

For $n = 3$, the characterization of Hermitian surfaces was completed in §19.5.

Now, a subset $\mathscr{K}$ of $PG(n, q)$ will be called *regular* if

(a) $\mathscr{K}$ is a $k_{r,n,q}$;
(b) $3 \le r \le q-1$;
(c) $\mathscr{K}$ has no plane section of type IV.

Lemma 23.5.8: *If $\mathscr{K}$ is a regular $k_{r,n,q}$ with $n \ge 4$, then $\mathscr{K}$ cannot have plane sections of type* I *and type* II.

Proof: Let π be a plane meeting $\mathscr{K}$ in a section $\mathscr{K}'$ of type I. Then, from Theorem 23.5.1, a solid through π meets $\mathscr{K}$ either in a cone $P\mathscr{K}'$ or in a Hermitian surface $\mathscr{U}_{3,q}$. If there are M solids of the latter type, then there are $\theta(n-3) - M$ of the former. Hence

$$\begin{aligned} k &= q\sqrt{q} + 1 + Mq(q\sqrt{q} + 1) + \{\theta(n-3) - M\}\{(q-1)(q\sqrt{q} + 1) + 1\} \\ &= q\sqrt{q}(q^{n-2} + M) + \theta(n-2). \end{aligned} \tag{23.24}$$

If π_0 is a plane meeting $\mathscr{K}$ in a plane section $\mathscr{K}_0$ of type II, then every solid through π_0 meets $\mathscr{K}$ in a cone $P\mathscr{K}_0$, by Theorem 23.5.1. Hence

$$\begin{aligned} k &= q + \sqrt{q} + 1 + \theta(n-3)\{(q-1)(q + \sqrt{q} + 1) + 1\} \\ &= \theta(n-1) + q^{n-2}\sqrt{q}. \end{aligned} \tag{23.25}$$

Equating (23.24) and (23.25) gives that

$$M = q^{n-3}(1 + \sqrt{q} - q) < 0,$$

a contradiction. □

Lemma 23.5.9: *If $\mathscr{K}$ is a regular $k_{r,n,q}$ with $n \geq 4$, then $\mathscr{K}$ cannot have plane sections of type* II *and type* III.

Proof: If π is a plane meeting $\mathscr{K}$ in a section $\mathscr{K}'$ of type III, then again by Theorem 23.5.1 any solid through π meets $\mathscr{K}$ in a cone $P\mathscr{K}'$. Hence

$$\begin{aligned} k &= (r-1)q + r + \theta(n-3)\{(q-1)[(r-1)q + r] + 1\} \\ &= (r-1)q^{n-1} + rq^{n-2} + \theta(n-3). \end{aligned} \tag{23.26}$$

If there is a section of type III and of type II, then $r = \sqrt{q} + 1$. So substituting this in (23.26) and equating it to (23.25) gives $q = 1$, a contradiction. □

Lemma 23.5.10: *If $\mathscr{K}$ is a regular $k_{r,n,q}$ with $n \geq 4$, and $\mathscr{K}$ has plane sections of type* I *and type* III, *then q^{n-3} solids through a plane of type* I *meet $\mathscr{K}$ in a Hermitian surface.*

Proof: Equating (23.24) and (23.26) with $r = \sqrt{q} + 1$ implies that $M = q^{n-3}$. □

Theorem 23.5.11: *If $\mathscr{K}$ is a regular $k_{r,n,q}$ with $n \geq 3$ and if there is a plane π meeting $\mathscr{K}$ in a section $\mathscr{K}'$ of type* II, *then $\mathscr{K} = \Pi_{n-3}\mathscr{K}'$.*

Proof: The result is true for $n = 3$ by Theorem 23.5.1. So let $n \geq 4$ and proceed by induction. Thus every prime through π meets $\mathscr{K}$ in a $k'_{r,n-1,q}$ which is a cone $\Pi_{n-4}\mathscr{K}'$. The theorem will follow if it is shown that the points of any such Π_{n-4} are singular for $\mathscr{K}$. For, considering a second prime through π, it then follows that $\mathscr{K}$ has at least a Π_{n-3} of singular points, and hence exactly a Π_{n-3} of singular points.

Let Π_{n-1} be one such prime and let P be a point of the vertex Π_{n-4}, that is the singular space of $\Pi_{n-1} \cap \mathscr{K}$. Suppose that P is non-singular for $\mathscr{K}$. Let l be an r-secant of $\mathscr{K}$ through P: necessarily $r = \sqrt{q} + 1$. The line l cannot belong to Π_{n-1} and so a plane through l meets Π_{n-1} in a line l' other than l, where l' is a line or a unisecant of $\mathscr{K} \cap \Pi_{n-1}$. The number of lines through P in $\mathscr{K} \cap \Pi_{n-1}$ is

$$\{\theta(n-4) - \theta(n-5)\}(q + \sqrt{q} + 1) + \theta(n-5) = q^{n-4}\sqrt{q} + \theta(n-3); \tag{23.27}$$

this is calculated by looking at the number of lines through P in each $(n-3)$-

space $\Pi_{n-4}Q$, where Q varies in $\mathscr{K}'$. Thus the number of lines through P in Π_{n-1} but not in $\mathscr{K}$ is

$$q^{n-4}(q^2 - \sqrt{q}). \tag{23.28}$$

Therefore the number of planes through l meeting Π_{n-1} in a line of $\mathscr{K}$ is given in (23.27) and the number meeting Π_{n-1} in a unisecant of $\mathscr{K}$ by (23.28).

By Lemmas 23.5.8 and 23.5.9, each of these planes meeting Π_{n-1} in a line of $\mathscr{K}$ must be a section of type V with $r = \sqrt{q} + 1$ and so has $r(q - 1) + 1 = q\sqrt{q} + q - \sqrt{q}$ points in common with $\mathscr{K}\backslash l$. Similarly the other planes meet $\mathscr{K}$ in a section of type II and so meet $\mathscr{K}\backslash l$ in q points. Using these numbers to find the size of $\mathscr{K}\backslash l$, we have

$$k - (\sqrt{q} + 1) = (q^{n-4}\sqrt{q} + \theta(n - 3))(q\sqrt{q} + q - \sqrt{q}) \\ + q^{n-4}(q^2 - \sqrt{q})q,$$

whence

$$k = \theta(n - 1) + q^{n-2}\sqrt{q} + q^{n-3}(q - 1). \tag{23.29}$$

Comparing this with (23.25) gives $q^{n-3}(q - 1) = 0$. This contradiction proves the result. □

Lemma 23.5.12: *Let $\mathscr{K}$ be a regular $k_{r,n,q}$ with $n \geq 3$ that contains a prime Π_{n-1}. Then $\mathscr{K}$ is one of the following*: (a) *r primes of a pencil*; (b) $\Pi_{n-3}\mathscr{K}_{\mathrm{III}}$, *where $\mathscr{K}_{\mathrm{III}}$ is a plane section of type* III.

Proof: Let π be any plane in Π_{n-1}, and P be any point of $\mathscr{K}\backslash\Pi_{n-1}$. The solid $P\pi$ contains a line l of $\mathscr{K}$ not in π and therefore not in Π_{n-1}: for, otherwise, the points of $\mathscr{K} \cap (P\pi\backslash\pi)$ constitute a k'-set $\mathscr{K}'$ with $k' = (r - 2)(q^2 + q + 1) + 1$ and of type $(0, r - 1)$ in $P\pi$. Such sets do not exist by Lemma 19.4.7.

Let $P' = l \cap \pi$ and let l' be any line through P' other than l. The plane $\pi' = ll'$ meets Π_{n-1} in a line m through P'. Thus π' contains the lines l and m of $\mathscr{K}$. Hence π' meets $\mathscr{K}$ in r lines of a pencil with centre P' or lies in $\mathscr{K}$. So l' is either a line of $\mathscr{K}$ or a unisecant at P'. Thus P' is singular. Therefore the singular space Π_d of $\mathscr{K}$ lies in Π_{n-1}, and every plane of Π_{n-1} meets Π_d. Hence $d \geq n - 3$ and the result follows from Theorem 23.5.3. □

Theorem 23.5.13: *If $\mathscr{K}$ is a regular $k_{r,n,q}$ with $n \geq 4$ and if there is a plane π meeting $\mathscr{K}$ in a section $\mathscr{K}'$ of type* III, *then $\mathscr{K} = \Pi_{n-3}\mathscr{K}'$.*

Proof: Let us start with the case $n = 4$. Take two solids Π_3 and Π'_3 through π. From Theorem 23.5.1, $\Pi_3 \cap \mathscr{K} = P\mathscr{K}'$ and $\Pi'_3 \cap \mathscr{K} = P'\mathscr{K}'$. The section $\mathscr{K}'$ is an $((r - 2)q + r - 1; r - 1)$-arc plus an external line l. The points P and P' are distinct as are the planes $\alpha = Pl$ and $\alpha' = P'l$. The line PP' is skew to π, since otherwise it would belong to both Π_3 and Π'_3, and hence to π. The planes α and α' lie in $\mathscr{K}$. If it is shown that the solid $\alpha\alpha'$ lies in $\mathscr{K}$, then the result follows from Lemma 23.5.12.

So let us suppose that the solid $\alpha\alpha'$ is not contained in $\mathscr{K}$. Since $\alpha\alpha'$ contains two planes in $\mathscr{K}$, it meets $\mathscr{K}$ in r planes through the line l, by Theorem 23.5.1; hence PP' is an r-secant of $\mathscr{K}$.

Any plane β through PP' does not lie in $\mathscr{K}$ and meets π in a point B. If $B \in \mathscr{K}'$ the lines BP and BP' lie in $\mathscr{K}$ and hence $\beta \cap \mathscr{K}$ is r lines of a pencil. If $B \notin \mathscr{K}'$ the lines BP and BP' are unisecants to $\mathscr{K}$ with contacts P and P'; then $\beta \cap \mathscr{K}$ can contain no lines, as otherwise such a line would have to pass through P and P'. So in this case, $\beta \cap \mathscr{K}$ is of type I by Lemma 23.5.9 and $r = \sqrt{q} + 1$.

The planes β of the first type number $(r-1)q + r$ and each meets $\mathscr{K}$ in $r(q-1)+1$ points off the line PP'. There are $\theta(2) - \{(r-1)q + r\}$ planes β of the second type, each of which contains $q\sqrt{q} + 1 - r$ points of $\mathscr{K}\backslash PP'$. Thus a count of the points of $\mathscr{K}\backslash PP'$ on the planes through PP' gives

$$k - r = \{(r-1)q + r\}\{r(q-1)+1\} + \{q^2 + q + 1 - (r-1)q - r\} \times \{q\sqrt{q} + 1 - r\};$$

hence, with $r = \sqrt{q} + 1$,

$$k = q\sqrt{q}(q^2 + q + 1) + q + 1. \tag{23.30}$$

However, with $r = \sqrt{q} + 1$ and $n = 4$, the number k is given by (23.26); namely,

$$k = q^3\sqrt{q} + q^2(\sqrt{q} + 1) + q + 1.$$

Thus $q\sqrt{q} = q^2$, a contradiction. This proves the result for $n = 4$.

Now let $n \geq 5$. Each solid χ through π meets $\mathscr{K}$ in a cone $Q\mathscr{K}'$. It suffices to show that Q is always singular. Let m be a line through Q. If it lies in the solid χ, it cannot be an r-secant. Suppose therefore that m is not in χ and so is skew to π. The 4-space $m\pi$ meets $\mathscr{K}$ in a $k'_{r,4,q}$ that is a cone $\Pi_1\mathscr{K}'$, by the previous part of the proof. Also $\chi \cap \Pi_1 = \{Q\}$. Hence Q is singular for $k'_{r,4,q}$; so the line m is not an r-secant of $k'_{r,4,q}$ and therefore not an r-secant of $\mathscr{K}$. Thus Q is singular and the result follows. □

The previous results allow a summary for sections of a non-singular regular set.

Theorem 23.5.14: *Let $\mathscr{K}$ be a regular, non-singular $k_{r,n,q}$ and let Π_2 and Π_3 be spaces not contained in $\mathscr{K}$. Then*

(i) *$\Pi_2 \cap \mathscr{K}$ is of type* I, V, *or* VI;

(ii) *$\Pi_3 \cap \mathscr{K}$ is a plane, r planes of a pencil, $\Pi_0\mathscr{K}_1$, or $\mathscr{U}_3$, where $\mathscr{K}_1$ is a section of type* I. □

Now we continue the study of regular sets $k_{r,n,q}$ for $n \geq 4$ and $q > 4$.

Lemma 23.5.15: *Let $\mathcal{K}$ be a non-singular, regular $k_{r,n,q}$ with $n \geq 4$ and $q > 4$. Then through any point P of $\mathcal{K}$ there passes a section of type* I, *whence q is a square and $r = \sqrt{q} + 1$.*

Proof: Suppose there is no unisecant through P. Let l_r be an r-secant through P and let π', π'' be distinct planes through l_r. By Theorem 23.5.14, $\pi' \cap \mathcal{K}$ is r lines through P' and $\pi'' \cap \mathcal{K}$ is r lines through P''; also $\pi'\pi'' \cap \mathcal{K}$ is r planes through $P'P''$. It follows that P' is singular for $\mathcal{K}$, which is a contradiction. Let l_1 be a unisecant through P. By Theorem 23.5.14, the plane $\pi = l_1 l_r$ meets $\mathcal{K}$ in a section of type I. □

For any point P of a regular, non-singular set $\mathcal{K}$, the *tangent space* T_P is the union of the unisecants and lines of $\mathcal{K}$ through P.

Lemma 23.5.16: (i) *T_P is a prime;*
(ii) *the singular space of $T_P \cap \mathcal{K}$ is $\{P\}$;*
(iii) $P \neq Q \Rightarrow T_P \neq T_Q$;
(iv) *a non-tangent prime meets $\mathcal{K}$ in a non-singular $k'_{r,n-1,q}$.*

Proof: Let $\mathcal{L}$ be the set consisting of the unisecants and lines of $\mathcal{K}$ through P. By the previous lemma, there is a plane π through P meeting $\mathcal{K}$ in a Hermitian arc $\mathcal{K}'$. Let l be the unisecant to $\mathcal{K}'$ in π at P. Each of the $\theta(n-3)$ solids Π_3 through π meets $\mathcal{K}$ either as a cone $Q\mathcal{K}'$ or $\mathcal{U}_3$. Let α be the tangent plane at P to $\Pi_3 \cap \mathcal{K}$; here α must contain l. The lines through P in α belong to $\mathcal{L}$ and so are the only lines of $\mathcal{L}$ in Π_3. Also, any line of $\mathcal{L}\backslash\{l\}$ is joined to π by some Π_3 and lies in the tangent plane at P to $\Pi_3 \cap \mathcal{K}$. Since distinct solids through π give distinct tangent planes, the number of lines in $\mathcal{L}$ is $q\theta(n-3) + 1 = \theta(n-2)$.

Now consider the pencil of lines containing two lines l_1 and l_2 of $\mathcal{L}$. The plane $l_1 l_2$ cannot meet $\mathcal{K}$ in a section of type I. Hence $l_1 l_2$ is of type V, VI, or VII, by Theorem 23.5.14. In each case all the lines of the pencil are in $\mathcal{L}$. Since $|\mathcal{L}| = \theta(n-2)$, the lines of $\mathcal{L}$ must be the set of lines through P in a prime; that is, T_P is a prime.

$T_P \cap \mathcal{K}$ is a $k'_{1,n-1,q}$ or a $k'_{r,n-1,q}$ for which P is singular. Suppose that $T_P \cap \mathcal{K}$ has another singular point P'. Then every point of PP' is singular. So every point Q of PP' has T_P as tangent prime. Hence every line through Q not in T_P is an r-secant.

Let α be a plane through PP' but not contained in T_P. Either α belongs to $\mathcal{K}$ or meets it in a line or in r lines of a pencil with centre Q_0, a point of PP'. This means that in α there is no r-secant through any point of PP' in the first two cases and through Q_0 in the third case. This contradicts the previous paragraph. Hence P is the only singular point of $T_P \cap \mathcal{K}$. This is (ii). Parts (iii) and (iv) now follow. □

Corollary: *For any point P in $\mathcal{K}$, the meet $T_P \cap \mathcal{K}$ is a cone $P\mathcal{K}'$ where $\mathcal{K}'$ is a regular, non-singular $k'_{r,n-2,q}$.*

Proof: Let $\Pi_{n-2} \subset T_P$ with $P \notin \Pi_{n-2}$. Then $\mathcal{K}' = \Pi_{n-2} \cap \mathcal{K}$ is non-singular and meets every line of $\mathcal{K}$ through P. Conversely, since P is singular in $T_P \cap \mathcal{K}$, the join QP is a line of $\mathcal{K}$ for every point Q of $\mathcal{K}'$. Hence $T_P \cap \mathcal{K} = P\mathcal{K}'$. □

Theorem 23.5.17: *Let $\mathcal{K}$ be a non-singular, regular $k_{r,n,q}$ with $n \geq 3$ and $q > 4$. Then, with $\mu_n = |\mathcal{U}_{n,q}|$,*

(i) $k = \mu_n$; (23.31)

(ii) *every section of type* I *is a Hermitian curve.*

Proof: (i) For $n = 3$, the theorem is part of Theorem 23.5.1. For $n \geq 4$, let P be a point of $\mathcal{K}$ and T_P the tangent prime to $\mathcal{K}$ at P. Thus $T_P \cap \mathcal{K} = P\mathcal{K}'$ where $\mathcal{K}'$ is a regular, non-singular $k'_{r,n-2,q}$. By the definition of T_P each of the q^{n-1} lines through P not in T_P is an r-secant of $\mathcal{K}$. Since $r = \sqrt{q} + 1$ by Lemma 23.5.15,

$$k = q^{n-1}\sqrt{q} + 1 + qk'.$$

This is the same recurrence relation as for μ_n in Theorem 23.2.4. Since a Hermitian arc has the same number of points as a Hermitian curve and the result is true for $n = 3$, the result is true for all n.

(ii) For any section by a plane π of type I, by (23.24),

$$k = q\sqrt{q}(q^{n-2} + M) + \theta(n-2),$$

where M is the number of solids through π meeting $\mathcal{K}$ in a $\mathcal{U}_{3,q}$. From (23.31), $k = \mu_n$; that is, from (23.9),

$$k = q^{n/2}\{q^{n/2} - (-1)^n\}/(\sqrt{q} + 1) + \theta(n-1).$$

Hence

$$M = \{q^{n-5/2} - (-1)^n q^{(n-3)/2}\}/(\sqrt{q} + 1).$$

Since $M \geq 1$, the section $\pi \cap \mathcal{K}$ is a Hermitian curve. □

As in the previous results let $\mathcal{K}$ be regular and non-singular, with $q > 4$. For P in $\mathcal{K}$, the meet $T_P \cap \mathcal{K}$ is the *tangent cone at* P. From the corollary to Lemma 23.5.16, the tangent cone is $P\mathcal{K}'$ with $\mathcal{K}' \subset \Pi_{n-2}$. From Theorem 23.5.17 when $n = 4$ and Theorem 23.5.1 when $n = 5$, the tangent cone is a Hermitian variety.

Consider a Hermitian variety $\mathcal{W}$ in $PG(n, q)$ with singular space Π_0. If

$\Pi_0 = \mathbf{U}_n$ and $\mathscr{W}$ is reduced to canonical form

$$\mathscr{W} = \mathbf{V}(x_0\bar{x}_0 + x_1\bar{x}_1 + \cdots + x_{n-1}\bar{x}_{n-1}),$$

then associate to any point $P = \mathbf{P}(a_0, a_1, \ldots, a_n)$ other than $\mathbf{U}_n$ the prime

$$\Pi_P = \mathbf{V}(x_0\bar{a}_0 + x_1\bar{a}_1 + \cdots + x_{n-1}\bar{a}_{n-1}).$$

As Π_P does not depend on a_n, the prime Π_P is associated to every point other than $\mathbf{U}_n$ on the line $P\mathbf{U}_n$. Hence the prime Π_P is associated to the line $P\mathbf{U}_n$. Conversely, associated to any prime $\Pi = \mathbf{V}(b_0x_0 + \cdots + b_{n-1}x_{n-1})$ is the line

$$l_\Pi = \{\mathbf{P}(\bar{b}_0, \bar{b}_1, \ldots, \bar{b}_{n-1}, t) | t \in \gamma^+\}.$$

If l_Π is associated to Π, with $\mathbf{U}_n$ in Π, and if Π' is any prime not through $\mathbf{U}_n$, then $\Pi \cap \Pi'$ is the polar prime of $l_\Pi \cap \Pi'$ with respect to $\mathscr{W} \cap \Pi'$.

Theorem 23.5.18: *Let $\mathscr{K}$ be a non-singular, regular $k_{r,n,q}$ with $n \geq 4$ and $q > 4$. If every tangent cone of $\mathscr{K}$ is a Hermitian variety, then $\mathscr{K}$ is a non-singular Hermitian variety.*

Proof: Let Π be a non-tangent prime, let P be a point of $\Pi \cap \mathscr{K} = \mathscr{K}_1$, and let $\Gamma = T_P \cap \mathscr{K} = P\mathscr{K}'$. The $(n-2)$-space $\beta_P = T_P \cap \Pi$ is tangent to $\mathscr{K}_1$ at P and so $\beta_P \cap \mathscr{K}_1$ is a cone $P\mathscr{K}''$, which is the same as $\Gamma \cap \beta_P$. Hence $\beta_P \cap \Gamma$ has only one singular point, namely P.

In the prime T_P, let l_P be the line associated with β_P as described before the theorem; it is a 1-secant to $\mathscr{K}$ with point of contact P. To each P in $\mathscr{K}_1$ is associated such a line l_P.

To see this when $n = 4$, consider the case that $\mathscr{K} = \mathscr{U}_4 = \mathbf{V}(x_0\bar{x}_0 + x_1\bar{x}_1 + x_2\bar{x}_3 + x_3\bar{x}_2 + x_4\bar{x}_4)$. Let $\Pi = \mathbf{u}_4$ and $P = \mathbf{U}_3$. Then $T_P = \mathbf{u}_2$, $\mathscr{K}_1 = \mathscr{U}_3 = \mathbf{V}(\bar{x}_0x_0 + \bar{x}_1x_1 + \bar{x}_2x_3 + \bar{x}_3x_2,\ x_4)$, $\mathscr{K}' = \mathscr{U}_2 = \mathbf{V}(x_0\bar{x}_0 + x_1\bar{x}_1 + x_4\bar{x}_4,\ x_2,\ x_3)$, $\beta_P = \mathbf{u}_2 \cap \mathbf{u}_4$, $\mathscr{K}'' = \mathscr{U}_1 = \mathbf{V}(x_0\bar{x}_0 + x_1\bar{x}_1,\ x_2,\ x_3,\ x_4)$, $\Gamma = \mathbf{U}_3\mathscr{K}'$, $\beta_P \cap \mathscr{K}_1 = \beta_P \cap \Gamma = \mathbf{U}_3\mathscr{K}'' = \mathbf{V}(x_0\bar{x}_0 + x_1\bar{x}_1,\ x_2,\ x_4)$, and $l_P = \mathbf{U}_3\mathbf{U}_4 = \mathbf{V}(x_0, x_1, x_2)$.

It will be shown that the lines l_P are concurrent at a point P_0. To do this it suffices to show that any two lines l_P intersect. So, let P_1 and P_2 be points of $\mathscr{K}_1$; for $i = 1$ and 2, let T_i be the tangent prime at P_i, let $\Gamma_i = \mathscr{K}_1 \cap T_i$, let $\beta_i = T_i \cap \Pi$, and let l_i be the line associated with P_i. To prove that l_1 and l_2 meet, two cases are distinguished: (a) $P_1P_2 \not\subset \mathscr{K}$; (b) $P_1P_2 \subset \mathscr{K}$.

(a) The $(n-2)$-space $T_1 \cap T_2$ contains neither P_1 nor P_2 and so $T_1 \cap T_2 \cap \mathscr{K}$ is a non-singular Hermitian variety. It follows that l_1 and l_2 are the lines joining P_1 and P_2 to the pole P_0 of the $(n-3)$-space $T_1 \cap T_2 \cap \Pi = \beta_1 \cap \beta_2$ with respect to the Hermitian variety $T_1 \cap T_2 \cap \mathscr{K}$.

(b) Both β_1 and β_2 contain P_1P_2. Let α be a plane in Π through P_1P_2 but in neither β_1 nor β_2. It does not lie in $\mathscr{K}$. If $\alpha \cap \mathscr{K} = P_1P_2$, then α would be in both T_1 and T_2 and so in $\beta_1 \cap \beta_2$. Also if α met $\mathscr{H}$ in r lines of a pencil with centre P_i, it would belong to β_i. Thus α meets $\mathscr{H}$ in r lines of a pencil through a point P of P_1P_2 with $P \neq P_1, P_2$. Let m_1 and m_2 be two of these lines other than P_1P_2; also let $Q_1 \in m_1\backslash\{P\}$ and let $Q_2 \in m_2\backslash\{P\}$ with Q_2 not on P_1Q_1 or P_2Q_1. Then the points P_1, P_2, Q_1, Q_2 of $\mathscr{H}_1$ have no three collinear and none of the lines P_1Q_1, P_1Q_2, P_2Q_1, P_2Q_2, Q_1Q_2 belong to $\mathscr{H}$. Hence, from (a), the lines l'_1 and l'_2 associated to Q_1 and Q_2 meet in a point P_0; they also must meet l_1 and l_2. Since P_i, Q_1, Q_2 are not collinear, so l_i, l'_1, l'_2 are not coplanar, for $i = 1, 2$. So l_1 and l_2 also contain P_0.

The point P_0 through which all the lines l_P pass as P varies in $\mathscr{H}_1$ is called the *pole* of Π. It is not in $\mathscr{H} \cup \Pi$.

From Theorem 23.5.17(i), the number of lines l_P is

$$k_1 = \mu_{n-1}. \tag{23.32}$$

On the other hand, let N denote the number of unisecants of $\mathscr{H}$ through P_0. Then counting the points of $\mathscr{H}$ on the lines through P_0 gives

$$k = N + r\{\theta(n-1) - N\}$$

where $r = \sqrt{q} + 1$. Hence

$$N = \{(\sqrt{q}+1)\theta(n-1) - k\}/\sqrt{q} = k_1$$

using (23.31) and (23.32). Thus every unisecant through P_0 is a line associated to a point of $\mathscr{H}_1$. Hence $\mathscr{H}_1$ is the set of points of contact of the unisecants through P_0 and these points generate Π, which can now be called the *polar prime of P_0 with respect to $\mathscr{H}$.*

From above, the correspondence which associates to a non-tangent prime Π its pole P_0 is bijective. Let $\mathfrak{S}$ be the bijection from the points to the primes of $PG(n, q)$ in which each point of $\mathscr{H}$ is mapped to its tangent prime and each point not on $\mathscr{H}$ is mapped to its polar prime.

It must be shown that $\mathfrak{S}$ is involutory; that is, if $\Pi = P\mathfrak{S}$ and $P' \in \Pi$, then the prime $\Pi' = P'\mathfrak{S}$ contains P. Let P and P' be off $\mathscr{H}$ and l be an r-secant of $\mathscr{H}$ through P' in Π. The r lines PQ for Q in $l \cap \mathscr{H}$ are unisecants of $\mathscr{H}$. So the plane $\alpha = Pl$ meets $\mathscr{H}$ in a Hermitian curve $\mathscr{U}_2$ by Theorems 23.5.14 and 23.5.17(ii). Also, l is the polar of P with respect to $\mathscr{U}_2$ and so the polar l' of P' with respect to $\mathscr{U}_2$ contains P. The line l' contains the r points of contact of the tangents to $\mathscr{U}_2$ through P'. Hence the polar prime Π' of P' contains l' and so P.

The other cases of P and P' are simpler. Thus the mapping $\mathfrak{S}$ is bijective and involutory, and transforms the points of a prime Π into the primes through the point $P = \Pi\mathfrak{S}^{-1}$ as well as vice versa. So $\mathfrak{S}$ is a polarity for which $\mathscr{H}$ is the set of self-polar points. Hence $\mathscr{H} = \mathscr{U}_n$. □

Theorem 23.5.19: *If $\mathscr{K}$ is a regular, non-singular $k_{r,n,q}$ with $n \geq 4$ and $q > 4$, then $\mathscr{K}$ is a non-singular Hermitian variety.*

Proof: From the previous result and Theorems 23.5.1 and 23.5.17(ii), the result is true for $n = 4$ and 5.

Let $n \geq 6$ and proceed by induction. The tangent cone Γ at a point P of $\mathscr{K}$ has base $\mathscr{K}'$ which is a regular, non-singular $k_{r,n-2,q}$. Since $n - 2 \geq 4$, the set $\mathscr{K}'$ is a non-singular Hermitian variety by the induction hypothesis. So Γ is a Hermitian variety and then Theorem 23.5.18 gives the result. □

23.6 The characterization of projections of quadrics

In §23.5, a description was given of sets $\mathscr{K} = k_{r,n,q}$ with one of the following properties:

(i) $r = 1$, 2 or q;
(ii) $\mathscr{K}$ is singular;
(iii) $\mathscr{K}$ is non-singular, $3 \leq r \leq q - 1$, $q > 4$, and $\mathscr{K}$ has no plane section of type IV.

This leaves the case that $\mathscr{K}$ is non-singular, $3 \leq r \leq q - 1$, and either $q = 4$ or $\mathscr{K}$ has a plane section of type IV.

When $q = 4$, we are dealing with sets $k_{3,n,4}$, that is sets of type (1, 3, 5) with at least one 3-secant. As explained in §19.6, the sets of type (1, 3, 5) form a vector space over $GF(2)$ of dimension $\frac{1}{3}(n + 1)(n^2 + 2n + 3)$. In $PG(3, 4)$ there are seven distinct types of non-singular $k_{3,3,4}$.

For the remainder of this section, although some notice is taken of the case $q = 4$, we deal mainly with the case that $\mathscr{K}$ is *non-singular*, $q > 4$, $n \geq 3$, $3 \leq r \leq q - 1$, *and $\mathscr{K}$ has a plane section of type* IV. It follows from Lemma 23.5.6 that $q = 2^h$ with $h \geq 2$ and $r = \frac{1}{2}q + 1$.

Theorem 23.6.1: *Let $\mathscr{K}$ be a non-singular $k_{r,n,q}$ with $3 \leq r \leq q - 1$ and having a plane section of type* IV. *Then*

(i) *if $q > 4$, $\mathscr{K}$ contains exactly one prime Π_{n-1};*
(ii) *if $q = 4$ and $\mathscr{K}$ contains no section of type* I *or* II, *it contains exactly one prime Π_{n-1}.*

Proof: For $n = 3$, the result is contained in Theorems 19.4.8 and 19.4.9, which show that $\mathscr{K}$ is a set $\mathscr{K}_1$, which contains precisely one plane. Subsequently, in Theorem 19.4.17, it is shown that $\mathscr{K}_1$ is $\mathscr{R}_3$, which is the projection of a quadric $\mathscr{P}_4$ onto a solid Π_3 from a point other than the nucleus. Now suppose $n > 3$.

Let l_1 be a unisecant of $\mathscr{K}$ (which exists by Theorem 19.4.7, Corollary) with point of contact P and let l_2 be an r-secant through P. So the plane $l_1 l_2 = \pi$ is of type III. Let Π_4 be a four-dimensional space containing π and consider the solids $\Pi_3^{(1)}, \ldots, \Pi_3^{(q+1)}$ in Π_4 containing π. Then $\Pi_3^{(i)} \cap \mathscr{K}$ is either a cone $P_i\mathscr{K}_{\text{III}}$ with vertex P_i and a section $\mathscr{K}_{\text{III}} = \pi \cap \mathscr{K}$ or $\Pi_3^{(i)} \cap \mathscr{K}$ is a set $\mathscr{R}_3^{(i)}$: this follows from Theorem 23.5.1 for $q > 4$ and from Theorem 19.6.8 for $q = 4$.

Suppose we have s cones with respective vertices $P_1, P_2, \ldots, P_s$ and $q + 1 - s$ sets $\mathscr{R}_3^{(1)}, \ldots, \mathscr{R}_3^{(q+1-s)}$. If $\Pi_4 \cap \mathscr{K}$ contains a solid Π_3, then Π_3 contains the line l_0 of $\mathscr{K}$ in π, the plane $l_0 P_i$, and the plane $\Pi_2^{(j)}$ of $\mathscr{K}$ in $\mathscr{R}_3^{(j)}$. It follows that Π_3 is unique, and is the union of the $q + 1$ planes $l_0 P_i$, $\Pi_2^{(j)}$.

Now suppose that the $q + 1$ planes $l_0 P_i$, $\Pi_2^{(j)}$ do not lie in a solid: they are in any case the only planes in $\Pi_4 \cap \mathscr{K}$ through l_0. Consider the solid Π'_3 containing two of the planes. Then $\Pi'_3 \cap \mathscr{K}$ is a cone with base type IV or V. If the base is of type IV, then $\Pi'_3 \cap \mathscr{K}$ contains exactly two of the $q + 1$ planes. Also, any other plane Π_2 through l_0 in Π'_3 is of type V. This plane Π_2 and π define a solid Π''_3 for which $\Pi''_3 \cap \mathscr{K}$ is an $\mathscr{R}_3$. Since Π_2 is of type V and since l_0 is in the plane on this $\mathscr{R}_3$, the line l_0 contains the exceptional point Q_0 of $\mathscr{R}_3$. So π is also of type V, a contradiction. Thus $\Pi'_3 \cap \mathscr{K}$ consists of $\frac{1}{2}q + 1$ planes through l_0. Hence the solids Π'_3 and the sections of type VII through l_0 in such solids form a $2 - (q + 1, \frac{1}{2}q + 1, 1)$ design, whence the number of such solids is $2(q + 1)/(\frac{1}{2}q + 1)$, which is not an integer. Thus the $q + 1$ planes $l_0 P_i$, $\Pi_2^{(j)}$ are the $q + 1$ planes through l_0 of a solid.

Consider now all the solids $\Pi_3^{(1)}, \ldots, \Pi_3^{(N)}$, $N = (q^{n-2} - 1)/(q - 1)$, which pass through π. Then $\Pi_3^{(i)} \cap \mathscr{K}$ is a cone $P_i\mathscr{K}_{\text{III}}$ or a set $\mathscr{R}_3^{(i)}$. Suppose we have t cones with respective vertices $P_1, \ldots, P_t$ and $N - t$ sets $\mathscr{R}_3^{(1)}, \ldots, \mathscr{R}_3^{(N-t)}$. If $\mathscr{K}$ contains a prime Π_{n-1}, then Π_{n-1} contains the line l_0, the plane $l_0 P_i$, and the plane $\Pi_2^{(j)}$ of $\mathscr{K}$ on $\mathscr{R}_3^{(j)}$. It follows that Π_{n-1} is unique and is the union of the N planes $l_0 P_i$, $\Pi_2^{(j)}$, which all contain l_0. However, the solid containing at least two of the planes $l_0 P_i$, $\Pi_2^{(j)}$ contains exactly $q + 1$, since we may argue as above for the Π_4 containing π and the two planes. Hence the planes $l_0 P_i$, $\Pi_2^{(j)}$ are all the planes through l_0 of a prime Π_{n-1}. So $\mathscr{K}$ contains exactly one prime Π_{n-1}. □

Corollary 1: *If $\mathscr{K}$ is a non-singular $k_{r,n,q}$ with $3 \leq r \leq q - 1$, then $\mathscr{K}$ contains at most one prime.*

Proof: Suppose $\mathscr{K}$ contains two primes Π_{n-1}, Π'_{n-1}: they intersect in a secundum Π_{n-2}. Let P be a point of Π_{n-2} and l be an r-secant of $\mathscr{K}$ through P; also let π be a plane through l such that $\pi \cap \Pi_{n-2} = \{P\}$. Now π contains l and meets Π_{n-1} and Π'_{n-1} in lines of $\mathscr{K}$, and so π is of type IV. For $q = 4$, since any plane contains at least one line of $\mathscr{K}$, no plane section is of type

I or II. Hence, by the theorem, $\mathscr{K}$ contains exactly one prime, a contradiction. □

Let $\mathscr{K}$ contain the unique prime Π_{n-1}. As in Theorem 19.4.9 for $PG(3, q)$, define $\mathscr{J}$, the *residual* of $\mathscr{K}$, to be

$$\mathscr{J} = (PG(n, q)\backslash\mathscr{K}) \cup \Pi_{n-1}.$$

This may also be written $\mathscr{J} = \mathscr{K} \nabla \Pi_{n-1}$, where $X \nabla Y$ is the complement of the symmetric difference of the two sets X and Y. For $q = 4$, this operation defines a vector space over $GF(2)$ on the sets $\mathscr{K}$ of type (1, 3, 5) as above; see §19.6.

Corollary 2: *If $\mathscr{K}$ is a non-singular $k_{r,n,q}$ with $3 \le r \le q - 1$ and contains a prime Π_{n-1}, then*

(i) *$\mathscr{K}$ contains a section of type* IV *and no section of type* I *or* II*;*
(ii) *$\mathscr{J} = (PG(n, q)\backslash\mathscr{K}) \cup \Pi_{n-1}$ is also a non-singular set of type $(1, \frac{1}{2}q + 1, q + 1)$ containing the same unique prime as $\mathscr{K}$.*

Proof: If $|l \cap \mathscr{K}| = i$ for a line l not in Π_{n-1}, then $|l \cap \mathscr{J}| = q + 2 - i$. So $\mathscr{J}$ is a set of type $(1, q + 2 - r, q + 1)$; note that $3 \le q + 2 - r \le q - 1$. Suppose P is a singular point of $\mathscr{J}$. If $P \notin \Pi_{n-1}$, then any line l through P contains two points of $\mathscr{J}$ and so $|l \cap \mathscr{J}| = q + 1$. Hence $\mathscr{J} = PG(n, q)$, a contradiction. If $P \in \Pi_{n-1}$, then any line through P contains one or $q + 1$ points of $\mathscr{K}$; that is, P is singular for $\mathscr{K}$, a contradiction. Hence $\mathscr{J}$ is a non-singular set of type $(1, q + 2 - r, q + 1)$ containing Π_{n-1}.

By Corollary 1, Π_{n-1} is the only prime of $\mathscr{K}$. Any plane section of $\mathscr{K}$ contains at least one line and is consequently not of type I or II. Similarly $\mathscr{J}$ contains no sections of type I or II. Let l_1 be a unisecant of $\mathscr{K}$ with point of contact P and let l_r be an r-secant through P. So the plane $l_1 l_r$ is of type III for $\mathscr{K}$ and hence of type IV for $\mathscr{J}$. Consequently $r = \frac{1}{2}q + 1$. In this argument, $\mathscr{K}$ and $\mathscr{J}$ can be interchanged. □

We now investigate the nature of sets $\mathscr{K}$ containing a prime in more detail. A non-empty subset $\mathscr{S}$ of $PG(d, q)$ is a *projective Shult space* with ambient space $PG(n, q)$ if $\mathscr{S}$ spans $PG(n, q)$ and there is a non-empty subset $\mathscr{L}$ of the set of lines in $\mathscr{S}$ such that, given a point S in $\mathscr{S}$ and a line l in $\mathscr{L}$ not containing S, then the line SQ is in $\mathscr{L}$ for exactly one or for all points Q of l. The Shult space $\mathscr{S}$ is *non-degenerate* if there is no point A in $\mathscr{S}$ such that AQ is in $\mathscr{L}$ for every point Q of $\mathscr{S}\backslash\{A\}$. Projective Shult spaces in $PG(n, q)$ are discussed in §26.3 and classified in Theorems 26.3.29 and 26.3.30.

If $\mathscr{K}$ in $PG(n, q)$ contains as prime Π_{n-1} and P is any point of $\mathscr{K}\backslash\Pi_{n-1}$, then the *support* of P is

$$\mathscr{S}_P = \{Q \in \Pi_{n-1} | PQ \subset \mathscr{K}\}.$$

In Theorem 23.6.5, it is in fact shown that $\mathscr{S}_P$ is a non-degenerate projective Shult space in Π_{n-1} or possibly, when $n = 4$, an elliptic quadric. When $n = 3$, it is proved in Lemma 19.4.16 that $\mathscr{S}_P$ is a conic.

The number of projectively distinct non-singular quadrics $\mathscr{Q}_n$ in $PG(n, q)$ is one or two according as n is even or odd. For n even, $\mathscr{Q}_n = \mathscr{P}_n$, and for n odd, $\mathscr{Q}_n = \mathscr{H}_n$ or $\mathscr{E}_n$. In the respective cases, the character w of $\mathscr{Q}_n$ is 1, 2, or 0; see §22.4. Also from (22.17),

$$|\mathscr{Q}_n| = q^{n-1} + q^{n-2} + \cdots + q + 1 + (w - 1)q^{(n-1)/2}.$$

It is also useful to assign the character $w = 0$ to a $(q^2 + 1)$-cap (of which $\mathscr{E}_3$ is an example).

Theorem 23.6.2: *Let $\mathscr{Q}_{n+1}$ be a non-singular quadric of character w in $PG(n+1, q)$, q even. Let Q be a point off $\mathscr{Q}_{n+1}$ other than the nucleus when $\mathscr{Q}_{n+1} = \mathscr{P}_{n+1}$ and let Π_n be a prime not containing Q. Then the projection of $\mathscr{Q}_{n+1}$ from Q onto Π_n is a non-singular set $\mathscr{R}_n$ of type $(1, \frac{1}{2}q + 1, q + 1)$ in Π_n containing a prime Π_{n-1} of Π_n with*

$$|\mathscr{R}_n| = \tfrac{1}{2}q^n + q^{n-1} + q^{n-2} + \cdots + q + 1 + \tfrac{1}{2}(w - 1)q^{n/2}.$$

Proof: Let l be a line in Π_n. The plane Ql meets $\mathscr{Q}_{n+1}$ in a point, a line, a line pair, or a conic. In the case that $Ql \cap \mathscr{Q}_{n+1} = \mathscr{P}_2$ either Q is the nucleus of $\mathscr{P}_2$, in which case the lines joining Q to the points of $\mathscr{P}_2$ are $q + 1$ distinct tangents, or the lines joining Q to $\mathscr{P}_2$ are $\frac{1}{2}q$ bisecants and one tangent.

Let $\mathscr{R}_n = \{P' = PQ \cap \Pi_n | P \in \mathscr{Q}_{n+1}\}$. Then we obtain the following table.

Table 23.3

$Ql \cap \mathscr{Q}_{n+1}$	Point	Line	Line pair	Conic
$\lvert l \cap \mathscr{R}_n\rvert$	1	$q + 1$	$q + 1$	$q + 1$ or $\frac{1}{2}q + 1$

Thus $\mathscr{R}_n$ is of type $(1, \frac{1}{2}q + 1, q + 1)$. The tangents to $\mathscr{Q}_{n+1}$ through Q meet $\mathscr{Q}_{n+1}$ in $\mathscr{P}_n$, $\mathscr{P}_n$, or $\Pi_0\mathscr{P}_{n-1}$ as $\mathscr{Q}_{n+1}$ is $\mathscr{H}_{n+1}$, $\mathscr{E}_{n+1}$, or $\mathscr{P}_{n+1}$; they meet Π_n in a prime Π_{n-1}. Since $\theta(n-1)$ is the number of tangent lines through Q to $\mathscr{Q}_{n+1}$,

$$k = |\mathscr{R}_n| = \theta(n-1) + \tfrac{1}{2}(|\mathscr{Q}_{n+1}| - \theta(n-1)),$$

whence k is as required.

To show that $\mathscr{R}_n$ is non-singular, it suffices to show that for any point P

of $\mathcal{Q}_{n+1}$, joined to Q by a line l, there exists a plane π through l meeting $\mathcal{Q}_{n+1}$ in a $\mathcal{P}_2$ for which Q is not the nucleus: if P projects to P' then π projects to a $(\frac{1}{2}q + 1)$-secant of $\mathcal{R}_n$ through P'.

First, let l be a bisecant of $\mathcal{Q}_{n+1}$. Then every plane through l meets $\mathcal{Q}_{n+1}$ in a conic $\mathcal{P}_2$ or a line pair $\Pi_0\mathcal{H}_1$. If there are b_0 of the former and b_1 of the latter,

$$b_0 + b_1 = \theta(n-1)$$

$$(q-1)b_0 + (2q-1)b_1 + 2 = \theta(n) + (w-1)q^{n/2},$$

whence

$$b_0 = q^{n-1} - (w-1)q^{(n-2)/2} > 0.$$

Now, let l be a tangent to $\mathcal{Q}_{n+1}$ through Q and let s_{n+1} be the number of planes through l meeting $\mathcal{Q}_{n+1}$ in a line pair $\Pi_0\mathcal{H}_1$. When $\mathcal{Q}_{n+1} = \mathcal{P}_{n+1}$ and the nucleus of $\mathcal{P}_{n+1}$ is on l, then $s_{n+1} = 0$; otherwise, $s_{n+1} = \frac{1}{2}(|\mathcal{Q}_{n-1}| - t_{n-1})$. It follows that there is a bisecant m through Q such that the plane ml does not meet $\mathcal{Q}_{n+1}$ in a line pair. So ml meets $\mathcal{Q}_{n+1}$ in a conic. □

When n is even, $\mathcal{R}_n$ is also denoted $\mathcal{R}_n^+$ or $\mathcal{R}_n^-$ as it is the projection of $\mathcal{H}_{n+1}$ or $\mathcal{E}_{n+1}$.

A description of $\mathcal{R}_n$ is required which does not depend on projection from higher space. So, let $F(x_0, x_1, \ldots, x_{n-1})$ be a non-degenerate quadratic form over $GF(q)$ and H be an additive subgroup of $GF(q)$ of index 2. Let

$$\mathcal{F}_\lambda = \mathbf{V}(F(x_0, x_1, \ldots, x_{n-1}) + \lambda x_n^2)$$

be a quadric in $PG(n, q)$; here $\mathcal{F}_\infty = \mathbf{V}(x_n) = \mathbf{u}_n$. Also we may assume that F is one of the forms

(i) $x_0^2 + x_1x_2 + \cdots + x_{n-2}x_{n-1}$, for n odd,
(ii) $x_0x_1 + x_2x_3 + \cdots + x_{n-2}x_{n-1}$ or $f(x_0, x_1) + x_2x_3 + \cdots + x_{n-2}x_{n-1}$ with f irreducible, for n even.

Theorem 23.6.3: $\mathcal{R}_n = \bigcup_{\lambda \in H \cup \{\infty\}} \mathcal{F}_\lambda$ *for* $n \geq 2$.

Proof: We consider in one go the cases that $\mathcal{Q}_{n+1} = \mathcal{P}_{n+1}, \mathcal{H}_{n+1}, \mathcal{E}_{n+1}$. Let us write $\mathcal{Q}_{n+1} = \mathbf{V}(G_{n+1})$, so that in each case $G_{n+1} = G_{n-1} + x_nx_{n+1}$ with $n \geq 1$. From the above canonical forms, we have $G_0 = x_0^2$ in the parabolic case, $G_1 = x_0x_1$ in the hyperbolic case, and $G_1 = f(x_0, x_1)$ in the elliptic case. Let $Q = \mathbf{P}(0, 0, \ldots, 0, 1, 1)$ and consider the pencil of primes through the secundum $\mathbf{V}(x_n, x_{n+1})$. Let

$$\mathcal{V}_t = \mathbf{V}(tx_n + x_{n+1}) \cap \mathcal{Q}_{n+1}.$$

For $t \neq \infty$,

$$\mathscr{V}_t = \mathbf{V}(G_{n-1} + tx_n^2, tx_n + x_{n+1}).$$

In particular, $\mathscr{V}_0 = \mathbf{V}(x_{n+1}) \cap \mathscr{Q}_{n+1}$, $\mathscr{V}_\infty = \mathbf{V}(x_n) \cap \mathscr{Q}_{n+1}$.

(i) For $\mathscr{Q}_{n+1} = \mathscr{P}_{n+1}$,

$$G_{n-1} + tx_n^2 = x_0^2 + x_1x_2 + \cdots + x_{n-2}x_{n-1} + tx_n^2.$$

So

$$\mathscr{V}_t = \Pi_0\mathscr{P}_{n-1}, \quad \text{for all } t.$$

(ii) For $\mathscr{Q}_{n+1} = \mathscr{H}_{n+1}$,

$$G_{n-1} + tx_n^2 = x_0x_1 + x_2x_3 + \cdots + x_{n-2}x_{n-1} + tx_n^2.$$

So

$$\mathscr{V}_t = \begin{cases} \mathscr{P}_n, & t \neq 0, \infty, \\ \Pi_0\mathscr{H}_{n-1}, & t = 0, \infty. \end{cases}$$

(iii) For $\mathscr{Q}_{n+1} = \mathscr{E}_{n+1}$,

$$G_{n-1} + tx_n^2 = f(x_0, x_1) + x_2x_3 + \cdots + x_{n-2}x_{n-1} + tx_n^2.$$

So

$$\mathscr{V}_t = \begin{cases} \mathscr{P}_n, & t \neq 0, \infty, \\ \Pi_0\mathscr{E}_{n-1}, & t = 0, \infty. \end{cases}$$

In each case, the $\mathscr{V}_t$ have a $\mathscr{Q}_{n-1}$ in common, of the same character w as $\mathscr{Q}_{n+1}$.

If $A = \mathbf{P}(a_0, a_1, \ldots, a_{n+1})$ lies in $\mathscr{Q}_{n+1}$, then QA meets $\mathscr{Q}_{n+1}$ again in $A' = \mathbf{P}(a_0, a_1, \ldots, a_{n-1}, a_{n+1}, a_n)$. If A also lies in $\mathscr{V}_t$, then

$$A = \mathbf{P}(a_0, \ldots, a_n, ta_n) \quad \text{and} \quad A' = \mathbf{P}(a_0, \ldots, a_{n-1}, ta_n, a_n),$$

whence A' lies in $\mathscr{V}_{1/t}$. Further, QA is a tangent when $t = 1$, and so the tangents through Q meet $\mathscr{Q}_{n+1}$ in $\mathscr{V}_1$.

Now project $\mathscr{Q}_{n+1}$ from Q to $\mathbf{u}_{n+1} = \mathbf{V}(x_{n+1})$:

$$A = \mathbf{P}(a_0, a_1, \ldots, a_n, a_{n+1}) \rightarrow \mathbf{P}(a_0, a_1, \ldots, a_{n-1}, a_n + a_{n+1}, 0).$$

Hence $\mathscr{V}_1 \rightarrow \mathbf{V}(x_n, x_{n+1})$ and, for $t \neq 1$,

$$\{\mathscr{V}_t, \mathscr{V}_{1/t}\} \rightarrow \mathscr{W}_t = \mathbf{V}\left(G_{n-1} + \frac{t}{t^2+1}x_n^2, x_{n+1}\right).$$

For, regarding the projection as $\mathbf{P}(X) \rightarrow \mathbf{P}(X')$, we have

$$x_0' = x_0, \ldots, x_{n-1}' = x_{n-1}, x_n' = x_n + x_{n+1}, x_{n+1}' = 0.$$

So, if $\mathbf{P}(X) \in \mathscr{V}_t$, $t \neq 1$, then $x_{n+1} = tx_n$, whence $x_n' = (t+1)x_n$. Also $\mathscr{W}_t$ is the same type of quadric as $\mathscr{V}_t$.

Let $K = \{t/(t^2+1) \mid t \in GF(q)\backslash\{1\}\}$; then $|K| = \frac{1}{2}q$. We show that K is a group. If we write $t = \lambda/\lambda'$, then $t/(t^2+1) = \lambda\lambda'/(\lambda^2+\lambda'^2)$. Now,

$$\frac{\lambda\lambda'}{\lambda^2+\lambda'^2} + \frac{\mu\mu'}{\mu^2+\mu'^2} = \frac{(\lambda\mu+\lambda'\mu')(\lambda\mu'+\lambda'\mu)}{(\lambda\mu+\lambda'\mu')^2+(\lambda\mu'+\lambda'\mu)^2}.$$

So K is a subgroup of $GF(q)$ of index 2. It has therefore been shown that

$$\mathscr{R}_n = \bigcup_{\lambda\in K} \mathscr{W}_\lambda \cup \mathbf{V}(x_n, x_{n+1}).$$

Select F as G_{n-1}. By Lemma 19.4.12, $K = \beta H$ for some β. So the projectivity $\mathbf{P}(X) \to \mathbf{P}(X')$ given by $x'_0 = x_0, x'_1 = x_1, \ldots, x'_{n-1} = x_{n-1}$, $x'_n = \sqrt{\beta}x_n$ transforms $\mathscr{R}_n$ to the required form. □

Now we prove a result on the characterization of elliptic quadrics in $PG(3, q)$ for q even that would more properly belong to Chapter 16 or 18. A weaker version will be required in the subsequent theorem. For q odd, or $q = 4$, a (q^2+1)-cap is an elliptic quadric, Theorem 16.1.7.

Lemma 23.6.4: *A (q^2+1)-cap $\mathscr{K}$ in $PG(3, q)$ containing $\frac{1}{2}(q^3-q^2+2q)$ conics is an elliptic quadric $\mathscr{E}_3$.*

Proof: From above, we can restrict our attention to q even with $q \geq 8$. First it is shown that there exist points P and Q on $\mathscr{K}$ such that

(a) through P there are $\frac{1}{2}q^2+1$ conics in $\mathscr{K}$;
(b) through both P and Q there are $\frac{1}{2}q+1$ conics in $\mathscr{K}$.

If there were no point P for which (a) holds, then every point of $\mathscr{K}$ would be on at most $\frac{1}{2}q^2$ conics. Counting the size of the set $\{(A, \mathscr{C}) | A \in \mathscr{K}, \mathscr{C}$ a conic in $\mathscr{K}$, $A \in \mathscr{C}\}$ in two ways, we obtain

$$\tfrac{1}{2}(q^3-q^2+2q)(q+1) \leq \tfrac{1}{2}q^2(q^2+1),$$

a contradiction.

If, given P, there is no point Q of $\mathscr{K}$ for which (b) holds, then through P and any one of the q^2 points of $\mathscr{K}\backslash\{P\}$ there can be at most $\frac{1}{2}q$ conics of $\mathscr{K}$. So, a count of the set $\{(A, \mathscr{C}) | A \in \mathscr{K}\backslash\{P\}, \mathscr{C}$ a conic in $\mathscr{K}$ containing A and $P\}$ gives

$$(\tfrac{1}{2}q^2+1)q \leq \tfrac{1}{2}q \cdot q^2,$$

another contradiction.

Since P and Q satisfy (a) and (b), let T_P and T_Q be the tangent planes to $\mathscr{K}$ at P and Q, let $\mathscr{C}_0, \mathscr{C}_1, \ldots, \mathscr{C}_{q/2}$ be distinct conics on $\mathscr{K}$ through P and Q, and let $\mathscr{C}$ be a conic on $\mathscr{K}$ containing P but not Q. The plane π of $\mathscr{C}$ meets T_P in a line which is tangent at P to at most one $\mathscr{C}_i$, say $\mathscr{C}_0$. Therefore

π meets the planes of $\mathscr{C}_1, \ldots, \mathscr{C}_{q/2}$ in bisecants of $\mathscr{K}$; if $P_1, \ldots, P_{q/2}$ denote the points of $\mathscr{K}$ other than P on these bisecants, where $P_i \in \mathscr{C}_i$, then $\mathscr{C}_i \cap \mathscr{C} = \{P, P_i\}$.

Let $\mathscr{Q}$ be the quadric containing $\mathscr{C}_1$, $\mathscr{C}_2$, and P_3; the nine conditions necessary for $\mathscr{C}_1$, $\mathscr{C}_2$, and P_3 to lie on a quadric ensure the existence of $\mathscr{Q}$ and it is impossible for two quadrics to meet in two conics plus a point. Since T_P contains the tangent lines at P to $\mathscr{C}_1$ and $\mathscr{C}_2$, it is also the tangent plane at P to $\mathscr{Q}$; it also contains the tangent line to $\mathscr{C}$ at P. As $\mathscr{Q}$ contains the four points P, P_1, P_2, P_3 of $\mathscr{C}$ and as the tangent plane T_P to $\mathscr{Q}$ at P contains the tangent to $\mathscr{C}$ at P, so $\mathscr{Q}$ contains $\mathscr{C}$. Also each conic $\mathscr{C}_i$, $i = 3, \ldots, \frac{1}{2}q$, lies on $\mathscr{Q}$, since $\mathscr{Q}$ contains the three points P, Q, P_i of $\mathscr{C}_i$ and the tangent planes T_P and T_Q to $\mathscr{Q}$ at P and Q contain the tangents of $\mathscr{C}_i$ at P and Q.

From (a), there exists a conic $\mathscr{D}$ on $\mathscr{K}$ which contains P but not Q and which does not touch $\mathscr{C}_0$ at P. If $\mathscr{D}$ is substituted for $\mathscr{C}$ in the above argument, then it follows that $\mathscr{C}_0$ lies on any quadric containing $\frac{1}{2}q - 1$ of the conics $\mathscr{C}_1, \ldots, \mathscr{C}_{q/2}$; hence $\mathscr{Q}$ also contains $\mathscr{C}_0$.

The number of points in $\mathscr{C}_0 \cup \mathscr{C}_1 \cup \cdots \cup \mathscr{C}_{q/2} \cup \mathscr{C}$ is at least

$$2 + (q-1)(\tfrac{1}{2}q + 1) + (\tfrac{1}{2}q - 1) = \tfrac{1}{2}(q^2 + 2q)$$
$$\geq \tfrac{1}{2}(q^2 + q + 4).$$

Hence, by the corollary to Lemma 18.1.8, $\mathscr{K}$ lies on $\mathscr{Q}$; so $\mathscr{K} = \mathscr{Q}$. □

Theorem 23.6.5: *In $PG(n, q)$ with $n \geq 4$, let $\mathscr{K}$ be a non-singular $k_{r,n,q}$ with $3 \leq r \leq q - 1$ containing a prime Π_{n-1} and let P be a point of $\mathscr{K} \backslash \Pi_{n-1}$. Then the support $\mathscr{S}_P$ of P is a non-singular quadric in Π_{n-1}. Thus, for n odd, $\mathscr{S}_P = \mathscr{P}_{n-1}$; for n even, $\mathscr{S}_P = \mathscr{H}_{n-1}$ or $\mathscr{E}_{n-1}$.*

Proof: If l is a line through P, then l contains a point of Π_{n-1}, whence $|l \cap \mathscr{K}| = \frac{1}{2}q + 1$ or $q + 1$. So consider all lines $l_1, l_2, \ldots, l_m$ through P which lie on $\mathscr{K}$, and let $l_i \cap \Pi_{n-1} = \{S_i\}$; that is, $\mathscr{S}_P = \{S_i | i = 1, 2, \ldots, m\}$. Suppose that three distinct lines l_1, l_2, l_3 lie in a plane Π_2 and consider a solid Π_3 containing Π_2. Since $\Pi_3 \cap \mathscr{K}$ contains a plane in Π_{n-1} as well as the lines l_1, l_2, l_3, it is singular by Lemma 19.4.10. In particular, it must be the join of a point to a plane section of type IV, V, or VII; that is, $\Pi_3 \cap \mathscr{K}$ consists of $q + 2$ concurrent planes no three of which have a line in common, $\frac{1}{2}q + 1$ planes through a line, or Π_3 itself. In each case, the plane Π_2 lies on $\mathscr{K}$ and hence every line through P in Π_2 is on $\mathscr{K}$. So $\mathscr{S}_P$ is a set of type $(0, 1, 2, q + 1)$ in Π_{n-1}.

We now show that $\mathscr{S}_P$ is a non-degenerate Shult space in some subspace Π_s of Π_{n-1}. Let l be a line of $\mathscr{S}_P$ and S_i a point of $\mathscr{S}_P \backslash \{l\}$. Consider the solid $\Pi_3 = l_i l$, where $l_i = PS_i$. Then $\Pi_3 \cap \mathscr{K}$ contains a plane of Π_{n-1}, the plane

Pl, and the line l_i skew to l. So $\Pi_3 \cap \mathcal{K}$ is Π_3 or a cone with base of type IV and vertex V on l. In the first case, the plane $S_i l$ is in $\mathcal{S}_P$; in the second case, there is just one line in $\mathcal{S}_P$ through S_i containing a point of l, namely VS_i, since the plane PVS_i lies in $\mathcal{K}$. Thus, if $\mathcal{S}_P$ contains at least one line l, it is a projective Shult space of type $(0, 1, 2, q + 1)$ in some subspace Π_s of Π_{n-1}; if $\mathcal{S}_P$ contains no line, it is a cap.

Now we show that the Shult space $\mathcal{S}_P$ is non-degenerate. If Π_3 is a solid containing P, then in each of the cases there is at least one line of $\Pi_3 \cap \mathcal{K}$ through P. So $\mathcal{S}_P$ is non-empty. Suppose $\mathcal{S}_P$ is degenerate with singular point A. Let Q be a point of $\mathcal{K} \backslash \Pi_{n-1}$ other than P, and let Π_3 contain A, P, Q. Then $\Pi_3 \cap \mathcal{S}_P$ consists of lines through A. If $\Pi_3 \cap \mathcal{K}$ is non-singular, then $\Pi_3 \cap \mathcal{S}_P$ is the support of P in $\Pi_3 \cap \mathcal{K}$ and so, by Lemma 19.4.10, is a $(q + 1)$-arc, a contradiction. Thus $\Pi_3 \cap \mathcal{K}$ is singular and is therefore the join of a point to a section of type III, IV, V, or VII.

(i) If the section is of type III, then $\Pi_3 \cap \mathcal{S}_P = \{A\}$ and A is the vertex of the cone. Hence AQ is on $\mathcal{K}$.
(ii) If the section is of type IV, then $\Pi_3 \cap \mathcal{S}_P$ is a pair of distinct lines which meet in the vertex V of the cone. Since V is the only singular point of $\Pi_3 \cap \mathcal{S}_P$, we see that $A = V$ and AQ is on $\mathcal{K}$.
(iii) If the section is of type V, then $\Pi_3 \cap \mathcal{S}_P$ is a line through A, namely the line of intersection of the $\frac{1}{2}q + 1$ planes of $\Pi_3 \cap \mathcal{K}$. Hence AQ is on $\mathcal{K}$.
(iv) If the section is of type VII, then $\Pi_3 \cap \mathcal{K} = \Pi_3$ and again AQ is on $\mathcal{K}$.

In each case AQ is on $\mathcal{K}$, whence A is a singular point of $\mathcal{K}$, a contradiction.

A *semi-quadric* in $PG(n, q)$ is a pair $(\mathcal{P}, \mathcal{L})$ where $\mathcal{P}$ is a set of points and $\mathcal{L}$ is a set of lines of $PG(n, q)$ such that one of the following holds:

(i) $\mathcal{P}$ is a non-singular quadric $\mathcal{Q}_n$ and $\mathcal{L}$ is the set of lines on $\mathcal{Q}_n$;
(ii) $\mathcal{P}$ is a non-singular Hermitian variety $\mathcal{U}_n$ and $\mathcal{L}$ is the set of lines of $\mathcal{U}_n$;
(iii) $\mathcal{P} = PG(n, q)$ and $\mathcal{L}$ is the set of lines of a linear complex.

If $\mathcal{S}_P$ contains no lines, it is a cap. If $\mathcal{S}_P$ has a projective index of at least one, then by Theorem 26.3.29 it is a semi-quadric in a subspace Π_s of Π_{n-1}. As $\mathcal{S}_P$ is of type $(0, 1, 2, q + 1)$, it cannot be a Hermitian variety. The symplectic case is also excluded as every line of Π_{n-1} in $\mathcal{S}_P$ is a line of it considered as a Shult space. Thus $\mathcal{S}_P$ is a cap or a non-singular quadric spanning a subspace Π_s of Π_{n-1}.

Suppose $s < n - 1$. Let l be a line of Π_s with $l \cap \mathcal{S}_P = \emptyset$, and let Π_3 be a solid containing P and l such that $\Pi_3 \cap \Pi_s = l$. Then, in $\Pi_3 \cap \mathcal{K}$, there is no line containing P, a contradiction. Hence $\mathcal{S}_P$ spans Π_{n-1}.

It remains to show that when $\mathcal{S}_P$ is a cap, then $n = 4$ and $\mathcal{S}_P = \mathcal{E}_3$. Let

Π_3 be the solid containing P and three points S_1, S_2, S_3 of $\mathscr{S}_P$. As $\Pi_3 \cap \mathscr{K}$ is necessarily non-singular, it follows that $\Pi_3 \cap \mathscr{K}$ is an $\mathscr{R}_3$ and $\Pi_3 \cap \mathscr{S}_P$ is a conic. So the points S_i and the conics $\Pi_3 \cap \mathscr{S}_P$ form a 3-$(m, q+1, 1)$ design with $m = |\mathscr{S}_P|$. However, if Π_3 is an arbitrary solid through P, S_1, S_2, then $\Pi_3 \cap \mathscr{K}$ is non-singular and $\Pi_3 \cap \mathscr{S}_P$ is again a conic. Thus every plane through two points of $\mathscr{S}_P$ meets $\mathscr{S}_P$ in a conic. The number of planes through a line in Π_{n-1} is $N = (q^{n-2} - 1)/(q - 1)$. Hence $m = N(q-1) + 2 = q^{n-2} + 1$. However, the maximum number M of points of a cap of $PG(d, q)$ with $q > 2$ satisfies $M = q^2 + 1$ for $d = 3$ and $M < q^{d-1} + 1$ for $d > 3$: see §27.3. So $n = 4$ and $m = q^2 + 1$. Since every plane of the Π_3 containing $\mathscr{S}_P$ intersects $\mathscr{S}_P$ in a conic or just one point, $\mathscr{S}_P = \mathscr{E}_3$ by Lemma 23.6.4. □

Theorem 23.6.6: *In $PG(n, q)$, let $\mathscr{K}$ be a non-singular $k_{r,n,q}$ with $3 \le r \le q - 1$ containing a prime Π_{n-1} and let P be a point of $\mathscr{K} \backslash \Pi_{n-1}$. If the support $\mathscr{S}_P$ of P has character w, then*

$$k = \tfrac{1}{2}q^n + q^{n-1} + q^{n-2} + \cdots + q + 1 + \tfrac{1}{2}(w-1)q^{n/2}.$$

Proof: For $n = 3$, this was proved in Theorem 19.4.9. For $d \ge 4$, Theorem 23.6.5 gives that

$$m = |\mathscr{S}_P| = |\mathscr{Q}_{n-1}| = q^{n-2} + q^{n-3} + \cdots + q + 1 + (w-1)q^{(n-2)/2}.$$

There are $(q^n - 1)/(q - 1)$ lines through P; of these, m are $(q+1)$-secants of $\mathscr{K}$ and the remainder are $(\frac{1}{2}q + 1)$-secants. Hence

$$\begin{aligned} k &= 1 + qm + \tfrac{1}{2}q[(q^n - 1)/(q-1) - m] \\ &= 1 + \tfrac{1}{2}qm + \tfrac{1}{2}q(q^n - 1)/(q-1) \\ &= 1 + \tfrac{1}{2}[q^{n-1} + q^{n-2} + \cdots + q^2 + q + (w-1)q^{n/2}] \\ &\quad + \tfrac{1}{2}[q^n + q^{n-1} + \cdots + q^2 + q] \\ &= \tfrac{1}{2}q^n + q^{n-1} + q^{n-2} + \cdots + q + 1 + \tfrac{1}{2}(w-1)q^{n/2}. \quad \square \end{aligned}$$

This theorem shows that k is the same as $|\mathscr{R}_n|$ in Theorem 23.6.2. We proceed to show that, if $\mathscr{K}$ is as in the previous two theorems, then $\mathscr{K} = \mathscr{R}_n$.

It is necessary to deal separately with the cases of n odd and even.

First a rather curious lemma is required. Consider the pencil $\mathscr{L}$ in $PG(2, q)$, q even, of plane quadrics $\mathscr{F}_\lambda$, where

$$\mathscr{F}_\lambda = \mathbf{V}(x_0^2 + bx_0x_1 + cx_1^2 + \lambda x_2^2);$$

λ varies in $GF(q) \cup \{\infty\}$ and $x^2 + bx + c$ is irreducible. So $\mathscr{F}_\infty = \mathbf{V}(x_2^2)$ is a line and $\mathscr{F}_0 = \mathbf{V}(x_0^2 + bx_0x_1 + cx_1^2) = \mathbf{P}(0, 0, 1)$, a point not lying on $\mathscr{F}_\infty$. The other $q - 1$ quadrics $\mathscr{F}_\lambda$ are all conics, no two of which have a point of intersection, since $\mathscr{F}_0 \cap \mathscr{F}_\infty = \emptyset$. If H is an additive subgroup of $GF(q)$, it is shown in Theorem 12.2.2 that $\mathscr{K}' = \bigcup_{\lambda \in H} \mathscr{F}_\lambda$ is a maximal arc.

However, implicit in the proof is the following result.

Lemma 23.6.7: *Let $H \subset GF(q)$ with $|H| = \frac{1}{2}q$ and let $\mathscr{K}' = \bigcup_{\lambda \in H} \mathscr{F}_\lambda$. If there is some line l other than $\mathscr{F}_\infty$ with $l \cap \mathscr{K}' = \emptyset$, then H is an additive group and $\mathscr{K}'$ is a maximal arc of type $(0, \frac{1}{2}q)$.*

Proof: Let $l = \mathbf{V}(a_0x_0 + a_1x_1 + x_2)$ with not both a_0 and a_1 zero (any line through $\mathbf{P}(0, 0, 1)$ meets every $\mathscr{F}_\lambda$). Let $\lambda \in H$ and so $l \cap \mathscr{F}_\lambda = \emptyset$. However, the intersection of l and $\mathscr{F}_\lambda$ is given by

$$x_0^2 + bx_0x_1 + cx_1^2 + \lambda(a_0^2x_0^2 + a_1^2x_1^2) = 0,$$

that is by

$$x_0^2(1 + \lambda a_0^2) + bx_0x_1 + (c + \lambda a_1^2)x_1^2 = 0,$$

or

$$x^2 + x + d = 0,$$

where

$$x = (1 + \lambda a_0^2)x_0/(bx_1), \qquad d = (1 + \lambda a_0^2)(c + \lambda a_1^2)/b^2.$$

So

$$d = e_0 + e_1\lambda + e_2\lambda^2,$$

where

$$e_0 = c/b^2, \qquad e_1 = (ca_0^2 + a_1^2)/b^2, \qquad e_2 = a_0^2a_1^2/b^2.$$

Also $e_1 + \sqrt{e_2} = (ca_0^2 + ba_0a_1 + a_1^2)/b^2 \neq 0$, since $x^2 + bx + c$ is irreducible.

Now, $x^2 + x + d = 0$ has two solutions or none in $GF(2^h)$ as $T(d) = 0$ or 1, where

$$T(d) = d + d^2 + d^4 + \cdots + d^{2^{h-1}};$$

T is an additive homomorphism, the *trace* function, from $GF(2^h)$ onto $GF(2)$ and satisfies $T(u + v) = T(u) + T(v)$, $T(u^2) = T(u)^2 = T(u)$. Also, since $x^2 + bx + c$ is irreducible, so is $x^2 + x + c/b^2$, whence $T(e_0) = 1$. Thus

$$\begin{aligned} T(d) &= T(e_0) + T(e_1\lambda) + T(e_2\lambda^2) = 1 + T(e_1\lambda) + T(\sqrt{e_2}\lambda) \\ &= 1 + T((e_1 + \sqrt{e_2})\lambda). \end{aligned}$$

Now $l \cap \mathscr{F}_\lambda = \emptyset$ for each λ in H, and so $T(d) = 1$ for each λ in H. Hence

$$T((e_1 + \sqrt{e_2})\lambda) = 0,$$

or

$$T(\mu) = 0,$$

where $\mu = (e_1 + \sqrt{e_2})\lambda$. The $\frac{1}{2}q$ solutions μ of this equation form an additive group, the kernel of the function T. So H is also an additive group. The rest of the lemma follows as in Theorem 12.2.2. □

It may be noted that in this case $\mathscr{K}'$ is the complement of the dual of a regular oval, where a regular oval is defined to be a conic plus its nucleus.

Theorem 23.6.8: *In $PG(n, q)$, n odd and $n \geq 5$, let $\mathscr{K}$ be a non-singular $k_{r,n,q}$ with $3 \leq r \leq q - 1$ containing a plane section of type* IV *and, for $q = 4$, also no section of type* I *or* II. *Then $\mathscr{K} = \mathscr{R}_n$, the projection of the quadric $\mathscr{P}_{n+1}$ in $PG(n + 1, q)$ onto $PG(n, q)$.*

Proof: By Theorem 23.6.3 we have to show that $\mathscr{K}$ comprises $\frac{1}{2}q + 1$ quadrics of a pencil, one being a prime Π_{n-1} and the others cones $\Pi_0^{(i)}\mathscr{P}_{n-1}$, $i = 1, 2, \ldots, \frac{1}{2}q$. From Theorem 23.6.1, $\mathscr{K}$ contains a unique prime Π_{n-1}. If P is any point of $\mathscr{K}\backslash\Pi_{n-1}$, then its support $\mathscr{S}_P$ in Π_{n-1} is a quadric $\mathscr{P}_{n-1}$, by Theorem 23.6.5. Let Q_0 be the nucleus of $\mathscr{P}_{n-1}$. The line PQ_0 is a $(\frac{1}{2}q + 1)$-secant of $\mathscr{K}$. So let P' be any point of $\mathscr{K} \cap PQ_0$ other than P and Q_0, and let S be any point of $\mathscr{S}_P$.

Suppose that $P'S$ is not a line of $\mathscr{K}$. Consider the plane $\pi = PP'S$. It contains the two $(\frac{1}{2}q + 1)$-secants PP' and $P'S$, and the two $(q + 1)$-secants PS and SQ_0. So π is a section of type IV. However, choose a solid Π_3 containing π in such a way that $\Pi_3 \cap \mathscr{P}_{n-1}$ is a conic $\mathscr{P}_2$. Then $\Pi_3 \cap \mathscr{K}$ is an $\mathscr{R}_3$. As π contains the nucleus Q_0 of the conic $\mathscr{P}_2$, it is of type V, Table 19.5: this contradicts that π is of type IV. So $P'S$ is a line of $\mathscr{K}$. Hence $\mathscr{S}_{P'} = \mathscr{S}_P = \mathscr{P}_{n-1}$.

Now we show that if, for $S, S' \in \mathscr{P}_{n-1}$, the lines PS and $P'S'$ intersect, then $S = S'$. Suppose that $S \neq S'$ and that $PS \cap P'S' = T$. The plane π containing PS and $P'S'$ contains Q_0 and therefore it is of type V. However, π contains the three lines PS, $P'S'$, and SQ_0, which are not concurrent, a contradiction. Hence $S = S'$.

This means that if P_1 and P_2 are any two points on $PQ_0 \cap \mathscr{K}$ other than Q_0, then the two cones $P_1\mathscr{P}_{n-1}$ and $P_2\mathscr{P}_{n-1}$, where $P_i\mathscr{P}_{n-1}$ comprises the points on all joins P_iQ for Q in $\mathscr{P}_{n-1}$, intersect exactly in $\mathscr{P}_{n-1}$. So, if $PQ_0 \cap \mathscr{K} = \{Q_0, P_1, P_2, \ldots, P_{q/2}\}$ where P is some P_i, say P_1, define

$$\mathscr{K}_0 = \Pi_{n-1} \cup \bigcup_{i=1}^{q/2} P_i\mathscr{P}_{n-1}.$$

Then $\mathscr{K}_0 \subset \mathscr{K}$ and

$$\begin{aligned} |\mathscr{K}_0| &= (q^n - 1)/(q - 1) + \tfrac{1}{2}q + \tfrac{1}{2}q(q - 1)|\mathscr{P}_{n-1}| \\ &= (q^n - 1)/(q - 1) + \tfrac{1}{2}q + \tfrac{1}{2}q(q - 1)(q^{n-1} - 1)/(q - 1) \\ &= \tfrac{1}{2}q^n + (q^n - 1)/(q - 1) \\ &= k, \end{aligned}$$

by Theorem 23.6.6. So $\mathscr{K}_0 = \mathscr{K}$.

We can now attach coordinates to $\mathscr{K}$. Let

$$\Pi_{n-1} = \mathbf{V}(x_n), \quad \mathscr{P}_{n-1} = \mathbf{V}(x_0^2 + x_1x_2 + \cdots + x_{n-2}x_{n-1}, x_n),$$
$$P_i = \mathbf{P}(\surd\lambda_i, 0, \ldots, 0, 1)$$

with $\lambda_1 = 0$. Then $Q_0 = \mathbf{P}(1, 0, \ldots, 0)$ and

$$P_i\mathscr{P}_{n-1} = \mathbf{V}(x_0^2 + x_1x_2 + \cdots + x_{n-2}x_{n-1} + \lambda_i x_n^2).$$

Now Lemma 23.6.7 can be used to show that $H = \{\lambda_i | i = 1, 2, \ldots, \frac{1}{2}q\}$ is an additive group. We select a plane Π_2 meeting Π_{n-1} in a line skew to $\mathscr{P}_{n-1}$, namely

$$\Pi_2 = \mathbf{V}(bx_0 + cx_1 + x_2, x_3, x_4, \ldots, x_{n-1}),$$

where $x^2 + bx + c$ is irreducible. Then

$$\Pi_2 \cap P_i\mathscr{P}_{n-1} = \mathbf{V}(x_0^2 + bx_0x_1 + cx_1^2 + \lambda_i x_n^2, bx_0 + cx_1 + x_2, x_3, \ldots, x_{n-1}).$$

So $\Pi_2 \cap \mathscr{K}$ consists of the line $\Pi_2 \cap \Pi_{n-1}$, $\frac{1}{2}q - 1$ conics of a pencil, and the point P_1, the nucleus of all the $\frac{1}{2}q - 1$ conics. So $\Pi_2 \cap \mathscr{K}$ is of type III containing a unisecant l, which is a 0-secant of $\mathscr{K}\backslash\Pi_{n-1}$ and is not the line $\Pi_2 \cap \Pi_{n-1}$ of the pencil. So, by the lemma, H is an additive group. Thus, by Theorem 23.6.3, $\mathscr{K} = \mathscr{R}_n$. □

It remains to consider the case that n is even.

Let $\mathscr{K}$ be a non-singular $k_{r,n,q}$ with $3 \leq r \leq q - 1$ containing a prime Π_{n-1} and let $P \in \mathscr{K}\backslash\Pi_{n-1}$. Then, from Theorem 23.6.5, $\mathscr{S}_P = \mathscr{H}_{n-1}$ or $\mathscr{E}_{n-1}$. Suppose it is shown when $\mathscr{S}_P = \mathscr{H}_{n-1}$ that $\mathscr{K}$ is projectively unique and so the projection of a quadric $\mathscr{H}_{n+1}$. Now take the other case and let $\mathscr{K}$ be such that $\mathscr{S}_P = \mathscr{E}_{n-1}$; then

$$k = \tfrac{1}{2}q^n + q^{n-1} + \cdots + q + 1 - \tfrac{1}{2}q^{n/2}$$

and, with $\mathscr{Q}$ the residual of $\mathscr{K}$,

$$|\mathscr{Q}| = (q^{n+1} - 1)/(q - 1) - k + (q^n - 1)/(q - 1)$$
$$= \tfrac{1}{2}q^n + q^{n-1} + \cdots + q + 1 + \tfrac{1}{2}q^{n/2}.$$

So, for $\mathscr{Q}$, the support of a point is an $\mathscr{H}_{n-1}$ and hence $\mathscr{Q}$ is the projection of $\mathscr{H}_{n+1}$. Thus $\mathscr{K}$ is projectively unique and so the projection of $\mathscr{E}_{n+1}$.

Theorem 23.6.9: *In $PG(n, q)$, n even and $n \geq 4$, let $\mathscr{K}$ be a non-singular $k_{r,n,q}$ with $3 \leq r \leq q - 1$ containing a plane section of type* IV *and, for $q = 4$, also no section of type* I *or* II. *Then $\mathscr{K}$ is the projection of either a hyperbolic quadric $\mathscr{H}_{n+1}$ or an elliptic quadric $\mathscr{E}_{n+1}$ in $PG(n + 1, q)$ onto $PG(n, q)$.*

Proof: By Theorem 23.6.1, $\mathscr{K}$ contains a unique prime Π_{n-1}. If P is any point of $\mathscr{K}\backslash\Pi_{n-1}$, then its support $\mathscr{S}_P$ in Π_{n-1} is a quadric $\mathscr{H}_{n-1}$ or $\mathscr{E}_{n-1}$, by Theorem 23.6.5. However, by the remark above, it suffices to consider the case that $\mathscr{S}_P = \mathscr{H}_{n-1}$. Now, by Theorem 23.6.3, we have to show that $\mathscr{K}$ comprises $\frac{1}{2}q + 1$ quadrics of a pencil, one being the prime Π_{n-1}, another the cone $P\mathscr{H}_{n-1}$, and the remainder parabolic quadrics $\mathscr{P}_n^{(i)}$, $i = 1, 2, \ldots, \frac{1}{2}q - 1$, each of which contains $\mathscr{H}_{n-1}$ and has P as its nucleus.

Each line l through P not joining P to $\mathscr{H}_{n-1}$ meets Π_{n-1} in a point of $\mathscr{K}$ and is therefore a $(\frac{1}{2}q + 1)$-secant of $\mathscr{K}$: it contains $\frac{1}{2}q - 1$ points other than P and the point of Π_{n-1}. We must construct the quadrics $\mathscr{P}_n^{(i)}$ by suitably selecting from every such line l one of the $\frac{1}{2}q - 1$ points for each quadric.

Let $\Pi_{n-1} = \mathbf{V}(x_n)$, let $P = \mathbf{P}(0, 0, \ldots, 0, 1)$, and let

$$\mathscr{S}_P = \mathscr{H}_{n-1} = \mathbf{V}(x_0x_1 + x_2x_3 + \cdots + x_{n-2}x_{n-1}, x_n).$$

In Π_{n-1}, consider the line $l = \mathbf{V}(x_0 + x_1, x_2 + x_3, x_4, \ldots, x_n)$. Then $l \cap \mathscr{S}_P = \{R\}$, where $R = \mathbf{P}(1, 1, 1, 1, 0, \ldots, 0)$. If l' is any line in Π_{n-1} not on $\mathscr{S}_P$, then in the plane Pl' there are through P two, one, or no lines of $\mathscr{K}$ and respectively $q - 1$, q, or $q + 1$ lines $(\frac{1}{2}q + 1)$-secant to $\mathscr{K}$ according as $|l' \cap \mathscr{S}_P| = 2, 1$, or 0; the plane Pl' meets $\mathscr{K}$ correspondingly in a section of type IV, V, or III. In particular, Pl meets $\mathscr{K}$ in a section of type V. The line l contains the point $Q = \mathbf{P}(0, 0, 1, 1, 0, \ldots, 0)$ in Π_{n-1}. So

$$PQ \cap \mathscr{K} = \{T_\lambda = \mathbf{P}(0, 0, \lambda, \lambda, 0, \ldots, 0, 1) \mid \lambda \in H \cup \{\infty\}\},$$

where $|H| = \frac{1}{2}q$ and $T_\infty = Q$. Consider, for $\lambda \neq \infty$, the line

$$l_\lambda = RT_\lambda = \{M_{\mu\lambda} = \mathbf{P}(\mu, \mu, \mu + \lambda, \mu + \lambda, 0, \ldots, 0, 1) \mid \mu \in GF(q) \cup \{\infty\}\}$$

of $\mathscr{K}$, where $M_{\infty\lambda} = R$. Define, for $\mu \neq \infty$ and $\lambda \neq 0, \infty$, the set

$$\mathscr{N}_{\mu\lambda} = \mathscr{S}_P \cap \mathscr{S}_{M_{\mu\lambda}}.$$

Then, for any point N in $\mathscr{N}_{\mu\lambda}$, the lines NP, $NM_{\mu\lambda}$, and $\Pi_{n-1} \cap NPM_{\mu\lambda}$ are all lines of $\mathscr{K}$. As $\pi = NPM_{\mu\lambda}$ contains the $(\frac{1}{2}q + 1)$-secant $PM_{\mu\lambda}$ and three concurrent lines of $\mathscr{K}$, it follows that $\pi \cap \mathscr{K}$ is of type V and $\Pi_{n-1} \cap \pi$ is a tangent to $\mathscr{S}_P$.

The point $S_{\mu\lambda} = \Pi_{n-1} \cap PM_{\mu\lambda} = \mathbf{P}(\mu, \mu, \mu + \lambda, \mu + \lambda, 0, \ldots, 0)$ lies on l. Not only is $NS_{\mu\lambda} = \Pi_{n-1} \cap \pi$ a tangent to $\mathscr{S}_P$, but conversely, if $S_{\mu\lambda}V$ is a tangent to $\mathscr{S}_P$ with point of contact V, then $V \in \mathscr{N}_{\mu\lambda}$. Thus $\mathscr{N}_{\mu\lambda}$ is the set of contact points of the tangents to $\mathscr{S}_P$ from $S_{\mu\lambda}$. Let $\mathscr{M}_{\mu\lambda}$ be the cone with vertex $M_{\mu\lambda}$ and base $\mathscr{N}_{\mu\lambda}$, and let $\Gamma_\lambda = \bigcup_{\mu \in GF(q)} \mathscr{M}_{\mu\lambda}$. Now

$$\mathscr{N}_{\mu\lambda} = \{\mathbf{P}(y_0, y_1, \ldots, y_{n-1}, 0) \mid F = G = 0\},$$

where

$$F = \mu(y_0 + y_1) + (\mu + \lambda)(y_2 + y_3), \qquad G = y_0y_1 + y_2y_3 + \cdots + y_{n-2}y_{n-1}.$$

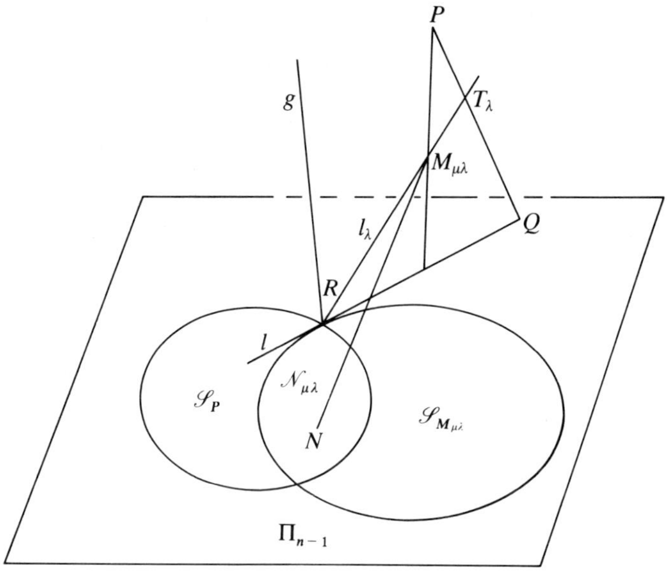

Figure 23.1

So

$$\mathcal{M}_{\mu\lambda} = \{\mathbf{P}(x_0, x_1, \ldots, x_n) | x_0 = \mu + ty_0, x_1 = \mu + ty_1, x_2 = \mu + \lambda + ty_2, \\ x_3 = \mu + \lambda + ty_3, x_4 = ty_4, \ldots, x_{n-1} = ty_{n-1}, \\ x_n = 1, F = G = 0, t \in GF(q) \cup \{\infty\}\}.$$

Elimination of μ, t, $y_0, \ldots, y_{n-1}$ from the equations for $\mathcal{M}_{\mu\lambda}$ and homogenization give $G_\lambda = 0$, where

$$G_\lambda = x_0x_1 + x_2x_3 + \cdots + x_{n-2}x_{n-1} + \lambda^2 x_n^2.$$

Thus Γ_λ is a subset of the quadric $\Delta_\lambda = \mathbf{V}(G_\lambda)$, where $\lambda \in H\backslash\{0\}$.

Each $\mathcal{N}_{\mu\lambda}$ is a $\mathcal{P}_{n-2}$, whence

$$|\mathcal{N}_{\mu\lambda}| = q^{n-3} + q^{n-4} + \cdots + 1.$$

Further,

$$\mathcal{N}_{\mu\lambda} \cap \mathcal{N}_{\rho\lambda} = \\ \mathbf{V}(x_0 + x_1, x_2 + x_3, x_n, x_0x_1 + x_2x_3 + \cdots + x_{n-2}x_{n-1}) = R\mathcal{P}_{n-4},$$

for $\mu \neq \rho$. So, if $\mathcal{N} = \bigcap_{\mu \in GF(q)} \mathcal{N}_{\mu\lambda}$, then $\mathcal{N} = R\mathcal{P}_{n-4}$ and

$$|\mathcal{N}| = q^{n-4} + q^{n-5} + \cdots + 1.$$

As $\mathcal{N}_{\mu\lambda}$ is a $\mathcal{P}_{n-2}$ and $\mathcal{M}_{\mu\lambda} = M_{\mu\lambda}\mathcal{N}_{\mu\lambda}$, it follows that

$$|\mathcal{M}_{\mu\lambda}| = q^{n-2} + q^{n-3} + \cdots + 1.$$

Also, for any point N in $\mathcal{N}\backslash\{R\}$, the plane $RT_\lambda N$ is on $\mathcal{K}$, whence $T_\lambda\mathcal{N} \subset \mathcal{M}_{\mu\lambda} \cap \mathcal{M}_{\rho\lambda}$, $\mu \neq \rho$. Conversely, if $M \in \mathcal{M}_{\mu\lambda} \cap \mathcal{M}_{\rho\lambda}$, $\mu \neq \rho$, and $M \notin RT_\lambda$, then $MRT_\lambda \cap \Pi_{n-1}$ is a line of $\mathcal{N}$, and so the plane MRT_λ is on $\mathcal{K}$, whence $M \in T_\lambda\mathcal{N}$. Hence, $\mathcal{M}_{\mu\lambda} \cap \mathcal{M}_{\rho\lambda} = T_\lambda\mathcal{N}$ (a fact which is also obtainable from the equations of $\mathcal{M}_{\mu\lambda}$, $\mathcal{M}_{\rho\lambda}$, and $T_\lambda\mathcal{N}$), from which

$$\mathcal{M} = \bigcap_{\mu \in GF(q)} \mathcal{M}_{\mu\lambda} = T_\lambda\mathcal{N} \quad \text{and} \quad |\mathcal{M}| = q^{n-3} + q^{n-4} + \cdots + 1.$$

Thus

$$|\Gamma_\lambda| = q(|\mathcal{M}_{\mu\lambda}| - |\mathcal{M}|) + |\mathcal{M}| = q^{n-1} + q^{n-3} + q^{n-4} + \cdots + 1.$$

In fact, $\Delta_\lambda\backslash\Gamma_\lambda$ consists of q^{n-3} lines through R. So let g be a line of Δ_λ through R other than l_λ. If $g \subset \Pi_{n-1}$, then g is on $\mathcal{K}$. Assume therefore that $g \not\subset \Pi_{n-1}$. If gl_λ is a plane of Δ_λ, then $gl_\lambda \cap \Pi_{n-1}$ is a line of $\mathcal{S}_P$ (note that $\mathcal{S}_P \subset \Delta_\lambda$). It now follows that $gl_\lambda \subset T_\lambda\mathcal{N}$ and so $g \subset \mathcal{K}$. If the plane gl_λ is not on Δ_λ, then let Π_3 be a solid containing gl_λ and intersecting Δ_λ in a hyperbolic quadric $\mathcal{H}_3$. The latter contains one line other than l_λ of each $\mathcal{M}_{\rho\lambda}$ for ρ in $GF(q)$. These q lines form with g a regulus on $\mathcal{H}_3$. Let g' be a line other than l_λ of the complementary regulus. Since $|g' \cap \mathcal{K}| \geq q$, we see that g' is a line of $\mathcal{K}$ and $g \cap g'$ is on $\mathcal{K}$, whence g is on $\mathcal{K}$. Hence $\Delta_\lambda \subset \mathcal{K}$.

Any two Δ_λ intersect in $\mathcal{S}_P$. Thus $\Pi_{n-1} \cup P\mathcal{S}_P \cup \bigcup_{\lambda \in H\backslash\{0\}}\Delta_\lambda$ is contained in $\mathcal{K}$ and has the same number of points as $\mathcal{K}$; so this set is $\mathcal{K}$. Therefore we may write

$$\mathcal{K} = \bigcup_{t \in H' \cup \{\infty\}} \mathcal{F}_t,$$

where $\mathcal{F}_t = \mathbf{V}(x_0x_1 + x_2x_3 + \cdots + x_{n-2}x_{n-1} + tx_n^2)$, $|H'| = \frac{1}{2}q$, and $0 \in H'$; here $\mathcal{F}_0 = P\mathcal{S}_P$, $\mathcal{F}_\infty = \Pi_{n-1}$.

Now, exactly as in the proof of Theorem 23.6.8, if we take a plane section of type III through P, Lemma 23.6.7 shows that H' is an additive subgroup of $GF(q)$; for example, we may use the plane

$$\Pi_2 = \mathbf{V}(x_0 + x_1, bx_1 + cx_2 + x_3, x_4, \ldots, x_{n-1}),$$

where $x^2 + bx + c$ is irreducible. Thus, by Theorem 23.6.3, $\mathcal{K}$ is the projection of $\mathcal{H}_{n+1}$; that is, $\mathcal{K} = \mathcal{R}_n^+$. □

Theorem 23.6.10: *In $PG(n, q)$ with $n \geq 3$ and $q > 4$, a non-singular $k_{r,n,q}$ with $3 \leq r \leq q - 1$ is either a non-singular Hermitian variety (with $r = \sqrt{q} + 1$) or the projection of a non-singular quadric in $PG(n + 1, q)$ (with $r = \frac{1}{2}q + 1$).*

Proof: The case $n = 3$ was summarized in Theorem 23.5.1. When $n \geq 4$ and the set has no plane section of type IV, the result is given by Theorem 23.5.19; when $n \geq 4$ and the set has a plane section of type IV, the result is given by Theorems 23.6.8 and 23.6.9. □

This theorem can be reworded as follows.

Corollary: *The projectively distinct non-singular sets* $k_{r,n,q}$ *in* $PG(n, q)$, *with* $n \geq 3$, $3 \leq r \leq q - 1$, *and* $q > 4$, *are given in Table* 23.4. □

Table 23.4

		n odd	n even
$q = p^h, p > 2$	h odd	–	–
	h even	$\mathcal{U}_n$	$\mathcal{U}_n$
$q = 2^h$	h odd	$\mathcal{R}_n$	$\mathcal{R}_n^+, \mathcal{R}_n^-$
	h even	$\mathcal{U}_n, \mathcal{R}_n$	$\mathcal{U}_n, \mathcal{R}_n^+, \mathcal{R}_n^-$

23.7 Notes and references

§§23.1–23.3. Although the theory of Hermitian forms over finite fields and their associated semi-linear groups is already contained in books such as Jordan (1870) and Dickson (1901), the first accounts with greater emphasis on the geometry rather than the group theory were given independently by Bose and Chakravarti (1966) and in the monumental paper of Segre (1965a). These sections follow in style the early sections on quadrics.

§23.4. The proof of Theorem 23.4.2, which is Witt's theorem, follows the treatment of Segre (1965a). The fundamental formula of Theorem 23.4.3 is due to Wan and Yang (1965) although they give a different proof.

§§23.5–23.6. These sections on the characterization of Hermitian varieties and sets of type $(1, r, q + 1)$ are an amalgamation of Tallini Scafati (1967a), Hirschfeld and Thas (1980a, b), and Glynn (1983a).

24

GRASSMANN VARIETIES

24.1 Plücker and Grassmann coordinates

Let Π_r be an r-space in $PG(n, K)$, $n \geq 3$, $1 \leq r \leq n-2$, and let $\mathbf{P}(X_0)$, $\mathbf{P}(X_1), \ldots, \mathbf{P}(X_r)$, with $X_i = (x_0^i, x_1^i, \ldots, x_n^i)$, be $r+1$ linearly independent points of Π_r. Write

$$T_x = \begin{bmatrix} x_0^0 & x_1^0 & \cdots & x_n^0 \\ x_0^1 & x_1^1 & \cdots & x_n^1 \\ \vdots & \vdots & & \vdots \\ x_0^r & x_1^r & \cdots & x_n^r \end{bmatrix}.$$

Also, write $(i_0 i_1 \cdots i_r)_x$, or $(i_0 i_1 \cdots i_r)$ if no confusion is possible, for the determinant of order $r+1$ whose columns coincide with the (i_0+1)th, (i_1+1)th, ..., (i_r+1)th column of the matrix T_x, with $i_0, i_1, \ldots, i_r \in \{0, 1, \ldots, n\}$. If two of the i_j are interchanged the sign of $(i_0 i_1 \cdots i_r)_x$ changes, and if two of the i_j are equal we have $(i_0 i_1 \cdots i_r)_x = 0$. Notice also that at least one of the determinants $(i_0 i_1 \cdots i_r)_x$ is not zero.

Lemma 24.1.1: *Let* $\mathbf{P}(X_0), \mathbf{P}(X_1), \ldots, \mathbf{P}(X_r)$ *and* $\mathbf{P}(Y_0), \mathbf{P}(Y_1), \ldots, \mathbf{P}(Y_r)$ *be two sets of* $r+1$ *linearly independent points of the r-space* Π_r *of* $PG(n, K)$, *where* $n \geq 3$ *and* $1 \leq r \leq n-2$. *Then* $(i_0 i_1 \cdots i_r)_y = t(i_0 i_1 \cdots i_r)_x$ *for some* $t \in K_0$ *which is independent of* $i_0, i_1, \ldots, i_r$.

Proof: Clearly $T_y = T T_x$ with $T = [t_{ij}]$ a non-singular $(r+1) \times (r+1)$ matrix over K. Hence if $t = |T| \neq 0$, then $(i_0 i_1 \cdots i_r)_y = t(i_0 i_1 \cdots i_r)_x$ for any $i_0, i_1, \ldots, i_r \in \{0, 1, \ldots, n\}$. □

Now choose $\mathbf{c}(n+1, r+1)$ ordered $(r+1)$-tuples $(i_0, i_1, \ldots, i_r)$ such that $i_0, i_1, \ldots, i_r$ are distinct elements of $\{0, 1, \ldots, n\}$ and such that the sets $\{i_0, i_1, \ldots, i_r\}$ are exactly all subsets of order $r+1$ of $\{0, 1, \ldots, n\}$. Further, order the set $\mathscr{V}$ of these $(r+1)$-tuples.

Consider again the $r+1$ linearly independent points $\mathbf{P}(X_0), \mathbf{P}(X_1), \ldots, \mathbf{P}(X_r)$ of Π_r. Then a *coordinate vector* of Π_r is

$$L = (l_0, l_1, l_2, \ldots, l_{\mathbf{c}(n+1, r+1)-1})$$

where the l_j are the elements $(i_0 i_1 \cdots i_r)_x$ with $(i_0, i_1, \ldots, i_r)$ in $\mathscr{V}$ in the given order. The elements $l_0, l_1, \ldots$ are called the *coordinates* of the r-space Π_r of $PG(n, K)$. By Lemma 24.1.1, L is determined by Π_r up to a factor of proportion. Write

$$\Pi_r = \mathbf{\Pi}_r(L).$$

For $n = 3$ and $r = 1$ these coordinates were introduced in §15.2. When $r = 1$ the coordinates are also called *Plücker coordinates.* In all other cases they are also called *Grassmann coordinates.*

Consider a projectivity ξ of $PG(n, K)$ with matrix A, and let $\Pi_r\xi = \Pi'_r$ with $\Pi_r = \mathbf{\Pi}_r(L)$ and $\Pi'_r = \mathbf{\Pi}_r(L')$. By a standard matrix manipulation as in Lemma 15.2.8,

$$tL' = L\tilde{A},$$

where $t \in K_0$ and where the elements of $\tilde{A}$ are, up to the sign, minors of order $r + 1$ of the matrix A. Also, $\tilde{A}$ is non-singular.

Next let $\pi(U_0), \pi(U_1), \ldots, \pi(U_{n-r-1})$ be $n - r$ linearly independent hyperplanes containing the r-space Π_r of $PG(n, K)$, where $U_i = (u_0^i, u_1^i, \ldots, u_n^i)$. Write

$$T_u = \begin{bmatrix} u_0^0 & u_1^0 & \cdots & u_n^0 \\ u_0^1 & u_1^1 & \cdots & u_n^1 \\ \vdots & \vdots & & \vdots \\ u_0^{n-r-1} & u_1^{n-r-1} & \cdots & u_n^{n-r-1} \end{bmatrix}.$$

Also, write $(j_0 j_1 \cdots j_{n-r-1})_u$, or $(j_0 j_1 \cdots j_{n-r-1})$ if no confusion is possible, for the determinant of order $n - r$ whose columns coincide with the $(j_0 + 1)$th, $(j_1 + 1)th, \ldots, (j_{n-r-1} + 1)$th column of the matrix T_u, with

$$j_0, j_1, \ldots, j_{n-r-1} \in \{0, 1, \ldots, n\}.$$

At least one of the determinants $(j_0 j_1 \cdots j_{n-r-1})_u$ is not zero.

For each element $(i_0, i_1, \ldots, i_r)$ of $\mathscr{V}$ we choose an ordered $(n - r)$-tuple $(j_0, j_1, \ldots, j_{n-r-1})$ such that $(i_0, i_1, \ldots, i_r, j_0, j_1, \ldots, j_{n-r-1})$ is an even permutation of $(0, 1, \ldots, n)$. The set of these $\mathbf{c}(n + 1, r + 1) = \mathbf{c}(n + 1, n - r)$ ordered $(n - r)$-tuples is denoted by $\mathscr{W}$. Clearly the ordering of $\mathscr{V}$ induces an ordering of $\mathscr{W}$.

Consider again the $n - r$ linearly independent hyperplanes containing Π_r. Then a *dual coordinate vector* of Π_r is

$$\hat{L} = (\hat{l}_0, \hat{l}_1, \ldots, \hat{l}_{\mathbf{c}(n+1,r+1)-1})$$

where the $\hat{l}_i$ are the elements $(j_0 j_1 \cdots j_{n-r-1})_u$, with $(j_0, j_1, \ldots, j_{n-r-1})$ in $\mathscr{W}$ in the given order. By the dual of Lemma 24.1.1, $\hat{L}$ is determined by Π_r up to a factor of proportion.

Lemma 24.1.2: $\rho l_i = \hat{l}_i$, $i = 0, 1, \ldots, \mathbf{c}(n+1, r+1) - 1$; *that is, up to a factor of proportion, $\hat{L}$ is L.*

Proof: Let $\mathbf{P}(X_0), \mathbf{P}(X_1), \ldots, \mathbf{P}(X_r)$ be $r+1$ linearly independent points of Π_r, with $X_i = (x_0^i, x_1^i, \ldots, x_n^i)$, $i = 0, 1, \ldots, r$. The l_j are of the form $(i_0 i_1 \cdots i_r)_x$. Consider $n - r$ points $\mathbf{P}(X_{r+1}), \mathbf{P}(X_{r+2}), \ldots, \mathbf{P}(X_n)$ of $PG(n, K)$ such that $\mathbf{P}(X_0), \mathbf{P}(X_1), \ldots, \mathbf{P}(X_n)$ are linearly independent, where $X_i = (x_0^i, x_1^i, \ldots, x_n^i)$ with $i = r+1, r+2, \ldots, n$. As hyperplanes $\pi(U_0), \pi(U_1), \ldots, \pi(U_{n-r-1})$ we choose the $(n-1)$-spaces

$$\Pi_r \mathbf{P}(X_{r+1})\mathbf{P}(X_{r+2}) \cdots \mathbf{P}(X_{r+i-1})\mathbf{P}(X_{r+i+1}) \cdots \mathbf{P}(X_n), \quad i = 1, 2, \ldots, n - r.$$

So for u_j^i we may take the cofactor of x_j^{r+i+1} in the matrix

$$D = \begin{bmatrix} x_0^0 & x_1^0 & \cdots & x_n^0 \\ x_0^1 & x_1^1 & \cdots & x_n^1 \\ \vdots & \vdots & & \vdots \\ x_0^n & x_1^n & \cdots & x_n^n \end{bmatrix},$$

$j = 0, 1, \ldots, n$ and $i = 0, 1, \ldots, n - r - 1$. If E is the matrix obtained from D by replacing each element x_j^i by its cofactor, then each minor of order $n - r$ of E is equal to the product of $|D|^{n-r-1}$ and the algebraic complement of the similarly placed minor in the matrix D. Hence

$$(j_0 j_1 \cdots j_{n-r-1})_u = |D|^{n-r-1}(-1)^d (i_0 i_1 \cdots i_r)_x,$$

where

$$\{j_0, j_1, \ldots, j_{n-r-1}\} \cup \{i_0, i_1, \ldots, i_r\} = \{0, 1, \ldots, n\},$$

and

$$d = i_0 + i_1 + \cdots + i_r + r + 1 + (r+2)(r+1)/2.$$

Now the result follows. □

If $n = 2r + 1$ then often the ordering in $\mathscr{W}$ is chosen in such a way that $\rho l_i = \hat{l}_{i+r}$, where indices are taken modulo $2r + 1$. Hence if $(i_0 i_1 \cdots i_r)_x$ is a coordinate of Π_r and $(j_0 j_1 \cdots j_r)_u$ is a dual coordinate of Π_r', where

$$|\{i_0, i_1, \ldots, i_r\} \cup \{j_0, j_1, \ldots, j_r\}| = 2r + 2,$$

then their positions differ by r.

Lemma 24.1.3: *Let Π_r be an r-space of $PG(n, K)$ with coordinates $(i_0 i_1 \cdots i_r)_x$, and let Π_{n-r-1} be an $(n-r-1)$-space with dual coordinates $(i_0 i_1 \cdots i_r)_u$, $1 \le r \le n - 2$ and $n \ge 3$. Then $\Pi_r \cap \Pi_{n-r-1} \ne \emptyset$ if and only if*

$$\sum (i_0 i_1 \cdots i_r)_x (i_0 i_1 \cdots i_r)_u = 0.$$

Proof: Choose $r+1$ linearly independent points $\mathbf{P}(X_0), \mathbf{P}(X_1), \ldots, \mathbf{P}(X_r)$ in Π_r and $n-r$ linearly independent points $\mathbf{P}(Y_0), \mathbf{P}(Y_1), \ldots, \mathbf{P}(Y_{n-r-1})$ in Π_{n-r-1}. Let

$$\Delta = \begin{vmatrix} x_0^0 & x_1^0 & \cdots & x_n^0 \\ x_0^1 & x_1^1 & \cdots & x_n^1 \\ \vdots & \vdots & & \vdots \\ x_0^r & x_1^r & \cdots & x_n^r \\ y_0^0 & y_1^0 & \cdots & y_n^0 \\ y_0^1 & y_1^1 & \cdots & y_n^1 \\ \vdots & \vdots & & \vdots \\ y_0^{n-r-1} & y_1^{n-r-1} & \cdots & y_n^{n-r-1} \end{vmatrix}.$$

Then $\Pi_r \cap \Pi_{n-r-1} \neq \varnothing$ if and only if $\Delta = 0$. By the Laplace expansion of Δ along the first $r+1$ rows and by Lemma 24.1.2 one immediately gets the required result. □

Corollary: *Consider an r-space Π_r of $PG(n, K)$ with coordinate vector $L = (l_0, l_1, \ldots)$, $1 \leq r \leq n-2$ and $n \geq 3$. If $l_k = (i_0 i_1 \cdots i_r)_x$, then $l_k = 0$ if and only if $\Pi_r \cap \mathbf{U}_{i_{r+1}} \mathbf{U}_{i_{r+2}} \cdots \mathbf{U}_{i_n} \neq \varnothing$, where $\{i_0, i_1, \ldots, i_n\} = \{0, 1, \ldots, n\}$.*

Proof: Let $\mathbf{U}_{i_{r+1}} \mathbf{U}_{i_{r+2}} \cdots \mathbf{U}_{i_n} = \Pi_{n-r-1}$. Since $\Pi_{n-r-1} = \mathbf{V}(x_{i_0}, x_{i_1}, \ldots, x_{i_r})$, the space Π_{n-r-1} has $(j_0 j_1 \cdots j_r)_u \neq 0$ if and only if $\{j_0, j_1, \ldots, j_r\} = \{i_0, i_1, \ldots, i_r\}$. By Lemma 24.1.3, $(i_0 i_1 \cdots i_r)_x = 0$ if and only if

$$\Pi_r \cap \Pi_{n-r-1} \neq \varnothing. \quad \square$$

Lemma 24.1.4: *Let Π_r be an r-space of $PG(n, K)$, $1 \leq r \leq n-2$ and $n \geq 3$, and assume that $\Pi_r \cap \mathbf{V}(x_{i_1}, x_{i_2}, \ldots, x_{i_r})$ is a point $\mathbf{P}(X)$, where $i_1, i_2, \ldots, i_r$ are distinct elements of $\{0, 1, \ldots, n\}$. If $\mathbf{P}(X_0), \mathbf{P}(X_1), \ldots, \mathbf{P}(X_r)$ are $r+1$ linearly independent points of Π_r,* where *$X_i = (x_0^i, x_1^i, \ldots, x_n^i)$, then up to a factor of proportion*

$$X = ((0\, i_1 i_2 \cdots i_r)_x, (1\, i_1 i_2 \cdots i_r)_x, \ldots, (n\, i_1 i_2 \cdots i_r)_x).$$

Proof: Since $\Pi_r \cap \mathbf{V}(x_{i_1}, x_{i_2}, \ldots, x_{i_r})$ is a point $\mathbf{P}(X)$, there is at least one hyperplane $\mathbf{V}(x_i)$ not containing $\mathbf{P}(X)$, $i \in \{0, 1, \ldots, n\}$. Then by the corollary of Lemma 24.1.3 we have $(i i_1 i_2 \cdots i_r)_x \neq 0$. Hence the matrix

$$D = \begin{bmatrix} x_{i_1}^0 & x_{i_2}^0 & \cdots & x_{i_r}^0 \\ x_{i_1}^1 & x_{i_2}^1 & \cdots & x_{i_r}^1 \\ \vdots & \vdots & & \vdots \\ x_{i_1}^r & x_{i_2}^r & \cdots & x_{i_r}^r \end{bmatrix}$$

has rank r. If $X = (x_0, x_1, \ldots, x_n)$ is the coordinate vector of P, then there are elements $t_0, \ldots, t_r$ determined up to a factor of proportion and not all zero such that

$$x_i = t_0 x_i^0 + t_1 x_i^1 + \cdots + t_r x_i^r,$$

$i = 0, 1, \ldots, n$. Since

$$t_0 x_{i_s}^0 + t_1 x_{i_s}^1 + \cdots + t_r x_{i_s}^r = 0,$$

$s = 1, 2, \ldots, r$, and since rank $D = r$, we may take for the t_j the minors of order r of D with alternating signs. Hence up to a factor of proportion x_i is $(i i_1 i_2 \cdots i_r)_x$, $i = 0, 1, \ldots, n$. □

Lemma 24.1.5: *Let* $\mathbf{P}(X_0), \mathbf{P}(X_1), \ldots, \mathbf{P}(X_r)$, $X_i = (x_i^0, x_i^1, \ldots, x_i^n)$, *be* $r + 1$ *linearly independent points of the r-space* Π_r *of* $PG(n, K)$, $1 \leq r \leq n - 2$ *and* $n \geq 3$. *The point* $\mathbf{P}(X)$, *where* $X = (x_0, x_1, \ldots, x_r)$, *is contained in* Π_r *if and only if*

$$x_{i_0}(i_1 i_2 \cdots i_{r+1})_x - x_{i_1}(i_0 i_2 \cdots i_{r+1})_x + \cdots + (-1)^{r+1} x_{i_{r+1}}(i_0 i_1 \cdots i_r)_x = 0, \tag{24.1}$$

for each choice of distinct $i_0, i_1, \ldots, i_{r+1}$ *in* $\{0, 1, \ldots, n\}$.

Proof: The point $\mathbf{P}(X)$ is contained in Π_r if and only if there exist elements $t_0, t_1, \ldots, t_r$ in K, not all zero, such that

$$X = t_0 X_0 + t_1 X_1 + \cdots + t_r X_r.$$

Equivalently,

$$x_i = t_0 x_i^0 + t_1 x_i^1 + \cdots + t_r x_i^r, \tag{24.2}$$

$i = 0, 1, \ldots, n$. Since the rank of the $(r + 1) \times (r + 1)$ matrix with elements x_j^i is equal to $r + 1$, the system (24.2) of $n + 1$ linear equations in $r + 1$ unknowns has at least one solution if and only if all minors of order $r + 2$ of the matrix

$$\begin{bmatrix} x_0 & x_1 & \cdots & x_n \\ x_0^0 & x_1^0 & \cdots & x_n^0 \\ \vdots & \vdots & & \vdots \\ x_0^r & x_1^r & \cdots & x_n^r \end{bmatrix}$$

are zero. Expanding these minors along the first row, we obtain the conditions (24.1). □

Theorem 24.1.6: *For any r-space* Π_r *of* $PG(n, K)$, $1 \leq r \leq n - 1$ *and* $n \geq 3$, *we have*

$$(i_0 i_1 \cdots i_r)(j_0 j_1 \cdots j_r) - (j_0 i_1 \cdots i_r)(i_0 j_1 \cdots j_r) + (j_1 i_1 \cdots i_r)(i_0 j_0 j_2 \cdots j_r) + \cdots + (-1)^{r+1}(j_r i_1 \cdots i_r)(i_0 j_1 \cdots j_{r-1}) = 0, \tag{24.3}$$

where $i_0, i_1, \ldots, j_r$ *are arbitrarily chosen elements in* $\{0, 1, \ldots, n\}$.

Proof: Assume that $\Pi_r \cap \mathbf{V}(x_{i_1}, x_{i_2}, \ldots, x_{i_r})$, with $i_1, i_2, \ldots, i_r$ distinct, is a point $\mathbf{P}(X)$. By Lemma 24.1.4, up to a factor of proportion

$$X = ((0\,i_1 i_2 \cdots i_r), (1\,i_1 i_2 \cdots i_r), \ldots, (n\,i_1 i_2 \cdots i_r)).$$

Since $\mathbf{P}(X)$ belongs to Π_r, by Lemma 24.1.5

$$\begin{aligned}(i_0 i_1 i_2 \cdots i_r)(j_0 j_1 \cdots j_r) - (j_0 i_1 \cdots i_r)(i_0 j_1 \cdots j_r) \\ + (j_1 i_1 \cdots i_r)(i_0 j_0 j_2 \cdots j_r) + \cdots \\ + (-1)^{r+1}(j_r i_1 \cdots i_r)(i_0 j_0 \cdots j_{r-1}) = 0, \quad (24.4)\end{aligned}$$

for any distinct $i_0, j_0, j_1, \ldots, j_r$.

If $\Pi_r \cap \mathbf{V}(x_{i_1}, x_{i_2}, \ldots, x_{i_r})$, with $i_1, i_2, \ldots, i_r$ distinct, is at least of projective dimension 1, then by the corollary of Lemma 24.1.3 we have $(i i_1 i_2 \cdots i_r) = 0$, $i = 0, 1, \ldots, n$. Hence also in this case (24.4) is satisfied.

If $i_1, i_2, \ldots, i_r$, or $i_0, j_0, \ldots, j_r$, are not all distinct, then (24.4) is trivially satisfied.

In conclusion (24.4) is satisfied whenever $i_0, i_1, \ldots, j_r$ are arbitrarily chosen in $\{0, 1, \ldots, n\}$. □

By Theorem 24.1.6 the coordinates of Π_r satisfy a number of *quadratic relations.* These relations (24.3) can also be written as follows:

$$(i_0 i_1 \cdots i_r)(j_0 j_1 \cdots j_r) = \sum_{s=0}^{r} (j_s i_1 i_2 \cdots i_r)(j_0 \cdots j_{s-1} i_0 j_{s+1} \cdots j_r). \quad (24.5)$$

In particular, putting $i_2 = j_2$, $i_3 = j_3, \ldots, i_r = j_r$, (24.5) becomes

$$\begin{aligned}(i_0 i_1 \cdots i_r)(j_0 j_1 i_2 \cdots i_r) = (j_0 i_1 i_2 \cdots i_r)(i_0 j_1 i_2 \cdots i_r) \\ + (j_1 i_1 i_2 \cdots i_r)(j_0 i_0 i_2 \cdots i_r). \quad (24.6)\end{aligned}$$

These are the *elementary quadratic relations.* If $r = 1$ or $r = n - 2$, it is easily seen that (24.5) and (24.6) are the same. For $r = 1$, (24.6) was also derived in Lemma 15.2.2.

Now let $1 < r < n - 2$. In this case $n \geq 5$. Suppose that $(k_0 k_1 \cdots k_r) \neq 0$ if and only if $\{k_0, k_1, \ldots, k_r\} = \{0, 1, 2, 6, 7, \ldots, r+3\}$ or $\{3, 4, 5, 6, 7, \ldots, r+3\}$, where $k_0, k_1, \ldots, k_r \in \{0, 1, \ldots, n\}$. Then the relations (24.6) are satisfied. Since $(0\,1\,2\,6\,7 \cdots r+3)(3\,4\,5\,6\,7 \cdots r+3) \neq 0$, the relations (24.5) are not satisfied. Hence elements $(k_0 k_1 \cdots k_r)$, $k_i = 0, 1, \ldots, n$ and $1 < r < n-2$, where $(k_0 k_1 k_2 \cdots k_r) = -(k_1 k_0 k_2 \cdots k_r)$, $(k_0 k_0 k_2 \cdots k_r) = 0$, etc., which satisfy the elementary quadratic relations and which are not all zero, do not necessarily correspond to some r-space Π_r of $PG(n, K)$.

Theorem 24.1.7: *Assume that the elements* $(k_0 k_1 \cdots k_r)$, *where* $k_i = 0, 1, \ldots, n$, $1 \leq r \leq n-2$ *and* $n \geq 3$, $(k_0 k_1 k_2 \cdots k_r) = -(k_1 k_0 k_2 \cdots k_r)$, $(k_0 k_0 k_2 \cdots k_r) = 0$, *etc., are not all zero and satisfy the quadratic relations* (24.5). *Then these elements* $(k_0 k_1 \cdots k_r)$ *correspond to exactly one r-space* Π_r *of* $PG(n, K)$.

Proof: Suppose, for example, that $(0\ 1\ \cdots\ r) \neq 0$. If there is an r-space Π_r corresponding to the (essentially $\mathbf{c}(n+1, r+1)$) given elements, then by the corollary of Lemma 24.1.3, $\Pi_r \cap \mathbf{V}(x_0, x_1, \ldots, x_r) = \varnothing$. Hence $\Pi_r \cap \mathbf{V}(x_0, x_1, \ldots, x_{s-1}, x_{s+1}, \ldots, x_r)$ is a point $\mathbf{P}(X_s)$, $s = 0, 1, \ldots, r$. By Lemma 24.1.4 up to a factor of proportion

$$X_s = (x_0^s, x_1^s, \ldots, x_n^s),$$

where $x_i^s = (0\ 1\ \cdots s-1\ i\ s+1\ \cdots\ r)$, $s = 0, 1, \ldots, r$, and $i = 0, 1, \ldots, n$. Calculating the determinant

$$\Delta = |x_i^s|, \quad i, s = 0, 1, \ldots, r,$$

we obtain $(0\ 1\ \cdots\ r)^{r+1} \neq 0$. Hence the points $\mathbf{P}(X_0), \mathbf{P}(X_1), \ldots, \mathbf{P}(X_r)$ are linearly independent. Now we show that up to a factor of proportion the given elements $(k_0 k_1 \cdots k_r)$ are the elements $(k_0 k_1 \cdots k_r)_x$ corresponding to the points $\mathbf{P}(X_s)$ of Π_r; more precisely we show that

$$(k_0 k_1 \cdots k_r)_x = (k_0 k_1 \cdots k_r)(0\ 1\ \cdots\ r)^r. \tag{24.7}$$

We already know that

$$(0\ 1\ \cdots\ r)_x = (0\ 1\ \cdots\ r)^{r+1}. \tag{24.8}$$

Further

$$\begin{aligned}(0\ 1\ \cdots\ s-1\ i\ s+1\ \cdots\ r)_x &= x_i^s(0\ 1\ \cdots\ r)^r \\ &= (0\ 1\ \cdots\ s-1\ i\ s+1\ \cdots\ r)(0\ 1\ \cdots\ r)^r,\end{aligned} \tag{24.9}$$

$s = 0, 1, \ldots, r$, and $i = 0, 1, \ldots, n$. We now proceed by induction on the number v of the k_j in the set $\{r+1, r+2, \ldots, n\}$. Equation (24.7) is satisfied if $v = 0$ or 1. Without loss of generality, it must be shown that

$$(j_1 j_2 \cdots j_v\, v\, v+1 \cdots r)_x = (j_1 j_2 \cdots j_v\, v\, v+1 \cdots r)(0\ 1\ \cdots\ r)^r, \tag{24.10}$$

where $j_1, j_2, \ldots, j_v \in \{r+1, r+2, \ldots, n\}$ and $v > 1$. Since both the elements $(k_0 k_1 \cdots k_r)$ and $(k_0 k_1 \cdots k_r)_x$ satisfy the quadratic relations (24.5), we have

$$\begin{aligned}&(0\ 1\ \cdots r)(j_1 j_2 \cdots j_v\, v\, v+1 \cdots r) \\ &= \sum_{s=1}^{v} (j_s\, 1\, 2 \cdots r)(j_1 \cdots j_{s-1}\, 0\, j_{s+1} \cdots j_v\, v\, v+1 \cdots r) \\ &\quad + \sum_{s=v}^{r} (s\, 1\, 2\ \cdots\ r)(j_1 \cdots j_v\, v \cdots s-1\, 0\, s+1 \cdots r) \\ &= \sum_{s=1}^{v} (j_s\, 1\, 2\ \cdots\ r)(j_1 \cdots j_{s-1}\, 0\, j_{s+1} \cdots j_v\, v\, v+1 \cdots r),\end{aligned} \tag{24.11}$$

and

$$(0\,1\,\cdots\,r)_x(j_1 j_2\cdots j_v\; v\; v+1\,\cdots\,r)_x$$
$$= \sum_{s=1}^{v} (j_s\,1\,2\,\cdots\,r)_x(j_1\cdots j_{s-1}\,0\,j_{s+1}\,\cdots\,j_v\; v\; v+1\,\cdots\,r)_x. \quad (24.12)$$

By (24.8) and (24.9),

$$(0\,1\,\cdots\,r)_x = (0\,1\,\cdots\,r)^{r+1}$$

and

$$(j_s\,1\,2\,\cdots\,r)_x = (j_s\,1\,2\,\cdots\,r)(0\,1\,\cdots\,r)^r.$$

By induction

$$(j_1\,\cdots\,j_{s-1}\,0\,j_{s+1}\,\cdots\,j_v\; v\; v+1\,\cdots\,r)_x$$
$$= (j_1\,\cdots\,j_{s-1}\,0\,j_{s+1}\,\cdots\,j_v\; v\; v+1\,\cdots\,r)(0\,1\,\cdots\,r)^r.$$

Hence (24.12) becomes

$$(0\,1\,\cdots\,r)^{r+1}(j_1\,j_2\,\cdots\,j_v\; v\; v+1\,\cdots\,r)_x$$
$$= \sum_{s=1}^{v} (0\,1\,\cdots\,r)^{2r}(j_s\,1\,2\,\cdots\,r)(j_1\,\cdots\,j_{s-1}\,0\,j_{s+1}\,\cdots\,j_v\; v\; v+1\,\cdots\,r).$$
$$(24.13)$$

Comparing (24.13) with (24.11), and since $(0\,1\,\cdots\,r) \neq 0$, we finally obtain

$$(j_1 j_2\cdots j_v\; v\; v+1\cdots r)_x = (0\,1\,\cdots\,r)^r(j_1 j_2\cdots j_v\; v\; v+1\,\cdots\,r),$$

which is the equality (24.10). □

24.2 Grassmann varieties

Let $PG^{(r)}(n, K)$ be the set of all r-spaces of $PG(n, K)$, $1 \leq r \leq n-2$ and $n \geq 3$. If

$$L = (l_0, l_1, \ldots, l_{\mathbf{c}(n+1,r+1)-1})$$

is a coordinate vector of $\Pi_r \in PG^{(r)}(n, K)$, then $\mathbf{P}(L)$ is a point of the projective space $PG(N, K)$, with $N = \mathbf{c}(n+1, r+1) - 1$.

The mapping which associates $\mathbf{P}(L)$ to Π_r is denoted by $\mathfrak{G}$. The algebraic variety $PG^{(r)}(n, K)\mathfrak{G}$ of $PG(N, K)$ is called the *Grassmannian* or the *Grassmann variety* of the r-spaces of $PG(n, K)$. It is denoted by $\mathcal{G}_{r,n,K}$ or $\mathcal{G}_{r,n}$. In the finite case it is also denoted by $\mathcal{G}_{r,n,q}$.

By Theorems 24.1.6 and 24.1.7, $\mathcal{G}_{r,n}$ is the intersection of the quadrics of $PG(N, K)$ represented by the equations (24.5). For $r = 1$ and $n = 3$ we have $N = 5$, and (24.5) represents only one quadric of $PG(5, K)$. In this case $\mathcal{G}_{1,3}$ is the hyperbolic quadric of §15.4. This quadric $\mathcal{G}_{1,3}$ is also called the *Klein quadric*.

Theorem 24.2.1: *The Grassmannian $\mathcal{G}_{r,n,q}$ has $\phi(r; n, q)$ points.*

Proof:
$$|\mathcal{G}_{r,n,q}| = |PG^{(r)}(n, q)| = \phi(r; n, q). \quad \square$$

Theorem 24.2.2: *No hyperplane of $PG(N, K)$ contains $\mathcal{G}_{r,n}$.*

Proof: Suppose that the hyperplane $\pi(U)$, where

$$U = (u_0, u_1, \ldots, u_N),$$

contains $\mathcal{G}_{r,n}$. Consider the r-space $\mathbf{U}_{i_0}\mathbf{U}_{i_1}\cdots\mathbf{U}_{i_r}$ of $PG(n, K)$, where $\{i_0, i_1, \ldots, i_r\} \subset \{0, 1, \ldots, n\}$. For this r-space we have $(k_0 k_1 \cdots k_r) = 0$ whenever $\{k_0, k_1, \ldots, k_r\} \neq \{i_0, i_1, \ldots, i_r\}$. If l_i is the coordinate of $\mathbf{U}_{i_0}\mathbf{U}_{i_1}\cdots\mathbf{U}_{i_r}$ corresponding to the set $\{i_0, i_1, \ldots, i_r\}$, then $l_i \neq 0$ while all other coordinates of this r-space are zero. Since $\mathcal{G}_{r,n}$ is contained in $\pi(U)$, we have $u_i = 0$. In this way we see that all coordinates of U are zero, a contradiction. $\square$

If $\mathbf{V}(F_1), \mathbf{V}(F_2), \ldots$ are the quadrics represented by (24.5), then $\mathcal{G}_{r,n,K} = \mathbf{V}_{N,K}(F_1, F_2, \ldots)$. Clearly $\overline{\mathcal{G}}_{r,n,K} = \mathbf{V}_{N,\bar{K}}(F_1, F_2, \ldots)$ is the Grassmannian $\mathcal{G}_{r,n,\bar{K}}$ of the r-spaces of $PG(n, \bar{K})$.

The following theorem will be stated without proof.

Theorem 24.2.3: *The algebraic variety $\mathcal{G}_{r,n}$ is absolutely irreducible and rational. All points of $\overline{\mathcal{G}}_{r,n}$ are simple. Finally, the dimension of $\mathcal{G}_{r,n}$ is $(r+1)(n-r)$ while its order is*

$$[(r+1)(n-r)]!\,\frac{((r))((n-r-1))}{((n))},$$

where $((m)) = 1!2!\cdots m!$. $\square$

Hence $\mathcal{G}_{1,3}$ has dimension 4 and order 2, $\mathcal{G}_{1,4}$ has dimension 6 and order 5, $\mathcal{G}_{1,n}$ has dimension $2(n-1)$ and order $[2(n-1)]!/\{(n-1)!n!\}$, $\mathcal{G}_{2,5}$ has dimension 9 and order 42, $\mathcal{G}_{r,2r+1}$ has dimension $(r+1)^2$ and order

$$(r+1)^2\,!\,\frac{1!2!\cdots r!}{(r+1)!(r+2)!\cdots(2r+1)!}.$$

Theorem 24.2.4: $\mathcal{G}_{r,n} \sim \mathcal{G}_{n-r-1,n}$, $1 \leq r \leq n-2$ *and* $n \geq 3$.

Proof: Suppose that the coordinate of $\Pi_r \in PG^{(r)}(n, K)$ in position $s+1$, and the dual coordinate of $\Pi_{n-r-1} \in PG^{(n-r-1)}(n, K)$ in the same position, correspond to the same ordered $(r+1)$-tuple $(i_0, i_1, \ldots, i_r)$, $s = 0, 1, \ldots, \mathbf{c}(n+1, r+1)$. If $\mathbf{P}(X_0), \mathbf{P}(X_1), \ldots, \mathbf{P}(X_r)$ are $r+1$ linearly independent points of Π_r, then the hyperplanes $\pi(X_0), \pi(X_1), \ldots, \pi(X_r)$ have a Π_{n-r-1} as intersection. Hence any coordinate vector of Π_r is also a dual coordinate vector of Π_{n-r-1}. By Lemma 24.1.2 any dual coordinate vector of Π_{n-r-1} is also a coordinate vector of Π_{n-r-1}. Now it is clear that $\mathscr{G}_{r,n} = \mathscr{G}_{n-r-1,n}$.

If we do not make the assumption at the beginning of the proof, then there is a projectivity of the form $\rho x_i' = \varepsilon_i x_j$, $\varepsilon_i \in \{+1, -1\}$ and $i, j = 0, 1, \ldots, \mathbf{c}(n+1, r+1) - 1$, of $PG(N, K)$ which takes $\mathscr{G}_{r,n}$ to $\mathscr{G}_{n-r-1,n}$. □

Now we shall consider in more detail the case $n = 2r + 1$. In this case we always assume that for any two coordinates $(i_0 i_1 \cdots i_r)_x$ and $(j_0 j_1 \cdots j_r)_x$ of Π_r, where $|\{i_0, i_1, \ldots, j_r\}| = 2r + 1$, the permutation $(i_0, i_1, \ldots, j_r)$ is even. Also we assume that the positions of any two such coordinates in the coordinate vector of Π_r differ by r. Let $\Pi_r = \mathbf{\Pi}_r(L)$ and $\Pi_r' = \mathbf{\Pi}_r(L')$ be r-spaces of $PG(2r+1, K)$ where

$$L = (l_0, l_1, \ldots) \quad \text{and} \quad L' = (l_0', l_1', \ldots).$$

By Lemmas 24.1.2 and 24.1.3 we have $\Pi_r \cap \Pi_r' \neq \varnothing$ if and only if

$$\sum_{i=0}^{r+1} (l_i l_{i+r}' + l_i' l_{i+r}) = 0 \quad \text{for } r \text{ odd},$$

and

$$\sum_{i=0}^{r+1} (l_i l_{i+r}' - l_i' l_{i+r}) = 0 \quad \text{for } r \text{ even}.$$

Suppose r odd. Then to $PG^{(r)}(2r+1, K)$ there is associated the polarity δ of $PG(N, K)$ with bilinear form

$$\sum_{i=0}^{r+1} (x_i x_{i+r}' + x_i' x_{i+r}). \tag{24.14}$$

If $K = GF(q)$ with q odd, then δ is the orthogonal polarity defined by the hyperbolic quadric

$$\mathbf{V}\left(\sum_{i=0}^{r+1} x_i x_{i+r}\right).$$

For $r = 1$ and $n = 3$ this quadric coincides with the Grassmannian $\mathscr{G}_{1,3}$. If $K = GF(q)$ with q even, then δ is a null polarity.

Suppose r even. Then to $PG^{(r)}(2r+1, K)$ there is associated the null polarity δ of $PG(N, K)$ with bilinear form

$$\sum_{i=0}^{r+1} (x_i x_{i+r}' - x_i' x_{i+r}). \tag{24.15}$$

The polarity δ is called the *fundamental polarity* associated to $\mathscr{G}_{r,2r+1}$.

Consider in $PG(2r+1, K)$ the correlation η represented by

$$\rho x_i = u_i, \qquad i = 0, 1, \ldots, 2r+1.$$

If $\Pi_r \eta = \Pi'_r$ and $\Pi_r = \mathbf{\Pi}_r(L)$, $\Pi'_r = \mathbf{\Pi}_r(L')$, where

$$L = (l_0, l_1, \ldots) \quad \text{and} \quad L' = (l'_0, l'_1, \ldots),$$

then by Lemma 24.1.2

$$\rho l'_i = l_{i+r}, \qquad i = 0, 1, \ldots, 2r+1$$

(indices being taken modulo $2r+1$). Hence to the correlation η there corresponds the projectivity

$$\rho x'_i = x_{i+r}, \qquad i = 0, 1, \ldots, 2r+1,$$

of $PG(N, K)$ which leaves $\mathscr{G}_{r,2r+1}$ invariant. This projectivity will be denoted by ζ.

Now we determine all subspaces of $PG(N, K)$ which are contained in $\mathscr{G}_{r,n}$. First of all the lines on $\mathscr{G}_{r,n}$ will be considered. We introduce the following term: the set of all r-spaces Π_r of $PG(n, K)$ contained in a given $(r+1)$-space Π_{r+1} and containing a given $(r-1)$-space $\Pi_{r-1} \subset \Pi_{r+1}$ will be called a *pencil of r-spaces*. This pencil will be denoted by (Π_{r-1}, Π_{r+1}).

Theorem 24.2.5: *The image of a pencil of r-spaces is a line of* $\mathscr{G}_{r,n}$. *Conversely, any line on* $\mathscr{G}_{r,n}$ *is the image of a pencil of r-spaces of* $PG(n, K)$.

Proof: Let Π_{r-1} be a given $(r-1)$-space of $PG(n, K)$; let $\Pi_{r+1} \supset \Pi_{r-1}$ be a given $(r+1)$-space of $PG(n, K)$. Further, let l be a line of Π_{r+1} which has no point in common with Π_{r-1}. Let $\mathbf{P}(X_0), \mathbf{P}(X_1), \ldots, \mathbf{P}(X_{r-1})$ be r linearly independent points of Π_{r-1}. Consider distinct elements Π_r^1, Π_r^2 of the pencil (Π_{r-1}, Π_{r+1}). The intersections of Π_r^1 and Π_r^2 with l are respectively denoted by $\mathbf{P}(X_r^1)$ and $\mathbf{P}(X_r^2)$. The coordinates of Π_r^i, i = 1, 2, are determined by the matrix

$$T_x^i = \begin{bmatrix} X_0 \\ X_1 \\ \vdots \\ X_{r-1} \\ X_r^i \end{bmatrix}.$$

Let Π_r be an arbitrary element of the pencil (Π_{r-1}, Π_{r+1}). A coordinate vector of the point $\Pi_r \cap l$ is of the form

$$t_1 X_r^1 + t_2 X_r^2,$$

$t_1, t_2 \in K$ and not both zero. Conversely, any vector of this type defines one element Π_r of (Π_{r-1}, Π_{r+1}). The coordinates of Π_r are determined by the matrix

$$T_x = \begin{bmatrix} X_0 \\ X_1 \\ \vdots \\ X_{r-1} \\ t_1 X_r^1 + t_2 X_r^2 \end{bmatrix}.$$

If

$$L_i = (l_0^i, l_1^i, \ldots), \qquad i = 1, 2,$$

is the coordinate vector defined by the matrix T_x^i, then

$$t_1 L_1 + t_2 L_2$$

is the coordinate vector defined by the matrix T_x. Hence $(\Pi_{r-1}, \Pi_{r+1})\mathfrak{G}$ is the line joining the points $\Pi_r^1\mathfrak{G}$ and $\Pi_r^2\mathfrak{G}$.

Conversely, let l be a line on $\mathscr{G}_{r,n}$ and let $\mathbf{P}(L_1), \mathbf{P}(L_2)$ be distinct points of l. Suppose that the intersection of $\mathbf{P}(L_1)\mathfrak{G}^{-1} = \Pi_r^1$ and $\mathbf{P}(L_2)\mathfrak{G}^{-1} = \Pi_r^2$ is an $(r-1)$-space Π_{r-1}. Then $\Pi_r^1\Pi_r^2$ is an $(r+1)$-space Π_{r+1}. Clearly then $(\Pi_{r-1}, \Pi_{r+1})\mathfrak{G} = \mathbf{P}(L_1)\mathbf{P}(L_2) = l$. So we show that $\Pi_r^1 \cap \Pi_r^2$ is an $(r-1)$-space.

Let $\Pi_r^1 \cap \Pi_r^2$ be a d-space Π_d, $-1 \le d \le r-1$, let $\mathbf{P}(X_0), \mathbf{P}(X_1), \ldots, \mathbf{P}(X_d)$ be $d+1$ linearly independent points of Π_d, and let $\mathbf{P}(X_0), \ldots, \mathbf{P}(X_d)$, $\mathbf{P}(X_{d+1}^i), \ldots, \mathbf{P}(X_r^i)$ be $r+1$ linearly independent points of Π_r^i, with $i = 1, 2$. Let ξ be a projectivity of $PG(n, K)$ with $\mathbf{P}(X_i)\xi = \mathbf{U}_i$, $i = 0, 1, \ldots, d$, $\mathbf{P}(X_i^1)\xi = \mathbf{U}_i$, $i = d+1, d+2, \ldots, r$, and $\mathbf{P}(X_{d+i}^2)\xi = \mathbf{U}_{r+i}$, $i = 1, 2, \ldots, r-d$. In §24.1, just after the definition of Grassmann coordinates, we have shown that ξ induces a projectivity $\tilde{\xi}$ of $PG(N, K)$ which leaves $\mathscr{G}_{r,n}$ invariant. Then

$$\mathbf{P}(L_1)\tilde{\xi} = (\mathbf{U}_0\mathbf{U}_1 \cdots \mathbf{U}_r)\mathfrak{G}$$

and

$$\mathbf{P}(L_2)\tilde{\xi} = (\mathbf{U}_0 \cdots \mathbf{U}_d\mathbf{U}_{r+1} \cdots \mathbf{U}_{2r-d})\mathfrak{G}.$$

The points $\mathbf{P}(L_1)\tilde{\xi}$ and $\mathbf{P}(L_2)\tilde{\xi}$ are on the line $l\tilde{\xi}$ of $\mathscr{G}_{r,n}$. For the space $\mathbf{U}_0\mathbf{U}_1 \cdots \mathbf{U}_r$ we have $(k_0 k_1 \cdots k_r) \neq 0$ if and only if $\{k_0, k_1, \ldots, k_r\} = \{0, 1, \ldots, r\}$; for the space $\mathbf{U}_0 \cdots \mathbf{U}_d\mathbf{U}_{r+1} \cdots \mathbf{U}_{2r-d}$ we have $(k_0 k_1 \cdots k_r) \neq 0$ if and only if $\{k_0, k_1, \ldots, k_r\} = \{0, \ldots, d, r+1, \ldots, 2r-d\}$. Since each point of $l\tilde{\xi}$ is on $\mathscr{G}_{r,n}$, for any two given elements $t_1, t_2 \in K$, not both zero, there is an r-space Π_r with $(0\,1 \cdots r) = t_1$, $(0 \cdots d\; r+1 \cdots 2r-d) = t_2$, and $(k_0 k_1 \cdots k_r) = 0$ in all other cases. Choose $t_1 \neq 0 \neq t_2$. Then

$(r\,r-1\cdots 0)(0\cdots d\,r+1\cdots 2r-d) \neq 0$. By (24.5)

$$\begin{aligned}(r\,r-1\cdots 0)&(0\cdots d\,r+1\cdots 2r-d)\\ &= \sum_{s=0}^{d} (s\,r-1\cdots 0)(0\cdots s-1\,r\,s+1\cdots d\,r+1\cdots 2r-d)\\ &\quad + \sum_{s=r+1}^{2r-d} (s\,r-1\cdots 0)(0\cdots d\,r+1\cdots s-1\,r\,s+1\cdots 2r-d)\\ &= \sum_{s=r+1}^{2r-d} (s\,r-1\cdots 0)(0\cdots d\,r+1\cdots s-1\,r\,s+1\cdots 2r-d).\end{aligned}$$

Hence for some $s \in \{r+1, r+2, \ldots, 2r-d\}$ we have $(s\,r-1\cdots 0) \neq 0$; that is, for some $s \in \{r+1, r+2, \ldots, 2r-d\}$ we have $\{0, 1, \ldots, r-1, s\} = \{0, \ldots, d, r+1, \ldots, 2r-d\}$. Consequently $d = r-1$, and so $\Pi_r^1 \cap \Pi_r^2$ is an $(r-1)$-space. □

Theorem 24.2.5 is equivalent to saying that the images $\Pi_r^1\mathfrak{G}$ and $\Pi_r^2\mathfrak{G}$ of two distinct r-spaces Π_r^1 and Π_r^2 of $PG(n, K)$ are on a common line of $\mathcal{G}_{r,n}$ if and only if the intersection $\Pi_r^1 \cap \Pi_r^2$ is an $(r-1)$-space, or equivalently if and only if $\Pi_r^1\Pi_r^2$ is an $(r+1)$-space.

Theorem 24.2.6: *The number of lines of $\mathcal{G}_{r,n,q}$ is equal to*

$$\prod_{i=3}^{n+1} (q^i-1)\Bigg/\left\{\prod_{i=1}^{r} (q^i-1) \prod_{i=1}^{n-r-1} (q^i-1)\right\}. \tag{24.16}$$

Proof: By Theorem 24.2.5 the number of lines of $\mathcal{G}_{r,n,q}$ is the number of pencils (Π_{r-1}, Π_{r+1}). Hence it is equal to the product of $|PG^{(r-1)}(n, q)|$ and the number of $(r+1)$-spaces containing a given Π_{r-1}. Consequently it is equal to

$$\begin{aligned}\phi(r-1; n, q)\chi(r-1, r+1; n, q)\\ &= \frac{[n-r+2, n+1]_-[3, n-r+1]_-}{[1, r]_-[1, n-r-1]_-}\\ &= \prod_{i=3}^{n+1} (q^i-1)\Bigg/\left\{\prod_{i=1}^{r} (q^i-1) \prod_{i=1}^{n-r-1} (q^i-1)\right\}. \quad \square\end{aligned}$$

Lemma 24.2.7: *The number of lines of $\mathcal{G}_{r,n,q}$ through a given point of $\mathcal{G}_{r,n,q}$ is equal to*

$$\theta(n-r-1)\theta(r). \tag{24.17}$$

Proof: The number of lines of $\mathcal{G}_{r,n,q}$ through a given point P of $\mathcal{G}_{r,n,q}$ is the number of pencils (Π_{r-1}, Π_{r+1}) with $\Pi_{r-1} \subset \Pi_r \subset \Pi_{r+1}$ and $\Pi_r = P\mathfrak{G}^{-1}$. Hence this number equals

$$\chi(r, r+1; n, q)\phi(r-1; r, q) = \theta(n-r-1)\theta(r). \quad \square$$

Lemma 24.2.8: *A line l of $PG(N, K)$ having at least three points in common with $\mathcal{G}_{r,n}$ is entirely contained in $\mathcal{G}_{r,n}$.*

Proof: Recall that $\mathcal{G}_{r,n}$ is the intersection of the quadrics of $PG(N, K)$ represented by the equations (24.5). The line l has at least three points in common with each of these quadrics, and hence is contained in all the quadrics. We conclude that l is contained in $\mathcal{G}_{r,n}$. $\square$

An s-space Π_s which is contained in $\mathcal{G}_{r,n}$, but in no $(s+1)$-space Π_{s+1} of $\mathcal{G}_{r,n}$, is called a *maximal space* or *maximal subspace* of $\mathcal{G}_{r,n}$. The next theorem describes all maximal spaces of $\mathcal{G}_{r,n}$.

Theorem 24.2.9: *The variety $\mathcal{G}_{r,n}$ contains two systems of maximal spaces. The first system consists of the $(n-r)$-spaces Π_{n-r} with $\Pi_{n-r}\mathfrak{G}^{-1}$ the set of all r-spaces through a common Π_{r-1}. The other system consists of the $(r+1)$-spaces Π_{r+1} with $\Pi_{r+1}\mathfrak{G}^{-1}$ the set of all r-spaces contained in a common $(r+1)$-space.*

Proof: Let Π_{r-1} be an $(r-1)$-space of $PG(n, K)$, and let $\mathcal{R}$ be the set of all r-spaces containing Π_{r-1}. If Π_r^1 and Π_r^2 are distinct elements of $\mathcal{R}$, then $\Pi_r^1 \cap \Pi_r^2$ is the $(r-1)$-space Π_{r-1} and so they belong to a pencil of r-spaces which is completely contained in $\mathcal{R}$. Hence if P_1 and P_2 are distinct points of $\mathcal{R}\mathfrak{G}$, then the line P_1P_2 is contained in $\mathcal{R}\mathfrak{G}$. This means that $\mathcal{R}\mathfrak{G}$ is a subspace of $PG(N, K)$. Let Π_{n-r} be a subspace of $PG(n, K)$ which is skew to Π_{r-1}. The bijection which maps each element of $\mathcal{R}$ onto its intersection with Π_{n-r} is denoted by δ. It is clear that $\delta^{-1}\mathfrak{G}$ is a bijection of Π_{n-r} onto $\mathcal{R}\mathfrak{G}$ which maps the lines of Π_{n-r} onto the lines of $\mathcal{R}\mathfrak{G}$. This means that $\delta^{-1}\mathfrak{G}$ is a collineation from Π_{n-r} onto $\mathcal{R}\mathfrak{G}$, and so $\mathcal{R}\mathfrak{G}$ has dimension $n-r$. If $\mathcal{R}\mathfrak{G}$ is properly contained in the subspace π of $\mathcal{G}_{r,n}$, then let $P \in \pi\backslash\mathcal{R}\mathfrak{G}$. The r-space $P\mathfrak{G}^{-1}$ of $PG(n, K)$ does not contain Π_{r-1} but has an $(r-1)$-space in common with each r-space through Π_{r-1}. This contradiction shows that the $(n-r)$-space $\mathcal{R}\mathfrak{G}$ is indeed a maximal space of $\mathcal{G}_{r,n}$.

Let Π_{r+1} be an $(r+1)$-space of $PG(n, K)$, and let $\mathcal{S}$ be the set of all r-spaces contained in Π_{r+1}. If Π_r^1 and Π_r^2 are distinct elements of $\mathcal{S}$, then $\Pi_r^1\Pi_r^2$ is the $(r+1)$-space Π_{r+1} and so they belong to a pencil of r-spaces which is completely contained in $\mathcal{S}$. Hence if P_1 and P_2 are distinct points of $\mathcal{S}\mathfrak{G}$, then the line P_1P_2 is contained in $\mathcal{S}\mathfrak{G}$. This means that $\mathcal{S}\mathfrak{G}$ is a subspace of $PG(N, K)$. Since the pencils of hyperplanes of Π_{r+1} are mapped by $\mathfrak{G}$ onto the lines of $\mathscr{S}\mathfrak{G}$, so $\mathfrak{G}$ induces a reciprocity from Π_{r+1} to $\mathscr{S}\mathfrak{G}$. Hence $\mathscr{S}\mathfrak{G}$ also has dimension $r + 1$. If $\mathscr{S}\mathfrak{G}$ is properly contained in the subspace π of $\mathscr{G}_{r,n}$, then let $P \in \pi \backslash \mathscr{S}\mathfrak{G}$. The r-space $P\mathfrak{G}^{-1}$ of $PG(n, K)$ is not contained in Π_{r+1} but has an $(r - 1)$-space in common with each r-space of Π_{r+1}, a contradiction. Consequently the $(r + 1)$-space $\mathscr{S}\mathfrak{G}$ is indeed a maximal space of $\mathscr{G}_{r,n}$.

Now consider an arbitrary maximal space π of $\mathscr{G}_{r,n}$. Clearly π has at least dimension 1. If l is a line of π, then each element of the pencil $(\Pi_{r-1}, \Pi_{r+1}) = l\mathfrak{G}^{-1}$ has at least an $(r - 1)$-space in common with each element $P\mathfrak{G}^{-1}$, where $P \in \pi$. So $P\mathfrak{G}^{-1}$ contains Π_{r-1} or is contained in Π_{r+1}. Suppose that for at least one point $P' \in \pi \backslash l$, the r-space $P'\mathfrak{G}^{-1}$ contains Π_{r-1}. For any element $P\mathfrak{G}^{-1}$, with $P \in \pi \backslash (l \cup \{P'\})$, we know that $P\mathfrak{G}^{-1} \subset \Pi_{r+1}$ or $\Pi_{r-1} \subset P\mathfrak{G}^{-1}$, and that $P\mathfrak{G}^{-1} \cap P'\mathfrak{G}^{-1}$ is an $(r - 1)$-space. Hence $\Pi_{r-1} \subset P\mathfrak{G}^{-1}$. So all elements of $\pi\mathfrak{G}^{-1}$ contain Π_{r-1}. Since π is maximal, it is the image of the set of all r-spaces containing Π_{r-1}. If for at least one point $P' \in \pi \backslash l$, the r-space $P'\mathfrak{G}^{-1}$ is contained in Π_{r+1}, then analogously one shows that π is the image of the set of all r-spaces contained in Π_{r+1}. □

The system of maximal spaces of $\mathscr{G}_{r,n}$ corresponding to the $(r - 1)$-spaces of $PG(n, K)$ will be called the *Latin system* and will be denoted by $\mathscr{S}_{\mathrm{L}}$; its elements are called the *Latin $(n - r)$-spaces.* The system of maximal spaces corresponding to the $(r + 1)$-spaces of $PG(n, K)$ will be called the *Greek system* and will be denoted by $\mathscr{S}_{\mathrm{G}}$; its elements are called the *Greek $(r + 1)$-spaces.* Notice that the Latin and the Greek spaces have the same dimension if and only if $n = 2r + 1$.

Let Π_{r-1} be an $(r - 1)$-space of $PG(n, K)$ and let Π_{n-r} be an $(n - r)$-space skew to Π_{r-1}. In the first part of the proof of Theorem 24.2.9, it was shown that $\mathfrak{G}$ induces a collineation ξ of Π_{n-r} onto the corresponding maximal space of $\mathscr{G}_{r,n}$. By the first part of the proof of Theorem 24.2.5, ξ preserves the cross-ratio of any four collinear points of Π_{n-r}. Hence ξ is a projectivity.

Similarly, let Π_{r+1} be an $(r + 1)$-space of $PG(n, K)$. In the second part of the proof of Theorem 24.2.9, it was shown that $\mathfrak{G}$ induces a reciprocity ξ of Π_{r+1} onto the corresponding maximal space of $\mathscr{G}_{r,n}$. Again by Theorem 24.2.5, ξ preserves the cross-ratio of any four hyperplanes of Π_{r+1} belonging to a common pencil in Π_{r+1}. Hence ξ is a correlation.

Lemma 24.2.10: *Any line l of $\mathscr{G}_{r,n}$ is contained in one Latin space and one Greek space. Hence two distinct Latin or two distinct Greek spaces have at most one point in common. If $\Pi_{n-r} \in \mathscr{S}_{\mathrm{L}}$ and $\Pi_{r+1} \in \mathscr{S}_{\mathrm{G}}$, then $\Pi_{n-r} \cap \Pi_{r+1}$ is the empty set or a line.*

Proof: Let $l\mathfrak{G}^{-1}$ be the pencil (Π_{r-1}, Π_{r+1}). Then the Latin space defined by Π_{r-1} is the only Latin space through l and the Greek space defined by Π_{r+1} is the only Greek space through l.

Let $\Pi_{n-r} \in \mathscr{S}_L$ and $\Pi_{r+1} \in \mathscr{S}_G$. The corresponding spaces of $PG(n, K)$ are respectively denoted by Π'_{r-1} and Π'_{r+1}. If $\Pi'_{r-1} \subset \Pi'_{r+1}$, then $\Pi_{n-r} \cap \Pi_{r+1}$ is a line; if $\Pi'_{r-1} \not\subset \Pi'_{r+1}$, then $\Pi_{n-r} \cap \Pi_{r+1} = \varnothing$. □

Let $\mathscr{T}_r(\Pi_g, \Pi_h)$ be the set of all r-spaces of $PG(n, K)$ through the g-space Π_g and contained in the h-space Π_h. In the previous notation $\mathscr{T}_r(\Pi_{r-1}, \Pi_{r+1})$, where $\Pi_{r-1} \subset \Pi_{r+1}$, is the pencil (Π_{r-1}, Π_{r+1}). Let $\mathscr{G}^{(s)}_{r,n}$ be the set of all s-spaces on $\mathscr{G}_{r,n}$.

Theorem 24.2.11: *For $0 \le s \le n - r$ the set $\mathscr{T}_r(\Pi_{r-1}, \Pi_{r+s})$, where $\Pi_{r-1} \subset \Pi_{r+s}$, is an element of $\mathscr{G}^{(s)}_{r,n}$; for $0 \le s \le r + 1$ the set $\mathscr{T}_r(\Pi_{r-s}, \Pi_{r+1})$, where $\Pi_{r-s} \subset \Pi_{r+1}$, is an element of $\mathscr{G}^{(s)}_{r,n}$. Also, any s-space of $\mathscr{G}_{r,n}$, where $0 \le s \le \max(n - r, r + 1)$, is obtained in such a way.*

Proof: (i) We first show that $\mathscr{T}_r(\Pi_{r-1}, \Pi_{r+s}) \in \mathscr{G}^{(s)}_{r,n}$, where $0 \le s \le n - r$. If Π_{n-r} is a space skew to Π_{r-1}, then $\Pi_{n-r} \cap \Pi_{r+s} = \Pi_s$. The maximal space of $\mathscr{G}_{r,n}$ defined by Π_{r-1} is denoted by Π'_{n-r}. Now, consider the projectivity $\xi: \Pi_{n-r} \to \Pi'_{n-r}$; then $\mathscr{T}_r(\Pi_{r-1}, \Pi_{r+s})\mathfrak{G} = \Pi_s\xi$ is an s-space Π'_s of Π'_{n-r}.

(ii) Similarly, $\mathscr{T}_r(\Pi_{r-s}, \Pi_{r+1}) \in \mathscr{G}^{(s)}_{r,n}$, where $0 \le s \le r + 1$. For, if the maximal space of $\mathscr{G}_{r,n}$ defined by Π_{r+1} is denoted by Π'_{r+1} and we consider the correlation $\xi: \Pi_{r+1} \to \Pi'_{r+1}$, then $\mathscr{T}_r(\Pi_{r-s}, \Pi_{r+1})\mathfrak{G}$ is an s-space Π'_s of Π'_{r+1}.

(iii) Conversely, consider Π'_s in $\mathscr{G}^{(s)}_{r,n}$, where $0 \le s \le \max(n - r, r + 1)$. The space Π'_s is contained in at least one element of $\mathscr{S}_L \cup \mathscr{S}_G$. So, either $\Pi'_s \subset \Pi'_{n-r}$ where $\Pi'_{n-r} \in \mathscr{S}_L$ or $\Pi'_s \subset \Pi'_{r+1}$ where $\Pi'_{r+1} \in \mathscr{S}_G$. The $(r - 1)$-space corresponding to Π'_{n-r} is denoted by Π_{r-1} and the $(r + 1)$-space corresponding to Π'_{r+1} is denoted by Π_{r+1}.

(a) Let Π_{n-r} be an $(n - r)$-space of $PG(n, K)$ skew to Π_{r-1}. From the projectivity $\xi: \Pi_{n-r} \to \Pi'_{n-r}$, we have that $\Pi'_s\xi^{-1}$ is an s-space Π_s of Π_{n-r}. Hence $\Pi'_s\mathfrak{G}^{-1}$ is the set $\mathscr{T}_r(\Pi_{r-1}, \Pi_s\Pi_{r-1} = \Pi_{r+s})$.

(b) From the correlation $\xi: \Pi_{r+1} \to \Pi'_{r+1}$, we see that $\Pi'_s\xi^{-1}$ is the set of all hyperplanes of Π_{r+1} containing a fixed Π_{r-s}. Hence $\Pi'_s\mathfrak{G}^{-1} = \mathscr{T}_r(\Pi_{r-s}, \Pi_{r+1})$. □

Remark: By Lemma 24.2.10, any s-space of $\mathscr{G}_{r,n}$, where $s > 1$, is contained in exactly one element of $\mathscr{S}_L \cup \mathscr{S}_G$.

Theorem 24.2.12: *Let $K = GF(q)$.*

(i) $|\mathscr{S}_L| = \phi(r - 1; n, q)$ *and* $|\mathscr{S}_G| = \phi(r + 1; n, q)$.

(ii) *For* $n < 2r + 1$ *and* $1 < s \leq n - r$,

$$|\mathscr{G}_{r,n}^{(s)}| = \phi(r - 1; n, q)\phi(s; n - r, q) + \phi(r + 1; n, q)\phi(s; r + 1, q).$$

(iii) *For* $n < 2r + 1$ *and* $n - r < s \leq r + 1$,

$$|\mathscr{G}_{r,n}^{(s)}| = \phi(r + 1; n, q)\phi(s; r + 1, q).$$

(iv) *For* $n > 2r + 1$ *and* $1 < s \leq r + 1$,

$$|\mathscr{G}_{r,n}^{(s)}| = \phi(r - 1; n, q)\phi(s; n - r, q) + \phi(r + 1; n, q)\phi(s; r + 1, q).$$

(v) *For* $n > 2r + 1$ *and* $r + 1 < s \leq n - r$,

$$|\mathscr{G}_{r,n}^{(s)}| = \phi(r - 1; n, q)\phi(s; n - r, q).$$

(vi) *For* $n = 2r + 1$ *and* $1 < s \leq r + 1$,

$$|\mathscr{G}_{r,n}^{(s)}| = 2\phi(r + 1; 2r + 1, q)\phi(s; r + 1, q).$$

(vii) *For* $n = 2r + 1$,

$$|\mathscr{S}_{\mathrm{L}}| = |\mathscr{S}_{\mathrm{G}}| = \phi(r + 1; 2r + 1, q).$$

Proof: We have

$$|\mathscr{S}_{\mathrm{L}}| = |PG^{(r-1)}(n, q)| = \phi(r - 1; n, q)$$

and

$$|\mathscr{S}_{\mathrm{G}}| = |PG^{(r+1)}(n, q)| = \phi(r + 1; n, q).$$

If $n = 2r + 1$, then $\phi(r - 1; 2r + 1, q) = \phi(r + 1; 2r + 1, q)$, whence $|\mathscr{S}_{\mathrm{L}}| = |\mathscr{S}_{\mathrm{G}}| = \phi(r + 1; 2r + 1, q)$.

Since any s-space of $\mathscr{G}_{r,n}$, with $s > 1$, is contained in exactly one element of $\mathscr{S}_{\mathrm{L}} \cup \mathscr{S}_{\mathrm{G}}$, the number of these s-spaces is equal to the sum of the number of all s-spaces in elements of $\mathscr{S}_{\mathrm{L}}$ and the number of all s-spaces in elements of $\mathscr{S}_{\mathrm{G}}$. This proves the theorem. □

Theorem 24.2.13: *Let* $1 \leq r \leq n - 2$, $\Pi_t \subset \Pi_s$, $-1 \leq t \leq r - 2$ *and* $r + 2 \leq s \leq n$. *Then* $\mathscr{T}_r(\Pi_t, \Pi_s)\mathfrak{G}$ *is projectively equivalent to a* $\mathscr{G}_{r-t-1,s-t-1}$ *and to a* $\mathscr{G}_{s-r-1,s-t-1}$.

Proof: By §24.1 we may assume, without loss of generality, that $\Pi_s = \mathbf{U}_0\mathbf{U}_1\cdots\mathbf{U}_s$ and $\Pi_t = \mathbf{U}_{s-t}\mathbf{U}_{s-t+1}\cdots\mathbf{U}_s$. For any $\Pi_r \in \mathscr{T}_r(\Pi_t, \Pi_s)$ we choose $r - t$ linearly independent points $\mathbf{P}(X_0), \mathbf{P}(X_1), \ldots, \mathbf{P}(X_{r-t-1})$ in $\Pi_r \cap \mathbf{U}_0\mathbf{U}_1\cdots\mathbf{U}_{s-t-1}$. Let

$$X_i = (x_0^i, x_1^i, \ldots, x_{s-t-1}^i, 0, \ldots, 0),$$

$i = 0, 1, \ldots, r - t - 1$. The coordinates of Π_r are determined by the $(r + 1) \times (n + 1)$ matrix

$$\begin{bmatrix} x_0^0 & x_1^0 & \cdots & x_{s-t-1}^0 & 0 & 0 & \cdots & 0 & 0 & 0 & \cdots & 0 \\ x_0^1 & x_1^1 & \cdots & x_{s-t-1}^1 & 0 & 0 & \cdots & 0 & 0 & 0 & \cdots & 0 \\ \vdots & \vdots & & \vdots & \vdots & \vdots & & \vdots & \vdots & \vdots & & \vdots \\ x_0^{r-t-1} & x_1^{r-t-1} & \cdots & x_{s-t-1}^{r-t-1} & 0 & 0 & \cdots & 0 & 0 & 0 & \cdots & 0 \\ 0 & 0 & \cdots & 0 & 1 & 0 & \cdots & 0 & 0 & 0 & \cdots & 0 \\ 0 & 0 & \cdots & 0 & 0 & 1 & \cdots & 0 & 0 & 0 & \cdots & 0 \\ \vdots & \vdots & & \vdots & \vdots & \vdots & & \vdots & \vdots & \vdots & & \vdots \\ 0 & 0 & \cdots & 0 & 0 & 0 & \cdots & 1 & 0 & 0 & \cdots & 0 \end{bmatrix}.$$

Clearly $(k_0 k_1 \cdots k_r)$, where $k_0, k_1, \ldots, k_r$ are distinct, is equal to zero if $\{s - t, s - t + 1, \ldots, s\} \not\subset \{k_0, k_1, \ldots, k_r\}$ or if $\{k_0, k_1, \ldots, k_r\} \not\subset \{0, 1, \ldots, s\}$. If

$$\{s - t, s - t + 1, \ldots, s\} \subset \{k_0, k_1, \ldots, k_r\} \subset \{0, 1, \ldots, s\}$$

and

$$\{l_0, l_1, \ldots, l_{r-t-1}\} = \{k_0, k_1, \ldots, k_r\} \backslash \{s - t, s - t + 1, \ldots, s\},$$

then up to the sign $(k_0 k_1 \cdots k_r)$ is equal to $(l_0 l_1 \cdots l_{r-t-1})$, where the latter is calculated with respect to the matrix

$$\begin{bmatrix} x_0^0 & x_1^0 & \cdots & x_{s-t-1}^0 \\ x_0^1 & x_1^1 & \cdots & x_{s-t-1}^1 \\ \vdots & \vdots & & \vdots \\ x_0^{r-t-1} & x_1^{r-t-1} & \cdots & x_{s-t-1}^{r-t-1} \end{bmatrix}.$$

From these considerations it is clear that $\mathscr{T}_r(\Pi_t, \Pi_s)\mathfrak{G}$ is projectively equivalent to the Grassmann variety $\mathscr{G}_{r-t-1,s-t-1}$. By Theorem 24.2.4 we have

$$\mathscr{G}_{r-t-1,s-t-1} \sim \mathscr{G}_{s-r-1,s-t-1}. \quad \square$$

Corollary 1: *The image of the set of all r-spaces of PG(n, K) containing a given t-space Π_t, where $1 \leq r \leq n - 2$ and $-1 \leq t \leq r - 2$, is projectively equivalent to a $\mathscr{G}_{r-t-1,n-t-1}$ and to a $\mathscr{G}_{n-r-1,n-t-1}$.*

Proof: This is Theorem 24.2.13 with $s = n$. □

Corollary 2: *The image of the set of all r-spaces of PG(n, K) contained in a given s-space Π_s, where $1 \leq r \leq n - 2$ and $r + 2 \leq s \leq n$, is projectively equivalent to a $\mathscr{G}_{r,s}$ and to a $\mathscr{G}_{s-r-1,s}$.*

Proof: This is Theorem 24.2.13 with $s = -1$. □

Corollary 3: *$\mathscr{G}_{r,n}$ contains a subvariety projectively equivalent to $\mathscr{G}_{1,n'}$ if and only if $3 \leq n' \leq \max(n - r + 1, r + 2)$.*

Proof: For $t = r - 2$, Theorem 24.2.13 gives us subvarieties projectively equivalent to $\mathscr{G}_{1,s-r+1}$, where $r + 2 \leq s \leq n$; for $s = r + 2$, Theorem 24.2.13 gives us subvarieties projectively equivalent to $\mathscr{G}_{1,r-t+1}$, where $-1 \leq t \leq r - 2$.

For $n' > \max(n - r + 1, r + 2)$ the variety $\mathscr{G}_{1,n'}$ contains r'-spaces with $r' = n' - 1$. The Latin spaces of $\mathscr{G}_{r,n}$ have dimension $n - r < n' - 1$, and the Greek spaces of $\mathscr{G}_{r,n}$ have dimension $r + 1 < n' - 1$. Hence in this case $\mathscr{G}_{r,n}$ cannot contain subvarieties equivalent to $\mathscr{G}_{1,n'}$. □

In the last part of this section we determine all projectivities of $PG(N, K)$ leaving $\mathscr{G}_{r,n}$ invariant. It will appear necessary to distinguish between the cases $n = 2r + 1$ and $n \neq 2r + 1$.

Theorem 24.2.14: *If ξ is a projectivity of $PG(N, K)$ leaving $\mathscr{G}_{r,n}$ invariant, then ξ leaves $\mathscr{S}_{\mathrm{L}}$ and $\mathscr{S}_{\mathrm{G}}$ invariant, or interchanges them. For $n \neq 2r + 1$, ξ always leaves $\mathscr{S}_{\mathrm{L}}$ and $\mathscr{S}_{\mathrm{G}}$ invariant.*

Proof: Let ξ be a projectivity of $PG(N, K)$ leaving $\mathscr{G}_{r,n}$ invariant. Evidently ξ maps a maximal space of $\mathscr{G}_{r,n}$ onto a maximal space of $\mathscr{G}_{r,n}$. For $n \neq 2r + 1$ the Greek and Latin spaces have different dimension, and so ξ leaves $\mathscr{S}_{\mathrm{L}}$ and $\mathscr{S}_{\mathrm{G}}$ invariant.

So assume that $n = 2r + 1$. Let $\Pi_{r+1} \in \mathscr{S}_{\mathrm{L}} \cup \mathscr{S}_{\mathrm{G}}$, for example $\Pi_{r+1} \in \mathscr{S}_{\mathrm{L}}$. Choose a space $\Pi'_{r+1} \in \mathscr{S}_{\mathrm{L}}$ which has exactly one point P in common with Π_{r+1}. Then $\Pi_{r+1}\xi \cap \Pi'_{r+1}\xi$ is the point $P\xi$. By Lemma 24.2.10 the maximal spaces $\Pi_{r+1}\xi$ and $\Pi'_{r+1}\xi$ both belong to $\mathscr{S}_{\mathrm{L}}$ or both belong to $\mathscr{S}_{\mathrm{G}}$. Next consider any space $\Pi''_{r+1} \in \mathscr{S}_{\mathrm{L}}$, $\Pi''_{r+1} \neq \Pi_{r+1}$. The $(r - 1)$-spaces of $PG(2r + 1, K)$ which correspond to Π_{r+1} and Π''_{r+1} are respectively denoted by Π_{r-1} and Π''_{r-1}. There exists a finite number of distinct $(r - 1)$-spaces $\Pi^0_{r-1}, \Pi^1_{r-1}, \ldots, \Pi^k_{r-1}$ such that

$$\Pi_{r-1}\Pi^0_{r-1}, \Pi^0_{r-1}\Pi^1_{r-1}, \ldots, \Pi^{k-1}_{r-1}\Pi^k_{r-1}, \Pi^k_{r-1}\Pi''_{r-1}$$

are r-spaces. Hence to $\Pi^0_{r-1}, \Pi^1_{r-1}, \ldots, \Pi^k_{r-1}$ there correspond Latin spaces $\Pi^0_{r+1}, \Pi^1_{r+1}, \ldots, \Pi^k_{r+1}$ of $\mathscr{G}_{r,2r+1}$ such that

$$\Pi_{r+1} \cap \Pi^0_{r+1}, \Pi^0_{r+1} \cap \Pi^1_{r+1}, \ldots, \Pi^{k-1}_{r+1} \cap \Pi^k_{r+1}, \Pi^k_{r+1} \cap \Pi''_{r+1}$$

are points. By repeating a previous argument, the maximal spaces $\Pi_{r+1}\xi$, $\Pi^0_{r+1}\xi, \ldots, \Pi^k_{r+1}\xi, \Pi''_{r+1}\xi$ all belong to $\mathscr{S}_{\mathrm{L}}$ or all belong to $\mathscr{S}_{\mathrm{G}}$. Hence $\Pi_{r+1}\xi$ and $\Pi''_{r+1}\xi$ both belong to $\mathscr{S}_{\mathrm{L}}$ or both belong to $\mathscr{S}_{\mathrm{G}}$. We have proved that

ξ maps all elements of $\mathscr{S}_L$ onto elements of one of $\mathscr{S}_L$ or $\mathscr{S}_G$. Analogously, ξ maps all elements of $\mathscr{S}_G$ onto elements of one of $\mathscr{S}_L$ or $\mathscr{S}_G$. Now it is clear that ξ leaves $\mathscr{S}_L$ and $\mathscr{S}_G$ invariant, or interchanges them. □

Recall that in the case $n = 2r + 1$ the correlation η of $PG(2r + 1, K)$ represented by

$$\rho x_i = u_i, \qquad i = 0, 1, \ldots, 2r + 1,$$

induces a projectivity ζ of $PG(N, K)$ which leaves $\mathscr{G}_{r,2r+1}$ invariant. Clearly ζ interchanges the systems $\mathscr{S}_L$ and $\mathscr{S}_G$ of $\mathscr{G}_{r,2r+1}$.

Let $G(\mathscr{G}_{r,n})$ be the subgroup of $PGL(N + 1, K)$ leaving $\mathscr{G}_{r,n}$ fixed.

Lemma 24.2.15: *Let $\theta\colon \xi \to \tilde{\xi}$ map each element of $PGL(n + 1, K)$ onto the corresponding element $\tilde{\xi}$ of $G(\mathscr{G}_{r,n})$. Then θ is a monomorphism of $PGL(n + 1, K)$ into $G(\mathscr{G}_{r,n})$. Also, distinct elements of $G(\mathscr{G}_{r,n})$ induce distinct permutations of $\mathscr{G}_{r,n}$.*

Proof: First we show that the identity mapping of $\mathscr{G}_{r,n}$ is induced only by the identity $\mathfrak{J}$ of $PGL(N + 1, K)$. Let $\mathfrak{J}' \in PGL(N + 1, K)$ fix $\mathscr{G}_{r,n}$ pointwise. Choose distinct points $P, P' \in \mathscr{G}_{r,n}$, and let $P\mathfrak{G}^{-1} = \Pi_r$ and $P'\mathfrak{G}^{-1} = \Pi'_r$. There is a finite number of elements $\Pi_r^0, \Pi_r^1, \ldots, \Pi_r^k \in PG^{(r)}(n, K)$ such that $\Pi_r \cap \Pi_r^0, \Pi_r^0 \cap \Pi_r^1, \ldots, \Pi_r^{k-1} \cap \Pi_r^k, \Pi_r^k \cap \Pi'_r$ are $(r - 1)$-spaces. This means that there is a finite number of points $P_0, P_1, \ldots, P_k$ on $\mathscr{G}_{r,n}$, such that $PP_0, P_0P_1, \ldots, P_{k-1}P_k, P_kP'$ are lines of $\mathscr{G}_{r,n}$. Hence all points of the subspace of $PG(N, K)$ generated by $\mathscr{G}_{r,n}$ are fixed by $\mathfrak{J}'$. By Theorem 24.2.2 all points of $PG(N, K)$ are fixed by $\mathfrak{J}'$; hence $\mathfrak{J}'$ is the identity of $PGL(N + 1, K)$.

From the preceding paragraph it follows that, if $\delta, \delta' \in G(\mathscr{G}_{r,n})$ coincide on $\mathscr{G}_{r,n}$, then $\delta = \delta'$.

Let $\xi, \xi' \in PGL(n + 1, K)$, with $\xi \neq \xi'$. Then the mappings induced by ξ and ξ' on $PG^{(r)}(n, K)$ are distinct. Hence $\tilde{\xi} \neq \tilde{\xi}'$. So θ is an injection of $PGL(n + 1, K)$ into $G(\mathscr{G}_{r,n})$. Again, consider elements $\xi, \xi' \in PGL(n + 1, K)$. Evidently $\tilde{\xi}\tilde{\xi}'$ and $\widetilde{\xi\xi'}$ coincide on $\mathscr{G}_{r,n}$. We have shown that then $\tilde{\xi}\tilde{\xi}' = \widetilde{\xi\xi'}$. We conclude that θ is a monomorphism of $PGL(n + 1, K)$ into $G(\mathscr{G}_{r,n})$. □

Theorem 24.2.16: *For $n \neq 2r + 1$, θ is an isomorphism of $PGL(n + 1, K)$ onto $G(\mathscr{G}_{r,n})$. For $n = 2r + 1$ we have*

$$G(\mathscr{G}_{r,n}) = PGL(n + 1, K)\theta \cup (PGL(n + 1, K)\theta)\zeta.$$

Proof: First suppose that $n \neq 2r + 1$. We must prove that $G(\mathscr{G}_{r,n}) = PGL(n + 1, K)\theta$. Consider an element $\delta \in G(\mathscr{G}_{r,n})$. By Theorem 24.2.14, δ leaves $\mathscr{S}_L$ and $\mathscr{S}_G$ invariant. Consider an $(r + 1)$-space $\bar{\Pi}_{r+1}^1 \in \mathscr{S}_G$. The $(r + 1)$-space $\bar{\Pi}_{r+1}^1\delta = \bar{\Pi}_{r+1}^{1'}$ also belongs to $\mathscr{S}_G$. The corresponding $(r + 1)$-spaces of $PG(n, K)$ are denoted by Π_{r+1}^1 and $\Pi_{r+1}^{1'}$ respectively. By the remark just preceding Lemma 24.2.10, $\mathfrak{G}$ induces a correlation of Π^1_{r+1} onto $\bar{\Pi}^1_{r+1}$ and of $\Pi^{1'}_{r+1}$ onto $\bar{\Pi}^{1'}_{r+1}$. Hence $\mathfrak{G}\delta\mathfrak{G}^{-1}$ induces a projectivity ξ_1 of Π^1_{r+1} onto $\Pi^{1'}_{r+1}$.

Next, consider an $(r + 1)$-space Π^2_{r+1} of $PG(n, K)$ for which $\Pi^1_{r+1} \cap \Pi^2_{r+1}$ is an r-space Π_r. To Π^2_{r+1} corresponds an $(r + 1)$-space $\bar{\Pi}^2_{r+1} \in \mathscr{S}_G$. Let $\bar{\Pi}^2_{r+1}\delta = \bar{\Pi}^{2'}_{r+1} \in \mathscr{S}_G$ and let $\Pi^{2'}_{r+1}$ be the corresponding $(r + 1)$-space of $PG(n, K)$. Again $\mathfrak{G}\delta\mathfrak{G}^{-1}$ induces a projectivity ξ_2 of Π^2_{r+1} onto $\Pi^{2'}_{r+1}$.

Since $\Pi^1_{r+1} \cap \Pi^2_{r+1}$ is an r-space, the intersection $\bar{\Pi}^1_{r+1} \cap \bar{\Pi}^2_{r+1}$ is a point P. Hence $\bar{\Pi}^{1'}_{r+1} \cap \bar{\Pi}^{2'}_{r+1}$ is a point P', whence $\Pi^{1'}_{r+1} \cap \Pi^{2'}_{r+1}$ is an r-space Π'_r. Since $P\delta = P'$, we have $\Pi_r\xi_1 = \Pi_r\xi_2 = \Pi'_r$. In Π_r we now choose an $(r - 1)$-space Π_{r-1}. If $\Pi^1_{r+1}\Pi^2_{r+1} = \Pi_{r+2}$ then $\mathscr{T}_r(\Pi_{r-1}, \Pi_{r+2})\mathfrak{G}$ is a plane Π_2. The intersections $\Pi_2 \cap \bar{\Pi}^1_{r+1}$ and $\Pi_2 \cap \bar{\Pi}^2_{r+1}$ are lines, which are respectively denoted by l_1 and l_2. It is evident that $l_1 \cap l_2 = \{P\}$. The plane $\Pi_2\delta = \Pi'_2$ intersects $\bar{\Pi}^{1'}_{r+1}$ and $\bar{\Pi}^{2'}_{r+1}$ in the respective lines $l'_1 = l_1\delta$ and $l'_2 = l_2\delta$; also $l'_1 \cap l'_2 = P'$. Clearly $l'_1\mathfrak{G}^{-1}$ is the pencil $(\Pi_{r-1}\xi_1, \Pi^{1'}_{r+1})$, and $l'_2\mathfrak{G}^{-1}$ is the pencil $(\Pi_{r-1}\xi_2, \Pi^{2'}_{r+1})$. These two pencils belong to $\Pi'_2\mathfrak{G}^{-1}$; hence by Theorem 24.2.11 all elements of the two pencils contain a common $(r - 2)$-space and are contained in a common $(r + 1)$-space, or contain a common $(r - 1)$-space and are contained in a common $(r + 2)$-space. Since the elements of the pencils generate $\Pi^{1'}_{r+1}\Pi^{2'}_{r+1}$, which is an $(r + 2)$-space, they all contain a common $(r - 1)$-space. Hence $\Pi_{r-1}\xi_1 = \Pi_{r-1}\xi_2 = \Pi'_{r-1}$. Consequently the actions of ξ_1 and ξ_2 on the set of all hyperplanes of Π_r coincide, which means that ξ_1 and ξ_2 also coincide on all points of Π_r.

Consider distinct $(r + 1)$-spaces Π^3_{r+1} and Π^4_{r+1} of $PG(n, K)$. Let their intersection be an s-space Π_s for some $s \in \{-1, 0, \ldots, r\}$. There exists a finite number of distinct $(r + 1)$-spaces $\Pi^3_{r+1} = \Pi^5_{r+1}, \Pi^6_{r+1}, \ldots, \Pi^k_{r+1} = \Pi^4_{r+1}$ in $PG(n, K)$ such that $\Pi^5_{r+1} \cap \Pi^6_{r+1}$, $\Pi^6_{r+1} \cap \Pi^7_{r+1}, \ldots, \Pi^{k-1}_{r+1} \cap \Pi^k_{r+1}$ are r-spaces which contain Π_s. As in the preceding paragraphs the spaces $\Pi^5_{r+1}, \Pi^6_{r+1}, \ldots, \Pi^k_{r+1}$ define projectivities $\xi_3 = \xi_5, \xi_6, \ldots, \xi_k = \xi_4$. Also ξ_i and ξ_{i+1} coincide on all points of $\Pi^i_{r+1} \cap \Pi^{i+1}_{r+1}$, $i = 5, 6, \ldots, k - 1$, and consequently coincide on all points of Π_s. Hence ξ_3 and ξ_4 coincide on all points of Π_s.

Define ξ as follows. Let P be an arbitrary point of $PG(n, K)$. Consider any $(r + 1)$-space Π^0_{r+1} containing P and the corresponding projectivity ξ_0. Then define $P\xi = P\xi_0$. By the foregoing $P\xi_0$ is independent of the choice of Π^0_{r+1} through P. If l is a line of $PG(n, K)$, then choosing Π^0_{r+1} through l we see that $l\xi$ is a line of $PG(n, K)$; if Π_r is an r-space of $PG(n, K)$, then choosing Π^0_{r+1} through Π_r we see that $\Pi_r\mathfrak{G}\delta = \Pi_r\xi\mathfrak{G}$. Hence ξ is a projectivity of $PG(n, K)$ and $\xi\theta = \tilde{\xi} = \delta$. This proves the theorem in the case $n \neq 2r + 1$.

Finally, suppose that $n = 2r + 1$. Clearly

$$PGL(2r + 2, K)\theta \cup (PGL(2r + 2, K)\theta)\zeta \subset G(\mathscr{G}_{r, 2r+1}).$$

Let $\delta \in G(\mathscr{G}_{r, 2r+1})$. If δ leaves $\mathscr{S}_L$ and $\mathscr{S}_G$ invariant, then as in the case

$n \neq 2r+1$ one shows that $\delta \in PGL(2r+2, K)\theta$. Now suppose that δ interchanges $\mathscr{S}_L$ and $\mathscr{S}_G$. Then $\delta\zeta^{-1} \in G(\mathscr{G}_{r,2r+1})$ and leaves $\mathscr{S}_L$ and $\mathscr{S}_G$ invariant. Consequently $\delta\zeta^{-1} \in PGL(2r+2, K)\theta$; that is, $\delta \in (PGL(2r+2, K)\theta)\zeta$. □

It follows from the proof that $(PGL(2r+2, K)\theta)\zeta$ is induced by the set of all correlations of $PG(2r+1, K)$.

Corollary: (i) *For* $n \neq 2r+1$,

$$|G(\mathscr{G}_{r,n,q})| = |PGL(n+1, q)|.$$

(ii) *For* $n = 2r+1$,

$$|G(\mathscr{G}_{r,2r+1,q})| = 2|PGL(2r+2, q)|.$$

Proof: This is immediate from Theorem 24.2.16. □

Theorem 24.2.17: *Let δ be a permutation of $\mathscr{G}_{r,n}$ such that δ and δ^{-1} fix $\mathscr{G}_{r,n}^{(1)}$. Then δ can be extended to an element of $G(\mathscr{G}_{r,n})$.*

Proof: The permutations δ and δ^{-1} map each s-space Π_s of $\mathscr{G}_{r,n}$ onto an s-space of $\mathscr{G}_{r,n}$. As in the proof of Theorem 24.2.14 one can show that, for $n \neq 2r+1$, both δ and δ^{-1} leave each of $\mathscr{S}_L$ and $\mathscr{S}_G$ invariant whereas, for $n = 2r+1$, both δ and δ^{-1} leave the pair $\{\mathscr{S}_L, \mathscr{S}_G\}$ invariant. Similarly to the proof of Theorem 24.2.16, when $n \neq 2r+1$ the permutation δ defines naturally a projectivity ξ of $PG(n, K)$; also, δ is the restriction of $\xi\theta = \tilde{\xi}$ to $\mathscr{G}_{r,n}$. Thus δ can be extended to $\tilde{\xi}$. When $n = 2r+1$, either δ or $\delta\zeta^{-1}$ defines the projectivity ξ and is correspondingly the restriction of $\xi\theta = \tilde{\xi}$ to $\mathscr{G}_{r,n}$; here δ can be extended to one of $\tilde{\xi}$ and $\tilde{\xi}\zeta$. In both cases δ can be extended to an element of $G(\mathscr{G}_{r,n})$. □

24.3 A characterization of Grassmann varieties

In this section it is always assumed that the objects considered are finite. However, all theorems stated can be generalized to the infinite case. Our main goal is to characterize the finite Grassmann varieties in terms of their subspaces.

Let $\mathscr{P}$ be a non-empty set whose elements are called *points*, and let $\mathscr{B}$ be a non-empty set consisting of subsets of $\mathscr{P}$. The elements of $\mathscr{B}$ are called *lines*. The pair $(\mathscr{P}, \mathscr{B})$ is said to be a *partial linear space* (PLS) if the following conditions are satisfied:

(i) any two distinct points in $\mathscr{P}$ belong to at most one line in $\mathscr{B}$;
(ii) any line in $\mathscr{B}$ contains at least two points of $\mathscr{P}$;
(iii) $\mathscr{B}$ is a *covering* of $\mathscr{P}$; that is, $\mathscr{P}$ is the union of all elements of $\mathscr{B}$.

Points P, P' of $\mathscr{P}$ are called *collinear* if there is a line l of $\mathscr{B}$ containing P and P'. In such a case we write $P \sim P'$; if P and P' are *non-collinear* we write $P \nsim P'$. Notice that $P \sim P$. If P and P' are distinct points on the line l, then l is also denoted by PP'.

If any two points of $\mathscr{P}$ are collinear, then we say that $(\mathscr{P}, \mathscr{B})$ is a *linear space* (LS). In the other case we say that $(\mathscr{P}, \mathscr{B})$ is a *proper partial linear space* (PPLS).

A subset $\mathscr{P}'$ of the PLS $(\mathscr{P}, \mathscr{B})$ is called a *subspace* of $(\mathscr{P}, \mathscr{B})$ if any two of its points are collinear and the line joining them is completely contained in $\mathscr{P}'$. The subspace $\mathscr{P}'$ is called a *maximal subspace* if it is not properly contained in any subspace of $(\mathscr{P}, \mathscr{B})$.

If each line l of the PLS $(\mathscr{P}, \mathscr{B})$ contains at least three points we say that $(\mathscr{P}, \mathscr{B})$ is *irreducible.* A PPLS $(\mathscr{P}, \mathscr{B})$ is said to be *connected* if for any two points P, P' there exist points $P_1, P_2, \ldots, P_k$ such that $P \sim P_1 \sim P_2 \sim \cdots \sim P_k \sim P'$.

Consider the Grassmann variety $\mathscr{G}_{r,n}$ of all r-spaces of $PG(n, q)$, $1 \le r \le n-2$. Let $\mathscr{P} = \mathscr{G}_{r,n}$ and let $\mathscr{B}$ be the set of all lines of $\mathscr{G}_{r,n}$. Then $(\mathscr{P}, \mathscr{B})$ is a PPLS whose subspaces are the subspaces of $PG(N, q)$ contained in $\mathscr{G}_{r,n}$, and whose maximal subspaces are the maximal spaces of $\mathscr{G}_{r,n}$. Also, $(\mathscr{P}, \mathscr{B})$ is irreducible and connected (cf. the proof of Lemma 24.2.15).

Lemma 24.3.1: *Let $(\mathscr{P}, \mathscr{B})$ be the PPLS corresponding to the Grassmann variety $\mathscr{G}_{r,n}$. Then $(\mathscr{P}, \mathscr{B})$ satisfies the following conditions*:

(i) *If $P, P', P'' \in \mathscr{P}$ with $P \sim P' \sim P'' \sim P$, then there is a subspace containing these points.*

(ii) *The set of maximal subspaces is partitioned into two families $\mathscr{S}_{\mathrm{L}}$ and $\mathscr{S}_{\mathrm{G}}$ with the further properties*:

I. *If $\pi \in \mathscr{S}_{\mathrm{L}}$ and $\pi' \in \mathscr{S}_{\mathrm{G}}$, then $\pi \cap \pi' = \varnothing$ or $\pi \cap \pi' \in \mathscr{B}$.*

II. *For each $l \in \mathscr{B}$ there is a unique $\pi \in \mathscr{S}_{\mathrm{L}}$ and a unique $\pi' \in \mathscr{S}_{\mathrm{G}}$ such that $\pi \cap \pi' = l$.*

III. *Let π, π', π'' be distinct elements of $\mathscr{S}_{\mathrm{L}}$ for which $\pi \cap \pi', \pi' \cap \pi'', \pi'' \cap \pi$ are distinct points. Then any element of $\mathscr{S}_{\mathrm{L}}$ other than π, π', π'' having distinct points in common with π and π' also has a point in common with π''.*

Similarly, let π, π', π'' be distinct elements of $\mathscr{S}_{\mathrm{G}}$ for which $\pi \cap \pi', \pi' \cap \pi''$, $\pi'' \cap \pi$ are distinct points. Then any element of $\mathscr{S}_{\mathrm{G}}$ other than π, π', π'' having distinct points in common with π and π' also has a point in common with π''.

(iii) *There exist distinct subspaces $\pi_1, \pi_2, \ldots, \pi_{r+1}$ such that $\pi_1 \in \mathscr{B}$, $\pi_{r+1} \in \mathscr{S}_{\mathrm{G}}, \pi_i \subset \pi_{i+1}$, and such that there is no subspace π with $\pi_i \subsetneq \pi \subsetneq \pi_{i+1}$.*

Similarly, there exist distinct subspaces $\pi_1, \pi_2, \ldots, \pi_{n-r}$ such that $\pi_1 \in \mathscr{B}$, $\pi_{n-r} \in \mathscr{S}_{\mathrm{L}}, \pi_i \subset \pi_{i+1}$, and such that there is no subspace π with $\pi_i \subsetneq \pi \subsetneq \pi_{i+1}$.

Proof: Let P, P', P'' be distinct points of $\mathscr{P}$, with $P \sim P' \sim P'' \sim P$. Clearly we may assume that P, P', P'' are not on a common line of $\mathscr{B}$. Recall that

$\mathcal{G}_{r,n}$ is the intersection of the quadrics of $PG(N, q)$ represented by the equations (24.5). The plane $PP'P''$ has three lines in common with each of these quadrics, hence is contained in all these quadrics. So the plane $PP'P''$ is contained in $\mathcal{G}_{r,n}$. Hence P, P', P'' are contained in a subspace of $(\mathcal{P}, \mathcal{B})$.

Properties (ii)I and (ii)II are proved in Theorem 24.2.9 and Lemma 24.2.10. Let π, π', π'' be distinct elements of $\mathcal{S}_L$ for which $\pi \cap \pi'$, $\pi' \cap \pi''$, and $\pi'' \cap \pi$ are distinct points. The $(r-1)$-spaces of $PG(n, q)$ which correspond to π, π', π'' are respectively denoted by $\Pi_{r-1}, \Pi'_{r-1}, \Pi''_{r-1}$. Since $\pi \cap \pi'$, $\pi' \cap \pi''$, and $\pi'' \cap \pi$ are distinct points, the spaces $\Pi_{r-1}, \Pi'_{r-1}, \Pi''_{r-1}$ contain a common $(r-2)$-space Π_{r-2}. If the space $\Pi''' \in \mathcal{S}_L \backslash \{\pi, \pi', \pi''\}$ has distinct points in common with π and π', then the $(r-1)$-space of $PG(n, q)$ corresponding to π''' contains Π_{r-2}. Hence π'' and π''' have a point in common. Similarly, let π, π', π'' be distinct elements of $\mathcal{S}_G$ for which $\pi \cap \pi'$, $\pi' \cap \pi''$, and $\pi'' \cap \pi$ are distinct points. The $(r+1)$-spaces of $PG(n, q)$ which correspond to π, π', π'' are respectively denoted by $\Pi_{r+1}, \Pi'_{r+1}, \Pi''_{r+1}$. Since $\pi \cap \pi'$, $\pi' \cap \pi''$, and $\pi'' \cap \pi$ are distinct points, the spaces $\Pi_{r+1}, \Pi'_{r+1}, \Pi''_{r+1}$ are contained in a common $(r+2)$-space Π_{r+2}. If the space $\pi''' \in \mathcal{S}_G \backslash \{\pi, \pi', \pi''\}$ has distinct points in common with π and π', then the $(r+1)$-space of $PG(n, q)$ corresponding to π''' is contained in Π_{r+2}. Hence π'' and π''' have a point in common.

Property (iii) is trivial. □

Let $(\mathcal{P}, \mathcal{B})$ be a connected irreducible PPLS. It will be said to be a *Grassmann space of index r* $(r \geq 1)$ if the following axioms are satisfied.

A1. If $P, P', P'' \in \mathcal{P}$ with $P \sim P' \sim P'' \sim P$, then there is a subspace of $(\mathcal{P}, \mathcal{B})$ containing these points.

A2. The set of maximal subspaces of $(\mathcal{P}, \mathcal{B})$ is partitioned into two families, say $\mathcal{S}$ and $\mathcal{T}$, with the following properties.

I. If $\pi \in \mathcal{S}$ and $\pi' \in \mathcal{T}$, then $\pi \cap \pi' = \varnothing$ or $\pi \cap \pi' \in \mathcal{B}$.

II. For each $l \in \mathcal{B}$ there is a unique $\pi \in \mathcal{S}$ and a unique $\pi' \in \mathcal{T}$ such that $l \subset \pi$ and $l \subset \pi'$.

III. Let π, π', π'' be distinct elements of $\mathcal{S}$, for which $\pi \cap \pi'$, $\pi' \cap \pi''$, and $\pi'' \cap \pi$ are distinct points. Then any element of $\mathcal{S} \backslash \{\pi, \pi', \pi''\}$ having distinct points in common with π and π' also has a point in common with π''.

A3. There exist distinct subspaces $\pi_1, \pi_2, \ldots, \pi_{r+1}$ such that $\pi_1 \subset \pi_2 \subset \cdots \subset \pi_{r+1}$, $\pi_1 \in \mathcal{B}$, $\pi_{r+1} \in \mathcal{T}$, and such that there is no subspace π with $\pi_i \subsetneqq \pi \subsetneqq \pi_{i+1}$ for $i = 1, 2, \ldots, r$.

Lemma 24.3.2: *Let $(\mathcal{P}, \mathcal{B})$ be the PPLS corresponding to the Grassmann variety $\mathcal{G}_{r,n}$. Then $(\mathcal{P}, \mathcal{B})$ is a Grassmann space of index r and of index $n - r - 1$.*

Proof: Putting $\mathscr{S} = \mathscr{S}_L$ and $\mathscr{T} = \mathscr{S}_G$ in Lemma 24.3.1 shows that $(\mathscr{P}, \mathscr{B})$ is a Grassmann space of index r; putting $\mathscr{S} = \mathscr{S}_G$ and $\mathscr{T} = \mathscr{S}_L$ in Lemma 24.3.1 shows that $(\mathscr{P}, \mathscr{B})$ is a Grassmann space of index $n - r - 1$. □

Let $(\mathscr{P}, \mathscr{B})$ and $(\mathscr{P}', \mathscr{B}')$ be two Grassmann spaces. A bijection ξ of $\mathscr{P}$ onto $\mathscr{P}'$ is called an *isomorphism* or *collineation* of $(\mathscr{P}, \mathscr{B})$ onto $(\mathscr{P}', \mathscr{B}')$, if $\mathscr{B}'$ is the set of all images under ξ of the elements of $\mathscr{B}$. In this case $(\mathscr{P}, \mathscr{B})$ and $(\mathscr{P}', \mathscr{B}')$ are called *isomorphic*. If ξ is an isomorphism of $(\mathscr{P}, \mathscr{B})$ onto $(\mathscr{P}', \mathscr{B}')$, then it is clear that ξ maps the subspaces and maximal subspaces of $(\mathscr{P}, \mathscr{B})$ onto the subspaces and maximal subspaces of $(\mathscr{P}', \mathscr{B}')$.

Let π and π' be distinct maximal subspaces belonging to the same system $\mathscr{D}$ of maximal subspaces of the Grassmann space $(\mathscr{P}, \mathscr{B})$. If $P \in \pi$, $P' \in \pi'$, $P \neq P'$, then by the connectivity of $(\mathscr{P}, \mathscr{B})$ there exist distinct points $P_1 = P, P_2, \ldots, P_k = P'$ such that $P_1 \sim P_2 \sim \cdots \sim P_k$. Let π_i be the element of $\mathscr{D}$ which contains the line P_iP_{i+1}, $i = 1, 2, \ldots, k - 1$. By A2.II, $\pi_i = \pi_{i+1}$ or $\pi_i \cap \pi_{i+1}$ is a point, $i = 1, 2, \ldots, k - 1$. If ξ is an isomorphism of the Grassmann space $(\mathscr{P}, \mathscr{B})$ onto the Grassmann space $(\mathscr{P}', \mathscr{B}')$, then clearly $\pi_i\xi = \pi_{i+1}\xi$ or $\pi_i\xi \cap \pi_{i+1}\xi$ is a point. Hence by A2.I, $\pi_1\xi, \pi_2\xi, \ldots, \pi_k\xi$ belong to the same system $\mathscr{D}'$ of maximal subspaces of $(\mathscr{P}', \mathscr{B}')$. Consequently $\pi\xi$ and $\pi'\xi$ belong to $\mathscr{D}'$. We conclude that ξ maps each system of maximal subspaces of $(\mathscr{P}, \mathscr{B})$ onto a system of maximal subspaces of $(\mathscr{P}', \mathscr{B}')$.

From now on we assume that $(\mathscr{P}, \mathscr{B})$ is a Grassmann space of index $r \geq 2$.

Lemma 24.3.3: *No line in $\mathscr{B}$ is a maximal subspace. Two distinct elements of the same system of maximal subspaces have at most one point in common.*

Proof: Let $l \in \mathscr{B}$ be a maximal subspace, say $l \in \mathscr{S}$. By A2.II, there is a $\pi \in \mathscr{T}$ which contains l. Since $\mathscr{T} \cap \mathscr{S} = \varnothing$, so l is properly contained in π, hence l is not maximal, a contradiction. Consequently no element of $\mathscr{B}$ is a maximal subspace.

Suppose that π and π' are distinct elements of $\mathscr{S} \cup \mathscr{T}$ which have distinct points P and P' in common. The line $l = PP'$ belongs to exactly one element of $\mathscr{S}$ and to exactly one element of $\mathscr{T}$. Hence π and π' belong to distinct families. □

Lemma 24.3.4: *Let $\pi, \pi' \in \mathscr{S}$, $\pi \neq \pi'$, and $P \in \pi \cap \pi'$. If $\pi'' \in \mathscr{T}$ and if $\pi \cap \pi'' = l$ and $\pi' \cap \pi'' = l'$ are lines, then $l \cap l' = \{P\}$ and so π'' contains P.*

Proof: By Lemma 24.3.3 we have $\pi \cap \pi' = \{P\}$; so $l \neq l'$. Let $Q \in l$, $Q' \in l'$, with P, Q, Q' distinct. Then $P \sim Q \sim Q' \sim P$. By A1 the points P, Q, Q' are contained in a subspace π^1, and π^1 is contained in a maximal subspace π^2. If $\pi^2 \in \mathscr{S}$, then by Lemma 24.3.3 we have $\pi = \pi^2$ and $\pi' = \pi^2$. Hence $\pi = \pi'$. a contradiction. So $\pi^2 \in \mathscr{T}$. Since π^2 contains the distinct points Q, Q' of π'', by Lemma 24.3.3, $\pi'' = \pi^2$. Consequently $P \in \pi''$. Finally, $P \in \pi \cap \pi'' = l$ and $P \in \pi' \cap \pi'' = l'$. □

Lemma 24.3.5: *Each element of $\mathcal{T}$ provided with its lines has the structure of points and lines of a projective space.*

Proof: If $\pi \in \mathcal{T}$, then π provided with its lines is a linear space. It is sufficient to show that in that linear space the Veblen–Wedderburn axiom holds.

Let l_1 and l_2 be distinct lines of π, with $l_1 \cap l_2 = \{P\}$; if l_3 and l_4 are distinct lines of π, each of them meeting both l_1 and l_2 at points distinct from P, then we must show that l_3 and l_4 meet at a point.

Let $l_1 \cap l_3 = \{P_1\}, l_2 \cap l_3 = \{P_2\}, l_1 \cap l_4 = \{Q_1\}$, and $l_2 \cap l_4 = \{Q_2\}$. Through l_i there is exactly one maximal subspace $\pi^i \in \mathcal{S}$, $i = 1, 2, 3, 4$. If $\pi^i = \pi^j$, $i \neq j$, then $\pi^i \cap \pi$ contains all points of $l_i \cup l_j$, in contradiction to A2.I. Hence the spaces $\pi^1, \pi^2, \pi^3, \pi^4$ are distinct. By Lemma 24.3.3 we have $\pi^1 \cap \pi^2 = \{P\}$, $\pi^1 \cap \pi^3 = \{P_1\}, \pi^2 \cap \pi^3 = \{P_2\}, \pi^1 \cap \pi^4 = \{Q_1\}, \pi^2 \cap \pi^4 = \{Q_2\}$. By A2.III, π^4 and π^3 have a common point Q. Since $Q \in \pi^3 \cap \pi^4$, $\pi \in \mathcal{T}$, $\pi^3 \cap \pi = l_3$, and $\pi^4 \cap \pi = l_4$, we have $l_3 \cap l_4 = \{Q\}$ by Lemma 24.3.4. □

Any projective space belonging to $\mathcal{T}$ contains projective planes. The set of all these projective planes is denoted by $\mathcal{C}$. Clearly any element of $\mathcal{C}$ is a subspace of $(\mathcal{P}, \mathcal{B})$, and is contained in exactly one element of $\mathcal{T}$.

Lemma 24.3.6: *Let π and π' be distinct elements of $\mathcal{T}$ which intersect in the point P. Assume that π^1, π^2, π^3 are distinct elements of $\mathcal{S}$ containing P, and let $\pi^i \cap \pi = l_i \in \mathcal{B}$, $\pi^i \cap \pi' = l'_i \in \mathcal{B}$, $i = 1, 2, 3$. If l_1, l_2, l_3 belong to a common plane in π, then l'_1, l'_2, l'_3 also belong to a common plane in π'.*

Proof: Notice that l_1, l_2, l_3 are distinct, and that l'_1, l'_2, l'_3 are distinct. Let Π_2 be the plane containing l_1, l_2, l_3, and let Π'_2 be the plane of π' containing l'_1 and l'_2. We must show that $l'_3 \subset \Pi'_2$.

Let l be a line of Π_2 which does not contain P, and let $\bar{\pi}$ be the element of $\mathcal{S}$ which contains l. If $\bar{\pi} \cap \pi'$ is not a line, then let l' be any line of Π'_2 which does not contain P; if $\bar{\pi} \cap \pi'$ is a line, then let l' be any line of Π'_2 which is distinct from $\bar{\pi} \cap \pi'$ and does not contain P. Let $l \cap l_i = \{Q_i\}$, $i = 1, 2, 3$, and let $l' \cap l'_i = \{Q'_i\}$, $i = 1, 2$. The five points $Q_1, Q_2, Q_3, Q'_1, Q'_2$ are distinct. Let $\bar{\pi}'$ be the element of $\mathcal{S}$ which contains l'. Then $\bar{\pi}, \bar{\pi}', \pi^1, \pi^2, \pi^3$ are distinct, $\bar{\pi} \cap \pi^i = \{Q_i\}$, $i = 1, 2, 3$, and $\bar{\pi}' \cap \pi^i = \{Q'_i\}$, $i = 1, 2$. Since $\pi^1, \pi^2, \bar{\pi}$ meet pairwise in distinct points and since $\bar{\pi}'$ has distinct points in common with π^1 and π^2, A2.III tells us that $\bar{\pi}$ and $\bar{\pi}'$ have a point in common. Let $\bar{\pi} \cap \bar{\pi}' = \{Q\}$. Clearly Q is distinct from Q_1 and Q'_1. Consequently the maximal subspaces $\bar{\pi}, \bar{\pi}', \pi^1$ meet pairwise in distinct points and π^3 meets $\bar{\pi}$ and π^1 at the distinct points Q_3 and P, respectively. Hence by A2.III the intersection of π^3 and $\bar{\pi}'$ is a point Q'. Since $\pi^3 \cap \pi' = l'_3$, $\bar{\pi}' \cap \pi' = l'$, and $\pi^3 \cap \bar{\pi}' = \{Q'\}$, by Lemma 24.3.4, $l' \cap l'_3 = \{Q'\}$. So l'_3 contains distinct points P and Q' of the plane Π'_2; that is, $l'_3 \subset \Pi'_2$. □

Let $P \in \mathscr{P}$ and $\pi \in \mathscr{C}$, with $P \in \pi$. Then the set consisting of all the elements of $\mathscr{S}$ meeting π at lines through P is denoted by $\mathscr{R}(P, \pi)$.

Lemma 24.3.7: *If $\pi \in \mathscr{C}$ with $P \in \pi$, then $|\mathscr{R}(P, \pi)| \geq 3$. If $\pi, \pi' \in \mathscr{C}$ with $P \in \pi$ and $P' \in \pi'$, then $|\mathscr{R}(P, \pi) \cap \mathscr{R}(P', \pi')| \geq 2$ implies $\mathscr{R}(P, \pi) = \mathscr{R}(P', \pi')$.*

Proof: Let $\pi \in \mathscr{C}$, $P \in \mathscr{P}$, where $P \in \pi$. Choose distinct lines l_1, l_2, l_3 in π through P. The elements of $\mathscr{S}$ containing l_1, l_2, l_3 are respectively denoted by π^1, π^2, π^3. Clearly π^1, π^2, π^3 belong to $\mathscr{R}(P, \pi)$. If $\pi^i = \pi^j$, $i \neq j$, then $\pi \cap \bar{\pi}$, with $\bar{\pi}$ the element of $\mathscr{T}$ containing π, contains all points of $l_i \cup l_j$. This contradicts A2.I. Hence $\pi^i \neq \pi^j$, $i \neq j$, and so $|\mathscr{R}(P, \pi)| \geq 3$.

Next, let $\pi, \pi' \in \mathscr{C}$, $P \in \pi$, $P' \in \pi'$, and $|\mathscr{R}(P, \pi) \cap \mathscr{R}(P', \pi')| \geq 2$. Choose distinct elements π^1 and π^2 in $\mathscr{R}(P, \pi) \cap \mathscr{R}(P', \pi')$. The points P and P' belong to both π^1 and π^2. By Lemma 24.3.3 we have $P = P'$. If $\pi = \pi'$, then obviously $\mathscr{R}(P, \pi) = \mathscr{R}(P', \pi')$. So assume that $\pi \neq \pi'$. Let $\bar{\pi}$ and $\bar{\pi}'$ be the elements of $\mathscr{T}$ which contain π and π', respectively. For at least one of π^1, π^2 the lines $\pi^i \cap \pi$ and $\pi^i \cap \pi'$ are distinct, say $\pi^1 \cap \pi \neq \pi^1 \cap \pi'$. If $\bar{\pi} = \bar{\pi}'$, then the distinct lines $\pi^1 \cap \pi$, $\pi^1 \cap \pi'$ belong to $\bar{\pi} \cap \pi^1$, in contradiction to A2.I. Hence $\bar{\pi} \neq \bar{\pi}'$ and $\bar{\pi} \cap \bar{\pi}' = \{P\}$. Now consider a subspace $\pi^3 \in \mathscr{R}(P, \pi) \backslash \{\pi^1, \pi^2\}$. Since $\pi^3 \cap \bar{\pi}' \neq \varnothing$, $\pi^3 \cap \bar{\pi}'$ is a line l'. By Lemma 24.3.6 the lines $\pi^1 \cap \bar{\pi}'$, $\pi^2 \cap \bar{\pi}'$, l' belong to a common plane. The lines $\pi^1 \cap \bar{\pi}'$, $\pi^2 \cap \bar{\pi}'$ belong to π'; hence $l' \subset \pi'$. This means that $\pi^3 \in \mathscr{R}(P, \pi')$. Consequently $\mathscr{R}(P, \pi) \subset \mathscr{R}(P, \pi')$. Analogously, $\mathscr{R}(P, \pi') \subset \mathscr{R}(P, \pi)$. So $\mathscr{R}(P, \pi') = \mathscr{R}(P, \pi)$, and the theorem is proved. □

The set whose elements are the subsets $\mathscr{R}(P, \pi)$ of $\mathscr{S}$ is denoted by $\mathscr{R}$.

Lemma 24.3.8: *The pair $(\mathscr{S}, \mathscr{R})$ is a connected irreducible PLS. Also, two elements in $\mathscr{S}$ are collinear in $(\mathscr{S}, \mathscr{R})$ if and only if they have a common point in $\mathscr{P}$.*

Proof: Let π and π' be two distinct elements of $\mathscr{S}$. If $\pi \cap \pi' = \{P\}$, then let l be a line in π through P. Through l there is a subspace $\bar{\pi} \in \mathscr{T}$ meeting π' at one point at least. Hence $\bar{\pi} \cap \pi'$ is a line l'. Let Π_2 be the plane of $\mathscr{T}$ containing the lines l and l'. It is clear that $\pi, \pi' \in \mathscr{R}(P, \Pi_2)$. Hence π and π' are collinear in $(\mathscr{S}, \mathscr{R})$. If $\pi \cap \pi' = \varnothing$, then clearly no element in $\mathscr{R}$ through π and π' exists.

Let $\pi \in \mathscr{S}$ and choose a line l in π. Through l there is a maximal subspace $\pi' \in \mathscr{T}$. Let $P \in l$ and let Π_2 be a plane of π' containing l. Then $\pi \in \mathscr{R}(P, \Pi_2)$, and so $\mathcal{R}$ is a covering of $\mathcal{S}$. By Lemma 24.3.7 any element of $\mathcal{R}$ contains at least three elements of $\mathcal{S}$, and two distinct elements of $\mathcal{S}$ belong to at most one element of $\mathcal{R}$. Hence $(\mathcal{S}, \mathcal{R})$ is an irreducible PLS.

Let π and π' be distinct elements of $\mathcal{S}$. Choose $P \in \pi$ and $P' \in \pi'$. Since $(\mathcal{P}, \mathcal{B})$ is connected, there are points $P = P_1, P_2, \ldots, P_k = P'$ with $P_1 \sim P_2 \sim \cdots \sim P_k$. Let π^i be the element of $\mathcal{S}$ containing the line P_iP_{i+1}, $i = 1, 2, \ldots, k-1$. Then $\pi \cap \pi^1 \neq \emptyset$, $\pi' \cap \pi^{k-1} \neq \emptyset$, and $\pi^i \cap \pi^{i+1} \neq \emptyset$, with $i = 1, 2, \ldots, k-1$. Hence π and π^1, π^i and π^{i+1}, π^{k-1} and π' are collinear in $(\mathcal{S}, \mathcal{R})$. This shows that $(\mathcal{S}, \mathcal{R})$ is connected. □

Let $P \in \mathcal{P}$, and let $\mathcal{S}_P$ be the set of all elements of $\mathcal{S}$ containing P. Then $\mathcal{S}_P$ is a subspace of the PLS $(\mathcal{S}, \mathcal{R})$.

Lemma 24.3.9: *Let P be a point of $\mathcal{P}$ and let π be an element of $\mathcal{T}$ through P. If π is the projective space $PG(s+1, q)$, then $\mathcal{S}_P$ provided with its lines is isomorphic to the linear space formed by the points and lines of $PG(s, q)$.*

Proof: Let $\mathcal{L}$ be the set of all lines of π through P. By A2.I each element of $\mathcal{S}_P$ meets π at a line in $\mathcal{L}$. Let ϕ be the mapping

$$\phi\colon \pi' \in \mathcal{S}_P \to \pi' \cap \pi \in \mathcal{L};$$

then ϕ is a bijection of $\mathcal{S}_P$ onto $\mathcal{L}$.

It is clear that ϕ^{-1} maps each pencil of lines in $\mathcal{L}$ onto a line of $(\mathcal{S}, \mathcal{R})$ which is contained in $\mathcal{S}_P$. Conversely, let us consider a line of the PLS $(\mathcal{S}, \mathcal{R})$ which is contained in $\mathcal{S}_P$. Such a line is of type $\mathcal{R}(P, \Pi_2)$. Let π^1 and π^2 be distinct elements of $\mathcal{R}(P, \Pi_2)$. Then $\pi^1 \cap \pi$ and $\pi^2 \cap \pi$ are distinct lines of π, which determine a plane Π_2'. By Lemma 24.3.7 we have $\mathcal{R}(P, \Pi_2) = \mathcal{R}(P, \Pi_2')$. Clearly $\mathcal{R}(P, \Pi_2)\phi = \mathcal{R}(P, \Pi_2')\phi$ consists of all lines of Π_2' through P. Hence $\mathcal{R}(P, \Pi_2)\phi$ is a pencil of lines in $\mathcal{L}$.

Notice that $\mathcal{L}$ provided with its pencils of lines is isomorphic to the structure of points and lines of $PG(s, q)$. Since ϕ is an isomorphism of the linear space formed by $\mathcal{S}_P$ and its lines onto the linear space formed by $\mathcal{L}$ and its pencils, we have proved that $\mathcal{S}_P$ provided with its lines is isomorphic to the linear space formed by the points and lines of $PG(s, q)$. □

Lemma 24.3.10: *Each π in $\mathcal{T}$ is an $(r+1)$-dimensional projective space over a common field $GF(q)$.*

Proof: By A3 there is a maximal subspace $\bar{\pi}$ in $\mathcal{T}$ which is an $(r+1)$-dimensional projective space over some field $GF(q)$. Let $\pi \in \mathcal{T} \backslash \{\bar{\pi}\}$. Choose a point P in $\bar{\pi}$ and a point P' in π, with $P \neq P'$. Since $(\mathcal{P}, \mathcal{B})$ is connected, there are distinct points $P = P_1, P_2, \ldots, P_k = P'$ such that $P_i \sim P_{i+1}$,

$i = 1, 2, \ldots, k - 1$. Let π^i be the element of $\mathscr{T}$ containing P_iP_{i+1}, $i = 1, 2, \ldots, k - 1$. Then π^i and π^{i+1} have a common point P_{i+1}, $i = 1, 2, \ldots, k - 1$. Also, $\bar{\pi}$ and π^1 both contain $P = P_1$, and π^{k-1} and π both contain $P' = P_k$. Suppose that π^1 is the projective space $PG(r' + 1, q')$. By Lemma 24.3.9, $\mathscr{S}_P$ provided with its lines is isomorphic to the linear space formed by the points and lines both of $PG(r, q)$ and of $PG(r', q')$. Hence $r = r'$ and $q = q'$. Repeating the argument, we finally see that π is the $(r + 1)$-dimensional projective space over $GF(q)$. □

Corollary: *Each $\mathscr{S}_P$, provided with its lines, is the linear space formed by the points and lines of $PG(r, q)$.*

Proof: Let $P \in \mathscr{P}$. Choose a line l containing P, and let π be the element of $\mathscr{T}$ containing l. By Lemma 24.3.10, π is the space $PG(r + 1, q)$. Now, by Lemma 24.3.9, $\mathscr{S}_P$ is the space $PG(r, q)$. □

Lemma 24.3.11: *Let π^1, π^2, π^3 be distinct elements of $\mathscr{S}$, which are pairwise collinear in $(\mathscr{S}, \mathscr{R})$. If π^1, π^2, π^3 contain a common point P of $\mathscr{P}$, then there exists a projective plane (over $GF(q)$) in $(\mathscr{S}, \mathscr{R})$ through them.*

Proof: By the corollary of Lemma 24.3.10 we know that $\mathscr{S}_P$ is the projective space $PG(r, q)$. That projective space contains π^1, π^2, π^3 as points. Hence there is a plane (over $GF(q)$) in $(\mathscr{S}, \mathscr{R})$ which contains π^1, π^2, π^3. □

Lemma 24.3.12: *Let π^1, π^2, π^3 be elements in $\mathscr{S}$, which are pairwise collinear in $(\mathscr{S}, \mathscr{R})$. Suppose that the subspaces π^1, π^2, π^3 of $(\mathscr{P}, \mathscr{B})$ do not contain a common point of $\mathscr{P}$. Then there exists a projective plane (over $GF(q)$) in $(\mathscr{S}, \mathscr{R})$ which contains π^1, π^2, π^3.*

Proof: Clearly π^1, π^2, π^3 are distinct and do not belong to a common line of $(\mathscr{S}, \mathscr{R})$. Let $\pi^2 \cap \pi^3 = \{P_1\}$, $\pi^3 \cap \pi^1 = \{P_2\}$, and $\pi^1 \cap \pi^2 = \{P_3\}$ (in $(\mathscr{P}, \mathscr{B})$). Since π^1, π^2, π^3 do not contain a common point we have $P_1 \neq P_3 \neq P_2 \neq P_1$. By A2.II the lines P_1P_2, P_2P_3, P_3P_1 of $\mathscr{B}$ are distinct. By A1 there is a subspace of $(\mathscr{P}, \mathscr{B})$ which contains P_1, P_2, P_3. That subspace contains a line of π^i, i = 1, 2, 3, and hence is contained in an element π of $\mathscr{T}$. Consequently P_1, P_2, P_3 generate a projective plane $PG(2, q) = \Pi_2$ of π. Now we prove that

$$\tilde{\Pi}_2 = \{\pi' \in \mathscr{S} \mid \pi' \cap \Pi_2 \in \mathscr{B}\}$$

is a projective plane of $(\mathscr{S}, \mathscr{R})$ which is isomorphic to Π_2.

Let π'_1 and π'_2 be any two distinct elements of $\tilde{\Pi}_2$. They meet Π_2 at two distinct lines l_1 and l_2 of $\mathscr{B}$, respectively. Let $l_1 \cap l_2 = \{P\}$. Hence π'_1 and π'_2 belong to the line $\mathscr{R}(P, \Pi_2)$ in $\mathscr{R}$. Also $\mathscr{R}(P, \Pi_2) \subset \tilde{\Pi}_2$. Now we let correspond to each line l in Π_2 the element of $\mathscr{S}$ containing l. This gives us an isomorphism from the dual of the plane Π_2 onto the linear space formed by the elements of $\tilde{\Pi}_2$ and lines of $(\mathscr{S}, \mathscr{R})$ contained in $\tilde{\Pi}_2$. Hence $\tilde{\Pi}_2$ is a projective plane isomorphic to the dual of $PG(2, q)$, and hence also to $PG(2, q)$. □

Lemma 24.3.13: *Each subspace of $(\mathscr{S}, \mathscr{R})$ is a projective space over $GF(q)$.*

Proof: This follows immediately from the previous two lemmas. □

Lemma 24.3.14: *Let P be a point in $\mathscr{P}$ and let π be an element of $\mathscr{S}$ which does not contain P. Then the set*

$$\mathscr{D} = \{\pi' \in \mathscr{S} \mid P \in \pi' \quad \text{and} \quad \pi \cap \pi' \neq \varnothing\}$$

is either a line in $\mathscr{R}$ or the empty set.

Proof: Assume that through P there are two distinct elements of $\mathscr{S}$, say π^1 and π^2, both meeting π at a point: $\pi \cap \pi^1 = \{P_1\}$, $\pi \cap \pi^2 = \{P_2\}$. By Lemma 24.3.3 we have $P_1 \neq P_2$. Since the points P_1, P_2, P are pairwise collinear, there is an element $\bar{\pi}$ in $\mathscr{T}$ containing these points. Call Π_2 the plane of $\bar{\pi}$ containing P_1, P_2, P. The line $\mathscr{R}(P, \Pi_2)$ in $(\mathscr{S}, \mathscr{R})$ consists of those elements in $\mathscr{S}$ containing P and any point of the line P_1P_2. Now assume that $\pi^3 \in \mathscr{S}$ contains P, that $\pi^3 \cap \pi \neq \varnothing$, and that $\pi^3 \notin \mathscr{R}(P, \Pi_2)$. Let $\pi^3 \cap \pi = \{P_3\}$; then $P_3 \notin P_1P_2$. Again P, P_1, P_3 are contained in some plane Π_2' contained in some element of $\mathscr{T}$. The intersection of the planes Π_2 and Π_2' is the line PP_1. Let $\bar{\pi}$ and $\bar{\pi}'$ be the elements of $\mathscr{T}$ containing Π_2 and Π_2', respectively. By Lemma 24.3.3 we have $\bar{\pi} = \bar{\pi}'$. Hence the points P_1, P_2, P_3 belong to $\bar{\pi}$, and so the plane $P_1P_2P_3$ belongs to $\pi \cap \bar{\pi}$, contradicting A2.I. This proves that $\mathscr{D} = \{\pi' \in \mathscr{S} \mid P \in \pi' \text{ and } \pi \cap \pi' \neq \varnothing\}$ is a line in $\mathscr{R}$.

Next, assume that through P there is at least one element of $\mathscr{S}$, say π', which has a point Q' in common with π. The element of $\mathscr{T}$ containing the line PQ' has a line l in common with π. Let $Q'' \in l \backslash \{Q'\}$. Then $P \sim Q''$, and the element of $\mathscr{S}$ through PQ'' belongs to $\mathscr{D}$. By the first part of the proof $\mathscr{D}$ is a line in $\mathscr{R}$.

We have proved that $\mathscr{D} = \varnothing$ or a line in $\mathscr{R}$. □

Corollary: *$(\mathscr{S}, \mathscr{R})$ is a proper PLS. More precisely, if $P \in \mathscr{P}$, $\pi \in \mathscr{S}$, $P \notin \pi$, then there is at least one element of $\mathscr{S}_P$ which has no point in common with π.*

Proof: Let $\pi \in \mathscr{S}$ and let P be a point not in π. We know that $\mathscr{S}_P$ is the projective space $PG(r, q)$, with $r \geq 2$. Since $\mathscr{D} = \varnothing$ or a line in $\mathscr{S}_P$, there are elements in $\mathscr{S}_P$ which do not belong to $\mathscr{D}$. Hence there are elements through P which have no points in common with π, which means that there exists at least one pair of non-collinear points in $(\mathscr{S}, \mathscr{R})$. □

Lemma 24.3.15: *For any $P \in \mathscr{P}$, the set $\mathscr{S}_P$ is a maximal subspace of $(\mathscr{S}, \mathscr{R})$.*

Proof: Suppose that $\mathscr{S}_P$ is not maximal. Then there exists an element $\pi \in \mathscr{S}$, with $\pi \notin \mathscr{S}_P$, such that π is collinear in $(\mathscr{S}, \mathscr{R})$ with each element of $\mathscr{S}_P$. By Lemma 24.3.8, π has a point in common with each element of $\mathscr{S}_P$. By the corollary of Lemma 24.3.14, $\mathscr{S}_P$ contains an element which has no point in common with π, a contradiction. □

The family consisting of the maximal subspaces $\mathscr{S}_P$ of $(\mathscr{S}, \mathscr{R})$ is denoted by $\tilde{\mathscr{T}}$. Note that each element of $\tilde{\mathscr{T}}$ is an r-dimensional projective space over $GF(q)$.

Lemma 24.3.16: *For distinct points $P, Q \in \mathscr{P}$ we have $|\mathscr{S}_P \cap \mathscr{S}_Q| \leq 1$; that is, two distinct maximal subspaces in $\tilde{\mathscr{T}}$ have at most one element of $\mathscr{S}$ in common.*

Proof: If $P \sim Q$ and $P \neq Q$, then $\mathscr{S}_P \cap \mathscr{S}_Q$ is the unique element of $\mathscr{S}$ containing the line PQ. If $P \nsim Q$ there is no element of $\mathscr{S}$ containing both P and Q. □

Lemma 24.3.17: *Let π, π' be distinct elements of $\mathscr{S}$ containing the point P. If π^1 and π^2 are distinct elements of $\mathscr{S}\backslash\{\pi, \pi'\}$ both meeting π and π' at distinct points, then π^1 and π^2 are collinear in $(\mathscr{S}, \mathscr{R})$. Also any element $\pi^3 \in \mathscr{S}$ belonging to the line $\pi^1\pi^2$ of $(\mathscr{S}, \mathscr{R})$ either meets both π and π' at distinct points or belongs to the line $\pi\pi'$ of $(\mathscr{S}, \mathscr{R})$.*

Proof: By A2.III the maximal subspaces π^1 and π^2 have a point in common; that is, they are collinear in $(\mathscr{S}, \mathscr{R})$. Let $\pi^1 \cap \pi^2 = \{Q\}$. Since $P \notin \pi^1 \cup \pi^2$, we have $P \neq Q$.

If $Q \in \pi$, then any $\pi^3 \in \mathscr{S}$ belonging to the line $\pi^1\pi^2$ of $(\mathscr{S}, \mathscr{R})$ has a point in common with π. Hence in $(\mathscr{S}, \mathscr{R})$ any such π^3 is collinear with π.

Now let $Q \notin \pi$. By Lemma 24.3.14 the set of all elements in $\mathscr{S}$ through Q and having a point in common with π is the line $\pi^1\pi^2$ of $(\mathscr{S}, \mathscr{R})$. Hence each element of the line $\pi^1\pi^2$ contains a point of π.

Similarly, each element of the line $\pi^1\pi^2$ contains a point of π'.

Next, assume that the element $\pi^3 \in \pi^1\pi^2$ contains P. We have to show that π^3 belongs to the line $\pi\pi'$ of $(\mathscr{S}, \mathscr{R})$. So $\pi^3 \in \mathscr{S}_P$ and $\pi^3 \cap \pi^1 = \{Q\}$. By Lemma 24.3.14 the set of all elements in $\mathscr{S}$ through P and having a point in common with π^1 is the line $\pi\pi'$ of $(\mathscr{S}, \mathscr{R})$. Consequently $\pi^3 \in \pi\pi'$, and now the lemma is completely proved. □

Let π, π' be distinct elements of $\mathscr{S}$ which are collinear in $(\mathscr{S}, \mathscr{R})$. Then π and π' have a common point P. Denote by $\mathscr{S}(\pi, \pi')$ the set consisting of all elements in $\mathscr{S}$ that either belong to the line $\pi\pi'$ of $(\mathscr{S}, \mathscr{R})$, or meet both π and π' at points of $\mathscr{P}\backslash\{P\}$.

Lemma 24.3.18: *$\mathscr{S}(\pi, \pi')$ is a subspace of $(\mathscr{S}, \mathscr{R})$, which properly contains the line $\pi\pi'$ of $(\mathscr{S}, \mathscr{R})$.*

Proof: Let $\pi \cap \pi' = \{P\}$ and let $\bar{\pi}$ be an element of $\mathscr{T}$ containing P. By A2.I we have $\pi \cap \bar{\pi} = l \in \mathscr{B}$ and $\pi' \cap \bar{\pi} = l' \in \mathscr{B}$. Let $Q \in l\backslash\{P\}$ and let $Q' \in l'\backslash\{P\}$. Then $Q \sim Q'$. The space $\pi'' \in \mathscr{S}$ which contains the line QQ' belongs to $\mathscr{S}(\pi, \pi')$, but not to $\pi\pi'$. Hence $\mathscr{S}(\pi, \pi')$ properly contains the line $\pi\pi'$ of $(\mathscr{S}, \mathscr{R})$.

We have still to show that $\mathscr{S}(\pi, \pi')$ is a subspace of $(\mathscr{S}, \mathscr{R})$. Three cases will be considered.

(i) Let π^1, π^2 be distinct elements of $\mathscr{S}(\pi, \pi')$ which both belong to the line $\pi\pi'$. Then π^1, π^2 are collinear in $(\mathscr{S}, \mathscr{R})$ and the line $\pi^1\pi^2 = \pi\pi'$ is completely contained in $\mathscr{S}(\pi, \pi')$.

(ii) Let π^1, π^2 be distinct elements of $\mathscr{S}(\pi, \pi')$, and suppose that $P \notin \pi^1$ and $P \notin \pi^2$. By Lemma 24.3.17, π^1 and π^2 are collinear in $(\mathscr{S}, \mathscr{R})$, and the line $\pi^1\pi^2$ is completely contained in $\mathscr{S}(\pi, \pi')$.

(iii) Let π^1, π^2 be distinct elements of $\mathscr{S}(\pi, \pi')$, where $P \notin \pi^1$ and $\pi^2 \in \pi\pi'$. By Lemma 24.3.14 the spaces π^1 and π^2 have a common point Q. Hence π^1 and π^2 are collinear in $(\mathscr{S}, \mathscr{R})$. If $Q \in \pi$, then it is trivial that each $\pi^3 \in \pi^1\pi^2$ contains a point of π; if $Q \notin \pi$, then by Lemma 24.3.14 each $\pi^3 \in \pi^1\pi^2$ contains a point of π. Similarly, each $\pi^3 \in \pi^1\pi^2$ contains a point of π'. Since π^2 is the only element of $\mathscr{S}$ containing P and Q, it is now clear that all spaces in $\pi^1\pi^2\backslash\{\pi^2\}$ belong to $\mathscr{S}(\pi, \pi')\backslash\pi\pi'$. Hence the line $\pi^1\pi^2$ of $(\mathscr{S}, \mathscr{R})$ is completely contained in $\mathscr{S}(\pi, \pi')$.

From (i), (ii), (iii) it follows that $\mathscr{S}(\pi, \pi')$ is a subspace of $(\mathscr{S}, \mathscr{R})$. □

Lemma 24.3.19: *Each $\mathscr{S}(\pi, \pi')$ is a maximal subspace of $(\mathscr{S}, \mathscr{R})$.*

Proof: Let $\pi \cap \pi' = \{P\}$, and suppose that $\mathscr{S}(\pi, \pi')$ is not maximal. Then there exists an element π'' in $\mathscr{S}$ not in $\mathscr{S}(\pi, \pi')$ such that in $(\mathscr{S}, \mathscr{R})$ the element π'' is collinear with each element of $\mathscr{S}(\pi, \pi')$. Hence π'' has a point in common with each element of $\mathscr{S}(\pi, \pi')$. If $P \notin \pi''$, then π'' meets both π and π' at points of $\mathscr{P}\backslash\{P\}$, and hence $\pi'' \in \mathscr{S}(\pi, \pi')$, a contradiction. So we have $P \in \pi''$. Let $\pi^1 \in \mathscr{S}(\pi, \pi')\backslash\pi\pi'$. Since π'' contains P and $\pi'' \cap \pi^1 \neq \varnothing$, by Lemma 24.3.14, $\pi'' \in \pi\pi'$; hence $\pi'' \in \mathscr{S}(\pi, \pi')$, again a contradiction. □

The family consisting of the maximal subspaces $\mathscr{S}(\pi, \pi')$ of $(\mathscr{S}, \mathscr{R})$ is denoted by $\tilde{\mathscr{S}}$. By Lemma 24.3.18 each element of $\tilde{\mathscr{S}}$ properly contains a line of $(\mathscr{S}, \mathscr{R})$.

Lemma 24.3.20: *Let π and π' be two distinct collinear elements of $(\mathscr{S}, \mathscr{R})$. Let $\pi \cap \pi' = \{P\}$. Then $\mathscr{S}_P$ and $\mathscr{S}(\pi, \pi')$ are the only maximal subspaces of $(\mathscr{S}, \mathscr{R})$ containing π and π', and $\mathscr{S}_P \cap \mathscr{S}(\pi, \pi') = \pi\pi'$.*

Proof: Clearly $\pi\pi' \subset \mathscr{S}_P$, $\pi\pi' \subset \mathscr{S}(\pi, \pi')$, and $\mathscr{S}_P \cap \mathscr{S}(\pi, \pi') = \pi\pi'$. Now let $\mathscr{H}$ be any subspace of $(\mathscr{S}, \mathscr{R})$ through π and π'. We must show that $\mathscr{H} \subset \mathscr{S}_P$ or $\mathscr{H} \subset \mathscr{S}(\pi, \pi')$.

If each element of $\mathscr{H}$ contains P, then $\mathscr{H} \subset \mathscr{S}_P$. So assume that $\pi'' \in \mathscr{H}$ and $P \notin \pi''$. Since π'' is collinear with π and π', it meets π and π' at distinct points. Hence $\pi'' \in \mathscr{S}(\pi, \pi')$. Each element π''' of $\mathscr{H}$ which contains P has a point in common with π'' and hence by Lemma 24.3.14 belongs to the line $\pi\pi'$ of $(\mathscr{S}, \mathscr{R})$. Again $\pi''' \in \mathscr{S}(\pi, \pi')$. So $\mathscr{H} \subset \mathscr{S}(\pi, \pi')$. □

Corollary: *The only maximal subspaces of $(\mathscr{S}, \mathscr{R})$ are the elements of $\tilde{\mathscr{S}} \cup \tilde{\mathscr{T}}$.*

Proof: This is immediate from Lemma 24.3.20. □

Remark: It is clear that $\tilde{\mathscr{S}} \cap \tilde{\mathscr{T}} = \varnothing$.

Lemma 24.3.21: *If π^1 and π^2 are distinct elements of $\mathscr{S}(\pi, \pi')$, then they have a common point, and $\mathscr{S}(\pi^1, \pi^2) = \mathscr{S}(\pi, \pi')$.*

Proof: Since $\pi^1, \pi^2 \in \mathscr{S}(\pi, \pi')$ they are collinear in $(\mathscr{S}, \mathscr{R})$; so they have a common point Q in $\mathscr{P}$. By Lemma 24.3.20, $\mathscr{S}_Q$ and $\mathscr{S}(\pi^1, \pi^2)$ are the only maximal subspaces containing the line $\pi^1\pi^2$ of $(\mathscr{S}, \mathscr{R})$. Since $\mathscr{S}(\pi, \pi')$ is a subspace containing the line $\pi^1\pi^2$, we have $\mathscr{S}(\pi, \pi') \subset \mathscr{S}_Q$ or $\mathscr{S}(\pi, \pi') \subset \mathscr{S}(\pi^1, \pi^2)$. It is clear that not all elements of $\mathscr{S}(\pi, \pi')$ have a common point. Hence $\mathscr{S}(\pi, \pi') \subset \mathscr{S}(\pi^1, \pi^2)$. Since $\mathscr{S}(\pi, \pi')$ is maximal, we finally have $\mathscr{S}(\pi, \pi') = \mathscr{S}(\pi^1, \pi^2)$. □

Lemma 24.3.22: *If $\mathscr{H}$, $\mathscr{H}'$, and $\mathscr{H}''$ are pairwise distinct elements in $\tilde{\mathscr{S}}$ such that $\mathscr{H} \cap \mathscr{H}' = \{\pi''\}$, $\mathscr{H}' \cap \mathscr{H}'' = \{\pi\}$, $\mathscr{H}'' \cap \mathscr{H} = \{\pi'\}$, with $\pi \neq \pi'$, then π, π', π'' have a point of $\mathscr{P}$ in common.*

Proof: If $\pi = \pi''$, then also $\pi = \pi'$, a contradiction. Hence $\pi \neq \pi''$, and analogously $\pi' \neq \pi''$. The spaces π, π', π'' are pairwise collinear in $(\mathscr{S}, \mathscr{R})$. By Lemmas 24.3.11 and 24.3.12 the elements π, π', π'' are contained in a subspace $\bar{\mathscr{H}}$ of $(\mathscr{S}, \mathscr{R})$. Denote by $\mathscr{H}'''$ the maximal subspace of $(\mathscr{S}, \mathscr{R})$ which contains $\bar{\mathscr{H}}$. Since $\pi, \pi' \in \mathscr{H}''' \cap \mathscr{H}''$, so $\mathscr{H}''' \in \tilde{\mathscr{T}}$ by Lemma 24.3.20. Hence there is some point $P \in \mathscr{P}$ for which $\mathscr{H}''' = \mathscr{S}_P$. We conclude that π, π', π'' contain the point P of $\mathscr{P}$. □

Lemma 24.3.23: $(\mathscr{S}, \mathscr{R})$ *is a Grassmann space of index $r - 1$.*

Proof: By Lemma 24.3.8 and the corollary of Lemma 24.3.14 we know that $(\mathcal{S}, \mathcal{R})$ is a connected irreducible PPLS.

Let π, π', π'' be pairwise collinear elements of $(\mathcal{S}, \mathcal{R})$. By Lemmas 24.3.11 and 24.3.12 the elements π, π', π'' are contained in a subspace of $(\mathcal{S}, \mathcal{R})$. This means that A1 is satisfied.

The set of all maximal subspaces of $(\mathcal{S}, \mathcal{R})$ is partitioned into the families $\tilde{\mathcal{S}}$ and $\tilde{\mathcal{T}}$. Consider an element $\mathcal{H} \in \tilde{\mathcal{T}}$ and an element $\mathcal{H}' \in \tilde{\mathcal{S}}$. For some point $P \in \mathcal{P}$ we have $\mathcal{H} = \mathcal{S}_P$. Assume that $\pi \in \mathcal{S}_P \cap \mathcal{H}'$ and let $\pi' \in \mathcal{H}' \backslash \{\pi\}$. By Lemma 24.3.21 we have $\mathcal{H}' = \mathcal{S}(\pi, \pi')$. If $\pi \cap \pi' = \{P\}$, then by Lemma 24.3.20 we have $\mathcal{S}_P \cap \mathcal{S}(\pi, \pi') = \pi\pi'$. Now assume that $P \notin \pi'$. By Lemma 24.3.14 the set of all elements of $\mathcal{S}_P$ which have a point in common with π' is a line $\mathcal{L}$ in $\mathcal{R}$. Let $\pi'' \in \mathcal{L} \backslash \{\pi\}$. Clearly $\pi'' \in \mathcal{S}(\pi, \pi')$. By Lemma 24.3.21, $\mathcal{S}(\pi, \pi') = \mathcal{S}(\pi, \pi'')$. Again by Lemma 24.3.20, $\mathcal{S}_P \cap \mathcal{S}(\pi, \pi'') = \pi\pi''$. Hence if $\mathcal{H} \cap \mathcal{H}' \neq \emptyset$, then $\mathcal{H} \cap \mathcal{H}'$ is a line of $(\mathcal{S}, \mathcal{R})$. This proves that A2.I is satisfied.

Let $\mathcal{L}$ be a line of $(\mathcal{S}, \mathcal{R})$. Suppose that $\pi, \pi' \in \mathcal{L}$, $\pi \neq \pi'$, and let $\pi \cap \pi' = \{P\}$. By Lemma 24.3.20, $\mathcal{S}_P$ in $\tilde{\mathcal{T}}$ and $\mathcal{S}(\pi, \pi')$ in $\tilde{\mathcal{S}}$ are the only maximal subspaces of $(\mathcal{S}, \mathcal{R})$ containing $\mathcal{L}$. Consequently A2.II is satisfied.

Next, let $\mathcal{H}, \mathcal{H}', \mathcal{H}''$ be distinct elements of $\tilde{\mathcal{S}}$, for which $\mathcal{H} \cap \mathcal{H}'$, $\mathcal{H}' \cap \mathcal{H}''$, and $\mathcal{H}'' \cap \mathcal{H}$ are distinct elements of $\mathcal{S}$. Let $\mathcal{H} \cap \mathcal{H}' = \{\pi''\}$, $\mathcal{H}' \cap \mathcal{H}'' = \{\pi\}$, and $\mathcal{H}'' \cap \mathcal{H} = \{\pi'\}$. Now consider an $\mathcal{H}''' \in \tilde{\mathcal{S}} \backslash \{\mathcal{H}, \mathcal{H}', \mathcal{H}''\}$ having distinct elements in common with $\mathcal{H}$ and $\mathcal{H}'$. Let $\mathcal{H}''' \cap \mathcal{H} = \{\bar{\pi}\}$ and $\mathcal{H}''' \cap \mathcal{H}' = \{\bar{\pi}'\}$. We shall show that $\mathcal{H}''' \cap \mathcal{H}'' \neq \emptyset$.

By Lemma 24.3.22 we know that π, π', π'' contain a common point $P \in \mathcal{P}$, and that $\bar{\pi}, \bar{\pi}', \pi''$ contain a common point $Q \in \mathcal{P}$.

First assume that $P = Q$. Since $\mathcal{H} \cap \mathcal{S}_P = \emptyset$ or a line of $(\mathcal{S}, \mathcal{R})$, the elements $\pi'', \pi', \bar{\pi}$ are collinear in $(\mathcal{S}, \mathcal{R})$; similarly $\pi'', \pi, \bar{\pi}'$ are collinear in $(\mathcal{S}, \mathcal{R})$. Hence in the projective space $\mathcal{S}_P$ the plane $\pi\pi'\pi''$ also contains $\bar{\pi}$ and $\bar{\pi}'$. Consequently in $(\mathcal{S}, \mathcal{R})$ the lines $\bar{\pi}\bar{\pi}'$ and $\pi\pi'$ meet at an element $\bar{\pi}'' \in \mathcal{S}$. Clearly $\bar{\pi}\bar{\pi}' \subset \mathcal{H}'''$ and $\pi\pi' \subset \mathcal{H}''$, so that $\mathcal{H}''' \cap \mathcal{H}'' \neq \emptyset$.

Next, assume that $P \neq Q$. Since $P \neq Q$, we have $\bar{\pi} \neq \pi'$. In $(\mathcal{S}, \mathcal{R})$, $\bar{\pi}$ and π' are collinear; so $\bar{\pi} \cap \pi' = \{P'\}$ for some $P' \in \mathcal{P}$. Similarly, $\bar{\pi}' \neq \pi$, and $\bar{\pi}' \cap \pi = \{Q'\}$. Clearly $P \neq Q$ implies that the points P, Q, P', Q' are distinct. In $(\mathcal{P}, \mathcal{B})$ the points P, Q, and P' are pairwise collinear; hence they are contained in a maximal subspace τ of $(\mathcal{S}, \mathcal{R})$. Since PQ and PP' are contained in the distinct elements $\pi'' \in \mathcal{S}$ and $\pi' \in \mathcal{S}$, respectively, we have $\tau \in \mathcal{T}$. Similarly, the points P, Q, and Q' are contained in a maximal subspace τ' in $\mathcal{T}$. The subspaces τ and τ' have at least two points in common, and so they coincide. Hence $P, Q, P', Q' \in \tau$; so the points P' and Q' are collinear. Let π''' be the element of $\mathcal{S}$ which contains P' and Q'. Since P' belongs to π''' and $\bar{\pi}$, and Q' belongs to π''' and $\bar{\pi}'$, we have $\pi''' \in \mathcal{S}(\bar{\pi}, \bar{\pi}') = \mathcal{H}'''$. Since P' belongs to π''' and π', and Q' belongs to π''' and π, we have $\pi''' \in \mathcal{S}(\pi, \pi') = \mathcal{H}''$. Consequently, $\mathcal{H}''' \cap \mathcal{H}'' \neq \emptyset$.

Since we have always $\mathscr{H}''' \cap \mathscr{H}'' \neq \varnothing$, also A2.III is satisfied.

By the corollary of Lemma 24.3.10 each $\mathscr{S}_P$ in $\tilde{\mathscr{T}}$ is an r-dimensional projective space over $GF(q)$. This means that for each $\mathscr{S}_P$ in $\tilde{\mathscr{T}}$ there exist distinct subspaces $\mathscr{H}_1, \mathscr{H}_2, \ldots, \mathscr{H}_r$ of $(\mathscr{S}, \mathscr{R})$, such that $\mathscr{H}_1$ is a line of $(\mathscr{S}, \mathscr{R})$, such that $\mathscr{H}_1 \subset \mathscr{H}_2 \subset \cdots \subset \mathscr{H}_r$, and such that there is no subspace $\mathscr{H}$ of $(\mathscr{S}, \mathscr{R})$ with $\mathscr{H}_i \subsetneqq \mathscr{H} \subsetneqq \mathscr{H}_{i+1}$ for $i = 1, 2, \ldots, r-1$. This proves that A3 is satisfied.

We conclude that $(\mathscr{S}, \mathscr{R})$ is a Grassmann space of index $r-1$. □

Lemma 24.3.24: *If $(\mathscr{S}, \mathscr{R})$ is isomorphic to the PPLS corresponding to the Grassmann variety $\mathscr{G}_{r-1,n}$, then $(\mathscr{P}, \mathscr{B})$ is isomorphic to the PPLS corresponding to the Grassmann variety $\mathscr{G}_{r,n}$.*

Proof: By hypothesis there exists a bijection ξ of $\mathscr{S}$ onto $\mathscr{G}_{r-1,n}$ such that the set of all lines of $\mathscr{G}_{r-1,n}$ consists of all images under ξ of the elements of $\mathscr{R}$. Then ξ maps the maximal subspaces of $(\mathscr{S}, \mathscr{R})$ onto the maximal subspaces of $\mathscr{G}_{r-1,n}$. By the observation preceding Lemma 24.3.3, ξ maps each of $\tilde{\mathscr{S}}$ and $\tilde{\mathscr{T}}$ onto a system of maximal spaces of $\mathscr{G}_{r-1,n}$. First let $n \neq 2r-1$. Then the elements of $\mathscr{S}_{\mathrm{L}}$, namely the Latin spaces of $\mathscr{G}_{r-1,n}$, and the elements of $\mathscr{S}_{\mathrm{G}}$, namely the Greek spaces of $\mathscr{G}_{r-1,n}$, have distinct dimension. By the corollary of Lemma 24.3.10 the elements of $\tilde{\mathscr{T}}$ have dimension r, which is the dimension of the elements of $\mathscr{S}_{\mathrm{G}}$. Hence ξ maps the elements of $\tilde{\mathscr{T}}$ onto the elements of $\mathscr{S}_{\mathrm{G}}$, and the elements of $\tilde{\mathscr{S}}$ onto the elements of $\mathscr{S}_{\mathrm{L}}$. Next, let $n = 2r-1$. In §24.2 it was shown that $G(\mathscr{G}_{r-1,2r-1})$ contains elements interchanging $\mathscr{S}_{\mathrm{L}}$ and $\mathscr{S}_{\mathrm{G}}$. It follows that also in this case we may assume that ξ maps the elements of $\tilde{\mathscr{T}}$ onto the elements of $\mathscr{S}_{\mathrm{G}}$, and the elements of $\tilde{\mathscr{S}}$ onto the elements of $\mathscr{S}_{\mathrm{L}}$. As in §24.2 the mapping which associates the points of $\mathscr{G}_{r-1,n}$ to the elements of $PG^{(r-1)}(n, q)$ is denoted by $\mathfrak{G}$.

Let $\Pi_r \in PG^{(r)}(n, q)$, and let $\Pi_r^{(r-1)}$ be the set of all $(r-1)$-dimensional subspaces of Π_r. Then $\Pi_r^{(r-1)} \subset PG^{(r-1)}(n, q)$ and $\Pi_r^{(r-1)}\mathfrak{G}\xi^{-1} = \mathscr{S}_P \in \tilde{\mathscr{T}}$. Now consider the mapping

$$\psi: PG^{(r)}(n, q) \to \mathscr{P},$$

defined by

$$\Pi_r\psi = P \quad \text{if and only if} \quad \Pi_r^{(r-1)}\mathfrak{G}\xi^{-1} = \mathscr{S}_P.$$

It is clear that ψ is a bijection of $PG^{(r)}(n, q)$ onto $\mathscr{P}$.

Let Π_r, Π_r', Π_r'' be elements of $PG^{(r)}(n, q)$, two of them at least being distinct, and let $\Pi_r\psi = P$, $\Pi_r'\psi = P'$, $\Pi_r''\psi = P''$. Then the following are equivalent:

(a) $\Pi_r \cap \Pi_r' \cap \Pi_r''$ is an element Π_{r-1} of $PG^{(r-1)}(n, q)$;
(b) $\Pi_r^{(r-1)} \cap \Pi_r'^{(r-1)} \cap \Pi_r''^{(r-1)} = \{\Pi_{r-1}\}$;
(c) $(\Pi_r^{(r-1)}\mathfrak{G}\xi^{-1}) \cap (\Pi_r'^{(r-1)}\mathfrak{G}\xi^{-1}) \cap (\Pi_r''^{(r-1)}\mathfrak{G}\xi^{-1}) = \{\Pi_{r-1}\mathfrak{G}\xi^{-1}\}$;
(d) $\mathscr{S}_P \cap \mathscr{S}_{P'} \cap \mathscr{S}_{P''} = \{\pi\}$ with $\pi = \Pi_{r-1}\mathfrak{G}\xi^{-1}$ in $\mathscr{S}$;
(e) $P, P', P'' \in \pi$.

Hence $\Pi_r \cap \Pi'_r \cap \Pi''_r$ is an element of $PG^{(r-1)}(n, q)$ if and only if P, P', P'' belong to a common element of $\mathscr{S}$.

Let Π_r and Π'_r be distinct elements of $PG^{(r)}(n, q)$, with $\Pi_r\psi = P$ and $\Pi'_r\psi = P'$. If $\Pi_r \cap \Pi'_r \in PG^{(r-1)}(n, q)$, then by the preceding paragraph the points P and P' belong to a common element of $\mathscr{S}$, and hence are collinear in $(\mathscr{P}, \mathscr{B})$. Conversely, if P and P' are collinear, then the line PP' belongs to an element of $\mathscr{S}$, and hence by the preceding paragraph $\Pi_r \cap \Pi'_r \in PG^{(r-1)}(n, q)$.

Assume again that $\Pi_r \cap \Pi'_r = \Pi_{r-1} \in PG^{(r-1)}(n, q)$. Then each element of $PG^{(r)}(n, q)$ through Π_{r-1} belongs to the unique π in $\mathscr{S}$ containing $P = \Pi_r\psi$ and $P' = \Pi'_r\psi$. Conversely, if $P'' \in \pi$, then $P''\psi^{-1}$ contains Π_{r-1}. So all elements of $PG^{(r)}(n, q)$ through Π_{r-1} are mapped by ψ onto all points of π. It also follows easily from the preceding paragraph that each π in $\mathscr{S}$ corresponds in this way to a Π_{r-1} in $PG^{(r-1)}(n, q)$.

Next, let $\bar{\pi} \in \mathscr{T}$ and let P, P', P'' be three points of $\bar{\pi}$, which do not belong to a common line of $(\mathscr{P}, \mathscr{B})$. Since P, P', and P'' are pairwise collinear, the spaces $P\psi^{-1} = \Pi_r$, $P'\psi^{-1} = \Pi'_r$, and $P''\psi^{-1} = \Pi''_r$ pairwise intersect at some element of $PG^{(r-1)}(n, q)$. If Π_r, Π'_r, Π''_r contain a common element of $PG^{(r-1)}(n, q)$, then P, P', P'' belong to a common π in $\mathscr{S}$. Hence $\pi \cap \bar{\pi}$ contains three non-collinear points, a contradiction. Consequently Π_r, Π'_r, Π''_r do not contain a common Π_{r-1}, whence they all lie in some Π_{r+1} in $PG^{(r+1)}(n, q)$. Now let P''' be any point of $\bar{\pi}$, and suppose that P, P', P'', say, do not belong to a common line of $(\mathscr{P}, \mathscr{B})$. Then the r-spaces $\Pi_r, \Pi'_r, \Pi'''_r = P'''\psi^{-1}$ belong to a common $(r + 1)$-space. Hence $\Pi'''_r \subset \Pi_{r+1}$. Conversely, let Π'''_r be any r-space contained in Π_{r+1}. Since $\Pi_r = \Pi'''_r$ or $\Pi_r \cap \Pi'''_r$ is an $(r - 1)$-space, the points P and $P''' = \Pi'''_r\psi$ are collinear. Similarly P''' and P' are collinear as are P''' and P''. If $P''' \notin \bar{\pi}$, then the maximal subspaces of $(\mathscr{P}, \mathscr{B})$ containing $PP'P'''$, $P'P''P'''$, and $PP''P'''$ are elements of $\mathscr{S}$, which easily gives a contradiction. Hence $P''' \in \bar{\pi}$. So we have shown that all elements of $PG^{(r)}(n, q)$ contained in Π_{r+1} are mapped by ψ onto all points of $\bar{\pi}$. Let us now consider any Π'_{r+1} in $PG^{(r+1)}(n, q)$. If $\Pi^1_r, \Pi^2_r \in \Pi'_{r+1}$, with $\Pi^1_r \neq \Pi^2_r$, then $\Pi^1_r\psi = P^1$ and $\Pi^2_r\psi = P^2$ are collinear. The points of the element of $\mathscr{T}$ through the line P^1P^2 are mapped by ψ^{-1} onto the r-spaces belonging to the $(r + 1)$-space Π'_{r+1} which contains Π^1_r and Π^2_r. Hence each $(r + 1)$-space of $PG(n, q)$ corresponds to some element of $\mathscr{T}$.

Next, let l be a line of $(\mathscr{P}, \mathscr{B})$. Then l is contained in a unique element π of $\mathscr{S}$ and a unique element $\bar{\pi}$ of $\mathscr{T}$; also $\pi \cap \bar{\pi} = l$. So we have $\pi\psi^{-1} \cap \bar{\pi}\psi^{-1} = l\psi^{-1}$. Since $l\psi^{-1} \neq \varnothing$, it is clear that $l\psi^{-1}$ is the pencil (Π_{r-1}, Π_{r+1}) of r-spaces, with Π_{r-1} the $(r - 1)$-space which corresponds to π and Π_{r+1} the $(r + 1)$-space which corresponds to $\bar{\pi}$. Conversely, let us consider a pencil (Π_{r-1}, Π_{r+1}) of r-spaces in $PG(n, q)$. Let π be the element of $\mathscr{S}$ which corresponds to Π_{r-1} and let $\bar{\pi}$ be the element of $\mathscr{T}$ which corresponds to Π_{r+1}. We have $(\Pi_{r-1}, \Pi_{r+1})\psi = \pi \cap \bar{\pi}$. Since $(\Pi_{r-1}, \Pi_{r+1})\psi \neq \varnothing$, clearly $\pi \cap \bar{\pi}$ is a line l of $(\mathscr{P}, \mathscr{B})$.

We have proved that ψ is a bijection of $PG^{(r)}(n, q)$ onto $\mathscr{P}$ such that $\mathscr{B}$ is the set of all images of the pencils of r-spaces of $PG(n, q)$. In other words, the Grassmann space $(\mathscr{P}, \mathscr{B})$ is isomorphic to the PPLS defined by the Grassmann variety $\mathscr{G}_{r,n}$. □

Theorem 24.3.25: *In a Grassmann space $(\mathscr{P}, \mathscr{B})$ of index $r = 1$ any two distinct elements of $\mathscr{S}$ have exactly one point in common.*

If $(\mathscr{P}, \mathscr{B})$ is a connected irreducible PPLS satisfying A1, A2.I, A2.II, *and* A2.III′: *any two distinct elements of $\mathscr{S}$ have exactly one point in common, then $(\mathscr{P}, \mathscr{B})$ is a Grassmann space of index* 1.

Proof: Let $(\mathscr{P}, \mathscr{B})$ be a Grassmann space of index $r = 1$. Then Lemmas 24.3.3 to 24.3.10 hold, so that each element of $\mathscr{T}$ is a projective plane over $GF(q)$. Let π and π' be distinct elements of $\mathscr{S}$. Suppose that $\pi \cap \pi' = \varnothing$. Let k be the minimum number for which there exist distinct points $P_1, P_2, \ldots, P_k$, with $P_1 \in \pi$, $P_k \in \pi'$, and $P_1 \sim P_2 \sim \cdots \sim P_k$. Then $P_2, P_3, \ldots, P_{k-1} \notin \pi \cap \pi'$. Notice that k exists by the connectivity of $(\mathscr{P}, \mathscr{B})$. Assume that $k > 2$. Let π'' be the element of $\mathscr{S}$ which contains the line P_2P_3, and let $\bar{\pi}$ be the element of $\mathscr{T}$ which contains the line P_1P_2. Since $\pi \cap \bar{\pi} \neq \varnothing$, we know that $\pi \cap \bar{\pi}$ is a line l of $(\mathscr{P}, \mathscr{B})$. Similarly $\pi'' \cap \bar{\pi}$ is a line l''. The space $\bar{\pi}$ is a projective plane; hence l and l'' have a common point P. The points $P, P_3, P_4, \ldots, P_k$ are distinct, and $P \sim P_3 \sim P_4 \sim \cdots \sim P_k$ with $P \in \pi$. This contradicts the assumption on the minimality of k. Hence $k = 2$, which means that there are points P_1 and P_2, with $P_1 \in \pi$, $P_2 \in \pi'$, and $P_1 \sim P_2$. Let $\bar{\pi}'$ be the element of $\mathscr{T}$ which contains the line P_1P_2. Since $\pi \cap \bar{\pi}' \neq \varnothing$, we know that $\pi \cap \bar{\pi}'$ is a line m. Similarly $\pi' \cap \bar{\pi}'$ is a line m'. Since $\bar{\pi}'$ is a projective plane, the lines m and m' have a point in common, whence $\pi \cap \bar{\pi}' \neq \varnothing$, a contradiction. We conclude that any two distinct elements of $\mathscr{S}$ have exactly one point in common.

Next, let $(\mathscr{P}, \mathscr{B})$ be a connected irreducible PPLS satisfying A1, A2.I, A2.II, and A2.III′. Then A2.III is trivially satisfied. Lemmas 24.3.3 to 24.3.10 are satisfied for a certain $r \geq 1$. Hence $(\mathscr{P}, \mathscr{B})$ is a Grassmann space of index r. Assume $r \geq 2$. By the corollary of Lemma 24.3.14 there exist disjoint elements in $\mathscr{S}$, in contradiction to A2.III′. We conclude that $r = 1$. □

Theorem 24.3.26: *Any Grassmann space of index $r \geq 1$ is isomorphic to the PPLS defined by some Grassmann variety $\mathscr{G}_{r,n}$.*

Proof: Let $(\mathscr{P}, \mathscr{B})$ be a Grassmann space of index r, where $r > 2$. Assume that each Grassmann space of index $r - 1$ is isomorphic to the PPLS defined by some Grassmann variety $\mathscr{G}_{r-1,n}$. Then, by Lemmas 24.3.23 and 24.3.24, $(\mathscr{P}, \mathscr{B})$ is isomorphic to the PPLS defined by the Grassmann variety $\mathscr{G}_{r,n}$. So we have only to show that any Grassmann space of index 1 is isomorphic to the PPLS defined by some Grassmann variety $\mathscr{G}_{1,n}$.

Let $(\mathscr{P}, \mathscr{B})$ be a Grassmann space of index 1. By Theorem 24.3.25 any two distinct elements of $\mathscr{S}$ have exactly one point in common. Since Lemmas 24.3.3 to 24.3.10 hold, each element of $\mathscr{T}$ is a projective plane over $GF(q)$. For any P in $\mathscr{P}$, let $\mathscr{S}_P$ denote the set of all elements of $\mathscr{S}$ through P. Further, let $\tilde{\mathscr{T}} = \{\mathscr{S}_P | P \in \mathscr{P}\}$. Consider a point P of $\mathscr{P}$ and a plane $\bar{\pi}$ of $\mathscr{T}$ through P. Through each line in $\bar{\pi}$ through P, there is a unique element of $\mathscr{S}_P$, and by A2.I we obtain in this way $q + 1$ distinct elements of $\mathscr{S}_P$. Now it is clear that $(\mathscr{S}, \tilde{\mathscr{T}})$ is an irreducible linear space. We shall prove that $(\mathscr{S}, \tilde{\mathscr{T}})$ is isomorphic to the linear space formed by the points and lines of some $PG(n, q)$. It is sufficient to show that in $(\mathscr{S}, \tilde{\mathscr{T}})$ the Veblen–Wedderburn axiom holds.

Let π, π', π'' be distinct elements of $\mathscr{S}$, let $\pi \cap \pi' = \{P''\}$, $\pi' \cap \pi'' = \{P\}$, $\pi'' \cap \pi = \{P'\}$, and assume that P, P', P'' are distinct. Further, let $\bar{\pi}$ and $\bar{\pi}'$ be distinct elements of $\mathscr{S} \backslash \{\pi''\}$, let $\bar{\pi} \cap \pi'' = \{P'\}$, $\bar{\pi}' \cap \pi'' = \{P\}$, $\bar{\pi} \cap \bar{\pi}' = \{\bar{P}''\}$, and assume again that $P, P', \bar{P}''$ are distinct. We must show that P'' and $\bar{P}''$ belong to a common element of $\mathscr{S}$; that is, we must show that $P'' \sim \bar{P}''$. By A1 the pairwise collinear points P, P', P'' belong to a maximal space π^1, which by A2.II is an element of $\mathscr{T}$. Similarly the pairwise collinear points $P, P', \bar{P}''$ belong to a maximal space π^2 in $\mathscr{T}$. As the line PP' is contained in both π^1 and π^2, we have $\pi^1 = \pi^2$ by A2.II. Hence $P'' \sim \bar{P}''$, and consequently $(\mathscr{S}, \tilde{\mathscr{T}})$ is isomorphic to the linear space formed by the points and lines of some $PG(n, q)$. Since the partial linear space $(\mathscr{P}, \mathscr{B})$ is proper, we can find disjoint lines in $(\mathscr{S}, \tilde{\mathscr{T}})$, and so $n \geq 3$.

Now we consider the mapping

$$\Phi: \tilde{\mathscr{T}} \to \mathscr{P},$$

defined by

$$\mathscr{S}_P \Phi = P.$$

It is clear that Φ is a bijection of $\tilde{\mathscr{T}}$ onto $\mathscr{P}$. Let $\mathscr{D}$ be a plane in $(\mathscr{S}, \tilde{\mathscr{T}})$ and let π, π', π'' be any three independent points of $\mathscr{D}$. Further, let $\pi \cap \pi' = \{P''\}$, $\pi' \cap \pi'' = \{P\}$, and $\pi'' \cap \pi = \{P'\}$. Then the points P, P', P'' are distinct. Since the points P, P', P'' are pairwise collinear in $(\mathscr{P}, \mathscr{B})$, there exists a maximal subspace $\bar{\pi}$ containing them; by A2.II, this subspace belongs to $\mathscr{T}$. Let π''' be an element of $\mathscr{S}$ containing a line l of $\bar{\pi}$. Suppose that l is distinct from the line PP'. If $\{Q\} = PP' \cap l$, then $\mathscr{S}_Q$ contains π'' and π''', and has an element in common with $\mathscr{S}_{P''}$. But $\mathscr{S}_{P''}$ is the line $\pi\pi'$ of $(\mathscr{S}, \tilde{\mathscr{T}})$, and hence π''' belongs to the plane $\mathscr{D}$. Conversely, let π''' be an element of the plane $\mathscr{D}$. Suppose, for example, that $\pi''' \neq \pi''$. Let $\pi'' \cap \pi''' = \{Q\}$. Then the lines $\mathscr{S}_{P''}$ and $\mathscr{S}_Q$ of $(\mathscr{S}, \tilde{\mathscr{T}})$ have an element in common; hence $P'' \sim Q$. Suppose that $Q \notin \bar{\pi}$. The pairwise collinear and distinct points P, P'', Q are contained in a maximal subspace $\bar{\pi}'$, which clearly belongs to $\mathscr{T}$. Since $\bar{\pi}$ and $\bar{\pi}'$ share the line PP'', we have $\bar{\pi} = \bar{\pi}'$. Hence $Q \in \bar{\pi}$, a contradiction. Consequently $Q \in \bar{\pi}$. By A2.I we know that $\pi''' \cap \bar{\pi}$ is a line of $(\mathscr{P}, \mathscr{B})$. We have proved that $\mathscr{D}$ consists of all elements of $\mathscr{S}$ meeting $\bar{\pi}$ at a line of $(\mathscr{P}, \mathscr{B})$. Notice that the lines of $\mathscr{D}$

are the elements $\mathcal{S}_P$ with $P \in \bar{\pi}$. Now it is easy to show that, for any plane $\bar{\pi}''$ of $\mathcal{T}$, the set of all spaces in $\mathcal{S}$ having a line in common with $\bar{\pi}''$ is a plane of $(\mathcal{S}, \tilde{\mathcal{T}})$.

Let $\pi \in \mathcal{S}$, let $\mathcal{D}$ be a plane of $(\mathcal{S}, \tilde{\mathcal{T}})$, and assume that $\pi \in \mathcal{D}$. By the preceding paragraph $\mathcal{D}$ consists of all elements of $\mathcal{S}$ containing a line of some plane $\bar{\pi}$ of $\mathcal{T}$. The lines of the pencil $(\pi, \mathcal{D})$ of $(\mathcal{S}, \tilde{\mathcal{T}})$ are the elements $\mathcal{S}_P$ with P in $\pi \cap \bar{\pi} = l$. Hence $(\pi, \mathcal{D})\Phi = l$. Conversely, let us consider any l' in $\mathcal{B}$. Let $l' \subset \pi' \in \mathcal{S}$ and $l' \subset \bar{\pi}' \in \mathcal{T}$. Then $l'\Phi^{-1}$ is the pencil $(\pi', \mathcal{D}')$, with $\mathcal{D}'$ the plane of $(\mathcal{S}, \tilde{\mathcal{T}})$ which consists of all elements of $\mathcal{S}$ having a line in common with $\bar{\pi}'$.

We have proved that the Grassmann space $(\mathcal{P}, \mathcal{B})$ of index 1 is isomorphic to the PPLS defined by the Grassmann variety $\mathcal{G}_{1,n}$. □

To end this section we give some examples of proper partial linear spaces which show that neither any of the axioms A2.I, A2.II, A2.III, nor the connectivity, nor the irreducibility, can be deleted in the characterization of Grassmann varieties.

Example 1: Let $\mathcal{P}$ consist of all 2-subsets of the set $\{1, 2, 3, 4\}$; let $\mathcal{B}$ consist of the elements of the form $\{\{a, b\}, \{a, c\}\}$ with a, b, c distinct. Then $(\mathcal{S}, \tilde{\mathcal{B}})$ is a connected PPLS which satisfies A1, A2.I, A2.II, A2.III. However, it is not irreducible.

Example 2: Let $(\mathcal{P}, \mathcal{B})$ and $(\mathcal{P}', \mathcal{B}')$ be Grassmann spaces with $\mathcal{P} \cap \mathcal{P}' = \emptyset$. Then $(\mathcal{P} \cup \mathcal{P}', \mathcal{B} \cup \mathcal{B}')$ is an irreducible PPLS which satisfies A1, A2.I, A2.II, A2.III, but which is not connected.

Example 3: Let $(\mathcal{P}, \mathcal{B})$ be a Grassmann space with lines having at least four points. Let $l \in \mathcal{B}$. Further, let $(l, \mathcal{L})$ be an irreducible linear space with $|\mathcal{L}| > 1$. Then $(\mathcal{P}, (\mathcal{B}\backslash\{l\})\cup\mathcal{L})$ is a connected irreducible PPLS which satisfies A1, A2.II, A2.III, but not A2.I.

Example 4: Let $(\mathcal{P}, \mathcal{B})$ be a Grassmann space. If the elements of $\mathcal{S}$ are projective spaces of dimension s over $GF(q)$, then we embed one element Π_s of $\mathcal{S}$ in a projective space Π_{s+1} of dimension $s+1$ over $GF(q)$ for which $\Pi_{s+1} \cap \mathcal{P} = \Pi_s$. Let $\mathcal{L}$ be the set of all lines of Π_{s+1}. Then $(\mathcal{P} \cup \Pi_{s+1}, \mathcal{B} \cup \mathcal{L})$ is a connected irreducible PPLS which satisfies A1, A2.I, A2.III, but not A2.II.

Example 5: Let $(\mathcal{P}, \mathcal{B})$ be a Grassmann space with lines having at least four points. Choose a point P in $\mathcal{P}$. Put $\mathcal{B}' = \{l\backslash\{P\} \mid l \in \mathcal{B}\}$, $\mathcal{P}' = \mathcal{P}\backslash\{P\}$. Then $(\mathcal{P}', \mathcal{B}')$ is a connected irreducible PPLS which satisfies A1, A2.I, and A2.II; for neither of the systems of maximal subspaces is A2.III satisfied.

We do not know whether or not A1 can be deleted in the characterization of Grassmann varieties.

24.4 Embedding of Grassmann spaces

Let $(\mathscr{P}, \mathscr{B})$ be a Grassmann space. If $\mathscr{P}$ is a point set of $PG(n, q)$ and $\mathscr{B}$ is a line set of $PG(n, q)$, then we say that $(\mathscr{P}, \mathscr{B})$ is *embedded* in $PG(n, q)$. Grassmann varieties are examples of embedded Grassmann spaces. In this section we will determine all embedded Grassmann spaces.

First we show that not every embedded Grassmann space is a Grassmann variety. Consider the Grassmann variety $\mathscr{G}_{1,7}$ of the lines of $PG(7, q)$. The number of points of $\mathscr{G}_{1,7}$ is a polynomial of degree 12 in q. The number of points on the lines having at least two points in common with $\mathscr{G}_{1,7}$ is a polynomial $a_0q^{25} + a_1q^{24} + \cdots + a_{25}$. By Theorem 24.2.2 the projective space generated by $\mathscr{G}_{1,7}$ has dimension 28. Hence for q large enough $PG(28, q)$ contains a point P, such that each line through it has at most one point in common with $\mathscr{G}_{1,7}$. Let Ω be a hyperplane of $PG(28, q)$, which does not contain P. By intersecting the cone $P\mathscr{G}_{1,7}$ with the hyperplane Ω, we obtain a variety which, together with the projections of the lines of $\mathscr{G}_{1,7}$, is a Grassmann space embedded in Ω. Assume that this Grassmann space $(\mathscr{P}, \mathscr{B})$ is a Grassmann variety. Since the maximal subspaces of $(\mathscr{P}, \mathscr{B})$ have dimensions 2 and 6, the only candidates for the Grassmann variety are $\mathscr{G}_{1,7}$ and $\mathscr{G}_{5,7}$. By Theorem 24.2.2 the projectively equivalent varieties $\mathscr{G}_{1,7}$ and $\mathscr{G}_{5,7}$ are not contained in a $PG(27, q)$, a contradiction. So we have here an example of a Grassmann space which is embedded in a projective space, but which is not a Grassmann variety.

In order to determine all embeddings of Grassmann spaces, it is necessary to introduce *homomorphisms* between projective spaces. Let ψ be the semi-linear transformation of the vector space $V(m + 1, K)$ into the vector space $V(n + 1, K)$ defined by the $(m + 1) \times (n + 1)$ matrix T over K and the automorphism σ of K. Clearly the kernel of ψ is a subspace of $V(m + 1, K)$, and the image of ψ is a subspace of $V(n + 1, K)$. Also, the sum of the dimensions of the kernel and the image of ψ is equal to $m + 1$. Let us now consider the projective spaces over K defined by $V(m + 1, K)$, $V(n + 1, K)$, the kernel of ψ, and the image of ψ. These spaces are respectively denoted by $PG(m, K)$, $PG(n, K)$, $P(\text{Ker}\,\psi)$, and $P(\text{Im}\,\psi)$. We have

$$\dim P(\text{Ker}\,\psi) + \dim P(\text{Im}\,\psi) = m - 1.$$

Now, the semi-linear transformation ψ induces a mapping ξ from $PG(m, K)\backslash P(\text{Ker}\,\psi)$ onto $P(\text{Im}\,\psi)$. Such a mapping ξ is called a *homomorphism of* $PG(m, K)$ *into* $PG(n, K)$ or *onto* $P(\text{Im}\,\psi)$. If ψ is bijective, that is if T is a non-singular $(n + 1) \times (n + 1)$ matrix over K, then ξ is a collineation between projective spaces.

Let us consider the Grassmann variety $\mathscr{G}_{r,n}$ of the r-spaces in $PG(n, q)$. The projective space generated by $\mathscr{G}_{r,n}$ is denoted by $PG(N, q)$. Suppose that ξ is a homomorphism of $PG(N, q)$ into $PG(N', q)$, where Ker $\psi \neq \{0\}$ and with the condition that any line of $PG(N, q)$ having at least two distinct points in common with $\mathscr{G}_{r,n}$ has no point in common with $P(\text{Ker } \psi)$. Then ξ maps the points and lines of $\mathscr{G}_{r,n}$ onto the points and lines of a Grassmann space $(\mathscr{P}, \mathscr{B})$ which is embedded in $P(\text{Im } \psi)$. Since the dimension of $P(\text{Im } \psi)$ is less than N, an argument of one of the previous paragraphs shows that $(\mathscr{P}, \mathscr{B})$ is not a Grassmann variety. The example given at the beginning of this section was constructed in such a way.

We will now prove that any Grassmann space embedded in a projective space can be obtained in the way described above. The proof is given for $GF(q)$, but is valid for any field K.

Theorem 24.4.1: *Let $(\mathscr{P}, \mathscr{B})$ be a Grassmann space of index r which is embedded in a projective space $PG(s, q)$ and let ξ be an isomorphism from the PPLS defined by $\mathscr{G}_{r,n}$ onto the PPLS $(\mathscr{P}, \mathscr{B})$. Then there is a unique homomorphism ψ from $PG(N, q)$, the space generated by $\mathscr{G}_{r,n}$, into $PG(s, q)$ which induces ξ on $\mathscr{G}_{r,n}$.*

Proof: Let $(\mathscr{P}, \mathscr{B})$ be a Grassmann space of index r, $r \geq 1$, which is embedded in $PG(s, q)$. We may assume that $\mathscr{P}$ generates $PG(s, q)$. By §24.3 we know that there is an isomorphism ξ from the PPLS corresponding to some Grassmann variety $\mathscr{G}_{r,n}$ onto $(\mathscr{P}, \mathscr{B})$. The projective space generated by $\mathscr{G}_{r,n}$ is denoted by $PG(N, q)$.

It is sufficient to prove the theorem for the pair (r, n), $1 \leq r \leq n-2$, under the assumption it is already proved for all pairs (r', n'), with $(r', n') \neq (r, n)$, $r' \leq r$, $n' \leq n$, and $1 \leq r' \leq n' - 2$.

Let π be a hyperplane of $PG(n, q)$. If $r \leq n-3$, then by Corollary 2 of Theorem 24.2.13, $\mathfrak{G}$ maps the r-spaces of π onto the points of a subvariety $\mathscr{G}_{r,n-1}$ of $\mathscr{G}_{r,n}$; if $r = n-2$, then $\mathfrak{G}$ maps the $(n-2)$-spaces of π onto the points of a maximal $(n-1)$-space π' of $\mathscr{G}_{n-2,n}$. For $r \leq n-3$, the images under ξ of the points and lines of $\mathscr{G}_{r,n-1}$ form a Grassmann space $(\mathscr{P}', \mathscr{B}')$ isomorphic to $\mathscr{G}_{r,n-1}$; for $r = n-2$ we obtain a maximal $(n-1)$-space $\pi'\xi$ of $(\mathscr{P}, \mathscr{B})$. Let $PG(N', q)$ be the projective space generated by $\mathscr{G}_{r,n-1}$. For $r \leq n-3$ the induction tells us that there is a unique homomorphism ψ_π from $PG(N', q)$ into $PG(s, q)$ which coincides with ξ on $\mathscr{G}_{r,n-1}$; if $r = n-2$ the restriction of ξ to π' is a collineation ψ_π of π' onto $\pi'\xi$. Notice that the semi-linear transformations $\bar{\psi}_\pi$ of $V(N'+1, q)$ into $V(s+1, q)$, which correspond to ψ_π, are determined by π up to a scalar multiple.

Next, let P be a point of $PG(n, q)$. If $r \geq 2$, then by Corollary 1 of Theorem 24.2.13, $\mathfrak{G}$ maps the r-spaces through P onto the points of a subvariety $\mathscr{G}_{r-1,n-1}$ of $\mathscr{G}_{r,n}$; if $r = 1$, then $\mathfrak{G}$ maps the lines through P onto the points of a maximal $(n-1)$-space π'' of $\mathscr{G}_{1,n}$. For $r \geq 2$, the images under ξ of the points and lines of $\mathscr{G}_{r-1,n-1}$ form a Grassmann space $(\mathscr{P}'', \mathscr{B}'')$ isomorphic to $\mathscr{G}_{r-1,n-1}$; for $r = 1$ we obtain a maximal $(n-1)$-space $\pi''\xi$ of $(\mathscr{P}, \mathscr{B})$. Let $PG(N'', q)$ be the projective space generated by $\mathscr{G}_{r-1,n-1}$. For $r \geq 2$ the induction tells us that there is a unique homomorphism ψ_P from $PG(N'', q)$ into $PG(s, q)$ which coincides with ξ on $\mathscr{G}_{r-1,n-1}$; if $r = 1$ the restriction of ξ to π'' is a collineation ψ_P of π'' onto $\pi''\xi$. Notice that the semi-linear transformations $\bar{\psi}_P$ of $V(N''+1, q)$ into $V(s+1, q)$ which correspond to ψ_P, are determined by P up to a scalar multiple.

We shall show that ψ_π and ψ_P have the same associated field automorphism. First suppose that $P \in \pi$. We use the notation of the preceding paragraphs. Let W be the image under $\mathfrak{G}$ of the set of all r-spaces of π containing P. If $r \neq 1, n-2$, then, by Theorem 24.2.13, $W = \mathscr{G}_{r,n-1} \cap \mathscr{G}_{r-1,n-1}$ is a Grassmann variety $\mathscr{G}_{r-1,n-2}$; if $r = n-2$ and $n \neq 3$, then $W = \pi' \cap \mathscr{G}_{n-3,n-1}$ is a maximal $(n-2)$-space $\bar{\pi}'$ of $\mathscr{G}_{n-3,n-1}$; if $r = 1$ and $n \neq 3$, then $W = \mathscr{G}_{1,n-1} \cap \pi''$ is a maximal $(n-2)$-space $\bar{\pi}''$ of $\mathscr{G}_{1,n-1}$; if $r = 1$ and $n = 3$, then $W = \pi' \cap \pi''$ is a line l. Let l be a line of W, and let P_1, P_2, P_3, P_4 be any four distinct points of l. Then $P_i\psi_P = P_i\psi_\pi = P_i\xi$, $i = 1, 2, 3, 4$. Let $P_i\xi = Q_i$, $i = 1, 2, 3, 4$. Then the cross-ratio $\{Q_1, Q_2; Q_3, Q_4\}$ is equal to $\{P_1, P_2; P_3, P_4\}^\sigma$ and to $\{P_1, P_2; P_3, P_4\}^{\sigma'}$, where σ and σ' are the field automorphisms associated to ψ_P and ψ_π. Hence we have $\sigma = \sigma'$. Next, let $P \notin \pi$. Consider a point $\bar{P}$ and a hyperplane $\bar{\pi}$ of $PG(n, q)$, where $P \in \bar{\pi}$, $\bar{P} \in \bar{\pi}$, $\bar{P} \in \pi$. The field automorphisms associated to $\psi_P, \psi_{\bar{\pi}}, \psi_{\bar{P}}, \psi_\pi$ are respectively denoted by $\sigma, \bar{\sigma}', \bar{\sigma}, \sigma'$. By a previous argument we have $\sigma = \bar{\sigma}'$, $\bar{\sigma}' = \bar{\sigma}$, and $\bar{\sigma} = \sigma'$. Hence $\sigma = \sigma'$. Next, let P and P' be distinct points of $PG(n, q)$. By considering a hyperplane through P and P', we see that the field automorphisms associated to ψ_P and $\psi_{P'}$ coincide. Similarly, the field automorphism associated to ψ_π, with π a hyperplane of $PG(n, q)$, is independent of the choice of π. This common field automorphism, associated to each ψ_P and each ψ_π, will be denoted by σ.

The following notation is required. For $i = 0, 1, \ldots, n$, let $\psi_i = \psi_{\mathbf{U}_i}$ and let $\psi^i = \psi_{\mathbf{u}_i}$. Recall that $\mathbf{U}_i \in \mathbf{u}_j$ for $i \neq j$. For the corresponding semi-linear transformations we write $\bar{\psi}_i$ and $\bar{\psi}^i$. Consider now homomorphisms ψ_i and ψ^j with $i \neq j$. Let W_i^j be the image under $\mathfrak{G}$ of the set of all r-spaces of $\mathbf{u}_j$ containing $\mathbf{U}_i$. On W_i^j the homomorphisms ψ_i and ψ^j coincide with ξ. First, let W_i^j be the Grassmann variety $\mathscr{G}_{r-1,n-2}$, and let $PG(N''', q)$ be the projective space generated by $\mathscr{G}_{r-1,n-2}$. Let $\bar{\Phi}_i$ be the restriction of $\bar{\psi}_i$ and $\bar{\Phi}^j$ the restriction of $\bar{\psi}^j$ to $V(N'''+1, q)$. By induction the semi-linear transformations $\bar{\Phi}_i$ and $\bar{\Phi}^j$ differ only by a scalar multiple. If W_i^j is a projective space Π_m, then again let $\bar{\Phi}_i$ and $\bar{\Phi}^j$ be the restrictions of $\bar{\psi}_i$ and $\bar{\psi}^j$ to $V(m+1, q)$. Here trivially $\bar{\Phi}_i$ and $\bar{\Phi}^j$ differ only by a scalar multiple. In both cases it is possible to choose $\bar{\psi}_i$ and $\bar{\psi}^j$ in such a way that $\bar{\Phi}_i$ and $\bar{\Phi}^j$ coincide; we call this process *normalization*.

Fix $\bar{\psi}^n$ and normalize $\bar{\psi}_0, \bar{\psi}_1, \ldots, \bar{\psi}_{n-1}$ each with respect to it. Next normalize $\bar{\psi}^0$ with respect to $\bar{\psi}_1$, and $\bar{\psi}_n$ with respect to $\bar{\psi}^0$. Finally, normalize $\bar{\psi}^1, \bar{\psi}^2, \ldots, \bar{\psi}^{n-1}$ with respect to $\bar{\psi}_n$. Now consider $\bar{\psi}_i$ and $\bar{\psi}_j$, $i \neq j$, and $i, j \in \{0, 1, \ldots, n-1\}$. The image under $\mathfrak{G}$ of the set of all r-spaces through $\mathbf{U}_i$ and $\mathbf{U}_j$ is denoted by $\mathscr{W}_{i,j}$; the image under $\mathfrak{G}$ of the set of all r-spaces of $\mathbf{u}_n$ through $\mathbf{U}_i$ and $\mathbf{U}_j$ is denoted by $\mathscr{W}^n_{i,j}$. Let $PG(M, q)$ be the projective space generated by $\mathscr{W}_{i,j}$, and let $\bar{\mathscr{W}}^n_{i,j}$ be the set of all vectors representing the points of $\mathscr{W}^n_{i,j}$. By previous arguments the restrictions of $\bar{\psi}_i$ and $\bar{\psi}_j$ to $V(M + 1, q)$ differ only by a scalar multiple. Since $\bar{\psi}_i$ and $\bar{\psi}_j$ coincide on $\bar{\mathscr{W}}^n_{i,j} \subset V(M + 1, q)$ and $\bar{\mathscr{W}}^n_{i,j}$ contains a non-zero vector, it is clear that the restrictions of $\bar{\psi}_i$ and $\bar{\psi}_j$ to $V(M + 1, q)$ coincide. Repeating the argument we finally see that any two elements of $\bar{\psi}_0, \bar{\psi}_1, \ldots, \bar{\psi}_n, \bar{\psi}^0, \bar{\psi}^1, \ldots, \bar{\psi}^n$ coincide on the intersection of their common domain and $\bar{\mathscr{G}}_{r,n}$, where $\bar{\mathscr{G}}_{r,n}$ is the set of all vectors representing the points of $\mathscr{G}_{r,n}$.

Let $V(N + 1, q)$ be the vector space generated by $\bar{\mathscr{G}}_{r,n}$. If E_i is the vector of $V(N + 1, q)$ with one in the $(i + 1)$th place and zeros elsewhere, then the vectors $E_0, E_1, \ldots, E_N$ are contained in $\bar{\mathscr{G}}_{r,n}$. Consider the vector E_i. It is in the common domain of $r + 1$ semi-linear transformations $\bar{\psi}_{i_0}, \bar{\psi}_{i_1}, \ldots, \bar{\psi}_{i_r}$. We have $\bar{\psi}_{i_0}E_i = \bar{\psi}_{i_1}E_i = \cdots = \bar{\psi}_{i_r}E_i$. Let us define as follows a semi-linear transformation $\bar{\psi}$ from $V(N + 1, q)$ into $V(s + 1, q)$ with associated field automorphism σ:

$$\bar{\psi}E_i = \bar{\psi}_{i_j}E_i, \quad \text{with } j \in \{0, 1, \ldots, r\},$$

$i = 0, 1, \ldots, N$. Consider now the basis $\{E_{i_0}, E_{i_1}, \ldots\}$ of the domain $\bar{\psi}_i$. Then we have $\bar{\psi}E_{i_j} = \bar{\psi}_i E_{i_j}$. Hence $\bar{\psi}$ agrees with $\bar{\psi}_0, \bar{\psi}_1, \ldots, \bar{\psi}_n$. Next, we consider the basis $\{E_{j_0}, E_{j_1}, \ldots\}$ of the domain of $\bar{\psi}^j$. Then, since $\bar{\psi}^j$ agrees with any $\bar{\psi}_i$, we easily see that $\bar{\psi}E_{j_i} = \bar{\psi}^j E_{j_i}$. Hence $\bar{\psi}$ agrees with $\bar{\psi}^0, \bar{\psi}^1, \ldots, \bar{\psi}^n$. Consequently the homomorphism ψ from $PG(N, q)$ into $PG(s, q)$, which corresponds with $\bar{\psi}$, agrees with $\psi_0, \psi_1, \ldots, \psi_n, \psi^0, \psi^1, \ldots, \psi^n$.

For a point P of $PG(n, q)$ let its *weight* wt(P) be the number of non-zero coordinates, and let the weight of a set of linearly independent points be the sum of its members' weights. The *weight* of an r-space of $PG(n, q)$ is the minimum of all the weights of its independent sets of size $r + 1$. Clearly the smallest possible weight of an r-space is $r + 1$. The weight of a point Q of $\mathscr{G}_{r,n}$ is the weight of the r-space $Q\mathfrak{G}^{-1}$ of $PG(n, q)$. Let Π_r be an r-space of weight at most $2r + 1$. Such a space must have a point of weight 1 in one of its independent sets of size $r + 1$ of minimal weight. Hence this space contains one of the points $\mathbf{U}_i$, and so ξ and ψ agree on the corresponding point of $\mathscr{G}_{r,n}$.

We prove by induction on the weight that ψ and ξ agree on all the points of $\mathscr{G}_{r,n}$. Assume that ψ and ξ agree on all points of $\mathscr{G}_{r,n}$ of weight $r + 1, r + 2, \ldots, h - 1$ and let $h \geq 2r + 2$. Let P be a point of weight h of $\mathscr{G}_{r,n}$. Suppose that $P_0, P_1, \ldots, P_r$ define the r-space $\Pi_r = P\mathfrak{G}^{-1}$ of $PG(n, q)$. We may assume that $h = \sum_{i=0}^{r} \text{wt}(P_i)$. Clearly there are indices i, j, with $i \neq j$, for which $\mathrm{wt}(P_i) \geq 2$ and $\mathrm{wt}(P_j) \geq 2$. For example, let $\mathrm{wt}(P_0) \geq 2$ and $\mathrm{wt}(P_1) \geq 2$. Further, let $P_i = \mathbf{P}(X_i)$, $i = 1, 2$. Now we choose points $Q_i = \mathbf{P}(Y_i)$, $i = 0, 1, 2, 3$, with $X_0 = Y_0 + Y_1$, $X_1 = Y_2 + Y_3$, $\mathrm{wt}(Q_0) < \mathrm{wt}(P_0)$, $\mathrm{wt}(Q_1) < \mathrm{wt}(P_0)$, $\mathrm{wt}(Q_2) < \mathrm{wt}(P_1)$, and $\mathrm{wt}(Q_3) < \mathrm{wt}(P_1)$. Clearly $P_0 \neq Q_0 \neq Q_1 \neq P_0$ and $P_1 \neq Q_2 \neq Q_3 \neq P_1$. Suppose that $Q_0 \in \Pi_r$; then also $Q_1 \in \Pi_r$. At most one of the points Q_0, Q_1 belongs to the $(r-1)$-space $\Pi_{r-1} = P_1 P_2 \cdots P_r$. So we may assume $Q_i \notin \Pi_{r-1}$, with $i \in \{0, 1\}$. Then the points $Q_i, P_1, P_2, \ldots, P_r$ define Π_r, and

$$\mathrm{wt}(Q_i) + \sum_{i=1}^{r} \mathrm{wt}(P_i) < \sum_{i=0}^{r} \mathrm{wt}(P_i) = h,$$

a contradiction. Hence $Q_0 \notin \Pi_r$. Similarly $Q_1, Q_2, Q_3 \notin \Pi_r$. Let

$$\Pi_r^1 = Q_0 P_1 P_2 \cdots P_r, \qquad \Pi_r^2 = Q_1 P_1 P_2 \cdots P_r,$$
$$\Pi_r^3 = P_0 Q_2 P_2 \cdots P_r, \qquad \Pi_r^4 = P_0 Q_3 P_2 \cdots P_r.$$

Evidently the r-spaces Π_r^1, Π_r^2 are distinct, and the r-spaces Π_r^3, Π_r^4 are distinct. Let $\Pi_r^i \mathfrak{G} = A_i$, $i = 1, 2, 3, 4$. Since the weight of A_i is less than h, by induction ψ and ξ agree on the points A_i. The r-spaces Π_r^1 and Π_r^2 have an $(r-1)$-space in common, so by Theorem 24.2.5, $A_1 A_2$ is a line of $\mathscr{G}_{r,n}$. Similarly, $A_3 A_4$ is a line of $\mathscr{G}_{r,n}$. Clearly, $\Pi_r^1 \cap \Pi_r^2 \neq \Pi_r^3 \cap \Pi_r^4$ implies that $A_1 A_2 \neq A_3 A_4$. And since Π_r belongs to the pencils defined both by Π_r^1 and Π_r^2, and by Π_r^3 and Π_r^4, we have $A_1 A_2 \cap A_3 A_4 = \{P\}$. Hence

$$\begin{aligned}\{P\psi\} &= (A_1 A_2)\psi \cap (A_3 A_4)\psi = (A_1\psi)(A_2\psi) \cap (A_3\psi)(A_4\psi)\\ &= (A_1\xi)(A_2\xi) \cap (A_3\xi)(A_4\xi) = (A_1 A_2)\xi \cap (A_3 A_4)\xi = \{P\xi\}.\end{aligned}$$

Therefore ξ and ψ agree on P. This shows that ξ and ψ agree on all points of $\mathscr{G}_{r,n}$.

Finally, let us prove that ψ is uniquely defined by ξ. To this end, let ψ' be a homomorphism from $PG(N, q)$ into $PG(s, q)$ which agrees with ξ on $\mathscr{G}_{r,n}$. A corresponding semi-linear transformation is denoted by $\bar{\psi}'$. Clearly for the restrictions $\bar{\psi}'_0, \bar{\psi}'_1, \ldots, \bar{\psi}'_n, \bar{\psi}'^0, \bar{\psi}'^1, \ldots, \bar{\psi}'^n$ of $\bar{\psi}'$, we have that $\bar{\psi}'^i$ is normalized with respect to $\bar{\psi}'_j$ for all $i \neq j$. From the uniqueness of the homomorphisms $\psi_0, \psi_1, \ldots, \psi_n, \psi^0, \psi^1, \ldots, \psi^n$ it follows that the transformations $\bar{\psi}'_0, \bar{\psi}'_1, \ldots, \bar{\psi}'^n$ are the transformations $\bar{\psi}_0, \bar{\psi}_1, \ldots, \bar{\psi}^n$, up to a common factor of proportion. Now we easily see that $\bar{\psi}$ and $\bar{\psi}'$ are equal up to a factor of proportion. We conclude that $\psi = \psi'$. □

24.5 Notes and references

§24.1. This is taken from Segre (1960a).

§24.2. For more details on Grassmann varieties, see for example Burau (1961), Hodge and Pedoe (1947). Theorem 24.2.3 is proved in Hodge and Pedoe (1947).

§24.3. This is taken from Tallini (1981b), Bichara and Tallini (1982, 1983). The final section on the independence of the axioms is due to Bichara and Mazzocca (1982b).

The finite Buekenhout incidence structures admitting the diagram

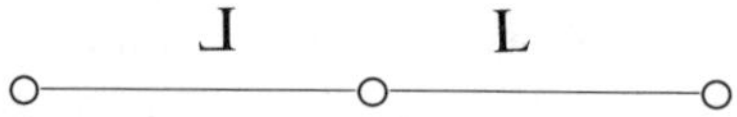

are determined by Sprague (1981b, 1985). Essentially, this is an alternative proof of the characterization theorem in §24.3. An infinite version of Sprague's theorem is proved in Shult (1983).

Characterizations of the Grassmann varieties $\mathcal{G}_{1,n}$ and $\mathcal{G}_{2,n}$ in terms of points and lines only are given by Lo Re and Olanda (1981), Biondi (1987). In these papers the authors prove that their axioms are equivalent to the axioms A1, A2, A3, and then apply the main theorem of §24.3.

The Grassmann varieties associated to an affine space are characterized by Bichara and Mazzocca (1982a, 1983).

Other characterizations of Grassmann varieties will be discussed in the notes and references of Chapter 26.

§24.4. This is taken from Wells (1983).

25

VERONESE AND SEGRE VARIETIES

25.1 Veronese varieties

The *Veronese variety* of all quadrics of $PG(n, K)$, $n \geq 1$, is the variety

$$\mathscr{V} = \{\mathbf{P}(x_0^2, x_1^2, \ldots, x_n^2, x_0x_1, x_0x_2, \ldots, x_0x_n, x_1x_2, \ldots, x_1x_n, \ldots, x_{n-1}x_n) \mid \mathbf{P}(x_0, x_1, \ldots, x_n) \text{ is a point of } PG(n, K)\}$$

of $PG(N, K)$ with $N = n(n + 3)/2$. Clearly $\mathscr{V}$ is a variety of dimension n. It is also called the *Veronesean* of quadrics of $PG(n, K)$, or simply the *quadric Veronesean* of $PG(n, K)$. It can be shown that the quadric Veronesean is absolutely irreducible and non-singular.

Let $PG(N, K)$ consist of all points $\mathbf{P}(Y)$ with

$$Y = (y_{00}, y_{11}, \ldots, y_{nn}, y_{01}, y_{02}, \ldots, y_{0n}, y_{12}, \ldots, y_{1n}, \ldots, y_{n-1,n}).$$

For y_{ij} we also write y_{ji}. Then $\mathscr{V}$ belongs to the intersection of the quadrics $\mathbf{V}(F_{ij})$ and $\mathbf{V}(F_{abc})$, with $i \neq j$ and $i, j \in \{0, 1, \ldots, n\}$, $a \neq b \neq c \neq a$ and $a, b, c \in \{0, 1, \ldots, n\}$, where

$$F_{ij} = y_{ij}^2 - y_{ii}y_{jj}, \qquad F_{abc} = y_{aa}y_{bc} - y_{ab}y_{ac}.$$

We shall now prove that the variety $\mathscr{V}$ is the intersection of these $(n + 1)n^2/2$ quadrics.

Lemma 25.1.1: *The quadric Veronesean $\mathscr{V}$ of $PG(n, K)$ is the intersection of all quadrics $\mathbf{V}(F_{ij})$ and $\mathbf{V}(F_{abc})$.*

Proof: Let $\mathbf{P}(Y)$, with $Y = (y_{00}, y_{11}, \ldots, y_{n-1,n})$, be a point of the intersection of the quadrics $\mathbf{V}(F_{ij})$ and $\mathbf{V}(F_{abc})$. Then $(y_{00}, y_{11}, \ldots, y_{nn}) \neq (0, 0, \ldots, 0)$, since $y_{00} = y_{11} = \cdots = y_{nn} = 0$ and $y_{ij}^2 - y_{ii}y_{jj} = 0$ imply that $y_{ij} = 0$ for all i, j. Assume for example that $y_{00} \neq 0$. Put $y_{00} = 1 = x_0$, and $y_{01} = x_1, y_{02} = x_2, \ldots, y_{0n} = x_n$. Let $i, j \in \{1, 2, \ldots, n\}$ with $i \neq j$. Then since $y_{00}y_{ij} - y_{0i}y_{0j} = 0$ we have $y_{ij} = x_ix_j$. Since $y_{0j}^2 - y_{00}y_{jj} = 0, j \neq 0$, we have $y_{jj} = x_j^2$. Hence $y_{ij} = x_ix_j$, for all $i, j \in \{0, 1, \ldots, n\}$; that is, P belongs to the quadric Veronesean $\mathscr{V}$. □

Theorem 25.1.2: *The quadric Veronesean $\mathscr{V}$ of $PG(n, K)$ consists of all points $\mathbf{P}(Y)$, $Y = (y_{00}, y_{11}, \ldots, y_{n-1,n})$, of $PG(N, K)$ for which rank $[y_{ij}] = 1$.*

Proof: Let $\mathbf{P}(Y)$ be a point for which rank $[y_{ij}] = 1$. Then $\mathbf{P}(Y)$ belongs to the intersection of the quadrics $\mathbf{V}(F_{ij})$ and $\mathbf{V}(F_{abc})$. By Lemma 25.1.1, $\mathbf{P}(Y)$ belongs to the quadric Veronesean $\mathscr{V}$. Conversely, let $\mathbf{P}(Y)$ be a point of the Veronesean $\mathscr{V}$. Then

$$y_{ij}y_{ab} - y_{ib}y_{aj} = x_i x_j x_a x_b - x_i x_b x_a x_j = 0,$$

for all $i, j, a, b \in \{0, 1, \ldots, n\}$. Hence rank $[y_{ij}] = 1$. □

Let ζ: $PG(n, K) \to PG(N, K)$, with $N = n(n+3)/2$ and $n \geq 1$, be defined by $\mathbf{P}(x_0, x_1, \ldots, x_n) \to \mathbf{P}(y_{00}, y_{11}, \ldots, y_{n-1,n})$, with $y_{ij} = x_i x_j$. It is an easy exercise to show that ζ is a bijection of $PG(n, K)$ onto the quadric Veronesean $\mathscr{V}$ of $PG(n, K)$. It then follows that the variety $\mathscr{V}$ is rational.

Theorem 25.1.3: *The quadrics of $PG(n, K)$ are mapped by ζ onto all hyperplane sections of $\mathscr{V}$.*

Proof: Let $\mathbf{V}(F)$, $F = \sum a_{ij}x_i x_j$, be a quadric $\mathscr{Q}$ of $PG(n, K)$. Then $\mathscr{Q}\zeta$ consists of all points of $\mathscr{V}$ for which $\sum a_{ij}y_{ij} = 0$, that is $\mathscr{Q}\zeta$ is a hyperplane section of $\mathscr{V}$. Conversely, let $\mathscr{H}$ be the intersection of $\mathscr{V}$ and the hyperplane $\sum a_{ij}y_{ij} = 0$ of $PG(N, K)$. Then $\mathscr{H}\zeta^{-1}$ consists of all points $\mathbf{P}(X)$ of $PG(n, K)$ satisfying $\sum a_{ij}x_i x_j = 0$; that is, $\mathscr{H}\zeta^{-1}$ is a quadric of $PG(n, K)$. □

Theorem 25.1.3 explains why $\mathscr{V}$ is called the Veronesean of quadrics of $PG(n, K)$.

Corollary: *No hyperplane of $PG(N, K)$ contains the quadric Veronesean $\mathscr{V}$.* □

Theorem 25.1.4: *The Veronese variety $\mathscr{V}$ of all quadrics of $PG(n, K)$, $n \geq 1$, has order 2^n.*

Proof: Let $\bar{K}$ be the algebraic closure of K, and let $\bar{\mathscr{V}}$ be the corresponding extension of $\mathscr{V}$. In $PG(N, \bar{K})$ we intersect $\bar{\mathscr{V}}$ with a $PG(N-n, \bar{K})$. Let $\Pi^1_{N-1}, \Pi^2_{N-1}, \ldots, \Pi^n_{N-1}$ be n linearly independent hyperplanes of $PG(N, \bar{K})$ through $PG(N-n, \bar{K})$. If $\mathscr{H}_i = \bar{\mathscr{V}} \cap \Pi^i_{N-1}$, then $|\bar{\mathscr{V}} \cap PG(N-n, \bar{K})|$ is equal to $|\mathscr{H}_1\zeta^{-1} \cap \mathscr{H}_2\zeta^{-1} \cap \cdots \cap \mathscr{H}_n\zeta^{-1}|$. Since $\mathscr{H}_1\zeta^{-1}, \mathscr{H}_2\zeta^{-1}, \ldots, \mathscr{H}_n\zeta^{-1}$ are n linearly independent quadrics of $PG(n, \bar{K})$, we have $|\bar{\mathscr{V}} \cap PG(N-n, \bar{K})| = 2^n$ for a general $PG(N-n, \bar{K})$ in $PG(N, \bar{K})$. This means that the quadric Veronesean $\mathscr{V}$ of $PG(n, K)$ has order 2^n. □

Notation: From now on the quadric Veronesean of $PG(n, K)$ will be denoted by $\mathscr{V}_n^{2^n}$ or simply $\mathscr{V}_n$. For $n = 1$, the Veronesean $\mathscr{V}_1^2$ is a conic of $PG(2, K)$; referring to the classification of plane quadrics in §7.2, it should be recalled that a conic is the same as a non-singular plane quadric. For $n = 2$, the

Veronesean is a surface $\mathscr{V}_2^4$ of order 4 in $PG(5, K)$. For $n = 3$, the Veronesean is a variety $\mathscr{V}_3^8$ of dimension 3 and order 8 of $PG(9, K)$.

It is easy to show that for $n = 1$ and any four points P_1, P_2, P_3, P_4 on $PG(1, K)$, we have $\{P_1, P_2; P_3, P_4\} = \{P_1\zeta, P_2\zeta; P_3\zeta, P_4\zeta\}$.

From now on it is assumed that K is the finite field $GF(q) = \gamma$, although many of the results will also hold in the case of a general field.

Theorem 25.1.5: $|\mathscr{V}_n| = \theta(n)$.

Proof: The variety $\mathscr{V}_n$ is bijectively mapped by ζ^{-1} onto $PG(n, q)$. Hence $\mathscr{V}_n$ has $\theta(n)$ points. □

Let ξ be a projectivity of $PG(n, q)$. Then ξ defines a permutation ξ' of the quadric Veronesean $\mathscr{V}_n$.

Lemma 25.1.6: *The permutation ξ' of $\mathscr{V}_n$ is induced by a projectivity $\tilde{\xi}$ of $PG(N, q)$.*

Proof: Let $\bar{\gamma}$ be the algebraic closure of γ, and let ζ' be the bijection which maps the quadric $\mathscr{Q} = \mathbf{V}(F)$, $F = \sum a_{ij}x_ix_j$, of $PG(n, \bar{\gamma})$ onto the hyperplane $\sum a_{ij}y_{ij} = 0$ of $PG(N, \bar{\gamma})$. For any hyperplane Π_{N-1} of $PG(N, \bar{\gamma})$, let $\Pi_{N-1}\bar{\eta}$ be the hyperplane $\Pi_{N-1}\zeta'^{-1}\bar{\xi}\zeta'$, with $\bar{\xi}$ the extension of ξ to $PG(n, \bar{\gamma})$. Clearly $\bar{\eta}$ is a permutation of the set of all hyperplanes of $PG(N, \bar{\gamma})$. Since $\bar{\xi}$ maps a pencil of quadrics onto a pencil of quadrics, the permutation $\bar{\eta}$ maps a pencil of hyperplanes onto a pencil of hyperplanes. Also, $\bar{\xi}$ leaves invariant the cross-ratio of any four elements of any pencil of quadrics of $PG(n, \bar{\gamma})$. It follows that $\bar{\eta}$ is a projectivity of the dual space of $PG(N, \bar{\gamma})$. Let $\bar{\tilde{\xi}}$ be the corresponding projectivity of $PG(N, \bar{\gamma})$, with $\tilde{\xi}$ the projectivity induced by $\bar{\tilde{\xi}}$ onto $PG(N, q)$. Now it is an easy exercise to show that $\tilde{\xi}$ leaves $\mathscr{V}_n$ invariant and induces ξ' on $\mathscr{V}_n$. This proves that ξ' is induced by at least one element $\tilde{\xi}$ of $PGL(N + 1, q)$. □

Theorem 25.1.7: *Let Π_s be any s-dimensional subspace of $PG(n, q)$. Then $\Pi_s\zeta$ is a quadric Veronesean $\mathscr{V}_s$, which is the complete intersection of $\mathscr{V}_n$ and the space $PG(s(s + 3)/2, q)$ containing $\mathscr{V}_s$.*

Proof: By Lemma 25.1.6 we may assume that Π_s contains the points $\mathbf{P}(E_0)$, $\mathbf{P}(E_1), \ldots, \mathbf{P}(E_s)$, where E_i is the vector with one in the $(i + 1)$th place and zeros elsewhere. So $\Pi_s\zeta$ consists of the points $\mathbf{P}(Y)$, where up to permutation the coordinates of the vector Y are $x_0^2, x_1^2, \ldots, x_s^2, x_0x_1, x_0x_2, \ldots, x_0x_s$, $x_1x_s, \ldots, x_{s-1}x_s, 0, 0, \ldots, 0$, where $x_i \in \gamma$ and $(x_0, x_1, \ldots, x_s) \neq (0, 0, \ldots, 0)$. Hence $\Pi_s\zeta$ is a quadric Veronesean $\mathscr{V}_s$.

The subspace π of $PG(N, q)$ which contains $\mathscr{V}_s$ is the intersection of the hyperplanes $y_{ij} = 0$ with i and j not both belonging to $\{0, 1, \ldots, s\}$. It is clear that the intersection of π and $\mathscr{V}_n$ corresponds to the set of all points of $PG(n, q)$ with $x_{s+1} = x_{s+2} = \cdots = x_n = 0$, that is $\pi \cap \mathscr{V}_n = \mathscr{V}_s$. □

As a particular case the lines of $PG(n, q)$ are mapped onto conics of $\mathscr{V}_n$.

Theorem 25.1.8: *The quadric Veronesean $\mathscr{V}_n$ is a $\theta(n)$-cap of $PG(N, q)$, $N = n(n+3)/2$.*

Proof: Suppose that P_1, P_2, P_3 are distinct collinear points of $\mathscr{V}_n$. Let π be a plane of $PG(n, q)$ containing the points $P_1\zeta^{-1}, P_2\zeta^{-1}, P_3\zeta^{-1}$. By Theorem 25.1.7, $\pi\zeta$ is a quadric Veronesean $\mathscr{V}_2^4$ which is contained in a subspace $PG(5, q)$ of $PG(N, q)$. The mapping ζ defines a bijection of the set of all plane quadrics of π containing $P_1\zeta^{-1}, P_2\zeta^{-1}, P_3\zeta^{-1}$ onto the set of all hyperplanes of $PG(5, q)$ containing P_1, P_2, P_3. There are $q^2 + q + 1$ plane quadrics of π through $P_1\zeta^{-1}, P_2\zeta^{-1}, P_3\zeta^{-1}$, and there are $q^3 + q^2 + q + 1$ hyperplanes of $PG(5, q)$ through P_1, P_2, P_3. This yields a contradiction; so $\mathscr{V}_n$ is a $\theta(n)$-cap of $PG(N, q)$. □

Now we prove a converse of Theorem 25.1.7.

Theorem 25.1.9: *Any quadric Veronesean $\mathscr{V}_s$ contained in $\mathscr{V}_n$ is of the form $\Pi_s\zeta$, with Π_s some s-dimensional subspace of $PG(n, q)$.*

Proof: We first prove the theorem for $s = 1$. So let $\mathscr{C}$ be a conic which is contained in $\mathscr{V}_n$. Let P_1, P_2, P_3 be three distinct points of $\mathscr{C}$, and let π be a plane containing $P_1\zeta^{-1}, P_2\zeta^{-1}, P_3\zeta^{-1}$. By Theorem 25.1.7, $\pi\zeta$ is a quadric Veronesean $\mathscr{V}_2^4$ in a subspace $PG(5, q)$ of $PG(N, q)$. Consider two distinct hyperplanes π', π'' of $PG(5, q)$ containing $\mathscr{C}$. Then $(\pi' \cap \mathscr{V}_2^4)\zeta^{-1} = \mathscr{C}'$ and $(\pi'' \cap \mathscr{V}_2^4)\zeta^{-1} = \mathscr{C}''$ are distinct plane quadrics of π which contain the set $\mathscr{C}\zeta^{-1}$ of order $q + 1$. Hence $\mathscr{C}\zeta^{-1}$ is necessarily a line of the plane π. It has therefore been shown that for any conic $\mathscr{C}$ on $\mathscr{V}_n$ the set $\mathscr{C}\zeta^{-1}$ is a line of $PG(n, q)$.

Next, consider any quadric Veronesean $\mathscr{V}_s$ contained in $\mathscr{V}_n$. Consider distinct points Q_1, Q_2 on $\mathscr{V}_s\zeta^{-1} = \mathscr{R}$. By the first part of the proof, the points $Q_1\zeta$ and $Q_2\zeta$ are contained in a unique conic $\mathscr{C}$ of $\mathscr{V}_s$ and in a unique conic $\mathscr{C}'$ of $\mathscr{V}_n$. Since $\mathscr{V}_s \subset \mathscr{V}_n$, we clearly have $\mathscr{C} = \mathscr{C}'$. Consequently the line $Q_1Q_2 = \mathscr{C}'\zeta^{-1} = \mathscr{C}\zeta^{-1}$ belongs to $\mathscr{R}$. Hence $\mathscr{R}$ is a subspace of $PG(n, q)$. Since $|\mathscr{R}| = |\mathscr{V}_s| = \theta(s)$, so $\mathscr{R}$ is an s-dimensional subspace of $PG(n, q)$. Thus any quadric Veronesean $\mathscr{V}_s$ contained in $\mathscr{V}_n$ is of the form $\Pi_s\zeta$, with Π_s some s-dimensional subspace of $PG(n, q)$. □

Corollary 1: *Any two points of $\mathscr{V}_n$ are contained in a unique conic of $\mathscr{V}_n$.*

Proof: Let P_1, P_2 be distinct points of $\mathscr{V}_n$. Then $P_1\zeta^{-1}$ and $P_2\zeta^{-1}$ are contained in a unique line of $PG(n, q)$; that is, P_1 and P_2 are contained in a unique $\mathscr{V}_1^2$. Since $\mathscr{V}_1^2$ is a conic, the result follows. □

Corollary 2: *The quadric Veronesean $\mathscr{V}_n$ contains $\phi(s; n, q)$ quadric Veroneseans $\mathscr{V}_s$.*

Proof: The number of Veroneseans $\mathscr{V}_s$ on the Veronesean $\mathscr{V}_n$ is the number of s-dimensional subspaces of $PG(n, q)$. □

Let ξ be a projectivity of $PG(n, q)$. Then ξ defines a permutation ξ' of $\mathscr{V}_n$, which by Lemma 25.1.6 is induced by a projectivity $\tilde{\xi}$ of $PG(N, q)$. Clearly $\xi_1 \neq \xi_2$ implies $\tilde{\xi}_1 \neq \tilde{\xi}_2$. Let $G(\mathscr{V}_n)$ be the subgroup of $PGL(N + 1, q)$ leaving $\mathscr{V}_n$ fixed.

Theorem 25.1.10: *For any projectivity ξ of $PG(n, q)$ the corresponding permutation ξ' of $\mathscr{V}_n$ is induced by a unique element $\tilde{\xi}$ of $G(\mathscr{V}_n)$. Also, the mapping θ: $\xi \to \tilde{\xi}$ is an isomorphism of $PGL(n + 1, q)$ onto $G(\mathscr{V}_n)$.*

Proof: Let $\tilde{\xi}$ be any element of $G(\mathscr{V}_n)$. The corresponding permutation of $\mathscr{V}_n$ is denoted by ξ'. We prove now that ξ' corresponds to a projectivity of $PG(n, q)$.

Let ξ be the permutation of the points of $PG(n, q)$ which corresponds to ξ'. Since ξ' maps conics of $\mathscr{V}_n$ onto conics of $\mathscr{V}_n$, the permutation ξ maps lines of $PG(n, q)$ onto lines of $PG(n, q)$. By a remark following Theorem 25.1.4 we also know that ξ leaves the cross-ratio of any four collinear points invariant. By the fundamental theorem of projective geometry the permutation ξ is a projectivity of $PG(n, q)$. Hence ξ' corresponds to a projectivity of $PG(n, q)$.

Next, consider any projectivity ξ of $PG(n, q)$ and also the corresponding permutation ξ' of $\mathscr{V}_n$. Assume that ξ' is induced by the projectivities $\tilde{\xi}$ and $\tilde{\xi}'$ of $PG(N, q)$. We shall prove that $\tilde{\xi} = \tilde{\xi}'$.

If $\tilde{\eta} = \tilde{\xi}'\tilde{\xi}^{-1}$, then $\tilde{\eta}$ induces the identity mapping of $\mathscr{V}_n$. Suppose that $\tilde{\eta}$ is not the identity mapping of $PG(N, q)$. By the corollary of Theorem 25.1.3 the Veronesean $\mathscr{V}_n$ generates $PG(N, q)$. Let $Q_0, Q_1, \ldots, Q_N$ be $N + 1$ linearly independent points on $\mathscr{V}_n$. Since $\tilde{\eta}$ is not the identity mapping, the points of $\mathscr{V}_n \backslash \{Q_0, Q_1, \ldots, Q_N\}$ are all contained in one of the $N + 1$ hyperplanes generated by N points of $\{Q_0, Q_1, \ldots, Q_N\}$. Hence there is a hyperplane containing $\theta(n) - 1$ points of $\mathscr{V}_n$; that is, there is a quadric of $PG(n, q)$ containing $\theta(n) - 1$ points. This contradiction shows that $\tilde{\eta}$ is the identity mapping of $PG(N, q)$; so $\tilde{\xi} = \tilde{\xi}'$.

Since $\xi \neq \eta$ clearly implies $\tilde{\xi} \neq \tilde{\eta}$, it has been shown that θ: $\xi \to \tilde{\xi}$ is an isomorphism of $PGL(n + 1, q)$ onto $G(\mathscr{V}_n)$. □

Corollary: $|G(\mathcal{V}_n)| = |PGL(n+1, q)|$. $\square$

Apart from the conic, the quadric Veronesean which is most studied and characterized is the surface $\mathcal{V}_2^4$ of $PG(5, q)$. In the second part of this section we prove some interesting properties of this surface, several of which can be generalized to all Veroneseans.

So let us consider the quadric Veronesean $\mathcal{V}_2^4$. By Corollary 2 of Theorem 25.1.9 the variety $\mathcal{V}_2^4$ contains $q^2 + q + 1$ conics, and by Corollary 1 of Theorem 25.1.9 any two points of $\mathcal{V}_2^4$ are contained in a unique one of these conics. Since the conics of $\mathcal{V}_2^4$ correspond to the lines of $PG(2, q)$, any two of these conics have a unique point in common.

To the quadrics of $PG(2, q)$ there correspond all hyperplane sections of $\mathcal{V}_2^4$. The hyperplane is uniquely determined by the plane quadric if and only if the latter is not a single point. If the quadric $\mathcal{C}$ of $PG(2, q)$ is a repeated line, then the corresponding prime Π_4 of $PG(5, q)$ meets $\mathcal{V}_2^4$ in a conic; if $\mathcal{C}$ is two distinct lines, then Π_4 meets $\mathcal{V}_2^4$ in two conics with exactly one point in common; if $\mathcal{C}$ is a conic, then Π_4 meets $\mathcal{V}_2^4$ in a rational quartic curve.

The planes of $PG(5, q)$ which meet $\mathcal{V}_2^4$ in a conic are called the *conic planes* of $\mathcal{V}_2^4$.

Theorem 25.1.11: *Any two conic planes π and π' of $\mathcal{V}_2^4$ have exactly one point in common, and this common point belongs to $\mathcal{V}_2^4$.*

Proof: Let $\pi \cap \mathcal{V}_2^4 = \mathcal{C}$ and $\pi' \cap \mathcal{V}_2^4 = \mathcal{C}'$. Then we know that $|\mathcal{C} \cap \mathcal{C}'| = 1$. Assume that $\pi \cap \pi'$ is a line and let $\pi \cup \pi'$ be contained in the distinct hyperplanes π'' and π''' of $PG(5, q)$. To the hyperplanes π'' and π''' there correspond two distinct quadrics of $PG(2, q)$, which both contain the distinct lines $\mathcal{C}\zeta^{-1}$ and $\mathcal{C}'\zeta^{-1}$. This contradiction proves the theorem. $\square$

Theorem 25.1.12: *The union of the conic planes of $\mathcal{V}_2^4$ is the primal $\mathcal{M}_4^3$ of order 3 with equation $F = 0$, where*

$$F = \begin{vmatrix} y_{00} & y_{01} & y_{02} \\ y_{01} & y_{11} & y_{12} \\ y_{02} & y_{12} & y_{22} \end{vmatrix}. \tag{25.1}$$

Proof: Let

$$u_0x_0 + u_1x_1 + u_2x_2 = 0$$

be the equation of any line l of $PG(2, q)$. Then, by multiplying this equation in turn by x_0, x_1, x_2, we see that the conic $\mathscr{C} = l\zeta$ is the section of $\mathscr{V}_2^4$ by the plane defined by the equations

$$\begin{aligned} u_0 y_{00} + u_1 y_{01} + u_2 y_{02} &= 0, \\ u_0 y_{01} + u_1 y_{11} + u_2 y_{12} &= 0, \\ u_0 y_{02} + u_1 y_{12} + u_2 y_{22} &= 0. \end{aligned} \tag{25.2}$$

A point $\mathbf{P}(y_{00}, y_{11}, y_{22}, y_{01}, y_{02}, y_{12})$ belongs to a conic plane, that is belongs to a plane having equations of the form (25.2), if and only if $F = 0$. □

Theorem 25.1.13: *The primal* $\mathscr{M}_4^3$ *has* $(q^2 + q + 1)(q^2 + 1)$ *points.*

Proof: The primal $\mathscr{M}_4^3$ is the union of the $q^2 + q + 1$ conic planes of $\mathscr{V}_2^4$. By Theorem 25.1.11 any two conic planes have exactly one point in common, which belongs to $\mathscr{V}_2^4$. Also each point of $\mathscr{V}_2^4$ belongs to at least one conic plane. Hence

$$|\mathscr{M}_4^3| = (q^2 + q + 1)q^2 + (q^2 + q + 1) = (q^2 + q + 1)(q^2 + 1). \quad \square$$

If the characteristic of γ is two, we have $\mathscr{M}_4^3 = \mathbf{V}(F)$ where

$$F = y_{00}y_{11}y_{22} + y_{00}y_{12}^2 + y_{11}y_{02}^2 + y_{22}y_{01}^2. \tag{25.3}$$

In this case $\mathscr{M}_4^3$ contains the plane $\mathbf{U}_3\mathbf{U}_4\mathbf{U}_5$. Notice that the plane $\mathbf{U}_3\mathbf{U}_4\mathbf{U}_5$ has no point in common with the Veronesean $\mathscr{V}_2^4$.

Lemma 25.1.14: *If the characteristic of* γ *is two, then the Veronesean* $\mathscr{V}_2^4$ *is the intersection of the quadrics* $\mathbf{V}(F_{01})$, $\mathbf{V}(F_{02})$, *and* $\mathbf{V}(F_{12})$, *where*

$$F_{01} = y_{01}^2 + y_{00}y_{11}, \quad F_{02} = y_{02}^2 + y_{00}y_{22}, \quad F_{12} = y_{12}^2 + y_{11}y_{22}. \tag{25.4}$$

Proof: Let $\mathbf{P}(Y)$, where $Y = (y_{00}, y_{11}, y_{22}, y_{01}, y_{02}, y_{12})$, be any point of $\mathbf{V}(F_{01}) \cap \mathbf{V}(F_{02}) \cap \mathbf{V}(F_{12})$. Then $y_{01}^2 = y_{00}y_{11}$ and $y_{02}^2 = y_{00}y_{22}$; so $y_{01}^2 y_{02}^2 = y_{00}^2 y_{11}y_{22}$. Since $y_{12}^2 = y_{11}y_{22}$, it follows that $y_{01}^2 y_{02}^2 = y_{00}^2 y_{12}^2$. Hence $y_{01}y_{02} = y_{00}y_{12}$; so $\mathbf{P}(Y)$ belongs to $\mathbf{V}(F_{012})$ with the notation of Lemma 25.1.1. Analogously, $\mathbf{P}(Y)$ belongs to $\mathbf{V}(F_{120})$ and $\mathbf{V}(F_{201})$. By Lemma 25.1.1 we have $\mathscr{V}_2^4 = \mathbf{V}(F_{01}) \cap \mathbf{V}(F_{02}) \cap \mathbf{V}(F_{12})$. □

Theorem 25.1.15: *The primal* $\mathscr{M}_4^3$ *has the Veronesean* $\mathscr{V}_2^4$ *as double surface.*

Proof: We have $\mathscr{M}_4^3 = \mathbf{V}(F)$, where

$$F = y_{00}y_{11}y_{22} + 2y_{01}y_{02}y_{12} - y_{00}y_{12}^2 - y_{11}y_{02}^2 - y_{22}y_{01}^2. \tag{25.5}$$

The partial derivatives of F are

$$\frac{\partial F}{\partial y_{00}} = y_{11}y_{22} - y_{12}^2 = -F_{12},$$

$$\frac{\partial F}{\partial y_{11}} = y_{00}y_{22} - y_{02}^2 = -F_{02},$$

$$\frac{\partial F}{\partial y_{22}} = y_{00}y_{11} - y_{01}^2 = -F_{01},$$

$$\frac{\partial F}{\partial y_{01}} = 2(y_{02}y_{12} - y_{22}y_{01}) = -F_{201},$$

$$\frac{\partial F}{\partial y_{02}} = 2(y_{01}y_{12} - y_{11}y_{02}) = -2F_{120},$$

$$\frac{\partial F}{\partial y_{12}} = 2(y_{01}y_{02} - y_{00}y_{12}) = -2F_{012}.$$

If the characteristic of γ is not two, then the singular points of $\mathcal{M}_4^3$ are the elements of $\mathcal{M}_4^3 \cap \mathbf{V}(F_{12}) \cap \mathbf{V}(F_{02}) \cap \mathbf{V}(F_{01}) \cap \mathbf{V}(F_{012}) \cap \mathbf{V}(F_{120}) \cap \mathbf{V}(F_{201}) = \mathcal{M}_4^3 \cap \mathcal{V}_2^4 = \mathcal{V}_2^4$.

If the characteristic of γ is two, then the singular points of $\mathcal{M}_4^3$ are the elements of $\mathcal{M}_4^3 \cap \mathbf{V}(F_{12}) \cap \mathbf{V}(F_{02}) \cap \mathbf{V}(F_{01})$. By Lemma 25.1.14 this set is again $\mathcal{V}_2^4$.

Finally, it is easy to check that all singular points of $\mathcal{M}_4^3$ are double points. □

The tangent lines of the conics of $\mathcal{V}_2^4$ are called the *tangents* or *tangent lines* of $\mathcal{V}_2^4$. Since no point of the surface $\mathcal{V}_2^4$ is singular, all tangent lines of $\mathcal{V}_2^4$ at the point P of $\mathcal{V}_2^4$ are contained in a plane $\pi(P)$. This plane $\pi(P)$ is called the *tangent plane* of $\mathcal{V}_2^4$ at P. Since P is contained in exactly $q + 1$ conics of $\mathcal{V}_2^4$ and since no two conic planes through P have a line in common, the tangent plane $\pi(P)$ is the union of the $q + 1$ tangent lines of $\mathcal{V}_2^4$ through P. Also $\pi(P) \cap \mathcal{V}_2^4 = \{P\}$.

Clearly all the tangent lines and tangent planes of the surface $\mathcal{V}_2^4$ belong to the primal $\mathcal{M}_4^3$. Since $\mathcal{M}_4^3$ is the union of the conic planes of $\mathcal{V}_2^4$, any point of $\mathcal{M}_4^3$ is on at least one tangent or bisecant of $\mathcal{V}_2^4$. As any two points of $\mathcal{V}_2^4$ are contained in a conic of $\mathcal{V}_2^4$, each bisecant of $\mathcal{V}_2^4$ is a line of $\mathcal{M}_4^3$. Hence $\mathcal{M}_4^3$ can also be described as the union of all tangents and bisecants of $\mathcal{V}_2^4$.

Theorem 25.1.16: *For any two distinct points P_1 and P_2 of $\mathcal{V}_2^4$, the tangent planes $\pi(P_1)$ and $\pi(P_2)$ have exactly one point in common.*

Proof: Let P_1 and P_2 be distinct points of $\mathscr{V}_2^4$, and let $\mathscr{C}$ be the conic of $\mathscr{V}_2^4$ through P_1 and P_2. The tangent l_i of $\mathscr{C}$ at P_i is contained in $\pi(P_i)$, $i = 1, 2$, and so $\pi(P_1)$ and $\pi(P_2)$ have the point $l_1 \cap l_2 = \{Q\}$ is common.

Assume that $Q' \in \pi(P_1) \cap \pi(P_2)$, with $Q \neq Q'$. Then P_iQ' is tangent to a conic $\mathscr{C}_i$ of $\mathscr{V}_2^4$, $i = 1, 2$. If $\mathscr{C}_1 = \mathscr{C}_2$, then $\mathscr{C}_1 = \mathscr{C}_2 = \mathscr{C}$ and so $Q = Q'$, a contradiction. So $\mathscr{C}_1 \neq \mathscr{C}_2$. If $\mathscr{C}_1 \cap \mathscr{C}_2 = \{P\}$, then the conic planes containing $\mathscr{C}_1$ and $\mathscr{C}_2$ have the distinct points P and Q' in common, contradicting Theorem 25.1.11. Hence we have proved that $\pi(P_1) \cap \pi(P_2) = \{Q\}$. □

Theorem 25.1.17: *Suppose that the characteristic of* γ *is two. Then each tangent plane of* $\mathscr{V}_2^4$ *meets the plane* $\mathbf{U}_3\mathbf{U}_4\mathbf{U}_5$ *in a line, each conic plane meets* $\mathbf{U}_3\mathbf{U}_4\mathbf{U}_5$ *in a point, and* $\mathbf{U}_3\mathbf{U}_4\mathbf{U}_5$ *consists of the nuclei of all conics on* $\mathscr{V}_2^4$.

Proof: It was observed above that in the case of characteristic two, the plane $\mathbf{U}_3\mathbf{U}_4\mathbf{U}_5$ belongs to the primal $\mathscr{M}_4^3$. If P is any point of $\mathscr{V}_2^4$ and Q is any point of the plane $\mathbf{U}_3\mathbf{U}_4\mathbf{U}_5$, then by Lemma 25.1.14 the line PQ has only the point P in common with $\mathscr{V}_2^4$.

Let π be any conic plane of $\mathscr{V}_2^4$, and let π be represented by the equations (25.2). Then $\pi \cap \mathbf{U}_3\mathbf{U}_4\mathbf{U}_5$ is determined by the following system of linear equations in y_{01}, y_{02}, y_{12}:

$$u_1 y_{01} + u_2 y_{02} = 0, \quad u_0 y_{01} + u_2 y_{12} = 0, \quad u_0 y_{02} + u_1 y_{12} = 0.$$

Since the rank of the matrix

$$\begin{bmatrix} u_1 & u_2 & 0 \\ u_0 & 0 & u_2 \\ 0 & u_0 & u_1 \end{bmatrix}$$

is equal to two, we have $|\pi \cap \mathbf{U}_3\mathbf{U}_4\mathbf{U}_5| = 1$. Let $\pi \cap \mathbf{U}_3\mathbf{U}_4\mathbf{U}_5 = \{Q\}$ and $\pi \cap \mathscr{V}_2^4 = \mathscr{C}$. Since $|PQ \cap \mathscr{C}| = 1$ for any point P of $\mathscr{C}$, the point Q is the nucleus of $\mathscr{C}$. Also, by Theorem 25.1.11 the nuclei of distinct conics of $\mathscr{V}_2^4$ are distinct. So any conic plane of $\mathscr{V}_2^4$ has exactly one point in common with $\mathbf{U}_3\mathbf{U}_4\mathbf{U}_5$, and $\mathbf{U}_3\mathbf{U}_4\mathbf{U}_5$ is the set of the nuclei of all conics on $\mathscr{V}_2^4$.

Finally, let $\pi(P)$ be the tangent plane of $\mathscr{V}_2^4$ at P. Since each line of $\pi(P)$ through P is tangent to some conic of $\mathscr{V}_2^4$, it contains the nucleus of this conic and hence contains a point of $\mathbf{U}_3\mathbf{U}_4\mathbf{U}_5$. Hence $\pi(P)$ and $\mathbf{U}_3\mathbf{U}_4\mathbf{U}_5$ have a line in common. □

In the case of characteristic two, the plane $\mathbf{U}_3\mathbf{U}_4\mathbf{U}_5$ is called the *nucleus* of the Veronesean $\mathscr{V}_2^4$.

Let $l = \mathbf{V}(F)$, where $F = u_0x_0 + u_1x_1 + u_2x_2$, be any line of $PG(2, q)$. If $\mathscr{C}$ is the plane quadric whose point set coincides with l, then to $\mathscr{C}$ there corresponds the hyperplane $\Pi_4 = \mathbf{V}(F')$ of $PG(5, q)$, where

$$F' = u_0^2 y_{00} + u_1^2 y_{11} + u_2^2 y_{22} + 2u_0 u_1 y_{01} + 2u_0 u_2 y_{02} + 2u_1 u_2 y_{12}. \quad (25.6)$$

Such a hyperplane Π_4 is called a *contact prime* of $\mathscr{V}_2^4$. The contact primes of $\mathscr{V}_2^4$ are the primes which have exactly one conic in common with $\mathscr{V}_2^4$.

First, let the characteristic of γ be two. Then the contact prime Π_4 always contains the nucleus $\mathbf{U}_3\mathbf{U}_4\mathbf{U}_5$ of $\mathscr{V}_2^4$. Let Π_2 be the conic plane containing $l\zeta = \Pi_4 \cap \mathscr{V}_2^4$. By Theorem 25.1.17 the contact prime Π_4 is generated by the conic plane Π_2 and the nucleus $\mathbf{U}_3\mathbf{U}_4\mathbf{U}_5$. By the same theorem the contact prime Π_4 contains the $q + 1$ planes tangent to $\mathscr{V}_2^4$ at the points of the conic $l\zeta$.

Next, let the characteristic of γ be different from two. Consider a point P of $PG(2, q)$, and also distinct lines l, l_1, l_2 through P. To the plane quadric $\mathscr{C}_i = l \cup l_i$ there corresponds the hyperplane Π_4^i of $PG(5, q)$, $i = 1, 2$. Clearly Π_4^i contains the tangent lines at $P\zeta$ of the conics $l\zeta$ and $l_i\zeta$, $i = 1, 2$. Hence Π_4^i contains the tangent plane $\pi(P\zeta)$ of $\mathscr{V}_2^4$ at $P\zeta$, $i = 1, 2$. The plane quadric $\mathscr{C}$ consisting of all points of the line l belongs to the pencil defined by $\mathscr{C}_1$ and $\mathscr{C}_2$; so the contact prime Π_4 of $\mathscr{V}_2^4$ which corresponds to $\mathscr{C}$ belongs to the pencil defined by Π_4^1 and Π_4^2. Hence Π_4 also contains $\pi(P\zeta)$. Thus also in this case, the contact prime Π_4 defined by the line l contains the $q + 1$ planes tangent to $\mathscr{V}_2^4$ at the points of the conic $l\zeta$.

From (25.6) it also follows that, if the characteristic of γ is not two, the set of all contact primes of $\mathscr{V}_2^4$ is nothing else than the dual of the Veronesean $\mathscr{V}_2^4$.

Theorem 25.1.18: *Suppose that the characteristic of γ is not two. Then $PG(5, q)$ admits a polarity which maps the set of all conic planes of $\mathscr{V}_2^4$ onto the set of all tangent planes of $\mathscr{V}_2^4$.*

Proof: Let

$$u_0 x_0 + u_1 x_1 + u_2 x_2 = 0$$

be the equation of the line l of $PG(2, q)$. By (25.2) the conic plane defined by the line l is given by

$$u_0 y_{00} + u_1 y_{01} + u_2 y_{02} = 0,$$
$$u_0 y_{01} + u_1 y_{11} + u_2 y_{12} = 0,$$
$$u_0 y_{02} + u_1 y_{12} + u_2 y_{22} = 0.$$

Next, let $\mathbf{P}(A) = Q$, where $A = (a_0, a_1, a_2)$, be a point of $PG(2, q)$. For any line m of $PG(2, q)$ through Q, the contact prime of $\mathscr{V}_2^4$ corresponding to m

contains the tangent plane of $\mathscr{V}_2^4$ at $Q\zeta$. Hence the tangent plane $\pi(Q\zeta)$ belongs to every prime with equation

$$u_0^2 y_{00} + u_1^2 y_{11} + u_2^2 y_{22} + 2u_0 u_1 y_{01} + 2u_0 u_2 y_{02} + 2u_1 u_2 y_{12} = 0,$$

where u_0, u_1, u_2 satisfy

$$u_0 a_0 + u_1 a_1 + u_2 a_2 = 0.$$

From these two equations the following points are elements of $\pi(Q\zeta)$:

$$\mathbf{P}(2a_0, 0, 0, a_1, a_2, 0), \quad \mathbf{P}(0, 2a_1, 0, a_0, 0, a_2), \quad \mathbf{P}(0, 0, 2a_2, 0, a_0, a_1).$$

Let η be the polarity of $PG(5, q)$ represented by

$$\rho v_{00} = y_{00}/2, \quad \rho v_{11} = y_{11}/2, \quad \rho v_{22} = y_{22}/2,$$

$$\rho v_{01} = y_{01}, \qquad \rho v_{02} = y_{02}, \qquad \rho v_{12} = y_{12},$$

where

$$v_{00} y_{00} + v_{11} y_{11} + v_{22} y_{22} + v_{01} y_{01} + v_{02} y_{02} + v_{12} y_{12} = 0$$

is the equation of a variable prime of $PG(5, q)$. Then it is clear that $\pi(Q\zeta)\eta$ is the conic plane defined by the line l of $PG(2, q)$ with equation

$$a_0 x_0 + a_1 x_1 + a_2 x_2 = 0.$$

Hence the polarity η maps the set of all conic planes of $\mathscr{V}_2^4$ onto the set of all tangent planes of $\mathscr{V}_2^4$. □

Corollary: *Let the characteristic of γ be different from two. Then for any three distinct points P_1, P_2, P_3 of $\mathscr{V}_2^4$, the intersection $\pi(P_1) \cap \pi(P_2) \cap \pi(P_3)$ of the tangent planes is empty.*

Proof: Assume that $\pi(P_1) \cap \pi(P_2) \cap \pi(P_3) \neq \varnothing$. By Theorem 25.1.16 we have $\pi(P_1) \cap \pi(P_2) \cap \pi(P_3) = \{Q\}$. So the prime $Q\eta$ contains the three distinct conic planes $\pi(P_i)\eta$, $i = 1, 2, 3$. Hence the quadric of $PG(2, q)$ which corresponds to $Q\eta$ has at least three distinct linear components, a contradiction. □

25.2 First characterizations of the Veronesean $\mathscr{V}_2^4$, with q odd

By Theorem 25.1.16, for any two distinct points P_1 and P_2 of the Veronesean $\mathscr{V}_2^4$ the tangent planes $\pi(P_1)$ and $\pi(P_2)$ have exactly one point in common. By a classical theorem, the Veronesean $\mathscr{V}_2^4$ over $\mathbf{C}$ is the only surface generating $PG(n, \mathbf{C})$, $n \geq 5$, which is not a cone (with non-trivial vertex) and which satisfies the property just mentioned. In this section it is our purpose to prove the corresponding characterization theorem in the case of a Galois field $\gamma = GF(q)$ with characteristic different from two.

So let γ be a finite field with characteristic different from two. By the corollary of Theorem 25.1.18, for any three distinct points P_1, P_2, P_3 of $\mathcal{V}_2^4$ we have $\pi(P_1) \cap \pi(P_2) \cap \pi(P_3) = \varnothing$. By Theorem 25.1.17 this property is not satisfied in the characteristic two case.

From now on let $\mathcal{F}$ be a set of $q^2 + q + 1$ planes of $PG(5, q)$, q odd, with the following properties:

(i) the elements of $\mathcal{F}$ generate $PG(5, q)$;
(ii) any two distinct elements of $\mathcal{F}$ have exactly one point in common;
(iii) any three distinct elements of $\mathcal{F}$ have an empty intersection.

It is our purpose to prove that $\mathcal{F}$ is the set of all tangent planes of the Veronesean surface $\mathcal{V}_2^4$.

Let π be a plane of $\mathcal{F}$. The other $q^2 + q$ planes of $\mathcal{F}$ intersect π in $q^2 + q$ distinct points. The remaining point of π is denoted by $T(\pi)$, and belongs to no other plane of $\mathcal{F}$; it is called the *contact point* of the plane π. In this way a set $\mathcal{V}$ of $q^2 + q + 1$ points $T(\pi)$ is obtained. We say that π is the *tangent plane of* $\mathcal{V}$ *at* $T(\pi)$.

Lemma 25.2.1: *Any prime* Π_4 *of* $PG(5, q)$ *contains at most* $q + 1$ *planes of* $\mathcal{F}$.

Proof: Let π be an element of $\mathcal{F}$ which does not belong to the prime Π_4. Each plane of $\mathcal{F}$ in Π_4 contains exactly one point of the line $\pi \cap \Pi_4 = l$, and distinct planes of $\mathcal{F}$ in Π_4 intersect l in distinct points. Hence the number of planes of $\mathcal{F}$ in Π_4 is at most the number of points on l, that is $q + 1$. □

Lemma 25.2.2: *Any prime* Π_4 *of* $PG(5, q)$ *contains* 0, 1, *or* $q + 1$ *planes of* $\mathcal{F}$.

Proof: Let Π_4 be a prime of $PG(5, q)$, and assume that Π_4 contains k planes of $\mathcal{F}$, where $2 \leq k \leq q$.

First of all we show that any plane of $\mathcal{F}$ in Π_4 belongs to at least one prime Π'_4, $\Pi'_4 \neq \Pi_4$, which contains k' elements of $\mathcal{F}$, where again $2 \leq k' \leq q$. If such is not the case, then there exists a plane π of $\mathcal{F}$ in Π_4, for which each prime (distinct from Π_4) through it contains 1 or $q + 1$ elements of $\mathcal{F}$. Let m be the number of primes containing π and q other elements of $\mathcal{F}$. Any of the $q^2 + q + 1 - k$ planes of $\mathcal{F}$ not in Π_4 belongs to one of these m primes. Counting in two ways the number of all ordered pairs (Π'_4, π'), with Π'_4 one of the m primes just introduced and π' any element of $\mathcal{F} \backslash \{\pi\}$ in Π'_4, we obtain:

$$mq = q^2 + q + 1 - k.$$

Hence q divides $k - 1$, contradicting $0 < k - 1 < q$.

Let π' be any plane of $\mathscr{F}$ which is not contained in Π_4. Suppose that the line $\pi' \cap \Pi_4 = l$ does not contain the point $T(\pi')$ of π'. Then any point of l is contained in a unique plane of $\mathscr{F} \backslash \{\pi'\}$: k points of l belong to some plane of $\mathscr{F}$ in Π_4, the remaining $q + 1 - k$ points of l belong to a plane of $\mathscr{F} \backslash \{\pi'\}$ not contained in Π_4. Let π'' be one of these $q + 1 - k$ planes. The plane π'' certainly exists as $q + 1 - k > 0$. The intersection of π'' and the prime Π_4 is the line l'. Any of the k planes of $\mathscr{F}$ in Π_4 contains a point Q of l and a point Q' of l'. Clearly $\{Q\} \neq l \cap l'$ and $\{Q'\} \neq l \cap l'$. Next, let π''' be one of the $q - k$ planes of $\mathscr{F} \backslash \{\pi', \pi''\}$ which does not belong to Π_4 but contains a point of l. Assume that the line $l'' = \pi''' \cap \Pi_4$ has no point in common with l'. Since each plane of $\mathscr{F}$ in Π_4 contains a point of each of l, l', l'', distinct from $l \cap l''$ and $l \cap l'$, it is clear that each such plane belongs to the solid defined by l, l', l''. Hence any two of the k planes of $\mathscr{F}$ in Π_4 have a line in common, a contradiction as $k \geq 2$. So the lines l' and l'' intersect. Let π^* be the plane containing l and l'. Now, the lines l, l' together with the k lines QQ' and the $q - k$ lines l'' form a set of $q + 2$ lines of π^* no three of which are concurrent, that is a dual $(q + 2)$-arc of the plane π^*. Since q is odd, this contradicts Theorem 8.1.3. We conclude that the line $l = \pi' \cap \Pi_4$ contains the point $T(\pi')$ of π'.

Since π' was any plane of $\mathscr{F}$ not contained in Π_4, the set $\mathscr{V}$ of all contact points belongs completely to Π_4. Similarly, the set $\mathscr{V}$ is contained in any prime which contains k' planes of $\mathscr{F}$, where $2 \leq k' \leq q$.

Now consider distinct planes π^1 and π^2 of $\mathscr{F}$ in Π_4. By the first part of the proof the plane π^i belongs to at least one prime Π_4^i, $\Pi_4^i \neq \Pi_4$, which contains k_i planes of $\mathscr{F}$, with $2 \leq k_i \leq q$, $i = 1, 2$. By the preceding section of the proof, the set $\mathscr{V}$ belongs also to the primes Π_4^1 and Π_4^2. Let $\Pi_3^i = \Pi_4 \cap \Pi_4^i$, $i = 1, 2$. Since $\pi^i \subset \Pi_3^i$, we have $\Pi_3^1 \neq \Pi_3^2$. So $\Pi_3^1 \cap \Pi_3^2$ is a plane π which contains $\mathscr{V}$. As $|\mathscr{V}| = |\pi|$, we have $\mathscr{V} = \pi$. The plane π has a line l^* in common with the plane π^1. Consequently π^1 contains $q + 1$ points of $\mathscr{V}$, that is $q + 1$ contact points, a contradiction. □

Lemma 25.2.3: *Let Π_4 be a prime of $PG(5, q)$ which contains $q + 1$ planes $\pi^0, \pi^1, \ldots, \pi^q$ of $\mathscr{F}$. If $\pi^0 \cap \pi^1 = \{Q_2\}$, $\pi^1 \cap \pi^2 = \{Q_0\}$, $\pi^2 \cap \pi^0 = \{Q_1\}$, then any line Q_iQ_j contains the contact point of the plane π^k, where $\{i, j, k\} = \{0, 1, 2\}$.*

Proof: Evidently the points Q_0, Q_1, Q_2 generate a plane π, since otherwise the planes π^0, π^1, π^2 share a line. Assume for example that the contact point of the plane π^2 is not on the line Q_0Q_1. Then Q_0Q_1 contains at least one point Q through which there passes a plane π' of $\mathscr{F} \backslash \{\pi^0, \pi^1, \ldots, \pi^q\}$. Let $\pi' \cap \Pi_4 = l$. Suppose that l is not a line of the plane π. The planes π^0 and π^1 have one point in common with π'; so π^0 and π^1 have one point in common with l. These points of π^0 and π^1 on l do not belong to π; hence the planes π^0 and π^1 are contained in the solid defined by π and l. It follows that π^0 and π^1 have a line in common, a contradiction. So l is a line of the plane π.

Any plane of $\mathscr{F}\backslash\{\pi^2\}$ in Π_4 which passes through a point of Q_0Q_1, also passes through a point of l. Since l belongs to π, any such plane of $\mathscr{F}$ has a line in common with π.

So we have shown that each of the $q+1$ planes of $\mathscr{F}\backslash\{\pi^2\}$ which have a non-empty intersection with Q_0Q_1 has a line in common with the plane π. Together with the line Q_0Q_1 there arise in this way $q+2$ lines of π, no three of which are concurrent. Hence π contains a dual $(q+2)$-arc, contradicting Theorem 8.1.3. □

Lemma 25.2.4: *Let Π_4 be a prime of $PG(5, q)$ which contains $q+1$ planes $\pi^0, \pi^1, \ldots, \pi^q$ of $\mathscr{F}$. Then the $q+1$ contact points $T(\pi^0)$, $T(\pi^1), \ldots, T(\pi^q)$ form a conic $\mathscr{C}$ of some plane $\bar{\pi}$ of Π_4. Also the tangent lines of $\mathscr{C}$ are the lines $\bar{\pi}\cap\pi^0, \bar{\pi}\cap\pi^1, \ldots, \bar{\pi}\cap\pi^q$, and $\mathscr{C}=\Pi_4\cap\mathscr{V}$.*

Proof: Let $\pi^0\cap\pi^1=\{Q_2\}$, $\pi^1\cap\pi^2=\{Q_0\}$, $\pi^2\cap\pi^0=\{Q_1\}$. The plane containing Q_0, Q_1, Q_2 is denoted by $\bar{\pi}$. Let $\{Q_i'\}=\pi^0\cap\pi^i$ and $\{Q_i''\}=\pi^1\cap\pi^i$, $i=3, 4, \ldots, q$. By applying Lemma 25.2.3 twice, we see that the contact point $T(\pi^0)$ belongs to the lines Q_2Q_1 and Q_2Q_i', and that the contact point $T(\pi^1)$ belongs to the lines Q_2Q_0 and Q_2Q_i''. Since $Q_2\neq T(\pi^0)$ and $Q_2\neq T(\pi^1)$, we have $Q_i'\in Q_1Q_2$ and $Q_i''\in Q_2Q_0$, with $Q_i', Q_i''\notin\{Q_0, Q_1, Q_2\}$. Hence π^i has the line $Q_i'Q_i''$ in common with the plane $\bar{\pi}$, $i=3, 4, \ldots, q$. Again by Lemma 25.2.3, the contact point $T(\pi^i)$ belongs to the line $Q_i'Q_i''$, $i=3, 4, \ldots, q$. So any plane π^i contains a line l_i of $\bar{\pi}$, and $T(\pi^i)$ belongs to l_i, $i=0, 1, \ldots, q$.

No three of the lines $l_0, l_1, \ldots, l_q$ are concurrent; so they form a dual $(q+1)$-arc. By Theorem 8.2.4 the set $\{l_0, l_1, \ldots, l_q\}$ is a dual conic Γ of $\bar{\pi}$, which means that it consists of the $q+1$ tangent lines of a conic $\mathscr{C}$. As the points of $\mathscr{C}$ are the unique elements of $l_0\cup l_1\cup\cdots\cup l_q$ which belong to exactly one line of Γ, it is clear that $\mathscr{C}$ consists of the contact points $T(\pi^0)$, $T(\pi^1), \ldots, T(\pi^q)$.

Finally, let π' be a plane of $\mathscr{F}\backslash\{\pi^0, \pi^1, \ldots, \pi^q\}$. The intersections of π' with the $q+1$ planes $\pi^0, \pi^1, \ldots, \pi^q$ all belong to $l'=\pi'\cap\Pi_4$, and so their union is l'. Consequently the contact point $T(\pi')$ is not in Π_4. So we have proved that $\Pi_4\cap\mathscr{V}=\mathscr{C}$. □

Corollary: *Let Π_4 be a prime of $PG(5, q)$ which contains $q+1$ planes $\pi^0, \pi^1, \ldots, \pi^q$ of $\mathscr{F}$. If $\bar{\pi}$ is the plane containing the contact points $T(\pi^0)$, $T(\pi^1), \ldots, T(\pi^q)$, then it also contains all points $\pi^i\cap\pi^j$, $i\neq j$ and $i, j=0, 1, \ldots, q$.*

Proof: Let $l_i = \bar{\pi} \cap \pi^i$ and $l_j = \bar{\pi} \cap \pi^j$, $i \neq j$. By Lemma 25.2.4 the sets l_i and l_j are lines; so $l_i \cap l_j = \pi^i \cap \pi^j$. Hence the point $\pi^i \cap \pi^j$ belongs to the plane $\bar{\pi}$. □

Lemma 25.2.5: *The set $\mathscr{V}$ is a $(q^2 + q + 1)$-cap of $PG(5, q)$.*

Proof: Let $\pi', \pi'' \in \mathscr{F}$ and $\pi' \cap \pi'' = \{Q\}$. Clearly the points Q, $T(\pi')$, and $T(\pi'')$ are linearly independent. By Lemma 25.2.2 the prime Π_4, defined by π' and π'', contains $q + 1$ planes of $\mathscr{F}$. By Lemma 25.2.4 and its corollary, the plane $\bar{\pi} = QT(\pi')T(\pi'')$ intersects $\mathscr{V}$ in a conic $\mathscr{C}$ having $QT(\pi')$ and $QT(\pi'')$ as tangent lines at the respective points $T(\pi')$ and $T(\pi'')$; also $\mathscr{V} \cap \Pi_4 = \mathscr{C}$. Hence no point of $\mathscr{V} \backslash \{T(\pi'), T(\pi'')\}$ is on the line $T(\pi')T(\pi'')$. Since $T(\pi')$ and $T(\pi'')$ are distinct arbitrarily chosen points of $\mathscr{V}$, it follows that $\mathscr{V}$ is a $(q^2 + q + 1)$-cap. □

The lines of $PG(5, q)$ having two distinct points in common with $\mathscr{V}$ are called the *bisecants* of $\mathscr{V}$; a *tangent* or *tangent line of $\mathscr{V}$ at Q* in $\mathscr{V}$ is any line of π' through Q, where $\pi' \in \mathscr{F}$ and $Q = T(\pi')$.

Let Π_4 be a prime of $PG(5, q)$ which contains $q + 1$ planes of $\mathscr{F}$, and let $\bar{\pi}$ be the plane defined by the contact points of these planes. The set of all these planes $\bar{\pi}$ will be denoted by $\mathscr{F}^*$.

Lemma 25.2.6: *The set $\mathscr{F}^*$ has $q^2 + q + 1$ elements.*

Proof: The number of elements of $\mathscr{F}^*$ is equal to the number of primes containing $q + 1$ planes of $\mathscr{F}$. Let $\pi', \pi'' \in \mathscr{F}$, $\pi' \neq \pi''$, and let Π_4 be the prime defined by π' and π''. By Lemma 25.2.2 the prime Π_4 contains $q + 1$ planes of $\mathscr{F}$, and so defines an element of $\mathscr{F}^*$. In this way every element of $\mathscr{F}^*$ is obtained exactly $\mathbf{c}(q + 1, 2)$ times. Hence $|\mathscr{F}^*| = \mathbf{c}(q^2 + q + 1, 2)/\mathbf{c}(q + 1, 2) = q^2 + q + 1$. □

Lemma 25.2.7: (i) *Each bisecant or tangent of $\mathscr{V}$ is contained in exactly one element of $\mathscr{F}^*$.*

(ii) *Any point $T(\pi)$ of $\mathscr{V}$ for π in $\mathscr{F}$ lies in exactly $q + 1$ planes $\bar{\pi}^0, \bar{\pi}^1, \ldots, \bar{\pi}^q$ of $\mathscr{F}^*$.*

(iii) *The $q + 1$ sets $\pi \cap \bar{\pi}^0, \pi \cap \bar{\pi}^1, \ldots, \pi \cap \bar{\pi}^q$ are the tangent lines of the conics $\mathscr{V} \cap \bar{\pi}^0, \mathscr{V} \cap \bar{\pi}^1, \ldots, \mathscr{V} \cap \bar{\pi}^q$ at $T(\pi)$.*

Proof: (i) Let l be a bisecant of $\mathscr{V}$, with $l \cap \mathscr{V} = \{Q_1, Q_2\}$. If the plane $\bar{\pi}$ of $\mathscr{F}^*$ contains l, then the prime Π_4 defining $\bar{\pi}$ contains the tangent planes of $\mathscr{V}$ at Q_1 and Q_2. It follows that Π_4 and $\bar{\pi}$ are uniquely defined by Q_1 and Q_2; that is, l is contained in exactly one element of $\mathscr{F}^*$.

Next, let t be a tangent line of $\mathscr{V}$ at $T(\pi)$, with π in $\mathscr{F}$. For any point Q

in $t\backslash\{T(\pi)\}$ there is a unique plane π' in $\mathscr{F}\backslash\{\pi\}$ containing it. The plane $QT(\pi)T(\pi')$ belongs to $\mathscr{F}^*$ and contains t. If $\bar{\pi}$ is any plane of $\mathscr{F}^*$ through t, then t is the tangent at $T(\pi)$ of the conic $\bar{\pi} \cap \mathscr{V}$ and any point of $t\backslash\{T(\pi)\}$ is contained in the tangent plane of $\mathscr{V}$ at some point of $\bar{\pi} \cap \mathscr{V}$. Hence the prime Π_4 defining $\bar{\pi}$ contains the planes π and π'. So Π_4 and $\bar{\pi}$ are uniquely defined by the line t.

(ii), (iii) Since the point $T(\pi)$ is on exactly $q + 1$ tangent lines of $\mathscr{V}$, the lemma is completely proved if we show that $T(\pi)$ is contained in exactly $q + 1$ planes of $\mathscr{F}^*$. The number of bisecants of $\mathscr{V}$ through $T(\pi)$ is equal to $q^2 + q$, and the number of bisecants of $\mathscr{V}$ through $T(\pi)$ in a given plane of $\mathscr{F}^*$ through $T(\pi)$ is equal to q. It follows that exactly $(q^2 + q)/q = q + 1$ elements of $\mathscr{F}^*$ contain $T(\pi)$. □

Lemma 25.2.8: *Any two distinct planes of $\mathscr{F}^*$ have exactly one point in common, and this point is on $\mathscr{V}$.*

Proof: Let $\bar{\pi} \in \mathscr{F}^*$ and $\bar{\pi} \cap \mathscr{V} = \mathscr{C}$. By Lemma 25.2.7 any point Q of $\mathscr{C}$ is contained in exactly q planes of $\mathscr{F}^*\backslash\{\bar{\pi}\}$, and none of these planes has a line in common with $\bar{\pi}$. Including $\bar{\pi}$, this gives $q(q + 1) + 1 = |\mathscr{F}^*|$ planes of $\mathscr{F}^*$. Since $\bar{\pi}$ was chosen arbitrarily, it follows that any two planes of $\mathscr{F}^*$ meet in exactly one point, which is in $\mathscr{V}$. □

Lemma 25.2.9: *Let $\bar{\pi}$ and $\bar{\pi}'$ be distinct planes in $\mathscr{F}^*$, with $\bar{\pi} \cap \mathscr{V} = \mathscr{C}$ and $\bar{\pi}' \cap \mathscr{V} = \mathscr{C}'$. If Π_4 is the prime containing $\bar{\pi}$ and $\bar{\pi}'$, then $\Pi_4 \cap \mathscr{V} = \mathscr{C} \cup \mathscr{C}'$. Hence at most two planes of $\mathscr{F}^*$ lie in a common prime.*

Proof: By Lemma 25.2.8, $\bar{\pi} \cap \bar{\pi}' = \mathscr{C} \cap \mathscr{C}' = \{Q\}$. So $\bar{\pi}$ and $\bar{\pi}'$ belong to a unique prime Π_4. Assume that $Q' \in \Pi_4 \cap \mathscr{V}$ and $Q' \notin \mathscr{C} \cup \mathscr{C}'$. Let Q'' be a point of $\mathscr{C}\backslash\{Q\}$, and let $\bar{\pi}''$ be the unique plane of $\mathscr{F}^*$ through Q'' and Q'. By Lemma 25.2.8 the plane $\bar{\pi}''$ and the conic $\mathscr{C}'$ have a unique point Q''' in common. The points Q', Q'', Q''' are linearly independent points of $\bar{\pi}''$; so $\bar{\pi}''$ is a plane of the prime Π_4. By Lemma 25.2.7 the tangent planes of $\mathscr{V}$ at Q'' and Q''' belong to Π_4. Hence $|\Pi_4 \cap \mathscr{V}| = q + 1$ by Lemma 25.2.4. This contradicts that $\mathscr{C} \cup \mathscr{C}' \subset \Pi_4$. □

Lemma 25.2.10: *Consider the plane π of $\mathscr{F}$, and also a plane π^* skew to π. Then the projection of $\mathscr{V}\backslash\{T(\pi)\}$ from π onto π^* is a conic $\mathscr{C}^*$ of π^*.*

Proof: Let Q be a point of $\mathscr{V}\backslash\{T(\pi)\}$, and let Π_3 be the solid containing π and Q. By Lemma 25.2.7 the unique plane $\bar{\pi}$ of $\mathscr{F}^*$ through $T(\pi)$ and Q contains a line of the plane π. Hence $\bar{\pi}$ lies in the solid Π_3. So the projection $\mathscr{C}^*$ of $\mathscr{V}\backslash\{T(\pi)\}$ from π onto π^* is the set consisting of the intersections of π^* with the solids defined by π and the planes of $\mathscr{F}^*$ through $T(\pi)$. By

Lemma 25.2.8, distinct planes of $\mathscr{F}^*$ through $T(\pi)$ define distinct solids Π_3. Hence $|\mathscr{C}^*| = q + 1$.

Suppose that $\mathscr{C}^*$ contains the three collinear points Q_1^*, Q_2^*, Q_3^*. Then the solids defined by π and the points Q_1^*, Q_2^*, Q_3^* belong to a common prime Π_4. Such a Π_4 contains three distinct planes of $\mathscr{F}^*$, which is in contradiction with Lemma 25.2.9. So $\mathscr{C}^*$ is a $(q + 1)$-arc of the plane π^*. By Theorem 8.2.4, $\mathscr{C}^*$ is a conic. □

Lemma 25.2.11: *Consider the bisecant l of $\mathscr{V}$, where $l \cap \mathscr{V} = \{Q, Q'\}$. Then the set of all planes projecting $\mathscr{V}\backslash l$ from l, together with all planes containing l and a tangent of $\mathscr{V}$ at Q or Q', is the family of all planes on a quadric $l\mathscr{H}_3$, where $\mathscr{H}_3$ is some hyperbolic quadric in a solid skew to l.*

Proof: Let π be the tangent plane of $\mathscr{V}$ at Q, and let π' be the tangent plane of $\mathscr{V}$ at Q'. The union of the planes containing l and a tangent of $\mathscr{V}$ at Q is the solid Π_3 defined by π and l, and the union of the planes containing l and a tangent of $\mathscr{V}$ at Q' is the solid Π'_3 defined by π' and l. By Lemma 25.2.7 the plane $\bar{\pi}$ of $\mathscr{F}^*$ through l contains a tangent of $\mathscr{V}$ at Q and a tangent of $\mathscr{V}$ at Q'; so $\bar{\pi}$ is the intersection of the solids Π_3 and Π'_3.

The elements of $\mathscr{F}^*\backslash\{\bar{\pi}\}$ through Q are denoted by $\bar{\alpha}_1, \bar{\alpha}_2, \ldots, \bar{\alpha}_q$; the elements of $\mathscr{F}^*\backslash\{\bar{\pi}\}$ through Q' are denoted by $\bar{\alpha}'_1, \bar{\alpha}'_2, \ldots, \bar{\alpha}'_q$. Further, let Π_3^i be the solid containing l and $\bar{\alpha}_i$, and $\Pi_3^{i\prime}$ the solid containing l and $\bar{\alpha}'_i$, $i = 1, 2, \ldots, q$. Since any two elements of $\mathscr{F}^*$ have exactly one point in common, these $2q$ solids are distinct. If Π_3^i and Π_3^j, $i \neq j$, meet in a plane, then the prime defined by Π_3^i and Π_3^j contains the conics $\bar{\alpha}_i \cap \mathscr{V} = \mathscr{C}_i$, $\bar{\alpha}_j \cap \mathscr{V} = \mathscr{C}_j$, and also the point $Q' \notin \mathscr{C}_i \cup \mathscr{C}_j$, in contradiction to Lemma 25.2.9. Hence Π_3^i and Π_3^j, $i \neq j$, have just the line l in common. Analogously, the intersection of $\Pi_3^{i\prime}$ and $\Pi_3^{j\prime}$, $i \neq j$, is the line l. Since Π_3^i and $\Pi_3^{j\prime}$, with $i, j \in \{1, 2, \ldots, q\}$, both contain the line l and also the point $\bar{\alpha}_i \cap \bar{\alpha}'_j$, they intersect in a plane. So we have two systems of q solids through l with the property that two solids of the same system intersect in the line l, and two solids of different systems intersect in a plane through l.

Each plane through l in Π_3^i contains a point of the set $\mathscr{C}_i\backslash\{Q\}$, with $\bar{\alpha}_i \cap \mathscr{V} = \mathscr{C}_i$, or contains the tangent of $\mathscr{C}_i$ at Q, and each plane through l in $\Pi_3^{j\prime}$ contains a point of the set $\mathscr{C}'_j\backslash\{Q'\}$, with $\bar{\alpha}'_j \cap \mathscr{V} = \mathscr{C}'_j$, or contains the tangent of $\mathscr{C}'_j$ at Q', $i, j = 1, 2, \ldots, q$. Also, applying Lemma 25.2.7, one deduces that any point of $\mathscr{V}\backslash\bar{\pi}$ is contained in one solid Π_3^i and one solid $\Pi_3^{j\prime}$, that any tangent of $\mathscr{V}$ at Q which does not belong to $\bar{\pi}$ is contained in exactly one solid Π_3^i, and that any tangent of $\mathscr{V}$ at Q' which does not belong to $\bar{\pi}$ is contained in exactly one solid $\Pi_3^{j\prime}$. Hence the union of all planes projecting $\mathscr{V}\backslash l$ from l, together with all planes containing l and a tangent of $\mathscr{V}$ at Q or Q', is also the union of the solids Π_3, Π'_3 and all solids Π_3^i, $\Pi_3^{j\prime}$.

The proof of the lemma is complete if it is shown that Π_3 and Π_3^i intersect in a plane, and that also Π_3' and $\Pi_3^{j\prime}$ intersect in a plane. Clearly $\Pi_3 \neq \Pi_3^i$, since otherwise $\bar{\alpha}_i$ and $\bar{\pi}$ have a line in common. By Lemma 25.2.7 the tangent t of $\bar{\alpha}_i \cap \mathscr{V}$ at Q belongs to the plane π; hence $\Pi_3 \cap \Pi_3^i$ is the plane tl. Analogously, $\Pi_3' \cap \Pi_3^{j\prime}$ is a plane. It follows that $\{\Pi_3', \Pi_3^1, \Pi_3^2, \ldots, \Pi_3^q\}$ and $\{\Pi_3, \Pi_3^{1\prime}, \Pi_3^{2\prime}, \ldots, \Pi_3^{q\prime}\}$ are the two systems of solids of a quadric $l\mathscr{H}_3$, where $\mathscr{H}_3$ is some hyperbolic quadric in a solid skew to l. □

Theorem 25.2.12: *The set $\mathscr{V}$ is the Veronesean $\mathscr{V}_2^4$, and $\mathscr{F}$ is the set of its tangent planes.*

Proof: Let Q_{00}, Q_{11}, Q_{22} be three distinct points of $\mathscr{V}$ which are not contained in a plane of $\mathscr{F}^*$. Further, let π^i be the tangent plane of $\mathscr{V}$ at Q_{ii}, $i = 0, 1, 2$. Put $\{Q_{ij}\} = \{Q_{ji}\} = \pi^i \cap \pi^j$, where $i \neq j$ and $i, j = 0, 1, 2$. By the corollary of Lemma 25.2.4 the plane $Q_{ii}Q_{jj}Q_{ij}$ belongs to $\mathscr{F}^*$ for all $i \neq j$. For all $i \neq j$ the lines $Q_{ii}Q_{ij}$ and $Q_{jj}Q_{ij}$ are the tangent lines of the conic $Q_{ii}Q_{jj}Q_{ij} \cap \mathscr{V} = \mathscr{C}_{ij} = \mathscr{C}_{ji}$ at the respective points Q_{ii} and Q_{jj}.

Suppose that the six points Q_{ij} belong to a prime Π_4. Then the three distinct planes $Q_{ii}Q_{jj}Q_{ij}$, $i \neq j$, of $\mathscr{F}^*$ lie in Π_4, a contradiction by Lemma 25.2.9. Hence the six points Q_{ij} are linearly independent. It follows that the coordinate system of $PG(5, q)$ can be chosen in such a way that $Q_{00} = \mathbf{P}(E_0)$, $Q_{11} = \mathbf{P}(E_1)$, $Q_{22} = \mathbf{P}(E_2)$, $Q_{01} = \mathbf{P}(E_3)$, $Q_{02} = \mathbf{P}(E_4)$, $Q_{12} = \mathbf{P}(E_5)$, where E_i is the vector with one in the $(i+1)$th place and zeros elsewhere. An arbitrary point U of $\mathscr{V}$, with $U \notin \mathscr{C}_{01} \cup \mathscr{C}_{02} \cup \mathscr{C}_{12}$, can be chosen as unit point. By Lemmas 25.2.4 and 25.2.9, any six of the seven points $U, Q_{00}, Q_{11}, Q_{22}, Q_{01}, Q_{02}, Q_{12}$ are linearly independent. An arbitrary point of $PG(5, q)$ is of the form $\mathbf{P}(Y)$, where $Y = (y_{00}, y_{11}, y_{22}, y_{01}, y_{02}, y_{12})$. For y_{ij} we also write y_{ji}.

The tangent plane π^k of $\mathscr{V}$ is skew to the plane $Q_{ii}Q_{jj}Q_{ij}$, where $\{i, j, k\} = \{0, 1, 2\}$. By Lemma 25.2.10 the projection of $\mathscr{V} \backslash \{Q_{kk}\}$ from π^k onto the plane $Q_{ii}Q_{jj}Q_{ij}$ is the conic $Q_{ii}Q_{jj}Q_{ij} \cap \mathscr{V} = \mathscr{C}_{ij}$. The cone $\pi^k\mathscr{C}_{ij}$ is the variety $\mathbf{V}(F)$ with $F = y_{ii}y_{jj} - ay_{ij}^2$. Since $U = \mathbf{U} \in \pi^k\mathscr{C}_{ij}$ we have $a = 1$, and so $F = y_{ii}y_{jj} - y_{ij}^2$.

By Lemma 25.2.11 the union $\mathscr{W}$ of all planes projecting $\mathscr{V} \backslash \{Q_{ii}, Q_{jj}\}$ from the line $Q_{ii}Q_{jj}$ and all planes containing the line $Q_{ii}Q_{jj} = l$ and a tangent of $\mathscr{V}$ at Q_{ii} or Q_{jj} is of the form $l\mathscr{H}_3$, where $\mathscr{H}_3$ is some hyperbolic quadric in a solid Π_3 skew to l. Let Π_3 be the solid containing the points $Q_{ij}, Q_{ik}, Q_{jk}, Q_{kk}$, with $\{i, j, k\} = \{0, 1, 2\}$. From the proof of Lemma 25.2.11 it follows that $\mathscr{H}_3$ contains the lines $Q_{ij}Q_{ik}$, $Q_{ij}Q_{jk}$, $Q_{kk}Q_{ik}$, and $Q_{kk}Q_{jk}$. Hence $\mathscr{W}$ is a variety $\mathbf{V}(F')$ with $F' = y_{kk}y_{ij} - by_{ik}y_{jk}$. Since $U = \mathbf{U} \in \mathscr{W}$ we have $b = 1$; so $F' = y_{kk}y_{ij} - y_{ik}y_{jk}$.

Thus $\mathscr{V}$ belongs to all primals

$$\mathbf{V}(y_{ii}y_{jj} - y_{ij}^2), \qquad i \neq j,$$
$$\mathbf{V}(y_{kk}y_{ij} - y_{ik}y_{jk}), \quad i \neq j \neq k \neq i.$$

By Lemma 25.1.1 we have $\mathcal{V} \subset \mathcal{V}_2^4$. Since $|\mathcal{V}| = |\mathcal{V}_2^4| = q^2 + q + 1$ this gives that $\mathcal{V} = \mathcal{V}_2^4$. Finally, by Lemma 25.2.7 the tangent planes of $\mathcal{V}$ coincide with the tangent planes of the Veronesean $\mathcal{V}_2^4$. □

Let $\mathcal{L}$ be any set of $q^2 + q + 1$ planes of $PG(5, q)$, q odd, with the following properties:

(i)′ there is no point belonging to all elements of $\mathcal{L}$;
(ii) any two distinct elements of $\mathcal{L}$ have exactly one point in common;
(iii)′ any three distinct elements of $\mathcal{L}$ generate $PG(5, q)$.

Theorem 25.2.13: *The set $\mathcal{L}$ is the set of all conic planes of a Veronesean $\mathcal{V}_2^4$. Hence the union of the elements of $\mathcal{L}$ is the variety $\mathcal{M}_4^3$.*

Proof: Let δ be any correlation of $PG(5, q)$. Then the set of planes $\mathcal{L}\delta$ satisfies the properties (i), (ii), and (iii). By Theorem 25.2.12, $\mathcal{L}\delta$ is the set of all tangent planes of a Veronesean of plane quadrics. Now, by Theorem 25.1.18, $\mathcal{L}$ is the set of all conic planes of a Veronesean $\mathcal{V}_2^4$. Hence the union of the planes of $\mathcal{L}$ is the variety $\mathcal{M}_4^3$ defined in §25.1. □

The next result is a generalization of Theorem 25.2.12.

Theorem 25.2.14: *Let $\mathcal{F}$ be a set of k planes of $PG(m, q)$, q odd, with the following properties:*

(i)″ *$k \geq q^2 + q + 1$, $m \geq 5$, and the elements of $\mathcal{F}$ generate $PG(m, q)$;*
(ii)′ *any two distinct elements of $\mathcal{F}$ have a non-empty intersection;*
(iii) *any three distinct elements of $\mathcal{F}$ have an empty intersection.*

Then $k = q^2 + q + 1$, $m = 5$, and $\mathcal{F}$ consists of all tangent planes of the Veronesean $\mathcal{V}_2^4$.

Proof: By Theorem 25.2.12 it must be shown that $k = q^2 + q + 1$ and $m = 5$, and that any two distinct elements of $\mathcal{F}$ meet in exactly one point.

We begin with the latter. Let $\pi, \pi' \in \mathcal{F}$, $\pi \neq \pi'$, and assume that $\pi \cap \pi'$ is a line l. If $\pi'' \in \mathcal{F}\backslash\{\pi, \pi'\}$ and $\pi \cap \pi''$ is also a line, then $\pi \cap \pi' \cap \pi''$ is non-empty, contradicting (iii). Hence $\pi \cap \pi'' = \{Q\}$, with Q a point not on the line l. Since distinct planes π'' define distinct points Q, the number of such planes π'' is at most q^2. Hence $k - 2 \leq q^2$, contradicting $k \geq q^2 + q + 1$. Consequently any two distinct planes of $\mathcal{F}$ have exactly one point in common.

Let π and π' be distinct planes of $\mathscr{F}$, and let $\pi \cap \pi' = \{Q''\}$. The four-dimensional space generated by π and π' is denoted by Π_4. Further, let π'' be a plane of $\mathscr{F}$ which is not contained in Π_4. The points $\pi'' \cap \pi$ and $\pi'' \cap \pi'$ are denoted by Q' and Q, respectively. By (iii) we have $Q \neq Q' \neq Q'' \neq Q$. Hence the space generated by Π_4 and π'' is a five-dimensional space Π_5. We now show that all planes of $\mathscr{F}$ are contained in Π_5.

So let $\pi^* \in \mathscr{F} \backslash \{\pi, \pi', \pi''\}$. None of the points Q, Q', Q'' is contained in π^*. Let $\pi \cap \pi^* = \{R\}$, $\pi' \cap \pi^* = \{R'\}$, $\pi'' \cap \pi^* = \{R''\}$. By (iii) the points R, R', R'' are distinct. If the points R, R', R'' are linearly independent, then the plane π^* belongs to Π_5. Suppose now that the points R, R', R'' lie on a common line l. If l does not meet the plane $QQ'Q''$, then the planes π, π', π'' belong to the four-dimensional space generated by $QQ'Q''$ and l, a contradiction. If the intersection of l and the plane $QQ'Q''$ is a point, then at least two of the planes π, π', π'' are contained in the solid generated by $QQ'Q''$ and l, contradicting the first part of the proof. Hence the line l is contained in the plane $QQ'Q''$. So for any plane $\pi^* \in \mathscr{F} \backslash \{\pi, \pi', \pi''\}$, the points $\pi^* \cap \pi$, $\pi^* \cap \pi'$, $\pi^* \cap \pi''$ are linearly independent or $\pi^* \cap QQ'Q''$ is a line l.

Assume that for any plane $\pi^* \in \mathscr{F} \backslash \{\pi, \pi', \pi''\}$ the intersection $\pi^* \cap QQ'Q''$ is a line l. By the first part of the proof we know that distinct planes π^* define distinct lines l, and by (iii) we know that any three distinct planes π^* define three non-concurrent lines l. Further, none of the lines QQ', $Q'Q''$, $Q''Q$ is concurrent with two lines l. Hence the $k - 3$ lines l, together with the lines QQ', $Q'Q''$, $Q''Q$, form a dual k-arc. By Theorem 8.1.3 we have $k \leq q + 1$, contradicting $k \geq q^2 + q + 1$. Consequently $\mathscr{F} \backslash \{\pi, \pi', \pi''\}$ contains at least one plane, say π^*, which intersects the respective planes π, π', π'' in linearly independent points R, R', R''. This plane π^* is contained in Π_5 and meets the plane $QQ'Q''$ in at most one point.

Assume that $QQ'Q'' \cap \pi^* = \{Q^*\}$ with $Q^* \notin QQ' \cup Q'Q'' \cup Q''Q$. The four-dimensional space generated by π^* and the plane $QQ'Q''$ is denoted by Π_4. Since none of the points R, R', R'' are in the set $QQ' \cup Q'Q'' \cup Q''Q$, the three planes π, π', π'' belong to Π_4, a contradiction. Hence if $QQ'Q'' \cap \pi^* = \{Q^*\}$, then $Q^* \in (QQ' \cup Q'Q'' \cup Q''Q) \backslash \{Q, Q', Q''\}$.

Let $\tilde{\pi}$ be any plane of $\mathscr{F} \backslash \{\pi, \pi', \pi''\}$, and assume that $\tilde{\pi} \not\subset \Pi_5$. Then $\tilde{\pi} \cap QQ'Q''$ is a line $\tilde{l}$. Let $\pi^* \cap \tilde{\pi} = \{\tilde{Q}\}$. If also $\{\tilde{Q}\} = \pi^* \cap QQ'Q''$, then $\tilde{Q} \in QQ' \cup Q'Q'' \cup Q''Q$; so $\tilde{Q}$ is contained in three planes of $\mathscr{F}$, a contradiction. Hence $\tilde{Q} \notin \pi^* \cap QQ'Q''$. If follows immediately that the plane $\tilde{\pi}$, which is generated by $\tilde{l}$ and $\tilde{Q}$, is contained in Π_5, a contradiction. We conclude that all planes of $\mathscr{F}$ are contained in Π_5; that is, $m = 5$.

Consider any plane π of $\mathscr{F}$. Each plane π' of $\mathscr{F} \backslash \{\pi\}$ has exactly one point Q in common with π and by (iii) distinct planes π' define distinct points Q. Hence $k - 1 \leq q^2 + q + 1$, and so $k = q^2 + q + 1$ or $k = q^2 + q + 2$. Assume that $k = q^2 + q + 2$. Since not all planes of $\mathscr{F} \backslash \{\pi\}$ are contained in a prime, the set $\mathscr{F} \backslash \{\pi\}$ satisfies the following properties:

(i) the elements of $\mathscr{F}\backslash\{\pi\}$ generate $PG(5, q)$;

(ii) any two distinct elements of $\mathscr{F}\backslash\{\pi\}$ have exactly one point in common;

(iii) any three distinct elements of $\mathscr{F}\backslash\{\pi\}$ have an empty intersection.

By Theorem 25.2.12 the set $\mathscr{F}\backslash\{\pi\}$ consists of all tangent planes of the Veronesean $\mathscr{V}_2^4$. Let π' be any plane of $\mathscr{F}\backslash\{\pi\}$. Then $\pi \cap \pi'$ is contained in no plane of $\mathscr{F}\backslash\{\pi, \pi'\}$; so $\pi \cap \pi'$ is the contact point of the tangent plane π' of $\mathscr{V}_2^4$. Since any point of π is contained in a plane $\pi' \in \mathscr{F}\backslash\{\pi\}$, we have $\pi = \mathscr{V}_2^4$, a contradiction. Consequently $k = q^2 + q + 1$.

Now Theorem 25.2.12 asserts that $\mathscr{F}$ consists of all tangent planes of the Veronesean $\mathscr{V}_2^4$. □

The characterization of the set of conic planes of $\mathscr{V}_2^4$ given in Theorem 25.2.13 can be strengthened.

Let $\mathscr{L}$ with $|\mathscr{L}| = k$ be any set of $(m-3)$-dimensional subspaces of $PG(m, q)$, q odd, with the following properties:

(i)‴ $k \geq q^2 + q + 1$, $m \geq 5$, and there is no point belonging to all elements of $\mathscr{L}$;

(ii)″ any two distinct elements of $\mathscr{L}$ belong to a common prime;

(iii)′ any three distinct elements of $\mathscr{L}$ generate $PG(m, q)$.

Theorem 25.2.15: *If $\mathscr{L}$ is any set satisfying* (i)‴, (ii)″, *and* (iii)′, *then $k = q^2 + q + 1$, $m = 5$, and $\mathscr{L}$ is the set of all conic planes of a Veronesean $\mathscr{V}_2^4$. Hence the union of the elements of $\mathscr{L}$ is the variety $\mathscr{M}_4^3$.*

Proof: Let δ be any correlation of $PG(m, q)$. Then the set of planes $\mathscr{L}\delta$ satisfies the properties (i)″, (ii)′, and (iii). By Theorem 25.2.14 we have $m = 5$, $k = q^2 + q + 1$, and $\mathscr{L}\delta$ consists of all tangent planes of the Veronesean of conics. Now by Theorem 25.1.18, $\mathscr{L}$ is the set of all conic planes of the Veronesean $\mathscr{V}_2^4$. Hence the union of the planes of $\mathscr{L}$ is the variety $\mathscr{M}_4^3$ defined in §25.1. □

Let us consider the Grassmannian $\mathscr{G}_{1,3}$ of the lines of $PG(3, q)$, q even or odd. If π is any plane of $\mathscr{G}_{1,3}$, then there are exactly $q^2 + q + 1$ planes $\pi^0, \pi^1, \ldots, \pi^{q^2+q}$ on $\mathscr{G}_{1,3}$ which intersect π in a line. These planes π^i generate $PG(5, q)$, and have in pairs exactly one point in common. If q is odd, then by the corollary of Theorem 25.1.18 the planes π^i are not the tangent planes of a Veronesean $\mathscr{V}_2^4$. Next, let q be even. Suppose that the planes π^i are the tangent planes of a Veronesean $\mathscr{V}_2^4$. Then $\mathscr{V}_2^4$ is a subset of $\mathscr{G}_{1,3}$. Let $\bar{\pi}$ be any conic plane of $\mathscr{V}_2^4$. Since $\bar{\pi}$ has a line in common with $q + 1$ tangent planes of $\mathscr{V}_2^4$, that is with $q + 1$ planes π^i, it is clear that $\bar{\pi}$ is a plane of $\mathscr{G}_{1,3}$. Any two points of $\mathscr{V}_2^4$ belong to a conic plane; hence any two points of $\mathscr{V}_2^4$

belong to a plane of $\mathcal{G}_{1,3}$. It follows that $\mathcal{V}_2^4$ belongs to a projective subspace of $\mathcal{G}_{1,3}$. Consequently $\mathcal{V}_2^4$ is a plane of $\mathcal{G}_{1,3}$, a contradiction. We conclude that also for q even, the planes π^i are not the tangent planes of some Veronesean $\mathcal{V}_2^4$.

Let us again consider the Grassmannian $\mathcal{G}_{1,3}$, q even or odd. If $\mathcal{K}$ is a k-cap or dual k-cap of $PG(3, q)$, then to the k elements of $\mathcal{K}$ there correspond k planes of $\mathcal{G}_{1,3}$. It is straightforward to show that for $k > 2$ the set $\mathcal{F}$ consisting of these k planes satisfies the conditions (i), (ii), and (iii) stated at the beginning of §25.2. By Theorems 16.1.1 and 16.1.4 we have $k \leq q^2 + 1$ for $q \neq 2$. Conversely, any set of k planes on $\mathcal{G}_{1,3}$ satisfying conditions (i), (ii), and (iii) can be obtained in this way.

25.3 The Veronesean $\mathcal{V}_2^4$ characterized by its number of common points with the planes and primes of $PG(5, q)$, for q odd

Since the hyperplane sections of $\mathcal{V}_2^4$ correspond to the quadrics of $PG(2, q)$, any prime Π_4 of $PG(5, q)$ has 1, $q + 1$, or $2q + 1$ points in common with $\mathcal{V}_2^4$. We now consider the intersections of $\mathcal{V}_2^4$ with the planes of $PG(5, q)$.

Lemma 25.3.1: *Any plane π of $PG(5, q)$ meets $\mathcal{V}_2^4$ in 0, 1, 2, 3, or $q + 1$ points.*

Proof: Suppose that the plane π contains at least four distinct points Q_1, Q_2, Q_3, Q_4 of $\mathcal{V}_2^4$. By Corollary 1 of Theorem 25.1.9, the points Q_i and Q_j, $i \neq j$, are contained in a unique conic of $\mathcal{V}_2^4$. Let $\mathcal{C}'$ be the conic defined by Q_1 and Q_2, and let $\mathcal{C}''$ be the conic defined by Q_2 and Q_3. Suppose that $\mathcal{C}' \neq \mathcal{C}''$. By Theorem 25.1.11 the conic planes π' and π'' containing $\mathcal{C}'$ and $\mathcal{C}''$ generate a prime Π_4. With the notation of §25.1 the set $(\Pi_4 \cap \mathcal{V}_2^4)\zeta^{-1}$ is a quadric of $PG(2, q)$. Hence $|\Pi_4 \cap \mathcal{V}_2^4| \leq 2q + 1$. Since $\pi \subset \Pi_4$, so $Q_4 \in \Pi_4$; since also $\mathcal{C}' \cup \mathcal{C}'' \subset \Pi_4$, it follows that $|\Pi_4 \cap \mathcal{V}_2^4| \geq |\mathcal{C}' \cup \mathcal{C}'' \cup \{Q_4\}| = 2q + 2$, a contradiction. Consequently $\mathcal{C}' = \mathcal{C}''$, and so $\mathcal{C}' = \pi \cap \mathcal{V}_2^4$. It follows that $|\pi \cap \mathcal{V}_2^4| = q + 1$. □

It is our purpose to characterize, in the odd case, the Veronesean $\mathcal{V}_2^4$ by the number of its common points with the primes and planes of $PG(5, q)$.

From now on let $\mathcal{K}$ be a set of k points of $PG(m, q)$, $m \geq 5$, with the following properties:

(i) $|\Pi_4 \cap \mathcal{K}| \leq 2q + 1$ for any four-dimensional subspace Π_4 of $PG(m, q)$ with equality for some Π_4;

(ii) any plane of $PG(m, q)$ meeting $\mathcal{K}$ in four points meets it in at least $q + 1$ points.

Lemma 25.3.2: *For any line l, either $l \subset \mathcal{K}$ or $|l \cap \mathcal{K}| \leq 3$.*

Proof: The lemma is trivial for $q \leq 3$. So we assume that $q \geq 4$. Let l be a line of $PG(m, q)$, where $l \not\subset \mathscr{K}$ and $|l \cap \mathscr{K}| = s$. Suppose that $4 \leq s \leq q$. Let Π_4 be a four-dimensional space containing the line l. By (ii) any plane π of Π_4 containing l has at least $q + 1$ points in common with $\mathscr{K}$. Consequently

$$|\Pi_4 \cap \mathscr{K}| \geq (q^2 + q + 1)(q + 1 - s) + s \geq q^2 + q + 1 + s \geq q^2 + q + 5.$$

By (i) we have $|\Pi_4 \cap \mathscr{K}| \leq 2q + 1$. So $q^2 + q + 5 \leq 2q + 1$, a contradiction. □

Lemma 25.3.3: *For the set $\mathscr{K}$ with $q \geq 5$, there is no pair (l_1, l_2) of skew lines in the following cases:* (a) *l_1 and l_2 are both lines of $\mathscr{K}$*; (b) *l_1 is a line of $\mathscr{K}$ and l_2 is a trisecant of $\mathscr{K}$*; (c) *l_1 and l_2 are both trisecants of $\mathscr{K}$.*

Proof: (a) Let l_1 and l_2 be distinct skew lines belonging to $\mathscr{K}$. Let Π_3 be the solid defined by l_1 and l_2. So $|\Pi_3 \cap \mathscr{K}| \geq 2q + 2$. Consequently any four-dimensional space of $PG(m, q)$ containing Π_3 has more than $2q + 1$ points in common with $\mathscr{K}$. This contradicts condition (i).

(b), (c) Here l_2 is a trisecant and l_1 is a trisecant or a line of $\mathscr{K}$. Then there exist distinct planes π, π', π'' containing l_2 and also a point of $l_1 \cap \mathscr{K}$. By (ii) we have $|\pi \cap \mathscr{K}| \geq q + 1$, $|\pi' \cap \mathscr{K}| \geq q + 1$, and $|\pi'' \cap \mathscr{K}| \geq q + 1$. Hence the solid $\Pi_3 = l_1 l_2$ contains at least $3(q + 1 - 3) + 3 = 3q - 3$ points of $\mathscr{K}$. Consequently any four-dimensional subspace of $PG(m, q)$ through Π_3 has at least $3q - 3$ points in $\mathscr{K}$. Since $q \geq 5$, we have $3q - 3 > 2q + 1$, contradicting condition (i). □

Lemma 25.3.4: *If $\mathscr{K}$ contains distinct lines l_1 and l_2, then $l_1 \cap l_2 \neq \varnothing$ and $\mathscr{K} = l_1 \cup l_2$.*

Proof: By the first part of the proof of Lemma 25.3.3 we know that $l_1 \cap l_2 \neq \varnothing$. Let $Q \in \mathscr{K}$, with $Q \notin l_1 \cup l_2$. Then any four-dimensional subspace containing l_1, l_2, and Q has at least $2q + 2$ points in common with $\mathscr{K}$, contradicting (i). Consequently $\mathscr{K} = l_1 \cup l_2$. □

Lemma 25.3.5: *If $\mathscr{K}$ lies in a plane π of $PG(m, q)$, then $|\mathscr{K}| = 2q + 1$ and there are the following possibilities*:

(a) *$|l \cap \mathscr{K}| \leq 3$ for any line l in π with equality for some line l*;
(b) *the set $\mathscr{K}$ consists of a line l and a q-arc $\mathscr{K}^*$ of π, where $l \cap \mathscr{K}^* = \varnothing$*;
(c) *the set $\mathscr{K}$ consists of two distinct lines of π.*

Proof: Any four-dimensional subspace containing π has at most $2q + 1$ points in $\mathscr{K}$. As $\mathscr{K} \subset \pi$, it follows that $|\mathscr{K}| \leq 2q + 1$. Since at least one four-dimensional subspace has exactly $2q + 1$ points in $\mathscr{K}$, we have $|\mathscr{K}| = 2q + 1$.

First, let us assume that $\mathscr{K}$ does not contain a line. By Lemma 25.3.2 any line of π has at most three distinct points in $\mathscr{K}$. If there is no line of π having exactly three points in $\mathscr{K}$, then $\mathscr{K}$ is a $(2q + 1)$-arc, contradicting Theorem 8.1.3. So π contains at least one trisecant of $\mathscr{K}$.

Next, let l be a line which lies on $\mathscr{K}$. If the q points of $\mathscr{K} \backslash l$ belong to a line l', then $\mathscr{K} = l \cup l'$ by Lemma 25.3.4. Assume now that there is no line containing $\mathscr{K} \backslash l$. Then by Lemma 25.3.2 the set $\mathscr{K} \backslash l$ is a q-arc of π. □

Lemma 25.3.6: *Let $q \geq 5$ and suppose that $\mathscr{K}$ generates a solid Π_3. Then $\mathscr{K} = l \cup \mathscr{K}^*$, with $\mathscr{K}^*$ a q-arc of some plane π, l a line not contained in π, and $l \cap \mathscr{K}^* = \varnothing$.*

Proof: Any four-dimensional subspace containing Π_3 has at most $2q + 1$ points in $\mathscr{K}$. As $\mathscr{K} \subset \Pi_3$, so $|\mathscr{K}| \leq 2q + 1$. Since at least one four-dimensional subspace has exactly $2q + 1$ points in $\mathscr{K}$, we have $|\mathscr{K}| = 2q + 1$.

Two cases are distinguished.

(a) *$\mathscr{K}$ does not contain a line*

By Lemma 25.3.2 any line has at most three points in $\mathscr{K}$.

First, suppose that $\mathscr{K}$ has at least one trisecant l. Let $Q \in \mathscr{K} \backslash l$. By condition (ii) the plane $\pi = Ql$ has at least $q + 1$ points in $\mathscr{K}$. Since $\mathscr{K}$ generates the solid Π_3, there is at least one point Q' in $\mathscr{K} \backslash \pi$. By (ii) the plane $\pi' = Q'l$ has at least $q + 1$ points in $\mathscr{K}$. If $Q'' \in \mathscr{K} \backslash (\pi \cup \pi')$, then the plane $\pi'' = Q''l$ has at least $q + 1$ points in $\mathscr{K}$; so $|\mathscr{K}| \geq 3(q + 1 - 3) + 3 = 3q - 3$. Hence $2q + 1 \geq 3q - 3$, whence $q \leq 4$, a contradiction. Consequently $\mathscr{K} \subset \pi \cup \pi'$. Suppose now that $\mathscr{K}$ has a trisecant l', with $l' \neq l$. Then l' lies in π or π', say $l' \subset \pi$. The plane $\pi^* = Q'l'$ has at least $q + 1$ points in $\mathscr{K}$, three of which at most belong to π'. Hence at least $q + 1 - 6$ points of $\pi^* \cap \mathscr{K}$ do not belong to $\pi \cup \pi'$. Since $\mathscr{K} \subset \pi \cup \pi'$, we necessarily have $q \leq 5$. Since it was assumed that $q \geq 5$, so $q = 5$. By the same argument the line $l'' = \pi^* \cap \pi'$ is a trisecant of $\mathscr{K}$, and the point $l' \cap l''$ of l is not in $\mathscr{K}$. On the lines l, l', l'' there are nine points of $\mathscr{K}$. Since $|\mathscr{K}| = 11$, there is a point $Q^* \in \mathscr{K} \backslash (l \cup l' \cup l'')$, say $Q^* \in \pi$. Since $q = 5$ at least one of the lines joining Q^* to a point of $l' \cap \mathscr{K}$ has a point in $l \cap \mathscr{K}$. Such a line l^* is a trisecant of $\mathscr{K}$. Then the plane $Q'l^*$ contains at least one point of $\mathscr{K} \backslash (\pi \cup \pi')$, a contradiction. So we have proved that l is the only trisecant of $\mathscr{K}$. Let $l \cap \mathscr{K} = \{Q_1, Q_2, Q_3\}$. Clearly $(\pi \cap \mathscr{K}) \backslash \{Q_i\}$ is a k_i-arc of the plane π, for all $i = 1, 2, 3$. Analogously $(\pi' \cap \mathscr{K}) \backslash \{Q_i\}$ is a k_i'-arc of the plane

π'. Since $|\mathscr{K}| = 2q + 1$ and $\mathscr{K} \subset \pi \cup \pi'$, we have $(k_i - 2) + (k_i' - 2) + 3 = 2q + 1$, for all $i = 1, 2, 3$. Hence $k_i + k_i' = 2q + 2$. Without loss of generality, let $k_1 \geq q + 1$. By §8.1 the k_1-arc $(\pi \cap \mathscr{K})\backslash\{Q_1\}$ has at least $(q-1)/2$ bisecants passing through Q_1. It follows that the plane π contains at least two trisecants of $\mathscr{K}$ through Q_1, a contradiction.

It has been shown that $\mathscr{K}$ has no trisecant; that is, $\mathscr{K}$ is a $(2q+1)$-cap of the solid Π_3. Assume that no four points of $\mathscr{K}$ are coplanar. Then $\mathscr{K}$ is a $(2q+1)$-arc of Π_3. By Theorems 21.2.4 and 21.3.8 we have $|\mathscr{K}| \leq q + 1$, which contradicts $|\mathscr{K}| = 2q + 1$. Hence Π_3 contains a plane π which has at least four points in common with $\mathscr{K}$. By condition (ii), $|\pi \cap \mathscr{K}| \geq q + 1$. Since $\pi \cap \mathscr{K}$ is a k'-arc of π, we have $|\pi \cap \mathscr{K}| \in \{q+1, q+2\}$; hence $|\mathscr{K}\backslash\pi| \geq q - 1$. Let l be a line containing at least two distinct points of $\mathscr{K}\backslash\pi$. Since $\mathscr{K}$ is a cap, the common point of l and π does not belong to $\mathscr{K}$. If $\pi \cap \mathscr{K}$ is a $(q+1)$-arc of π and q is even, then, since $q - 1 > 2$, we may assume without loss of generality that l does not contain the nucleus of $\pi \cap \mathscr{K}$. Since $q \geq 5$ the point $l \cap \pi$ is on at least two bisecants l_1 and l_2 of $\pi \cap \mathscr{K}$. By (ii) the planes ll_1 and ll_2 have at least $q + 1$ points in $\mathscr{K}$. Hence $|\mathscr{K}| \geq 2(q + 1 - 2) + 2 + (q + 1 - 4) = 3q - 3$. So $2q + 1 \geq 3q - 3$, whence $q \leq 4$, a contradiction. We conclude that case (a) cannot occur.

(b) *$\mathscr{K}$ contains a line l*

Since $\mathscr{K}$ generates the solid Π_3, Lemma 25.3.4 tells us that l is the only line on $\mathscr{K}$. Assume that $\mathscr{K}\backslash l$ contains distinct collinear points Q_1, Q_2, Q_3. By Lemma 25.3.2 the line Q_1Q_2 is a trisecant of $\mathscr{K}$, and by Lemma 25.3.3 the line Q_1Q_2 has a point on the line l. Consequently the line Q_1Q_2 has at least four distinct points in $\mathscr{K}$, a contradiction.

Hence $\mathscr{K}\backslash l$ is a q-cap of Π_3. On $\mathscr{K}\backslash l$ the distinct points P_1, P_2, P_3 can be chosen such that the plane $\pi = P_1P_2P_3$ does not contain l. Since P_1, P_2, P_3, and $\pi \cap l$ are four distinct points of $\mathscr{K}$, the plane π contains at least $q + 1$ points of $\mathscr{K}$. But $|\mathscr{K}| = 2q + 1$, whence $|\pi \cap \mathscr{K}| = q + 1$ and $\mathscr{K}$ consists of l and the q-arc $\mathscr{K}^* = (\pi \cap \mathscr{K})\backslash l$. □

Lemma 25.3.7: *Let $q \geq 5$ and suppose that $\mathscr{K}$ generates a $PG(s, q)$ with $s \geq 4$. Then any line l which is not contained in $\mathscr{K}$ has at most two points in $\mathscr{K}$.*

Proof: Let l be a line which is not contained in $\mathscr{K}$ and suppose that $|l \cap \mathscr{K}| > 2$. By Lemma 25.3.2 we have $|l \cap \mathscr{K}| = 3$. Since $\mathscr{K}$ generates a $PG(s, q)$ with $s \geq 4$, there are points P_1, P_2, P_3 in $\mathscr{K}$ such that l, P_1, P_2, and P_3 generate a $PG(4, q)$. Let Π_2^i be the plane containing l and P_i, $i = 1, 2, 3$. Since $|\Pi_2^i \cap \mathscr{K}| \geq 4$, we have $|\Pi_2^i \cap \mathscr{K}| \geq q + 1$. Hence $|PG(4, q) \cap \mathscr{K}| \geq 3q - 3$. By (i) it follows that $3q - 3 \leq 2q + 1$; so $q \leq 4$, a contradiction. □

Corollary: *Let $q \geq 5$ and suppose that $\mathcal{K}$ generates a $PG(s, q)$ with $s \geq 4$. Then $\mathcal{K}$ is a k-cap, or $\mathcal{K}$ contains a unique line l and any other line has at most two points in common with $\mathcal{K}$.*

Proof: This follows immediately from Lemmas 25.3.3, 25.3.4, and 25.3.7. □

From now on let $\mathcal{K}$ be a set of k points of $PG(5, q)$ satisfying the following:

(i)′ $|\Pi_4 \cap \mathcal{K}| = 1, q + 1$, or $2q + 1$ for any prime Π_4 of $PG(5, q)$ and $|\Pi_4 \cap \mathcal{K}| = 2q + 1$ for some prime Π_4;
(ii) any plane of $PG(5, q)$ with four points in $\mathcal{K}$ has at least $q + 1$ points in $\mathcal{K}$.

Lemma 25.3.8: *For $q \geq 5$ any set $\mathcal{K}$* satisfying (i)′ *and* (ii) *is a $(q^2 + q + 1)$-cap which generates $PG(5, q)$.*

Proof: By (i)′ the set $\mathcal{K}$ does not generate a line. Assume that $\mathcal{K}$ generates a plane Π_2. By Lemma 25.3.5 there is a line l of Π_2 with $|l \cap \mathcal{K}| \in \{2, 3\}$. Let Π_4 be a prime of $PG(5, q)$ which contains l but not Π_2. Then $|\Pi_4 \cap \mathcal{K}| \in \{2, 3\}$, contradicting (i)′. Next, assume that $\mathcal{K}$ generates a solid Π_3. By Lemma 25.3.6 we have $\mathcal{K} = l \cup \mathcal{K}^*$, with $\mathcal{K}^*$ a q-arc of some plane Π_2, l a line not contained in Π_2, and $l \cap \mathcal{K}^* = \emptyset$. Let Π_2' be a plane containing two points of $\mathcal{K}^*$ and one point of $l \backslash \Pi_2$; then $|\Pi_2' \cap \mathcal{K}| = 3$. Hence any prime containing Π_2' but not Π_3 has exactly three points in $\mathcal{K}$, contradicting (i)′. Finally, assume that $\mathcal{K}$ generates a prime Π_4. By (i)′ we have $|\mathcal{K}| = 2q + 1$ and each solid Π_3 of Π_4 has 1 or $q + 1$ points in $\mathcal{K}$. Let l be a line having at least two points in $\mathcal{K}$, and let Π_2 be a plane of Π_4 containing l. Further, let $|l \cap \mathcal{K}| = a$ and $|\Pi_2 \cap \mathcal{K}| = b$. Counting the points of $\mathcal{K}$ in the solids of Π_4 containing Π_2, we obtain

$$(q + 1 - b)(q + 1) + b = 2q + 1.$$

Hence $b = q$. Counting the points of $\mathcal{K}$ in the planes of Π_4 containing l, we obtain

$$(q - a)(q^2 + q + 1) + a = 2q + 1.$$

Hence

$$q^3 + q^2 - aq^2 - aq - q = 1.$$

Consequently q divides 1, a contradiction. So it has been proved that $\mathcal{K}$ generates $PG(5, q)$.

Now let us show that $\mathcal{K}$ is a k-cap. By Lemma 25.3.7 it is sufficient to prove that $\mathcal{K}$ does not contain a line. Hence assume that $\mathcal{K}$ contains some line l. By the corollary of Lemma 25.3.7 any plane through l has at most one point in $\mathcal{K} \backslash l$. Let Π_3 be a solid skew to l. By projecting $\mathcal{K} \backslash l$ from l

onto Π_3 we obtain a set $\mathscr{K}'$ of Π_3 of size $k - (q + 1)$. By (i)′ any plane of Π_3 has 0 or q points in $\mathscr{K}'$. Let $\Pi_2^1, \Pi_2^2, \ldots$ be the planes of Π_3, and let t_i be the number of points of $\mathscr{K}'$ in Π_2^i. Counting in two ways the number of ordered pairs (P, Π_2^i), with $P \in \mathscr{K}'$ and $P \in \Pi_2^i$, we obtain

$$\sum t_i = (k - q - 1)(q^2 + q + 1). \tag{25.7}$$

Counting in two ways the number of ordered triples (P, P', Π_2^i), with $P \in \mathscr{K}'$, $P' \in \mathscr{K}'$, $P \neq P'$, $P \in \Pi_2^i$, and $P' \in \Pi_2^i$, we obtain

$$\sum t_i(t_i - 1) = (k - q - 1)(k - q - 2)(q + 1). \tag{25.8}$$

Since $t_i \in \{0, q\}$ for all i, we have $\sum t_i(t_i - q) = 0$. Hence

$$\sum t_i(t_i - 1) - (q - 1) \sum t_i = 0. \tag{25.9}$$

By (25.7), (25.8), and (25.9),

$$(k - q - 1)(k - q - 2)(q + 1) - (q - 1)(k - q - 1)(q^2 + q + 1) = 0.$$

Since $k \neq q + 1$, this yields

$$(k - q - 2)(q + 1) - (q - 1)(q^2 + q + 1) = 0.$$

Hence

$$k = (q^3 + q^2 + 3q + 1)/(q + 1).$$

Consequently $q + 1$ divides $q^3 + q^2 + 3q + 1 = q^2(q + 1) + 3(q + 1) - 2$; that is, $q + 1$ divides 2, a contradiction. So we have proved that $\mathscr{K}$ is a k-cap.

Finally, it must be shown that $k = q^2 + q + 1$. Let $\Pi_4^1, \Pi_4^2, \ldots$ be the primes of $PG(5, q)$, and let s_i be the number of points of $\mathscr{K}$ in Π_4^i. Counting in two ways the number of ordered pairs (P, Π_4^i), with $P \in \mathscr{K}$ and $P \in \Pi_4^i$, we obtain

$$\sum s_i = k(q^4 + q^3 + q^2 + q + 1). \tag{25.10}$$

Counting in two ways the number of ordered triples (P, P', Π_4^i), with P, $P' \in \mathscr{K}$, P, $P' \in \Pi_4^i$, and $P \neq P'$, we obtain

$$\sum s_i(s_i - 1) = k(k - 1)(q^3 + q^2 + q + 1). \tag{25.11}$$

The set $\mathscr{K}$ is a k-cap; so counting in two ways the number of ordered 4-tuples (P, P', P'', Π_4^i), with $P, P', P'' \in \mathscr{K}$, $P, P', P'' \in \Pi_4^i$, and $P \neq P' \neq P'' \neq P$, we obtain

$$\sum s_i(s_i - 1)(s_i - 2) = k(k - 1)(k - 2)(q^2 + q + 1). \tag{25.12}$$

Since $s_i \in \{1, q + 1, 2q + 1\}$ for all i,

$$\sum (s_i - 1)(s_i - q - 1)(s_i - 2q - 1) = 0, \tag{25.13}$$

which expands to

$$\sum s_i(s_i - 1)(s_i - 2) - 3q \sum s_i(s_i - 1) \\ + (q+1)(2q+1) \sum s_i - (q+1)(2q+1)\theta(5) = 0. \quad (25.14)$$

By (25.10), (25.11), (25.12), and (25.14) we have

$$k(k-1)(k-2)(q^2+q+1) - 3qk(k-1)(q^3+q^2+q+1) \\ + (q+1)(2q+1)k(q^4+q^3+q^2+q+1) - (q+1)(2q+1)\theta(5) = 0.$$

Hence

$$(q^2+q+1)k^3 - 3(q^4+q^3+2q^2+2q+1)k^2 \\ + (2q^6+5q^5+9q^4+9q^3+11q^2+9q+3)k \\ - (2q^7+5q^6+6q^5+6q^4+6q^3+6q^2+4q+1) = 0.$$

It follows that, if $k \neq q^2+q+1$, then

$$(q^2+q+1)k^2 - (2q^4+q^3+3q^2+4q+2)k \\ + (2q^5+3q^4+q^3+2q^2+3q+1) = 0. \quad (25.15)$$

If for at least one prime Π_4^i we have $s_i = 1$, then there exists at least one solid having exactly one point in $\mathscr{K}$. Now suppose that there is at least one prime Π_4^i with $s_i = q+1$. If $P \in \Pi_4^i \cap \mathscr{K}$, then it is easy to show that we successively can find a line l, a plane Π_2, and a solid Π_3 of Π_4^i, with $P \in l \subset \Pi_2 \subset \Pi_3 \subset \Pi_4^i$ and $|l \cap \mathscr{K}| = |\Pi_2 \cap \mathscr{K}| = |\Pi_3 \cap \mathscr{K}| = 1$. So also in this case there exists a solid having exactly one point in $\mathscr{K}$. Next, assume that for all primes Π_4^i we have $s_i = 2q+1$. Then by (25.10),

$$\theta(5)(2q+1) = k\theta(4).$$

Hence

$$k = q(2q+1) + (2q+1)/\theta(4).$$

It follows that $\theta(4) \leq 2q+1$, a contradiction. So we conclude that there is always a solid Π_3 having exactly one point P in common with $\mathscr{K}$. Now counting the points of $\mathscr{K}$ in all primes containing Π_3, we see that

$$k \equiv 1 \pmod{q}. \quad (25.16)$$

Assume that $k \neq q^2+q+1$; that is, k satisfies (25.15). Let

$$F(x) = (q^2+q+1)x^2 - (2q^4+q^3+3q^2+4q+2)x \\ + (2q^5+3q^4+q^3+2q^2+3q+1).$$

Then

$$F(q+1) = q^4 - q^2 > 0 \quad \text{and} \quad F(q+2) = -q^4 + q^3 + q^2 + q + 1 < 0.$$

Consequently $F(x)$ has a root $k' = q + a$, with $1 < a < 2$. Since $k \geq 2q + 1$, we have $k \neq k'$. The sum of the roots of $F(x)$ is equal to

$$k + k' = k + q + a = (2q^4 + q^3 + 3q^2 + 4q + 2)/(q^2 + q + 1).$$

Hence

$$k = 2q^2 - 2q + 2 - a + 3q/(q^2 + q + 1). \tag{25.17}$$

By (25.16), q divides $2q^2 - 2q + 1 - a + 3q/(q^2 + q + 1)$. Hence q divides $1 - a + 3q/(q^2 + q + 1)$. Since $1 < a < 2$, we have

$$|1 - a + 3q/(q^2 + q + 1)| < q,$$

and so

$$1 - a + 3q/(q^2 + q + 1) = 0. \tag{25.18}$$

Now by (25.17) and (25.18),

$$k = 2q^2 - 2q + 1.$$

Hence

$$F(2q^2 - 2q + 1) = q^2(q - 2)(q - 4) = 0,$$

a contradiction since $q \geq 5$.

We conclude that $k = q^2 + q + 1$. □

Lemma 25.3.9: *In $PG(5, 3)$ any set $\mathscr{K}$ satisfying* (i)′ *is a 13-cap which generates $PG(5, 3)$.*

Proof: By (i)′ the set $\mathscr{K}$ does not generate a line. Assume that $\mathscr{K}$ generates a plane Π_2. By Lemma 25.3.5 there is a line l of Π_2 with $|l \cap \mathscr{K}| \in \{2, 3\}$. Let Π_4 be a prime of $PG(5, 3)$ which contains l but not Π_2. Then $|\Pi_4 \cap \mathscr{K}| \in \{2, 3\}$, contradicting (i)′. Next, assume that $\mathscr{K}$ generates a solid Π_3. Then $|\mathscr{K}| = 7$ and each plane of Π_3 has one or four points in $\mathscr{K}$. Let P and P' be distinct points of $\mathscr{K}$. The line PP' has b points in $\mathscr{K}$. Counting the points of $\mathscr{K}$ in the planes of Π_3 through the line $PP' = l$, we obtain $4(4 - b) + b = 7$, whence $b = 3$. Let $l \cap \mathscr{K} = \{P, P', P''\}$ and let $\Pi_2 \cap \mathscr{K} = \{P, P', P'', P'''\}$ with Π_2 some plane of Π_3 through l. Then the line PP''' has only two points in $\mathscr{K}$, a contradiction. Finally, assume that $\mathscr{K}$ generates a prime Π_4. By (i)′ we have $|\mathscr{K}| = 7$ and each solid Π_3 of Π_4 has one or four points in $\mathscr{K}$. Let l be a line having at least two points in $\mathscr{K}$, and let Π_2 be a plane of Π_4 containing l. Further, let $|l \cap \mathscr{K}| = a$ and $|\Pi_2 \cap \mathscr{K}| = b$. Counting the points of $\mathscr{K}$ in the solids of Π_4 containing Π_2, we obtain

$4(4-b)+b=7$; hence $b=3$. Counting the points of $\mathscr{K}$ in the planes of Π_4 containing l, we obtain $13(3-a)+a=7$. Consequently $a=8/3$, a contradiction. So it has been shown that $\mathscr{K}$ generates $PG(5,3)$.

Now we show that $\mathscr{K}$ is a k-cap. First suppose that there is a line l which contains exactly three points P, P', P'' of $\mathscr{K}$. Let R_1, R_2, R_3 be points of $\mathscr{K}\backslash l$, such that the planes lR_1, lR_2, lR_3 generate a prime Π_4 of $PG(5,3)$; then $|\Pi_4 \cap \mathscr{K}| = 7$. Clearly $|lR_i \cap \mathscr{K}| \in \{4, 5\}$, with $i = 1, 2, 3$. If the plane lR_1 contains five points of $\mathscr{K}$, then counting the points of $\mathscr{K}$ in the primes through the solid lR_1R_2 we obtain $|\mathscr{K}| = 4\cdot(7-6)+6 = 10$. Counting the points of $\mathscr{K}$ in the primes through the solid lR_2R_3, we have $|\mathscr{K}| = 4\cdot(7-5)+5 = 13$. This contradiction shows that the plane lR_1 contains exactly four points of $\mathscr{K}$. Analogously the planes lR_2 and lR_3 contain exactly four points of $\mathscr{K}$. Let $(\mathscr{K}\cap\Pi_4)\backslash l = \{R_1, R_2, R_3, R_4\}$. By a previous argument we have $|\mathscr{K}\cap lR_1R_2| = |\mathscr{K}\cap lR_2R_3| = |\mathscr{K}\cap lR_1R_3|$. It follows that R_4 does not belong to any of the solids $lR_1R_2, lR_2R_3, lR_1R_3$. Counting the points of $\mathscr{K}$ in the primes through the solid lR_1R_2 gives $|\mathscr{K}| = 4\cdot(7-5)+5 = 13$. Let Π_3 be a solid skew to l and let $\mathscr{K}\backslash l = \mathscr{K}'$. Then $l\mathscr{K}'\cap\Pi_3 = \mathscr{K}''$ is a set of 10 points of Π_3. No three points of $\mathscr{K}''$ are collinear, and any plane containing at least three points of $\mathscr{K}''$ contains exactly four points of $\mathscr{K}''$. Hence $\mathscr{K}''$ is an ovaloid of Π_3. Now it is also clear that $\{P, P', P''\}$ is the only set consisting of three collinear points of $\mathscr{K}$.

Let $\Pi_4^1, \Pi_4^2, \ldots$ be the primes of $PG(5,3)$, and let s_i be the number of points of $\mathscr{K}$ in Π_4^i. Counting in two ways the number of ordered pairs (R, Π_4^i), with $R\in\mathscr{K}$ and $R\in\Pi_4^i$, we obtain

$$\sum s_i = 13\cdot 121 = 1573. \tag{25.19}$$

Counting in two ways the number of ordered triples (R, R', Π_4^i), with $R, R'\in\mathscr{K}$, $R, R'\in\Pi_4^i$, and $R\neq R'$, we obtain

$$\sum s_i(s_i-1) = 13\cdot 12\cdot 40 = 6240. \tag{25.20}$$

Counting in two ways the number of ordered 4-tuples (R, R', R'', Π_4^i), with $R, R', R''\in\mathscr{K}$, $R, R', R''\in\Pi_4^i$, and $R\neq R'\neq R''\neq R$, we obtain

$$\sum s_i(s_i-1)(s_i-2) = (13\cdot 12\cdot 11 - 6)\cdot 13 + 6\cdot 40 = 22\,470. \tag{25.21}$$

By (25.19), (25.20), and (25.21) we have

$$\sum s_i = 1573, \qquad \sum s_i^2 = 7813, \qquad \sum s_i^3 = 42\,763. \tag{25.22}$$

By (25.22),

$$\sum (s_i-1)(s_i-4)(s_i-7) = 162. \tag{25.23}$$

Since (i)′ is satisfied, we have $s_i\in\{1, 4, 7\}$ for all i, and so

$$\sum (s_i-1)(s_i-4)(s_i-7) = 0,$$

contradicting (25.23).

Hence there is no line which contains exactly three points of $\mathcal{K}$.

Assume now that there is a line l which is contained in $\mathcal{K}$. Let R_1, R_2, R_3 be points of $\mathcal{K}\backslash l$, such that the planes lR_1, lR_2, lR_3 generate a prime Π_4 of $PG(5, 3)$; then $|\Pi_4 \cap \mathcal{K}| = 7$. Hence the planes lR_1, lR_2, lR_3 contain exactly five points of $\mathcal{K}$. Also, the solids lR_1R_2, lR_2R_3, lR_1R_3 contain exactly six points of $\mathcal{K}$. Counting the points of $\mathcal{K}$ in the primes through the solid lR_1R_2, we obtain $|\mathcal{K}| = 4\cdot(7 - 6) + 6 = 10$. Let Π_3 be a solid skew to l and let $\mathcal{K}\backslash l = \mathcal{K}'$. Then $l\mathcal{K}' \cap \Pi_3 = \mathcal{K}''$ is a set of six points of Π_3. No four of these six points are coplanar. Hence $\mathcal{K}''$ is a 6-arc of Π_3. By Theorem 21.2.1 an arc of $PG(3, 3)$ has at most five points, giving a contradiction. Thus it has been proved that $\mathcal{K}$ is a k-cap.

Finally, it must be shown that $k = 13$. Let $\Pi_4^1, \Pi_4^2, \ldots$ be the primes of $PG(5, 3)$, and let s_i be the number of points of $\mathcal{K}$ in Π_4^i. Counting in two ways the number of ordered pairs (P, Π_4^i), with $P \in \mathcal{K}$ and $P \in \Pi_4^i$, we obtain

$$\sum s_i = 121k. \tag{25.24}$$

Counting in two ways the number of ordered triples (P, P', Π_4^i), with $P, P' \in \mathcal{K}$, $P, P' \in \Pi_4^i$, and $P \neq P'$, we obtain

$$\sum s_i(s_i - 1) = 40k(k - 1). \tag{25.25}$$

The set $\mathcal{K}$ is a k-cap; so counting in two ways the number of ordered 4-tuples (P, P', P'', Π_4^i), with $P, P', P'' \in \mathcal{K}$, $P, P', P'' \in \Pi_4^i$, and $P \neq P' \neq P'' \neq P$, we obtain

$$\sum s_i(s_i - 1)(s_i - 2) = 13k(k - 1)(k - 2). \tag{25.26}$$

Since $s_i \in \{1, 4, 7\}$ for all i, we have

$$\sum (s_i - 1)(s_i - 4)(s_i - 7) = 0. \tag{25.27}$$

By (25.27),

$$\sum s_i(s_i - 1)(s_i - 2) - 9\sum s_i(s_i - 1) + 28\sum s_i - 10\,192 = 0. \tag{25.28}$$

By (25.24), (25.25), (25.26), and (25.28) we have

$$13k(k - 1)(k - 2) - 360k(k - 1) + 3388k - 10\,192 = 0.$$

Hence

$$13k^3 - 399k^2 + 3774k - 10\,192 = 0.$$

It follows that $k = 13$ or $13k^2 - 230k + 784 = 0$.

Since no integer satisfies the equation $13x^2 - 230x + 784 = 0$, we have $k = 13$, completing the proof of the lemma. □

Lemma 25.3.10: *Any solid Π_3 of $PG(5, q)$, $q = 3$ or $q \geq 5$, meets $\mathcal{K}$ in at most $q + 2$ points.*

Proof: Let $|\Pi_3 \cap \mathscr{K}| = m$ and suppose that $m \geq q + 2$. Counting the points of $\mathscr{K}$ in the primes through Π_3 gives

$$k = (q+1)(2q+1-m) + m.$$

By Lemmas 25.3.8 and 25.3.9 we have $k = q^2 + q + 1$. Hence $m = q + 2$. □

Lemma 25.3.11: *When $q = 3$ or $q \geq 5$, suppose that the plane Π_2 meets $\mathscr{K}$ in more than three points. Then $\Pi_2 \cap \mathscr{K}$ is a $(q+1)$-arc and so, for q odd, is a conic.*

Proof: Let $|\Pi_2 \cap \mathscr{K}| = n$. By (ii) we have $n \geq q + 1$. Since $\mathscr{K}$ is a cap, $n \leq q + 2$ by Theorem 8.1.3. If $n = q + 2$, then any solid containing Π_2 and a point of $\mathscr{K} \backslash \Pi_2$ has at least $q + 3$ points in common with $\mathscr{K}$, contradicting Lemma 25.3.10. Hence $n = q + 1$, and $\Pi_2 \cap \mathscr{K}$ is a $(q+1)$-arc of Π_2. By Theorem 8.2.4, $\Pi_2 \cap \mathscr{K}$ is a conic when q is odd. □

Lemma 25.3.12: *For $q = 3$ or $q \geq 5$, any two distinct points of $\mathscr{K}$ are contained in a unique plane meeting $\mathscr{K}$ in a $(q+1)$-arc.*

Proof: Let P and P' be distinct points of $\mathscr{K}$. Suppose that no plane through the line PP' meets $\mathscr{K}$ in a $(q+1)$-arc. Then by Lemma 25.3.11 any plane through PP' has at most three points in $\mathscr{K}$. Now we project the set $\mathscr{K} \backslash \{P, P'\}$ from the line PP' onto a solid Π_3 skew to PP'. This gives a set $\mathscr{K}'$ of $q^2 + q - 1$ points in Π_3. Since any prime of $PG(5, q)$ through PP' meets $\mathscr{K}$ in $q + 1$ or $2q + 1$ points, any plane of Π_3 meets $\mathscr{K}'$ in $q - 1$ or $2q - 1$ points. Let $\Pi_2^1, \Pi_2^2, \ldots$ be the planes of Π_3, and let s_i be the number of points of $\mathscr{K}'$ in Π_2^i. Counting in two ways the number of ordered pairs (R, Π_2^i), with $R \in \mathscr{K}'$ and $R \in \Pi_2^i$, we obtain

$$\sum s_i = (q^2+q-1)(q^2+q+1). \tag{25.29}$$

Counting in two ways the number of ordered triples (R, R', Π_2^i), with $R, R' \in \mathscr{K}'$, $R, R' \in \Pi_2^i$, and $R \neq R'$, we obtain

$$\sum s_i(s_i - 1) = (q^2+q-1)(q^2+q-2)(q+1). \tag{25.30}$$

Since $s_i \in \{q-1, 2q-1\}$ for all i, we have

$$\sum (s_i - q + 1)(s_i - 2q + 1) = 0. \tag{25.31}$$

Hence

$$\sum s_i(s_i - 1) + 3(1-q)\sum s_i + (q-1)(2q-1)(q^3+q^2+q+1) = 0. \tag{25.32}$$

By (25.29), (25.30), and (25.32),

$$(q^2 + q - 1)(q^2 + q - 2)(q + 1) + 3(1 - q)(q^2 + q - 1)(q^2 + q + 1) + (q - 1)(2q - 1)(q^3 + q^2 + q + 1) = 0;$$

that is,

$$q^2(q - 1)(q - 2) = 0,$$

contradicting $q > 2$. Consequently there exists a plane through PP' which meets $\mathscr{K}$ in a $(q + 1)$-arc.

Now assume that PP' is contained in distinct planes Π_2 and Π'_2 meeting $\mathscr{K}$ in $(q + 1)$-arcs. Then the solid defined by Π_2 and Π'_2 has at least $2q$ points in common with $\mathscr{K}$, contradicting Lemma 25.3.10.

We conclude that the points P, P' are contained in a unique plane meeting $\mathscr{K}$ in a $(q + 1)$-arc. □

Lemma 25.3.13: *Let $q = 3$ or $q \geq 5$. The number of planes meeting $\mathscr{K}$ in a $(q + 1)$-arc is $q^2 + q + 1$, and any two distinct planes meeting $\mathscr{K}$ in a $(q + 1)$-arc have exactly one point in common.*

Proof: Let b be the number of planes meeting $\mathscr{K}$ in a $(q + 1)$-arc. By Lemma 25.3.12 we have $|\mathscr{K}|(|\mathscr{K}| - 1)/\{(q + 1)q\} = b$. Since $|\mathscr{K}| = q^2 + q + 1$, it follows that $b = q^2 + q + 1$. Let Π_2 and Π'_2 be distinct planes meeting $\mathscr{K}$ in a $(q + 1)$-arc. If Π_2 and Π'_2 have a line in common, then the solid defined by Π_2 and Π'_2 meets $\mathscr{K}$ in at least $2q$ points, contradicting Lemma 25.3.10. Now suppose that $\Pi_2 \cap \Pi'_2 = \varnothing$. For any point P in $\Pi_2 \cap \mathscr{K}$ and any point P' in $\Pi'_2 \cap \mathscr{K}$, there is exactly one plane containing PP' and meeting $\mathscr{K}$ in a $(q + 1)$-arc. Hence there are at least $2 + (q + 1)^2$ planes meeting $\mathscr{K}$ in a $(q + 1)$-arc. This contradicts the first part of the proof. So we conclude that Π_2 and Π'_2 have exactly one point in common. □

Theorem 25.3.14: *If $\mathscr{K}$ is a set of k points of $PG(5, q)$, q odd, which satisfies* (i)′ *and* (ii), *then $\mathscr{K}$ is a Veronesean $\mathscr{V}_2^4$.*

Proof: Let $\mathscr{L}$ be the set of all planes intersecting $\mathscr{K}$ in a $(q + 1)$-arc. By Lemma 25.3.13 we know that $|\mathscr{L}| = q^2 + q + 1$ and that any two distinct elements of $\mathscr{L}$ meet in exactly one point. Hence condition (ii) of §25.2 is satisfied.

Now we show that condition (i)′ of §25.2 is also satisfied. Let $\Pi_2 \in \mathscr{L}$, let $P \in \Pi_2$, let $P' \in \Pi_2 \cap \mathscr{K}$ with $P \neq P'$, and let $P'' \in \mathscr{K} \backslash \Pi_2$. Then the element Π'_2 of $\mathscr{L}$ containing P' and P'' has only the point P' in common with Π_2. Hence $P \notin \Pi'_2$. Since P was arbitrarily chosen in Π_2, this means that there is no point belonging to all elements of $\mathscr{L}$.

Let Π_2, Π_2', Π_2'' be three distinct elements of $\mathscr{L}$. If Π_2, Π_2', Π_2'' generate a prime Π_4, then $|\Pi_4 \cap \mathscr{K}| \geq 3q$, contradicting that $|\Pi_4 \cap \mathscr{K}| \in \{1, q+1, 2q+1\}$. Hence Π_2, Π_2', Π_2'' generate $PG(5, q)$. This is condition (iii)′ of §25.2.

Now by Theorem 25.2.13 the set $\mathscr{K}$ is a Veronesean $\mathscr{V}_2^4$. □

For $q = 3$ any set $\mathscr{K}$ satisfies condition (ii). Hence any set $\mathscr{K}$ of $PG(5, 3)$ which satisfies (i)′ is a Veronesean $\mathscr{V}_2^4$.

25.4 Characterization of the quadric Veronesean $\mathscr{V}_n$

First we prove two properties which hold for the quadric Veronesean $\mathscr{V}_n$. The planes of $PG(N, q)$, $N = n(n+3)/2$, meeting $\mathscr{V}_n$ in a conic are called the *conic planes* of $\mathscr{V}_n$.

Lemma 25.4.1: *Two distinct conic planes of $\mathscr{V}_n$ with non-empty intersection meet in exactly one point, and this point lies in $\mathscr{V}_n$.*

Proof: Let Π_2' and Π_2'' be distinct conic planes of $\mathscr{V}_n$, and let $P \in \Pi_2' \cap \Pi_2''$. Assume that $P \notin \mathscr{V}_n$.

First, let q be odd. The point P belongs to at least one bisecant l of $\Pi_2' \cap \mathscr{V}_n = \mathscr{C}'$. By Corollary 1 of Theorem 25.1.9 the line l is not contained in the plane Π_2''. Let $\mathscr{C}' \cap l = \{P_1, P_2\}$ and $\Pi_2'' \cap \mathscr{V}_n = \mathscr{C}''$. The plane of $PG(n, q)$ containing the line $\mathscr{C}''\zeta^{-1}$ and the point $P_1\zeta^{-1}$ is denoted by Π_2, with ζ as in §25.1. Then $\Pi_2\zeta$ is a Veronesean $\mathscr{V}_2^4$ containing $\mathscr{C}''$ and P_1. The point P_2 belongs to the space $PG(5, q)$ containing $\mathscr{V}_2^4$. By Theorem 25.1.7 the Veronesean $\mathscr{V}_2^4$ is the complete intersection of $PG(5, q)$ and $\mathscr{V}_n$, and so P_2 belongs to $\mathscr{V}_2^4$. The points P_1 and P_2 are contained in a unique conic of $\mathscr{V}_2^4$ and a unique conic of $\mathscr{V}_n$. Consequently $\mathscr{C}'$ is a conic of $\mathscr{V}_2^4$. Now by Theorem 25.1.11 we have $P \in \mathscr{V}_2^4$, contradicting $P \notin \mathscr{V}_n$. Hence $P \in \mathscr{V}_n$. Since $\mathscr{V}_n$ is a cap it is now also clear that P is the unique common point of Π_2' and Π_2''.

Next, let q be even. If P is not the nucleus of $\mathscr{C}' = \Pi_2' \cap \mathscr{V}_n$ or $\mathscr{C}'' = \Pi_2'' \cap \mathscr{V}_n$, then the argument of the preceding paragraph shows again that $P \in \mathscr{V}_n$. So assume that P is the nucleus of both $\mathscr{C}'$ and $\mathscr{C}''$. Let $P' \in \mathscr{C}' \backslash \mathscr{C}''$. Then the line $PP' = l$ is a tangent of the conic $\mathscr{C}'$, and hence is a tangent of the algebraic variety $\mathscr{V}_n$. The plane of $PG(n, q)$ containing the line $\mathscr{C}''\zeta^{-1}$ and the point $P'\zeta^{-1}$ is denoted by Π_2. Then $\Pi_2\zeta$ is a Veronesean $\mathscr{V}_2^4$ containing $\mathscr{C}''$ and P'. The line l belongs to the space $PG(5, q)$ containing $\mathscr{V}_2^4$. Hence l is a tangent of the algebraic variety $\mathscr{V}_2^4$. By §25.1 the Veronesean $\mathscr{V}_2^4$ contains a unique conic $\mathscr{C}$ which is tangent to l at P'. Since P is the nucleus of $\mathscr{C}''$, by Theorem 25.1.17 it belongs to the nucleus of $\mathscr{V}_2^4$. Then again by Theorem 25.1.17 the point P is the nucleus of the conic $\mathscr{C}$. Hence the conics $\mathscr{C}$ and $\mathscr{C}''$ of $\mathscr{V}_2^4$ have a common nucleus. This contradicts Theorem 25.1.17. This proves that again $P \in \mathscr{V}_n$. Since $\mathscr{V}_n$ is a cap it now follows that P is the unique common point of Π_2' and Π_2''. □

Lemma 25.4.2: *Let $\mathscr{C}$ be a conic of $\mathscr{V}_n$ and let P be a point of $\mathscr{V}_n \backslash \mathscr{C}$. Then the union of the tangents at P of the conics of $\mathscr{V}_n$ which pass through P and a point of $\mathscr{C}$ is a plane.*

Proof: Let Π_2 be the plane of $PG(n, q)$ which contains the line $\mathscr{C}\zeta^{-1}$ and the point $P\zeta^{-1}$. Then $\Pi_2\zeta$ is a Veronesean $\mathscr{V}_2^4$ which contains $\mathscr{C}$ and P. By Corollary 1 of Theorem 25.1.9 the $q + 1$ conics of $\mathscr{V}_n$ which contain P and a point of $\mathscr{C}$ are precisely the $q + 1$ conics of $\mathscr{V}_2^4$ through P. From the remarks preceding Theorem 25.1.16 the union of the tangents at P to the $q + 1$ conics of $\mathscr{V}_2^4$ through P is the tangent plane of $\mathscr{V}_2^4$ at P. □

In this section we shall prove that Corollary 1 of Theorem 25.1.9, together with Lemmas 25.4.1 and 25.4.2, characterize the Veronesean $\mathscr{V}_n$.

From now on let $\mathscr{K}$ be a set of k points of some projective space $PG(N, q)$, $N > 2$, where $\mathscr{K}$ generates $PG(N, q)$. Further, let Γ be a set of conics on $\mathscr{K}$, where each element of Γ is the complete intersection of $\mathscr{K}$ and some plane. The elements of Γ will be called Γ-*conics*, and the planes of the Γ-conics will be called Γ-*planes*. The tangents of the Γ-conics will be called Γ-*tangents*. The Γ-tangent which is tangent to $\mathscr{C} \in \Gamma$ at $P \in \mathscr{C}$ will be denoted by $t(P, \mathscr{C})$.

Suppose that $\mathscr{K}$ satisfies the following:

(i) any two distinct points of $\mathscr{K}$ belong to a Γ-plane;
(ii) any two distinct intersecting Γ-planes meet on $\mathscr{K}$;
(iii) if $\mathscr{C}$ is a Γ-conic and P belongs to $\mathscr{K} \backslash \mathscr{C}$, then the tangents at P of the Γ-conics passing through P and a point of $\mathscr{C}$ are coplanar.

Lemma 25.4.3: (a) *The set $\mathscr{K}$ is a k-cap.*
(b) *Any two distinct points of $\mathscr{K}$ are contained in a unique Γ-conic.*
(c) *A Γ-tangent $t(P, \mathscr{C})$ is tangent to a unique Γ-conic through P.*

Proof: Let $P, P' \in \mathscr{K}$ and let Π_2 be a Γ-plane containing P and P'. Since $\Pi_2 \cap \mathscr{K}$ is a conic, it is clear that $\mathscr{K} \cap PP' = \{P, P'\}$. Hence $\mathscr{K}$ is a k-cap. Let Π_2' be another Γ-plane containing the line PP'. By (ii) the line $\Pi_2 \cap \Pi_2'$ is a line of $\mathscr{K}$, contradicting the first part of the proof. Finally, consider a Γ-tangent $t(P, \mathscr{C})$. Assume that $t(P, \mathscr{C}) = t(P, \mathscr{C}')$, with $\mathscr{C} \neq \mathscr{C}'$. By (ii) the Γ-planes containing $\mathscr{C}$ and $\mathscr{C}'$ meet on $\mathscr{K}$. Since $t(P, \mathscr{C})$ is the intersection of these Γ-planes, the set $\mathscr{K}$ contains a line, a contradiction. □

The unique Γ-conic containing the points P, P', $P \neq P'$, of $\mathscr{K}$ will be denoted by $[P, P']$. For $P \in \mathscr{K}$ and $\mathscr{C}$ a Γ-conic not passing through P, the plane containing the tangents at P of the Γ-conics passing through P and a point of $\mathscr{C}$ will be denoted by $\pi(P, \mathscr{C})$. By Lemma 25.4.3 there are exactly $q + 1$ Γ-conics containing P and a point of $\mathscr{C}$, and $\pi(P, \mathscr{C})$ is the union of the corresponding $q + 1$ Γ-tangents.

Lemma 25.4.4: *The incidence structure $\mathscr{S}$ formed by the points of $\mathscr{K}$ and the conics of Γ is the incidence structure of points and lines of some projective space.*

Proof: Let P, P', P'' be points of $\mathscr{K}$ not on the same Γ-conic. It is sufficient to prove that the substructure of $\mathscr{S}$ generated by P, P', P'' is a projective plane. Let $A \in [P, P']\backslash\{P\}$ and $B \in [P, P'']\backslash\{P\}$. By Lemma 25.4.3 the Γ-tangents to $[P, P']$ and $[P, P'']$ at P are distinct, and by (iii) belong to the intersection of $\pi(P, [P', P''])$ and $\pi(P, [A, B])$. Hence $\pi(P, [P', P'']) = \pi(P, [A, B]) = \Pi_2$. Suppose that $R \in [P', P'']$ with $[P, R] \cap [A, B] = \varnothing$. Then the line $t(P, [P, R])$ is tangent to some conic $[P, R']$ with $R' \in [A, B]$. Hence $[P, R']$ and $[P, R]$ are distinct Γ-conics tangent to $t(P, [P, R])$ at P. This contradicts Lemma 25.4.3. Consequently $[P, R] \cap [A, B] \neq \varnothing$. It easily follows that the Γ-conic through any two distinct points of $\mathscr{V} = \bigcup [P, X]$, where X varies in $[P', P'']$, is completely contained in $\mathscr{V}$. Consequently $\mathscr{V}$ together with all Γ-conics contained in $\mathscr{V}$ is the substructure of $\mathscr{S}$ generated by P, P', P''. Since each Γ-conic contains $q + 1$ points and $\mathscr{V}$ contains $q^2 + q + 1$ points, this substructure is a projective plane of order q. Hence $\mathscr{S}$ is the incidence structure of points and lines of some projective space. □

The *dimension* of $\mathscr{K}$ is defined to be the dimension of $\mathscr{K}$ considered as a projective space. Suppose $\mathscr{K}$ has dimension n. Then $\mathscr{K}$ has $\theta(n)$ points. Since $\mathscr{K}$ generates $PG(N, q)$, with $N > 2$, it is clear that $n \geq 2$.

For any point P in $\mathscr{K}$, let $T(P) = \bigcup t(P, \mathscr{C})$, with the union taken over all Γ-conics containing P.

Lemma 25.4.5: *For any point P the set $T(P)$ is an n-dimensional projective subspace of $PG(N, q)$.*

Proof: Let R, R' be distinct points of $T(P)$, where R, R', P are not collinear. Let $PR = t(P, \mathscr{C})$ and $PR' = t(P, \mathscr{C}')$. For $A \in \mathscr{C}\backslash\{P\}$ and $B \in \mathscr{C}'\backslash\{P\}$, the plane $\pi(P, [A, B])$ coincides with the plane PRR'. Since $\pi(P, [A, B]) \subset T(P)$ and $\pi(P, [A, B]) = PRR'$, the line RR' belongs to $T(P)$. Hence it follows that $T(P)$ is a projective subspace of $PG(N, q)$.

Since the projective space $\mathscr{K}$ has dimension n, the number of Γ-conics through P is equal to $\theta(n - 1)$. Hence the number of tangents $t(P, \mathscr{C})$ in $T(P)$ is equal to $\theta(n - 1)$. Thus the projective space $T(P)$ is n-dimensional. □

For any point P of $\mathcal{K}$ the space $T(P)$ will be called the *tangent space* of $\mathcal{K}$ at P.

Lemma 25.4.6: *Let $\mathcal{C}$ and $\mathcal{C}'$ be Γ-conics, with $P \in \mathcal{C}$ and $P' \in \mathcal{C}'$, $P \neq P'$. If $t(P, \mathcal{C}) \cap t(P', \mathcal{C}') \neq \varnothing$, then $\mathcal{C} = \mathcal{C}'$.*

Proof: If $t(P, \mathcal{C}) = t(P', \mathcal{C}')$, then $P = P'$, a contradiction. So let $t(P, \mathcal{C}) \cap t(P', \mathcal{C}') = \{R\}$. Suppose that $\mathcal{C} \neq \mathcal{C}'$. The planes Π_2 and Π_2' containing $\mathcal{C}$ and $\mathcal{C}'$, respectively, are distinct and have R in common. By (ii) the point R belongs to $\mathcal{K}$. Since $t(P, \mathcal{C}) \cap \mathcal{K} = \{P\}$ and $t(P', \mathcal{C}') \cap \mathcal{K} = \{P'\}$, we have $P = R$ and $P' = R$. Hence $P = P'$, a contradiction. We conclude that $\mathcal{C} = \mathcal{C}'$. □

Lemma 25.4.7: *Two distinct tangent spaces of $\mathcal{K}$ meet in a unique point, and this point is not in $\mathcal{K}$.*

Proof: Let P, P' be distinct points of $\mathcal{K}$. The lines $t(P, [P, P'])$ and $t(P', [P, P'])$ have a point R in common, which clearly belongs to $(T(P) \cap T(P'))\backslash\mathcal{K}$. Now, let $R' \in T(P) \cap T(P')$. Then the lines PR' and $P'R'$ are Γ-tangents. Let PR' be tangent to the Γ-conic $\mathcal{C}$, and let $P'R'$ be tangent to the Γ-conic $\mathcal{C}'$. By Lemma 25.4.6 we have $\mathcal{C} = \mathcal{C}'$. Now by Lemma 25.4.3, $\mathcal{C} = [P, P']$; hence $R' = R$. Consequently $T(P)$ and $T(P')$ have a unique point R in common, with $R \notin \mathcal{K}$. □

Lemma 25.4.8: *Let $\mathcal{K}$ have dimension 2. If $P \in \mathcal{K}$, then the tangent plane $T(P)$ is skew to any Γ-plane not through P.*

Proof: Let Π_2 be a Γ-plane not through P and suppose that Π_2 and $T(P)$ have a point R in common. Then $R \neq P$ and the line PR is tangent to a Γ-conic $\mathcal{C}$ at P. The plane containing $\mathcal{C}$ and the plane Π_2 have the point R in common; hence $R \in \mathcal{K}$ by (ii). Since P is the only point of $T(P)$ in $\mathcal{K}$, we have $P = R$, a contradiction. □

Lemma 25.4.9: *If $\mathcal{K}$ has dimension 2, then $N = 5$.*

Proof: Let $P \in \mathcal{K}$ and let Π_2 be a Γ-plane which does not contain P. By Lemma 25.4.8 the planes Π_2 and $T(P)$ generate a projective space $PG(5, q)$ of dimension 5. Hence $N \geq 5$. Let P' be any point of $\mathcal{K}\backslash\{P\}$. By Lemma 25.4.4 the incidence structure $\mathcal{S}$ is a projective plane; hence the Γ-conic in Π_2 and the Γ-conic $[P, P']$ have a unique point R in common. Further, the Γ-tangent $t(P, [P, P'])$ belongs to $T(P)$. Consequently the Γ-plane containing $[P, P']$ belongs to $PG(5, q)$. Hence P' is a point of $PG(5, q)$, and so $\mathcal{K}$ is contained in $PG(5, q)$. We conclude that $\mathcal{K}$ generates $PG(5, q)$. □

Lemma 25.4.10: *Let $\mathscr{K}$ have dimension 2 and suppose that $P_1, P_2, P_3 \in \mathscr{K}$ do not belong to the same Γ-conic. If $\{P_{ij}\} = T(P_i) \cap T(P_j)$, $i \neq j$, then the points $P_1, P_2, P_3, P_{12}, P_{13}, P_{23}$ generate $PG(5, q)$.*

Proof: Suppose that the points P_1, P_{12}, and P_{13} are collinear. The line P_1P_{12} is tangent to the conic $[P_1, P_2]$ at P_1, and the line P_1P_{13} is tangent to the conic $[P_1, P_3]$ at P_1. By Lemma 25.4.3 we have $[P_1, P_2] = [P_1, P_3]$, a contradiction. Hence P_1, P_{12}, and P_{13} are not collinear. Analogously P_2, P_{12}, P_{23} are not collinear, and P_3, P_{13}, P_{23} are not collinear. By Lemma 25.4.8 the plane $T(P_1)$ and the plane of the conic $[P_2, P_3]$ are skew. The line P_2P_{23} is tangent to $[P_2, P_3]$ at P_2. Hence the plane $T(P_1)$ and the line P_2P_{23} are skew. This means that the points $P_1, P_2, P_{12}, P_{13}, P_{23}$ are linearly independent. Let Π_4 be the prime spanned by these points. Suppose that P_3 belongs to Π_4. Then the non-collinear points P_3, P_{13}, P_{23} belong to $T(P_3)$, and hence $T(P_3)$ is a plane of Π_4. Consequently the plane $T(P_3)$ and the Γ-plane $P_1P_2P_{12}$ containing $[P_1, P_2]$ have a non-empty intersection, contradicting Lemma 25.4.8. We conclude that $P_3 \notin \Pi_4$; that is, the points $P_1, P_2, P_3, P_{12}, P_{13}, P_{23}$ generate $PG(5, q)$. □

Lemma 25.4.11: *Let $\mathscr{K}$ have dimension 2. Consider a tangent plane $T(P)$ of $\mathscr{K}$ and also a plane Π_2 skew to $T(P)$. Then the projection of $\mathscr{K}\backslash\{P\}$ from $T(P)$ onto Π_2 is a conic of Π_2.*

Proof: Let $\mathscr{C}$ be a Γ-conic not containing the point P. By Lemma 25.4.8 the $q+1$ points R of $\mathscr{C}$ define $q+1$ solids $T(P)R$. For R on $\mathscr{C}$ the line $t(P, [P, R])$ belongs to $T(P)$; hence the Γ-conic $[P, R]$ belongs to $T(P)R$. From Lemmas 25.4.8 and 25.4.3 it follows that $\mathscr{K} \cap T(P)R = [P, R]$. Let $P' \in \mathscr{K}\backslash\{P\}$. Since the incidence structure $\mathscr{S}$ is a projective plane, the Γ-conic $[P', P]$ has a unique point R' in $\mathscr{C}$. As $[P, P'] = [P, R']$ belongs to $T(P)R'$, we have $T(P)P' \cap \Pi_2 = T(P)R' \cap \Pi_2$. Hence the projection of $\mathscr{K}\backslash\{P\}$ from $T(P)$ onto Π_2 coincides with the projection of $\mathscr{C}$ from $T(P)$ onto Π_2. This proves the lemma. □

Lemma 25.4.12: *Let $\mathscr{K}$ have dimension 2. Any plane having at least four distinct points in $\mathscr{K}$ is a Γ-plane.*

Proof: Consider a plane Π_2 having at least four distinct points P_1, P_2, P_3, P_4 in $\mathscr{K}$. Suppose that $[P_1, P_2] \neq [P_3, P_4]$. The Γ-planes of the respective conics $[P_1, P_2]$ and $[P_3, P_4]$ both contain the common point R of the lines P_1P_2 and P_3P_4. By (ii) R belongs to $\mathscr{K}$. Hence $\mathscr{K}$ contains the collinear points P_1, P_2, and R. This contradicts Lemma 25.4.3. Hence $[P_1, P_2] = [P_3, P_4]$; that is, P_1, P_2, P_3, P_4 are contained in a common Γ-plane. □

Corollary: *Let $\mathscr{K}$ have dimension 2 and let P, P' be distinct points of $\mathscr{K}$. If Π_2 is the Γ-plane through PP', then there are exactly q^2 planes through PP', but distinct from Π_2, which meet $\mathscr{K}\backslash\{P, P'\}$. Each of these q^2 planes contains exactly one point of $\mathscr{K}\backslash\{P, P'\}$.*

Proof: This is straightforward. □

Lemma 25.4.13: *Let $\mathscr{K}$ have dimension 2 and let P, P' be distinct points of $\mathscr{K}$. Then the set of all planes projecting $\mathscr{K}\backslash\{P, P'\}$ from $PP' = l$, together with all planes containing l and a Γ-tangent of $\mathscr{K}$ at P or P', is the family of all planes on a quadric $l\mathscr{H}_3$, where $\mathscr{H}_3$ is some hyperbolic quadric in a solid skew to l.*

Proof: The union of the planes containing l and a Γ-tangent of $\mathscr{K}$ at P is the solid $T(P)P'$, and the union of the planes containing l and a Γ-tangent of $\mathscr{K}$ at P' is the solid $T(P')P$. The intersection of the solids $T(P)P'$ and $T(P')P$ is the Γ-plane containing the conic $[P, P']$.

The Γ-conics of $\mathscr{K}$ through P are denoted by $[P, P']$, $\mathscr{C}_1, \mathscr{C}_2, \ldots, \mathscr{C}_q$, and the Γ-conics of $\mathscr{K}$ through P' are denoted by $[P, P']$, $\mathscr{C}'_1, \mathscr{C}'_2, \ldots, \mathscr{C}'_q$. Further, let Π_3^i be the solid containing l and $\mathscr{C}_i$, and let $\Pi_3^{i\prime}$ be the solid containing l and $\mathscr{C}'_i$, $i = 1, 2, \ldots, q$. From (ii) and Lemma 25.4.3 it follows that the $2(q+1)$ solids $\Pi_3^1, \Pi_3^2, \ldots, \Pi_3^q, \Pi_3^{1\prime}, \Pi_3^{2\prime}, \ldots, \Pi_3^{q\prime}, T(P)P', T(P')P$ are distinct. If Π_3^i and Π_3^j, $i \neq j$, have a plane in common, then by the corollary of Lemma 25.4.12 this plane is in the solid $T(P)P'$. Hence $\mathscr{C}_i$ and $\mathscr{C}_j$ have a common tangent at P or $T(P)P' \cap \mathscr{K} \neq [P, P']$, a contradiction. Consequently Π_3^i and Π_3^j, $i \neq j$, have just the line l in common. Analogously, the intersection of $\Pi_3^{i\prime}$ and $\Pi_3^{j\prime}$, $i \neq j$, is the line l. Since $\mathscr{S}$ is a projective plane, the conics $\mathscr{C}_i$ and $\mathscr{C}'_j$, with $i, j \in \{1, 2, \ldots, q\}$, have exactly one point in common. So the solids Π_3^i and $\Pi_3^{j\prime}$ have a plane in common. Since each Π_3^i contains a line of the plane $T(P)$, the solids Π_3^i and $T(P)P'$ have a plane in common. Analogously, each solid $\Pi_3^{i\prime}$ contains a plane of the solid $T(P')P$. Further, we have $\Pi_3^i \cap T(P')P = l$ and $\Pi_3^{i\prime} \cap T(P)P' = l$, for all $i = 1, 2, \ldots, q$.

So there are two systems of $q + 1$ solids through l with the property that two solids of the same system intersect in the line l, and two solids of different systems intersect in a plane through l.

Each plane through l in at least one of the $2(q + 1)$ solids contains a point of $\mathscr{K}\backslash\{P, P'\}$, or a Γ-tangent of $\mathscr{K}$ at P, or a Γ-tangent of $\mathscr{K}$ at P'. Conversely, any plane containing l and a point of $\mathscr{K}\backslash\{P, P'\}$, or l and a Γ-tangent of $\mathscr{K}$ at P, or l and a Γ-tangent of $\mathscr{K}$ at P', belongs to at least one of the $2(q + 1)$ solids. Hence the union of all planes projecting $\mathscr{K}\backslash\{P, P'\}$ from l, together with all planes containing l and a Γ-tangent of $\mathscr{K}$ at P or P', is also the union of the solids $T(P)P'$, $T(P')P$, and all solids Π_3^i, $\Pi_3^{j\prime}$.

From the preceding paragraphs it now follows that the set of all planes projecting $\mathscr{K}\backslash\{P, P'\}$ from l, together with all planes containing l and a Γ-tangent of $\mathscr{K}$ at P or P', is the family of all planes on a quadric $l\mathscr{H}_3$, with $\mathscr{H}_3$ some hyperbolic quadric in a solid skew to l. □

Theorem 25.4.14: *If* $\mathscr{K}$ *satisfies* (i), (ii), (iii), *and has dimension* 2, *then* $\mathscr{K}$ *is the Veronesean* $\mathscr{V}_2^4$, *whose conic planes are the* Γ*-planes of* $\mathscr{K}$.

Proof: Let Q_{00}, Q_{11}, Q_{22} be three distinct points of $\mathscr{K}$ not contained in the same Γ-plane. Suppose that $\{Q_{ij}\} = \{Q_{ji}\} = T(Q_{ii}) \cap T(Q_{jj})$, where $i \neq j$ and $i, j = 0, 1, 2$. By Lemma 25.4.10 the points Q_{00}, Q_{11}, Q_{22}, Q_{01}, Q_{02}, Q_{12} generate $PG(5, q)$. Further, let U be a point of $\mathscr{K}$ which is not in $[Q_{00}, Q_{11}] \cup [Q_{11}, Q_{22}] \cup [Q_{00}, Q_{22}]$. We shall prove that no six of the seven points U, Q_{00}, Q_{11}, Q_{22}, Q_{01}, Q_{02}, Q_{12} are linearly dependent. It is sufficient to show that U, Q_{00}, Q_{11}, Q_{22}, Q_{01}, Q_{02} and U, Q_{00}, Q_{11}, Q_{01}, Q_{02}, Q_{12} are linearly independent. First suppose that U, Q_{00}, Q_{11}, Q_{22}, Q_{01}, Q_{02} are contained in a prime Π_4. Then the plane $T(Q_{00}) = Q_{00}Q_{01}Q_{02}$ belongs to Π_4. The solids $T(Q_{00})Q_{11}$, $T(Q_{00})Q_{22}$, and $T(Q_{00})U$ are three distinct solids in Π_4. Let Π_2 be a plane which is skew to $T(Q_{00})$. Then the points common to Π_2 and the solids $T(Q_{00})Q_{11}$, $T(Q_{00})Q_{22}$, and $T(Q_{00})U$ are collinear. This contradicts Lemma 25.4.11. Next, suppose that U, Q_{00}, Q_{11}, Q_{01}, Q_{02}, Q_{12} are contained in a prime Π'_4. Then $\Pi'_4 = T(Q_{00})T(Q_{11})$. Since the plane $T(Q_{00})$ and the conic $[U, Q_{11}]$ belong to Π'_4, Lemma 25.4.8 is contradicted.

If follows that the coordinate system of $PG(5, q)$ can be chosen in such a way that $Q_{00} = \mathbf{P}(E_0)$, $Q_{11} = \mathbf{P}(E_1)$, $Q_{22} = \mathbf{P}(E_2)$, $Q_{01} = \mathbf{P}(E_3)$, $Q_{02} = \mathbf{P}(E_4)$, $Q_{12} = \mathbf{P}(E_5)$, where E_i is the vector with one in the $(i + 1)$th place and zeros elsewhere. The point U can be chosen as unit point. An arbitrary point of $PG(5, q)$ is of the form $\mathbf{P}(Y)$, where $Y = (y_{00}, y_{11}, y_{22}, y_{01}, y_{02}, y_{12})$. For y_{ij} we also write y_{ji}.

The tangent plane $T(Q_{kk})$ is skew to the Γ-plane $Q_{ii}Q_{jj}Q_{ij}$, where $\{i, j, k\} = \{0, 1, 2\}$. By Lemma 25.4.11 the projection of $\mathscr{K}\backslash\{Q_{kk}\}$ from $T(Q_{kk})$ onto the plane $Q_{ii}Q_{jj}Q_{ij}$ is the conic $[Q_{ii}, Q_{jj}] = \mathscr{C}_{ij}$. The cone $T(Q_{kk})\mathscr{C}_{ij}$ is the variety $\mathbf{V}(F)$ with $F = y_{ii}y_{jj} - ay_{ij}^2$. Since $U = \mathbf{U} \in T(Q_{kk})\mathscr{C}_{ij}$ we have $a = 1$, and so $F = y_{ii}y_{jj} - y_{ij}^2$.

By Lemma 25.4.13 the union $\mathscr{W}$ of all planes projecting $\mathscr{K}\backslash\{Q_{ii}, Q_{jj}\}$ from the line $Q_{ii}Q_{jj}$, and all planes containing the line $Q_{ii}Q_{jj} = l$ and a Γ-tangent of $\mathscr{K}$ at Q_{ii} or Q_{jj}, is of the form $l\mathscr{H}_3$, where $\mathscr{H}_3$ is some hyperbolic quadric in a solid Π_3 skew to l. Let Π_3 be the solid containing the points Q_{ij}, Q_{ik}, Q_{jk}, Q_{kk}, with $\{i, j, k\} = \{0, 1, 2\}$. From the proof of Lemma 25.4.13 it follows that $\mathscr{H}_3$ contains the lines $Q_{ij}Q_{ik}$, $Q_{ij}Q_{jk}$, $Q_{kk}Q_{ik}$, and $Q_{kk}Q_{jk}$. Hence $\mathscr{W}$ is a variety $\mathbf{V}(F')$ with $F' = y_{kk}y_{ij} - by_{ik}y_{jk}$. Since $U = \mathbf{U} \in \mathscr{W}$ we have $b = 1$, whence $F' = y_{kk}y_{ij} - y_{ik}y_{jk}$.

So $\mathscr{K}$ belongs to all primals

$$\mathbf{V}(y_{ii}y_{jj} - y_{ij}^2), \qquad i \neq j,$$
$$\mathbf{V}(y_{kk}y_{ij} - y_{ik}y_{jk}), \quad i \neq j \neq k \neq i.$$

By Lemma 25.1.1, $\mathscr{K} \subset \mathscr{V}_2^4$. Since $|\mathscr{K}| = |\mathscr{V}_2^4| = q^2 + q + 1$ we have the result $\mathscr{K} = \mathscr{V}_2^4$.

By Corollary 1 of Theorem 25.1.9 the conic planes of $\mathscr{V}_2^4$ are the Γ-planes of $\mathscr{K}$. □

Lemma 25.4.15: *If $\mathscr{K}$ generates $PG(N, q)$ and has dimension n, then $N \leq n(n+3)/2$.*

Proof: When $n = 2$ then $N = 5$ by Lemma 25.4.9. Now let $n > 2$, and proceed by induction on n. By Lemma 25.4.4 the incidence structure $\mathscr{S}$ is the incidence structure of points and lines of a projective space of dimension n. Let $P \in \mathscr{K}$ and let $\mathscr{K}_{n-1}$ be a prime of the projective space $\mathscr{K}$ which does not pass through P. Further, let $PG(N', q)$ be the subspace of $PG(N, q)$ generated by $\mathscr{K}_{n-1}$. As $n - 1 \geq 2$, we have $N' > 2$. The set $\mathscr{K}_{n-1}$ together with the Γ-conics on $\mathscr{K}_{n-1}$ and the Γ-planes of these Γ-conics satisfy conditions (i), (ii), and (iii). Since the dimension of $\mathscr{K}_{n-1}$ is $n - 1$, by the induction hypothesis we have $N' \leq (n-1)(n+2)/2$. Let R be any point of $\mathscr{K} \backslash \{P\}$. The plane of the Γ-conic $[P, R]$ contains the line $t(P, [P, R])$ of $T(P)$. Since $[P, R]$ is a line of the projective space $\mathscr{K}$, it contains a unique point of the prime $\mathscr{K}_{n-1}$. Hence the plane of $[P, R]$ is contained in the subspace $PG(N'', q)$ of $PG(N, q)$ generated by $T(P)$ and $PG(N', q)$. Consequently $\mathscr{K}$ is contained in $PG(N'', q)$, and so $N'' = N$. Thus

$$N = N'' \leq N' + n + 1 \leq (n-1)(n+2)/2 + n + 1.$$

We conclude that $N \leq n(n+3)/2$. □

From now on assume that $\mathscr{K}$ satisfies

(iv) $N = n(n+3)/2$.

Lemma 25.4.16: *Let $P \in \mathscr{K}$ and let $\mathscr{K}_{n-1}$ be a prime of the projective space $\mathscr{K}$, with $P \notin \mathscr{K}_{n-1}$. If $\mathscr{K}_{n-1}$ generates the subspace $PG(N', q)$ of $PG(N, q)$, then $N' = (n-1)(n+2)/2$ and $T(P) \cap PG(N', q) = \varnothing$.*

Proof: By Lemma 25.4.8 the lemma is satisfied for $n = 2$. Now let $n > 2$. From the proof of Lemma 25.4.15 we have

$$N = n(n+3)/2 \leq N' + n + 1 \leq (n-1)(n+2)/2 + n + 1.$$

Hence $N' = (n-1)(n+2)/2$. Also, since $PG(N, q)$ is generated by $T(P)$ and $PG(N', q)$, we have $T(P) \cap PG(N', q) = \emptyset$. □

Lemma 25.4.17: *Let $Q_{00}, Q_{11}, \ldots, Q_{nn}$ be $n+1$ points of $\mathscr{K}$ which generate the n-dimensional projective space $\mathscr{K}$. Then the tangent spaces $T(Q_{00})$, $T(Q_{11}), \ldots, T(Q_{nn})$ generate the projective space $PG(N, q)$.*

Proof: By Lemma 25.4.10 the lemma is true for $n = 2$. Now suppose that $n > 2$ and proceed by induction on n. Let $\mathscr{K}_{n-1}$ be the prime of the projective space $\mathscr{K}$ which is generated by $Q_{11}, Q_{22}, \ldots, Q_{nn}$. By Lemma 25.4.16 the set $\mathscr{K}_{n-1}$ generates a subspace $PG(N', q)$ of $PG(N, q)$, with $N' = (n-1) \times (n+2)/2$. By the induction hypothesis the space $PG(N', q)$ is generated by the tangent spaces $T'(Q_{11}), T'(Q_{22}), \ldots, T'(Q_{nn})$ of $\mathscr{K}_{n-1}$. Clearly $T'(Q_{ii}) \subset T(Q_{ii})$, with $i = 1, 2, \ldots, n$. Since $PG(N, q)$ is generated by $T(Q_{00})$ and $PG(N', q)$, it now follows that $PG(N, q)$ is generated by the tangent spaces $T(Q_{00}), T(Q_{11}), \ldots, T(Q_{nn})$. □

Lemma 25.4.18: *Let $Q_{00}, Q_{11}, \ldots, Q_{nn}$ be $n+1$ points of $\mathscr{K}$ which generate the n-dimensional projective space $\mathscr{K}$. If $\{Q_{ij}\} = T(Q_{ii}) \cap T(Q_{jj})$, $i \neq j$, then for any i the $n+1$ points $Q_{i0}, Q_{i1}, \ldots, Q_{in}$ generate the tangent space $T(Q_{ii})$.*

Proof: By the first part of the proof of Lemma 25.4.10 the lemma is satisfied for $n = 2$. So assume that $n > 2$. Let Π_{n-1} be a prime of the tangent space $T(Q_{ii})$. Further, let $\mathscr{K}_{n-1}$ be the prime of $\mathscr{K}$ containing $\{Q_{00}, Q_{11}, \ldots, Q_{nn}\} \backslash \{Q_{ii}\}$. For any point R of $\mathscr{K}_{n-1}$ let $\bar{R}$ be the intersection of Π_{n-1} and the line $t(Q_{ii}, [Q_{ii}, R])$. Let $\eta\colon \mathscr{K}_{n-1} \to \Pi_{n-1}$ be defined by $R \to \bar{R}$. Then η is a bijection of $\mathscr{K}_{n-1}$ onto Π_{n-1}. By condition (iii) the bijection η is a collineation of the projective space $\mathscr{K}_{n-1}$ onto the projective space Π_{n-1}. Hence the n points $Q_{jj}\eta$, $j \neq i$, generate Π_{n-1}. Since $Q_{jj}\eta$, $j \neq i$, is the common point of Π_{n-1} and the line $Q_{ii}Q_{ij}$, the n lines $Q_{ii}Q_{ij}$ generate $T(Q_{ii})$. So we have proved that the $n+1$ points $Q_{i0}, Q_{i1}, \ldots, Q_{in}$ generate $T(Q_{ii})$. □

Lemma 25.4.19: *Let $Q_{00}, Q_{11}, \ldots, Q_{nn}$ be $n+1$ points of $\mathscr{K}$ which generate the n-dimensional projective space $\mathscr{K}$. If $\{Q_{ij}\} = T(Q_{ii}) \cap T(Q_{jj})$, $i \neq j$, then the $n(n+3)/2 + 1$ points Q_{ij}, with $i, j = 0, 1, \ldots, n$, generate $PG(N, q)$.*

Proof: This follows immediately from Lemmas 25.4.17 and 25.4.18. □

Lemma 25.4.20: *Let $Q_{00}, Q_{11}, \ldots, Q_{nn}$ be $n+1$ linearly independent points of the n-dimensional projective space $\mathscr{K}$, and let $T(Q_{ii}) \cap T(Q_{jj}) = \{Q_{ij}\}$ for all $i \neq j$. Further, let $\mathscr{K}^{i}_{n-1}$ be the prime of $\mathscr{K}$ generated by $\{Q_{00}, Q_{11}, \ldots, Q_{nn}\} \backslash \{Q_{ii}\}$ and let $U \in \mathscr{K} \backslash (\mathscr{K}^{0}_{n-1} \cup \mathscr{K}^{1}_{n-1} \cup \cdots \cup \mathscr{K}^{n}_{n-1})$.*

Then any $N + 1 = n(n+3)/2 + 1$ *of the points* $Q_{00}, Q_{11}, \ldots, Q_{nn}, Q_{01}, Q_{02}, \ldots, Q_{0n}, \ldots, Q_{n-1,n}, U$ *are linearly independent in* $PG(N, q)$.

Proof: By the first part of the proof of Lemma 25.4.14, the lemma is satisfied for $n = 2$. So assume that $n > 2$ and proceed by induction on n. Consider the point Q_{ii} and let U_i be the unique point common to the conic $[Q_{ii}, U]$ and $\mathscr{K}^i_{n-1}$. The point Q_{jl}, $i \neq j \neq l \neq i$, is the meet of the tangent spaces $T_i(Q_{jj})$ and $T_i(Q_{ll})$ of $\mathscr{K}^i_{n-1}$. Let $\mathscr{K}^{ij}_{n-2}$, with $i \neq j$, be the prime of $\mathscr{K}^i_{n-1}$ generated by $\{Q_{00}, Q_{11}, \ldots, Q_{nn}\} \backslash \{Q_{ii}, Q_{jj}\}$. Then $U_i \in \mathscr{K}^i_{n-1} \backslash (\bigcup_{j \neq i} \mathscr{K}^{ij}_{n-2})$. By the induction hypothesis any $(n-1)(n+2)/2 + 1$ of the points Q_{jj}, Q_{jl}, U_i, with $j \neq i$ and $j \neq l \neq i$, are linearly independent in the projective space $PG^i(N', q)$ generated by $\mathscr{K}^i_{n-1}$. Then, by Lemmas 25.4.16 and 25.4.18, any $(n-1)(n+2)/2 + 1$ of the points Q_{jj}, Q_{jl}, U_i, with $j \neq i$ and $j \neq l \neq i$, together with the $n + 1$ points $Q_{i0}, Q_{i1}, \ldots, Q_{in}$, generate $PG(N, q)$. The point U_i is in the plane defined by the line $t(Q_{ii}, [Q_{ii}, U])$ and the point U. Hence U_i is a point of the projective space generated by $Q_{i0}, Q_{i1}, \ldots, Q_{in}, U$.

Let us consider the points $Q_{11}, Q_{22}, \ldots, Q_{nn}, Q_{01}, Q_{02}, \ldots, Q_{0n}, \ldots, Q_{n-1,n}, U$. The point U_1 is a point of the projective space generated by $Q_{10}, Q_{11}, \ldots, Q_{1n}, U$. Now by the preceding paragraph the $n(n+3)/2 + 1$ given points generate $PG(N, q)$. Next let us consider the points $Q_{00}, Q_{11}, \ldots, Q_{nn}, Q_{02}, Q_{03}, \ldots, Q_{0n}, Q_{12}, \ldots, Q_{n-1,n}, U$. The point U_2 is a point of the projective space generated by $Q_{20}, Q_{21}, \ldots, Q_{2n}, U$. Again by the first paragraph the $n(n+3)/2 + 1$ given points generate $PG(N, q)$. Further, recall that by Lemma 25.4.19 also the points $Q_{00}, Q_{11}, \ldots, Q_{nn}, Q_{01}, \ldots, Q_{0n}, \ldots, Q_{n-1,n}$ generate $PG(N, q)$. Now it is clear that any $n(n+3)/2 + 1$ of the points Q_{jj}, Q_{jl}, U, with $j \neq l$, are linearly independent. □

Theorem 25.4.21: *If* $\mathscr{K}$ *has dimension n and satisfies* (i), (ii), (iii), *and* (iv), *then* $\mathscr{K}$ *is the Veronesean* $\mathscr{V}_n$. *Also the conic planes of* $\mathscr{V}_n$ *are the* Γ*-planes of* $\mathscr{K}$.

Proof: Let $Q_{00}, Q_{11}, \ldots, Q_{nn}$ be $n + 1$ linearly independent points of the n-dimensional projective space $\mathscr{K}$, and let $T(Q_{ii}) \cap T(Q_{jj}) = \{Q_{ij}\}$ for all $i \neq j$. Further, let $\mathscr{K}^i_{n-1}$ be the prime of $\mathscr{K}$ generated by $\{Q_{00}, Q_{11}, \ldots, Q_{nn}\} \backslash \{Q_{ii}\}$ and let $U \in \mathscr{K} \backslash (\mathscr{K}^0_{n-1} \cup \mathscr{K}^1_{n-1} \cup \cdots \cup \mathscr{K}^n_{n-1})$. By Lemma 25.4.20 the coordinate system of $PG(5, q)$ can be chosen in such a way that $Q_{00} = \mathbf{P}(E_0)$, $Q_{11} = \mathbf{P}(E_1), \ldots$, $Q_{nn} = \mathbf{P}(E_n)$, $Q_{01} = \mathbf{P}(E_{n+1})$, $Q_{02} = \mathbf{P}(E_{n+2}), \ldots, Q_{0n} = \mathbf{P}(E_{2n}), \ldots, Q_{n-1,n} = \mathbf{P}(E_N)$, with $N = n(n+3)/2$ and where E_i is the vector with one in the $(i+1)$th place and zeros elsewhere, and the point U can be chosen as unit point. An arbitrary point of $PG(N, q)$ is of the form $\mathbf{P}(Y)$, where

$$Y = (y_{00}, y_{11}, \ldots, y_{nn}, y_{01}, y_{02}, \ldots, y_{0n}, \ldots, y_{n-1,n}).$$

For y_{ij} we also write y_{ji}. We shall show that with respect to this coordinate system the set $\mathscr{K}$ is the intersection of all primals

$$\mathbf{V}(y_{ii}y_{jj} - y_{ij}^2), \qquad i \neq j, \tag{25.33}$$

$$\mathbf{V}(y_{kk}y_{ij} - y_{ik}y_{jk}), \quad i \neq j \neq k \neq i. \tag{25.34}$$

By the proof of Theorem 25.4.14, for $n = 2$ the set $\mathscr{K}$ is indeed the intersection of all primals (25.33) and (25.34). So assume $n > 2$ and let us proceed by induction on n. Consider the point Q_{ii} and let U_i be the unique common point of the conic $[Q_{ii}, U]$ and $\mathscr{K}^i_{n-1}$. Let $\mathscr{K}^{ij}_{n-2}$, with $i \neq j$, be the prime of $\mathscr{K}^i_{n-1}$ generated by $\{Q_{00}, Q_{11}, \ldots, Q_{nn}\}\backslash\{Q_{ii}, Q_{jj}\}$. Then $U_i \in \mathscr{K}^i_{n-1}\backslash(\bigcup_{j\neq i}\mathscr{K}^{ij}_{n-2})$. The point U_i is also the unique common point of the space $T(Q_{ii})U$ and the space $PG^i(N', q)$ generated by $\mathscr{K}^i_{n-1}$. Also, the point Q_{jl}, $i \neq j \neq l \neq i$, is the common point of the tangent spaces $T_i(Q_{jj})$ and $T_i(Q_{ll})$ of $\mathscr{K}^i_{n-1}$. By the induction hypothesis the set $\mathscr{K}^i_{n-1}$ is the intersection of all primals

$$\mathbf{V}(y_{ij}), \qquad i \neq j;$$

$$\mathbf{V}(y_{jj}y_{ll} - y_{jl}^2), \qquad i \neq j \neq l \neq i;$$

$$\mathbf{V}(y_{kk}y_{jl} - y_{jk}y_{lk}),$$

with j, l, k distinct elements of $\{0, 1, \ldots, n\}\backslash\{i\}$. Since $\mathscr{K}^i_{n-1}$ is the projection of $\mathscr{K}\backslash\{Q_{ii}\}$ from $T(Q_{ii})$ onto $PG^i(N', q)$, for any given i the set $\mathscr{K}$ lies in all the primals

$$\mathbf{V}(y_{jj}y_{ll} - y_{jl}^2),\ i \neq j \neq l \neq i, \quad \text{and} \quad \mathbf{V}(y_{kk}y_{jl} - y_{jk}y_{lk}),$$

with j, l, k distinct elements of $\{0, 1, \ldots, n\}\backslash\{i\}$. Since $n > 2$, it now follows that $\mathscr{K}$ is in all the primals

$$\mathbf{V}(y_{jj}y_{ll} - y_{jl}^2), \qquad j \neq l,$$

$$\mathbf{V}(y_{kk}y_{jl} - y_{jk}y_{lk}), \quad k \neq l \neq j \neq k.$$

By Lemma 25.1.1, $\mathscr{K} \subset \mathscr{V}_n$. Since $|\mathscr{K}| = |\mathscr{V}_n| = \theta(n)$ we have $\mathscr{K} = \mathscr{V}_n$.

By Corollary 1 of Theorem 25.1.9 the conic planes of $\mathscr{V}_n$ are the Γ-planes of $\mathscr{K}$. $\square$

In the statement of the theorem, saying that $\mathscr{K}$ has dimension n is equivalent to saying that $|\mathscr{K}| = \theta(n)$.

Finally, by Theorem 25.4.15, for dimension $n = 2$ condition (iv) may be deleted in the statement of the theorem.

25.5 Segre varieties

Consider k projective spaces $PG(n_1, K)$, $PG(n_2, K), \ldots, PG(n_k, K)$, with $n_i \geq 1$, over the field K, and suppose that $PG(n_i, K)$ consists of all points $\mathbf{P}(X^i)$ with

$$X^i = (x_0^{(i)}, x_1^{(i)}, \ldots, x_{n_i}^{(i)}).$$

Let $\bar{\mathbf{N}}_r = \{0, 1, \ldots, r\}$ for any $r \geq 1$, and let η be a bijection of $\bar{\mathbf{N}}_{n_1} \times \bar{\mathbf{N}}_{n_2} \times \cdots \times \bar{\mathbf{N}}_{n_k}$ onto $\bar{\mathbf{N}}_m$, with $m + 1 = (n_1 + 1)(n_2 + 1) \cdots (n_k + 1)$. Then the *Segre variety* of the k given projective spaces is the variety

$$\mathscr{S}_{n_1;n_2;\ldots;n_k} = \{\mathbf{P}(x_0, x_1, \ldots, x_m) \mid x_j = x_{(i_1, i_2, \ldots, i_k)\eta} = x_{i_1}^{(1)} x_{i_2}^{(2)} \cdots x_{i_k}^{(k)}$$
$$\text{with } \mathbf{P}(x_0^{(i)}, x_1^{(i)}, \ldots, x_{n_i}^{(i)}) \text{ a point of } PG(n_i, K)\}$$

of $PG(m, K)$. Since $(x_0^{(i)}, x_1^{(i)}, \ldots, x_{n_i}^{(i)}) \neq (0, 0, \ldots, 0)$ for all i, we have $(x_0, x_1, \ldots, x_m) \neq (0, 0, \ldots, 0)$. The integers $n_1, n_2, \ldots, n_k$ are called the *indices* of the variety. This variety has dimension $n_1 n_2 \cdots n_k$, and it can be shown that $\mathscr{S}_{n_1;n_2;\ldots;n_k}$ is absolutely irreducible and non-singular, with order equal to

$$\frac{(n_1 + n_2 + \cdots + n_k)!}{n_1!\, n_2! \cdots n_k!}.$$

Any point $\mathbf{P}(x_0, x_1, \ldots, x_m)$ of the Segre variety satisfies the equations

$$x_{(i_1, \ldots, i_k)\eta} x_{(j_1, \ldots, j_k)\eta} - x_{(i_1, \ldots, i_{s-1}, j_s, i_{s+1}, \ldots, i_k)\eta} x_{(j_1, \ldots, j_{s-1}, i_s, j_{s+1}, \ldots, j_k)\eta} = 0. \tag{25.35}$$

Theorem 25.5.1: *The Segre variety $\mathscr{S}_{n_1;n_2;\ldots;n_k}$ is the intersection of all quadrics of $PG(m, K)$ defined by the equations* (25.35), *and conversely any point of $PG(m, K)$ satisfying the equations* (25.35) *corresponds to a unique element of $PG(n_1, K) \times PG(n_2, K) \times \cdots \times PG(n_k, K)$.*

Proof: Let $\mathbf{P}(x_0', x_1', \ldots, x_m')$ be a point satisfying the equations (25.35). Without loss of generality we may assume that $x'_{(0,0,\ldots,0)\eta} = 1$. If the points $\mathbf{P}(x_0^{(i)}, x_1^{(i)}, \ldots, x_{n_i}^{(i)}) = P_i$ of $PG(n_i, K)$ define the given point of $PG(m, K)$, then $x_0^{(1)} x_0^{(2)} \cdots x_0^{(k)} \neq 0$. Hence we may assume that $x_0^{(1)} = x_0^{(2)} = \cdots = x_0^{(k)} = 1$. Then we have $x_{(i_1, 0, \ldots, 0)\eta} = x_{i_1}^{(1)}, x_{(0, i_2, 0, \ldots, 0)\eta} = x_{i_2}^{(2)}, \ldots, x_{(0,0,\ldots,0,i_k)\eta} = x_{i_k}^{(k)}$, with $i_s = 0, 1, \ldots, n_s$. Consequently the given point of $PG(m, K)$ corresponds to at most one element of $PG(n_1, K) \times PG(n_2, K) \times \cdots \times PG(n_k, K)$.

Consider the k points $P_i = \mathbf{P}(x_0^{(i)}, x_1^{(i)}, \ldots, x_{n_i}^{(i)})$, with $x'_{(i_1, 0, \ldots, 0)\eta} = x_{i_1}^{(1)}$,

$x'_{(0,i_2,0,\dots,0)\eta} = x_{i_2}^{(2)}, \dots, x'_{(0,0,\dots,0,i_k)\eta} = x_{i_k}^{(k)}$. For these points,

$$\begin{aligned} x'_{(i_1,i_2,\dots,i_k)\eta} &= x'_{(i_1,i_2,\dots,i_k)\eta} x'_{(0,0,\dots,0)\eta} \\ &= x'_{(i_1,i_2,\dots,i_{k-1},0)\eta} x'_{(0,0,\dots,0,i_k)\eta} \\ &= x'_{(i_1,i_2,\dots,i_{k-1},0)\eta} x_{i_k}^{(k)} \\ &= x'_{(i_1,i_2,\dots,i_{k-1},0)\eta} x'_{(0,0,\dots,0)\eta} x_{i_k}^{(k)} \\ &= x'_{(i_1,i_2,\dots,i_{k-2},0,0)\eta} x'_{(0,0,\dots,0,i_{k-1},0)\eta} x_{i_k}^{(k)} \\ &= x'_{(i_1,i_2,\dots,i_{k-2},0,0)\eta} x_{i_{k-1}}^{(k-1)} x_{i_k}^{(k)} \\ &= \cdots = x_{i_1}^{(1)} x_{i_2}^{(2)} \cdots x_{i_k}^{(k)} = x_{(i_1,i_2,\dots,i_k)\eta}. \end{aligned}$$

Hence to the element $(P_1, P_2, \dots, P_k) \in PG(n_1, K) \times PG(n_2, K) \times \cdots \times PG(n_k, K)$ there corresponds the given point of $PG(m, K)$. □

Let

$$\delta\colon PG(n_1, K) \times PG(n_2, K) \times \cdots \times PG(n_k, K) \to \mathscr{S}_{n_1;n_2;\dots;n_k}$$

be defined by

$$(\mathbf{P}(x_0^{(1)}, x_1^{(1)}, \dots, x_{n_1}^{(1)}), \dots, \mathbf{P}(x_0^{(k)}, x_1^{(k)}, \dots, x_{n_k}^{(k)})) \to \mathbf{P}(x_0, x_1, \dots, x_m),$$

with

$$x_j = x_{(i_1,i_2,\dots,i_k)\eta} = x_{i_1}^{(1)} x_{i_2}^{(2)} \cdots x_{i_k}^{(k)}.$$

By Theorem 25.5.1 the mapping δ is a bijection.

Theorem 25.5.2: *For given points $P_1, P_2, \dots, P_{i-1}, P_{i+1}, \dots, P_k$ of the respective spaces $PG(n_1, K)$, $PG(n_2, K), \dots,$ $PG(n_{i-1}, K)$, $PG(n_{i+1}, K), \dots,$ $PG(n_k, K)$, the set of all points $(P_1, P_2, \dots, P_k)\delta$, with $P_i \in PG(n_i, K)$, is an n_i-dimensional projective space.*

Proof: Up to order, the coordinates of $(P_1, P_2, \dots, P_k)\delta$ are of the form $x_0^{(i)} r_1$, $x_1^{(i)} r_1, \dots, x_{n_i}^{(i)} r_1, x_0^{(i)} r_2, x_1^{(i)} r_2, \dots, x_{n_i}^{(i)} r_2, \dots, x_0^{(i)} r_{(m+1)/(n_i+1)}, \dots, x_{n_i}^{(i)} r_{(m+1)/(n_i+1)}$, with $r_1, r_2, \dots, r_{(m+1)/(n_i+1)}$ constants which are not all zero. Since $\mathbf{P}(x_0^{(i)}, x_1^{(i)}, \dots, x_{n_i}^{(i)})$ is a variable point of $PG(n_i, K)$, it easily follows that the set of all points $(P_1, P_2, \dots, P_k)\delta$ is an n_i-dimensional projective space on $\mathscr{S}_{n_1;n_2;\dots;n_k}$. □

For $i = 1, 2, \dots, k$, the variation of $(P_1, P_2, \dots, P_{i-1}, P_{i+1}, \dots, P_k)$ gives a system Σ_i of n_i-dimensional projective spaces on $\mathscr{S}_{n_1;n_2;\dots;n_k}$.

Theorem 25.5.3: *Any two distinct elements of Σ_i are skew and each point of $\mathscr{S}_{n_1;n_2;\dots;n_k}$ is contained in exactly one element of each Σ_i. Also, an element of Σ_i meets an element of Σ_j, $i \neq j$, in at most one point.*

Proof: Let $\Pi_{n_i} \in \Sigma_i$ correspond to the points $P_1, P_2, \ldots, P_{i-1}, P_{i+1}, \ldots, P_k$, and let $\Pi'_{n_i} \in \Sigma_i$ correspond to the points $P'_1, P'_2, \ldots, P'_{i-1}, P'_{i+1}, \ldots, P'_k$, where $P_j, P'_j \in PG(n_j, K)$ and $(P_1, P_2, \ldots, P_{i-1}, P_{i+1}, \ldots, P_k) \neq (P'_1, P'_2, \ldots, P'_{i-1}, P'_{i+1}, \ldots, P'_k)$. For any points $P_i, P'_i \in PG(n_i, K)$ we have $(P_1, P_2, \ldots, P_k) \neq (P'_1, P'_2, \ldots, P'_k)$; so $(P_1, P_2, \ldots, P_k)\delta \neq (P'_1, P'_2, \ldots, P'_k)\delta$. This means that $\Pi_{n_i} \cap \Pi'_{n_i} = \varnothing$.

Let $(P_1, P_2, \ldots, P_k)\delta$ be any point of $\mathscr{S}_{n_1;n_2;\ldots;n_k}$; this point lies in the space Π_{n_i} of Σ_i corresponding to the points $P_1, P_2, \ldots, P_{i-1}, P_{i+1}, \ldots, P_k$.

Finally consider the space $\Pi_{n_i} \in \Sigma_i$ corresponding to the points P_1, $P_2, \ldots, P_{i-1}, P_{i+1}, \ldots, P_k$, and the space $\Pi_{n_j} \in \Sigma_j$ corresponding to the points $P'_1, P'_2, \ldots, P'_{j-1}, P'_{j+1}, \ldots, P'_k$, where $i \neq j$. If $|\Pi_{n_i} \cap \Pi_{n_j}| \neq \varnothing$, then $P_s = P'_s$ for all s with $s \neq i$ and $s \neq j$. If $P_s = P'_s$ for all s with $s \neq i$ and $s \neq j$, then $(P_1, P_2, \ldots, P_{i-1}, P'_i, P_{i+1}, \ldots, P_k)\delta$ is the unique common point of Π_{n_i} and Π_{n_j}. □

From now on we assume that K is the finite field $GF(q) = \gamma$, although many of the results also hold in the case of a general field.

Theorem 25.5.4: (i) $|\mathscr{S}_{n_1;n_2;\ldots;n_k}| = \theta(n_1)\theta(n_2)\cdots\theta(n_k)$.
(ii) $|\Sigma_i| = \theta(n_1)\theta(n_2)\cdots\theta(n_k)/\theta(n_i)$.

Proof: Since δ is a bijection from $PG(n_1, K) \times PG(n_2, K) \times \cdots \times PG(n_k, K)$ onto $\mathscr{S}_{n_1;n_2;\ldots;n_k}$ we have $|\mathscr{S}_{n_1;n_2;\ldots;n_k}| = \theta(n_1)\theta(n_2)\cdots\theta(n_k)$. By Theorem 25.5.3 the elements of Σ_i form a partition of $\mathscr{S}_{n_1;n_2;\ldots;n_k}$, and so $|\Sigma_i| = |\mathscr{S}_{n_1;n_2;\ldots;n_k}|/\theta(n_i)$. □

We give some examples. For $n_1 = n_2 = \cdots = n_k = 1$ we have $m = 2^k - 1$, and $\mathscr{S}_{n_1;n_2;\ldots;n_k}$ has order equal to $k!$. In this case the elements of Σ_i are lines, $|\mathscr{S}_{1;1;\ldots;1}| = (q+1)^k$, and $|\Sigma_i| = (q+1)^{k-1}$. Hence $\mathscr{S}_{1;1}$ is a hyperbolic quadric of $PG(3, q)$. For $k = 3$ we have $m = 7$, $\mathscr{S}_{1;1;1}$ has order 6, $|\mathscr{S}_{1;1;1}| = (q+1)^3$, and $|\Sigma_i| = (q+1)^2$.

Next, let us assume that $n_1 = n_2 = \cdots = n_k = n$. Then we have $m = (n+1)^k - 1$, and $\mathscr{S}_{n_1;n_2;\ldots;n_k}$ has order equal to $(kn)!/(n!)^k$. Further, $|\mathscr{S}_{n;n;\ldots;n}| = \theta(n)^k$ and $|\Sigma_i| = \theta(n)^{k-1}$. As an example, for $k = n = 2$, we have $m = 8$, $\mathscr{S}_{2;2}$ has order 6, $|\mathscr{S}_{2;2}| = (q^2+q+1)^2$, and $|\Sigma_i| = q^2+q+1$.

Now assume that $k = 2$. Then we have $m = (n_1+1)(n_2+1) - 1 = n_1n_2 + n_1 + n_2$, and $\mathscr{S}_{n_1;n_2}$ has order $(n_1+n_2)!/(n_1!n_2!)$. Further, $|\mathscr{S}_{n_1;n_2}| = \theta(n_1)\theta(n_2)$, $|\Sigma_1| = \theta(n_2)$, and $|\Sigma_2| = \theta(n_1)$. For $n_1 = n_2 = n$, we have $m = n(n+2)$, $\mathscr{S}_{n;n}$ has order $(2n)!/(n!)^2$, $|\mathscr{S}_{n;n}| = \theta(n)^2$, and $|\Sigma_i| = \theta(n)$. For $n_1 = 1$ and $n_2 = n$, we have $m = 2n+1$, $\mathscr{S}_{1;n}$ has order $n+1$, $|\mathscr{S}_{1;n}| = (q+1)\theta(n)$, $|\Sigma_1| = \theta(n)$, and $|\Sigma_2| = q+1$. As an example, for $n_1 = 1$ and $n_2 = 2$ we have $m = 5$, $\mathscr{S}_{1;3}$ has order 3, $|\mathscr{S}_{1;3}| = (q+1)(q^2+q+1)$, $|\Sigma_1| = q^2+q+1$, and $|\Sigma_2| = q+1$.

The Segre variety most studied is the variety $\mathscr{S}_{n_1;n_2}$. Hence we restrict ourselves to the case $k = 2$. However, several properties of $\mathscr{S}_{n_1;n_2}$ can be generalized to all Segre varieties.

Theorem 25.5.5: *On the Segre variety $\mathscr{S}_{n_1;n_2}$ each element of Σ_1 has exactly one point in common with each element of Σ_2.*

Proof: Let Π_{n_1} correspond to the point P_2 of $PG(n_2, q)$, and let Π_{n_2} correspond to the point P_1 of $PG(n_1, q)$; then $\Pi_{n_1} \cap \Pi_{n_2} = \{(P_1, P_2)\delta\}$. □

Theorem 25.5.6: *No hyperplane of $PG(m, q)$ contains the Segre variety $\mathscr{S}_{n_1;n_2}$.*

Proof: Suppose that the prime $\Pi_{m-1} = \mathbf{V}(F)$, with $F = \sum_j a_j x_j$, contains the Segre variety $\mathscr{S}_{n_1;n_2}$. Let $a_j = a_{(i_1,i_2)\eta} = b_{i_1 i_2}$. Then

$$\sum_{i_1=0}^{n_1} \sum_{i_2=0}^{n_2} b_{i_1 i_2} x_{i_1}^{(1)} x_{i_2}^{(2)} = 0$$

for all $x_0^{(1)}, x_1^{(1)}, \ldots, x_{n_1}^{(1)}$ and all $x_0^{(2)}, x_1^{(2)}, \ldots, x_{n_2}^{(2)}$. Fix the elements $x_0^{(1)}$, $x_1^{(1)}, \ldots, x_{n_1}^{(1)}$. Since

$$\sum_{i_2=0}^{n_2} \left(\sum_{i_1=0}^{n_1} b_{i_1 i_2} x_{i_1}^{(1)} \right) x_{i_2}^{(2)} = 0$$

for all $x_0^{(2)}, x_1^{(2)}, \ldots, x_{n_2}^{(2)}$, it follows that for any $i_2 \in \{0, 1, \ldots, n_2\}$ we have

$$\sum_{i_1=0}^{n_1} b_{i_1 i_2} x_{i_1}^{(1)} = 0$$

for all $x_0^{(1)}, x_1^{(1)}, \ldots, x_{n_1}^{(1)}$. Hence $b_{i_1 i_2} = 0$ for all $i_1 = 0, 1, \ldots, n_1$ and all $i_2 = 0, 1, \ldots, n_2$. So $a_0 = a_1 = \cdots = a_m = 0$, a contradiction. □

Let us now introduce the following notation: $x_0^{(1)} = y_0, x_1^{(1)} = y_1, \ldots, x_{n_1}^{(1)} = y_{n_1}$; $x_0^{(2)} = z_0$, $x_1^{(2)} = z_1, \ldots, x_{n_2}^{(2)} = z_{n_2}$; $x_j = x_{(i_1,i_2)\eta}$ will be denoted by $x_{i_1 i_2}$. Further, let $(i_1, i_2)\eta = i_1(n_2 + 1) + i_2$. The equations (25.35) become

$$x_{i_1 i_2} x_{j_1 j_2} - x_{j_1 i_2} x_{i_1 j_2} = 0. \tag{25.36}$$

Theorem 25.5.7: *The Segre variety $\mathscr{S}_{n_1;n_2}$ consists of all points $\mathbf{P}(X)$, $X = (x_{00}, x_{01}, \ldots, x_{0n_2}, x_{10}, \ldots, x_{1n_2}, \ldots, x_{n_1 n_2})$, of $PG(m, q)$ for which rank $[x_{ij}] = 1$.*

Proof: By Theorem 25.5.1 the Segre variety $\mathscr{S}_{n_1;n_2}$ consists of all points $\mathbf{P}(X)$, $X = (x_{00}, x_{01}, \ldots, x_{0n_2}, x_{10}, \ldots, x_{1n_2}, \ldots, x_{n_1 n_2})$, satisfying equations (25.36). Hence $\mathscr{S}_{n_1;n_2}$ consists of all points $\mathbf{P}(X)$ for which rank $[x_{ij}] = 1$. □

Theorem 25.5.8: *The intersection of the Segre variety* $\mathscr{S}_{n;n}$ *and the projective space* $\Pi_{n(n+3)/2} = \mathbf{V}(x_{01} - x_{10}, x_{02} - x_{20}, \ldots, x_{n-1,n} - x_{n,n-1})$ *is the quadric Veronesean* $\mathscr{V}_n$ *of all quadrics of* $PG(n, q)$.

Proof: This follows immediately from Theorems 25.5.7 and 25.1.2. □

Let ξ_i be a projectivity of $PG(n_i, q)$, $i = 1, 2$, and let ξ be defined by $(P_1, P_2)\delta\xi = (P_1\xi_1, P_2\xi_2)\delta$ for all P_1 in $PG(n_1, q)$ and all P_2 in $PG(n_2, q)$. Then ξ is a permutation of the Segre variety $\mathscr{S}_{n_1;n_2}$, for which $\Sigma_i\xi = \Sigma_i$, $i = 1, 2$.

If $n_1 = n_2 = n$, then let ψ_1 be a projectivity from $PG(n_1, q) = \Pi_n^1$ onto $PG(n_2, q) = \Pi_n^2$, and let ψ_2 be a projectivity from Π_n^2 onto Π_n^1. Let ψ be defined by $(P_1, P_2)\delta\psi = (P_2\psi_2, P_1\psi_1)\delta$ for all P_1 in Π_n^1 and all P_2 in Π_n^2. Then ψ is a permutation of the Segre variety $\mathscr{S}_{n;n}$ which interchanges Σ_1 and Σ_2.

Now, define $G(\mathscr{S}_{n_1;n_2})$ to be the subgroup of $PGL(m + 1, q)$, with $m = n_1n_2 + n_1 + n_2$, leaving $\mathscr{S}_{n_1;n_2}$ fixed.

Theorem 25.5.9: *The permutation* ξ *of* $\mathscr{S}_{n_1;n_2}$ *is induced by a unique element* $\tilde{\xi}$ *of* $G(\mathscr{S}_{n_1;n_2})$. *If* $n_1 = n_2 = n$, *then also the permutation* ψ *of* $\mathscr{S}_{n;n}$ *is induced by a unique element* $\tilde{\psi}$ *of* $G(\mathscr{S}_{n;n})$.

Proof: Let ξ_i be the projectivity of $PG(n_i, q)$ with matrix $A_i = [a_{jk}^{(i)}]$, $i = 1, 2$. If $(x'_{00}, x'_{01}, \ldots, x'_{n_1n_2}) = (x_{00}, x_{01}, \ldots, x_{n_1n_2})\xi$, then

$$x'_{i_1i_2} = y'_{i_1}z'_{i_2} = \sum_{r=0}^{n_1} a_{ri_1}^{(1)}y_r \sum_{s=0}^{n_2} a_{si_2}^{(2)}z_s = \sum_{r=0}^{n_1}\sum_{s=0}^{n_2} a_{ri_1}^{(1)}a_{si_2}^{(2)}y_rz_s$$

$$= \sum_{r=0}^{n_1}\sum_{s=0}^{n_2} a_{ri_1}^{(1)}a_{si_2}^{(2)}x_{rs}. \tag{25.37}$$

Hence ξ is induced by the element $\tilde{\xi}$ of $G(\mathscr{S}_{n_1;n_2})$ with matrix

$$A = \begin{bmatrix} a_{00}^{(1)}A_2 & a_{01}^{(1)}A_2 & \cdots & a_{0n_1}^{(1)}A_2 \\ a_{10}^{(1)}A_2 & a_{11}^{(1)}A_2 & \cdots & a_{1n_1}^{(1)}A_2 \\ \vdots & \vdots & & \vdots \\ a_{n_10}^{(1)}A_2 & a_{n_11}^{(1)}A_2 & \cdots & a_{n_1n_1}^{(1)}A_2 \end{bmatrix}.$$

This matrix A is simply the Kronecker product $A_1 \otimes A_2$ of the matrices A_1 and A_2. One consequence is that $|A| = |A_1 \otimes A_2| = |A_1|^{n_2+1}|A_2|^{n_1+1}$.

Assume that ξ is also induced by the element $\tilde{\xi}'$ of $G(\mathscr{S}_{n_1;n_2})$, where $\tilde{\xi}'$ has matrix $A' = [a'_{jk}]$. Then the projectivity $\tilde{\xi}'\tilde{\xi}^{-1}$ with matrix $A'A^{-1} = B$ induces the identity mapping of $\mathscr{S}_{n_1;n_2}$. Let $B = [b_{jk}]$ and put $b_{jk} = b_{(j_1j_2)(k_1k_2)}$, where $(j_1, j_2)\eta = j$ and $(k_1, k_2)\eta = k$. Then

$$ty_{k_1}z_{k_2} = \sum_{j_1=0}^{n_1} \sum_{j_2=0}^{n_2} b_{(j_1 j_2)(k_1 k_2)} y_{j_1} z_{j_2},$$

some t in γ, for all $\mathbf{P}(Y)$, $Y = (y_0, y_1, \ldots, y_{n_1})$, of $PG(n_1, q)$ and all $\mathbf{P}(Z)$, $Z = (z_0, z_1, \ldots, z_{n_2})$, of $PG(n_2, q)$. Putting $Y = (1, 0, \ldots, 0)$ and $Z = (1, 0, \ldots, 0)$, we find $b_{(00)(k_1 k_2)} = 0$ if $(k_1, k_2) \neq (0, 0)$. Proceeding in this way we obtain $b_{(j_1 j_2)(k_1 k_2)} = 0$ for $(k_1, k_2) \neq (j_1, j_2)$. So,

$$ty_{k_1}z_{k_2} = b_{(k_1 k_2)(k_1 k_2)} y_{k_1} z_{k_2},$$

for all $\mathbf{P}(Y)$ of $PG(n_1, q)$ and all $\mathbf{P}(Z)$ of $PG(n_2, q)$. Consequently, $b_{(k_1 k_2)(k_1 k_2)}$ is independent of k_1 and k_2. Hence $B = tI$ and so $\tilde{\xi}' = \tilde{\xi}$. It has therefore been shown that the permutation ξ of $\mathscr{S}_{n_1; n_2}$ is induced by a unique element $\tilde{\xi}$ of $G(\mathscr{S}_{n_1; n_2})$.

Now assume that $n_1 = n_2 = n$. Let ψ_1 be the projectivity from $PG(n_1, q) = \Pi_n^1$ onto $PG(n_2, q) = \Pi_n^2$ defined by the matrix D_1, and let ψ_2 be the projectivity from Π_n^2 onto Π_n^1 defined by the matrix D_2. Further, let ψ be defined by $(P_1, P_2)\delta\psi = (P_2\psi_2, P_1\psi_1)\delta$ for all P_1 in Π_n^1 and all P_2 in Π_n^2. Put $D_i = [d_{jk}^{(i)}]$, $i = 1, 2$, and $(x'_{00}, x'_{01}, \ldots, x'_{nn}) = (x_{00}, x_{01}, \ldots, x_{nn})\psi$. Then

$$\begin{aligned} x'_{i_1 i_2} = y'_{i_1} z'_{i_2} &= \sum_{r=0}^{n} d_{ri_1}^{(2)} z_r \sum_{s=0}^{n} d_{si_2}^{(1)} y_s = \sum_{r=0}^{n} \sum_{s=0}^{n} d_{ri_1}^{(2)} d_{si_2}^{(1)} z_r y_s \\ &= \sum_{s,r=0}^{n} d_{ri_1}^{(2)} d_{si_2}^{(1)} x_{sr}. \end{aligned} \tag{25.38}$$

Since (25.38) represents an element $\tilde{\psi}$ of $PGL(m+1, q)$, the permutation ψ is induced by the element $\tilde{\psi} \in G(\mathscr{S}_{n;n})$. Let $\tilde{\zeta}$ be the element of $PGL(m+1, q)$, with $m = n^2 + 2n$, defined by $x'_{sr} = x_{rs}$ for all $r, s = 0, 1, \ldots, m$. For any point P of $\mathscr{S}_{n;n}$, the coordinates of $P\tilde{\zeta}$ satisfy (25.36); so by Theorem 25.5.1 the point $P\tilde{\zeta}$ is in $\mathscr{S}_{n;n}$. Hence $\tilde{\zeta} \in G(\mathscr{S}_{n;n})$, and $\tilde{\psi}\tilde{\zeta} \in G(\mathscr{S}_{n;n})$ has matrix $D_1 \otimes D_2$. So by the first part of the proof the projectivity $\tilde{\psi}\tilde{\zeta}$ corresponds to (ξ_1, ξ_2) with ξ_i the projectivity of Π_n^i with matrix D_i, $i = 1, 2$.

Assume that ψ is also induced by the element $\tilde{\psi}'$ of $G(\mathscr{S}_{n;n})$. Then $\tilde{\psi}\tilde{\zeta}$ and $\tilde{\psi}'\tilde{\zeta}$ induce the same permutation of $\mathscr{S}_{n;n}$. Now by the preceding paragraph and the first part of the proof it follows that $\tilde{\psi}\tilde{\zeta} = \tilde{\psi}'\tilde{\zeta}$; hence $\tilde{\psi} = \tilde{\psi}'$. We conclude that ψ is induced by a unique element $\tilde{\psi}$ of $G(\mathscr{S}_{n;n})$. □

Since $\mathscr{S}_{n_1; n_2}$ is the intersection of the quadrics (25.36), any line l of $PG(m+1, q)$ meets $\mathscr{S}_{n_1; n_2}$ in 0, 1, 2, or $q+1$ points. In the next lemma we show that the lines of the elements of Σ_1 and Σ_2 are the only lines which are completely contained in $\mathscr{S}_{n_1; n_2}$.

Lemma 25.5.10: *Any line l of $\mathscr{S}_{n_1; n_2}$ is contained in an element of Σ_1 or Σ_2.*

Proof: Let l be a line of $\mathscr{S}_{n_1;n_2}$ and let P and P' be distinct points of l. Further, let $(P_1, P_2)\delta = P$ and $(P'_1, P'_2)\delta = P'$. Assume that $P_1 \neq P'_1$ and $P_2 \neq P'_2$. Now let ξ_1 be a projectivity of $PG(n_1, q)$ for which $P_1\xi_1 = U'_0$ and $P'_1\xi_1 = U'_1$, with $U'_0 = \mathbf{P}(1, 0, \ldots, 0)$ and $U'_1 = \mathbf{P}(0, 1, 0, \ldots, 0)$, and let ξ_2 be a projectivity of $PG(n_2, q)$ for which $P_2\xi_2 = U''_0$ and $P'_2\xi_2 = U''_1$, with $U''_0 = \mathbf{P}(1, 0, \ldots, 0)$ and $U''_1 = \mathbf{P}(0, 1, 0, \ldots, 0)$. The projectivity of $G(\mathscr{S}_{n_1;n_2})$ which corresponds to (ξ_1, ξ_2) is denoted by $\tilde{\xi}$. Then $l\tilde{\xi}$ is also a line of $\mathscr{S}_{n_1;n_2}$. The line $l\tilde{\xi}$ contains the points $(U'_0, U''_0)\delta = \mathbf{U}_0$ and $(U'_1, U''_1)\delta = \mathbf{U}_{n_2+2}$, where $\mathbf{U}_i = \mathbf{P}(E_i)$ with E_i the vector with one in the $(i + 1)$th place and zeros elsewhere. Hence $l\tilde{\xi}$ also contains the point $\mathbf{P}(E_0 + E_{n_2+2}) = R$. But the coordinates of R do not satisfy (25.36); hence $R \notin \mathscr{S}_{n_1;n_2}$, yielding a contradiction. Therefore, either $P_1 = P'_1$ or $P_2 = P'_2$; suppose the former.

Let P'' be any point of l, and let $(P''_1, P''_2)\delta = P''$. If $P''_1 \neq P_1$, then by the preceding paragraph we have $P''_2 = P_2$ and $P''_2 = P'_1$, a contradiction. Consequently $P''_1 = P_1$. We conclude that l is a line of the element of Σ_2 which corresponds to the point P_1 of $PG(n_1, q)$. □

An s-space Π_s which is contained in $\mathscr{S}_{n_1;n_2}$, but in no $(s + 1)$-space Π_{s+1} of $\mathscr{S}_{n_1;n_2}$, is called a *maximal space* or *maximal subspace* of $\mathscr{S}_{n_1;n_2}$. The next theorem describes all maximal spaces of $\mathscr{S}_{n_1;n_2}$.

Theorem 25.5.11: *The maximal spaces of the Segre variety $\mathscr{S}_{n_1;n_2}$ are the elements of Σ_1 and Σ_2.*

Proof: Let Π_s be a maximal subspace of $\mathscr{S}_{n_1;n_2}$, and assume that Π_s is not contained in an element of $\Sigma_1 \cup \Sigma_2$. Choose a point P in Π_s, and also a line l of Π_s through P. By Theorem 25.5.3 and Lemma 25.5.10 the line l is contained in a unique element π' of $\Sigma_1 \cup \Sigma_2$. Since Π_s is not contained in π', there exists a line l' through P which is contained in Π_s but not in π'. Let π'' be the unique element of $\Sigma_1 \cup \Sigma_2$ which contains l'; then $\pi' \cap \pi'' = \{P\}$. Since $\Pi_s \neq (\Pi_s \cap \pi') \cup (\Pi_s \cap \pi'')$, there exists a line l'' through P not in $\pi' \cup \pi''$. The line l'' is contained in a unique element π''' of $\Sigma_1 \cup \Sigma_2$. Thus P is contained in at least three distinct elements of $\Sigma_1 \cup \Sigma_2$, a contradiction.

Hence Π_s is contained in an element of $\Sigma_1 \cup \Sigma_2$. Since Π_s is maximal, it is an element of $\Sigma_1 \cup \Sigma_2$. □

Corollary 1: *Each s-space of $\mathscr{S}_{n_1;n_2}$, with $s > 0$, is contained in a unique element of $\Sigma_1 \cup \Sigma_2$.*

Proof: Let Π_s be an s-space of $\mathscr{S}_{n_1;n_2}$, with $s > 0$. This space is contained in a maximal subspace of $\mathscr{S}_{n_1;n_2}$, and hence, by the previous theorem, in an element π of $\Sigma_1 \cup \Sigma_2$. By Theorem 25.5.3 the space π is uniquely determined by Π_s. □

Corollary 2: *Let* $n_1 \leq n_2$. *The number of s-spaces contained in* $\mathscr{S}_{n_1;n_2}$ *is*

(i) $\theta(n_1)\phi(s; n_2, q) + \theta(n_2)\phi(s; n_1, q)$, *for* $0 < s \leqslant n_1$;

(ii) $\theta(n_1)\phi(s; n_2, q)$, *for* $n_1 < s \leqslant n_2$.

Proof: This is immediate from Corollary 1. □

Theorem 25.5.12: *Let* $P_i \in PG(n_i, q)$ *and let* $PG(d_i, q)$ *be a* d_i*-space of* $PG(n_i, q)$, $i = 1, 2$. *Then*

(i) $(\{P_1\} \times PG(d_2, q))\delta$ *is a* d_2*-subspace and* $(PG(d_1, q) \times \{P_2\})\delta$ *is a* d_1*-subspace of* $\mathscr{S}_{n_1;n_2}$;

(ii) *all subspaces of* $\mathscr{S}_{n_1;n_2}$ *are obtained as in* (i);

(iii) *when* $d_i > 0$, $i = 1, 2$, $(PG(d_1, q) \times PG(d_2, q))\delta$ *is a Segre variety* $\mathscr{S}_{d_1;d_2}$ *on* $\mathscr{S}_{n_1;n_2}$;

(iv) $\mathscr{S}_{d_1;d_2} = \mathscr{S}_{n_1;n_2} \cap PG(m', q)$, *where* $m' = d_1 d_2 + d_1 + d_2$ *and* $PG(m', q)$ *is the* m'*-space generated by* $\mathscr{S}_{d_1;d_2}$;

(v) *all Segre subvarieties of* $\mathscr{S}_{n_1;n_2}$ *are obtained as in* (iii).

Proof: Let $PG(d_i, q)$, $d_i \geq 0$, be a d_i-space of $PG(n_i, q)$, with $i = 1, 2$. By Theorem 25.5.9, coordinates can be chosen in such a way that $PG(d_1, q)$ contains the points $U'_0 = \mathbf{P}(E'_0)$, $U'_1 = \mathbf{P}(E'_1), \ldots, U'_{d_1} = \mathbf{P}(E'_{d_1})$, and that $PG(d_2, q)$ contains the points $U''_0 = \mathbf{P}(E''_0)$, $U''_1 = \mathbf{P}(E''_1), \ldots, U''_{d_2} = \mathbf{P}(E''_{d_2})$, with E'_i and E''_i vectors with one in the $(i + 1)$th place and zeros elsewhere. Then $(PG(d_1, q) \times PG(d_2, q))\delta = \mathscr{V}$ is the set of all points $(y_0 z_0, y_0 z_1, \ldots, y_0 z_{n_2}, y_1 z_0, \ldots, y_1 z_{n_2}, \ldots, y_{n_1} z_{n_2})$ with $z_{d_2+1} = z_{d_2+2} = \cdots = z_{n_2} = 0$ and $y_{d_1+1} = y_{d_1+2} = \cdots = y_{n_1} = 0$. If $d_1 = 0$, then $\mathscr{V}$ is a d_2-space of $\mathscr{S}_{n_1;n_2}$. If $d_2 = 0$, then $\mathscr{V}$ is a d_1-space of $\mathscr{S}_{n_1;n_2}$. If $d_1 > 0$ and $d_2 > 0$, then $\mathscr{V}$ clearly is a Segre variety $\mathscr{S}_{d_1;d_2}$. The subspace $PG(m', q)$, $m' = d_1 d_2 + d_1 + d_2$, of $PG(m, q)$ generated by $\mathscr{S}_{d_1;d_2}$ is the intersection of the primes $\mathbf{V}(x_{i_1 i_2})$, where we have at least one of $i_1 > d_1$ or $i_2 > d_2$. For any point of $\mathscr{S}_{n_1;n_2}$ in the intersection of these primes, $y_{i_1} = 0$ for $i_1 > d_1$ and $z_{i_2} = 0$ for $i_2 > d_2$. Hence $\mathscr{S}_{d_1;d_2} = \mathscr{S}_{n_1;n_2} \cap PG(m', q)$.

If $d_1 = 0$, then δ defines a projectivity from $PG(d_2, q)$ onto the d_2-space $(PG(d_1, q) \times PG(d_2, q))\delta$; if $d_2 = 0$, then δ defines a projectivity from $PG(d_1, q)$ onto the d_1-space $(PG(d_1, q) \times PG(d_2, q))\delta$.

Conversely, let Π_{d_2} be a d_2-space contained in $\mathscr{S}_{n_1;n_2}$. By Corollary 1 of Theorem 25.5.11, the space Π_{d_2} is contained in an element of $\Sigma_1 \cup \Sigma_2$. Assume for example that Π_{d_2} is contained in an element Π_{n_2} of Σ_1. Let $\Pi_{n_2}\delta^{-1} = \{P_1\} \times PG(n_2, q)$. Since δ defines a projectivity from $PG(n_2, q)$ onto Π_{n_2}, we have $\Pi_{d_2}\delta^{-1} = \{P_1\} \times PG(d_2, q)$, where $PG(d_2, q)$ is a d_2-dimensional subspace of $PG(n_2, q)$.

Next, let $\mathscr{S}_{d_1;d_2}$ be a Segre subvariety of $\mathscr{S}_{n_1;n_2}$. The systems of maximal subspaces of $\mathscr{S}_{d_1;d_2}$ are denoted by Σ'_1 and Σ'_2, where the elements of Σ'_i are contained in elements of Σ_i, $i = 1, 2$. Let $P \in \mathscr{S}_{d_1;d_2}$ and $P\delta^{-1} = (P_1, P_2)$. The element of Σ_1' containing P is denoted by Π_{d_2}, the element of Σ_2' containing P is denoted by Π_{d_1}. Let $\Pi_{d_2}\delta^{-1} = \{P_1\} \times PG(d_2, q)$, with $PG(d_2, q)$ a d_2-space of $PG(n_2, q)$, and $\Pi_{d_1}\delta^{-1} = PG(d_1, q) \times \{P_2\}$ with $PG(d_1, q)$ a d_1-space of $PG(n_1, q)$. The points of $\mathscr{S}_{d_1;d_2}$ are the points P', where $\{P'\} = \Pi'_{d_1} \cap \Pi'_{d_2}$ with $\Pi'_{d_1} \in \Sigma_2'$ and $\Pi'_{d_2} \in \Sigma_1'$. It follows that $\mathscr{S}_{d_1;d_2}$ consists of the points P', where $\{P'\} = \Pi_{n_1} \cap \Pi_{n_2}$ with Π_{n_1} any space of Σ_2 containing a point of Π_{d_2} and Π_{n_2} any space of Σ_1 containing a point of Π_{d_1}. Hence $\mathscr{S}_{d_1;d_2}\delta^{-1}$ consists of all ordered pairs (P_1', P_2') with P_1' in $PG(d_1, q)$ and P_2' in $PG(d_2, q)$. We conclude that $\mathscr{S}_{d_1;d_2} = (PG(d_1, q) \times PG(d_2, q))\delta$. □

Corollary 1: *Let $n_1 \leq n_2$. For given d_1 and d_2, with $0 < d_1 \leq n_1$, $0 < d_2 \leq n_2$, and $d_1 \leq d_2$, the number of Segre subvarieties $\mathscr{S}_{d_1;d_2}$ of the Segre variety $\mathscr{S}_{n_1;n_2}$ is equal to*

(i) $\phi(d_1; n_1, q)\phi(d_2; n_2, q) + \phi(d_1; n_2, q)\phi(d_2; n_1, q)$ *for* $d_1 < d_2 \leq n_1$;
(ii) $\phi(d_1; n_1, q)\phi(d_2; n_2, q)$ *for* $d_1 = d_2 \leq n_1$;
(iii) $\phi(d_1; n_1, q)\phi(d_2; n_2, q)$ *for* $d_1 \leq n_1 < d_2$.

Proof: This is immediate from Theorem 25.5.12. □

Corollary 2: *Let Π_s be an s-space of $\mathscr{S}_{n_1;n_2}$, $s \geq 1$, contained in an element Π_{n_1} of Σ_2. Then the elements of Σ_1 meeting Π_s in a point are the elements of a system of maximal spaces of a Segre subvariety $\mathscr{S}_{s;n_2}$ of $\mathscr{S}_{n_1;n_2}$.*

Proof: Let $\Pi_s\delta^{-1} = PG(s, q) \times \{P_2\}$, with P_2 a point of $PG(n_2, q)$ and $PG(s, q)$ an s-space of $PG(n_1, q)$. Then the elements of Σ_1 having a point in common with Π_s are the elements of a system of maximal spaces of the Segre subvariety $(PG(s, q) \times PG(n_2, q))\delta = \mathscr{S}_{s;n_2}$ of $\mathscr{S}_{n_1;n_2}$. □

Let ξ_i be a projectivity of $PG(n_i, q)$, $i = 1, 2$, and let $\tilde{\xi}$ be the corresponding element of $G(\mathscr{S}_{n_1;n_2})$. By Theorem 25.5.9 the map θ: $(\xi_1, \xi_2) \to \tilde{\xi}$ defines a monomorphism from $PGL(n_1 + 1, q) \times PGL(n_2 + 1, q)$ into $G(\mathscr{S}_{n_1;n_2})$. Now let $n_1 = n_2 = n$, let ζ_1 be the projectivity from $PG(n_1, q) = \Pi_n^1$ onto $PG(n_2, q) = \Pi_n^2$ with matrix I, and let ζ_2 be the projectivity from Π_n^2 onto Π_n^1 with matrix I. By (25.38) the element of $G(\mathscr{S}_{n;n})$ which corresponds to (ζ_1, ζ_2) is the element $\tilde{\zeta} \in PGL(m + 1, q)$, $m = n^2 + 2n$, defined by $x'_{sr} = x_{rs}$ for all $r, s = 0, 1, \ldots, m$.

Theorem 25.5.13: *For $n_1 \neq n_2$ the mapping θ: $(\xi_1, \xi_2) \to \tilde{\xi}$ is an isomorphism from $PGL(n_1 + 1, q) \times PGL(n_2 + 1, q)$ onto $G(\mathscr{S}_{n_1;n_2})$. For $n_1 = n_2 = n$,*

$$G(\mathscr{S}_{n;n}) = (PGL(n + 1, q) \times PGL(n + 1, q))\theta \cup ((PGL(n + 1, q) \times PGL(n + 1, q))\theta)\tilde{\zeta}.$$

Proof: Let $n_1 \neq n_2$, and let $\tilde{\xi} \in G(\mathscr{S}_{n_1;n_2})$. By Theorem 25.5.11 we have $\Sigma_1\tilde{\xi} = \Sigma_1$ and $\Sigma_2\tilde{\xi} = \Sigma_2$. Hence $\tilde{\xi}$ defines a permutation ξ_1 of $PG(n_1, q)$, and a permutation ξ_2 of $PG(n_2, q)$. Let l be any line of $PG(n_1, q)$. To the points of l there correspond the elements $\Pi^0_{n_2}, \Pi^1_{n_2}, \ldots, \Pi^q_{n_2}$ of a system of maximal n_2-spaces of the Segre subvariety $\mathscr{S}_{1;n_2} = (l \times PG(n_2, q))\delta$ of $\mathscr{S}_{n_1;n_2}$. By the second paragraph of the proof of Theorem 25.5.12, δ defines a projectivity from l onto any line $(l \times \{P_2\})\delta = m$, with $P_2 \in PG(n_2, q)$. The n_2-spaces $\Pi^0_{n_2}\tilde{\xi}$, $\Pi^1_{n_2}\tilde{\xi}, \ldots, \Pi^q_{n_2}\tilde{\xi}$ are the elements of a system of maximal n_2-spaces of the Segre subvariety $\mathscr{S}_{1;n_2}\tilde{\xi} = \mathscr{S}'_{1;n_2}$ of $\mathscr{S}_{n_1;n_2}$. By Theorem 25.5.12, $\mathscr{S}'_{1;n_2} = (l' \times PG(n_2, q))\delta$, with l' some line of $PG(n_1, q)$. Again by the proof of Theorem 25.5.12, δ defines a projectivity from l' onto the line $m\tilde{\xi}$. Hence $l\xi_1 = l'$, and ξ_1 induces a projectivity from l onto l'. It follows that ξ_1 is a projectivity of $PG(n_1, q)$. Analogously ξ_2 is a projectivity of $PG(n_2, q)$. Hence $(\xi_1, \xi_2)\theta$ is the given $\tilde{\xi}$. Therefore $(PGL(n_1 + 1, q) \times PGL(n_2 + 1, q))\theta = G(\mathscr{S}_{n_1;n_2})$, which proves the first part of the theorem.

Next, let $n_1 = n_2 = n$, let $PG(n_1, q) = \Pi^1_n$, and let $PG(n_2, q) = \Pi^2_n$. Consider any element $\tilde{\eta}$ in $G(\mathscr{S}_{n;n})$. We have $\Sigma_1\tilde{\eta} = \Sigma_1$ and $\Sigma_2\tilde{\eta} = \Sigma_2$, or $\Sigma_1\tilde{\eta} = \Sigma_2$ and $\Sigma_2\tilde{\eta} = \Sigma_1$. If $\Sigma_i\tilde{\eta} = \Sigma_i$, $i = 1, 2$, then as in the first part of the proof we show that there exists a projectivity ξ_i of Π^i_n, $i = 1, 2$, such that $(\xi_1, \xi_2)\theta = \tilde{\eta}$. Now suppose that we have $\Sigma_2\tilde{\eta} = \Sigma_1$ and $\Sigma_1\tilde{\eta} = \Sigma_2$. Then $\Sigma_2\tilde{\eta}\tilde{\zeta} = \Sigma_2$ and $\Sigma_1\tilde{\eta}\tilde{\zeta} = \Sigma_1$. Hence there exists a projectivity η_i of Π^i_n, $i = 1, 2$, such that $(\eta_1, \eta_2)\theta = \tilde{\eta}\tilde{\zeta}$. Since $\tilde{\zeta}^{-1} = \tilde{\zeta}$, we have $(\eta_1, \eta_2)\theta\tilde{\zeta} = \tilde{\eta}$. Consequently $G(\mathscr{S}_{n;n}) = (PGL(n + 1, q) \times PGL(n + 1, q))\theta \cup ((PGL(n + 1, q) \times PGL(n + 1, q))\theta)\tilde{\zeta}$. □

Corollary: (i) *For* $n_1 \neq n_2$,

$$|G(\mathscr{S}_{n_1;n_2})| = |PGL(n_1 + 1, q)||PGL(n_2 + 1, q)|.$$

(ii) *For* $n_1 = n_2 = n$,

$$|G(\mathscr{S}_{n;n})| = 2|PGL(n + 1, q)|^2.$$

Proof: This is immediate from Theorem 25.5.13. □

Theorem 25.5.14: *Let* $n_1 = n_2 = n$ *and let* ψ_1 *be a projectivity from* $PG(n_1, q) = \Pi^1_n$ *onto* $PG(n_2, q) = \Pi^2_n$. *Then the set of all points* $(P_1, P_1\psi_1)\delta$, *with* $P_1 \in \Pi^1_n$, *is a quadric Veronesean* $\mathscr{V}_n$.

Proof: Coordinates can be chosen in such a way that ψ_1 has matrix I. Then $(P_1, P_1\psi_1)\delta$ is the set of all points $(y_0^2, y_0y_1, \ldots, y_0y_n, y_1y_0, y_1^2, \ldots, y_1y_n, \ldots, y_n^2)$, with $(y_0, y_1, \ldots, y_n) \neq (0, 0, \ldots, 0)$. Hence $(P_1, P_1\psi_1)\delta$ is a quadric Veronesean $\mathscr{V}_n$. Since $(P_1, P_1\psi_1)\delta$ is the intersection of $\mathscr{S}_{n;n}$ and the projective space $\mathbf{V}(x_{01} - x_{10}, x_{02} - x_{20}, \ldots, x_{n-1,n} - x_{n,n-1})$, it is the quadric Veronesean $\mathscr{V}_n$ described in Theorem 25.5.8. □

Let ψ_1 be a projectivity from $PG(n_1, q) = \Pi_n^1$ onto $PG(n_2, q) = \Pi_n^2$ and let $\psi_2 = \psi_1^{-1}$. Further, let $\tilde{\psi}$ be the element of $G(\mathscr{S}_{n;n})$ which corresponds to the ordered pair (ψ_1, ψ_2). Then the points $P = (P_1, P_2)\delta$ of $\mathscr{S}_{n;n}$ which are fixed by $\tilde{\psi}$ are determined by $(P_2\psi_1^{-1}, P_1\psi_1)\delta = (P_1, P_2)\delta$. Hence these points are of the form $(P_1, P_1\psi_1)\delta$, with $P_1 \in \Pi_n^1$. By Theorem 25.5.14 the set of all these fixed points is a quadric Veronesean $\mathscr{V}_n$.

Finally, the coordinates of the maximal spaces of $\mathscr{S}_{n_1;n_2}$ and the images of these spaces on the corresponding Grassmannians merit a brief description. Consider the element $(PG(n_1, q) \times \{P_2\})\delta = \Pi_{n_1}$ of Σ_1, where $P_2 = \mathbf{P}(Z)$ and $Z = (z_0, z_1, \ldots, z_{n_2})$. The projective space Π_{n_1} is generated by the independent points $R_i = \mathbf{P}(X_i)$, $i = 0, 1, \ldots, n_1$, where $X_0 = (z_0, z_1, \ldots, z_{n_2}, 0, 0, \ldots, 0)$, $X_1 = (0, 0, \ldots, 0, z_0, z_1, \ldots, z_{n_2}, 0, 0, \ldots, 0)$ with z_0 in the $(n_2 + 2)$th place, $\ldots$, $X_{n_1} = (0, 0, \ldots, 0, z_0, z_1, \ldots, z_{n_2})$. The coordinates of Π_{n_1} are denoted by $(i_0 i_1 \cdots i_{n_1})$, with $i_0 < i_1 < \cdots < i_{n_1}$ and $\{i_0, i_1, \ldots, i_{n_1}\}$ a subset of order $n_1 + 1$ of $\{0, 1, \ldots, n_1 n_2 + n_1 + n_2 = m\}$. Let $V_k = \{k(n_2 + 1), k(n_2 + 1) + 1, \ldots, k(n_2 + 1) + n_2\}$, $k = 0, 1, \ldots, n_1$, and $i_{ks} = k(n_2 + 1) + s$. Then $(i_0 i_1 \cdots i_{n_1}) = 0$ if $(i_0, i_1, \ldots, i_{n_1}) \notin V_0 \times V_1 \times \cdots \times V_{n_1}$. Now let $(i_0, i_1, \ldots, i_{n_1}) \in V_0 \times V_1 \times \cdots \times V_{n_1}$, with $i_0 = i_{0s_0}$, $i_1 = i_{1s_1}, \ldots, i_{n_1} = i_{n_1 s_{n_1}}$. Then we have $(i_0 i_1 \cdots i_{n_1}) = z_{s_0} z_{s_1} \cdots z_{s_{n_1}}$. Analogous remarks can be made about the elements of Σ_2.

Let $\Sigma_1\mathfrak{G}$ be the image of Σ_1 onto the Grassmannian $\mathscr{G}_{n_1,m}$, with $m = n_1 n_2 + n_1 + n_2$. If $n_1 = 1$, then from the observations in the preceding paragraph it follows that $\Sigma_1\mathfrak{G}$ is the Veronesean of quadrics of $PG(n_2, q)$. If $n_2 = 1$, then $\Sigma_1\mathfrak{G}$ is a normal rational curve of order $n_1 + 1$. In particular, if $n_1 = n_2 = 1$ then $\Sigma_1\mathfrak{G}$ and $\Sigma_2\mathfrak{G}$ are conics of the quadric $\mathscr{G}_{1,3}$.

25.6 Regular n-spreads and Segre varieties $\mathscr{S}_{1;n}$

In §17.1 regular spreads of lines in $PG(3, q)$ were studied in detail. In particular great attention was paid to the reguli contained in such a spread. Some of the results will be extended here to regular spreads of n-spaces in $PG(2n + 1, q)$.

A spread of n-spaces in $PG(2n + 1, q)$ will be called an *n-spread* of $PG(2n + 1, q)$, and a system of maximal n-spaces of a Segre variety $\mathscr{S}_{1;n}$ will be called an *n-regulus*. Hence a regulus is simply a 1-regulus.

Theorem 25.6.1: *If Π_n, Π_n', Π_n'' are mutually skew n-spaces in $PG(2n + 1, q)$, $n \geq 1$, then the set of all lines having a non-empty intersection with Π_n, Π_n', and Π_n'' is a system of maximal spaces of a Segre variety $\mathscr{S}_{1;n}$.*

Proof: Coordinates are chosen in such a way that Π_n contains the points $\mathbf{U}_0 = \mathbf{P}(E_0)$, $\mathbf{U}_1 = \mathbf{P}(E_1), \ldots, \mathbf{U}_n = \mathbf{P}(E_n)$, where E_i is the vector with one in the $(i + 1)$th place and zeros elsewhere. Through each point $\mathbf{U}_i$, with

$i = 0, 1, \ldots, n$, there is exactly one line l_i meeting Π'_n and Π''_n in a point. Assume that the common points of the lines $l_0, l_1, \ldots, l_n$ and the space Π'_n generate a projective space $\Pi'_{n'}$, and that the common points of these lines with the space Π''_n generate a projective space $\Pi''_{n''}$. The $(n' + n'' + 1)$-space generated by $\Pi'_{n'}$ and $\Pi''_{n''}$ contains the points $\mathbf{U}_0, \mathbf{U}_1, \ldots, \mathbf{U}_n$ and hence contains Π_n. Since $\Pi_n \cap \Pi'_{n'} = \varnothing$ we necessarily have $n = n''$; analogously $n = n'$. Hence take $l_i \cap \Pi'_n = \{\mathbf{U}_{i+n+1}\}$ with $\mathbf{U}_{i+n+1} = \mathbf{P}(E_{i+n+1})$, $i = 0, 1, \ldots, n$. Let $l_i \cap \Pi''_n = \{Q_i\}$, $i = 0, 1, \ldots, n$, and let U be a point of Π''_n contained in none of the $(n-1)$-spaces generated by n of the points $Q_0, Q_1, \ldots, Q_n$. Then U may be taken as $\mathbf{P}(E)$ with $E = (1, 1, \ldots, 1)$. Now it is easy to verify that $Q_i = \mathbf{P}(E_{i,i+n+1})$, $i = 0, 1, \ldots, n$, with $E_{i,i+n+1}$ the vector with one in the $(i+1)$th and $(i+n+2)$th places and zeros elsewhere.

Let $P = \mathbf{P}(Z)$, with $Z = (z_0, z_1, \ldots, z_n, 0, 0, \ldots, 0)$, be any point of the space Π_n, and let l be the line through $\mathbf{P}(Z)$ having a non-empty intersection with Π'_n and Π''_n. Let $l \cap \Pi'_n = \{P'\}$, with $P' = \mathbf{P}(Z')$ and $Z' = (0, 0, \ldots, 0, z'_0, z'_1, \ldots, z'_n)$, and $l \cap \Pi''_n = \{P''\}$, with $P'' = \mathbf{P}(Z'')$. Then $Z'' = (r_0 z_0, r_0 z_1, \ldots, r_0 z_n, r_1 z'_0, r_1 z'_1, \ldots, r_1 z'_n)$ with $r_0 \neq 0 \neq r_1$. Since Z'' is a linear combination of the vectors $E_{0,n+1}, E_{1,n+2}, \ldots, E_{n,2n+1}$, it follows that $r_0 z_i = r_1 z'_i$, $i = 0, 1, \ldots, n$. Hence take $z_i = z'_i$, $i = 0, 1, \ldots, n$. Then any point of the line l is of the form $\mathbf{P}(X)$, where $X = (y_0 z_0, y_0 z_1, \ldots, y_0 z_n, y_1 z_0, y_1 z_1, \ldots, y_1 z_n)$ with $(y_0, y_1) \neq (0, 0)$. Now it is clear that all lines l form a system of maximal spaces of a Segre variety $\mathscr{S}_{1;n}$. □

Corollary: *If Π_n, Π'_n, Π''_n are mutually skew n-spaces in $PG(2n+1, q)$, $n \geq 1$, then there is exactly one n-regulus which contains Π_n, Π'_n, and Π''_n.*

Proof: A Segre variety $\mathscr{S}_{1;n}$ containing the spaces Π_n, Π'_n, Π''_n is necessarily the union of the $\theta(n)$ lines having a point in common with Π_n, Π'_n, and Π''_n. In Theorem 25.6.1 it was proved that this union is indeed a Segre variety $\mathscr{S}_{1;n}$. The system of maximal spaces containing Π_n, Π'_n, Π''_n of this unique Segre variety $\mathscr{S}_{1;n}$ is the unique n-regulus containing Π_n, Π'_n, Π''_n. □

The n-regulus containing Π_n, Π'_n, and Π''_n will be denoted by $\mathscr{R}(\Pi_n, \Pi'_n, \Pi''_n)$.

Theorem 25.6.2: *The number of all n-reguli of $PG(2n+1, q)$ is*

$$|PGL(2n+2, q)|/(|PGL(2, q)||PGL(n+1, q)|).$$

Proof: The number of all n-reguli of $PG(2n+1, q)$, $n > 1$, is the number of all Segre varieties $\mathscr{S}_{1;n}$ in $PG(2n+1, q)$, which is $|PGL(2n+2, q)|/|G(\mathscr{S}_{1;n})|$. So by the corollary of Theorem 25.5.13 this number is equal to $|PGL(2n+2, q)|/(|PGL(2, q)||PGL(n+1, q)|)$.

The number of all reguli of $PG(3, q)$ is twice the number of Segre varieties

$\mathscr{S}_{1;1}$ in $PG(3, q)$. The number of Segre varieties $\mathscr{S}_{1;1}$ is equal to $|PGL(4, q)|/|G(\mathscr{S}_{1;1})|$. By the corollary of Theorem 25.5.13, this number of varieties $\mathscr{S}_{1;1}$ is equal to $|PGL(4, q)|/(2|PGL(2, q)|^2)$. Hence the number of all reguli equals $|PGL(4, q)|/|PGL(2, q)|^2$. □

Let $l_0, l_1, \ldots, l_n$ be $n + 1$ lines of $PG(2n + 1, q^{n+1})$ which generate the space and are conjugate with respect to the $(n + 1)$th extension $GF(q^{n+1})$ of $GF(q)$. Further, let P_0 be any point of l_0 and let $P_1, P_2, \ldots, P_n$ be the points conjugate to P_0. Then $P_i \in l_i$, with $i = 0, 1, \ldots, n$, and the points $P_0, P_1, \ldots, P_n$ generate an n-space $\bar{\Pi}_n$ of $PG(2n + 1, q^{n+1})$. The intersection of $\bar{\Pi}_n$ and $PG(2n + 1, q)$ is an n-space Π_n of $PG(2n + 1, q)$. The set of these $q^{n+1} + 1$ spaces Π_n is denoted by $\mathscr{S}(l_0, l_1, \ldots, l_n)$. By Lemma 17.1.2 the set $\mathscr{S}(l_0, l_1)$ is an elliptic congruence of $PG(3, q)$, or equivalently a regular spread of lines of $PG(3, q)$.

Lemma 25.6.3: *The set* $\mathscr{S}(l_0, l_1, \ldots, l_n)$ *is an n-spread of* $PG(2n + 1, q)$.

Proof: Let Π_n and Π'_n correspond to distinct points P_0 and P'_0 of l_0. Then $\bar{\Pi}_n \cap l_i \neq \bar{\Pi}'_n \cap l_i$, $i = 0, 1, \ldots, n$, with $\bar{\Pi}_n$ the extension of Π_n and $\bar{\Pi}'_n$ the extension of Π'_n. Assume that $\Pi_n \cap \Pi'_n \neq \varnothing$; the space generated by Π_n and Π'_n has dimension less than $2n + 1$ and its extension contains the lines $l_0, l_1, \ldots, l_n$, a contradiction. Hence $\Pi_n \cap \Pi'_n = \varnothing$. Since $|\mathscr{S}(l_0, l_1, \ldots, l_n)| = |l_0| = q^{n+1} + 1$, the set $\mathscr{S}(l_0, l_1, \ldots, l_n)$ is a partition of $PG(2n + 1, q)$. We conclude that $\mathscr{S}(l_0, l_1, \ldots, l_n)$ is an n-spread of $PG(2n + 1, q)$. □

For $n > 1$ an n-spread $\mathscr{S}$ of $PG(2n + 1, q)$ is called *regular* if there exist lines $l_0, l_1, \ldots, l_n$ of $PG(2n + 1, q^{n+1})$ for which $\mathscr{S} = \mathscr{S}(l_0, l_1, \ldots, l_n)$.

Theorem 25.6.4: *The following are equivalent*:

(i) *if* Π_n, Π'_n, Π''_n *are any three distinct elements of the n-spread* $\mathscr{S}$ *of* $PG(2n + 1, q)$, *then the whole n-regulus* $\mathscr{R}(\Pi_n, \Pi'_n, \Pi''_n)$ *is contained in* $\mathscr{S}$;

(ii) $\mathscr{S}$ *is an n-spread of* $PG(2n + 1, q)$ *such that the n-spaces of* $\mathscr{S}$ *meeting any line not in an element of* $\mathscr{S}$ *form an n-regulus.*

Finally, any regular n-spread $\mathscr{S}$ *satisfies* (i) *and* (ii).

Proof: (i) ⇒ (ii). Suppose that $\mathscr{S}$ satisfies (i), and let l be a line which is not contained in an element of the n-spread $\mathscr{S}$. Further, let P, P', P'' be distinct points of l and let Π_n, Π'_n, Π''_n be the n-spaces of $\mathscr{S}$ containing these points. Then $\mathscr{R}(\Pi_n, \Pi'_n, \Pi''_n) \subset \mathscr{S}$. The elements of $\mathscr{R}(\Pi_n, \Pi'_n, \Pi''_n)$ are the n-spaces of $\mathscr{S}$ containing a point of l. Hence the n-spaces of $\mathscr{S}$ meeting the line l form an n-regulus.

(ii) ⇒ (i). Suppose that $\mathscr{S}$ satisfies (ii), and let Π_n, Π'_n, Π''_n be distinct elements of $\mathscr{S}$. Further, let l be a line meeting Π_n, Π'_n, and Π''_n. The n-regulus consisting of the $q+1$ n-spaces of $\mathscr{S}$ meeting l contains Π_n, Π'_n, and Π''_n, and hence coincides with $\mathscr{R}(\Pi_n, \Pi'_n, \Pi''_n)$. Consequently $\mathscr{R}(\Pi_n, \Pi'_n, \Pi''_n) \subset \mathscr{S}$.

Finally, let $\mathscr{S}$ be the regular n-spread $\mathscr{S}(l_0, l_1, \ldots, l_n)$. Consider a line l which is not contained in an element of $\mathscr{S}$. The elements of $\mathscr{S}$ meeting l are denoted by $\Pi_n^0, \Pi_n^1, \ldots, \Pi_n^q$. The $q+1$ spaces Π_n of $\mathscr{R}(\Pi_n^0, \Pi_n^1, \Pi_n^2)$ meet l, $l_0, l_1, \ldots, l_n$ in the extension $PG(2n+1, q^{n+1})$ of $PG(2n+1, q)$. Hence $\mathscr{R}(\Pi_n^0, \Pi_n^1, \Pi_n^2) = \{\Pi_n^0, \Pi_n^1, \ldots, \Pi_n^q\}$, and so $\{\Pi_n^0, \Pi_n^1, \ldots, \Pi_n^q\}$ is an n-regulus. Consequently (ii) is satisfied. By the previous section of the proof, also (i) is satisfied. □

The following interesting theorem was proved by Bruck and Bose.

Theorem 25.6.5: *For $q > 2$ any n-spread $\mathscr{S}$ of $PG(2n+1, q)$ which satisfies* (i) *or* (ii) *in the statement of Theorem* 25.6.4 *is regular.* □

For $q = 2$ conditions (i) and (ii) are trivially satisfied. Many examples of non-regular n-spreads in $PG(2n+1, 2)$ are known.

Theorem 25.6.6: (i) *The number of n-reguli contained in a regular n-spread of $PG(2n+1, q)$ is*

$$q^n(q^{2n+2}-1)/(q^2-1) = q^n(q^{2n} + q^{2n-2} + q^{2n-4} + \cdots + q^2 + 1).$$

(ii) *The number of regular n-spreads of $PG(2n+1, q)$ is*

$$q^{2n(n+1)}[1, 2n+1]_-/\{(q^{n+1}-1)(n+1)\}.$$

Proof: (i) Let $\mathscr{S}$ be a regular n-spread of $PG(2n+1, q)$. By Theorem 25.6.4 the number of n-reguli contained in $\mathscr{S}$ is the number of subsets of order 3 of $\mathscr{S}$ divided by the number of subsets of order 3 of an n-regulus. Hence this number is

$$(q^{n+1}+1)q^{n+1}(q^{n+1}-1)/\{(q+1)q(q-1)\} = q^n(q^{2n+2}-1)/(q^2-1).$$

(ii) Let $\Pi_n^0, \Pi_n^1, \Pi_n^2$ be mutually skew n-spaces of $PG(2n+1, q)$. For any point $P \in \Pi_n^0$, let l be the line containing P and meeting the n-spaces Π_n^1 and Π_n^2, and let $l \cap \Pi_n^1 = \{P'\}$ and $l \cap \Pi_n^2 = \{P''\}$. We show that $\psi: P' \to P''$ is a projectivity from Π_n^1 onto Π_n^2. By the proof of Theorem 25.6.1 coordinates can be chosen in such a way that the Segre variety $\mathscr{S}_{1;n}$ containing $\mathscr{R}(\Pi_n^0, \Pi_n^1, \Pi_n^2)$ consists of all points $\mathbf{P}(X)$ with $X = (y_0z_0, y_0z_1, \ldots, y_0z_n, y_1z_0, y_1z_1, \ldots, y_1z_n)$ and that $\psi: P' = \mathbf{P}(0, 0, \ldots, 0, z_0, z_1, \ldots, z_n) \to P'' = \mathbf{P}(z_0, z_1, \ldots, z_n, 0, 0, \ldots, 0)$ for all $(z_0, z_1, \ldots, z_n) \neq (0, 0, \ldots, 0)$. Hence ψ is a projectivity from Π_n^1 onto Π_n^2.

Next, let $\mathscr{S}$ be a regular n-spread of $PG(2n+1, q)$, where $\mathscr{S} = \mathscr{S}(l_0,$

$l_1, \ldots, l_n) = \mathscr{S}(m_0, m_1, \ldots, m_n)$. Consider elements Π_n^0, Π_n^1, Π_n^2, Π_n^3 of $\mathscr{S}$ which do not belong to the same n-regulus. For any point P of Π_n^0, let l be the line containing P and meeting the n-spaces Π_n^1 and Π_n^2, and let l' be the line containing P and meeting the n-spaces Π_n^1 and Π_n^3. Further, let $l \cap \Pi_n^1 = \{P'\}$ and $l' \cap \Pi_n^1 = \{P''\}$. Then $\psi: P \to P'$ and $\psi': P \to P''$ are projectivities from Π_n^0 onto Π_n^1. Hence $\psi^{-1}\psi' = \delta: P' \to P''$ is a projectivity of Π_n^1 onto itself. Since $\Pi_n^0, \Pi_n^1, \Pi_n^2, \Pi_n^3$ do not belong to the same n-regulus, δ is not the identity. In the extension $PG(2n+1, q^{n+1})$ of $PG(2n+1, q)$, let $l_i \cap \bar{\Pi}_n^1 = \{P_i\}$ and $m_i \cap \bar{\Pi}_n^1 = \{Q_i\}$, $i = 0, 1, \ldots, n$, where $\bar{\Pi}_n^1$ is the extension of Π_n^1. In $PG(2n+1, q^{n+1})$ the points $P_0, P_1, \ldots, P_n, Q_0, Q_1, \ldots, Q_n$ are fixed by δ. Since the conjugate points $P_0, P_1, \ldots, P_n$, and also $Q_0, Q_1, \ldots, Q_n$, are linearly independent we necessarily have $\{P_0, P_1, \ldots, P_n\} = \{Q_0, Q_1, \ldots, Q_n\}$. It follows that $\{l_0, l_1, \ldots, l_n\} = \{m_0, m_1, \ldots, m_n\}$.

Now we calculate the number of all regular n-spreads of $PG(2n+1, q)$ containing a given n-regulus $\mathscr{R}$. Let $\Pi_n \in \mathscr{R}$ and let $\mathscr{S}$ be a regular n-spread containing $\mathscr{R}$. Then $\mathscr{S} = \mathscr{S}(l_0, l_1, \ldots, l_n)$ where $\{l_0, l_1, \ldots, l_n\}$ is uniquely defined by $\mathscr{S}$. The points P_i with $\{P_i\} = l_i \cap \bar{\Pi}_n$ and with $\bar{\Pi}_n$ the extension of Π_n, $i = 0, 1, \ldots, n$, are linearly independent in $\bar{\Pi}_n$. Conversely, let us consider $n+1$ linearly independent and conjugate points $P_0, P_1, \ldots, P_n$ of $\bar{\Pi}_n$. If $\bar{\mathscr{S}}_{1;n}$ is the extension of the Segre variety $\mathscr{S}_{1;n}$ defined by $\mathscr{R}$, then $\bar{\mathscr{S}}_{1;n}$ contains exactly one line l_i through P_i which meets all elements of $\mathscr{R}$, $i = 0, 1, \ldots, n$. The lines $l_0, l_1, \ldots, l_n$ are conjugate and generate $PG(2n+1, q^{n+1})$, and so they define a unique regular n-spread $\mathscr{S} = \mathscr{S}(l_0, l_1, \ldots, l_n)$. Hence the number of regular n-spreads containing $\mathscr{R}$ is the number of sets $\{P_0, P_1, \ldots, P_n\}$. Let $\Pi_n = \mathbf{V}(x_{n+1}, x_{n+2}, \ldots, x_{2n+1})$ and let $P_0 = \mathbf{P}(X_0)$,

$$X_0 = (a_{00} + a_{01}\alpha + \cdots + a_{0n}\alpha^n, a_{10} + a_{11}\alpha + \cdots + a_{1n}\alpha^n, \ldots, a_{n0} + a_{n1}\alpha + \cdots + a_{nn}\alpha^n, 0, 0, \ldots, 0),$$

with $a_{ij} \in GF(q)$ and $GF(q^{n+1}) = \{a_0 + a_1\alpha + \cdots + a_n\alpha^n \mid a_i \in GF(q)\}$. The points P_i, $i = 1, 2, \ldots, n$, which are conjugate to P_0 are the points $\mathbf{P}(X_i)$,

$$X_i = (a_{00} + a_{01}\alpha^{q^i} + \cdots + a_{0n}\alpha^{nq^i}, \ldots, a_{n0} + a_{n1}\alpha^{q^i} + \cdots + a_{nn}\alpha^{nq^i}, 0, \ldots, 0).$$

The points $P_0, P_1, \ldots, P_n$ are linearly independent in $\bar{\Pi}_n$ if and only if $\Delta\Delta' \neq 0$, with

$$\Delta = \begin{vmatrix} 1 & \alpha & \alpha^2 & \cdots & \alpha^n \\ 1 & \alpha^q & (\alpha^q)^2 & \cdots & (\alpha^q)^n \\ \vdots & \vdots & \vdots & & \vdots \\ 1 & \alpha^{q^n} & (\alpha^{q^n})^2 & \cdots & (\alpha^{q^n})^n \end{vmatrix} \quad \text{and} \quad \Delta' = \begin{vmatrix} a_{00} & a_{01} & \cdots & a_{0n} \\ a_{10} & a_{11} & \cdots & a_{1n} \\ \vdots & \vdots & & \vdots \\ a_{n0} & a_{n1} & \cdots & a_{nn} \end{vmatrix}.$$

Since

$$\Delta = \prod_{\substack{i,j=0 \\ i>j}}^{n} (\alpha^{q^i} - \alpha^{q^j}),$$

we have $\Delta \neq 0$. Consequently the points $P_0, P_1, \ldots, P_n$ are linearly independent if and only if $\Delta' \neq 0$. So the number of such sets $\{P_0, P_1, \ldots, P_n\}$ is equal to $|GL(n+1, q)|/\{(q^{n+1} - 1)(n+1)\}$. This is also the number of all regular n-spreads containing $\mathscr{R}$.

From Theorem 25.6.2 and the first part of Theorem 25.6.6, it now follows that the number of all regular n-spreads of $PG(2n+1, q)$ is

$$\frac{|PGL(2n+2, q)|}{|PGL(2, q)||PGL(n+1, q)|} \cdot \frac{|GL(n+1, q)|}{(q^{n+1} - 1)(n+1)} \cdot \frac{q^2 - 1}{q^n(q^{2n+2} - 1)}$$

$$= q^{2n(n+1)} \prod_{i=1}^{2n+1} (q^i - 1)/\{(q^{n+1} - 1)(n+1)\}. \quad \square$$

Corollary: *The number of lines l_0 of $PG(2n+1, q^{n+1})$, for which l_0 together with its conjugates $l_1, l_2, \ldots, l_n$ generate $PG(2n+1, q^{n+1})$, is*

$$q^{2n(n+1)}[1, 2n+1]_-/(q^{n+1} - 1).$$

Proof: To each such line l_0 there corresponds one regular n-spread $\mathscr{S}(l_0, l_1, \ldots, l_n)$ of $PG(2n+1, q)$, and to each regular n-spread there correspond $n+1$ of these lines. Hence the number of such lines l_0 is equal to the number of regular n-spreads times $n+1$. $\square$

Theorem 25.6.7: *The group $PGL(2n+2, q)$ acts transitively on the set of all regular n-spreads. The subgroup $G(\mathscr{S})$ of $PGL(2n+2, q)$ consisting of all projectivities fixing a given regular n-spread $\mathscr{S}$ has order*

$$(n+1)q^{n+1}(q^{n+1} - 1)(q^{2n+2} - 1)/(q - 1).$$

Proof: Let $\mathscr{S} = \mathscr{S}(l_0, l_1, \ldots, l_n)$ and $\mathscr{S}' = \mathscr{S}'(l'_0, l'_1, \ldots, l'_n)$ be regular n-spreads of $PG(2n+1, q)$. Further let $\Pi_n^0, \Pi_n^1, \Pi_n^2$ be distinct elements of $\mathscr{S}$, and let $\Pi_n^{0\prime}, \Pi_n^{1\prime}, \Pi_n^{2\prime}$ be distinct elements of $\mathscr{S}'$. There is an element ξ in $PGL(2n+2, q)$ for which $\Pi_n^{i\prime}\xi = \Pi_n^i$, $i = 0, 1, 2$. By Theorem 25.6.1 we have $\mathscr{R}(\Pi_n^{0\prime}, \Pi_n^{1\prime}, \Pi_n^{2\prime})\xi = \mathscr{R}(\Pi_n^0, \Pi_n^1, \Pi_n^2)$. Let $l'_0\xi \cap \bar{\Pi}_n^0 = \{P'_0\}$ and $l_0 \cap \bar{\Pi}_n^0 = \{P_0\}$, with $\bar{\Pi}_n^0$ the extension of Π_n^0. Coordinates are chosen so that $\Pi_n^0 = \mathbf{V}(x_{n+1}, x_{n+2}, \ldots, x_{2n+1})$ and $P_0 = \mathbf{P}(X_0)$, $P'_0 = \mathbf{P}(X'_0)$,

$$X_0 = (a_{00} + a_{01}\alpha + \cdots + a_{0n}\alpha^n, a_{10} + a_{11}\alpha + \cdots + a_{1n}\alpha^n, \ldots, a_{n0} + a_{n1}\alpha + \cdots + a_{nn}\alpha^n, 0, 0, \ldots, 0),$$

$$X'_0 = (a'_{00} + a'_{01}\alpha + \cdots + a'_{0n}\alpha^n, a'_{10} + a'_{11}\alpha + \cdots + a'_{1n}\alpha^n, \ldots, a'_{n0} + a'_{n1}\alpha + \cdots + a'_{nn}\alpha^n, 0, 0, \ldots, 0),$$

with $a_{ij}, a'_{ij} \in GF(q)$ and $GF(q^{n+1}) = \{a_0 + a_1\alpha + \cdots + a_n\alpha^n | a_i \in GF(q)\}$. By the proof of Theorem 25.6.6 the matrices $A = [a_{ij}]$ and $A' = [a'_{ij}]$ are non-singular. Also, from the proof of Theorem 25.6.1, we may assume that $\Pi_n^1 = \mathbf{V}(x_0, x_1, \ldots, x_n)$ and $\Pi_n^2 = \mathbf{V}(x_0 - x_{n+1}, x_1 - x_{n+2}, \ldots, x_n - x_{2n+1})$. Then, for the projectivity η of $PG(2n + 1, q)$ with matrix

$$\begin{bmatrix} A'A^{-1} & 0 \\ 0 & A'A^{-1} \end{bmatrix}^*,$$

we have $\Pi_n^i\eta = \Pi_n^i$, $i = 0, 1, 2$, and consequently $\mathscr{R}(\Pi_n^0, \Pi_n^1, \Pi_n^2)\eta = \mathscr{R}(\Pi_n^0, \Pi_n^1, \Pi_n^2)$. Extended to $PG(2n + 1, q^{n+1})$, this projectivity η maps P_0 onto P'_0. Now we have $l_0\eta = l'_0\xi$. Hence $l_0\eta\xi^{-1} = l'_0$, and consequently $\{l_0, l_1, \ldots, l_n\}\eta\xi^{-1} = \{l'_0, l'_1, \ldots, l'_n\}$. It follows that $\mathscr{S}\eta\xi^{-1} = \mathscr{S}'$.

Since $PGL(2n + 2, q)$ acts transitively on the set of all regular n-spreads of $PG(2n + 1, q)$, the order of $G(\mathscr{S})$, with $\mathscr{S}$ a regular n-spread, is equal to $|PGL(2n + 2, q)|$ divided by the number of all regular n-spreads. So, from Theorem 25.6.6,

$$|G(\mathscr{S})| = (n + 1)q^{n+1}(q^{n+1} - 1)(q^{2n+2} - 1)/(q - 1). \quad \square$$

Lemma 25.6.8: *Let $PG(2n + 1, q^2)$ be an extension of the projective space $PG(2n + 1, q)$. In $PG(2n + 1, q^2)$, let $\Pi_{n,q}$ be an n-space over $GF(q)$ skew to $PG(2n + 1, q)$. If $P \in \Pi_{n,q}$ and if $\tilde{P}$ is the conjugate point of P with respect to the extension $GF(q^2)$ of $GF(q)$, then the intersection of the line $P\tilde{P}$ and the space $PG(2n + 1, q)$ is a line l of $PG(2n + 1, q)$. These lines l form a system of maximal spaces of a Segre variety $\mathscr{S}_{1;n}$ of $PG(2n + 1, q)$.*

Proof: The intersection of the line $P\tilde{P}$ and the space $PG(2n + 1, q)$ is a line l of $PG(2n + 1, q)$. Let $P_0, P_1, \ldots, P_{n+1}$ be $n + 2$ points of $\Pi_{n,q}$, such that any $n + 1$ of them are linearly independent in $\Pi_{n,q}$. If $l_0, l_1, \ldots, l_{n+1}$ are the corresponding lines of $PG(2n + 1, q)$, then any $n + 1$ of them generate $PG(2n + 1, q)$. Let Q_0, Q_1, Q_2 be three distinct points of the line l_0. Through Q_i there is exactly one n-space $\Pi_{n,q}^i$ of $PG(2n + 1, q)$ which has a point in common with each of the lines $l_1, l_2, \ldots, l_{n+1}$, $i = 0, 1, 2$. Let $\mathscr{S}_{1;n}$ be the Segre variety of $PG(2n + 1, q)$ defined by the n-regulus $\mathscr{R}(\Pi_{n,q}^0, \Pi_{n,q}^1, \Pi_{n,q}^2) = \mathscr{R}$. The extensions of $\Pi_{n,q}, \mathscr{R}, \mathscr{S}_{1;n}$ to $PG(2n + 1, q^2)$ are respectively denoted by $\bar{\Pi}_{n,q}, \bar{\mathscr{R}}, \bar{\mathscr{S}}_{1;n}$. The space $\bar{\Pi}_{n,q}$ is the unique n-space of $PG(2n + 1, q^2)$ containing P_0 and meeting the lines $P_1\tilde{P}_1, P_2\tilde{P}_2, \ldots, P_{n+1}\tilde{P}_{n+1}$

in a point; hence $\bar{\Pi}_{n,q}$ belongs to $\bar{\mathscr{R}}$. Let l be any line of $\mathscr{S}_{1;n}$ meeting all elements of $\mathscr{R}$. The extension $\bar{l}$ of l has a point P in common with $\bar{\Pi}_{n,q}$. Since l is a line of $PG(2n+1, q)$, the line $\bar{l}$ also contains the conjugate point $\tilde{P}$ of P. The set of all points P is projectively equivalent to $\Pi^0_{n,q}$ and hence is a projective n-space $\Pi'_{n,q}$ over $GF(q)$. Since the points $P_0, P_1, \ldots, P_{n+1}$ are contained in a unique n-space over $GF(q)$, we necessarily have $\Pi'_{n,q} = \Pi_{n,q}$. Now the theorem is completely proved. □

Application: Construction method for n-spreads of PG(2n + 1, q)

Consider a projective space $PG(2m, q^2)$, and let $\mathscr{P}$ be a partition of it by projective $2m$-spaces $\Pi^i_{2m,q}$ over $GF(q)$, $i = 1, 2, \ldots, (q^{2m+1}+1)/(q+1)$. By Theorem 4.3.6 such a partition $\mathscr{P}$ exists. Embed $PG(2m, q^2)$ in the extension $PG(4m+1, q^2)$ of $PG(4m+1, q)$, and assume that $PG(2m, q^2)$ contains no point of $PG(4m+1, q)$. By Lemma 25.6.8 the $2m$-space $\Pi^i_{2m,q}$ defines a Segre variety $\mathscr{S}^i_{1;2m}$ of $PG(4m+1, q)$. These $(q^{2m+1}+1)/(q+1)$ Segre varieties $\mathscr{S}^i_{1;2m}$ form a partition of $PG(4m+1, q)$. Hence the $q^{2m+1}+1$ maximal $2m$-spaces of these Segre varieties form a $2m$-spread $\mathscr{S}$ of $PG(4m+1, q)$. We conclude that each partition $\mathscr{P}$ of $PG(2m, q^2)$ by $2m$-spaces over $GF(q)$ defines a $2m$-spread $\mathscr{S}$ of $PG(4m+1, q)$.

Next, consider a projective space $PG(2m+1, q^2)$, $m \geq 0$. Let $\mathscr{P}$ be a partition of $PG(2m+1, q^2)$ consisting of α spaces $\Pi^i_{2m+1,q}$ of dimension $2m+1$ over $GF(q)$ and β spaces Π^j_{m,q^2} of dimension m over $GF(q^2)$; then $\alpha(q+1) + \beta = q^{2m+2} + 1$. By Theorem 4.1.1 such a partition always exists for $\alpha = 0$. Embed $PG(2m+1, q^2)$ in the extension $PG(4m+3, q^2)$ of $PG(4m+3, q)$, and assume that $PG(2m+1, q^2)$ does not contain a point of $PG(4m+3, q)$. By Lemma 25.6.8 the $(2m+1)$-space $\Pi^i_{2m+1,q}$ defines a Segre variety $\mathscr{S}^i_{1;2m+1}$ of $PG(4m+3, q)$. The m-space Π^j_{m,q^2} and its conjugate $\tilde{\Pi}^j_{m,q^2}$ generate a $(2m+1)$-space Π^j_{2m+1,q^2} of $PG(4m+3, q^2)$, and $\Pi^j_{2m+1,q^2} \cap PG(4m+3, q)$ is a $(2m+1)$-space $\Pi^j_{2m+1,q}$ of $PG(4m+3, q)$. The α Segre varieties $\mathscr{S}^i_{1;2m+1}$ and the β spaces $\Pi^j_{2m+1,q}$ form a partition of $PG(4m+3, q)$. Let Σ^i be the system of maximal $(2m+1)$-spaces of $\mathscr{S}^i_{1;2m+1}$ for $m \neq 0$, and let Σ^i be a system of lines of $\mathscr{S}^i_{1;2m+1}$ for $m = 0$. Then the elements of $\Sigma^1 \cup \Sigma^2 \cup \cdots \cup \Sigma^\alpha$ together with the β spaces $\Pi^1_{2m+1,q}, \Pi^2_{2m+1,q}, \ldots, \Pi^\beta_{2m+1,q}$ form a $(2m+1)$-spread $\mathscr{S}$ of $PG(4m+3, q)$.

25.7 Notes and references

§25.1. For more details on quadric Veroneseans and their projections, see for example Bertini (1923), Godeaux (1948), Semple and Roth (1949), Burau (1961), Herzer (1982).

The following important theorem on Veroneseans over **C** is due to Kronecker and Castelnuovo. Any surface of $PG(m, \mathbf{C})$ which contains ∞^2 plane quadrics is the Veronesean $\mathscr{V}_2^4$ or one of its projections; any surface of $PG(3, \mathbf{C})$ having ∞^2 reducible plane sections is either the projection of a Veronesean $\mathscr{V}_2^4$ or a scroll.

Consider the Veronesean $\mathscr{V}_2^4$ in $PG(5, \mathbf{C})$, and let l be a line meeting $\mathscr{M}_4^3$ in three distinct points. By Theorem 25.1.15 the Veronesean $\mathscr{V}_2^4$ is a double surface of $\mathscr{M}_4^3$, and so l has no point in common with $\mathscr{V}_2^4$. The projection of $\mathscr{V}_2^4$ from l onto a solid $PG(3, \mathbf{C})$ of $PG(5, \mathbf{C})$ having no point in common with l is a surface $\mathscr{F}_2^4$ of order 4 of $PG(3, \mathbf{C})$ and is called a Steiner surface. It has three double lines which meet in a triple point of the surface. In a suitable coordinate system $\mathscr{F}_2^4 = \mathbf{V}(x_0x_1x_2x_3 - x_2^2x_3^2 - x_3^2x_1^2 - x_1^2x_2^2)$.

§25.2. The characterization Theorems 25.2.12 and 25.2.14 are taken from Tallini (1958a).

§25.3. For q odd and $q > 3$ the characterization Theorem 25.3.14 is taken from Ferri (1976). The proof of Lemma 25.3.8 has been modified from the latter. Lemma 25.3.9 was essential for the characterization of $\mathscr{V}_2^4$ in the case $q = 3$.

§25.4. For the odd case, Theorem 25.4.14 is due to Mazzocca and Melone (1984). Since Proposition 2.7 of Mazzocca and Melone (1984) does not hold for q even, the final step of the proof of Theorem 25.4.14 was revised to hold for all q. It was necessary to add axiom (iv) to obtain the characterization Theorem 25.4.21.

Let $\mathscr{M}$ be the algebraic variety formed by the tangent spaces of the algebraic variety $\mathscr{V}_n$. Melone (1983) gives a characterization of $\mathscr{M}$ in terms of its points and the lines contained in the tangent spaces of $\mathscr{V}_n$.

§25.5. For more details on Segre varieties, see for example Segre (1891), Godeaux (1948), Burau (1961), Melone and Olanda (1981).

For $n_2 = 1$, the normal rational curve $\Sigma_1\mathfrak{G}$ on the Grassmannian $\mathscr{G}_{n_1, 2n_1+1}$ is also considered by Herzer (1984).

§25.6. Theorem 25.6.5 is due to Bruck and Bose (1966). Its proof depends on deep theorems concerning translation planes. In André (1954), in Segre (1964b), and in Bruck and Bose (1966), it is shown that the study of n-spreads in $PG(2n + 1, q)$ is equivalent to the study of finite translation planes. The n-spread is regular if and only if the corresponding translation plane is Desarguesian. To a translation plane of order 2^{n+1} there always corresponds an n-spread of $PG(2n + 1, 2)$. Since there are many non-Desarguesian translation planes of order 2^{n+1} (see for example Dembowski (1968)), it follows that there are many non-regular n-spreads in $PG(2n + 1, 2)$. This explains a remark following Theorem 25.6.5.

In Bruen and Thas (1976) there is a construction of translation planes which is equivalent to the second construction in the last part of the section. In this connection Lemma 17.6.6, Corollary 2 shows that $PG(3, 4)$ can be partitioned into 14 lines and one $PG(3, 2)$. The corresponding translation plane of order 16 is non-Desarguesian and can be shown to be isomorphic to the plane discovered by Lorimer (1974).

26

EMBEDDED GEOMETRIES

26.1 Polar spaces

A *polar space* $\mathcal{S}$ of (finite) *rank* n or *projective index* $n - 1$, $n \geq 3$, is a set $\mathcal{P}$ of elements called *points* together with distinguished subsets called *subspaces* with the following properties.

(i) A subspace together with the subspaces it contains is a d-dimensional projective space with $-1 \leq d \leq n - 1$.

(ii) The intersection of any two subspaces is a subspace.

(iii) Given a subspace π of dimension $n - 1$ and a point P in $\mathcal{P} \backslash \pi$, there exists a unique subspace π' containing P such that the dimension of $\pi \cap \pi'$ is $n - 2$. The subspace π' contains all points of π which are joined to P by some subspace of dimension 1.

(iv) There exist disjoint subspaces of dimension $n - 1$.

A *polar space* of *rank* 2 or *projective index* 1 is an *incidence structure* $\mathcal{S} = (\mathcal{P}, \mathcal{B}, \mathrm{I})$ in which $\mathcal{P}$ and $\mathcal{B}$ are disjoint (non-empty) sets of objects called *points* and *lines* respectively, and for which I is a symmetric point–line *incidence relation* satisfying the following axioms.

(i) Each point is incident with $1 + t$ lines ($t \geq 1$) and two distinct points are incident with at most one line.

(ii) Each line is incident with $1 + s$ points ($s \geq 1$) and two distinct lines are incident with at most one point.

(iii) If P is a point and l is a line not incident with P, then there is a unique pair $(P', l') \in \mathcal{P} \times \mathcal{B}$ for which $P \mathrm{I} l' \mathrm{I} P' \mathrm{I} l$; see fig. 26.1.

Polar spaces of rank 2 are usually called *generalized quadrangles.* The integers s and t are the *parameters* of the generalized quadrangle and $\mathcal{S}$ is said to have *order* (s, t); if $s = t$, $\mathcal{S}$ is said to have *order* s. There is a point–line duality for generalized quadrangles (of order (s, t)) for which in any definition or theorem the words 'point' and 'line' are interchanged and the parameters s and t are interchanged. Normally, we assume without further remark that the dual of a given theorem or definition has also been given.

The main reason for the difference between the axioms in the cases $n = 2$ and $n \geq 3$ is that, in the latter case, the axioms applied to $n = 2$ do not imply that each line contains a constant number of points and similarly that each point is on a constant number of lines.

Isomorphisms and automorphisms of polar spaces are defined in the usual way; similarly for isomorphisms (or collineations), anti-isomorphisms (or

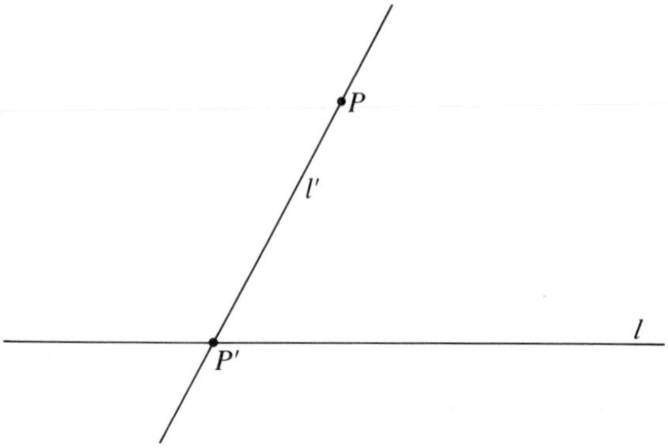

Figure 26.1

reciprocities), automorphisms, anti-automorphisms, involutions, and polarities of generalized quadrangles.

Examples: (a) Let $\mathcal{Q}$ be a non-singular quadric of $PG(d, q)$ of projective index $n - 1$ with $n \geq 2$. Then $\mathcal{Q}$ together with the projective subspaces lying on it is a polar space of rank n.

(b) Let $\mathcal{U}$ be a non-singular Hermitian variety of $PG(d, q^2)$, $d \geq 3$. Then $\mathcal{U}$ together with the subspaces lying on it is a polar space. The projective index of this polar space is the maximum dimension of subspaces lying on $\mathcal{U}$.

(c) Let ζ be a null polarity of $PG(d, q)$, with d odd. Then $PG(d, q)$ together with all subspaces of the self-polar $(d - 1)/2$-dimensional spaces is a polar space of projective index $(d - 1)/2$.

(d) Let $\mathcal{P} = \{P_{ij} | i, j = 0, 1, \ldots, s\}$, $s > 0$, $\mathcal{B} = \{l_0, l_1, \ldots, l_s, m_0, m_1, \ldots, m_s\}$, $P_{ij} \mathrm{I}\, l_k$ if and only if $i = k$, and $P_{ij} \mathrm{I}\, m_k$ if and only if $j = k$. Then $(\mathcal{P}, \mathcal{B}, \mathrm{I})$ is a generalized quadrangle of order $(s, 1)$. Up to an isomorphism, there is only one generalized quadrangle of order $(s, 1)$, for any given $s > 0$. The generalized quadrangles with $t = 1$ are called *grids*.

(e) Let π be a plane of $PG(3, q)$, q even, and let $\mathcal{O}$ be an oval in π. Further, let $\mathcal{P} = PG(3, q)\backslash\pi$, let $\mathcal{B}$ be the set of all lines of $PG(3, q)$ not contained in π but containing a point of $\mathcal{O}$, and let I be the incidence of $PG(3, q)$. Then $(\mathcal{P}, \mathcal{B}, \mathrm{I})$ is a generalized quadrangle of order $(q - 1, q + 1)$ and is denoted by $\mathcal{T}_2^*(\mathcal{O})$.

A complete classification of the polar spaces of rank at least three has been obtained by Tits. We state his result (without proof) in the finite case.

Theorem 26.1.1: *If $\mathcal{S}$ is a finite polar space of rank at least three, then $\mathcal{S}$ is isomorphic to one of* (a), (b), (c). □

The examples (d) and (e) show that this theorem is not valid in the rank 2 case. In fact many other examples of generalized quadrangles are known.

A *Shult space* $\mathcal{S}$ is a non-empty set $\mathcal{P}$ of *points* together with distinguished subsets of cardinality at least two called *lines* such that for each line l of $\mathcal{S}$ and each point P of $\mathcal{P}\backslash l$, the point P is collinear (or adjacent) with either one or all points of l; two points P_1 and P_2 (not necessarily distinct) being called *collinear* (or *adjacent*), with the notation $P_1 \sim P_2$, if there is at least one line of $\mathcal{S}$ containing P_1 and P_2. A Shult space is called *non-degenerate* if no point of $\mathcal{S}$ is collinear with all other points. A *subspace* X of a Shult space $\mathcal{S}$ is a set of pairwise collinear points such that any line meeting X in more than one point is contained in X. We say that $\mathcal{S}$ has *rank* n or *projective index* $n - 1$ $(n \geq 1)$ if n is the largest integer for which there is a chain $X_0 \subset X_1 \subset \cdots \subset X_n$ of distinct subspaces $X_0 = \varnothing, X_1, X_2, \ldots, X_n$.

From Theorem 26.1.1 it follows that, for any finite polar space $\mathcal{S}$ of rank n, $n \geq 3$, the point set $\mathcal{P}$ together with the subspaces of dimension 1 is a Shult space of rank n. In fact this result also holds for infinite polar spaces. Next, let $\mathcal{S}$ be a generalized quadrangle of order (s, t). If each line of $\mathcal{S}$ is identified with the set of its points, then a Shult space of rank 2 is obtained.

The following converse, which we state without proof, is due to Buekenhout and Shult.

Theorem 26.1.2: *A non-degenerate Shult space of rank n, $n \geq 3$, all of whose lines have cardinality at least three, together with its subspaces, is a polar space of rank n. A Shult space of rank 2, all of whose lines have cardinality at least three and all of whose points are contained in at least three lines, is a generalized quadrangle.* □

Finally, let $\mathcal{S}$ be a degenerate Shult space. The point set consisting of all points of $\mathcal{S}$ which are collinear with each point of $\mathcal{S}$ is denoted by $\mathcal{R}$ and is called the *radical* of $\mathcal{S}$; it is a subspace of $\mathcal{S}$. An equivalence relation ρ is defined on the point set $\mathcal{P}$ of $\mathcal{S}$ by putting $P \,\rho\, P'$ if and only if the set of all points collinear with P coincides with the set of all points collinear with P'. Let $\rho(P)$ denote the equivalence class containing the point P for the relation ρ; then $\rho(P) = \mathcal{R}$ for all P in $\mathcal{R}$.

Lemma 26.1.3: *Let $\mathcal{S}$ be a degenerate Shult space with radical $\mathcal{R}$. If the set of all points collinear with the point P is contained in the set of all points collinear with the point P', then either $P \,\rho\, P'$ or $P' \in \mathcal{R}$.*

Proof: Assume that $P' \notin \rho(P)$ and $P' \notin \mathcal{R}$. Then P and P' are distinct collinear points, and there exists a point T which is not collinear with P'. Let l be a line containing P and P', and let T' be a point of l collinear with T. Since $P' \notin \rho(P)$ there exists a point Z which is collinear with P' but not with P. Let

m be a line containing Z and P'. On $m\backslash\{P'\}$ there is a point Z' which is collinear with T. Since Z is not collinear with P, also Z' is not collinear with P. On a line m' through T and Z' there is a point W which is collinear with P. Hence W is collinear with P'. Since W and Z' are collinear with P', also T is collinear with P', a contradiction. We conclude that $P' \in \rho(P)$ or $P' \in \mathscr{R}$. □

Let P be a point of $\mathscr{S}$ which is not contained in the radical $\mathscr{R}$. A corollary of Lemma 26.1.3 is that $\mathscr{R} \cup \rho(P)$ is a subspace of $\mathscr{S}$. Now we introduce a new structure $\mathscr{S}'$ with point set $\mathscr{P}'$ and line set $\mathscr{B}'$. Let $\mathscr{P}'$ be the set of classes $\rho(P)$ with $P \notin \mathscr{R}$, and call a line of $\mathscr{S}'$ any set $\{\rho(P) | P \in l$ with l a line of $\mathscr{S}$ not contained in a subspace of the form $\rho(T) \cup \mathscr{R}\}$. Then the following result is readily obtained.

Theorem 26.1.4: *If $\mathscr{R} \neq \mathscr{P}$, then the structure $\mathscr{S}'$ is a non-degenerate Shult space.* □

Finally, for a non-degenerate Shult space the radical $\mathscr{R}$ is defined to be the empty set.

26.2 Generalized quadrangles

Only finite generalized quadrangles are considered.

A start is made by giving a brief description of three families of examples known as the *classical* generalized quadrangles, all of which are associated with classical groups.

(a) Consider a non-singular quadric $\mathscr{Q}$ of projective index 1 in the projective space $PG(d, q)$, with $d = 3$, 4, or 5. Then the points of $\mathscr{Q}$ together with the lines of $\mathscr{Q}$ (which are the subspaces of maximal dimension on $\mathscr{Q}$) form a generalized quadrangle $\mathscr{Q}(d, q)$ with the following parameters:

$$s = q, t = 1, \quad \text{when } d = 3,$$

$$s = q, t = q, \quad \text{when } d = 4,$$

$$s = q, t = q^2, \quad \text{when } d = 5.$$

Since $\mathscr{Q}(3, q)$ is a grid, its structure is trivial. From §5.1, the quadric $\mathscr{Q}$ has the following canonical form:

$$\mathscr{Q} = \mathscr{H}_3 = \mathbf{V}(x_0x_1 + x_2x_3), \qquad \text{when } d = 3;$$

$$\mathscr{Q} = \mathscr{P}_4 = \mathbf{V}(x_0^2 + x_1x_2 + x_3x_4), \qquad \text{when } d = 4;$$

$$\mathscr{Q} = \mathscr{E}_5 = \mathbf{V}(f(x_0, x_1) + x_2x_3 + x_4x_5), \quad \text{when } d = 5,$$

where $f(x_0, x_1)$ is an irreducible binary quadratic form.

(b) Let $\mathscr{U}$ be a non-singular Hermitian variety in the projective space $PG(d, q^2)$, $d = 3$ or 4. Then the points of $\mathscr{U}$ together with the lines on $\mathscr{U}$ form a generalized quadrangle $\mathscr{U}(d, q^2)$ with parameters as follows:

$$s = q^2, t = q, \quad \text{when } d = 3,$$

$$s = q^2, t = q^3, \quad \text{when } d = 4.$$

From §5.1, $\mathscr{U}$ has the following canonical form

$$\mathscr{U} = \mathbf{V}(x_0^{q+1} + x_1^{q+1} + \cdots + x_d^{q+1}).$$

(c) The points of $PG(3, q)$, together with the self-polar lines of a null polarity ζ, form a generalized quadrangle $\mathscr{W}(q)$ with parameters

$$s = q, t = q.$$

From Chapter 15, the lines of $\mathscr{W}(q)$ are the elements of a general linear complex of lines of $PG(3, q)$. Further, a null polarity of $PG(3, q)$ has the canonical bilinear form

$$x_0 y_1 - x_1 y_0 + x_2 y_3 - x_3 y_2.$$

The examples (d) and (e) of §26.1 show that there exist generalized quadrangles other than the classical ones and their duals. The order of each known generalized quadrangle is one of the following:

$(s, 1)$ with $s \geq 1$;
$(1, t)$ with $t \geq 1$;
(q, q) with q a prime power;
(q, q^2), (q^2, q) with q a prime power;
(q^2, q^3), (q^3, q^2) with q a prime power;
$(q - 1, q + 1)$, $(q + 1, q - 1)$ with q a prime power.

Let $\mathscr{S} = (\mathscr{P}, \mathscr{B}, \mathrm{I})$ be a generalized quadrangle of order (s, t). Given two (not necessarily distinct) points P, P' of $\mathscr{S}$, we write $P \sim P'$ and say that P and P' are *collinear*, provided that there is some line l for which $P \,\mathrm{I}\, l \,\mathrm{I}\, P'$; hence $P \nsim P'$ means that P and P' are not collinear. Dually, for $l, l' \in \mathscr{B}$, we write $l \sim l'$ or $l \nsim l'$ according as l and l' are *concurrent* or non-concurrent. When $P \sim P'$ we also say that P is *orthogonal* or *perpendicular* to P'; similarly for $l \sim l'$. The line incident with distinct collinear points P and P' is denoted PP', and the point incident with distinct concurrent lines l and l' is denoted $l \cap l'$.

For $P \in \mathscr{P}$ put $P^\perp = \{P' \in \mathscr{P} | P \sim P'\}$, and note that $P \in P^\perp$. The *trace* of a pair $\{P, P'\}$ of distinct points is defined to be the set $P^\perp \cap P'^\perp$ and is denoted $\mathrm{tr}(P, P')$ or $\{P, P'\}^\perp$; then $|\{P, P'\}^\perp| = s + 1$ or $t + 1$ according as $P \sim P'$ or $P \nsim P'$. More generally, if $\mathscr{A} \subset \mathscr{P}$, $\mathscr{A}$ 'perp' is defined by $\mathscr{A}^\perp = \bigcap \{P^\perp | P \in \mathscr{A}\}$.

For $P \neq P'$, the *span* of the pair $\{P, P'\}$ is $\text{sp}\{P, P'\} = \{P, P'\}^{\perp\perp} = \{Y \in \mathscr{P} \mid Y \in Z^{\perp}$ for all $Z \in P^{\perp} \cap P'^{\perp}\}$. When $P \not\sim P'$, then $\{P, P'\}^{\perp\perp}$ is also called the *hyperbolic line* defined by P and P', and $|\{P, P'\}^{\perp\perp}| = s + 1$ or $|\{P, P'\}^{\perp\perp}| \leq t + 1$ according as $P \sim P'$ or $P \not\sim P'$.

A *triad* (of points) is a triple of pairwise non-collinear points. Given a triad $\mathscr{T} = \{P, P', P''\}$, a *centre* of $\mathscr{T}$ is just a point of $\mathscr{T}^{\perp}$.

These definitions will be illustrated by some examples.

Let $P \not\sim P'$ in $\mathscr{W}(q)$, and let ζ be the null polarity defining $\mathscr{W}(q)$. If l is the polar line of the line PP' of $PG(3, q)$, then $\{P, P'\}^{\perp} = l$ and $\{P, P'\}^{\perp\perp} = PP'$. Hence each hyperbolic line of $\mathscr{W}(q)$ contains $q + 1$ points.

Let $P \not\sim P'$ in $\mathscr{Q}(4, q)$; then $\{P, P'\}^{\perp}$ is a conic. For q odd, $\{P, P'\}^{\perp\perp} = \{P, P'\}$, and for q even, $\{P, P'\}^{\perp\perp}$ is the intersection of $\mathscr{Q}$ and the plane $PP'N$, where N is the nucleus of the quadric $\mathscr{Q}$. In the even case $\{P, P'\}^{\perp\perp}$ is a conic, and each hyperbolic line contains $q + 1$ points.

Let $P \not\sim P'$ in $\mathscr{Q}(5, q)$, and let ζ be the polarity defined by $\mathscr{Q}$. If π is the polar solid of the line PP' of $PG(5, q)$, then $\{P, P'\}^{\perp}$ is the elliptic quadric $\pi \cap \mathscr{Q}$ of π, and $\{P, P'\}^{\perp\perp} = \{P, P'\}$.

Let $P \not\sim P'$ in $\mathscr{U}(3, q^2)$, and let ζ be the unitary polarity defined by $\mathscr{U}$. If l is the polar line of the line PP' of $PG(3, q^2)$, then $\{P, P'\}^{\perp} = \mathscr{U} \cap l$ and $\{P, P'\}^{\perp\perp} = PP' \cap \mathscr{U}$. Hence each hyperbolic line has $q + 1$ points.

Let $P \not\sim P'$ in $\mathscr{U}(4, q^2)$, and let ζ be the unitary polarity defined by $\mathscr{U}$. If π is the polar plane of the line PP' of $PG(4, q^2)$, then $\{P, P'\}^{\perp}$ is the non-singular Hermitian curve $\pi \cap \mathscr{U}$, and $\{P, P'\}^{\perp\perp} = PP' \cap \mathscr{U}$. Hence each hyperbolic line has $q + 1$ points.

Consider again $\mathscr{W}(q)$ and its defining polarity ζ. If $\mathscr{T} = \{P, P', P''\}$ is a triad of $\mathscr{W}(q)$ for which P, P', P'' are collinear in $PG(3, q)$, then $\mathscr{T}^{\perp} = \{P, P'\}^{\perp}$ and so $|\mathscr{T}^{\perp}| = q + 1$. If $\mathscr{T} = \{P, P', P''\}$ is a triad for which P, P', P'' are not collinear, then the pole of the plane $PP'P''$ is the unique centre of $\mathscr{T}$.

Finally consider again $\mathscr{Q}(5, q)$ and the corresponding polarity ζ. If $\mathscr{T} = \{P, P', P''\}$ is a triad of $\mathscr{Q}\{5, q\}$, then $\mathscr{T}^{\perp}$ is the conic $\mathscr{Q} \cap \pi$, where π is the polar plane of the plane $PP'P''$. Hence in this case any triad has $q + 1$ centres.

Let $\mathscr{S} = (\mathscr{P}, \mathscr{B}, \mathrm{I})$ be a generalized quadrangle of order (s, t), and put $|\mathscr{P}| = v$ and $|\mathscr{B}| = b$.

Theorem 26.2.1: (i) $v = (s + 1)(st + 1)$; (ii) $b = (t + 1)(st + 1)$.

Proof: Let l be a fixed line of $\mathscr{S}$ and count in different ways the number of ordered pairs $(P, m) \in \mathscr{P} \times \mathscr{B}$ with $P \not\mathrm{I}\, l$, $P \,\mathrm{I}\, m$, and $l \sim m$. Then $v - s - 1 = (s + 1)ts$, whence $v = (s + 1)(st + 1)$. Dually $b = (t + 1)(st + 1)$. □

Theorem 26.2.2: *The integer $s + t$ divides $st(s + 1)(t + 1)$.*

Proof: If $\mathscr{E} = \{\{P, P'\} \mid P, P' \in \mathscr{P}$ and $P \sim P'\}$, then it is evident that $(\mathscr{P}, \mathscr{E})$

is a strongly regular graph with parameters $v = (s+1)(st+1)$, k (or n_1) $= st + s$, λ (or p^1_{11}) $= s - 1$, μ (or p^2_{11}) $= t + 1$. The graph $(\mathscr{P}, \mathscr{E})$ is called the *point graph* of the generalized quadrangle. Let $\mathscr{P} = \{P_1, P_2, \ldots, P_v\}$ and let $A = [a_{ij}]$ be the $v \times v$ matrix over $\mathbf{R}$ for which $a_{ij} = 0$ if $i = j$ or $P_i \not\sim P_j$, and $a_{ij} = 1$ if $i \neq j$ and $P_i \sim P_j$; that is, A is an adjacency matrix of the graph $(\mathscr{P}, \mathscr{E})$.

If $A^2 = [c_{ij}]$, then (a) $c_{ii} = (t+1)s$; (b) $i \neq j$ and $P_i \not\sim P_j$ imply $c_{ij} = t + 1$; (c) $i \neq j$ and $P_i \sim P_j$ imply $c_{ij} = s - 1$. Consequently

$$A^2 - (s - t - 2)A - (t+1)(s-1)I = (t+1)J;$$

here I is the $v \times v$ identity matrix and J is the $v \times v$ matrix with each entry equal to one. Evidently $(t+1)s$ is an eigenvalue of A, and J has eigenvalues $0, v$ with multiplicities $v - 1, 1$ respectively. Since

$$((t+1)s)^2 - (s-t-2)(t+1)s - (t+1)(s-1) = (t+1)(st+1)(s+1) = (t+1)v,$$

the eigenvalue $(t+1)s$ of A corresponds to the eigenvalue v of J, and so $(t+1)s$ has multiplicity 1. The other eigenvalues of A are roots of the equation

$$x^2 - (s-t-2)x - (t+1)(s-1) = 0.$$

Denote the multiplicities of these eigenvalues θ_1, θ_2 by m_1, m_2. Then we have $\theta_1 = -t - 1$, $\theta_2 = s - 1$, $v = 1 + m_1 + m_2$, and $s(t+1) - m_1(t+1) + m_2(s-1) = \operatorname{tr} A = 0$. Hence

$$m_1 = (st+1)s^2/(s+t) \quad \text{and} \quad m_2 = st(s+1)(t+1)/(s+t).$$

Since m_1, m_2 are positive integers, the integer $s + t$ divides the integers $(st+1)s^2$ and $st(s+1)(t+1)$. Finally, we notice that $s + t$ divides $(st+1)s^2$ if and only if it divides $st(s+1)(t+1)$. □

Theorem 26.2.3 (Higman's inequality): *If $s > 1$ and $t > 1$, then $t \leq s^2$, and dually $s \leq t^2$.*

Proof: Let P, P' be two non-collinear points of $\mathscr{S}$. Put

$$\mathscr{V} = \{T \in \mathscr{P} \mid P \not\sim T \quad \text{and} \quad P' \not\sim T\};$$

so $|\mathscr{V}| = d = (s+1)(st+1) - 2 - 2(t+1)s + (t+1)$. Denote the elements of $\mathscr{V}$ by $T_1, T_2, \ldots, T_d$ and let

$$t_i = |\{Z \in \{P, P'\}^\perp \mid Z \sim T_i\}|.$$

Count in different ways the number of ordered pairs $(T_i, Z) \in \mathscr{V} \times \{P, P'\}^\perp$ with $Z \sim T_i$ to obtain

$$\sum_i t_i = (t+1)(t-1)s. \tag{26.1}$$

Next count in different ways the number of ordered triples $(T_i, Z, Z') \in \mathscr{V} \times \{P, P'\}^\perp \times \{P, P'\}^\perp$ with $Z \neq Z'$, $Z \sim T_i$, $Z' \sim T_i$, to obtain

$$\sum_i t_i(t_i - 1) = (t + 1)t(t - 1). \qquad (26.2)$$

From (26.1) and (26.2) it follows that

$$\sum_i t_i^2 = (t + 1)(t - 1)(s + t).$$

With $d\bar{t} = \sum_i t_i$, the inequality $0 \leq \sum_i (\bar{t} - t_i)^2$ simplifies to

$$d\sum_i t_i^2 - \left(\sum_i t_i\right)^2 \geq 0,$$

which implies

$$d(t + 1)(t - 1)(s + t) \geq (t + 1)^2(t - 1)^2 s^2,$$

or

$$t(s - 1)(s^2 - t) \geq 0,$$

completing the proof. □

There is an immediate corollary of the proof.

Corollary: *When $s > 1$ and $t > 1$, the following are equivalent:*

(i) $s^2 = t$;
(ii) $d\sum t_i^2 - (\sum t_i)^2 = 0$ *for any pair $\{P, P'\}$ of non-collinear points;*
(iii) $t_i = \bar{t}$ *for $i = 1, 2, \ldots, d$ and any pair $\{P, P'\}$ of non-collinear points;*
(iv) *each triad of points has a constant number of centres, in which case this number is $s+1$.* □

Theorem 26.2.4: *If $s \neq 1$, $t \neq 1$, $s \neq t^2$, and $t \neq s^2$, then $t \leq s^2 - s$ and dually $s \leq t^2 - t$.*

Proof: Suppose $s \neq 1$ and $t \neq s^2$. By Theorem 26.2.3 we have $t = s^2 - x$ with $x > 0$, and by Theorem 26.2.2 the integer $s + s^2 - x$ divides $s(s^2 - x) \times (s + 1)(s^2 - x + 1)$. Hence modulo $s + s^2 - x$ we have $0 \equiv x(-s)(-s + 1) \equiv x(x - 2s)$. If $x < 2s$, then $s + s^2 - x \leq x(2s - x)$ forces $x \in \{s, s + 1\}$. Consequently $x = s$, $x = s + 1$, or $x \geq 2s$, from which it follows that $t \leq s^2 - s$. □

The only classical generalized quadrangle which has $t = s^2$ is the generalized quadrangle $\mathscr{Q}(5, q)$; the only classical example with $s = t^2$ is the generalized quadrangle $\mathscr{U}(3, q^2)$. In one of the examples preceding Theorem 26.2.1 it was verified that any triad of $\mathscr{Q}(5, q)$ has $q + 1$ centres.

In the next two theorems isomorphisms and anti-isomorphisms between the classical generalized quadrangles are described.

Theorem 26.2.5: (i) *$\mathcal{Q}(4, q)$ is isomorphic to the dual of $\mathcal{W}(q)$;*
(ii) *$\mathcal{Q}(4, q)$ (or $\mathcal{W}(q)$) is self-dual if and only if q is even.*

Proof: Let $\mathcal{H}_5 = \mathcal{G}_{1,3}$ be the Klein quadric, that is the Grassmannian of the lines of $PG(3, q)$. The image of $\mathcal{W}(q)$ on $\mathcal{H}_5$ is the intersection of $\mathcal{H}_5$ with a non-tangent prime $PG(4, q)$ of $PG(5, q)$ (cf. §15.4). The non-singular quadric $\mathcal{H}_5 \cap PG(4, q)$ of $PG(4, q)$ is denoted by $\mathcal{Q}$. The lines of $\mathcal{W}(q)$ which are incident with a given point form a flat pencil of lines, hence their images on $\mathcal{H}_5$ form a line of $\mathcal{Q}$. Now it easily follows that $\mathcal{W}(q)$ is anti-isomorphic with $\mathcal{Q}(4, q)$.

In Theorem 16.4.13 it was shown that $\mathcal{W}(q)$ is self-dual if and only if q is even. By the first part of the proof, also $\mathcal{Q}(4, q)$ is self-dual if and only if q is even. □

In §16.4 it was shown that $\mathcal{W}(q)$ admits a polarity if and only if $q = 2^{2h+1}$ with $h \geq 0$.

An algebraic proof of the existence of an anti-isomorphism between $\mathcal{Q}(5, q)$ and $\mathcal{U}(3, q^2)$ can be found in §19.2. Here it is shown in a purely geometrical way that $\mathcal{Q}(5, q)$ and $\mathcal{U}(3, q^2)$ are anti-isomorphic.

Theorem 26.2.6: *The generalized quadrangle $\mathcal{Q}(5, q)$ is isomorphic to the dual of $\mathcal{U}(3, q^2)$.*

Proof: Let $\mathcal{Q}$ be an elliptic quadric in $PG(5, q)$. Extend $PG(5, q)$ to $PG(5, q^2)$. Then the extension of $\mathcal{Q}$ is a hyperbolic quadric $\mathcal{H}_5$ in $PG(5, q^2)$. Hence $\mathcal{H}_5$ is the Klein quadric of the lines of $PG(3, q^2)$. So to $\mathcal{Q}$ in $\mathcal{H}_5$ there corresponds a set $\mathcal{V}$ of lines in $PG(3, q^2)$. To a given line l of the generalized quadrangle $\mathcal{Q}(5, q)$ there correspond $q + 1$ lines of $PG(3, q^2)$ that all lie in a plane and pass through a point P. Let $\mathcal{U}$ be the set of points on the lines of $\mathcal{V}$. Then to each point of $\mathcal{Q}(5, q)$ there corresponds a line of $\mathcal{V}$, and to each line l of $\mathcal{Q}(5, q)$ there corresponds a point P of $\mathcal{U}$. To distinct lines l, l' of $\mathcal{Q}(5, q)$ correspond distinct points P, P' of $\mathcal{U}$ as a plane of $\mathcal{H}_5$ contains at most one line of $\mathcal{Q}$. Since a point T of $\mathcal{Q}(5, q)$ is on $q^2 + 1$ lines of $\mathcal{Q}(5, q)$, these $q^2 + 1$ lines are mapped onto the $q^2 + 1$ points of the image of T. Hence an anti-isomorphism is obtained from $\mathcal{Q}(5, q)$ onto the structure $(\mathcal{U}, \mathcal{V}, \mathrm{I})$, where I is the natural incidence relation. So $(\mathcal{U}, \mathcal{V}, \mathrm{I})$ is a generalized quadrangle of order (q^2, q) embedded in $PG(3, q^2)$. But now by a celebrated result of Buekenhout and Lefèvre, which is part of Theorem 26.3.29, the generalized quadrangle $(\mathcal{U}, \mathcal{V}, \mathrm{I})$ must be $\mathcal{U}(3, q^2)$. □

26.3 Embedded Shult spaces

A *projective Shult space* $\mathcal{S}$ is a Shult space for which the point set $\mathcal{P}$ is a subset of the point set of some projective space $PG(d, K)$, and for which the line set $\mathscr{B}$ is a non-empty set of lines of $PG(d, K)$. In such a case we also say that the Shult space $\mathscr{S}$ is *embedded* in $PG(d, K)$. If $PG(d', K)$ is the subspace of $PG(d, K)$ generated by all points of $\mathscr{P}$, then we say that $PG(d', K)$ is the *ambient space* of $\mathscr{S}$.

Examples (a), (b), (c) of §26.1 yield projective Shult spaces. It is our goal to prove that these are the only non-degenerate Shult spaces embedded in a Galois space; a direct proof is given, without relying on Theorems 26.1.1 and 26.1.2.

Theorem 26.3.1: *A non-degenerate Shult space $\mathscr{S}$ of rank 2 embedded in $PG(d, q)$ is a generalized quadrangle.*

Proof: Let $\mathscr{P}$ be the point set of $\mathscr{S}$, and let $\mathscr{B}$ be the line set of $\mathscr{S}$. On each line of $\mathscr{B}$ there are exactly $q + 1$ points. Let $P \in \mathscr{P}$, $l \in \mathscr{B}$, and $P \notin l$. By the definition of Shult space, $\mathscr{B}$ contains one or $q + 1$ lines through P which are concurrent with l. If $\mathscr{B}$ contains $q + 1$ lines through P and concurrent with l, then the plane Pl is a subspace of $\mathscr{S}$. This yields a contradiction since $\mathscr{S}$ has rank 2. So there is exactly one line of $\mathscr{S}$ through P which is concurrent with l.

Let P, P' be distinct points of $\mathscr{P}$ for which PP' is not a line of $\mathscr{B}$. If l is any line of $\mathscr{B}$ through P, then there is exactly one line l' of $\mathscr{B}$ through P' which is concurrent with l. It follows that the number of lines of $\mathscr{B}$ through P is equal to the number of lines of $\mathscr{B}$ through P'. Now we show that any point P of $\mathscr{P}$ is contained in at least two lines of $\mathscr{B}$. Since $\mathscr{S}$ is non-degenerate there is a line l in $\mathscr{B}$ which does not contain P. Let l' be the line of $\mathscr{B}$ through P which is concurrent with l. The common point of l and l' is denoted by P'. Since $\mathscr{S}$ is non-degenerate there is a point $P'' \neq P'$ in $\mathscr{P}$ such that $P'P''$ is not a line of $\mathscr{B}$. The line of $\mathscr{B}$ which contains P'' and is concurrent with l is denoted by l''. Clearly $P \notin l''$ and $l' \cap l'' = \emptyset$. So the line of $\mathscr{B}$ which contains P and is concurrent with l'' is distinct from l'. Consequently P is contained in at least two distinct lines of $\mathscr{B}$. Now consider distinct points T, T' of $\mathscr{P}$, where TT' is in $\mathscr{B}$. Let m be a line of $\mathscr{B}$ through T which is distinct from TT', and let m' be a line of $\mathscr{B}$ through T' which is distinct from TT'; then $m \cap m' = \emptyset$. Let $M \in m\backslash\{T\}$, $M' \in m'\backslash\{T'\}$, with MM' not in $\mathscr{B}$; then $T'M \notin \mathscr{B}$ and $TM' \notin \mathscr{B}$. By a previous argument the number of lines of $\mathscr{B}$ through T is equal to the number of lines of $\mathscr{B}$ through M', which equals the number of lines of $\mathscr{B}$ through M, which in turn equals the number of lines of $\mathscr{B}$ through T'. Therefore each point of $\mathscr{P}$ is contained in a constant number $t + 1$, $t \geq 1$, of lines of $\mathscr{B}$.

We conclude that $\mathscr{S}$ is a generalized quadrangle of order (q, t). □

Theorem 26.3.2: *Let $\mathscr{S}$ be a non-degenerate projective Shult space of rank 2 with ambient space $PG(d, q)$. If some point of $\mathscr{S}$ is contained in exactly two lines of $\mathscr{S}$, then $d = 3$ and $\mathscr{S}$ is the generalized quadrangle $\mathscr{Q}(3, q)$.*

Proof: By Lemma 26.3.1, $\mathcal{S}$ is a generalized quadrangle of order $(q, 1)$. Since $\mathcal{S}$ is a grid it is clear that $d = 3$ and $\mathcal{S} = \mathcal{Q}(3, q)$. □

Let $\mathcal{S}$ be a Shult space embedded in $PG(d, q)$. Let $\mathcal{P}$ be the point set of $\mathcal{S}$ and let $\mathcal{B}$ be the line set of $\mathcal{S}$. If $P, P' \in \mathcal{P}$ and if there is at least one line of $\mathcal{B}$ through P and P', then we say that P and P' are *adjacent* and write $P \sim P'$. Thus there will be no confusion between collinearity in $PG(d, q)$ and adjacency in $\mathcal{S}$. For the subspace of $PG(d, q)$ generated by the point sets or points $\mathcal{V}_1, \mathcal{V}_2, \ldots, \mathcal{V}_k$, we shall frequently use the notation $\langle \mathcal{V}_1, \mathcal{V}_2, \ldots, \mathcal{V}_k \rangle$. If π is a subspace of $PG(d, q)$, then $\pi \cap \mathcal{S}$ (or $\mathcal{S} \cap \pi$) will denote the structure with point set $\pi \cap \mathcal{P}$ and line set the set of all lines of $\mathcal{S}$ contained in π. If $PG(d, q)$ is the ambient space of $\mathcal{S}$, then $\langle \pi \cap \mathcal{P} \rangle$ does not necessarily coincide with π.

We assume that, from now on, $\mathcal{S}$ is a projective Shult space with point set $\mathcal{P}$, line set $\mathcal{B}$, and ambient space $PG(d, q)$.

Theorem 26.3.3: *The radical $\mathcal{R}$ of a degenerate Shult space $\mathcal{S}$ is a subspace of $PG(d, q)$. Let $\mathcal{R}$ have dimension r $(r \neq -1)$, and let $PG(d - r - 1, q) = \pi$ be a subspace of $PG(d, q)$ which is skew to $\mathcal{R}$. If $\mathcal{R} \neq PG(d, q)$, then $\pi \cap \mathcal{S}$ is a non-degenerate Shult space, $\mathcal{P}$ is the union of all lines joining every point of $\mathcal{R}$ to every point of $\pi \cap \mathcal{S}$, and two points of $\mathcal{P} \backslash \mathcal{R}$ are adjacent if and only if their projections from $\mathcal{R}$ onto π are adjacent.*

Proof: In §26.1 it was already mentioned that $\mathcal{R}$ is a subspace of the Shult space $\mathcal{S}$. Hence $\mathcal{R}$ has the property that the line joining any two distinct points of $\mathcal{R}$ is completely contained in $\mathcal{R}$. Hence $\mathcal{R}$ is a subspace of $PG(d, q)$. Let $\mathcal{R}$ have dimension r, $r < d$, and let $PG(d - r - 1, q) = \pi$ be a subspace of $PG(d, q)$ which is skew to $\mathcal{R}$. In §26.1 we introduced the equivalence relation ρ on $\mathcal{P}$ by writing $P \rho P'$ if and only if the set of all points collinear with P coincides with the set of all points collinear with P'. Let $\rho(P)$ denote the equivalence class containing the point P. If $P \in \mathcal{P} \backslash \mathcal{R}$, then it was shown in §26.1 that $\mathcal{R} \cup \rho(P)$ is a subspace of $\mathcal{S}$ and hence a projective subspace of $PG(d, q)$. Now we prove that $\mathcal{R} \cup \rho(P)$ is the projective $(r + 1)$-space $\mathcal{R}P$.

Let $P' \in \mathcal{R}P$, with $P' \notin \mathcal{R} \cup \{P\}$, and let P'' be the common point of PP' and $\mathcal{R}$. If $T \sim P$, then since $T \sim P''$ we have also $T \sim P'$. Analogously $T \sim P'$ implies $T \sim P$. Hence $P' \in \rho(P)$, and so $\mathcal{R}P \subset \mathcal{R} \cup \rho(P)$. Next, let $P_1 \in \rho(P) \backslash \{P\}$ and suppose that $PP_1 \cap \mathcal{R} = \varnothing$; then $PP' \subset \rho(P)$. Since $P \notin \mathcal{R}$ there is a point T not adjacent to P. On PP' there is at least one point T' adjacent to T. Since $T' \in \rho(P)$, we have $T \sim P$, a contradiction. Hence $PP_1 \cap \mathcal{R} \neq \varnothing$, and so $P_1 \in \mathcal{R}P$. Consequently $\mathcal{R} \cup \rho(P) \subset \mathcal{R}P$. It follows that $\mathcal{R} \cup \rho(P) = \mathcal{R}P$.

The set $\mathcal{P}$ is the union of all lines joining every point of $\mathcal{R}$ to every point of $\pi \cap \mathcal{S}$. Let $\rho(P) \neq \rho(P')$, with $P, P' \notin \mathcal{R}$, let $T \in \rho(P)$, $T' \in \rho(P')$, and let $P \sim P'$. Since $T \in \rho(P)$ we have $T \sim P'$, and since $T' \in \rho(P')$ we have $T \sim T'$.

It now follows that two points of $\mathscr{P}\backslash\mathscr{R}$ are adjacent if and only if their projections from $\mathscr{R}$ onto π are adjacent. Finally, by Theorem 26.1.4, $\pi \cap \mathscr{S}$ is a non-degenerate Shult space. □

The non-degenerate Shult space $\pi \cap \mathscr{S}$ will be called a *basis* of $\mathscr{S}$.

Lemma 26.3.4: *Let $\mathscr{S}$ be a non-degenerate Shult space in $PG(d, q)$ and let P, P' be adjacent points of $\mathscr{S}$. Then there is a point T in $\mathscr{S}$ such that $T \nsim P$ and $T \nsim P'$. Further, each point of $\mathscr{S}$ is contained in a constant number $t + 1$ $(t \geq 1)$ of lines of $\mathscr{B}$.*

Proof: Assume the contrary. As $\mathscr{S}$ is non-degenerate, there is a point P_1 with $P_1 \nsim P'$; so $P_1 \sim P$. There is also a point P_2 for which $P_2 \nsim P$ and $P_2 \sim P'$. Let l be a line through P_1 intersecting $P'P_2$ in P_2'. Then $P_2' \neq P'$ and $P \nsim P_2'$. If T is a point on l distinct from P_1 and P_2', then $T \nsim P$ and $T \nsim P'$.

Let $M, M' \in \mathscr{P}$, with $M \nsim M'$. If m is any line of $\mathscr{B}$ through M, then there is a unique line of $\mathscr{B}$ through M' concurrent with m, and conversely. It follows that the number of lines of $\mathscr{B}$ through M is equal to the number of lines of $\mathscr{B}$ through M'. Next, let $M, M' \in \mathscr{P}$, with $M \neq M'$ and $M \sim M'$. There is a point T in $\mathscr{S}$ such that $T \nsim M$ and $T \nsim M'$. The number of lines of $\mathscr{B}$ through T is equal to the number of lines of $\mathscr{B}$ through M and to the number of lines of $\mathscr{B}$ through M'. Again the number of lines of $\mathscr{B}$ through M is equal to the number of lines of $\mathscr{B}$ through M'. Hence the number of lines of $\mathscr{B}$ through the point $M \in \mathscr{P}$ is a constant $t + 1$. Let $m \in \mathscr{B}$ with $M \notin m$. On m there is a point M' which is collinear with M; so M' is contained in at least two lines of $\mathscr{B}$. We conclude that $t \geq 1$. □

Lemma 26.3.5: *If π is a subspace of $PG(d, q)$ for which $\pi \cap \mathscr{P}$ is non-empty, then $\pi \cap \mathscr{S}$ is a Shult space. If $\mathscr{S}$ is non-degenerate and π is a hyperplane, then $\pi \cap \mathscr{P}$ generates π.*

Proof: Let π be a subspace of $PG(d, q)$. If $\pi \cap \mathscr{P}$ is non-empty, then it is immediate that $\pi \cap \mathscr{S}$ is a Shult space. Now let $\mathscr{S}$ be non-degenerate and let π be a hyperplane. Since $PG(d, q)$ is the ambient space of $\mathscr{S}$, there is a point P in $\mathscr{P}\backslash(\pi \cap \mathscr{P})$. It suffices to show that an arbitrary line l of $\mathscr{B}$ is in $\langle \pi \cap \mathscr{P}, P \rangle$. We may suppose that l meets π in some point P'. If $P \in l$, the required conclusion is obvious. So suppose $P \notin l$. If $P \sim P''$ with $P'' \in l\backslash\{P'\}$, there is a line l' of $\mathscr{B}$ through P and P'' which is in $\langle \pi \cap \mathscr{P}, P \rangle$. Hence $P'' \in \langle \pi \cap \mathscr{P}, P \rangle$, and consequently l is in $\langle \pi \cap \mathscr{P}, P \rangle$. Finally, suppose $P \notin l$ and suppose that P' is the only point of l which is adjacent to P. Let R be a point not adjacent to P'. The line PP' contains a point R' for which $R' \sim R$. As $R' \in \langle \pi \cap \mathscr{P}, P \rangle$, we have $RR' \subset \langle \pi \cap \mathscr{P}, P \rangle$. Let $T \in RR'\backslash\{R'\}$ with $T \notin \pi$. Then $T \in \langle \pi \cap \mathscr{P}, P \rangle$. Let $T' \in l$ with $T \sim T'$. Then $T' \neq P'$, since otherwise $R \sim P'$. The line TT' is contained in $\langle \pi \cap \mathscr{P}, P \rangle$; hence $T' \in \langle \pi \cap \mathscr{P}, P \rangle$. We conclude that $P'T' = l$ is contained in $\langle \pi \cap \mathscr{P}, P \rangle$. $\square$

For P in $\mathscr{P}$ put $P^\perp = \{P' \in \mathscr{P} | P' \sim P\}$. Then $P^\perp$ is the union of all lines of $\mathscr{B}$ through P. A *tangent to* $\mathscr{S}$ *at* $P \in \mathscr{P}$ is any line l through P such that either $l \in \mathscr{B}$ or $l \cap \mathscr{P} = \{P\}$. The union of all tangents to $\mathscr{S}$ at P will be called the *tangent set of* $\mathscr{S}$ *at* P and is denoted by $\mathscr{S}(P)$. The relation between $\mathscr{S}(P)$ and $P^\perp$ is that $P^\perp = \mathscr{P} \cap \mathscr{S}(P)$. A line l of $PG(d, q)$ is a *secant* of $\mathscr{S}$ if l intersects $\mathscr{P}$ in at least two points but is not a member of $\mathscr{B}$.

Lemma 26.3.6: *For each* $P \in \mathscr{P}$ *we have* $\langle P^\perp \rangle \subset \mathscr{S}(P)$.

Proof: We must show that for each line l through P in $\langle P^\perp \rangle$ either $l \in \mathscr{B}$ or l intersects $\mathscr{P}$ exactly in P. So suppose that $P \in l \notin \mathscr{B}$, $l \subset \langle P^\perp \rangle$. First, suppose that there is some line l_1 of $\mathscr{B}$ through P and a second tangent l_2 to $\mathscr{S}$ at P for which the plane $\pi = \langle l_1, l_2 \rangle$ contains l. If l were not a tangent at P it would contain some point P', where $P \neq P' \in \mathscr{P}$. There would be a unique line $m \in \mathscr{B}$ through P' and intersecting l_1 in P_1, with $P_1 \neq P$. As m is contained in π, so m meets l_2 in a point P_2, with $P_2 \neq P$. Then $P, P_2 \in l_2$ implies $l_2 \in \mathscr{B}$, since l_2 is a tangent to $\mathscr{S}$ containing at least two points of $\mathscr{S}$. But then l_1 and l_2 are two lines of $\mathscr{S}$ through P intersecting m, contradicting $P \nsim P'$ and the assumption that $\mathscr{S}$ is a Shult space. Hence l must be a tangent.

Now, suppose there is an integer k such that $\langle P^\perp \rangle$ is generated by k lines $l_1, l_2, \ldots, l_k$ of $\mathscr{S}$ through P. Let $\pi^{(i)} = \langle l_1 \cup l_2 \cup \cdots \cup l_i \rangle$, $i = 2, 3, \ldots, k$. From the first case we know that $\pi^{(2)} \in \mathscr{S}(P)$. Now we use induction on i. Assume $\pi^{(i)} \subset \mathscr{S}(P)$, and let l be some line of $\pi^{(i+1)}$ through P. We may suppose $l \neq l_{i+1}$ and $l \not\subset \pi^{(i)}$. Then the plane $\pi = \langle l, l_{i+1} \rangle$ intersects $\pi^{(i)}$ along a line l'. By the induction hypothesis l' is tangent to $\mathscr{S}$ at P, so that $\pi = \langle l', l_{i+1} \rangle$ satisfies the hypothesis of the first case. Hence l is a tangent to $\mathscr{S}$ at P, and it follows that $\pi^{(i+1)} \subset \mathscr{S}(P)$. $\square$

We assume that from now on $\mathscr{S}$ is non-degenerate.

Lemma 26.3.7: *For any point* P *in* $\mathscr{P}$ *we have* $\langle P^\perp \rangle \neq PG(d, q)$.

Proof: Otherwise $\langle P^\perp \rangle = PG(d, q)$ and, by Lemma 26.3.6, $\mathscr{S}(P) = PG(d, q)$. But then P is adjacent to all points of $\mathscr{S}$, and so $\mathscr{S}$ is degenerate, a contradiction. $\square$

Lemma 26.3.8: *The dimension of* $\langle P^\perp \rangle$ *is independent of* P *in* $\mathscr{P}$.

Proof: If $P \nsim P'$, with $P, P' \in \mathscr{P}$, then we show that $\langle P^\perp \rangle \cap \langle P'^\perp \rangle$ is a hyperplane in $\langle P^\perp \rangle$. We have $\langle P^\perp \cap P'^\perp \rangle \subset \langle P^\perp \rangle \cap \langle P'^\perp \rangle$ and, since $\mathscr{S}$ is a Shult space, P and $P^\perp \cap P'^\perp$ generate $\langle P^\perp \rangle$. Hence $\langle P^\perp \rangle \cap \langle P'^\perp \rangle$ is a hyperplane of $\langle P^\perp \rangle$, or $\langle P^\perp \rangle \cap \langle P'^\perp \rangle = \langle P^\perp \rangle$. If $\langle P^\perp \rangle = \langle P^\perp \rangle \cap \langle P'^\perp \rangle$, then $\langle P'^\perp \rangle \subset \langle P^\perp \rangle \subset \mathscr{S}(P)$, contradicting $P \nsim P'$. Consequently $\langle P^\perp \rangle \cap \langle P'^\perp \rangle$ is a hyperplane of $\langle P^\perp \rangle$. Analogously $\langle P^\perp \rangle \cap \langle P'^\perp \rangle$ is a hyperplane of $\langle P'^\perp \rangle$. It follows that $\langle P^\perp \rangle$ and $\langle P'^\perp \rangle$ have the same dimension.

Next, let $P \sim P'$ with P, P' distinct points of $\mathscr{P}$. By Lemma 26.3.4 there is a point T in $\mathscr{S}$ such that $T \nsim P$ and $T \nsim P'$. Now by the first part of the proof, $\dim\langle T^\perp \rangle = \dim\langle P^\perp \rangle$ and $\dim\langle T^\perp \rangle = \dim\langle P'^\perp \rangle$. Hence $\dim\langle P^\perp \rangle = \dim\langle P'^\perp \rangle$.

We conclude that the dimension of $\langle P^\perp \rangle$ is independent of P. □

Lemma 26.3.9: *The point P in $\mathscr{P}$ is the unique point of $\mathscr{S}$ adjacent to all points of $P^\perp$.*

Proof: Let $P' \in \mathscr{P} \backslash \{P\}$ be adjacent to all points of $P^\perp$. Clearly $P' \in P^\perp$. Since $\mathscr{S}$ is a Shult space, all points of the line PP' are adjacent to all points of $P^\perp$. Now take a point $T \in \mathscr{P} \backslash P^\perp$. Since $\langle P^\perp \rangle \subset \mathscr{S}(P)$, we have $T \notin \langle P^\perp \rangle$. There is a line m through T intersecting PP' in a point P'', and so $\langle P''^\perp \rangle$ contains $\langle P^\perp \rangle$ properly. This contradicts Lemma 26.3.8. □

Lemma 26.3.10: *For each P in $\mathscr{P}$ the projective subspace $\langle P^\perp \rangle$ is a prime of $PG(d, q)$.*

Proof: Consider a point P' in $\mathscr{P} \backslash \langle P^\perp \rangle$. By Lemma 26.3.5, $\langle P^\perp, P' \rangle \cap \mathscr{S}$ is a Shult space $\mathscr{S}'$. Assume that $\mathscr{S}'$ is degenerate with radical $\mathscr{R}'$. If $R \in \mathscr{R}'$, then R is adjacent to all points of $P^\perp$. By Lemma 26.3.9 we have $R = P$. Hence $P' \sim P$, a contradiction. Consequently $\mathscr{S}'$ is non-degenerate. Now assume that $\langle P^\perp, P' \rangle \neq PG(d, q)$, and let $P'' \in PG(d, q) \backslash \langle P^\perp, P' \rangle$. Consider a point T in $\mathscr{S}'$, with $T \sim P''$. If T is contained in $t + 1$ lines of $\mathscr{S}$ and $t' + 1$ lines of $\mathscr{S}'$, then $t \geq t' + 1$. But P is contained in exactly $t + 1$ lines of $\mathscr{S}'$. Hence $t = t'$, contradicting $t > t'$. We conclude that $\langle P^\perp, P' \rangle = PG(d, q)$, so that $\langle P^\perp \rangle$ is a prime of $PG(d, q)$. □

Lemma 26.3.11: *The hyperplane $\langle P^\perp \rangle$, for P in $\mathscr{P}$, is the tangent set $\mathscr{S}(P)$ of $\mathscr{S}$ at P.*

Proof: By Lemmas 26.3.6 and 26.3.10, $\langle P^\perp \rangle$ is a hyperplane contained in $\mathscr{S}(P)$. If equality did not hold, there would be some tangent l at P not in $\langle P^\perp \rangle$.

Let l_1 be a line of $\mathscr{S}$ through P and let Π_2 be the plane $\langle l, l_1 \rangle$. If there were a point $P' \in \Pi_2 \cap \mathscr{P}$ with $P' \notin l_1$, there would be a line m of $\mathscr{B}$ through P' meeting l_1 in a point other than P. But m would be in Π_2 and hence meet l in a point $(\neq P)$ of $\mathscr{P}$, an impossibility. Hence each line of Π_2 through P is a tangent of $\mathscr{S}$ at P.

Let l_2 be a line of $\mathscr{S}$ through P, with $l_2 \neq l_1$. If m, with $m \neq l_1$, is a line of Π_2 through P, then by the previous paragraph each line of $\langle l_2, m\rangle$ through P is a tangent of $\mathscr{S}$ at P. Hence each line of $\Pi_3 = \langle l, l_1, l_2\rangle$ through P is a tangent of $\mathscr{S}$ at P.

Let l_3 be a line of $\mathscr{S}$ through P, with l_3 not in the plane $\langle l_1, l_2\rangle$. If m', with m' not in the plane $\langle l_1, l_2\rangle$, is a line of Π_3 through P, then by the previous paragraph each line of $\langle l_3, m'\rangle$ through P is a tangent of $\mathscr{S}$ at P. Hence each line of $\Pi_4 = \langle l, l_1, l_2, l_3\rangle$ through P is a tangent of $\mathscr{S}$ at P.

Proceeding in this way, we finally obtain that each line of $\langle P^{\perp}, l\rangle = PG(d, q)$ through P is a tangent of $\mathscr{S}$ at P. Consequently P belongs to the radical of $\mathscr{S}$, a contradiction as $\mathscr{S}$ is non-degenerate.

So we conclude that $\langle P^{\perp}\rangle = \mathscr{S}(P)$. □

The tangent set $\mathscr{S}(P)$ of $\mathscr{S}$ at P, for P in $\mathscr{P}$, is also called the *tangent hyperplane of $\mathscr{S}$ at P.*

Lemma 26.3.12: *Let l be a secant of $\mathscr{S}$ containing three distinct points P, A, A' of $\mathscr{P}$. Then the perspectivity σ of $PG(d, q)$ with centre P and axis $\mathscr{S}(P)$ mapping A onto A' leaves $\mathscr{P}$ invariant.*

Proof: The map σ fixes all points of $\mathscr{S}(P)$ and thus fixes $P^{\perp}$. Let $P' \in \mathscr{P}\backslash P^{\perp}$. First suppose P' is not on l and let π be the plane $\langle P, A, P'\rangle$. Consider the line $m = \langle A, P'\rangle$. Then m intersects $\mathscr{S}(P)$ at a point P'', fixed by σ. Hence $m\sigma = \langle A', P''\rangle$.

If m is a line of $\mathscr{S}$, then $P'' \in \mathscr{P}$ and so the tangent $\langle P, P''\rangle$ is a line of $\mathscr{S}$. Thus the plane $\langle P, A, P''\rangle = \pi$ is in the tangent hyperplane $\mathscr{S}(P'')$. Hence, since $A' \in \pi$, it follows that $A' \sim P''$, that $m\sigma$ is a line of $\mathscr{S}$, and that $P'\sigma$ is a point of $\mathscr{S}$.

If m is not a line of $\mathscr{S}$, suppose there is a point $D \in \mathscr{P}\backslash\mathscr{S}(P)$ with $D \in A^{\perp} \cap P'^{\perp}$. The argument of the previous paragraph, with D in the role of P', shows that $D\sigma \in \mathscr{P}$. Then with D and $D\sigma$ playing the roles of A and A', respectively, it follows that $P'\sigma \in \mathscr{P}$. On the other hand, suppose $A^{\perp} \cap P'^{\perp} \subset \mathscr{S}(P)$. Consider a line l_1 of $\mathscr{B}$ containing P, and let P_1 be defined by $P_1 \in l_1$ and $A \sim P_1 \sim P'$. By Lemma 26.3.4 there is a point $T \in \mathscr{P}$ with $T \sim P$ and $T \nsim P_1$. The line of $\mathscr{B}$ through T which is concurrent with l_1 is denoted by m_1. Let D, D' be defined by $D, D' \in m_1$ and $A \sim D$, $D' \sim P'$. Then D, D' are distinct points of $\mathscr{P}\backslash\mathscr{S}(P)$. Consecutive applications of the previous paragraph show that $D\sigma$, $D'\sigma$, and finally $P'\sigma$ are all in $\mathscr{P}$.

Secondly, suppose P' is on l, and use the fact that if D is any point of $\mathscr{P}$ not on l then $D\sigma \in \mathscr{P}$. It follows readily that $P'\sigma \in \mathscr{P}$. □

Lemma 26.3.13: *If A, B, C are three collinear, distinct points of $\mathscr{P}$, then the intersections $\mathscr{S}(A) \cap \mathscr{S}(B)$, $\mathscr{S}(B) \cap \mathscr{S}(C)$, and $\mathscr{S}(C) \cap \mathscr{S}(A)$ coincide.*

Proof: We show that $\mathcal{S}(A) \cap \mathcal{S}(B) \subset \mathcal{S}(C)$.

First, suppose that A, B, C are on a line l of $\mathcal{S}$. Let $P \in \mathcal{S}(A) \cap \mathcal{S}(B)$. If $P \in \mathcal{P}\backslash l$, then $P \sim A$ and $P \sim B$, and so $P \sim C$. If $P \in l$, then $P \in \mathcal{S}(C)$. Now let $P \notin \mathcal{P}$. Suppose that PC is a secant of $\mathcal{S}$, and let $C' \in PC \cap \mathcal{P}$ with $C \neq C'$. The plane $\langle A, P, B \rangle$ is in $\mathcal{S}(A)$; hence $AC' \in \mathcal{B}$. Analogously $BC' \in \mathcal{B}$. Consequently $C \sim C'$, a contradiction. So PC is a tangent of $\mathcal{S}$ at C.

Secondly, suppose that A, B, C are on a secant l of $\mathcal{S}$. Let $P \in \mathcal{S}(A) \cap \mathcal{S}(B)$. If $P \in \mathcal{P}$, then $P \sim A$ and $P \sim B$, and so the line AB is in $\mathcal{S}(P)$. Hence $P \sim C$ and so $P \in \mathcal{S}(C)$. Now let $P \notin \mathcal{P}$, and suppose that $P \notin \mathcal{S}(C)$. Then there is a second point C' of $\mathcal{P}$ on the line PC. Consider the perspectivity σ with centre A and axis $\mathcal{S}(A)$ mapping C onto B. By Lemma 26.3.12 we have $\mathcal{P}\sigma = \mathcal{P}$. This perspectivity σ fixes $P \in \mathcal{S}(A)$ and so σ maps the line PC onto the line PB. Hence $C'\sigma$ is a point of $PB\backslash\{B\}$ on $\mathcal{P}$, a contradiction since $P \in \mathcal{S}(B)$. Consequently $P \in \mathcal{S}(C)$.

Hence $\mathcal{S}(A) \cap \mathcal{S}(B) \subset \mathcal{S}(C)$. Analogously $\mathcal{S}(A) \cap \mathcal{S}(C) \subset \mathcal{S}(B)$ and $\mathcal{S}(B) \cap \mathcal{S}(C) \subset \mathcal{S}(A)$. We conclude that $\mathcal{S}(A) \cap \mathcal{S}(B)$, $\mathcal{S}(B) \cap \mathcal{S}(C)$, and $\mathcal{S}(C) \cap \mathcal{S}(A)$ coincide. □

Lemma 26.3.14: *All secant lines contain the same number of points of $\mathcal{S}$.*

Proof: Let l and l' be secant lines of $\mathcal{S}$. First suppose l and l' have a point P of $\mathcal{P}$ in common, and let m be any secant line through P. If some m contains more than two points in $\mathcal{P}$, by Lemma 26.3.12 we may consider the non-trivial group G of all perspectivities with centre P and axis $\mathcal{S}(P)$ leaving $\mathcal{P}$ invariant. The group G is regular on the set of points of m in $\mathcal{P}$ other than P. Hence each secant through P has $1 + |G|$ points of $\mathcal{P}$, so that l and l' have the same number of points of $\mathcal{S}$. If no m is incident with more than two points of $\mathcal{P}$, then l and l' contain two points of $\mathcal{S}$.

Secondly, suppose l and l' do not have any point of $\mathcal{P}$ in common, and choose points P, P' of $\mathcal{P}$ on l, l', respectively. If $P \nsim P'$, then PP' is a secant, and so meets $\mathcal{P}$ in the same number of points as do l and l', by the previous paragraph. If $P \sim P'$, then by Lemma 26.3.4 there is a point T in $\mathcal{S}$ such that $P \nsim T \nsim P'$. Now apply the previous paragraph to the secants l, PT, TP', l'. □

For a point P of $PG(d, q)$, the *collar* $\mathcal{S}_P$ *of* $\mathcal{S}$ *for* P is the set of all points P' of $\mathcal{S}$ such that $P = P'$ or the line $\langle P, P' \rangle$ is a tangent to $\mathcal{S}$ at P. For example, if $P \in \mathcal{P}$, $\mathcal{S}_P$ is just $P^\perp$. If $P \notin \mathcal{P}$, the collar $\mathcal{S}_P$ is the set of points P' of $\mathcal{P}$ such that $\langle P, P' \rangle \cap \mathcal{P} = \{P'\}$.

For all $P \in PG(d, q)$ the *polar* $P\zeta$ of P with respect to $\mathcal{S}$ is the subspace of $PG(d, q)$ generated by the collar $\mathcal{S}_P$. In particular, if $P \in \mathcal{P}$, then $P\zeta = \langle P^\perp \rangle = \mathcal{S}(P)$.

Lemma 26.3.15: *For any point P, let P_1 and P_2 be distinct points of $\mathcal{S}_P$. Then $\mathcal{P} \cap \langle P_1, P_2 \rangle \subset \mathcal{S}_P$.*

Proof: Suppose $P' \in \mathcal{P} \cap \langle P_1, P_2 \rangle$, $P_1 \neq P' \neq P_2$. Since $P \in \mathcal{S}(P_1) \cap \mathcal{S}(P_2)$, by Lemma 26.3.13 the point P is also in $\mathcal{S}(P')$ and hence $P' \in \mathcal{S}_P$. □

Lemma 26.3.16: *Each line l of $\mathcal{S}$ intersects the collar $\mathcal{S}_P$, with $P \in PG(d, q)$, in exactly one point, unless each point of l is in $\mathcal{S}_P$.*

Proof: When $P \in \mathcal{P}$, the result is immediate. So suppose that $P \notin \mathcal{P}$ and put $\pi = \langle l, P \rangle$. If $\pi \cap \mathcal{P} = l$, then each point of l is in $\mathcal{S}_P$. So suppose $P' \in \pi \cap \mathcal{P}$, $P' \notin l$. Let $\langle P, P' \rangle \cap l = \{T\}$; then $T \nsim P'$. Hence P' is adjacent to a unique point T' of l. By Lemma 26.3.11 each line of π through T' is tangent to $\mathcal{S}$ at T', and hence $T' \in \mathcal{S}_P$. Also, by Lemma 26.3.15, T' is the unique point of l in $\mathcal{S}_P$ unless each point of l is in $\mathcal{S}_P$. □

Lemma 26.3.17: *Either $P\zeta = \langle \mathcal{S}_P \rangle$ is a prime or $P\zeta = PG(d, q)$.*

Proof: Again assume that $P \notin \mathcal{P}$. If the assertion is false for some point P, then $P\zeta$ is contained in some $(d-2)$-space π of $PG(d, q)$. By Lemma 26.3.16 each line l of $\mathcal{B}$ intersects π. Therefore if P' is a point of $\mathcal{S}$ not in π, then $\mathcal{S}_{P'} = P'^{\perp}$ is contained in $\langle \pi, P' \rangle$. As $\langle \mathcal{S}_{P'} \rangle$ is a hyperplane, $\langle \pi, P' \rangle = \langle \mathcal{S}_{P'} \rangle = \mathcal{S}(P')$. Any line l' of $\mathcal{S}$ through P' must contain a second point P'' of $\mathcal{P}$ not in π. Then $\mathcal{S}(P') = \langle \pi, P' \rangle = \langle \pi, P'' \rangle = \mathcal{S}(P'')$. This contradicts Lemma 26.3.9. □

Lemma 26.3.18: *If $P\zeta$ is a prime, then $\mathcal{S}_P = \mathcal{P} \cap P\zeta$.*

Proof: Clearly $\mathcal{S}_P \subset \mathcal{P} \cap P\zeta$. Suppose there were a point P' of $\mathcal{P} \cap P\zeta$ not in $\mathcal{S}_P$. Then either some line l of $\mathcal{S}$ through P' does not lie in $P\zeta$, or $P\zeta = \mathcal{S}(P')$. In the first case l intersects $P\zeta$ exactly in P'. As $P' \notin \mathcal{S}_P$, so l is on no point of $\mathcal{S}_P$, contradicting Lemma 26.3.16. In the second case, as $P' \notin \mathcal{S}_P$, each line of $\mathcal{B}$ through P' has exactly one point in $\mathcal{S}_P$. So on any line of $\mathcal{B}$ through P' there is a point P'', $P'' \neq P'$, of $\mathcal{S}(P') \backslash \mathcal{S}_P$. By Lemma 26.3.9 we have $\mathcal{S}(P') \neq \mathcal{S}(P'')$; so there is a line of $\mathcal{B}$ through P'' but not in $P\zeta = \mathcal{S}(P')$, leading back to the first case. □

Lemma 26.3.19: *Let P be a point of $PG(d, q)$ and A, A' distinct points of $\mathcal{P} \backslash \{P\}$ collinear with P but not in $P\zeta$. Then the perspectivity σ of $PG(d, q)$ with centre P and axis $P\zeta$ mapping A onto A' leaves $\mathcal{P}$ invariant.*

Proof: Since $A, A' \notin P\zeta$ we know that $P\zeta$ is a prime. First, let $P \in \mathcal{P}$. Since $A, A' \notin P\zeta$ the line $\langle P, A, A' \rangle$ is a secant of $\mathcal{S}$. In this case the result is known by Lemma 26.3.12. Now let $P \notin \mathscr{P}$. Note that σ fixes all points of $\mathscr{P} \cap P\zeta$. Let P' be a point of $\mathscr{P} \backslash P\zeta$ not on $\langle A, A' \rangle$. Let π be the plane $\langle P, A, P' \rangle$ and m the line $\langle A, P' \rangle$. If $m \cap P\zeta = \{P''\}$, then $m\sigma = \langle A', P'' \rangle$.

If m is a line of $\mathscr{S}$, then $P'' \in \mathscr{P} \cap P\zeta = \mathscr{S}_P$ by Lemma 26.3.18. So $\langle P'', P \rangle$ and m, and hence π, are in the tangent hyperplane $\mathscr{S}(P'')$. Then $A' \in \pi \subset \mathscr{S}(P'')$, forcing $m\sigma = \langle A', P'' \rangle \in \mathscr{B}$; that is, $P'\sigma \in \mathscr{P}$.

If m is not a line of $\mathscr{S}$, suppose there is a point $D \in \mathscr{P} \backslash P\zeta$ with $D \in A^{\perp} \cap P'^{\perp}$. The argument of the previous paragraph, with D in the role of P', shows that $D\sigma \in \mathscr{P}$. Then with D and $D\sigma$ playing the roles of A and A' respectively, it follows that $P'\sigma \in \mathscr{P}$. On the other hand, suppose $A^{\perp} \cap P'^{\perp} \subset \mathscr{S}_P$. Let $T \in A^{\perp} \cap P'^{\perp}$. Now assume that each point of $A^{\perp} \cap P'^{\perp}$ is adjacent to T. Since each line of $\mathscr{B}$ through P' has a point in common with $\mathscr{S}_P$, it follows that T is adjacent to all points of $P'^{\perp}$, contradicting Lemma 26.3.9. Hence there exists a point $T' \in A^{\perp} \cap P'^{\perp}$, with $T \nsim T'$. Let $D \in \langle A, T \rangle \backslash \{A, T\}$, and let $D' \in \langle P', T' \rangle$ with $D \sim D'$. If $D' = P'$, then $A \sim P'$, a contradiction. If $D = T'$, then $T \sim T'$, a contradiction. However, $D \neq D'$, which means that D and D' are distinct points of $\mathscr{P} \backslash P\zeta$. Consecutive applications of the previous paragraph show that $D\sigma$, $D'\sigma$, and finally $P'\sigma$ are all in $\mathscr{P}$.

Finally, suppose that P' is on $\langle A, A' \rangle$, and use the fact that if D is any point of $\mathscr{P}$ not on $\langle A, A' \rangle$ then $P'\sigma \in \mathscr{P}$. It follows readily that $P'\sigma \in \mathscr{P}$. □

By Lemma 26.3.14 all secant lines of $\mathscr{S}$ contain the same number α of points of $\mathscr{S}$; note that $\alpha \geq 2$.

Lemma 26.3.20: *If $\alpha \neq 2$, there is no point in all tangent hyperplanes of $\mathscr{S}$.*

Proof: Assume that P belongs to all tangent hyperplanes of $\mathscr{S}$. If $P \in \mathscr{P}$, then $\mathscr{P} = P^{\perp}$, contradicting the non-degeneracy of $\mathscr{S}$; so $P \notin \mathscr{P}$. Let $P', P'' \in \mathscr{P}$, with $P' \nsim P''$. Then the plane $\pi = \langle P, P', P'' \rangle$ contains no line of $\mathscr{S}$, since otherwise at least one of the tangents $\langle P, P' \rangle$, $\langle P, P'' \rangle$ contains at least two distinct points of $\mathscr{S}$. Since $P' \nsim P''$ we have $\pi \not\subset \mathscr{S}(P')$; so $\langle P, P' \rangle$ is the only line of π which is tangent to $\mathscr{S}$ at P'. Counting the points of $\pi \cap \mathscr{P}$ on the lines of π through P', we obtain $q(\alpha - 1) + 1$. For each point $T \in \pi \cap \mathscr{P}$ the line $\langle T, P \rangle$ is a tangent of $\mathscr{S}$ at T; so the number of points in $\pi \cap \mathscr{P}$ is at most the number of lines of π through P. Consequently $q(\alpha - 1) + 1 \leq q + 1$. Hence $\alpha = 2$. □

Theorem 26.3.21: *When $\alpha = 2$, then $\mathscr{S}$ is formed by the points and lines of a non-singular quadric of $PG(d, q)$.*

Proof: Each line of $PG(d, q)$ contains 0, 1, 2, or $q + 1$ points of $\mathscr{S}$. Since the union of all tangent lines at any point of $\mathscr{P}$ is a prime, $\mathscr{P}$ is a non-singular quadratic set in the sense of §22.10. By Theorem 22.10.23, $\mathscr{P}$ is a non-singular quadric of $PG(d, q)$. Clearly all lines in $\mathscr{B}$ are lines of the quadric $\mathscr{P}$. Conversely, let l be a line of the quadric $\mathscr{P}$. Since l contains more than $\alpha = 2$ points of $\mathscr{P}$, it is a line of $\mathscr{B}$. Hence $\mathscr{B}$ is the set of all lines of the quadric $\mathscr{P}$. □

From now on we assume that $\alpha > 2$. By Lemma 26.3.20 there is no point in all tangent hyperplanes of $\mathscr{S}$.

Lemma 26.3.22: *If $\mathscr{S}$ has rank at least three, then $P\zeta$ is a prime for any P in $PG(d, q)$.*

Proof: Assume that $\mathscr{S}$ contains a plane π. If $P \in \mathscr{P}$, then $P\zeta$ is the prime $\langle P^{\perp} \rangle = \mathscr{S}(P)$. By way of contradiction assume that $P \notin \mathscr{P}$ and $P\zeta = PG(d, q)$. By Lemma 26.3.16 each line of π has at least one point in $\mathscr{S}_P$. Hence consider in π two points P_1 and P_2 of $\mathscr{S}_P$. By Lemma 26.3.15 the line $\langle P_1, P_2 \rangle$ is contained in $\mathscr{S}_P$. Hence $\mathscr{S}_P$ contains at least one line of $\mathscr{B}$. Again by Lemma 26.3.15 it is clear that $\mathscr{S}_P$, together with the lines of $\mathscr{S}$ in $\mathscr{S}_P$, forms a projective Shult space $\mathscr{S}'$ with ambient space $\langle \mathscr{S}_P \rangle = P\zeta = PG(d, q)$. The Shult space $\mathscr{S}'$ cannot be degenerate, since otherwise $\mathscr{S}_P$ would be contained in a tangent hyperplane of $\mathscr{S}$, contradicting $\langle \mathscr{S}_P \rangle = PG(d, q)$. Consequently the lines of $\mathscr{S}'$ through $T \in \mathscr{S}_P$ generate a prime π' of $PG(d, q)$. Hence $\pi' = \langle T^{\perp} \rangle$, and so the tangent hyperplanes of $\mathscr{S}'$ are tangent hyperplanes of $\mathscr{S}$. Let $T \in \mathscr{S}_P$, and consider the secant lines of $\mathscr{S}'$ through T. As the tangent hyperplanes $\mathscr{S}'(T)$ and $\mathscr{S}(T)$ coincide, these secant lines are also the secant lines of $\mathscr{S}$ through T. Hence, by Lemma 26.3.15, all points of $\mathscr{P}$ non-adjacent to T are in $\mathscr{S}_P$. Next, let $T' \in T^{\perp} \backslash \{T\}$. By Lemma 26.3.9 there is a point T'' in $T'^{\perp}$ which is not adjacent to T. The line $\langle T', T'' \rangle$ contains q points not adjacent to T. Hence these q points are in $\mathscr{S}_P$. By Lemma 26.3.15, T' is also in $\mathscr{S}_P$. Consequently $\mathscr{S}_P = \mathscr{P}$. So P is in all tangent hyperplanes of $\mathscr{S}$, contradicting Lemma 26.3.20. □

The next few lemmas show that Lemma 26.3.22 also holds in the rank 2 case, that is in the case of a generalized quadrangle.

Lemma 26.3.23: *If $\mathscr{S}$ is a generalized quadrangle, then $\alpha = (t/q^{d-3}) + 1$ and $d = 3$ or 4.*

Proof: If $\mathscr{S}$ is a generalized quadrangle, then $\mathscr{S}$ has order (q, t). Since $\mathscr{S}$ is non-degenerate it is clear that $d \geq 3$. The secant lines through a point P of $\mathscr{P}$ are the q^{d-1} lines of $PG(d, q)$ through P which do not lie in the tangent hyperplane $\mathscr{S}(P)$. Hence the total number of points of $\mathscr{S}$ is $(\alpha - 1)q^{d-1} + |P^{\perp}|$. By Theorem 26.2.1 we have $|\mathscr{P}| = (1 + q)(1 + qt)$. Hence $\alpha = (t/q^{d-3}) + 1$.

By Theorem 26.2.3 we know that $t \le q^2$, so that $2 < \alpha \le (q^2/q^{d-3}) + 1$, implying $d = 3$ or $d = 4$. $\square$

A subset $\mathscr{C}$ of $\mathscr{P}$ is called *linearly closed* in $\mathscr{P}$ (or $\mathscr{S}$) if for all $P, P' \in \mathscr{C}$, $P \neq P'$, the intersection $\langle P, P' \rangle \cap \mathscr{P}$ is contained in $\mathscr{C}$. Thus any subset $\mathscr{D}$ of $\mathscr{P}$ generates a *linear closure* $\bar{\mathscr{D}}$ in $\mathscr{P}$ (or $\mathscr{S}$).

Lemma 26.3.24: *Let $\mathscr{S}$ be a generalized quadrangle with ambient space $PG(3, q)$. If $P_1, P_2, P_3 \in \mathscr{P}$ are non-collinear in $PG(3, q)$, then $\overline{\{P_1, P_2, P_3\}} = \mathscr{P} \cap \langle P_1, P_2, P_3 \rangle$.*

Proof: Let $\mathscr{S}$ have order (q, t). If the plane $\pi = \langle P_1, P_2, P_3 \rangle$ contains a line of $\mathscr{S}$, then $\mathscr{P} \cap \langle P_1, P_2, P_3 \rangle$ consists of $t + 1$ distinct concurrent lines of $\mathscr{B}$. In this case the lemma is trivial. Hence suppose π contains no lines of $\mathscr{B}$. As $d = 3$, by Lemma 26.3.23 any secant line intersects $\mathscr{P}$ in exactly $t + 1$ points. Take a point P, with $P \neq P_1$, on $\langle P_1, P_2 \rangle \cap \mathscr{P}$. The $t + 1$ secant lines $\langle P, T \rangle$, where T is a point of $\mathscr{P} \cap \langle P_1, P_3 \rangle$, intersect $\mathscr{P}$ in points which are in the linear closure $\overline{\{P_1, P_2, P_3\}}$. As each of these lines $\langle P, T \rangle$ intersects $\mathscr{P}$ in $t + 1$ points, there are $t(t + 1) + 1$ points of $\mathscr{P}$ on these lines. Hence $|\overline{\{P_1, P_2, P_3\}}| \ge t^2 + t + 1$. If Lemma 26.3.24 were false, there would be a point T' in $(\mathscr{P} \cap \pi) \backslash \overline{\{P_1, P_2, P_3\}}$. Then every line of π through T' contains at most one point of $\overline{\{P_1, P_2, P_3\}}$; so there are at least $t^2 + t + 1$ lines of π through T'. Hence $t^2 + t + 1 \le q + 1$, and so $t^2 < q$. By Theorem 26.2.3 it follows that $t = 1$. Consequently $\alpha = 2$, a contradiction. We conclude that $\overline{\{P_1, P_2, P_3\}} = \mathscr{P} \cap \langle P_1, P_2, P_3 \rangle$. $\square$

Lemma 26.3.25: *Let $\mathscr{S}$ be a generalized quadrangle with ambient space $PG(4, q)$. If $P_1, P_2, P_3 \in \mathscr{P}$ are non-collinear in $PG(4, q)$, then $\overline{\{P_1, P_2, P_3\}} = \mathscr{P} \cap \langle P_1, P_2, P_3 \rangle$.*

Proof: Let $\mathscr{S}$ have order (q, t). As before we may suppose that $\pi = \langle P_1, P_2, P_3 \rangle$ contains no line of $\mathscr{S}$. Fix a point $P \in \mathscr{P} \cap \pi$ and a line $l \in \mathscr{B}$ incident with P. Put $\pi' = \langle P_1, P_2, P_3, l \rangle$. By Lemma 26.3.5, $\pi' \cap \mathscr{S}$ is a Shult space with ambient space π'. For $\pi' \cap \mathscr{S}$ there are the following possibilities: (a) $\pi' \cap \mathscr{S}$ is non-degenerate and then, by Theorem 26.3.1, $\pi' \cap \mathscr{S}$ is a projective subquadrangle of $\mathscr{S}$; (b) $\pi' \cap \mathscr{S}$ is degenerate, and the lines of $\pi' \cap \mathscr{S}$ contain a distinguished point of π'. If (a) holds, then by Lemma 26.3.24, $\mathscr{P} \cap \pi$ is the linear closure of $\{P_1, P_2, P_3\}$ in $\mathscr{S}$. If (b) holds, two cases are possible. (i) There exists a line l' in $\mathscr{B}$ through a point of π such that $\langle \pi, l' \rangle$ intersects $\mathscr{S}$ in a subquadrangle. Then Lemma 26.3.24 still applies. (ii) For each line l' of $\mathscr{B}$ intersecting π, the lines of $\langle \pi, l' \rangle \cap \mathscr{S}$ all contain a point T_i of l' not in π. Here the prime $\langle \pi, l' \rangle$ is the tangent hyperplane of $\mathcal{S}$ at T_i. Hence π contains $1 + t$ points of $\mathcal{P}$: these are $P_1, P_2, \ldots, P_{t+1}$, and $P_j \sim T_i$ for all i and j. Further, by the definition of the points T_i, the lines of $\mathcal{S}$ through a given point P_j are the lines $\langle P_j, T_i \rangle$. Hence there are exactly $1 + t$ points T_i. This means $\mathcal{S}$ has two (disjoint) sets $\{P_j\}, \{T_i\}$ of $1 + t$ (pairwise non-adjacent) points with $P_j \sim T_i$, $0 \leq i, j \leq t$. If $T' \in \langle P_j, T_i \rangle \cap \langle P_{j'}, T_{i'} \rangle$, with $j \neq j'$ and $i \neq i'$, then a triangle $P_j T_{i'} T'$ is obtained in $\mathcal{S}$, a contradiction since $\mathcal{S}$ is a generalized quadrangle. Hence there are exactly $(q - 1)(t + 1)^2 + 2(t+1)$ points of $\mathcal{S}$ on the lines $\langle P_j, T_i \rangle$. Since $(q-1)(t+1)^2 + 2(t+1) \leq |\mathcal{P}| = (q+1)(qt+1)$, we have $(t-q)t(q-1) \leq 0$, whence $t \leq q$. Then $\alpha = (t/q) + 1 \leq 2$, contradicting $\alpha > 2$. □

Lemma 26.3.26: *Let $\{P_1, P_2, \ldots, P_k\}$ be a set of points of the generalized quadrangle $\mathcal{S}$. Then the linear closure of $\{P_1, P_2, \ldots, P_k\}$ in $\mathcal{S}$ is $\mathcal{P} \cap \langle P_1, P_2, \ldots, P_k \rangle$.*

Proof: First note that if $q = 2$ then, since $\alpha > 2$, any line of $PG(d, q)$ containing at least two points of $\mathcal{P}$ is entirely contained in $\mathcal{P}$. Hence $\mathcal{P}$ is a projective subspace of $PG(d, q)$ and $\mathcal{P} = PG(d, q)$. Consequently in that case the lemma is trivial.

Hence we assume $q > 2$. The result holds trivially when $k = 1$ or $k = 2$. By Lemmas 26.3.24 and 26.3.25 the result also holds if $\langle P_1, P_2, \ldots, P_k \rangle$ is a plane. Further, we may assume that the points P_i are linearly independent in $PG(d, q)$. Now we apply induction; so suppose the result is true for $k - 1$ points $P_1, P_2, \ldots, P_{k-1}$, $3 \leq k - 1$. We show that the result holds for $\{P_1, P_2, \ldots, P_k\}$, with $P_k \in \mathcal{P} \backslash \langle P_1, P_2, \ldots, P_{k-1} \rangle$. Indices can be chosen so that $\langle P_1, P_2, \ldots, P_{k-1} \rangle$ is not contained in $\mathcal{S}(P_1)$. Put $l_i = \langle P_1, P_i \rangle$, $i = 2, 3, \ldots, k$, and let π be any plane through l_k contained in $\langle l_2, l_3, \ldots, l_k \rangle = \langle P_1, P_2, \ldots, P_k \rangle$. Clearly π intersects $\langle l_2, l_3, \ldots, l_{k-1} \rangle$ in a line l. If it is shown that $\mathcal{P} \cap \langle l_k, l \rangle \subset \overline{\{P_1, P_2, \ldots, P_k\}}$, the desired result follows immediately. Suppose l contains at least two points of $\mathcal{P}$. By the induction hypothesis the points of $\mathcal{P}$ on l are all in $\overline{\{P_1, P_2, \ldots, P_{k-1}\}}$. Lemmas 26.3.24 and 26.3.25 then show that $\mathcal{P} \cap \langle l_k, l \rangle$ is in $\overline{\{P_1, P_2, \ldots, P_k\}}$. Now suppose that l is a tangent line whose points are not all in $\mathcal{P}$. If $\langle l_k, l \rangle$ contains no point of $\mathcal{P}$ not on l_k, there is nothing more to show. So suppose P is a point of $\mathcal{P} \cap \langle l_k, l \rangle$ but not on l_k. Consider the plane π' generated by l and a secant through P_1 in the space $\langle l_2, l_3, \ldots, l_{k-1} \rangle$; such a secant exists since $\mathcal{S}(P_1)$ does not contain $\langle P_1, P_2, \ldots, P_{k-1} \rangle$. This plane π' is not in the tangent hyperplane $\mathcal{S}(P_1)$; so l is the unique tangent at P_1 in π'. Hence there are two secants m_1, m_2 in π' and through P_1. Each of the planes $\langle l_k, m_1 \rangle, \langle l_k, m_2 \rangle$ is not in $\mathcal{S}(P_1)$ and hence contains exactly one tangent at P_1. Consider in $\langle l_k, m_1 \rangle$ a secant m, with $m \neq l_k$, such that the plane $\langle m, P \rangle$ intersects $\langle l_k, m_2 \rangle$ in a secant m'.

Note that m exists, because $\langle l_k, m_1 \rangle$ has at least four lines through P_1. By the induction hypothesis the points of $\mathscr{P}$ on m_1 and m_2 belong to $\overline{\{P_1, P_2, \ldots, P_{k-1}\}}$. Hence by Lemmas 26.3.24 and 26.3.25 the points of $\mathscr{P}$ on m and m' belong to $\overline{\{P_1, P_2, \ldots, P_k\}}$. But as $P \in \langle m, m' \rangle$, again by Lemmas 26.3.24 and 26.3.25, $P \in \overline{\{P_1, P_2, \ldots, P_k\}}$. □

Lemma 26.3.27: *If $\mathscr{S}$ is a generalized quadrangle, then $P\zeta$ is a prime for any P in $PG(d, q)$.*

Proof: If $P \in \mathscr{P}$, then $P\zeta$ is the prime $\langle P^\perp \rangle = \mathscr{S}(P)$. So suppose $P \notin \mathscr{P}$. Consider the intersection $P\zeta \cap \mathscr{P}$. By Lemmas 26.3.15 and 26.3.26 all points of $P\zeta \cap \mathscr{P}$ are in $\mathscr{S}_P$, implying $\mathscr{S}_P = P\zeta \cap \mathscr{P}$. If $P\zeta$ were not a hyperplane, then, by Lemma 26.3.17, $P\zeta = PG(d, q)$, implying $\mathscr{S}_P = \mathscr{P}$. Hence P belongs to all hyperplanes of $\mathscr{S}$, contradicting Lemma 26.3.20. We conclude that $P\zeta$ is a hyperplane. □

Theorem 26.3.28: (i) *The mapping $P \to P\zeta$ is a polarity of $PG(d, q)$;*
(ii) *$\mathscr{P}$ is the set of all self-conjugate points of ζ;*
(iii) *$\mathscr{B}$ consists of all lines l of $PG(d, q)$ such that $l \subset l\zeta$.*

Proof: First of all we show that $P \to P\zeta$ defines a bijection from the set of all points of $PG(d, q)$ onto the set of all primes of $PG(d, q)$. By Lemmas 26.3.22 and 26.3.27, $P\zeta$ is a prime for any $P \in PG(d, q)$. Assume that $P \neq P'$ and $P\zeta = P'\zeta$. Let $X \in \langle P, P' \rangle$ and $Y \in \mathscr{S}_P = \mathscr{P} \cap P\zeta = \mathscr{P} \cap P'\zeta = \mathscr{S}_{P'}$, with $Y \neq X, P, P'$. The lines $\langle Y, P \rangle$ and $\langle Y, P' \rangle$ are tangents of $\mathscr{S}$; hence the line $\langle Y, X \rangle$ is also a tangent of $\mathscr{S}$. If $Y = P$ and $Y \neq X$, then the line $\langle Y, X \rangle = \langle Y, P' \rangle$ is a tangent of $\mathscr{S}$. Analogously, if $Y = P'$ and $Y \neq X$, then the line $\langle Y, X \rangle = \langle Y, P \rangle$ is a tangent of $\mathscr{S}$. Hence $\mathscr{S}_P \subset \mathscr{S}_X$ for any $X \in \langle P, P' \rangle$. Consequently $P\zeta = \langle \mathscr{S}_P \rangle \subset \langle \mathscr{S}_X \rangle = X\zeta$; so $P\zeta = X\zeta$ since $P\zeta$ and $X\zeta$ are primes. Let $P'' \in \mathscr{P} \backslash P\zeta$; then $P \neq P'' \neq P'$. Let X be the common point of $\langle P, P' \rangle$ and the prime $P''\zeta$. Then $X = P''$, or $\langle X, P'' \rangle$ is a tangent; so $P'' \in X\zeta$. Hence $P'' \in P\zeta = X\zeta$, a contradiction. It follows that $P \neq P'$ implies $P\zeta \neq P'\zeta$. Since $PG(d, q)$ is finite, we conclude that $P \to P\zeta$ defines a bijection of the set of all points of $PG(d, q)$ onto the set of all primes of $PG(d, q)$.

Next, suppose that $P' \in P\zeta$, with $P \neq P'$. Let $A \in \mathscr{P} \backslash P\zeta$, and let $A' \in \langle P, A \rangle \cap \mathscr{P}$ with $A' \neq P$ and $A' \neq A$. By Lemma 26.3.19 the perspectivity σ with centre P and axis $P\zeta$ mapping A onto A' leaves $\mathscr{P}$ invariant. Since P' is on the axis of σ we have $P'\sigma = P'$; hence $P'\zeta\sigma = P'\zeta$. By the previous paragraph $P\zeta \neq P'\zeta$; so $P'\zeta$ contains the centre P of σ. Consequently it has been proved that $P' \in P\zeta$ implies $P \in P'\zeta$. This means that ζ is a polarity of $PG(d, q)$.

Let $\mathcal{S}'$ be the Shult space defined by the polarity ζ. The point set of $\mathcal{S}'$ is denoted by $\mathcal{P}'$, and the line set of $\mathcal{S}'$ is denoted by $\mathcal{B}'$. If $P \in \mathcal{P}$, then $P \in P\zeta$ and hence $P \in \mathcal{P}'$; so $\mathcal{P} \subset \mathcal{P}'$. It is also clear that $\mathcal{B} \subset \mathcal{B}'$. Assume that $P' \in \mathcal{P}' \backslash \mathcal{P}$ and let T, T' be distinct points of $\mathcal{S}_{P'}$. First, suppose that the plane $\langle P', T, T' \rangle = \pi$ contains no line of $\mathcal{B}$. Since $T \nsim T'$ we have $\pi \not\subset \mathcal{S}(T)$; so $\langle P', T \rangle$ is the only line of π tangent to $\mathcal{S}$ at T. Counting the points of $\pi \cap \mathcal{P}$ on the lines of π through T, we obtain $q(\alpha - 1) + 1$. Since $P' \in \mathcal{P}'$ the lines $\langle P', T \rangle$ and $\langle P', T' \rangle$ are contained in $P'\zeta$; so for each point $A \in \pi \cap \mathcal{P}$ the line $\langle P', A \rangle$ is contained in $P'\zeta$. Consequently $\langle P', A \rangle$ is tangent to $\mathcal{S}$ at A. So the number of points in $\pi \cap \mathcal{P}$ is at most the number of lines of π through P'. Therefore $q(\alpha - 1) + 1 \leq q + 1$, whence $\alpha = 2$, a contradiction. Secondly, suppose that the plane π contains a line l of $\mathcal{B}$. Since $\langle P', T \rangle$ and $\langle P', T' \rangle$ are tangents of $\mathcal{S}$, it follows that $l = \langle T, T' \rangle$. Hence any two points of $\mathcal{S}_{P'}$ are adjacent in $\mathcal{S}$. By Lemma 26.3.15 we have $T^{\perp} \supset \mathcal{S}_{P'}$. Hence $T\zeta \supset P'\zeta$ and so $T\zeta = P'\zeta$, giving $T = P'$, a contradiction. Thus $\mathcal{P}' = \mathcal{P}$. Next, let $l' \in \mathcal{B}'$. If D, D' are distinct points of l', then D' belongs to the tangent hyperplane $D\zeta$ of $\mathcal{S}'$ at D. As $D' \in D\zeta$ and $D, D' \in \mathcal{P}$, the line $l' = \langle D, D' \rangle$ belongs to $\mathcal{B}$. Consequently also $\mathcal{B} = \mathcal{B}'$. We conclude that $\mathcal{S} = \mathcal{S}'$, which means that $\mathcal{P}$ is the set of all self-conjugate points of the polarity ζ and that $\mathcal{B}$ consists of all lines l with $l \subset l\zeta$. □

Theorem 26.3.29: *Let $\mathcal{S}$ be a non-degenerate projective Shult space with ambient space $PG(d, q)$. Then one of the following holds:*

(a) *$\mathcal{S}$ is formed by the points and lines of a non-singular quadric of $PG(d, q)$;*

(b) *q is a square and $\mathcal{S}$ is formed by the points and lines of a non-singular Hermitian variety of $PG(d, q)$;*

(c) *d is odd, the points of $\mathcal{S}$ are the points of $PG(d, q)$, and the lines of $\mathcal{S}$ are the lines of $PG(d, q)$ in the self-polar $(d - 1)/2$-dimensional spaces with respect to some null polarity ζ of $PG(d, q)$.*

Proof: If $\alpha = 2$, then by Theorem 26.3.21 we have case (a). If $\alpha > 2$, then by Theorem 26.3.28 we have one of the cases (b) and (c). □

Theorem 26.3.30: *let $\mathcal{S}$ be a projective Shult space with ambient space $PG(d, q)$. Then $\mathcal{S}$ is one of the following types.*

(a) *$\mathcal{S}$ is formed by the points and the lines of $PG(d, q)$, $d \geq 1$. The radical $\mathcal{R}$ of $\mathcal{S}$ is $PG(d, q)$.*

(b) *The point set of $\mathcal{S}$ is the union of k spaces $PG(r + 1, q)$ through a $PG(r, q)$, where $k > 1$ and $r \geq 0$. The line set of $\mathcal{S}$ is the set of all lines in these $(r + 1)$-spaces. The radical $\mathcal{R}$ of $\mathcal{S}$ is $PG(r, q)$.*

(c) *$\mathcal{S}$ is formed by the points and lines of a quadric $\mathcal{Q}$ (of projective index at least one) of $PG(d, q)$, $d \geq 3$. The radical $\mathcal{R}$ of $\mathcal{S}$ is the space of all singular points of $\mathcal{Q}$.*

(d) *q is a square and $\mathscr{S}$ is formed by the points and lines of a Hermitian variety $\mathscr{U}$ (of projective index at least one) of $PG(d, q)$, $d \geq 3$. The radical $\mathscr{R}$ of $\mathscr{S}$ is the space of all singular points of $\mathscr{U}$.*

(e) *The points of $\mathscr{S}$ are the points of $PG(d, q)$, $d \geq 3$. In $PG(d, q)$ there is a space $PG(r, q)$, $r \geq -1$, and a space $PG(d - r - 1, q)$, $d - r - 1$ odd and at least three, skew to $PG(r, q)$ such that the lines of $\mathscr{S}$ are all the lines of $PG(d, q)$ in the $(r + 2)$-spaces joining $PG(r, q)$ to the lines of $PG(d - r - 1, q)$ in the self-polar $(d - r - 2)/2$-spaces of some null polarity ζ in $PG(d - r - 1, q)$. The radical $\mathscr{R}$ of $\mathscr{S}$ is $PG(r, q)$.*

Proof: First, let $\mathscr{S}$ be non-degenerate. By Theorem 26.3.29 we have either case (c) with $\mathscr{R} = \varnothing$, or case (d) with $\mathscr{R} = \varnothing$, or case (e) with $\mathscr{R} = \varnothing$.

Next, let $\mathscr{S}$ be degenerate with radical $\mathscr{R}$ and let $\mathscr{R} = PG(d, q)$. Then case (a) holds.

Finally, let $\mathscr{S}$ be degenerate with radical $\mathscr{R} = PG(r, q)$ and $-1 < r < d$. Let $PG(d - r - 1, q) = \pi$ be a subspace of $PG(d, q)$ skew to $\mathscr{R}$. By Theorem 26.3.3, $\pi \cap \mathscr{S}$ is a non-degenerate Shult space, the point set of $\mathscr{S}$ is the union of all lines joining every point of $\mathscr{R}$ to every point of $\pi \cap \mathscr{S}$, and two points of $\mathscr{P}\backslash\mathscr{R}$ are adjacent if and only if their projections from $\mathscr{R}$ onto π are adjacent. If $\pi \cap \mathscr{S}$ contains at least one line, then $\pi \cap \mathscr{S}$ is a non-degenerate projective Shult space with ambient space π. In this case, by Theorem 26.3.29, we have one of the cases (c), (d), (e) with $\mathscr{R} \neq \varnothing$. Finally, suppose that $\pi \cap \mathscr{S}$ contains no line; this gives case (b). □

Consider a pair $\mathscr{S} = (\mathscr{P}, \mathscr{B})$, where $\mathscr{P}$ is a non-empty point set of $PG(d, q)$ and $\mathscr{B}$ is a (possibly empty) line set of $PG(d, q)$. If $\mathscr{B} \neq \varnothing$, then assume that $\mathscr{P}$ is the union of all lines of $\mathscr{B}$. The subspace $PG(d', q)$ of $PG(d, q)$ generated by all points of $\mathscr{P}$ will be called the *ambient space* of $\mathscr{S}$. A *tangent to $\mathscr{S}$ at* $P \in \mathscr{P}$ is any line l through P such that either $l \in \mathscr{B}$ or $l \cap \mathscr{P} = \{P\}$. The union of all tangents to $\mathscr{S}$ at P will be called the *tangent set of $\mathscr{S}$ at P*, and we denote it by $\mathscr{S}(P)$. We say that $\mathscr{S}$ is a *semi-quadratic* set of $PG(d, q)$, $d \geq 2$, if $PG(d, q)$ is the ambient space of $\mathscr{S}$ and if for each $P \in \mathscr{P}$ the tangent set $\mathscr{S}(P)$ is either a hyperplane or $PG(d, q)$. If $\mathscr{S}(P) = PG(d, q)$, then P is called a *singular* point of $\mathscr{S}$. The set of all singular points of $\mathscr{S}$ is the *radical* $\mathscr{R}$ of $\mathscr{S}$. A semi-quadratic set $\mathscr{S}$ of $PG(d, q)$ is a *semi-ovaloid* if $\mathscr{B} = \varnothing$; in this case $\mathscr{R} = \varnothing$.

Theorem 26.3.31: *The pair $\mathscr{S} = (\mathscr{P}, \mathscr{B})$ is a semi-quadratic set of $PG(d, q)$, $d \geq 2$, if and only if one of the following holds.*

(a) *$\mathscr{S}$ is of type* (a), (c), (d), *or* (e) *in the statement of Theorem* 26.3.30. *In each of these cases the radical of the Shult space $\mathscr{S}$ coincides with the radical of the semi-quadratic set $\mathscr{S}$.*

(b)′ *The point set of $\mathscr{S}$ is the union of k spaces $PG(r + 1, q)$ through a*

$PG(r, q)$, where $k > 1$ and $d - 3 \geq r \geq -1$. The line set of $\mathscr{S}$ is the set of all lines in these $(r+1)$-spaces. If $PG(d-r-1, q)$ is skew to $PG(r, q)$, then $\mathscr{P} \cap PG(d-r-1, q)$ is a semi-ovaloid of $PG(d-r-1, q)$. The radical $\mathscr{R}$ of $\mathscr{S}$ is $PG(r, q)$.

Proof: Let $\mathscr{S} = (\mathscr{P}, \mathscr{B})$ be a semi-quadratic set of $PG(d, q)$, $d \geq 2$. We show that $\mathscr{S}$ is a Shult space. Let $P \in \mathscr{P}$, $l \in \mathscr{B}$, and $P \notin l$. Since the tangent set $\mathscr{S}(P)$ is a prime or $PG(d, q)$ itself, we have $|l \cap \mathscr{S}(P)| = 1$ or $l \subset \mathscr{S}(P)$. If $P' \in l \cap \mathscr{S}(P)$, then $\langle P, P' \rangle$ is a line of $\mathscr{B}$. Hence P is adjacent to one point or to all points of l. It follows that $\mathscr{S}$ is a Shult space.

First, assume that $\mathscr{B} \neq \varnothing$. Then $\mathscr{S}$ is a projective Shult space with ambient space $PG(d, q)$. So we have one of the cases (a) to (e) in the statement of Theorem 26.3.30. Conversely, each Shult space of type (a), (c), (d), or (e) is a semi-quadratic set. Now let us consider a Shult space of type (b), and let $PG(d-r-1, q)$ be skew to $PG(r, q)$, $r \geq 0$. Then $\mathscr{S}$ is a semi-quadratic set if and only if $\mathscr{P} \cap PG(d-r-1, q)$ is a semi-ovaloid of $PG(d-r-1, q)$ with $d-r-1 \geq 2$. In each of these cases the radical of the Shult space $\mathscr{S}$ coincides with the radical of the semi-quadratic set $\mathscr{S}$.

Secondly, assume that $\mathscr{B} = \varnothing$. Then, the semi-quadratic set $\mathscr{S}$ is a semi-ovaloid of $PG(d, q)$. This gives case (b)$'$ with $r = -1$. □

Let $\mathscr{S} = (\mathscr{P}, \varnothing)$ be a semi-ovaloid of $PG(d, q)$, $d \geq 2$. Any tangent of $\mathscr{S}$ has exactly one point in common with $\mathscr{S}$, and the tangent set $\mathscr{S}(P)$ of $\mathscr{S}$ at $P \in \mathscr{P}$ is always a prime. In the next theorem all semi-ovaloids are classified.

Theorem 26.3.32: *If $\mathscr{S} = (\mathscr{P}, \varnothing)$ is a semi-ovaloid of $PG(d, q)$, then there are only two possibilities.*

(i) $d = 2$ and $q + 1 \leq |\mathscr{P}| \leq q\sqrt{q} + 1$. If $|\mathscr{P}| = q + 1$, then $\mathscr{P}$ is a $(q+1)$-arc of $PG(2, q)$; if $|\mathscr{P}| = q\sqrt{q} + 1$, then $\mathscr{P}$ is a Hermitian arc of $PG(2, q)$.

(ii) $d = 3$ and $|\mathscr{P}| = q^2 + 1$. For $q > 2$, $\mathscr{P}$ is an ovaloid of $PG(3, q)$; for $q = 2$, $\mathscr{P}$ is an elliptic quadric of $PG(3, 2)$.

Proof: Let $\mathscr{S} = (\mathscr{P}, \varnothing)$ be a semi-ovaloid of $PG(2, q)$. Let $P \in \mathscr{P}$ and let l be the tangent of $\mathscr{S}$ at P. If the line m contains P and $m \neq l$, then $|m \cap \mathscr{P}| > 1$. Counting the points of $\mathscr{P}$ on the lines through P, we obtain $|\mathscr{P}| \geq q + 1$. If $|\mathscr{P}| = q + 1$, then each non-tangent contains exactly zero or two points of $\mathscr{P}$, whence $\mathscr{P}$ is a $(q+1)$-arc of $PG(2, q)$. Conversely, any $(q+1)$-arc of $PG(2, q)$ is a semi-ovaloid of $PG(2, q)$.

Next, let $\mathscr{S} = (\mathscr{P}, \varnothing)$ be a semi-ovaloid of $PG(d, q)$, $d > 2$. Let l be a tangent of $\mathscr{S}$ at $P \in \mathscr{P}$. If π is a plane through l which does not belong to the tangent set $\mathscr{S}(P)$, then $\pi \cap \mathscr{S}$ is a semi-ovaloid of π. Counting the points of $\mathscr{S}$ in the planes through l we obtain

$$|\mathscr{P}| = \sum_{\pi} (|\mathscr{P} \cap \pi| - 1) + 1 \geq q \cdot q^{d-2} + 1 = q^{d-1} + 1.$$

Let $\mathscr{S} = (\mathscr{P}, \varnothing)$ be a semi-ovaloid of $PG(d, q)$, $d \geq 2$. From the first part of the proof we have

$$|\mathscr{P}| \geq q^{d-1} + 1. \tag{26.3}$$

Let $\mathscr{P} = \{P_1, P_2, \ldots, P_\alpha\}$ and $PG(d, q)\backslash\mathscr{P} = \{T_1, T_2, \ldots, T_\beta\}$, with $\alpha + \beta = (q^{d+1} - 1)/(q - 1)$. Further, let t_i be the number of tangents of $\mathscr{S}$ through T_i, $i = 1, 2, \ldots, \beta$. Now, count in different ways the number of ordered pairs (T_i, P_j), where $\langle T_i, P_j\rangle$ is a tangent of $\mathscr{S}$; this gives

$$\sum_{i=1}^{\beta} t_i = \alpha q \sum_{k=0}^{d-2} q^k. \tag{26.4}$$

Next, count in different ways the number of ordered triples $(T_i, P_j, P_{j'})$, with $P_j \neq P_{j'}$ and $\langle T_i, P_j\rangle$, $\langle T_i, P_{j'}\rangle$ tangents of $\mathscr{S}$. Hence

$$\sum_{i=1}^{\beta} t_i(t_i - 1) = \alpha(\alpha - 1) \sum_{k=0}^{d-2} q^k. \tag{26.5}$$

From (26.4) and (26.5) follows that

$$\sum_{i=1}^{\beta} t_i^2 = \alpha(\alpha + q - 1) \sum_{k=0}^{d-2} q^k.$$

With $\beta\bar{t} = \sum_{i=1}^{\beta} t_i$, $0 \leq \sum_{i=1}^{\beta} (\bar{t} - t_i)^2$ simplifies to

$$\beta \sum_{i=1}^{\beta} t_i^2 - \left(\sum_{i=1}^{\beta} t_i\right)^2 \geq 0,$$

which implies

$$\beta\alpha(\alpha + q - 1) \sum_{k=0}^{d-2} q^k - \alpha^2 q^2 \left(\sum_{k=0}^{d-2} q^k\right)^2 \geq 0.$$

As $\beta = \sum_{k=0}^{d} q^k - \alpha$, manipulation gives

$$(\alpha - 1)^2 \leq q^{d+1}. \tag{26.6}$$

First, let $d = 2$. From (26.3) and (26.6) it follows that

$$q + 1 \leq \alpha \leq q\sqrt{q} + 1.$$

If $q + 1 = \alpha$, then in the first part of the proof it was shown that $\mathscr{P}$ is a $(q + 1)$-arc. Let $\alpha = q\sqrt{q} + 1$. Then

$$0 = \sum_{i=1}^{\beta} (\bar{t} - t_i)^2;$$

so $t_i = \bar{t} = (\sum_{i=1}^{\beta} t_i)/\beta = \alpha q/\beta = \sqrt{q} + 1$, $i = 1, 2, \ldots, \beta$. From §§12.2 and 12.3, the set of all tangents of $\mathscr{S}$ forms a dual Hermitian arc of $PG(2, q)$, and hence $\mathscr{P}$ is a Hermitian arc of $PG(2, q)$.

Next, let $d = 3$. From (26.3) and (26.6) it follows that $\alpha = q^2 + 1$. Any line l through $P_i \in \mathscr{P}$, but not in $\mathscr{S}(P_i)$, contains at least one point of $\mathscr{P}\backslash\{P_i\}$. Since there are q^2 such lines l through P_i and $|\mathscr{P}\backslash\{P_i\}| = q^2$, it follows that l contains exactly two points of $\mathscr{P}$. Hence $\mathscr{P}$ is a $(q^2 + 1)$-cap of $PG(3, q)$. If $q > 2$, then $\mathscr{P}$ is an ovaloid; if $q = 2$, then the 5-cap $\mathscr{P}$ is an elliptic quadric of $PG(3, 2)$.

Finally, let $d > 3$; then (26.3) contradicts (26.6). □

Let $\mathscr{S} = (\mathscr{P}, \varnothing)$ be a semi-ovaloid of the plane $PG(2, q)$. If $P \in \mathscr{P}$ and all non-tangents through P intersect $\mathscr{P}$ in more than two points, then $\mathscr{S}' = (\mathscr{P}\backslash\{P\}, \varnothing)$ is still a semi-ovaloid of $PG(2, q)$. In $PG(2, 3)$ there is a class of semi-ovaloids with six points: take the vertices of a quadrangle together with two of its diagonal points.

26.4 Characterizations of the classical generalized quadrangles

In this section we review the most important characterizations of the classical generalized quadrangles. Apart from a few exceptions the proofs are long, complicated, and technical. So, proofs are given only in the simpler cases. First some new ideas are introduced.

Let $\mathscr{S} = (\mathscr{P}, \mathscr{B}, \mathrm{I})$ be a finite generalized quadrangle of order (s, t). If $P \sim P'$, $P \neq P'$, or if $P \nsim P'$ and $|\{P, P'\}^{\perp\perp}| = t + 1$, we say the pair $\{P, P'\}$ is *regular*. The point P is *regular* provided $\{P, P'\}$ is regular for all $P' \in \mathscr{P}$, $P' \neq P$. A point P is *coregular* provided each line incident with P is regular. The pair $\{P, P'\}$, $P \nsim P'$, is *anti-regular* provided $|P''^{\perp} \cap \{P, P'\}^{\perp}| \leq 2$ for all $P'' \in \mathscr{P}\backslash\{P, P'\}$. A point P is *anti-regular* provided $\{P, P'\}$ is anti-regular for all $P' \in \mathscr{P}\backslash P^{\perp}$.

The *closure* of the pair $\{P, P'\}$ is $\mathrm{cl}(P, P') = \{P'' \in \mathscr{P} | P''^{\perp} \cap \{P, P'\}^{\perp\perp} \neq \varnothing\}$.

Theorem 26.4.1: *Let $\mathscr{S} = (\mathscr{P}, \mathscr{B}, \mathrm{I})$ be a generalized quadrangle of order $s > 1$.*

(a) *For a regular point P, the incidence structure π_P with point set $P^{\perp}$, with line set the set of spans $\{P', P''\}^{\perp\perp}$, where $P', P'' \in P^{\perp}$ with $P' \neq P''$, and with the natural incidence is a projective plane of order s.*

(b) *For an anti-regular point P and a point P' in $P^{\perp}\backslash\{P\}$, the incidence structure $\pi(P, P')$ with point set $P^{\perp}\backslash\{P, P'\}^{\perp}$, with lines the sets $\{P, P_1\}^{\perp\perp}\backslash\{P\}$ with $P \sim P_1 \nsim P'$ and the sets $\{P, P_2\}^{\perp}\backslash\{P'\}$ with $P' \sim P_2 \nsim P$, and with the natural incidence is an affine plane of order s.*

Proof: Both parts are straightforward verifications of the axioms. □

An *ovoid* of $\mathscr{S}$ is a set $\mathscr{O}$ of points of $\mathscr{S}$ such that each line of $\mathscr{S}$ is incident with a unique point of $\mathscr{O}$. A *spread* of $\mathscr{S}$ is a set $\mathscr{R}$ of lines of $\mathscr{S}$ such that each point of $\mathscr{S}$ is incident with a unique line of $\mathscr{R}$. It follows that any ovoid or spread of $\mathscr{S}$ has exactly $1 + st$ elements.

Now let $s^2 = t > 1$; then, by the corollary of Theorem 26.2.3, for any triad $\{P, P', P''\}$ we have $|\{P, P', P''\}^\perp| = s + 1$. Thus $|\{P, P', P''\}^{\perp\perp}| \le s + 1$ and we say that $\{P, P', P''\}$ is 3-*regular* provided $|\{P, P', P''\}^{\perp\perp}| = s + 1$. Finally, the point P is called 3-*regular* if and only if each triad containing P is 3-regular.

These definitions will be illustrated by some examples.

In §26.2 it was observed that each hyperbolic line of $\mathscr{W}(q)$ contains $q + 1$ points. Hence all points of $\mathscr{W}(q)$ are regular. Dually, all lines of $\mathscr{Q}(4, q)$ are regular. By Theorem 26.2.5 all lines of $\mathscr{W}(q)$, with q even, are regular. Dually, all points of $\mathscr{Q}(4, q)$, q even, are regular. From the examples in §26.2 it follows that each point of $\mathscr{Q}(4, q)$, q odd, is anti-regular, and, dually, that each line of $\mathscr{W}(q)$, q odd, is anti-regular. Further, each hyperbolic line of $\mathscr{U}(3, q^2)$ contains $q + 1$ points; hence each point of $\mathscr{U}(3, q^2)$ is regular. Dually, each line of $\mathscr{Q}(5, q)$ is regular.

Consider $\mathscr{Q}(5, q)$ and the corresponding polarity ζ. If $\mathscr{T} = \{P, P', P''\}$ is a triad of $\mathscr{Q}(5, q)$, then $\mathscr{T}^\perp$ is the conic $\mathscr{Q} \cap \pi$, where π is the polar plane of the plane $PP'P''$, and $\mathscr{T}^{\perp\perp}$ is the conic $\mathscr{Q} \cap PP'P''$. So $|\mathscr{T}^{\perp\perp}| = q + 1$, and consequently each point of $\mathscr{Q}(5, q)$ is 3-regular. Dually, each line of $\mathscr{U}(3, q^2)$ is 3-regular.

The generalized quadrangle $\mathscr{Q}(4, q)$ always has ovoids, and has spreads if and only if q is even; see §§AVI.3 and AVI.4. Further, $\mathscr{Q}(5, q)$ has spreads but no ovoids, and $\mathscr{U}(4, q^2)$ has no ovoids. For $q = 2$, $\mathscr{U}(4, q^2)$ has no spreads; for $q > 2$, the existence of a spread is an open problem.

Historically, the next result is probably the oldest combinatorial characterization of a class of generalized quadrangles. A proof is essentially contained in a paper by Singleton (although he erroneously thought he had proved a stronger result), but the first satisfactory treatment may have been given by Benson. Undoubtedly it was discovered independently by several authors; see §26.9.

Theorem 26.4.2: *A generalized quadrangle $\mathscr{S}$ of order s, $s \neq 1$, is isomorphic to $\mathscr{W}(s)$ if and only if all its points are regular.*

Proof: All points of $\mathscr{W}(s)$ are regular.

Conversely, let us assume that $\mathscr{S} = (\mathscr{P}, \mathscr{B}, \mathrm{I})$ is a generalized quadrangle of order s, $s \neq 1$, for which all points are regular. Now we introduce the incidence structure $\mathscr{S}' = (\mathscr{S}', \mathscr{B}', \mathrm{I}')$, with $\mathscr{P}' = \mathscr{P}$, $\mathscr{B}'$ the set of spans of all point pairs of $\mathscr{P}$, and I' the natural incidence. Then $\mathscr{S}$ is isomorphic to the substructure of $\mathscr{S}'$ formed by all points and the spans of all pairs of points collinear in $\mathscr{S}$.

Let $\mathscr{T} = \{P, P', P''\}$ be a triad of points in $\mathscr{S}$. Counting all points on the lines joining a point of $\{P, P'\}^\perp$ and a point of $\{P, P'\}^{\perp\perp}$, we obtain

$$(s + 1)^2(s - 1) + 2(s + 1) = (s + 1)(s^2 + 1) = |\mathscr{P}|.$$

Hence P'' is on at least one line joining a point of $\{P, P'\}^{\perp}$ to a point of $\{P, P'\}^{\perp\perp}$, and so $\mathcal{T}^{\perp} \neq \emptyset$. Now by Theorem 26.4.1(a), it follows that any three non-collinear points of $\mathcal{S}'$ generate a projective plane of order s. Since $|\mathcal{P}'| = s^3 + s^2 + s + 1$, $\mathcal{S}'$ is the design of points and lines of $PG(3, s)$. All spans (in $\mathcal{S}$) of collinear point pairs containing a given point P form a pencil of lines in $PG(3, s)$. By Theorem 15.2.13 the set of all spans of collinear point pairs is a general linear complex of lines of $PG(3, q)$ or, equivalently, is the set of all self-polar lines of a null polarity ζ. Thus $\mathcal{S} \cong \mathcal{W}(s)$. □

The next result is a slight generalization of the preceding theorem and is stated without proof.

Theorem 26.4.3: *A generalized quadrangle $\mathcal{S}$ of order (s, t), $s \neq 1$, is isomorphic to $\mathcal{W}(s)$ if and only if each hyperbolic line has at least $s + 1$ points.* □

Theorem 26.4.4: *Up to isomorphism there is only one generalized quadrangle of order 2.*

Proof: Let $\mathcal{S}$ be a generalized quadrangle of order 2. Consider two points P, P' with $P \nsim P'$, and let $\{P, P'\}^{\perp} = \{Z_1, Z_2, Z_3\}$. Let $\{Z_1, Z_2\}^{\perp} = \{P, P', U\}$, let Y be the unique point of PZ_3 collinear with U, and let Y' be the unique point of $P'Z_3$ collinear with U. If $Y \neq Z_3 \neq Y'$, then U is incident with the four distinct lines UZ_1, UZ_2, UY, UY', a contradiction since $t = 2$. Consequently $Y = Z_3 = Y'$, and so $U \sim Z_3$. Hence $|\{P, P'\}^{\perp\perp}| = 3$. So every point of $\mathcal{S}$ is regular, and now by Theorem 26.4.2 we have $\mathcal{S} \cong \mathcal{W}(2)$. □

The following four characterizations are stated without proof.

Theorem 26.4.5: *A generalized quadrangle $\mathcal{S}$ of order s, $s \neq 1$, is isomorphic to $\mathcal{W}(2^h)$ if and only if it has an ovoid $\mathcal{O}$ each triad of which has at least one centre.* □

Theorem 26.4.6: *A generalized quadrangle $\mathcal{S}$ of order s, $s \neq 1$, is isomorphic to $\mathcal{W}(2^h)$ if and only if it has an ovoid $\mathcal{O}$ each point of which is regular.* □

Theorem 26.4.7: *A generalized quadrangle $\mathcal{S}$ of order s, $s \neq 1$, is isomorphic to $\mathcal{W}(2^h)$ if and only if it has a regular pair $\{l_1, l_2\}$ of non-concurrent lines with the property that any triad of points lying on lines of $\{l_1, l_2\}^{\perp}$ has at least one centre.* □

Theorem 26.4.8: *Let $\mathcal{S}$ be a generalized quadrangle of order s, $s \neq 1$, having an anti-regular point P. Then $\mathcal{S}$ is isomorphic to $\mathcal{Q}(4, s)$ if and only if there is a point P' in $P^{\perp}\backslash\{P\}$, for which the associated affine plane $\pi(P, P')$ is Desarguesian.* □

Corollary: *Let $\mathcal{S}$ be a generalized quadrangle of order s, $s \neq 1$, having an anti-regular point P. If $s \leq 8$, then $\mathcal{S}$ is isomorphic to $\mathcal{Q}(4, s)$.*

Proof: Since each plane of order s, $s \leq 8$, is Desarguesian, the result follows. □

The following characterization theorem is very important, not only for the theory of generalized quadrangles, but also for other areas in combinatorics. The proof is again very long, and so it will not be given.

Theorem 26.4.9: *Let $\mathcal{S}$ be a generalized quadrangle of order (s, s^2).*

(i) *When $s > 1$, then $\mathcal{S} \cong \mathcal{Q}(5, s)$ if and only if all points of $\mathcal{S}$ are 3-regular.*

(ii) *When s is odd and $s > 1$, then $\mathcal{S} \cong \mathcal{Q}(5, s)$ if and only if it has a 3-regular point.*

(iii) *When s is even, then $\mathcal{S} \cong \mathcal{Q}(5, s)$ if and only if it has at least one 3-regular point not incident with some regular line.* □

Corollary: *Up to isomorphism there is only one generalized quadrangle of the following orders:*

(a) (2, 4); (b) (3, 9).

Proof: (a) Let $\mathcal{S}$ be a generalized quadrangle of order (2, 4). If $\{P_1, P_2, P_3\}$ is a triad of points, then $\{P_1, P_2, P_3\}^{\perp\perp} = \{P_1, P_2, P_3\}$. Hence $|\{P_1, P_2, P_3\}^{\perp\perp}| = 1 + s$, every point is 3-regular, and by Theorem 26.4.9(i) we have $\mathcal{S} \cong \mathcal{Q}(5, 2)$.

(b) Let $\mathcal{S}$ be a generalized quadrangle of order (3, 9). Let $\{P_1, P_2, P_3\}$ be a triad of points, let $\{P_1, P_2, P_3\}^{\perp} = \{U_1, U_2, U_3, U_4\}$, and let $\{U_1, U_2, U_3\}^{\perp} = \{P_1, P_2, P_3, P_4\}$. The number of points collinear with U_4 and also with at least two points of $\{U_1, U_2, U_3\}$ is at most six, and the number of points collinear with U_4 and incident with some line P_4U_i, $i = 1, 2, 3$, is at most three. Since $3 + 6 < 10 = t + 1$, there is a line l incident with U_4, but not concurrent with P_4U_i, $i = 1, 2, 3$, and not incident with an element of $\{U_i, U_j\}^{\perp}$, $i \neq j$, $1 \leq i \leq 3$, $1 \leq j \leq 3$. The point incident with l and collinear with U_i is denoted by Z_i, $i = 1, 2, 3$; the points Z_1, Z_2, Z_3 are distinct. Since $\mathcal{S}$ has no triangles, the point P_4 is not collinear with any Z_i, forcing $P_4 \sim U_4$. Hence the triad $\{P_1, P_2, P_3\}$ is 3-regular, and so every point of $\mathcal{S}$ is 3-regular. Now by Theorem 26.4.9 we have $\mathcal{S} \cong \mathcal{Q}(5, 3)$. □

The generalized quadrangle $\mathcal{S}' = (\mathcal{P}', \mathcal{B}', \mathrm{I}')$ of order (s', t') is called a *subquadrangle* of the generalized quadrangle $\mathcal{S} = (\mathcal{P}, \mathcal{B}, \mathrm{I})$ of order (s, t) if $\mathcal{P}' \subset \mathcal{P}$, $\mathcal{B}' \subset \mathcal{B}$, and if I′ is the restriction of I to $(\mathcal{P}' \times \mathcal{B}') \cup (\mathcal{B}' \times \mathcal{P}')$. If $\mathcal{S}' \neq \mathcal{S}$, then $\mathcal{S}'$ is a *proper* subquadrangle of $\mathcal{S}$. If $|\mathcal{P}| = |\mathcal{P}'|$ it follows that $s = s'$ and $t = t'$; hence if $\mathcal{S}'$ is a proper subquadrangle then $\mathcal{P} \neq \mathcal{P}'$, and dually $\mathcal{B} \neq \mathcal{B}'$.

We give some examples.

Consider $\mathcal{Q}(5, q)$, with $\mathcal{Q}$ a non-singular quadric of projective index 1 in $PG(5, q)$. Intersect $\mathcal{Q}$ with a non-tangent prime $PG(4, q)$. Then the points and lines of $\mathcal{Q}' = \mathcal{Q} \cap PG(4, q)$ form the generalized quadrangle $\mathcal{Q}'(4, q)$. Here $s^2 = t = q^2$, $s = s' = t'$. Since all lines of $\mathcal{Q}(5, q)$ and of $\mathcal{Q}'(4, q)$ are regular, both have subquadrangles of order (s'', t'') with $t'' = 1$ and $s'' = s' = s$, each of which is a hyperbolic quadric in some solid of $PG(5, q)$.

Similarly, consider $\mathcal{U}(4, q^2)$, with $\mathcal{U}$ a non-singular Hermitian variety of $PG(4, q^2)$. Intersect $\mathcal{U}$ with a non-tangent prime $PG(3, q^2)$. Then the points and lines of $\mathcal{U}' = \mathcal{U} \cap PG(3, q^2)$ form the generalized quadrangle $\mathcal{U}'(3, q^2)$. Here $t = s^{3/2} = q^3$, $s = s'$, $t' = \sqrt{s}$. Since all points of $\mathcal{U}'(3, q^2)$ are regular, $\mathcal{U}'(3, q^2)$ has subquadrangles with $t'' = t' = \sqrt{s}$ and $s'' = 1$.

Now consider $\mathcal{Q}(4, q)$ and extend $GF(q)$ to $GF(q^2)$. Then $\mathcal{Q}$ extends to $\bar{\mathcal{Q}}$ and $\mathcal{Q}(4, q)$ to $\bar{\mathcal{Q}}(4, q^2)$. Here $\mathcal{Q}(4, q)$ is a subquadrangle of $\bar{\mathcal{Q}}(4, q^2)$, and we have $t = s = q^2$ and $t' = s' = q$.

Next we consider the role of subquadrangles in characterizing $\mathcal{Q}(5, q)$. Proofs are again omitted.

Theorem 26.4.10: *Let $\mathcal{S}$ be a generalized quadrangle of order (s, t) with $s > 1$ and $t > 1$. Then $\mathcal{S}$ is isomorphic to $\mathcal{Q}(5, s)$ if and only if either* (i) *or* (ii) *holds*:

(i) *every triad of lines with at least one centre is contained in a proper subquadrangle of order (s, t');*

(ii) *for each triad $\{P, P', P''\}$ with distinct centres Z, Z', the five points P, P', P'', Z, Z' are contained in a proper subquadrangle of order (s, t').* □

Let $\mathcal{S}$ be a generalized quadrangle of order (s, t), and let $\{l_1, l_2, l_3\}$ and $\{m_1, m_2, m_3\}$ be two triads of lines for which $l_i \not\sim m_j$ if and only if $\{i, j\} = \{1, 2\}$. Let P_i be the point defined by $l_i \mathrm{I} P_i \mathrm{I} m_i$, $i = 1, 2$. This configuration $\mathcal{T}$ of seven distinct points and six distinct lines is called a *broken grid* with *carriers* P_1 and P_2 (fig. 26.2). We say that $\mathcal{T}$ satisfies *axiom (D)* with respect to the

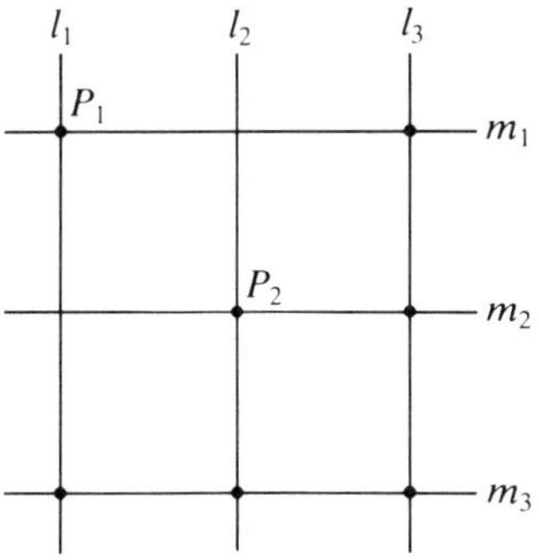

Figure 26.2

pair $\{l_1, l_2\}$ provided the following holds: if $l_4 \in \{m_1, m_2\}^\perp$ with $l_4 \not\sim l_i$, $i = 1, 2, 3$, then $\{l_1, l_2, l_4\}$ has at least one centre. Interchanging l_i and m_i gives the definition of axiom (D) for $\mathcal{T}$ with respect to the pair $\{m_1, m_2\}$. Further, $\mathcal{T}$ is said to satisfy *axiom* (D) provided it satisfies axiom (D) with respect to both pairs $\{l_1, l_2\}$ and $\{m_1, m_2\}$.

Let P be any point of $\mathcal{S}$. Then $\mathcal{S}$ is said to satisfy *axiom* $(D)'_P$ if the broken grid $\mathcal{T}$ satisfies axiom (D) with respect to $\{l_1, l_2\}$ whenever $P \mathrm{I} l_1$; it satisfies *axiom* $(D)''_P$ if $\mathcal{T}$ satisfies axiom (D) with respect to $\{m_1, m_2\}$ whenever $P \mathrm{I} l_1$.

Now another interesting characterization of $\mathcal{Q}(5, s)$ is the following.

Theorem 26.4.11: *Let $\mathcal{S}$ be a generalized quadrangle of order (s, t), with $s \neq t$, $s > 1$, $t > 1$.*

(i) *If s is odd, then $\mathcal{S} \cong \mathcal{Q}(5, s)$ if and only if $\mathcal{S}$ contains a coregular point P for which $(D)'_P$ or $(D)''_P$ is satisfied.*

(ii) *If s is even, then $\mathcal{S} \cong \mathcal{Q}(5, s)$ if and only if all lines of $\mathcal{S}$ are regular and $\mathcal{S}$ contains a point P for which $(D)'_P$ or $(D)''_P$ is satisfied.* □

In order to conclude this section dealing with characterizations of $\mathcal{Q}(5, s)$, we introduce one more basic concept. Let $\mathcal{S} = (\mathcal{P}, \mathcal{B}, \mathrm{I})$ be a generalized quadrangle of order (s, t). If $\mathcal{B}^{\perp\perp}$ is the set of all spans $\{P, P'\}^{\perp\perp}$ with $P \not\sim P'$, then let $\mathcal{S}^{\perp\perp} = (\mathcal{P}, \mathcal{B}^{\perp\perp}, \in)$. For $P \in \mathcal{P}$, say that $\mathcal{S}$ satisfies *property* $(A)_P$ if for any $m = \{P', P''\}^{\perp\perp} \in \mathcal{B}^{\perp\perp}$ with $P \in \{P', P''\}^\perp$, and $U \in \mathrm{cl}(P', P'') \cap (P^\perp \backslash \{P\})$ with $U \notin m$, the substructure of $\mathcal{S}^{\perp\perp}$ generated by m and U is a dual affine plane. The generalized quadrangle $\mathcal{S}$ is said to satisfy *property* (A) if it satisfies $(A)_P$ for all $P \in \mathcal{P}$. So $\mathcal{S}$ satisfies (A) if for any $m = \{P', P''\}^{\perp\perp} \in \mathcal{B}^{\perp\perp}$ and any $U \in \mathrm{cl}(P', P'') \backslash (\{P', P''\}^\perp \cup \{P', P''\}^{\perp\perp})$ the substructure of $\mathcal{S}^{\perp\perp}$ generated by m and U is a dual affine plane. The duals of $(A)_P$ and (A) are denoted by $(\hat{A})_l$ and $(\hat{A})$, respectively.

Again the proof of the following theorem will not be given.

Theorem 26.4.12: *Let $\mathcal{S}$ be a generalized quadrangle of order (s, t), $s \neq t$, $t > 1$.*

(i) *If $s > 1$, s odd, then $\mathcal{S}$ is isomorphic to $\mathcal{Q}(5, s)$ if and only if $(\hat{A})_l$ is satisfied for all lines l incident with some coregular point P.*

(ii) *If s is even, then $\mathcal{S}$ is isomorphic to $\mathcal{Q}(5, s)$ if and only if all lines of $\mathcal{S}$ are regular and $(\hat{A})_l$ is satisfied for all lines l incident with some point P.* □

Let $\mathcal{S} = (\mathcal{P}, \mathcal{B}, \mathrm{I})$ be a generalized quadrangle of order (s, t), and let $\mathcal{B}^* = \{\{P, P'\}^{\perp\perp} | P, P' \in \mathcal{P}, P \neq P'\}$. Then $\mathcal{S}^* = (\mathcal{P}, \mathcal{B}^*, \in)$ is a *linear space*; see §26.9, p. 283. So as to have no confusion between collinearity in $\mathcal{S}$ and collinearity in $\mathcal{S}^*$, points $P_1, P_2, \ldots$ *of* $\mathcal{P}$ which are on a line of $\mathcal{S}^*$ will be called $\mathcal{S}^*$*-collinear*. A *linear variety* of $\mathcal{S}^*$ is a subset $\mathcal{P}' \subset \mathcal{P}$ such that $P, P' \in \mathcal{P}'$, $P \neq P'$, implies $\{P, P'\}^{\perp\perp} \subset \mathcal{P}'$. If $\mathcal{P}' \neq \mathcal{P}$ and $|\mathcal{P}'| > 1$, the linear variety is *proper*; if $\mathcal{P}'$ is generated by three points which are not $\mathcal{S}^*$-collinear, $\mathcal{P}'$ is said to be a *plane* of $\mathcal{S}^*$.

Now we state a fundamental characterization of the generalized quadrangle $\mathcal{U}(3, s)$.

Theorem 26.4.13: *Let $\mathcal{S} = (\mathcal{P}, \mathcal{B}, \mathrm{I})$ be a generalized quadrangle of order (s, t), with $s \neq t$, $s > 1$, and $t > 1$. Then $\mathcal{S}$ is isomorphic to $\mathcal{U}(3, s)$ if and only if*

(i) *all points of $\mathcal{S}$ are regular, and*

(ii) *if the lines l and l' of $\mathcal{B}^*$ are contained in a proper linear variety of $\mathcal{S}^*$, then also the lines $l^\perp$ and $l'^\perp$ of $\mathcal{B}^*$ are contained in a proper linear variety of $\mathcal{S}^*$.* □

A beautiful characterization theorem, but with a long and complicated proof, is the following.

Theorem 26.4.14: *A generalized quadrangle $\mathcal{S}$ of order (s, t), $s^3 = t^2$ and $s \neq 1$, is isomorphic to $\mathcal{U}(4, s)$ if and only if every hyperbolic line has at least $\sqrt{s} + 1$ points.* □

Relying on the preceding theorem one can prove the next characterization of $\mathcal{U}(4, s)$.

Theorem 26.4.15: *Let $\mathcal{S}$ have order (s, t) with $1 < s^3 \leq t^2$. Then $\mathcal{S}$ is isomorphic to $\mathcal{U}(4, s)$ if and only if each trace $\{P, P'\}^\perp$, with $P \nsim P'$, is a plane of $\mathcal{S}^*$ which is generated by any three non-$\mathcal{S}^*$-collinear points in it.* □

Next, conditions are given which simultaneously characterize several classical generalized quadrangles. But first two new definitions are required. A point U of $\mathcal{S}$ is called *semi-regular* provided that $P'' \in \mathrm{cl}(P, P')$ whenever U is the unique centre of the triad $\{P, P', P''\}$. A point U has *property* (H) provided $P'' \in \mathrm{cl}(P, P')$ if and only if $P \in \mathrm{cl}(P', P'')$, whenever $\{P, P', P''\}$ is a triad consisting of points in $U^\perp$. It follows easily that any semi-regular point has property (H).

We give some examples.

In $\mathcal{W}(q)$, $\mathcal{Q}(4, q)$, $\mathcal{Q}(5, q)$, and $\mathcal{U}(3, q^2)$ all points and lines are semi-regular and have property (H). In $\mathcal{U}(4, q^2)$ all points are semi-regular and have property (H); all lines have property (H). Finally, we show that no line of $\mathcal{U}(4, q^2)$ is semi-regular. Consider three distinct lines l, m, n of $\mathcal{U}(4, q^2)$ with $l \sim n \sim m \nsim l$. Further, let r be a line of $\mathcal{U}(4, q^2)$ for which $r \sim n$, $l \nsim r \nsim m$, and which is not contained in the solid $PG(3, q^2)$ defined by l and m. Then n is the unique centre of the triad $\{l, m, r\}$, but $r \notin \mathrm{cl}(l, m)$ since $\mathrm{cl}(l, m)$ consists of all lines concurrent with at least one of l, m. Hence n is not semi-regular. So property (H) does not imply semi-regularity.

Now we state, without proof, six characterizations, most of them involving more than one classical generalized quadrangle.

Theorem 26.4.16: *Let $\mathscr{S}$ have order (s, t) with $s \neq 1$. Then $|\{P, P'\}^{\perp\perp}| \geq (s^2/t) + 1$ for all P, P', with $P \not\sim P'$, if and only if one of the following occurs:*

(i) $t = s^2$;
(ii) $\mathscr{S} \cong \mathscr{W}(s)$;
(iii) $\mathscr{S} \cong \mathscr{U}(4, s)$. □

Theorem 26.4.17: *In the generalized quadrangle $\mathscr{S}$ of order (s, t) each point has property (H) if and only if one of the following holds:*

(i) *each point is regular;*
(ii) *each hyperbolic line has exactly two points;*
(iii) $\mathscr{S} \cong \mathscr{U}(4, s)$. □

Theorem 26.4.18: *Let $\mathscr{S}$ be a generalized quadrangle of order (s, t). Then each point is semi-regular if and only if one of the following occurs:*

(i) *$s > t$ and each point is regular;*
(ii) *$s = t$ and $\mathscr{S} \cong \mathscr{W}(s)$;*
(iii) *$s = t$ and each point is anti-regular;*
(iv) *$s < t$, each hyperbolic line has exactly two points, and no triad of points has a unique centre;*
(v) $\mathscr{S} \cong \mathscr{U}(4, s)$. □

Theorem 26.4.19: *In a generalized quadrangle $\mathscr{S}$ of order (s, t) all triads $\{P, P', P''\}$ with $P'' \notin \mathrm{cl}(P, P')$ have a constant number of centres if and only if one of the following occurs:*

(i) *all points are regular;*
(ii) $s^2 = t$;
(iii) $\mathscr{S} \cong \mathscr{U}(4, s)$. □

Theorem 26.4.20: *The generalized quadrangle $\mathscr{S}$ of order (s, t), $s > 1$, is isomorphic to one of $\mathscr{W}(s)$, $\mathscr{Q}(5, s)$, or $\mathscr{U}(4, s)$ if and only if, for each triad $\{P, P', P''\}$ with $P \notin \mathrm{cl}(P', P'')$, the set $\{P\} \cup \{P', P''\}^{\perp}$ is contained in a proper subquadrangle of order (s, t').* □

Theorem 26.4.21: *Let $\mathscr{S}$ be a generalized quadrangle of order (s, t) with not all points regular. Then $\mathscr{S}$ is isomorphic to $\mathscr{Q}(4, s)$, with s odd, to $\mathscr{Q}(5, s)$, or to $\mathscr{U}(4, s)$ if and only if each set $\{P\} \cup \{P', P''\}^{\perp}$, where $\{P, P', P''\}$ is a triad with at least one centre and $P \notin \mathrm{cl}(P', P'')$, is contained in a proper subquadrangle of order (s, t').* □

Next, we give a characterization in terms of matroids. A finite *matroid* is an ordered pair $(\mathscr{P}, \mathfrak{M})$ where $\mathscr{P}$ is a finite set of elements called *points*, and $\mathfrak{M}$ is a closure operator which associates to each subset $\mathscr{X}$ of $\mathscr{P}$ a subset $\bar{\mathscr{X}}$ (the *closure* of $\mathscr{X}$) of $\mathscr{P}$, such that the following conditions are satisfied:

(i) $\bar{\varnothing} = \varnothing$, and $\overline{\{P\}} = \{P\}$ for all $P \in \mathscr{P}$;
(ii) $\mathscr{X} \subset \bar{\mathscr{X}}$ for all $\mathscr{X} \subset \mathscr{P}$;
(iii) $\mathscr{X} \subset \bar{\mathscr{Y}} \Rightarrow \bar{\mathscr{X}} \subset \bar{\mathscr{Y}}$ for all $\mathscr{X}, \mathscr{Y} \subset \mathscr{P}$;
(iv) $P' \in \overline{\mathscr{X} \cup \{P\}}$, $P' \notin \bar{\mathscr{X}} \Rightarrow P \in \overline{\mathscr{X} \cup \{P'\}}$ for all $P, P' \in \mathscr{P}$ and $\mathscr{X} \subset \mathscr{P}$.

The sets $\bar{\mathscr{X}}$ are called the *closed sets* of the matroid $(\mathscr{P}, \mathfrak{M})$. It is immediate that the intersection of closed sets is always closed. A closed set $\bar{\mathscr{X}}$ has *dimension h* if $h + 1$ is the minimum number of points in any subset of $\bar{\mathscr{X}}$ whose closure coincides with $\bar{\mathscr{X}}$. The closed sets of dimension 1 are the *lines* of the matroid.

Theorem 26.4.22: *Suppose that $\mathscr{S} = (\mathscr{P}, \mathscr{B}, \mathrm{I})$ is a generalized quadrangle of order (s, t), $s > 1$ and $t > 1$. Then $\mathscr{P}$ is the point set and $\mathscr{B}^* = \{\{P, P'\}^{\perp\perp} | P, P' \in \mathscr{P}$ and $P \neq P'\}$ is the line set of some matroid $(\mathscr{P}, \mathfrak{M})$ having all sets $P^{\perp}$, $P \in \mathscr{P}$, as closed sets, if and only if one of the following occurs:*

(i) $\mathscr{S} \cong \mathscr{W}(s)$;
(ii) $\mathscr{S} \cong \mathscr{Q}(4, s)$;
(iii) $\mathscr{S} \cong \mathscr{U}(4, s)$;
(iv) $\mathscr{S} \cong \mathscr{Q}(5, s)$;
(v) *all points of $\mathscr{S}$ are regular, $s = t^2$, and every three non-$\mathscr{S}^*$-collinear points are contained in a proper linear variety of the linear space $\mathscr{S}^* = (\mathscr{P}, \mathscr{B}^*, \in)$.* □

Now we give a characterization of $\mathscr{Q}(4, q)$ and $\mathscr{Q}(5, q)$ which utilizes Theorem 26.4.28 on Moufang generalized quadrangles. The statement of the theorem, however, is purely combinatorial.

Let $\mathscr{S}$ be a generalized quadrangle of order (s, t), $s > 1$ and $t > 1$. A *quadrilateral* of $\mathscr{S}$ is just a subquadrangle of order (1, 1). A quadrilateral $\mathscr{V}$ is said to be *opposite* a line l if the lines of $\mathscr{V}$ are not concurrent with l. If $\mathscr{V}$ is opposite l, the four lines incident with the points of $\mathscr{V}$ and concurrent with l are called the *lines of perspectivity of $\mathscr{V}$ from l*. Two quadrilaterals $\mathscr{V}$ and $\mathscr{V}'$ are in *perspective from l* if either

(a) $\mathscr{V} = \mathscr{V}'$ and $\mathscr{V}$ is opposite l; or
(b) (i) $\mathscr{V} \neq \mathscr{V}'$,
(ii) $\mathscr{V}$ and $\mathscr{V}'$ are both opposite l,
(iii) the lines of perspectivity of $\mathscr{V}$, and of $\mathscr{V}'$, from l are the same.

Theorem 26.4.23: *The generalized quadrangle* $\mathcal{S} = (\mathcal{P}, \mathcal{B}, \mathrm{I})$ *of order* (s, t), $s > 1$ *and* $t > 2$, *is isomorphic to* $\mathcal{Q}(4, s)$ *or* $\mathcal{Q}(5, s)$ *if and only if given a quadrilateral* $\mathcal{V}$ *opposite a line* l *and a point* P', $P' \not\mathrm{I}\, l$, *incident with a line of perspectivity of* $\mathcal{V}$ *from* l, *there is a quadrilateral* $\mathcal{V}'$ *containing* P' *and in perspective with* $\mathcal{V}$ *from* l. □

Remark: If $t = 2$ and $s > 1$, then by Theorems 26.2.2 and 26.2.3 we have $s \in \{2, 4\}$. Now by Theorems 26.4.4 and 26.4.9, $\mathcal{S} \cong \mathcal{Q}(4, 2)$ or $\mathcal{S} \cong \mathcal{U}(3, 4)$. One can check that in these two cases the quadrilateral condition of the preceding theorem is satisfied.

We conclude this section on characterization theorems of purely combinatorial type with a fundamental characterization of all classical and dual classical generalized quadrangles with $s > 1$ and $t > 1$. The reader is reminded of properties (A) and $(\hat{A})$ introduced in the paragraph preceding Theorem 26.4.12. Let $\mathcal{B}^{\perp\perp}$ be the set of all hyperbolic lines of the generalized quadrangle $\mathcal{S} = (\mathcal{P}, \mathcal{B}, \mathrm{I})$, and let $\mathcal{S}^{\perp\perp} = (\mathcal{P}, \mathcal{B}^{\perp\perp}, \in)$. We say that $\mathcal{S}$ satisfies *property* (A) if, for any $m = \{P, P'\}^{\perp\perp} \in \mathcal{B}^{\perp\perp}$ and any $U \in \mathrm{cl}(P, P')\backslash(\{P, P'\}^{\perp} \cup \{P, P'\}^{\perp\perp})$, the substructure of $\mathcal{S}^{\perp\perp}$ generated by m and U is a dual affine plane. The dual of (A) is denoted by $(\hat{A})$.

Theorem 26.4.24: *Let* $\mathcal{S} = (\mathcal{P}, \mathcal{B}, \mathrm{I})$ *be a generalized quadrangle of order* (s, t), *with* $s > 1$ *and* $t > 1$. *Then* $\mathcal{S}$ *is a classical or a dual classical generalized quadrangle if and only if it satisfies either condition* (A) *or* $(\hat{A})$. □

In the last part of this section we give, without proof, some important characterizations of classical generalized quadrangles, formulated in terms of automorphisms. First some new ideas are introduced, always with the trivial cases $s = 1$ and $t = 1$ excluded.

Let θ be an automorphism of the generalized quadrangle $\mathcal{S} = (\mathcal{P}, \mathcal{B}, \mathrm{I})$ of order (s, t). If θ fixes each point of $P^{\perp}$, $P \in \mathcal{P}$, then θ is a *symmetry about* P. If θ is the identity or if θ fixes each line incident with P but no point of $\mathcal{P}\backslash P^{\perp}$, then θ is an *elation about* P. It is possible to prove that any symmetry about P is automatically an elation about P. We say that $\mathcal{S}$ is an *elation generalized quadrangle* with *elation group* G and *base point* P, if there is a group G of elations about P acting regularly on $\mathcal{P}\backslash P^{\perp}$; briefly, one says that $(\mathcal{S}^{(P)}, G)$ or $\mathcal{S}^{(P)}$ is an elation generalized quadrangle. If the group G is Abelian, then we say that the elation generalized quadrangle $(\mathcal{S}^{(P)}, G)$ is a *translation generalized quadrangle* with *translation group* G and *base point* P. One can prove that the base point P of a translation generalized quadrangle is coregular.

Let $P, P' \in \mathcal{P}$, $P \not\sim P'$. A *generalized homology* with *centres* P, P' is an automorphism θ of $\mathcal{S}$ which fixes all lines incident with P and all lines incident with P'. The group of all generalized homologies with centres P, P'

is denoted $H(P, P')$. The generalized quadrangle $\mathcal{S}$ is said to be (P, P')-*transitive* if for each $P'' \in \{P, P'\}^{\perp}$ the group $H(P, P')$ is transitive on $\{P, P''\}^{\perp\perp}\backslash\{P, P''\}$ and on $\{P', P''\}^{\perp\perp}\backslash\{P', P''\}$.

Finally, $\mathcal{S}$ is a *Moufang generalized quadrangle* if the following condition and its dual are satisfied: for any point P and any two distinct lines l and m incident with P, the group of automorphisms of $\mathcal{S}$ fixing l and m pointwise and P linewise is transitive on the lines ($\neq l$) incident with a given point P' on l ($P' \neq P$).

Theorem 26.4.25: *Let $(\mathcal{S}^{(P)}, G)$ be a translation generalized quadrangle of order (s, t), $s > 1$, $t > 1$.*

(i) *If s is prime, then $\mathcal{S} \cong \mathcal{Q}(4, s)$ or $\mathcal{S} \cong \mathcal{Q}(5, s)$.*

(ii) *If all lines are regular, then $\mathcal{S} \cong \mathcal{Q}(4, s)$ or $t = s^2$.*

(iii) *Let $t = s^2$ with s odd. Then $(\mathcal{S}^{(P)}, G)$ is isomorphic to $\mathcal{Q}(5, s)$ if and only if for a fixed point P', with $P' \nsim P$, the group $H(P, P')$ has order $s - 1$.*

(iv) *Let $t = s^2$ with s even. Then $(\mathcal{S}^{(P)}, G)$ is isomorphic to $\mathcal{Q}(5, s)$ if and only if all lines are regular and for a fixed point P', with $P' \nsim P$, the group $H(P, P')$ has order $s - 1$.*

(v) *If $t = s^2$, with $s = p^2$ and p prime, and if all lines are regular, then $(\mathcal{S}^{(P)}, G)$ must be isomorphic to $\mathcal{Q}(5, s)$.* □

The reader may verify that $\mathcal{Q}(4, q)$ and $\mathcal{Q}(5, q)$ are translation generalized quadrangles for any choice of the base point P.

Theorem 26.4.26: *The group of symmetries about each point of the generalized quadrangle $\mathcal{S}$ of order (s, t), $s > 1$, $t > 1$, has even order if and only if one of the following holds:*

(i) $\mathcal{S} \cong \mathcal{W}(s)$;
(ii) $\mathcal{S} \cong \mathcal{U}(3, s)$;
(iii) $\mathcal{S} \cong \mathcal{U}(4, s)$. □

Theorem 26.4.27: *Let $\mathcal{S} = (\mathcal{P}, \mathcal{B}, \mathrm{I})$ be a generalized quadrangle of order (s, t), $s \neq 1$ and $t \neq 1$. Then $\mathcal{S}$ is classical if and only if $\mathcal{S}$ is (P, P')-transitive for all $P, P' \in \mathcal{P}$ with $P \nsim P'$.* □

Theorem 26.4.28: *A generalized quadrangle $\mathcal{S}$ of order (s, t), $s \neq 1$ and $t \neq 1$, is Moufang if and only if $\mathcal{S}$ is classical or dual classical.* □

26.5 Partial geometries

A (finite) *partial geometry* is an incidence structure $\mathcal{S} = (\mathcal{P}, \mathcal{B}, \mathrm{I})$ in which $\mathcal{P}$ and $\mathcal{B}$ are disjoint (non-empty) sets of objects called *points* and *lines*, and for which I is a symmetric point–line incidence relation satisfying the following axioms:

(i) each point is incident with $1 + t$ lines ($t \geq 1$) and two distinct points are incident with at most one line;

(ii) each line is incident with $1 + s$ points ($s \geq 1$) and two distinct lines are incident with at most one point;

(iii) if P is a point and l is a line not incident with P, then there are exactly α ($\alpha \geq 1$) points $P_1, P_2, \ldots, P_\alpha$ and α lines $l_1, l_2, \ldots, l_\alpha$ such that $P \mathrm{I} l_i \mathrm{I} P_i \mathrm{I} l$, with $i = 1, 2, \ldots, \alpha$.

From the axioms, the partial geometries with $\alpha = 1$ are the generalized quadrangles.

The integers s, t, and α are the *parameters* of the partial geometry. Given two (not necessarily distinct) points P and P' of $\mathcal{S}$, they are *collinear*, written $P \sim P'$, if there is some line l for which $P \mathrm{I} l \mathrm{I} P'$; so $P \nsim P'$ means that P and P' are not collinear. Dually, for $l, l' \in \mathcal{B}$, we write $l \sim l'$ or $l \nsim l'$ according as l and l' are *concurrent* or non-concurrent. The line incident with distinct collinear points P and P' is denoted PP'; the point incident with distinct concurrent lines l and l' is denoted $l \cap l'$.

Let $\mathcal{S} = (\mathcal{P}, \mathcal{B}, \mathrm{I})$ be a partial geometry with parameters s, t, and α. Put $|\mathcal{P}| = v$ and $|\mathcal{B}| = b$.

Theorem 25.5.1: (i) $v = (s + 1)(st + 1)/\alpha$; (ii) $b = (t + 1)(st + 1)/\alpha$.

Proof: Let l be a fixed line of $\mathcal{S}$ and count in different ways the number of ordered pairs $(P, m) \in \mathcal{P} \times \mathcal{B}$ with $P \not\mathrm{I}\, l$, $P \mathrm{I} m$, and $l \sim m$. This gives $(v - s - 1)\alpha = (s + 1)ts$ or $v = (s + 1)(st + 1)/\alpha$. Dually $b = (t + 1)(st + 1)/\alpha$. □

Corollary: *The elements $st(s + 1)/\alpha$ and $st(t + 1)/\alpha$ are integers.* □

Theorem 26.5.2: *The integer $\alpha(s + t + 1 - \alpha)$ divides $st(s + 1)(t + 1)$.*

Proof: This is analogous to the proof of Theorem 26.2.2. □

Theorem 26.5.3 (The Krein inequalities): *The integers s, t, and α satisfy the inequalities*

$$(s + 1 - 2\alpha)t \leq (s - 1)(s + 1 - \alpha)^2 \tag{26.7}$$

and

$$(t + 1 - 2\alpha)s \leq (t - 1)(t + 1 - \alpha)^2. \tag{26.8}$$

When equality holds in (26.7), the number of points collinear with three points P_1, P_2, P_3 depends only on the number of collinearities in $\{P_1, P_2, P_3\}$; when equality holds in (26.8), the number of lines concurrent with three lines l_1, l_2, l_3 depends only on the number of concurrencies in $\{l_1, l_2, l_3\}$.

Proof: See §26.9. □

Partial geometries $\mathscr{S}$ can be divided into four (non-disjoint) classes.

(a) $\mathscr{S}$ has $\alpha = s + 1$ or, dually, $\alpha = t + 1$; when $\alpha = s + 1$, then $\mathscr{S}$ is a $2 - (v, s + 1, 1)$ design.

(b) $\mathscr{S}$ has $\alpha = s$ or, dually, $\alpha = t$; when $\alpha = t$, then $\mathscr{S}$ is a *net* of *order* $s + 1$ and *degree* $t + 1$.

(c) When $\alpha = 1$, $\mathscr{S}$ is a generalized quadrangle.

(d) When $1 < \alpha < \min(s, t)$, then $\mathscr{S}$ is *proper*.

Now we give some examples. It is easy to construct $2 - (v, s + 1, 1)$ designs. For example, the points and lines of the projective space $PG(n, q)$, $n \geq 2$, form a $2 - (\theta(n), q + 1, 1)$ design with $\theta(n) = (q^{n+1} - 1)/(q - 1)$. The points and lines of the affine space $AG(n, q)$, $n \geq 2$, form a $2 - (q^n, q, 1)$ design. Let $\mathscr{K}$ be a maximal $(qn - q + n; n)$-arc in $PG(2, q)$, with $n \geq 2$ (cf. §12.2). Then the points of $\mathscr{K}$ together with the non-empty intersections $l \cap \mathscr{K}$, where l is any line of $PG(2, q)$, form a $2 - (qn - q + n, n, 1)$ design. If $n < q$, then the points of $PG(2, q)\backslash\mathscr{K}$ together with the lines having an empty intersection with $\mathscr{K}$, form a dual design with parameters $s = q$, $t = (q/n) - 1$, $\alpha = q/n$.

If we delete d classes, $0 \leq d \leq q - 1$, of parallel lines of $AG(2, q)$, then the remaining structure is a net of order q and degree $q + 1 - d$. Let $\mathscr{P}$ be the set of all points of $PG(n, q)$ which are not contained in a fixed subspace $PG(n - 2, q)$, with $n \geq 2$. Let $\mathscr{B}$ be the set of all lines of $PG(n, q)$ having no point in common with $PG(n - 2, q)$. Finally, let I be the natural incidence. Then $(\mathscr{P}, \mathscr{B}, \mathrm{I})$ is a partial geometry with parameters $s = q$, $t = q^{n-1} - 1$, $\alpha = q$. This dual net will be denoted by H_q^n.

Now we give two classes of proper partial geometries. Let $\mathscr{K}$ be a maximal $(qn - q + n; n)$-arc of $PG(2, q)$, $2 \leq n < q$. Let $\mathscr{P}$ be the set of all points of $PG(2, q)\backslash\mathscr{K}$, let $\mathscr{B}$ be the set of all lines of $PG(2, q)$ having a non-empty intersection with $\mathscr{K}$, and let I be the natural incidence. Then $\mathscr{S}(\mathscr{K}) = (\mathscr{P}, \mathscr{B}, \mathrm{I})$ is a partial geometry with parameters

$$s = q - n, \qquad t = q(n - 1)/n, \qquad \alpha = (q - n)(n - 1)/n. \tag{26.9}$$

Now we embed the plane $PG(2, q)$ of $\mathscr{K}$ in the projective space $PG(3, q)$. Let $\mathscr{P}' = PG(3, q)\backslash PG(2, q)$, let $\mathscr{B}'$ consist of all lines of $PG(3, q)$ having a unique point in common with $\mathscr{K}$, and let I$'$ be the natural incidence. Then $\mathscr{T}_2^*(\mathscr{K}) = (\mathscr{P}', \mathscr{B}', \mathrm{I}')$ is a partial geometry with parameters

$$s = q - 1, \qquad t = (q + 1)(n - 1), \quad \alpha = n - 1. \tag{26.10}$$

By Theorem 12.2.2 there exist maximal $(2^{m+h} - 2^h + 2^m; 2^m)$-arcs in $PG(2, 2^h)$ for any m with $1 \leq m < h$. Hence there exist partial geometries with parameters as follows:

(a) $s = 2^h - 2^m$, $t = 2^h - 2^{h-m}$, $\alpha = (2^m - 1)(2^{h-m} - 1)$ with $1 \leq m < h$; (26.11)

(b) $s = 2^h - 1$, $t = (2^h + 1)(2^m - 1)$, $\alpha = 2^m - 1$ with $1 \leq m < h$. (26.12)

Such a partial geometry has $\alpha = 1$ or is proper. A partial geometry of type (a) is a generalized quadrangle if and only if $h = 2$ and $m = 1$. This gives the following model of the unique generalized quadrangle $\mathscr{S}$ with 15 points and 15 lines: points of $\mathscr{S}$ are the 15 points of $PG(2, 4)\backslash\mathscr{K}$ with $\mathscr{K}$ a given oval of $PG(2, 4)$, lines of $\mathscr{S}$ are the 15 bisecants of $\mathscr{K}$, and incidence is the natural one. A partial geometry of type (b) is a generalized quadrangle if and only if $m = 1$. In this case, $\mathscr{K}$ is an oval of $PG(2, q)$, $q = 2^h$, and $\mathscr{T}_2^*(\mathscr{K})$ is the generalized quadrangle described in §26.1.

Up to duality, the parameters of the known proper partial geometries are the following:

$s = 2^h - 2^m, t = 2^h - 2^{h-m}, \alpha = (2^m - 1)(2^{h-m} - 1)$, with $h \neq 2$ and $1 \leq m < h$;

$s = 2^h - 1$, $t = (2^h + 1)(2^m - 1)$, $\alpha = 2^m - 1$, with $1 < m < h$;

$s = 2^{2h-1} - 1$, $t = 2^{2h-1}$, $\alpha = 2^{2h-2}$, with $h > 1$;

$s = 26$, $t = 27$, $\alpha = 18$;

$s = t = 5$, $\alpha = 2$;

$s = 4$, $t = 17$, $\alpha = 2$.

26.6 Embedded partial geometries

A *projective partial geometry* $\mathscr{S} = (\mathscr{P}, \mathscr{B}, \mathrm{I})$ is a partial geometry for which the point set $\mathscr{P}$ is a subset of the point set of some projective space $PG(n, q)$, and for which the line set $\mathscr{B}$ is a set of lines of $PG(n, q)$. In such a case we also say that the partial geometry $\mathscr{S}$ is *embedded* in $PG(n, q)$. If $PG(n', q)$ is the subspace of $PG(n, q)$ generated by all points of $\mathscr{P}$, then we say that $PG(n', q)$ is the *ambient space* of $\mathscr{S}$.

Theorem 26.6.1: *If $\mathscr{S} = (\mathscr{P}, \mathscr{B}, \mathrm{I})$ is a partial geometry with parameters s, t, α which is projective with ambient space $PG(n, s)$, then one of the following holds:*

(a) *$\alpha = s + 1$ and $\mathscr{S}$ is the $2 - (\theta(n), s + 1, 1))$ design formed by all points and all lines of $PG(n, s)$;*

(b) *$\alpha = 1$ and $\mathscr{S}$ is a classical generalized quadrangle;*

(c) *$\alpha = t + 1$, $n = 2$, $PG(2, s)\backslash\mathscr{P}$ is a maximal $(sd - s + d; d)$-arc $\mathscr{K}$ of $PG(2, s)$ with $d = s/\alpha$ and $2 \leq d < s$, and $\mathscr{B}$ consists of all lines of $PG(2, s)$ having an empty intersection with $\mathscr{K}$;*

(d) *$\alpha = s$, $n \geq 2$ and $\mathscr{S} = H_s^n$.*

Proof: If $\alpha = s + 1$, then $\mathscr{S}$ is a $2-(v, s+1, 1)$ design. Hence $\mathscr{S}$ consists of all points and all lines of some subspace $PG(m, s)$ of $PG(n, s)$. Since $PG(n, s)$ is the ambient space of $\mathscr{S}$, we have $m = n$. Consequently $\mathscr{S}$ is the design formed by all points and all lines of $PG(n, s)$.

If $\alpha = 1$, then by Theorem 26.3.29 the partial geometry $\mathscr{S}$ is a classical generalized quadrangle.

Now let $\alpha = t + 1$. Since any two lines of $\mathscr{S}$ have a non-empty intersection, the ambient space of $\mathscr{S}$ is a plane $PG(2, s)$. Each line of $PG(2, s)$ not in $\mathscr{B}$ has exactly s/α points in common with $PG(2, s)\backslash\mathscr{P}$. If $\alpha = s$, then $\mathscr{S}$ is the dual affine plane H_s^2. If $2 \le d < s$ with $d = s/\alpha$, then $PG(2, s)\backslash\mathscr{P}$ is a maximal $(sd - s + d; d)$-arc $\mathscr{K}$ and $\mathscr{B}$ is the set of all lines of $PG(2, s)$ having an empty intersection with $\mathscr{K}$.

Let us now suppose that $1 < \alpha < s$ and $\alpha \neq t + 1$. In this case we have $n \geq 3$.

First, let $n = 3$. Suppose that l is a line of $\mathscr{S}$ and that P is a point of $\mathscr{S}$ with $P \notin l$. Let π be the plane Pl of $PG(3, s)$. The points and lines of $\mathscr{S}$ in π constitute a partial geometry $\mathscr{S}_\pi$ with parameters $t_\pi = \alpha - 1$, $s_\pi = s$, and $\alpha_\pi = \alpha$. Hence the points of π which do not belong to $\mathscr{S}_\pi$ form a maximal $(s(s\alpha^{-1} + \alpha^{-1} - 1); s\alpha^{-1})$-arc of π. Let m be any line of $PG(3, s)$ which contains at least two points P', P'' of $\mathscr{S}$. Further, let $m' \in \mathscr{B}$, with $m \neq m'$ and $P' \in m'$. Since $P''m'\backslash\mathscr{P}$ is an $(s(s\alpha^{-1} + \alpha^{-1} - 1); s\alpha^{-1})$-arc of the plane $P''m'$, we have $|m \cap \mathscr{P}| \in \{s + 1, s + 1 - s\alpha^{-1}\}$. Hence each line of $PG(3, s)$ intersects $\mathscr{P}$ in $0, 1, s + 1 - s\alpha^{-1}$, or $s + 1$ points, which means that $\mathscr{P}$ is a set of type $(0, 1, s + 1 - s\alpha^{-1}, s + 1)$ in $PG(3, s)$. The $(s + 1)$-secants of $\mathscr{P}$ are the lines of $\mathscr{S}$. Now we show that $\mathscr{P}$ has no 1-secant.

Suppose that l is a 1-secant of $\mathscr{P}$, and let P be the common point of l and $\mathscr{P}$. The lines of $\mathscr{S}$ through P are denoted by $m_1, m_2, \ldots, m_{t+1}$. First, let us assume that each plane lm_i contains exactly $s + 1$ points of $\mathscr{P}$, with $i = 1, 2, \ldots, t + 1$. Since $n = 3$, each line m of $\mathscr{B}$ contains at least one point of $\mathscr{P} \cap lm_i = m_i$, $i = 1, 2, \ldots, t + 1$. It follows that all lines of $\mathscr{S}$ belong to a common plane; so $PG(3, s)$ is not the ambient space of $\mathscr{S}$, a contradiction. Consequently for at least one index i we have $|\mathscr{P} \cap lm_i| > s + 1$, say $P' \in (\mathscr{P} \cap lm_i)\backslash m_i$. By the previous paragraph the line l contains $s + 1 - s\alpha^{-1}$ or $s + 1$ points of the partial geometry $\mathscr{S}_{\pi_i}$, with $\pi_i = P'm_i = lm_i$. Hence $1 \in \{s + 1 - s\alpha^{-1}, s + 1\}$, a contradiction. Therefore $\mathscr{P}$ has no 1-secant; that is, $\mathscr{P}$ is of type $(0, s + 1 - s\alpha^{-1}, s + 1)$. Next, we show that such a set with $1 < \alpha < s$ cannot exist.

Counting the points of $\mathscr{P}$ on all lines of $PG(3, s)$ containing a fixed point of $\mathscr{P}$, we obtain

$$|\mathscr{P}| = 1 + (t + 1)s + (s^2 + s - t)(s - s\alpha^{-1}). \qquad (26.13)$$

By Theorem 26.5.1 we also have

$$|\mathscr{P}| = (s + 1)(st + \alpha)/\alpha. \qquad (26.14)$$

From (26.13) and (26.14) it follows that $t = (s+1)(\alpha - 1)$. Since $\alpha \neq s+1$ we have $\mathscr{P} \neq PG(3, s)$. Let $A \in PG(3, s)\backslash\mathscr{P}$. Counting the points of $\mathscr{P}$ on all lines of $PG(3, s)$ containing A, we see that $s + 1 - s\alpha^{-1}$ divides $|\mathscr{P}|$. Hence $s\alpha + \alpha - s$ divides

$$(s+1)(s^2\alpha + s\alpha - s^2 - s + \alpha) = (\alpha s + \alpha - s)(s^2 + s + 1) - s^2.$$

Consequently $s\alpha + \alpha - s$ divides s^2. Let $s = p^h$, $\alpha = p^{h'}$, with $0 < h' < h$. Then $p^h + 1 - p^{h-h'}$ divides $p^{2h-h'}$. Since $(p^h + 1 - p^{h-h'}, p^{2h-h'}) = 1$ we necessarily have $p^h + 1 - p^{h-h'} = 1$, clearly a contradiction.

It has been shown that, for $1 < \alpha < s$ and $\alpha \neq t+1$, we necessarily have $n > 3$.

So let $1 < \alpha < s$, $\alpha \neq t+1$, and $n > 3$. Let l be a line of $\mathscr{S}$, let π be the plane defined by l and a point P in $\mathscr{P}\backslash l$, and let $PG(3, s)$ be the solid defined by π and a point P' in $\mathscr{P}\backslash\pi$. Let P_1 and P_2 be distinct points of $\mathscr{P}$ in $PG(3, s)$. Counting the number of pairs (l_1, l_2), with $l_1, l_2 \in \mathscr{B}$, l_1 and l_2 in $PG(3, s)$, $P_1 \in l_1$, $P_2 \in l_2$, and $l_1 \sim l_2$, in different ways, it appears that the number of lines of $\mathscr{B}$ in $PG(3, s)$ which contain P_1 is equal to the number of lines of $\mathscr{B}$ in $PG(3, s)$ which contain P_2. It follows that the points and lines of $\mathscr{S}$ in $PG(3, s)$ constitute a partial geometry $\mathscr{S}'$ with parameters t', $s' = s$, $\alpha' = \alpha$. Since $1 < \alpha' < s'$ and $\alpha' \neq t' + 1$ ($\mathscr{S}'$ is not contained in a plane), such a geometry cannot exist.

So the only possibilities are $\alpha = 1$, $\alpha = s+1$, $\alpha = t+1$, and $\alpha = s$.

Finally, assume that $\alpha = s$ with $\alpha \neq 1$ and $\alpha \neq t+1$. In this case we have $n \geq 3$. Let l be a line of $\mathscr{S}$, and suppose that the point P of $\mathscr{S}$ is not on l. The points and lines of $\mathscr{S}$ in the plane $\pi = Pl$ form a partial geometry with parameters $s' = s$, $t' = \alpha - 1 = s' - 1$, and $\alpha' = \alpha = s'$, that is a dual affine plane of order s. If the line m of $PG(n, s)$ contains at least two point of $\mathscr{P}$, then m is contained in at least one plane π in which $\mathscr{S}$ induces a dual affine plane of order s. Hence $\mathscr{P}$ is a set of type $(0, 1, s, s+1)$ in $PG(n, s)$, and a line m of $PG(n, s)$ contains $s+1$ points of $\mathscr{P}$ if and only if m belongs to $\mathscr{B}$. If $\mathscr{P}$ has no 1-secant, then all planes of $PG(n, s)$ through a fixed line l of $\mathscr{S}$ contain a point of $\mathscr{P}\backslash l$; hence the points of $\mathscr{S}$ in such a plane are the points of a dual affine plane of order s. It follows that

$$|\mathscr{P}| = s + 1 + (s^2 - 1)(s^{n-1} - 1)/(s - 1) = s^n + s^{n-1}. \qquad (26.15)$$

Conversely, if $|\mathscr{P}| = s^n + s^{n-1}$ then $\mathscr{P}$ admits no 1-secant. Now we show that $\mathscr{P}$ has no 1-secant.

First, let $n = 3$. By a reasoning analogous to that in a preceding paragraph, it follows that $\mathscr{P}$ has no 1-secant; so $|\mathscr{P}| = s^3 + s^2$ for $n = 3$. Now we use induction, and assume that any projective partial geometry with $\alpha = s$, $\alpha \neq 1$, $\alpha \neq t+1$, and ambient space $PG(n-1, s)$, $n \geq 4$, has no 1-secant. Next, assume that $\mathscr{S} = (\mathscr{P}, \mathscr{B}, \mathrm{I})$ is a projective partial geometry with $\alpha = s$, $\alpha \neq 1$, $\alpha \neq t+1$, and ambient space $PG(n, s)$, $n \geq 4$, which has at least one 1-secant l.

Let $l \cap \mathscr{P} = \{P\}$. Consider a line m of $\mathscr{B}$ and $n - 2$ points $P_1, P_2, \ldots, P_{n-2}$ of $\mathscr{P}$ such that $m, P_1, \ldots, P_{n-2}$ generate a prime $PG(n-1, s)$. The geometry induced by $\mathscr{S}$ in $PG(n-1, s)$ is a partial geometry $\bar{\mathscr{S}}$ with parameters $\bar{t}$, $\bar{s} = s = \bar{\alpha}$ and ambient space $PG(n-1, s)$. Clearly $\bar{\alpha} \neq 1$, and $\bar{\alpha} \neq \bar{t} + 1$ since $n - 1 > 2$. By the induction hypothesis we have

$$|\bar{\mathscr{P}}| = |\mathscr{P} \cap PG(n-1, s)| = s^{n-1} + s^{n-2}. \tag{26.16}$$

Let $m_1, m_2, \ldots, m_{t+1}$ be the lines of $\mathscr{B}$ through P. The plane lm_1 contains $s + 1$ points of $\mathscr{S}$. If the intersection of $\mathscr{P}$ and the solid lm_1m_i, with $i > 1$, generates lm_1m_i, then the set $\mathscr{P} \cap lm_1m_i$ has no 1-secant, a contradiction. Hence for all $i > 1$, $\mathscr{P} \cap lm_im_1$ is contained in the plane m_im_1, and consequently $|\mathscr{P} \cap lm_im_1| = s^2 + s$. Let P' be any point of $\mathscr{P} \backslash m_1$. Then the plane $P'm_1$ contains s lines through P' which also belong to $\mathscr{B}$. Therefore P' belongs to at least one of the solids lm_im_1, with $i > 1$. It follows that

$$|\mathscr{P}| \leq \theta(n-3)(s^2 - 1) + s + 1 = s^{n-1} + s^{n-2}, \tag{26.17}$$

where $\theta(n-3)$ is the number of solids containing the plane lm_1. From (26.17) and (26.16) it now follows that $\mathscr{P} = \mathscr{P} \cap PG(n-1, s)$; hence $PG(n, s)$ is not the ambient space of $\mathscr{S}$, a contradiction. So $\mathscr{P}$ has no 1-secant.

So $|\mathscr{P}| = s^n + s^{n-1}$ and $\mathscr{P}$ is of type $(0, s, s+1)$. Consequently $PG(n, s) \backslash \mathscr{P}$ is of type $(0, 1, s+1)$, and so $PG(n, s) \backslash \mathscr{P}$ is a projective subspace of $PG(n, s)$. Since $|\mathscr{P}| = s^n + s^{n-1}$, the subspace $PG(n, s) \backslash \mathscr{P}$ has dimension $n - 2$. Let $PG(n, s) \backslash \mathscr{P} = PG(n-2, s)$. The lines of $\mathscr{B}$ are exactly the lines of $PG(n, s)$ having an empty intersection with $PG(n-2, s)$. We conclude that $\mathscr{S} = H_s^n$, and now the theorem is completely proved. □

Any projective partial geometry $\mathscr{S} = (\mathscr{P}, \mathscr{B}, \mathrm{I})$ satisfies the axiom of Pasch:

(Pa) If $l_1 \,\mathrm{I}\, P \,\mathrm{I}\, l_2$, $l_1 \neq l_2$, $m_1 \,\not{\mathrm{I}}\, P \,\not{\mathrm{I}}\, m_2$, $l_i \sim m_j$ for all $i, j \in \{1, 2\}$, then $m_1 \sim m_2$.

The known partial geometries satisfying (Pa) are (i) all known generalized quadrangles, (ii) all known partial geometries having $\alpha = t + 1$, (iii) the partial geometries isomorphic to the design formed by the points and lines of some $PG(n, q)$, (iv) the partial geometries isomorphic to some H_q^n.

Theorem 26.6.2: *Let $\mathscr{S}$ be a dual net of order $s + 1$ and degree $t + 1$, with $s < t + 1$. If $\mathscr{S}$ satisfies (Pa), then $\mathscr{S}$ is isomorphic to a partial geometry H_q^n ($s = q$ and $t = q^{n-1} - 1$).*

Proof: See §26.9. □

26.7 (0, α)-geometries and semi-partial geometries

A (finite) $(0, \alpha)$-*geometry*, $\alpha > 1$, is an incidence structure $\mathscr{S} = (\mathscr{P}, \mathscr{B}, \mathrm{I})$ in which $\mathscr{P}$ and $\mathscr{B}$ are disjoint (non-empty) sets of objects called *points* and *lines*, and for which I is a symmetric point–line incidence relation satisfying the following axioms:

(i) two distinct points are incident with at most one line;

(ii) if a point P and a line l are not incident, then there are 0 or α points which are collinear with P and incident with l;

(iii) each line is incident with at least two points, and each point is incident with at least two lines;

(iv) $\mathcal{S}$ is connected (which means that for any two elements P and P' of $\mathcal{P} \cup \mathcal{B}$, there exists a set of elements $P_1, P_2, \ldots, P_r \in \mathcal{P} \cup \mathcal{B}$ such that $P \mathrm{I} P_1 \mathrm{I} P_2 \mathrm{I} \cdots \mathrm{I} P_r \mathrm{I} P'$).

Terms such as 'collinear' and 'concurrent', and notations such as $\sim$ and $\nsim$, are defined as for generalized quadrangles and partial geometries.

Theorem 26.7.1: *Each point is incident with a constant number* $1 + t$ $(t \geq 1)$ *of lines, and each line is incident with a constant number* $1 + s$ $(s \geq 1)$ *of points.*

Proof: Let P and P', $P \neq P'$, be collinear points of $\mathcal{S}$; let $1 + t$ and $1 + t'$ be the respective numbers of lines incident with P and P'. Counting in different ways the number of ordered pairs (l, l'), with $P \mathrm{I} l$, $P' \mathrm{I} l'$, $l \neq l'$, and $l \sim l'$, we obtain $t(\alpha - 1) = t'(\alpha - 1)$; hence $t = t'$. By the connectedness of $\mathcal{S}$ we now see that each point of $\mathcal{S}$ is incident with $1 + t$ lines. Dually, each line is incident with a constant number $1 + s$ of points. □

The integers s, t, and α are called the *parameters* of the $(0, \alpha)$-geometry. Let $|\mathcal{P}| = v$ and $|\mathcal{B}| = b$. It should be noted that v and b are not uniquely determined by s, t, and α.

A (finite) *semi-partial geometry* is an incidence structure $\mathcal{S} = (\mathcal{P}, \mathcal{B}, \mathrm{I})$ in which $\mathcal{P}$ and $\mathcal{B}$ are disjoint (non-empty) sets of objects called *points* and *lines*, and for which I is a symmetric point–line incidence relation satisfying the axioms that follow:

(i) each point is incident with $1 + t$ $(t \geq 1)$ lines and two distinct points are incident with at most one line;

(ii) each line is incident with $1 + s$ $(s \geq 1)$ points and two distinct lines are incident with at most one point;

(iii) if a point P and a line l are not incident, then there are 0 or α $(\alpha \geq 1)$ points which are collinear with P and incident with l;

(iv) if two points are not collinear, then there are μ $(\mu > 0)$ points collinear with both.

The integers s, t, α, and μ are called the *parameters* of the semi-partial geometry.

Clearly any semi-partial geometry with $\alpha > 1$ is a $(0, \alpha)$-geometry. The semi-partial geometries with $\alpha = 1$ are also called *partial quadrangles*. A semi-partial geometry is a partial geometry if and only if the zero in axiom

(iii) does not occur, which is equivalent to the condition $\mu = (t + 1)\alpha$. If we write '$\rightarrow$' for 'generalizes to', then we have the following scheme:

$$\begin{array}{ccc} \text{generalized quadrangle} & \longrightarrow & \text{partial geometry} \\ \downarrow & & \downarrow \\ \text{partial quadrangle} & \longrightarrow & \text{semi-partial geometry} \end{array}$$

Let $\mathcal{S} = (\mathcal{P}, \mathcal{B}, \mathrm{I})$ be a semi-partial geometry with parameters s, t, α, and μ. Further, let $|\mathcal{P}| = v$ and $|\mathcal{B}| = b$.

Theorem 26.7.2: *The parameters of a semi-partial geometry satisfy*

(i) $v(t + 1) = b(s + 1)$;
(ii) $v = 1 + (1 + t)s(1 + t(s - \alpha + 1)/\mu)$.

Proof: Count in different ways the number of ordered pairs (P, l), with $P \in \mathcal{P}$, $l \in \mathcal{B}$, and $P \,\mathrm{I}\, l$. We obtain $v(t + 1) = b(s + 1)$.

Now we count in different ways the number of ordered triples (P, P', P''), where $P, P', P'' \in \mathcal{P}$, $P \sim P'$, $P \nsim P''$, and $P' \sim P''$. This gives

$$v(t + 1)st(s + 1 - \alpha) = v(v - (t + 1)s - 1)\mu,$$

and hence $v = 1 + (1 + t)s(1 + t(s - \alpha + 1)/\mu)$. □

Corollary : *The elements* $st(1 + t)(s - \alpha + 1)/\mu$ *and* $st(\mu + (1 + t)^2(s - \alpha + 1))/(\mu(s + 1))$ *are integers.*

Proof: This follows from the fact that v and b are integers. □

Theorem 26.7.3: *For* $\alpha \neq s + 1$,

(a) α *divides* $st(t + 1)$ *and* $st(s + 1)$;
(b) α *divides* μ;
(c) α^2 *divides* μst;
(d) α^2 *divides* $t(\alpha(t + 1) - \mu)$;
(e) $\alpha^2 \leq \mu \leq (t + 1)\alpha$.

Proof: For any non-incident point–line pair (P, l), we define $[P, l]$ to be the number of points incident with l and collinear with P. For any line l, let v_l be the number of points P for which $[P, l] = \alpha$; for any point P, let b_P be the number of lines l for which $[P, l] = \alpha$.

For a fixed line l we now compute in different ways the number of ordered pairs (P, l'), with $P \in \mathcal{P}$, $l' \in \mathcal{B}$, $P \,\not\mathrm{I}\, l$, $P \,\mathrm{I}\, l'$, and $l \sim l'$. We obtain $v_l\alpha = (s + 1)ts$, and so α divides $st(s + 1)$.

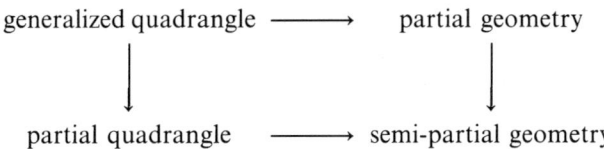

For a fixed point P we compute in different ways the number of ordered pairs (P', l), with $P' \in \mathscr{P}$, $l \in \mathscr{B}$, $P \not\mathrm{I}\, l$, $P' \mathrm{I}\, l$, and $P \sim P'$. We obtain $b_P\alpha = (t + 1)st$, and so α divides $st(t + 1)$.

Let P and P' be two non-collinear points. Then the number of lines l for which $P \mathrm{I}\, l$ and $[P', l] = \alpha$ is given by μ/α. Consequently α divides μ.

Consider again two non-collinear points P and P'. Let β be the number of lines l, with $P \not\mathrm{I}\, l$, $P' \not\mathrm{I}\, l$, $[P, l] = \alpha$, and $[P', l] = 0$. Now we count in different ways the number of ordered pairs (l, l'), with $l, l' \in \mathscr{B}$, $P \not\mathrm{I}\, l$, $P' \not\mathrm{I}\, l$, $P \mathrm{I}\, l'$, $l \sim l'$, and $[P', l] = 0$. Since there are μ/α lines l' with $P \mathrm{I}\, l'$ and $[P', l'] = \alpha$, we obtain

$$\begin{aligned}\beta\alpha &= (t + 1 - \mu/\alpha)s(t - \mu/\alpha) + (\mu/\alpha)(s - \alpha)(t + 1 - \mu/\alpha)\\ &= ((t + 1)\alpha - \mu)(st - \mu)/\alpha.\end{aligned}$$

Since α divides μ and also $st(t + 1)$, we see that α^2 divides μst.

Now consider distinct collinear points P and P'. Then the number of lines l, with $P \not\mathrm{I}\, l$, $P' \not\mathrm{I}\, l$, $[P, l] = \alpha$, and $[P', l] = 0$, is given by

$$t(s + 1 - \alpha)(t + 1 - \mu/\alpha)/\alpha.$$

Since α divides μ, $st(t + 1)$, and μst, we see that α^2 divides $t(\alpha(t + 1) - \mu)$.

Let P and P' be non-collinear points. The number of lines l with $P \mathrm{I}\, l$ and $[P', l] = \alpha$ equals μ/α. Hence $\mu/\alpha \le t + 1$. Equality occurs if and only if $\mathscr{S}$ is a partial geometry. Let $P \mathrm{I}\, m$, $P' \mathrm{I}\, m'$, $m \sim m'$. Since $[P, m'] = \alpha$ there are at least α lines l with $P \mathrm{I}\, l$ and $[P', l] = \alpha$. Hence $\alpha \le \mu/\alpha$. □

If $\alpha = s + 1$ then $\mathscr{S}$ is a $2 - (v, s + 1, 1)$ design.

Theorem 26.7.4: *If $\mathscr{S} = (\mathscr{P}, \mathscr{B}, \mathrm{I})$ is a semi-partial geometry with parameters s, t, α, and μ, with $\alpha \ne s + 1$, then*

(i) $D = (t(\alpha - 1) + s - 1 - \mu)^2 + 4((t + 1)s - \mu)$ *is a square, except in the case $s = t = \alpha = \mu = 1$ where $\mathscr{S}$ is the pentagon and $D = 5$;*

(ii) $[2(t + 1)s + (v - 1)(t(\alpha - 1) + s - 1 - \mu + D^{1/2})]/(2D^{1/2})$ *is an integer.*

Proof: If $\mathscr{E} = \{\{P, P'\} | P, P' \in \mathscr{P}$ and $P \sim P'\}$, then $(\mathscr{P}, \mathscr{E})$ is a strongly regular graph with parameters $v = 1 + (1 + t)s(1 + t(s - \alpha + 1)/\mu)$, k (or $n_1) = st + s$, λ (or $p^1_{11}) = t(\alpha - 1) + s - 1$, μ (or $p^2_{11}) = \mu$. The graph $(\mathscr{P}, \mathscr{E})$ is called the *point graph* of the semi-partial geometry. Let $\mathscr{P} = \{P_1, P_2, \ldots, P_v\}$ and let $A = [a_{ij}]$ be the $v \times v$ matrix over $\mathbf{R}$ for which $a_{ij} = 0$ if $i = j$ or $P_i \nsim P_j$, and $a_{ij} = 1$ if $i \ne j$ and $P_i \sim P_j$; that is, A is an adjacency matrix of the graph $(\mathscr{P}, \mathscr{E})$.

If $A^2 = [c_{ij}]$, then (a) $c_{ii} = (t + 1)s$; (b) $i \ne j$ and $P_i \nsim P_j$ imply $c_{ij} = \mu$; (c) $i \ne j$ and $P_i \sim P_j$ imply $c_{ij} = t(\alpha - 1) + s - 1$. So

$$A^2 - (t(\alpha - 1) + s - 1 - \mu)A - (s(t + 1) - \mu)I = \mu J; \qquad (26.18)$$

as before, I is the $v \times v$ identity matrix and J is the $v \times v$ matrix with each entry equal to one. Evidently $(t+1)s$ is an eigenvalue of A, and J has eigenvalues $0, v$ with multiplicities $v-1, 1$ respectively. Since

$$((t+1)s)^2 - (t(\alpha-1)+s-1-\mu)(t+1)s - (s(t+1)-\mu) = \mu v,$$

the eigenvalue $(t+1)s$ of A corresponds to the eigenvalue v of J, and so $(t+1)s$ has multiplicity 1. The other eigenvalues of A are roots of the equation

$$x^2 - (t(\alpha-1)+s-1-\mu)x - (s(t+1)-\mu) = 0. \tag{26.19}$$

Denote the multiplicities of these eigenvalues θ_1, θ_2 by m_1, m_2 respectively.

Put

$$D = (t(\alpha-1)+s-1-\mu)^2 + 4((t+1)s-\mu).$$

If $D = 0$, then from $\mu \le (t+1)\alpha \le (t+1)s$ it follows that $t(\alpha-1)+s-1-\mu = 0$ and $(t+1)s-\mu = 0$, and so $t(\alpha-(s+1)) = 1$, a contradiction. Hence $D \neq 0$, and so $\theta_1 \neq \theta_2$. From (26.19),

$$\theta_1 = (t(\alpha-1)+s-1-\mu+D^{1/2})/2,$$

and

$$\theta_2 = (t(\alpha-1)+s-1-\mu-D^{1/2})/2.$$

Since $1+m_1+m_2 = v$ and $s(t+1)+m_1\theta_1+m_2\theta_2 = \operatorname{tr} A = 0$,

$$m_2 = \frac{2(t+1)s+(v-1)(t(\alpha-1)+s-1-\mu+D^{1/2})}{2D^{1/2}}.$$

It follows that

$$\frac{2(t+1)s+(v-1)(t(\alpha-1)+s-1-\mu+D^{1/2})}{2D^{1/2}}$$

is an integer.

Now suppose that D is not a square. Then, as m_2 is an integer,

$$2(t+1)s+(v-1)(t(\alpha-1)+s-1-\mu) = 0. \tag{26.20}$$

Since $v-1 > (t+1)s$, we have

$$0 < \mu - t(\alpha-1) - s + 1 < 2.$$

Hence $\mu = s + t(\alpha-1)$. From (26.20),

$$v = 2(t+1)s + 1. \tag{26.21}$$

By Theorem 26.7.2,

$$v = 1 + (1+t)s\left(1 + \frac{t(s-\alpha+1)}{s+t(\alpha-1)}\right). \tag{26.22}$$

From (26.21) and (26.22) it follows that

$$\alpha = 1 + s(t - 1)/(2t). \tag{26.23}$$

Hence

$$\mu = s(t + 1)/2. \tag{26.24}$$

From (26.23), $2t$ divides $s(t - 1)$; so t divides s. Since $\mu \leq \alpha(t + 1)$, by (26.23) and (26.24) we have $s \leq 2t$. Since t divides s, we necessarily have $s \in \{t, 2t\}$.

First, suppose $s = 2t$. Then $\alpha = t$ and $\mu = t(t + 1)$. But then $D = (1 + 2t)^2$, a contradiction since D is not a square.

So $s = t$, $\alpha = (t + 1)/2$, and $\mu = t(t + 1)/2$. By Theorem 26.7.3, α^2 divides μst, and so $t + 1$ divides $2t^3$; hence $t = 1$. Consequently we have $s = t = \alpha = \mu = 1$, $v = b = 5$, and $D = 5$. This means that $\mathcal{S}$ is the pentagon. □

Theorem 26.7.5: *If $\mathcal{S}$ is a semi-partial geometry, but not a partial geometry, then $b \geq v$.*

Proof: Let $\mathcal{S} = (\mathcal{P}, \mathcal{B}, \mathrm{I})$ be a semi-partial geometry with parameters s, t, α, and μ. Assume that $\mathcal{S}$ is not a partial geometry; so $\mu < \alpha(t + 1)$.

Let $\mathcal{P} = \{P_1, P_2, \ldots, P_v\}$ and $\mathcal{B} = \{l_1, l_2, \ldots, l_b\}$. The corresponding adjacency matrix of the point graph of $\mathcal{S}$ is denoted by $A = [a_{ij}]$. Let $M = [m_{ij}]$ be the $v \times b$ matrix over $\mathbf{R}$ for which $m_{ij} = 0$ if $P_i \not\mathrm{I}\, l_j$, and $m_{ij} = 1$ if $P_i \mathrm{I}\, l_j$; that is, M is an incidence matrix of the geometry $\mathcal{S}$.

It is straightforward that

$$MM^* = (t + 1)I + A. \tag{26.25}$$

Suppose that $v > b$. We have rank $M \leq b$; hence rank $MM^* \leq b$. Therefore the $v \times v$ matrix MM^* is singular, whence $-(t + 1)$ is an eigenvalue of A. Then by (26.19),

$$(t + 1)^2 + (t(\alpha - 1) + s - 1 - \mu)(t + 1) - (s(t + 1) - \mu) = 0.$$

This equation is equivalent to $t(\mu - \alpha(t + 1)) = 0$. Hence $\mu = \alpha(t + 1)$, and so $\mathcal{S}$ is a partial geometry, a contradiction.

Hence, if $\mathcal{S}$ is not a partial geometry, then $b \geq v$. □

Theorem 27.7.6: *The dual of a semi-partial geometry $\mathcal{S}$ is a semi-partial geometry if and only if $\mathcal{S}$ is a partial geometry or $v = b$. If $v = b$, then $\mathcal{S}$ and its dual $\bar{\mathcal{S}}$ have the same parameters.*

Proof: Let $\mathcal{S} = (\mathcal{P}, \mathcal{B}, \mathrm{I})$ be a semi-partial geometry, and let $\bar{\mathcal{S}}$ be the dual geometry of $\mathcal{S}$.

Suppose that $\bar{\mathcal{S}}$ is also a semi-partial geometry. If $\mathcal{S}$ is not a partial geometry then $\bar{\mathcal{S}}$ is not a partial geometry, and by Theorem 26.7.5 we have $b \geq v$ and $v \geq b$. Hence $\mathcal{S}$ is a partial geometry or $v = b$.

If $\mathcal{S}$ is a partial geometry, then $\bar{\mathcal{S}}$ is a partial geometry and consequently also a semi-partial geometry. Now suppose that $\mathcal{S}$ is not a partial geometry but let $b = v$. We use the notation introduced in the proof of the preceding theorem. Since $\mu \neq \alpha(t+1)$ we know that $-(t+1)$ is not an eigenvalue of A. Hence MM^* is not singular. Consequently the $v \times v$ matrix M is not singular. By (26.25),

$$M^* = (s+1)M^{-1} + M^{-1}A$$

(since $b = v$ we have $s = t$). Hence

$$M^*M = (s+1)I + M^{-1}AM.$$

Let $M^{-1}AM = B = [b_{ij}]$. Then $b_{ii} + s + 1$ is the number of points incident with l_i; so $b_{ii} = 0$. Further, for $i \neq j$, b_{ij} is the number of points incident with l_i and l_j. Hence B is an adjacency matrix of the point graph of $\bar{\mathcal{S}}$. Also $B^2 = M^{-1}A^2M$; hence by (26.18),

$$B^2 = (s\alpha - 1 - \mu)B + (s(s+1) - \mu)I + \mu M^{-1}JM.$$

Since $MJ = JM = (s+1)J$, it follows that

$$B^2 = (s\alpha - 1 - \mu)B + (s(s+1) - \mu)I + \mu J. \qquad (26.26)$$

Let $B^2 = [d_{ij}]$. By (26.26), for $l_i \nsim l_j$ we have $d_{ij} = \mu$. But

$$d_{ij} = \sum_{r=1}^{v} b_{ir}b_{jr},$$

whence d_{ij} is the number of lines l_r with $l_i \sim l_r \sim l_j$. It follows that $\bar{\mathcal{S}}$ is also a semi-partial geometry, with the same parameters as $\mathcal{S}$. □

Now we list some of the many known $(0, \alpha)$-geometries and semi-partial geometries.

Examples: (a) Let $\mathcal{O}$ be an ovaloid of $PG(3, q)$, $q > 2$, or an elliptic quadric of $PG(3, 2)$. Suppose that $PG(3, q)$ is a prime of $PG(4, q)$. Points of $\mathcal{S}$ are the points of $PG(4, q)\backslash PG(3, q)$. Lines of $\mathcal{S}$ are the lines of $PG(4, q)$ which contain a point of $\mathcal{O}$ but are not contained in $PG(3, q)$. Here I is the incidence of $PG(4, q)$. Then $\mathcal{S}$ is a partial quadrangle with parameters

$$s = q - 1, \quad t = q^2, \quad \alpha = 1, \quad \text{and} \quad \mu = q^2 - q.$$

(b) Consider a subplane $PG(2, q)$ of the projective plane $PG(2, q^2)$. Suppose that $PG(2, q^2)$ is a plane of $PG(3, q^2)$. Points of $\mathcal{S}$ are the points of $PG(3, q^2)\backslash PG(2, q^2)$. Lines of $\mathcal{S}$ are the lines of $PG(3, q^2)$ which contain a point of $PG(2, q)$ but are not contained in $PG(2, q^2)$. Here I is the incidence of $PG(3, q^2)$. Then $\mathcal{S}$ is a semi-partial geometry with parameters

$$s = q^2 - 1, \quad t = q(q+1), \quad \alpha = q, \quad \text{and} \quad \mu = q(q+1).$$

(c) Consider a Hermitian arc $\mathscr{U}$ of the plane $PG(2, q^2)$. Proceeding as in example (b) we obtain a semi-partial geometry with parameters

$$s = q^2 - 1, \quad t = q^3, \quad \alpha = q, \quad \text{and} \quad \mu = q^2(q^2 - 1).$$

(d) Let Π_{n-2} be an $(n - 2)$-dimensional subspace of $PG(n, q)$, $n \geq 3$. The points of $\mathscr{S}$ are the lines of $PG(n, q)$ which have no point in common with Π_{n-2}. The lines of $\mathscr{S}$ are the planes of $PG(n, q)$ which have exactly one point in common with Π_{n-2}. Incidence is here inclusion. Then $\mathscr{S}$ is a semi-partial geometry with parameters

$$s = q^2 - 1, \quad t = q^{n-2} + q^{n-3} + \cdots + q, \quad \alpha = q, \quad \text{and} \quad \mu = q(q + 1).$$

(e) Let $\mathscr{P}$ be the set of all lines of $PG(n, q)$, $n \geq 3$, let $\mathscr{B}$ be the set of all planes of $PG(n, q)$, and let I be inclusion. Then $\mathscr{S} = (\mathscr{P}, \mathscr{B}, \mathrm{I})$ is a semi-partial geometry with parameters

$$s = q(q + 1), \quad t = q^{n-2} + q^{n-3} + \cdots + q, \quad \alpha = q + 1, \quad \text{and} \quad \mu = (q + 1)^2.$$

(f) Let $\mathscr{U}$ be a Hermitian variety of $PG(3, q^2)$. Points of $\mathscr{S}$ are the points of $PG(3, q^2)\backslash\mathscr{U}$. Lines of $\mathscr{S}$ are the 1-secants of $\mathscr{U}$. Incidence is the natural one. Then $\mathscr{S}$ is a semi-partial geometry with parameters

$$s = q^2 - 1, \quad t = q^3, \quad \alpha = q + 1, \quad \text{and} \quad \mu = q(q + 1)(q^2 - 1).$$

(g) Let $\mathscr{Q}$ be an elliptic quadric of the projective space $PG(5, q)$. Let $P \notin \mathscr{Q}$, and let Π_4 be a prime not containing P. The projection of $\mathscr{Q}$ from P onto Π_4 is denoted by $\mathscr{Q}'$. Further, let $\mathscr{Q}''$ be the set of all points P' of $\mathscr{Q}'$ for which PP' is a 1-secant of $\mathscr{Q}$. If q is odd, then $\mathscr{Q}''$ is a non-singular quadric of Π_4; if q is even, then $\mathscr{Q}''$ is a solid of Π_4. Let $\mathscr{P} = \mathscr{Q}'\backslash\mathscr{Q}''$, let $\mathscr{B}$ be the set of all lines of Π_4 which are contained in $\mathscr{Q}'$ but not in $\mathscr{Q}''$, and let I be the incidence of Π_4. Then $\mathscr{S} = (\mathscr{P}, \mathscr{B}, \mathrm{I})$ is a semi-partial geometry with parameters

$$s = q - 1, \quad t = q^2, \quad \alpha = 2, \quad \text{and} \quad \mu = 2q(q - 1).$$

(For q even, see also §23.6.)

(h) Let $\mathscr{U}$ be a Hermitian variety of $PG(3, q^2)$ and let l be a fixed line of $\mathscr{U}$. Points of $\mathscr{S}$ are the points of $\mathscr{U}\backslash l$, lines of $\mathscr{S}$ are the lines of $\mathscr{U}$ having no point in common with l, and incidence is containment. Then the dual of $\mathscr{S}$ is a partial quadrangle with parameters

$$s = q - 1, \quad t = q^2, \quad \alpha = 1, \quad \text{and} \quad \mu = q^2 - q.$$

(i) Let $\mathscr{V}$ be a set with h elements ($h \geq 4$), let $\mathscr{V}_2 = \{T \subset \mathscr{V} \text{ with } |T| = 2\}$, and let $\mathscr{V}_3 = \{T \subset \mathscr{V} \text{ with } |T| = 3\}$. If I is inclusion, then $\mathscr{S}_h = (\mathscr{V}_2, \mathscr{V}_3, \mathrm{I})$ is a semi-partial geometry with

$$s = \alpha = 2, \quad t = m - 3, \quad \text{and} \quad \mu = 4.$$

(j) Let M be an $(n+1) \times (n+1)$ skew-symmetric matrix over $GF(q)$, $n \geq 2$. Then rank M is even; let rank $M = 2k$, with $k > 0$. The (possibly singular) polarity ζ of $PG(n, q)$ defined by M is called a *null polarity*. The subspace of $PG(n, q)$ consisting of all points having no image with respect to ζ is called the *radical* of ζ and is denoted by $\mathcal{R}$; it has dimension $n - 2k$.

The points of $\mathcal{S}$ are the points of $PG(n, q)\backslash\mathcal{R}$. The lines of $\mathcal{S}$ are the lines l of $PG(n, q)\backslash\mathcal{R}$ for which $l \not\subset l\zeta$ if $n \geq 3$, and $l\zeta \notin l$ if $n = 2$. The incidence is the natural one. Then $\mathcal{S} = (\mathcal{P}, \mathcal{B}, \mathrm{I})$ is a $(0, \alpha)$-geometry with parameters

$$s = \alpha = q, \quad t = q^{n-1} - 1.$$

This geometry $\mathcal{S}$ is also denoted by $\overline{W(n, 2k, q)}$.

If $k = 1$, then $\overline{W(n, 2, q)}$ is the dual net H_q^n introduced in §26.5.

If $2k = n + 1$, so n is odd, then the null polarity ζ is non-singular. In this case the geometry $\overline{W(n, n+1, q)}$ is a semi-partial geometry with $\mu = q^{n-1}(q-1)$, and is also denoted by $\overline{W(n, q)}$.

In all other cases $\overline{W(n, 2k, q)}$ is not a semi-partial geometry; that is, it is a *proper* $(0, \alpha)$-geometry.

(k) Take a quadric $\mathcal{Q}$ in $PG(n, 2)$, $n \geq 3$. Suppose that $\mathcal{Q}$ does not consist of one or two primes, that $\mathcal{Q}$ is not the hyperbolic quadric of $PG(3, 2)$, and that $\mathcal{Q}$ is not the cone with vertex $PG(n-4, 2)$, $n \geq 4$, and base a hyperbolic quadric of a solid skew to $PG(n-4, 2)$. Let $\mathcal{B}$ be the set of lines skew to $\mathcal{Q}$, let $\mathcal{P}$ be the set of points of $PG(n, 2)$ on at least one line of $\mathcal{B}$, and let I be the incidence of $PG(n, 2)$. Then $\mathcal{S} = (\mathcal{P}, \mathcal{B}, \mathrm{I})$ is a $(0, 2)$-geometry.

If $n = 2d - 1$ and $\mathcal{Q}$ is (non-singular) elliptic, then $\mathcal{S}$ is a semi-partial geometry, denoted by $NQ^-(2d-1, 2)$, with parameters

$$s = \alpha = 2, \quad t = 2^{2d-3} + 2^{d-2} - 1, \quad \text{and} \quad \mu = 2^{2d-3} + 2^{d-1}.$$

If $n = 2d - 1$ and $\mathcal{Q}$ is (non-singular) hyperbolic, then $\mathcal{S}$ is a semi-partial geometry, denoted by $NQ^+(2d-1, 2)$, with parameters

$$s = \alpha = 2, \quad t = 2^{2d-3} - 2^{d-2} - 1, \quad \text{and} \quad \mu = 2^{2d-3} - 2^{d-1}.$$

If $n = 2d$ and $\mathcal{Q}$ is (non-singular) parabolic, then $\mathcal{S}$ is a semi-partial geometry, denoted by $NQ(2d, 2)$, with parameters

$$s = \alpha = 2, \quad t = 2^{2d-2} - 1, \quad \text{and} \quad \mu = 2^{2d-1} - 2^{2d-2}.$$

In all other cases $\mathcal{S}$ is a proper $(0, 2)$-geometry.

(l) Let $\mathcal{Q}$ be a hyperbolic quadric of $PG(3, 2^h)$, $h \geq 2$. Let $\mathcal{B}$ be the set of lines skew to $\mathcal{Q}$, let $\mathcal{P}$ be the set of points of $PG(3, 2^h)\backslash\mathcal{Q}$, and let I be the incidence of $PG(3, 2^h)$. Then $\mathcal{S} = (\mathcal{P}, \mathcal{B}, \mathrm{I})$ is a proper $(0, 2^{h-1})$-geometry $NQ^+(3, 2^h)$ with

$$s = 2^h \quad \text{and} \quad t = 2^{2h-1} - 2^{h-1} - 1.$$

Also for the following parameter sets at least one semi-partial geometry is known:

$$s = 1, \quad t = r - 1, \quad \alpha = 1, \quad \text{and} \quad \mu = 1, \quad \text{for } r = 2, 3, 7;$$

$$s = t = \alpha = 6 \quad \text{and} \quad \mu = 36;$$

$$s = 1, \quad t = 9, \quad \alpha = 1, \quad \text{and} \quad \mu = 2;$$

$$s = 1, \quad t = 15, \quad \alpha = 1, \quad \text{and} \quad \mu = 4;$$

$$s = 1, \quad t = 21, \quad \alpha = 1, \quad \text{and} \quad \mu = 6;$$

$$s = 2, \quad t = 10, \quad \alpha = 1, \quad \text{and} \quad \mu = 2;$$

$$s = 2, \quad t = 55, \quad \alpha = 1, \quad \text{and} \quad \mu = 20;$$

$$s = q^{m+1} - 1, \quad t = q^{m+2}, \quad \alpha = q^m, \quad \text{and} \quad \mu = q^{m+1}(q^{m+1} - 1),$$

with q any prime power and $m \geq 2$.

26.8 Embedded $(0, \alpha)$-geometries and semi-partial geometries

A *projective* $(0, \alpha)$-geometry or semi-partial geometry $\mathcal{S} = (\mathcal{P}, \mathcal{B}, \mathrm{I})$ is a $(0, \alpha)$-geometry or semi-partial geometry whose point set $\mathcal{P}$ is a subset of the point set of some $PG(n, q)$ and whose line set $\mathcal{B}$ is a set of lines of $PG(n, q)$; one also says that $\mathcal{S}$ is *embedded* in $PG(n, q)$. If $PG(n', q)$ is the subspace of $PG(n, q)$ generated by the points of $\mathcal{P}$, then $PG(n', q)$ is the *ambient space* of $\mathcal{S}$.

A $(0, \alpha)$-geometry or semi-partial geometry embedded in $PG(2, q)$ is a partial geometry. As the latter were classified in §26.6, the dimension of the ambient space is henceforth taken to be at least three.

Let $\mathcal{S} = (\mathcal{P}, \mathcal{B}, \mathrm{I})$ be a $(0, \alpha)$-geometry or a semi-partial geometry with $\alpha > 1$, with parameters s, t, α which is projective with ambient space $PG(3, s)$. Consider distinct concurrent lines l, l' of $\mathcal{S}$, and let π be the plane defined by l and l'. The points and lines of $\mathcal{S}$ in π constitute a partial geometry $\mathcal{S}(\pi)$ with parameters $s(\pi) = s$, $t(\pi) = \alpha - 1$, $\alpha(\pi) = \alpha$, together with $m(\pi)$ *isolated points.* A point P of $\mathcal{S}$ in π is *isolated* if and only if no line of $\mathcal{S}$ through P is contained in π. By Theorem 26.6.1 there are the following possibilities for $\mathcal{S}(\pi)$:

(a) $\alpha = s + 1$ and $\mathcal{S}(\pi)$ is the $2 - (s^2 + s + 1, s + 1, 1)$ design formed by all points and all lines of π;

(b) the points of π not in $\mathcal{S}(\pi)$ form a maximal $(sd - s + d; d)$-arc $\mathcal{K}(\pi)$ of π, with $d = s/\alpha$ and $2 \leq d < s$, and the lines of $\mathcal{S}(\pi)$ are the lines of π having empty intersection with $\mathcal{K}(\pi)$;

(c) $\alpha = s$, π contains exactly one point $P(\pi)$ which is not in $\mathcal{S}(\pi)$, and the lines of $\mathcal{S}(\pi)$ are the lines of π which do not contain $P(\pi)$.

Lemma 26.8.1: *The number $m(\pi)$ of isolated points in π is independent of the choice of π. The number of lines of $\mathscr{S}$, where $m = m(\pi)$, is*

$$b = \frac{(s\alpha - s + \alpha)((s+1)(t+1) - \alpha s)}{\alpha} + m(t+1), \tag{26.27}$$

and the number of points of $\mathscr{S}$ is equal to

$$v = \frac{(s+1)(s\alpha - s + \alpha)((s+1)(t+1) - \alpha s)}{\alpha(t+1)} + m(s+1). \tag{26.28}$$

Proof: The number of points of $\mathscr{S}(\pi)$ is equal to $(s+1)(s\alpha - s + \alpha)/\alpha$ and the number of lines of $\mathscr{S}(\pi)$ is equal to $s\alpha - s + \alpha$. The number of lines of $\mathscr{S}$ containing exactly one point of $\mathscr{S}(\pi)$ is equal to $(t + 1 - \alpha)(s+1)(s\alpha - s + \alpha)/\alpha$, and the number of lines of $\mathscr{S}$ containing an isolated point in π is equal to $(t+1)m(\pi)$. Consequently

$$|\mathscr{B}| = b = s\alpha - s + \alpha + \frac{(t+1-\alpha)(s+1)(s\alpha - s + \alpha)}{\alpha} + (t+1)m(\pi),$$

and so

$$b = \frac{(s\alpha - s + \alpha)((s+1)(t+1) - \alpha s)}{\alpha} + (t+1)m(\pi).$$

It immediately follows that $m(\pi) = m$ is independent of the choice of π.

Counting in different ways the number of ordered pairs (P, l), with $P \in \mathscr{P}$, $l \in \mathscr{B}$, and $P \mathrm{I} l$, we obtain $|\mathscr{P}|(t+1) = |\mathscr{B}|(s+1)$. Hence

$$|\mathscr{P}| = v = \frac{(s+1)(s\alpha - s + \alpha)((s+1)(t+1) - \alpha s)}{\alpha(t+1)} + (s+1)m. \quad \square$$

Lemma 26.8.2: *With respect to $\mathscr{S}$ three types of planes are possible in the ambient space $PG(3, s)$:*

(a) *those containing $s\alpha - s + \alpha$ lines of $\mathscr{S}$ and*

$$\rho_a = \frac{(s+1)(s\alpha - s + \alpha)}{\alpha} + m \tag{26.29}$$

points of $\mathscr{S}$;

(b) *those containing exactly one line of $\mathscr{S}$ and*

$$\rho_b = s + 1 + \frac{s(s+1)(t+1-\alpha)(\alpha - 1)}{\alpha(t+1)} + m \tag{26.30}$$

points of $\mathscr{S}$;

(c) *those containing no line of $\mathscr{S}$ and*

$$\rho_c = \frac{(s\alpha - s + \alpha)((s+1)(t+1) - \alpha s)}{\alpha(t+1)} + m \tag{26.31}$$

points of $\mathscr{S}$.

Proof: Let π be a plane of the ambient space $PG(3, s)$.

If π contains at least two lines of $\mathcal{S}$, then the points and lines of $\mathcal{S}$ in π constitute a partial geometry $\mathcal{S}(\pi)$ with parameters $s(\pi) = s$, $t(\pi) = \alpha - 1$, $\alpha(\pi) = \alpha$, together with m isolated points. Hence π contains $s\alpha - s + \alpha$ lines of $\mathcal{S}$ and

$$\rho_a = \frac{(s + 1)(s\alpha - s + \alpha)}{\alpha} + m$$

points of $\mathcal{S}$.

Next, suppose that π contains exactly one line of $\mathcal{S}$. If ρ_b is the number of points of $\mathcal{S}$ in π, then

$$b = 1 + (s + 1)t + (\rho_b - s - 1)(t + 1).$$

So, by Lemma 26.8.1,

$$\rho_b = s + 1 + \frac{s(s + 1)(t + 1 - \alpha)(\alpha - 1)}{\alpha(t + 1)} + m.$$

Finally, suppose that π contains no line of $\mathcal{S}$. If ρ_c is the number of points of $\mathcal{S}$ in π, then

$$b = \rho_c(t + 1).$$

Again, by Lemma 26.8.1,

$$\rho_c = \frac{(s\alpha - s + \alpha)((s + 1)(t + 1) - \alpha s)}{\alpha(t + 1)} + m. \quad \square$$

Corollary : *If there is at least one plane of type* (b) *and at least one plane of type* (c), *then* $t + 1$ *divides* s.

Proof: From the theorem $(t + 1)(\rho_b - \rho_c) = s$. Hence, the result follows. $\square$

Theorem 26.8.3: *If* $\mathcal{S} = (\mathcal{P}, \mathcal{B}, \mathrm{I})$ *is a semi-partial geometry with parameters* s, t, α, μ *which is projective with ambient space* $PG(3, s)$, *then one of the following holds*:

(a) $\alpha = s + 1$ *and* $\mathcal{S}$ *is the* $2 - ((s^2 + 1)(s + 1), s + 1, 1)$ *design formed by all points and all lines of* $PG(3, s)$;

(b) $\alpha = 1$ *and* $\mathcal{S}$ *is a classical generalized quadrangle*;

(c) $\alpha = s$ *and* $\mathcal{S} = \overline{H_s^3}$;

(d) $\alpha = s$ *and* $\mathcal{S} = W(3, s)$;

(e) $\alpha = s = 2$ *and* $\mathcal{S} = NQ^-(3, 2)$.

Proof: If $\mathcal{S}$ is a partial geometry, then by Theorem 26.6.1 we have one of (a), (b), (c).

From now on we assume that $\mathcal{S}$ is not a partial geometry. Then $\mu < (t + 1)\alpha$, and by Theorems 26.7.3 and 26.7.5 we have $\alpha^2 \leq \mu$ and

$|\mathscr{B}| = b \geq |\mathscr{P}| = v$. So $b(s+1) = v(t+1)$ implies that $t \geq s$. By Theorem 26.7.3, α divides μ, and so $\mu < (t+1)\alpha$ implies $\mu \leq \alpha t$.

Let $\alpha = 1$. Then by Theorem 26.7.2,

$$v = 1 + (1+t)s(1 + ts/\mu).$$

Hence $v \leq s^3 + s^2 + s + 1$ implies $\mu(s^2 + s - t) \geq st(t+1)$. Since $\mu \leq t$, it follows that $st(t+1) \leq t(s^2 + s - t)$ and so $t \leq s - 1 + 1/(s+1) < s$, a contradiction. Hence $\alpha \neq 1$.

Next, let $\alpha = s + 1$. From the connectedness of $\mathscr{S}$ it follows that $\mathscr{S}$ is a $2-(v, s+1, 1)$ design, a contradiction since $\mathscr{S}$ is not a partial geometry. Hence $\alpha \neq s + 1$.

Assume that there exists a plane π of type (c). The number of points of $\mathscr{S}$ in π is equal to $\rho_c = v/(s+1)$. Let P be a fixed point of $\mathscr{S}$ in π. Now we count in different ways the number of ordered pairs (P', P''), with $P', P'' \in \mathscr{P}\backslash\{P\}$, $P'' \in \pi$, and $PP', P'P'' \in \mathscr{B}$. We obtain $(\rho_c - 1)\mu = (t+1)st$. Hence $\rho_c = 1 + (t+1)st/\mu$. From these two expressions for ρ_c, it follows that

$$v = s + 1 + (s+1)(t+1)st/\mu. \tag{26.32}$$

By Theorem 26.7.2 we have

$$v = 1 + (1+t)s(1 + t(s - \alpha + 1)/\mu). \tag{26.33}$$

From (26.32) and (26.33) it follows that $\mu = \alpha(t+1)$, a contradiction. Consequently there are no planes of type (c).

By (26.28) we have

$$m = \frac{v}{s+1} - \frac{(s\alpha - s + \alpha)((s+1)(t+1) - \alpha s)}{\alpha(t+1)}, \tag{26.34}$$

with v given by (26.33).

Let $P, P' \in \mathscr{P}$ with $P \neq P'$ and $PP' \notin \mathscr{B}$. Further let π be a plane containing PP'. If π is of type (b), or if π is of type (a) with at least one of P, P' isolated in π, then π does not contain a point $P'' \in \mathscr{P}\backslash\{P, P'\}$ with PP'' and $P'P''$ belonging to $\mathscr{B}$. If π is of type (a) and neither P nor P' is isolated in π, then π contains exactly α^2 points $P'' \in \mathscr{P}\backslash\{P, P'\}$ for which PP'' and $P'P''$ are lines of $\mathscr{B}$. Considering all planes through PP' it follows that α^2 divides μ.

Now suppose that all planes of $PG(3, s)$ are of type (a). Counting all points of $\mathscr{P}$ in all planes through a given line l of $\mathscr{B}$, we obtain

$$v = s + 1 + (s+1)(\rho_a - s - 1), \tag{26.35}$$

with ρ_a given by (26.29). Hence

$$m = \frac{v}{s+1} - \frac{(s+1)(s\alpha - s + \alpha)}{\alpha} + s. \tag{26.36}$$

From (26.34) and (26.36) it follows that

$$t = (s + 1)(\alpha - 1). \tag{26.37}$$

Counting in different ways the number of pairs (P, π), with $P \in \mathscr{P}$ and π a plane containing P, we obtain

$$v(s^2 + s + 1) = (s^3 + s^2 + s + 1)\rho_a. \tag{26.38}$$

Eliminating ρ_a from (26.35) and (26.38) gives

$$v = (s^2 + 1)(s + 1),$$

and so $\mathscr{P} = PG(3, s)$. Now from Theorem 26.7.2 and (26.37),

$$\mu = (\alpha - 1)(s\alpha - s + \alpha).$$

Since α divides μ, it divides s. Let $s = p^h$ and $\alpha = p^r$, with p prime and $r \leq h$. Since α^2 divides μ, so p^{2r} divides $p^{h+r} - p^h + p^r$, whence p^r divides $p^h - p^{h-r} + 1$. Hence $h = r$, which means that $s = \alpha$. Consequently we have

$$s = \alpha, \quad t = \alpha^2 - 1, \quad \mu = \alpha^2(\alpha - 1), \quad m = 1.$$

So any plane π contains exactly one point $P(\pi)$ not in $\mathscr{P}$, and the lines of $\mathscr{B}$ in π are the lines of π not containing $P(\pi)$. Therefore the structure $\mathscr{S}'$ consisting of all points of $PG(3, s)$ and all lines of $PG(3, s)$ not in $\mathscr{B}$ is a projective generalized quadrangle. By Theorem 26.3.29, $\mathscr{S}'$ is a classical generalized quadrangle; so, since $PG(3, s)$ is the point set of $\mathscr{S}'$, the structure $\mathscr{S}'$ is the generalized quadrangle $\mathscr{W}(s)$ arising from a non-singular null polarity of $PG(3, s)$. Thus $\mathscr{S}$ is the semi-partial geometry $\overline{W(3, s)}$.

Next, we suppose that there is at least one plane π of type (b). Let l be the unique line of $\mathscr{B}$ in π. Fix a point P in l, and count in different ways the number of ordered pairs (P', P''), with $P', P'' \in \mathscr{P}\backslash\{P\}$, $P'' \in \pi\backslash l$, $P' \notin \pi$, and $PP', P'P'' \in \mathscr{B}$. We obtain

$$(\rho_b - s - 1)\mu = ts(t + 1 - \alpha). \tag{26.39}$$

Now fix a point P_1 in $\pi\backslash l$, and count in different ways the number of ordered pairs (P'_1, P''_1), with $P'_1, P''_1 \in \mathscr{P}\backslash\{P_1\}$, $P''_1 \in \pi\backslash\{P_1\}$, $P'_1 \notin \pi$, and $P_1P'_1, P'_1P''_1 \in \mathscr{B}$. We obtain

$$(\rho_b - 1)\mu = (t + 1)st. \tag{26.40}$$

From (26.39) and (26.40) it follows that $\mu = \alpha t$. Hence

$$\rho_b = \frac{(t + 1)s}{\alpha} + 1,$$

and from Theorem 26.7.2,

$$v = 1 + (1 + t)s(s + 1)/\alpha.$$

Eliminating t gives

$$v = 1 + (\rho_b - 1)(s + 1). \tag{26.41}$$

Counting in different ways the number of pairs (P^*, l^*), with $P^* \in \mathscr{P} \cap \pi$, $l^* \in \mathscr{B}\backslash\{l\}$, and $P^* \in l^*$, we obtain

$$b - 1 = (s + 1)t + (\rho_b - s - 1)(t + 1).$$

Since $v(t + 1) = b(s + 1)$ it follows that

$$v = \frac{(s + 1)(st + t + 1)}{t + 1} + (\rho_b - s - 1)(s + 1). \tag{26.42}$$

Now, from (26.41) and (26.42) it is deduced that $s = t$. By Theorem 26.7.4 either $D = 1 + 4s(s + 1 - \alpha)$ is a square or $D = 5$. Since $1 < \alpha < s + 1$, we have $1 + 4s(s + 1 - \alpha) > 5$, and so D is a square. Consequently there exists a positive integer g for which $s(s + 1 - \alpha) = g(g + 1)$. As s is a prime power, it divides either g or $g + 1$. Hence

$$g + 1 \geq s > s + 1 - \alpha \geq g.$$

It follows that $s + 1 - \alpha = g$ and $s = g + 1$, and so $\alpha = 2$. Therefore $s = t$, $\alpha = 2$, $\mu = 2t$, and $D = (2s - 1)^2$. Also, since α^2 divides μ, we have $t = s = 2^h$. By Theorem 26.7.4, $2\sqrt{D}$ divides

$$2(t + 1)s + (v - 1)(t(\alpha - 1) + s - 1 - \mu + \sqrt{D}),$$

and so $2^{h+1} - 1$ divides

$$(2^h + 1)(2^{2h} + 1)2^{h-1}.$$

Hence $h \in \{1, 3\}$. So either $s = t = \alpha = 2$ and $\mu = 4$, or $s = t = 8$, $\alpha = 2$, and $\mu = 16$.

First, let $s = t = \alpha = 2$ and $\mu = 4$. Then $v = 10$; by (26.34), $m = 0$; by (26.29), $\rho_a = 6$; and by (26.30), $\rho_b = 4$. Since there are no planes of type (c), no three points of $PG(3, 2)\backslash\mathscr{P}$ are collinear. Hence $PG(3, 2)\backslash\mathscr{P}$ is a 5-cap of $PG(3, 2)$. Since $\rho_a = 6$ and $\rho_b = 4$, any plane of $PG(3, 2)$ contains either one or three points of $PG(3, 2)\backslash\mathscr{P}$. Now it is clear that the complement of $\mathscr{P}$ is an elliptic quadric $\mathscr{Q}$ of $PG(3, 2)$ and that $\mathscr{B}$ consists of the 10 external lines of $\mathscr{Q}$. So we have proved that $\mathscr{S} = NQ^-(3, 2)$.

Finally, let $s = t = 8$, $\alpha = 2$, and $\mu = 16$. Then $v = 325$; by (26.34), $m = 0$; by (26.29), $\rho_a = 45$; and by (26.30), $\rho_b = 37$. Let π be a plane of type (b), let l be the line of $\mathscr{B}$ in π, and let $\mathscr{K}$ be the set of the 28 points of $\mathscr{P}$ in $\pi\backslash l$. If P, P' are distinct points of $\mathscr{K}$, then from $\mu > 0$ it follows that there is at least one plane π' of type (a) through PP'. Since π' is of type (a) and $m = 0$, the line PP' contains five points of $\pi' \cap \mathscr{P}$. Since PP' contains exactly one point of l, it contains four points of $\mathscr{K}$. Hence $\mathscr{K}$ is a maximal $(28; 4)$-arc of π. Therefore each line of $PG(3, 8)$ has 1, 5, or 9 points in common with $\mathscr{P}$. In fact, by §19.4, $\mathscr{P}$ is a non-singular $28_{5,3,8}$ of $PG(3, 8)$. Then, by Theorems 19.4.8 and 19.4.9, we have $v = (8^3/2) + 8^2 + 8 + 1 = 329$, contradicting $v = 325$. We conclude that the case $s = t = 8$, $\alpha = 2$, and $\mu = 16$ cannot occur. □

Theorem 26.8.4: *Let $\mathscr{S} = (\mathscr{P}, \mathscr{B}, \mathrm{I})$ be a $(0, \alpha)$-geometry with parameters s, t, α which is projective with ambient space $PG(3, s)$. If $m = 0$, then one of the following holds:*

(a) *$\alpha = s + 1$ and $\mathscr{S}$ is the $2 - ((s^2 + 1)(s + 1), s + 1, 1)$ design formed by all points and all lines of $PG(3, s)$;*
(b) *$\alpha = s$ and $\mathscr{S} = H_s^3$;*
(c) *$\alpha = s = 2$ and $\mathscr{S} = NQ^-(3, 2)$.*

Proof: By the definition of a $(0, \alpha)$-geometry we have $\alpha > 1$.

First, suppose that all lines of some plane π of type (a) belong to $\mathscr{B}$; then $\alpha = s + 1$. It follows that all lines of each plane of type (a) belong to $\mathscr{B}$. Let l be a line of $\mathscr{B}$ not contained in π. If $\{P\} = l \cap \pi$ and if $l_1, l_2, \ldots, l_{s+1}$ are the lines of π through P, then all lines of ll_i through P belong to $\mathscr{B}$, $i = 1, 2, \ldots, s + 1$; hence $t = s^2 + s$. By (26.27) we now have $b = (s^2 + 1)(s^2 + s + 1)$ and, by (26.28), $v = (s^2 + 1)(s + 1)$. So $\mathscr{S}$ is the $2 - ((s^2 + 1)(s + 1), s + 1, 1)$ design formed by all points and all lines of $PG(3, s)$.

Now suppose that in each plane of type (a) there is at least one line not in $\mathscr{B}$. Let π be a plane of type (a) and let l be a line of π which does not belong to $\mathscr{B}$. Since $m = 0$ we have $|l \cap \mathscr{P}| = (s\alpha - s + \alpha)/\alpha$. Let τ be the number of planes of type (a) through l. The number of lines of $\mathscr{B}$ containing a point of l and contained in a plane of type (a) through l is $\tau(s\alpha - s + \alpha)$. Hence the number of planes of type (b) through l is

$$(t + 1)(s\alpha - s + \alpha)/\alpha - \tau(s\alpha - s + \alpha) = (s\alpha - s + \alpha)(t + 1 - \tau\alpha)/\alpha. \qquad (26.43)$$

By (26.43) we have $t + 1 \geq \tau\alpha$. Since the number of planes of type (b) through l is at most $s + 1 - \tau$, we have

$$(s\alpha - s + \alpha)(t + 1 - \tau\alpha)/\alpha \leq s + 1 - \tau. \qquad (26.44)$$

The inequality (26.44) is equivalent to

$$t + 1 - \tau\alpha \leq 1 + \frac{1}{\alpha - 1} - \frac{t + 1}{(\alpha - 1)(s + 1)}. \qquad (26.45)$$

Hence $0 \leq t + 1 - \tau\alpha \leq 1$, and so either $t + 1 = \tau\alpha$ or $t = \tau\alpha$.

Let $t + 1 = \tau\alpha$. Then the number of planes of type (b) through l is zero. By way of contradiction assume that $\mathscr{S}$ contains at least one plane π' of type (b). Let l' be the line of $\mathscr{B}$ in π'. Further, let π'' be a plane of type (a) not containing l' and let $\pi'' \cap \pi' = l''$. Since $l'' \notin \mathscr{B}$ it is contained in no plane of type (b), a contradiction since π' is of type (b). Consequently there are no planes of type (b). It follows that $\mathscr{S}$ is a partial geometry and now, by Theorem 26.6.1, $\alpha = s$ and $\mathscr{S} = H_s^3$.

Next, let $t = \tau\alpha$. Then by (26.44) we have $s \geq t$. Let π be a plane of type (a) and let P be any point of $\mathscr{S}$ in π. Now count in different ways the number η of ordered pairs (π', l'), with π' a plane of type (b) having its line of $\mathscr{S}$ through P and with $l' = \pi \cap \pi'$ not in $\mathscr{B}$. Since there are $t/(\alpha - 1)$ planes of type (a) through a given line of $\mathscr{S}$, and consequently $s+1 - t/(\alpha - 1)$ planes of type (b) through that line, we have

$$\eta = (t + 1 - \alpha)\left(s + 1 - \frac{t}{\alpha - 1}\right). \tag{26.46}$$

For a given line $l' \notin \mathscr{B}$ of π through P, the number of lines l'' of $\mathscr{B}$ through P for which $l'l''$ is of type (b) is $t + 1 - \tau\alpha = 1$. Hence

$$\eta = s + 1 - \alpha. \tag{26.47}$$

From (26.46) and (26.47) it follows that

$$s + 1 - \alpha = (t + 1 - \alpha)\left(s + 1 - \frac{t}{\alpha - 1}\right). \tag{26.48}$$

Since $s \geq t$,

$$s + 1 - \alpha \geq (t + 1 - \alpha)\left(s + 1 - \frac{s}{\alpha - 1}\right),$$

which is equivalent to

$$t + 1 - \alpha \leq 1 + \frac{\alpha - \alpha^2 + s}{\alpha s + \alpha - 2s - 1}. \tag{26.49}$$

Since $t = \tau\alpha$ we have $t + 1 - \alpha \geq 1$, and so $\alpha - \alpha^2 + s \geq 0$. An easy calculation shows that $(\alpha - \alpha^2 + s)/(\alpha s + \alpha - 2s - 1)$ is at least one if and only if $\alpha = 2$ and $s \geq 3$. This means that $t = \alpha$ whenever $\alpha \neq 2$. But since there are $t/(\alpha - 1)$ planes of type (a) through a given line of $\mathscr{S}$, the case $\alpha \neq 2$ and $t = \alpha$ cannot occur. So we always have $\alpha = 2$. Then (26.48) becomes $(s - t)(t - 2) = 0$. Consequently $s = t$ or $t = 2$.

First, suppose that $\alpha = 2$ and $s = t$. Since $m = 0$, the dual of $\mathscr{S}$ is a semi-partial geometry $\mathscr{S}^*$ with parameters

$$s^* = t, \quad t^* = s, \quad \alpha^* = \alpha, \quad \text{and} \quad \mu^* = \alpha t/(\alpha - 1).$$

Here $s^* = t^*$ and so, by Theorem 26.7.6, $\mathscr{S}$ is also a semi-partial geometry. Then, by Theorem 26.8.3 and since $s = t$ and $\alpha = 2$, we have $\mathscr{S} = NQ^-(3, 2)$.

Finally assume that $\alpha = 2$ and $t = 2$; then $\tau = 1$. Let π be a plane of type (a) and let l be a line of π which does not belong to $\mathscr{B}$. It was shown that the number of planes of type (b) through l is $1 + s/2$; hence $s = 2^h$. If there is at least one plane of type (c), then, by the corollary of Theorem 26.8.2, 2^h is divisible by three, a contradiction. So there are no planes of type (c). It follows that each of the planes through l is either of type (a) or of type (b), and so $2 + s/2 = s + 1$, implying $s = 2$. Again $s = t$ and $\alpha = 2$, and now by the previous paragraph $\mathscr{S} = NQ^-(3, 2)$. □

Theorem 26.8.5: *Let $\mathscr{S} = (\mathscr{P}, \mathscr{B}, \mathrm{I})$ be a $(0, \alpha)$-geometry with parameters s, t, α which is projective with ambient space $PG(3, s)$. If $m \neq 0$, then there is no plane of type* (b).

Proof: Suppose that there is at least one plane of each type.

The total number of planes of type (a) is equal to

$$\frac{bt}{(\alpha - 1)(s\alpha - s + \alpha)}. \tag{26.50}$$

By (26.27) the number (26.50) becomes

$$\left(\frac{\delta((s + 1)(t + 1) - \alpha s)}{\alpha} + m(t + 1)\right)\frac{t}{(\alpha - 1)\delta}, \tag{26.51}$$

with $\delta = s\alpha - s + \alpha$. The total number of planes of type (b) is equal to

$$b\left(s + 1 - \frac{t}{\alpha - 1}\right). \tag{26.52}$$

By (26.27), the number (26.52) becomes

$$\left(\frac{\delta((s + 1)(t + 1) - \alpha s)}{\alpha} + m(t + 1)\right)\left(s + 1 - \frac{t}{\alpha - 1}\right). \tag{26.53}$$

The total number of planes of type (a) and (b) is less than or equal to $(s^2 + 1)(s + 1)$, and so, adding (26.51) and (26.53), we obtain

$$\left(\frac{\delta((s+1)(t+1) - \alpha s)}{\alpha} + m(t+1)\right)\left(s+1 - \frac{t}{\alpha - 1} + \frac{t}{(\alpha - 1)\delta}\right) \leq (s^2 + 1)(s + 1);$$

that is,

$$\left(\frac{\delta((s + 1)(t + 1) - \alpha s)}{\alpha} + m(t + 1)\right)(s + 1)(\delta - t)/\delta \leq (s^2 + 1)(s + 1). \tag{26.54}$$

The inequality (26.54) is equivalent to

$$\frac{((s + 1)(t + 1) - \alpha s)}{\alpha}(s\alpha - s + \alpha - t) + m(t + 1)\left(1 - \frac{t}{s\alpha - s + \alpha}\right) \leq s^2 + 1. \tag{26.55}$$

If l is a line of the plane π of type (a) and l does not belong to $\mathscr{B}$, then $|l \cap \mathscr{P}| = (s\alpha - s + \alpha)/\alpha$ (since $m \neq 0$ such a line l exists). Hence α divides s. By the corollary of Theorem 26.8.2, $t + 1$ divides s. Since $\alpha \leq t + 1$ and s is a prime power, it follows that α divides $t + 1$.

By assumption there is at least one plane of type (b), and so by (26.52) we have

$$t < (\alpha - 1)(s + 1). \tag{26.56}$$

Hence

$$m(t + 1)\left(1 - \frac{t}{s\alpha - s + \alpha}\right) > \frac{m(t + 1)}{s\alpha - s + \alpha} > 0. \tag{26.57}$$

By (26.55) and (26.57),

$$\frac{((s + 1)(t + 1) - \alpha s)}{\alpha}(s\alpha - s + \alpha - t) < s^2 + 1,$$

which is seen to be equivalent to

$$\alpha + \alpha s - t - 1 - \frac{\alpha s(\alpha + s)}{t + s} + \frac{\alpha s(t + 1)}{(s + 1)(t + s)} < 0. \tag{26.58}$$

From (26.58) it follows that

$$t + 1 + \frac{\alpha s(\alpha + s)}{t + s} > \alpha s,$$

or equivalently

$$t(t + 1) + s(t + \alpha^2 - \alpha t + 1) > 0. \tag{26.59}$$

We have $t + \alpha^2 - \alpha t + 1 < 0$ if and only if $t > \alpha + (\alpha + 1)/(\alpha - 1)$. Let l, l', l'' be distinct non-coplanar lines of $\mathscr{B}$ through P in $\mathscr{P}$. The α lines of $\mathscr{B}$ in ll' through P are denoted by $l = l_1$, $l' = l_2, l_3, \ldots, l_\alpha$. Counting the lines of $\mathscr{B}$ through P in the α planes $l''l_i$, with $i = 1, 2, \ldots, \alpha$, we obtain $t \geq \alpha^2 - \alpha$. If $\alpha \geq 3$ then

$$\alpha^2 - \alpha > \alpha + \frac{\alpha + 1}{\alpha - 1}.$$

Hence for $\alpha \geq 3$ we have $t + \alpha^2 - \alpha t + 1 < 0$ and so, by (26.59),

$$s < \frac{t(t + 1)}{\alpha t - \alpha^2 - t - 1}. \tag{26.60}$$

Since $t + 1$ divides s, we have $t + 1 \leq s$, and consequently from (26.60) it follows that

$$(\alpha - 2)t < \alpha^2 + 1. \tag{26.61}$$

In a previous paragraph we showed that α divides $t+1$; so $t \geq \alpha^2 - \alpha$ implies $t \geq \alpha^2 - 1$. Now from (26.61) it follows that $(\alpha - 2)(\alpha^2 - 1) < \alpha^2 + 1$, and therefore $\alpha \leq 3$. So we conclude that $\alpha \in \{2, 3\}$.

Suppose that $\alpha = 3$. By (26.61) we have $t < 10$, and since $t \geq \alpha^2 - \alpha$ we have $t \geq 6$. But α and $t + 1$ divide s, and so $t = 8$. Now by (26.59) we have $s < 12$, and so $s = 9$. This contradicts the inequality (26.58).

Next we assume that $\alpha = 2$ and $t = 3$. By (26.58) we then have $s < 3$, a contradiction since $t + 1$ divides s.

Finally, we assume that $\alpha = 2$ and $t \neq 3$. Since α divides $t + 1$, we have $t > 3$. The integer α divides s; so $s = 2^h$. As $t + 1$ divides s, we have $t \geq 7$. Putting $s = (t + 1)r$, the inequality (26.59) implies $(r - 1)(t - 5) < 5$. Since $t \geq 7$ and $r = 2^{h'}$, it follows that $r \in \{1, 2\}$. If $r = 1$, so $s = t + 1$; then (26.58) gives $t(t - 1) < 4$, contradicting $t \geq 7$. If $r = 2$, so $s = 2(t + 1)$; then (26.59) gives $t < 10$. Hence $s = 16$ and $t = 7$. Now again (26.58) yields a contradiction.

Thus planes of type (b) and (c) cannot both occur.

Suppose that there are no planes of type (c), but at least one plane of type (b). Then by (26.52) we have

$$t < (\alpha - 1)(s + 1). \tag{26.62}$$

The total number of planes of type (a) and planes of type (c) is $(s^2 + 1)(s + 1)$. Hence by (26.50) and (26.52),

$$b\left(1 - \frac{t}{s\alpha - s + \alpha}\right) = s^2 + 1. \tag{26.63}$$

From (26.63) and (26.27) it follows that

$$m(t + 1) = \frac{t(s^2 + 1)}{s\alpha - s + \alpha - t} + 1 - (s + 1)s(t + 1 - \alpha) \\ + (s + 1)(t + 1)\left(\frac{s}{\alpha} - 1\right). \tag{26.64}$$

If all points of $PG(3, s)$ are elements of $\mathscr{P}$, then $\rho_{\mathrm{a}} = s^2 + s + 1$, and so by (26.29), $m = (s + 1 - \alpha)(s/\alpha)$. Since $v = (s^2 + 1)(s + 1)$, (26.28) gives $t = (s + 1)(\alpha - 1)$, contradicting (26.62). Hence $PG(3, s)$ contains at least one point which does not belong to $\mathscr{P}$.

Let $P \in PG(3, s)\backslash\mathscr{P}$, and let τ_{b} denote the number of planes of type (b) through P. Counting in different ways the number of pairs (π, l), with $l \in \mathscr{B}$ and $\pi = lP$, we obtain

$$b = \tau_{\mathrm{b}} + (s^2 + s + 1 - \tau_{\mathrm{b}})(s\alpha - s + \alpha),$$

and so, by (26.27),

$$m(t + 1) = -\tau_{\mathrm{b}}(s + 1)(\alpha - 1) + (s\alpha - s + \alpha)(s + 1)\left(s - \frac{t + 1}{\alpha} + 1\right). \tag{26.65}$$

By (26.65), $s+1$ divides $m(t+1)$ and now, by (26.64), $s+1$ divides $1+[t(s^2+1)/(s\alpha-s+\alpha-t)]$. It follows that $s+1$ divides $t+1$. Since $t+1 \leq (s+1)(\alpha-1)$ we have

$$\text{either } t+1=(s+1)(\alpha-1), \quad \text{or } t+1 \leq (s+1)(\alpha-2). \tag{26.66}$$

Let $P' \in \mathscr{P}$. The number of planes of type (a) containing at least one line of $\mathscr{B}$ through P' is

$$\frac{(t+1)t}{\alpha(\alpha-1)}.$$

The number of planes of type (b) having its line of $\mathscr{B}$ through P' is

$$(t+1)\left(s+1-\frac{t}{\alpha-1}\right).$$

Since P' is contained in exactly s^2+s+1 planes, we have

$$\frac{(t+1)t}{\alpha(\alpha-1)}+(t+1)\left(s+1-\frac{t}{\alpha-1}\right) \leq s^2+s+1.$$

This inequality can be written as

$$t^2-t(\alpha s+\alpha-1)+s^2\alpha \geq 0. \tag{26.67}$$

The corresponding discriminant is equal to $D=(\alpha s+\alpha-1)^2-4s^2\alpha$. One easily verifies that $D<0$ if and only if either $\alpha=3$ and $s>4$, or $\alpha=2$. For $\alpha=s=3$ and $\alpha \geq 4$ we have $D>0$. Since $m \neq 0$ we cannot have $\alpha=s+1=3$, and since $\alpha<s+1$ the integer α divides s and so we cannot have $\alpha=3$ and $s=4$.

Now we consider the following six cases.

Case 1: $t+1 \leq (s+1)(\alpha-2)$, $D>0$, *and* $t \geq (\alpha s+\alpha-1+\sqrt{D})/2$

From these inequalities

$$s\alpha-2s+\alpha-2 \geq t+1 \geq (\alpha s+\alpha+1+\sqrt{D})/2. \tag{26.68}$$

Therefore

$$s^2(4-\alpha)+2s(5-2\alpha)+2(3-\alpha) \geq 0. \tag{26.69}$$

If $\alpha \geq 4$, then $s^2(4-\alpha) \leq 0$, $2s(5-2\alpha)<0$, and $2(3-\alpha)<0$, a contradiction. For $s=\alpha=3$, (26.68) becomes $4 \geq t+1 \geq 9$, again a contradiction.

Consequently $D>0$ implies either $t+1>(s+1)(\alpha-2)$ or $t<(\alpha s+\alpha-1+\sqrt{D})/2$. Recall that, by (26.66), $t+1>(s+1)(\alpha-2)$ is equivalent to $t+1=(s+1)(\alpha-1)$ and, by (26.67), $D>0$ together with $t<(\alpha s+\alpha-1+\sqrt{D})/2$ implies $t \leq (\alpha s+\alpha-1-\sqrt{D})/2$.

Case 2: $D > 0$, $t \le (\alpha s + \alpha - 1 - \sqrt{D})/2$, *and* $t \ge 2s + 1$

From these inequalities

$$4s + 2 \le 2t \le \alpha s + \alpha - 1 - \sqrt{D}. \tag{26.70}$$

Therefore

$$s^2(4 - \alpha) + 3s(2 - \alpha) + 2 - \alpha \ge 0. \tag{26.71}$$

If $\alpha \ge 4$, then $s^2(4 - \alpha) \le 0$, $3s(2 - \alpha) < 0$, and $2 - \alpha < 0$, a contradiction. For $s = \alpha = 3$, (26.71) also yields a contradiction.

Thus $D > 0$ together with $t \le (\alpha s + \alpha - 1 - \sqrt{D})/2$ implies $t < 2s + 1$. Since $s + 1$ divides $t + 1$, $t < 2s + 1$ is equivalent to $t = s$.

Case 3: $D > 0$, $t \le (\alpha s + \alpha - 1 - \sqrt{D})/2$, *and* $s = t$

By (26.64) we have

$$m(s + 1) = \frac{s(s^2 + 1)}{s\alpha - 2s + \alpha} + 1 - (s + 1)s(s + 1 - \alpha) + (s + 1)^2\left(\frac{s}{\alpha} - 1\right). \tag{26.72}$$

This is equivalent to

$$m\alpha(s\alpha - 2s + \alpha) = -s(\alpha - 2)((\alpha - 1)(s - \alpha)(s + 1) + s\alpha). \tag{26.73}$$

Since $D > 0$ we have $\alpha > 2$. Hence $m\alpha(s\alpha - 2s + \alpha) > 0$ and $-s(\alpha - 2)((\alpha - 1)(s - \alpha)(s + 1) + s\alpha) < 0$, contradicting (26.73).

Case 4: $D > 0$ *and* $t + 1 = (s + 1)(\alpha - 1)$

Counting the number of planes of type (a) through a line of $\mathscr{B}$ we deduce that $\alpha - 1$ divides t. Since $t + 1 = (s + 1)(\alpha - 1)$, so $\alpha - 1$ also divides $t + 1$. Hence $\alpha = 2$, contradicting $D > 0$.

Case 5: $D < 0$ *and* $\alpha = 2$

By (26.62) we have $t < s + 1$. Since $s + 1$ divides $t + 1$, we necessarily have $s = t$. Then from (26.64) it follows that $m = 0$, a contradiction.

Case 6: $D < 0$, $\alpha = 3$, *and* $s > 4$

By (26.62) we have $t < 2s + 2$. Since $s + 1$ divides $t + 1$, we necessarily have either $s = t$ or $t = 2s + 1$. First, let $s = t$. Then (26.64) is equivalent to (26.73), giving $3m(s + 3) = -s(2(s - 3)(s + 1) + 3s)$. Since $3m(s + 3) > 0$ and

$-s(2(s-3)(s+1)+3s) < 0$, we again have a contradiction. Finally, let $t = 2s+1$. Now (26.64) becomes $2m = -(2s^2+5s+3)/6$. So $m < 0$, again a contradiction.

We conclude that there are no planes of type (b). □

Corollary 1: *Let $\mathcal{S} = (\mathcal{P}, \mathcal{B}, \mathrm{I})$ be a $(0, \alpha)$-geometry with parameters s, t, α which is projective with ambient space $PG(3, s)$. If $m \neq 0$, then $t = (\alpha - 1)(s+1)$; if also there are no planes of type* (c), *then $m = s(s-\alpha+1)/\alpha$ and $\mathcal{P} = PG(3, s)$.*

Proof: Suppose $m \neq 0$. By Theorem 26.8.5 there are no planes of type (b). Now from (26.52) it follows that $t = (\alpha - 1)(s+1)$.

If there are also no planes of type (c), then every plane is of type (a); so, from (26.50),

$$\frac{bt}{(\alpha-1)(s\alpha - s + \alpha)} = (s^2+1)(s+1).$$

So $b = (s^2+1)(s\alpha - s + \alpha)$, and $v = (s^2+1)(s+1)$. Now, by (26.28), $m = s(s-\alpha+1)/\alpha$. □

Corollary 2: *Let $\mathcal{S} = (\mathcal{P}, \mathcal{B}, \mathrm{I})$ be a $(0, \alpha)$-geometry with parameters s, t, α which is projective with ambient space $PG(3, s)$. If there is at least one plane of type* (b), *then $s = \alpha = 2$ and $\mathcal{S} = NQ^-(3, 2)$.*

Proof: By Theorem 26.8.5 we have $m = 0$. Now by Theorem 26.8.4, $\mathcal{S}$ is either the design formed by all points and all lines of $PG(3, s)$, or $\mathcal{S} = H_s^3$, or $\mathcal{S} = NQ^-(3, 2)$. But only $NQ^-(3, 2)$ admits planes of type (b). □

Theorem 26.8.6: *Let $\mathcal{S} = (\mathcal{P}, \mathcal{B}, \mathrm{I})$ be a projective $(0, \alpha)$-geometry with ambient space $PG(3, s)$ and $m = 1$. Then one of the following holds:*

(a) *$\alpha = 2$ and $\mathcal{S} = \overline{W(3, s)}$;*
(b) *$\alpha = s/2$, $s = 2^h$, and $\mathcal{S} = NQ^+(3, s)$.*

Proof: Let $\alpha \neq s$. We count the number τ of lines of $PG(3, s)$ containing exactly $(s\alpha - s + \alpha)/\alpha$ points of $\mathcal{P}$. In a plane of type (a) there are $s^2 + s + 1 - (s\alpha - s + \alpha) - (s+1) = (s-\alpha)(s+1)$ such lines. By (26.50) and Corollary 1 of Theorem 26.8.5, there are $b(s+1)/(s\alpha - s + \alpha)$ planes of type (a). Now by (26.27) the number of planes of type (a) is equal to

$$(s+1)\left((s+1)\left(s - \frac{s}{\alpha}\right) + 2\right).$$

If a line $l \notin \mathcal{B}$ of $PG(3, s)$ contains ρ points of $\mathcal{P}$, then l is contained in

$\rho(t+1)/(s\alpha - s + \alpha) = \rho$ planes of type (a). It follows that

$$\tau = (s-\alpha)(s+1)^2\left((s+1)\left(s-\frac{s}{\alpha}\right)+2\right)\alpha/(s\alpha - s+\alpha). \qquad (26.74)$$

Hence $s\alpha - s + \alpha$ divides $(s-\alpha)(s+1)^2((s+1)s(\alpha-1)+2\alpha)$, so divides $(s-\alpha)(s+1)^2\alpha(s-1) = (s^2 - \alpha s + s - \alpha)(\alpha s + \alpha)(s-1)$, and therefore divides $s^3(s-1)$. Let $s = p^h$ with p a prime. Then $\alpha = p^k$, with $0 < k < h$. Consequently $p^{h+k} - p^h + p^k$ divides $p^{3h}(p^h - 1)$, and so $p^h - p^{h-k} + 1$ divides $p^{3h-k}(p^h - 1)$. Since $(p^h - p^{h-k} + 1, p^{3h-k}) = 1$, so $p^h - p^{h-k} + 1$ divides $p^h - 1$; therefore $p^h - p^{h-k} + 1$ divides $p^{h-k} - 2$. If $p^{h-k} \neq 2$, then $p^h - p^{h-k} + 1 \leq p^{h-k} - 2$, implying $p^h - 2p^{h-k} < 0$; so $\alpha = p^k < 2$, a contradiction. Hence $p^{h-k} = 2$, which means that $p = 2$ and $\alpha = s/2$. We conclude that either $\alpha = s$ or $\alpha = s/2$.

Assume that $\alpha = s$; then $t = s^2 - 1$. By (26.27) and (26.28) we have $b = s^4 + s^2$ and $v = s^3 + s^2 + s + 1$. Hence $\mathscr{P} = PG(3, s)$. By (26.50) the number of planes of type (a) is $(s^2+1)(s+1)$. Consequently there are no planes of type (c). In any plane π of type (a) there is exactly one point P which does not belong to $\mathscr{P}$, and the lines of π not in $\mathscr{B}$ form the pencil of π through P. Now by Theorem 15.2.13 the lines of $PG(3, s)$ which do not belong to $\mathscr{B}$ are the lines of a general linear complex. Hence $\mathscr{S} = \overline{W(3, s)}$.

Next, assume that $\alpha = s/2$; then $t = (s-2)(s+1)/2$. By (26.27) and (26.28) we have $b = s^2(s-1)^2/2$ and $v = s(s^2-1)$. A plane of type (a) contains s^2 points of $\mathscr{P}$ and $s(s-1)/2$ lines of $\mathscr{B}$, and a plane of type (c) contains $s(s-1)$ points of $\mathscr{P}$. By (26.50) there are $s(s^2-1)$ planes of type (a), and so $(s+1)^2$ planes of type (c). Recall that if a line $l \notin \mathscr{B}$ contains ρ points of $\mathscr{P}$, then l is in ρ planes of type (a). Hence if $l \notin \mathscr{B}$ contains at least one point of $\mathscr{P}$, then l is in at least one plane of type (a). For any line l of a plane π of type (a) we have $|l \cap \mathscr{P}| \in \{s-1, s, s+1\}$. Hence for any line l of $PG(3, s)$ we have $|l \cap \mathscr{P}| \in \{0, s-1, s, s+1\}$. Let $\mathscr{P}' = PG(3, s)\backslash\mathscr{P}$. Then $|\mathscr{P}'| = (s+1)^2$, and any line with at least three points in $\mathscr{P}'$ lies entirely in $\mathscr{P}'$. Now by Theorem 16.2.2, $\mathscr{P}'$ is a hyperbolic quadric or consists of a plane and a line or consists of lines joining an oval to a vertex. Clearly $\mathscr{P}'$ does not contain a plane. If $\mathscr{P}'$ consists of lines joining an oval to a vertex, then there are planes through the vertex which contain exactly $s^2 + s$ points of $\mathscr{P}$. Such planes cannot be of type (a) nor of type (c), a contradiction. Hence $\mathscr{P}'$ is a hyperbolic quadric. Clearly no line of $\mathscr{B}$ has a point in common with $\mathscr{P}'$. Since $b = s^2(s-1)^2/2$, the set $\mathscr{B}$ consists of all lines having an empty intersection with the quadric $\mathscr{P}'$. We conclude that $\mathscr{S} = NQ^+(3, s)$. □

Theorem 26.8.7: *For any projective* $(0, \alpha)$*-geometry* $\mathscr{S} = (\mathscr{P}, \mathscr{B}, \mathrm{I})$ *with ambient space* $PG(3, s)$ *the number* m *of isolated points satisfies* $m \neq 2$.

Proof: Suppose that $m = 2$.

We count the number τ of lines of $PG(3, s)$ containing exactly $(s\alpha - s + 3\alpha)/\alpha$ points of $\mathscr{P}$. In any plane of type (a) there is exactly one such line. By (26.50) and Corollary 1 of Theorem 26.8.5, there are $b(s+1)/(s\alpha - s + \alpha)$ planes of type (a). Now by (26.27) the number of planes of type (a) is $(s+1) \times (s(s+1)(\alpha - 1) + 3\alpha)/\alpha$. Recall that if a line l not in $\mathscr{B}$ contains ρ points of $\mathscr{P}$, then l is in ρ planes of type (a). Hence

$$\tau = (s+1)(s(s+1)(\alpha - 1) + 3\alpha)/(s\alpha - s + 3\alpha).$$

Consequently $s\alpha - s + 3\alpha$ divides $(s+1)(s(s+1)(\alpha - 1) + 3\alpha)$, and so $s - (s/\alpha) + 3$ divides $(s - (s/\alpha) + 3 + (s/\alpha) - 2)\{(s - (s/\alpha) + 3)(s+1) - 3(s+1) + 3\}$; hence $s - (s/\alpha) + 3$ divides $3s((s/\alpha) - 2)$. Let $s = p^h$ with p a prime; then $\alpha = p^k$, with $0 < k \leq h$. If $\alpha = s$, then for any plane of type (a) the union of its $s\alpha - s + \alpha = s^2$ lines of $\mathscr{B}$ is a set of order $s^2 + s$; so $m \leq 1$, a contradiction. Hence $k < h$. We have shown that $p^h - p^{h-k} + 3$ divides $3p^h(p^{h-k} - 2)$.

First, let $p = 3$. Then $3^{h-1} - 3^{h-k-1} + 1$ divides $3^h(3^{h-k} - 2)$. Suppose that $h \neq k + 1$. Then $(3^{h-1} - 3^{h-k-1} + 1, 3^h) = 1$ and so $3^{h-1} - 3^{h-k-1} + 1$ divides $3^{h-k} - 2$. Hence $3^{h-1} - 3^{h-k-1} + 1 \leq 3^{h-k} - 2$, and so $3^{h-1} - 3^{h-k-1} - 3^{h-k} < 0$; therefore $3^k - 1 - 3 < 0$, whence $k = 1$ and $\alpha = 3$. Thus, either $s/\alpha = 3$ or $\alpha = 3$. Now consider a plane π of type (a). The lines of $\mathscr{B}$ in π form either a maximal $(2s + 3; 3)$-arc, or a maximal $(s(s-2)/3; s/3)$-arc. By Theorem 12.2.5 and its corollary we necessarily have $h = 1$, a contradiction.

Next, let $p \neq 3$. Since $(p^h - p^{h-k} + 3, p^h) = 1$, the integer $p^h - p^{h-k} + 3$ divides $3(p^{h-k} - 2)$. If $p^{h-k} = 2$, then $p = 2$ and $h = k + 1$. Now let $p^{h-k} \neq 2$. Then $p^h - p^{h-k} + 3 \leq 3(p^{h-k} - 2)$, and so $p^h - 4p^{h-k} + 9 \leq 0$, whence $p^h - 4p^{h-k} < 0$. Hence $p^k < 4$, implying $p = 2$ and $k = 1$. Consequently $2^{h-1} + 3$ divides $3(2^{h-1} - 2)$; so $2^{h-1} + 3$ divides $2^{h-1} - 2$. Hence $h = 2$, and $p^{h-k} = 2$, a contradiction. We conclude that $p = 2$ and $h = k + 1$.

Let π be a plane of type (a) and let P be a point of π not in $\mathscr{P}$. Since $s/\alpha = 2$ and $m = 2$, the set $\pi \backslash \mathscr{P}$ is an s-arc $\mathscr{K}$ of π. However, $\mathscr{K}$ contains P, and $\mathscr{K}$ together with the two isolated points of π forms an oval of π. It follows that π contains exactly two lines of $\mathscr{B}$ through P having s points in common with $\mathscr{P}$. The number of planes of type (a) through P is $b/(s\alpha - s + \alpha) = s^2 - s + 1$. Therefore the number of lines through P having exactly s points in $\mathscr{P}$ is $(s^2 - s + 1)2/s$. Hence s divides 2; so $s = 2$ and $\alpha = 1$, a contradiction. □

Conjecture: *If $\mathscr{S}$ is a projective $(0, \alpha)$-geometry with ambient space $PG(3, s)$, then it is one of the following types:*

(a) $m = 0$ *and $\mathscr{S}$ is either the design formed by all points and all lines of $PG(3, s)$, or $\mathscr{S} = H_s^3$, or $\mathscr{S} = NQ^-(3, 2)$;*

(b) $m = 1$ *and either $\mathscr{S} = W(3, s)$, or s is even and $\mathscr{S} = NQ^+(3, s)$.*

Lemma 26.8.8: *Suppose that $\mathcal{S} = (\mathcal{P}, \mathcal{B}, \mathrm{I})$ is a projective $(0, \alpha)$-geometry with ambient space $PG(n, s)$, $n \geq 3$. If P is any point of $\mathcal{P}$, then the $t + 1$ lines of $\mathcal{B}$ through P do not lie in the same prime of $PG(n, s)$.*

Proof: Suppose that the $t + 1$ lines of $\mathcal{B}$ through P are contained in the prime Π_{n-1}. If l is any of these lines, then a point P' on l, $P' \neq P$, lies on $t + 1$ lines $l, l_1, l_2, \ldots, l_t$ of $\mathcal{B}$. On $l_i \backslash \{P'\}$ there are $\alpha - 1$ (≥ 1) points which are joined to P by a line of $\mathcal{B}$. It follows that l_i is contained in Π_{n-1}, $i = 1, 2, \ldots, t$. Since $\mathcal{S}$ is connected, repeated application of this argument shows that $\mathcal{S}$ is contained in Π_{n-1}, a contradiction. □

Suppose that $(\mathcal{P}, \mathcal{B}, \mathrm{I})$ is a projective $(0, \alpha)$-geometry with ambient space $PG(n, s)$, $n \geq 3$. Let $P \in \mathcal{P}$ and let $l_1, l_2, \ldots, l_r$, $r \geq 2$, be distinct lines of $\mathcal{B}$ through P which generate a $PG(n', s)$, with $2 \leq n' \leq n$. Further, let $\mathcal{S} \cap PG(n', s)$ denote the structure formed by all points and all lines of $\mathcal{S}$ in $PG(n', s)$. By definition, the *connected component* of $\mathcal{S} \cap PG(n', s)$ through P is the structure $\mathcal{S}'$ formed by all elements T of $\mathcal{S} \cap PG(n', s)$, for which there exist elements $T_1, T_2, \ldots, T_u$ of $\mathcal{S} \cap PG(n', s)$ with $T \mathrm{I} T_1 \mathrm{I} \cdots \mathrm{I} T_u \mathrm{I} P$. Then $\mathcal{S}'$ is a projective $(0, \alpha)$-geometry with ambient space $PG(n', s)$. In particular each point of $\mathcal{S}'$ is incident with the same number $t' + 1$, $t' \geq 1$, of lines of $\mathcal{S}'$.

Theorem 26.8.9: *Suppose that $\mathcal{S} = (\mathcal{P}, \mathcal{B}, \mathrm{I})$ is a projective $(0, \alpha)$-geometry with ambient space $PG(n, s)$, with $s > 2$ and $n \geq 4$. Then $\mathcal{S}$ is either the design formed by all points and all lines of $PG(n, s)$, or $\mathcal{S} = \overline{W(n, 2k, s)}$ with $2k \in \{2, 4, \ldots, n + 1\}$.*

Proof: We shall prove that $\alpha = s$ or $\alpha = s + 1$.

(a) $n = 4$. Let P be a point of $\mathcal{S}$. By Lemma 26.8.8 there exist lines l_1, l_2, l_3, l_4 of $\mathcal{B}$ through P which are not contained in a solid. Let $PG(3, s)$ be the solid containing l_1, l_2, and l_3. The connected component $\mathcal{S}'$ of $\mathcal{S} \cap PG(3, s)$ through P is a projective $(0, \alpha)$-geometry with ambient space $PG(3, s)$. Since $s > 2$, by Corollary 2 of Theorem 26.8.5 no plane of $PG(3, s)$ contains exactly one line of $\mathcal{S}'$. Considering all planes of $PG(3, s)$ through l_1 one sees that the parameter t' of $\mathcal{S}'$ is equal to $(\alpha - 1)(s + 1)$. Hence the parameter t of $\mathcal{S}$ satisfies $t > (\alpha - 1)(s + 1)$.

Let l be any line of $\mathcal{S}$ through P, and assume that l is contained in a plane π such that l is the only line of $\mathcal{S}$ in π. The other lines of $\mathcal{B}$ through P are denoted by $l'_1, l'_2, \ldots, l'_t$. The solid defined by π and l'_i is denoted by π_i, $i = 1, 2, \ldots, t$. Clearly P belongs to at least $\alpha - 2$ lines l'_j, $j \neq i$, of π_i. Now suppose that P is on exactly α lines of $\mathcal{S} \cap \pi_i$, for all $i = 1, 2, \ldots, t$. In other words suppose that for all i the lines of $\mathcal{S} \cap \pi_i$ through P lie in a plane. Considering all solids of $PG(4, s)$ through π, we then obtain $t \leq (\alpha - 1)(s + 1)$, a contradiction. Hence there exist lines l'_i, l'_j, with $i \neq j$, such that $\pi_i = \pi_j$ and with l'_i, l'_j, l not coplanar. Since the plane π of π_i contains exactly one line of the connected component of $\mathscr{S} \cap \pi_i$ through P, we have a contradiction by Corollary 2 of Theorem 26.8.5. So we conclude that each plane π through l contains exactly $s\alpha - s + \alpha$ lines of $\mathscr{S}$.

Now let $PG(3, s)$ be a prime of $PG(4, s)$ not containing P. The $t + 1$ lines of $\mathscr{B}$ through P intersect $PG(3, s)$ in the $t + 1$ points $P_0, P_1, \ldots, P_t$. By the preceding paragraph each line of $PG(3, s)$ contains either 0 or α points of the set $\mathscr{V} = \{P_0, P_1, \ldots, P_t\}$. Considering all lines of $PG(3, s)$ through P_0 gives $t = (s^2 + s + 1)(\alpha - 1)$. Let π be any plane of $PG(3, s)$ through P_0. Then $\pi \cap \mathscr{V}$ is a maximal $(s\alpha - s + \alpha; \alpha)$-arc of π. Let $\alpha \neq s + 1$ and let l' be a line of π having no point in common with that arc. Considering all planes of $PG(3, s)$ through l', one sees that $s\alpha - s + \alpha$ divides $t + 1$. Hence $s\alpha - s + \alpha$ divides $s(s\alpha - s + \alpha) - s + \alpha$, and so $s\alpha - s + \alpha$ divides $s - \alpha$. Since $s\alpha - s + \alpha > s - \alpha$ we necessarily have $s = \alpha$. We conclude that either $\alpha = s + 1$ or $\alpha = s$.

(b) $n > 4$. Let P be a point of $\mathscr{S}$. By Lemma 26.8.8 there exist lines l_1, l_2, l_3, l_4 of $\mathscr{B}$ through P which generate a subspace $PG(4, s)$ of $PG(n, s)$. The connected component $\mathscr{S}'$ of $\mathscr{S} \cap PG(4, s)$ through P is a projective $(0, \alpha)$-geometry with ambient space $PG(4, s)$. Now by (a) we necessarily have either $\alpha = s + 1$ or $\alpha = s$.

Suppose that $\mathscr{S}$ is a projective $(0, \alpha)$-geometry with ambient space $PG(n, s)$, with $s > 2$, $\alpha = s + 1$, and $n \geq 4$. Let P be a point of $\mathscr{S}$, and let l and l' be two lines of $\mathscr{S}$ through P. Since $\alpha = s + 1$, every line of the plane ll' is a line of $\mathscr{S}$. It follows that the union of all lines of $\mathscr{S}$ through P is a subspace $PG(n', s)$ of $PG(n, s)$. Now by Lemma 26.8.8 we have $n' = n$. Hence $\mathscr{S}$ is the design formed by all points and all lines of $PG(n, s)$.

Next, suppose that $\mathscr{S}$ is a projective $(0, \alpha)$-geometry with ambient space $PG(n, s)$, with $s > 2$, $\alpha = s$, and $n \geq 4$. We shall prove that $t = s^{n-1} - 1$. Considering all planes containing a given line l of $\mathscr{B}$ we see that this is equivalent to proving that no plane through l contains exactly one line of $\mathscr{B}$. By (a) this holds for $n = 4$, and in this case $t = s^3 - 1$. Let us now proceed by induction on n. For any line l of $\mathscr{S}$, assume that it is contained in a plane π and that no other line of $\mathscr{S}$ is in π. Let $P \in l$ and let the other lines of $\mathscr{B}$ through P be denoted by $l_1, l_2, \ldots, l_t$. The solid through π and l_i is denoted by π_i, $i = 1, 2, \ldots, t$. Clearly P belongs to at least $\alpha - 2$ lines l_j, $j \neq i$, of π_i. Now suppose that P is on exactly α lines of $\mathscr{S} \cap \pi_i$, for all $i = 1, 2, \ldots, t$. Considering all solids of $PG(n, s)$ through π, we then obtain $t \leq s^{n-2} - 1$. By Lemma 26.8.8 we may assume that $l, l_1, l_2, \ldots, l_{n-1}$ generate $PG(n, s)$. Let $PG(n - 1, s) = ll_1 \cdots l_{n-2}$ and let $\mathscr{S}'$ be the connected component of $\mathscr{S} \cap PG(n - 1, s)$ through P. By the induction hypothesis the parameter t' of $\mathscr{S}'$ is equal to $s^{n-2} - 1$. Since l_{n-1} does not belong to $PG(n - 1, s)$, we necessarily have $t > s^{n-2} - 1$, contradicting $t \leq s^{n-2} - 1$. Consequently each plane through l contains exactly s^2 lines of $\mathscr{B}$, which is equivalent to $t = s^{n-1} - 1$.

Let P be any point of $\mathscr{S}$ and let $l_0, l_1, \ldots, l_t$ be the lines of $\mathscr{B}$ through P. Further, let $PG(n-1, s)$ be a prime not through P and let $l_i \cap PG(n-1, s) = \{P_i\}$, $i = 0, 1, \ldots, t$. By the preceding paragraph any line of $PG(n-1, s)$ has 0 or s points in common with the set $\mathscr{V} = \{P_0, P_1, \ldots, P_t\}$. Hence $\mathscr{V}$ is the complement of a prime of $PG(n-1, s)$. Therefore the union of the lines $l_0, l_1, \ldots, l_t$ is the complement of a prime of $PG(n, s)$.

Now we consider the incidence structure $\mathscr{S}'$ formed by all the points of $PG(n, s)$ and all the lines of $PG(n, s)$ not in $\mathscr{B}$. By the preceding paragraph the union of all lines of $\mathscr{S}'$ through any point of $PG(n, s)$ is a prime of $PG(n, s)$. It follows immediately that $\mathscr{S}'$ is a projective Shult space. Since $PG(n, s)$ is the point set of $\mathscr{S}'$, we necessarily have either case (b) or case (e) of Theorem 26.3.30. First, suppose we are in case (b). Since $PG(n, s)$ is the point set of $\mathscr{S}'$ and since the union of all lines of $\mathscr{S}'$ through any point of $PG(n, s)$ is a prime, we then have (with the notation of Theorem 26.3.30) $r = n - 2$ and $k = s + 1$. In that case $\mathscr{S} = H_s^n = \overline{W(n, 2, s)}$. If we are in case (e), then $\mathscr{S} = \overline{W(n, 2k, s)}$ with $2k \in \{4, 6, \ldots, n + 1)$. □

Theorem 26.8.10: *Suppose that $\mathscr{S} = (\mathscr{P}, \mathscr{B}, \mathrm{I})$ is a projective semi-partial geometry with ambient space $PG(n, s)$, $n \geq 4$. If $\alpha > 1$ and $s > 2$, then it is one of the following types:*

(a) $\alpha = s + 1$ *and $\mathscr{S}$ is the design formed by all points and all lines of* $PG(n, s)$;

(b) $\alpha = s$ *and* $\mathscr{S} = H_s^n$;

(c) $\alpha = s$, *n is odd, and* $\mathscr{S} = \overline{W(n, s)}$.

Proof: $\mathscr{S}$ is a projective $(0, \alpha)$-geometry with ambient space $PG(n, s)$, where $n \geq 4$ and $s > 2$. By Theorem 26.8.9, $\mathscr{S}$ is either the design formed by all points and all lines of $PG(n, s)$, or $\mathscr{S} = \overline{W(n, 2k, s)}$ with $2k \in \{2, 4, \ldots, n + 1\}$. In §26.7 it was observed that $\overline{W(n, 2k, s)}$ is a semi-partial geometry if and only if either $2k = n + 1$, in which case n is odd, or $k = 1$. If $k = 1$, then $\mathscr{S} = \overline{W(n, 2, s)} = H_s^n$; if $2k = n + 1$, then $\mathscr{S} = \overline{W(n, n + 1, s)} = \overline{W(n, s)}$. □

Theorem 26.8.11: *Suppose that $\mathscr{S} = (\mathscr{P}, \mathscr{B}, \mathrm{I})$ is a projective dual semi-partial geometry with ambient space $PG(n, s)$, $n \geq 3$ and $\alpha > 1$. Then it is one of the following types:*

(a) $\alpha = s + 1$ *and $\mathscr{S}$ is the design formed by all points and all lines of* $PG(n, s)$;

(b) $\alpha = s$ *and* $\mathscr{S} = H_s^n$;

(c) $\alpha = s = 2$ *and* $\mathscr{S} = NQ^-(3, 2)$.

Proof: For any plane π containing at least two lines of $\mathscr{B}$, let $\bar{\pi}$ be the set of all lines of $\mathscr{B}$ in $\bar{\pi}$; then $|\bar{\pi}| = s\alpha - s + \alpha$. The set of all these sets $\bar{\pi}$ will be denoted by $\bar{\mathscr{P}}$. Now we consider the incidence structure $\bar{\mathscr{S}} = (\bar{\mathscr{P}}, \mathscr{B}, \bar{\mathrm{I}})$, with $\bar{\pi}\,\bar{\mathrm{I}}\,l$ ($\bar{\pi} \in \mathscr{P}$ and $l \in \mathscr{B}$) if and only if $l \in \bar{\pi}$. Then $\bar{\mathscr{S}}$ is a $(0, \bar{\alpha})$-geometry with $\bar{\alpha} = \alpha$, $\bar{t} = (\alpha - 1)(s + 1)$, and $\bar{s} = t/(\alpha - 1) - 1$; two lines of $\mathscr{B}$ are concurrent in $\bar{\mathscr{S}}$ if and only if they are concurrent in $\mathscr{S}$. Hence $\bar{\mathscr{S}}$ is also a dual semi-partial geometry.

First, suppose that $\mathscr{S}$ is a partial geometry. Then by Theorem 26.6.1 we have either case (a) or case (b).

Now suppose that $\mathscr{S}$ is not a partial geometry. By Theorem 26.7.5 we have $s \geq t$. It follows immediately that $\bar{s} < \bar{t}$. Again by Theorem 26.7.5, $\bar{\mathscr{S}}$ is a partial geometry. Consequently for any line $l \in \mathscr{B}$ and any element $\bar{\pi} \in \bar{\mathscr{P}}$, with $l \notin \bar{\pi}$, the line l is concurrent with α lines of $\mathscr{B}$ in π. So for any point $P \in \mathscr{P}$, with $P \notin \pi$, the lines of $\mathscr{B}$ through $\mathscr{P}$ are contained in the solid $P\pi$. By Lemma 26.8.8 we necessarily have $n = 3$. Now assume that $m \neq 0$ and let P' be a point of $\mathscr{P}$ in π which is on none of the $s\alpha - s + \alpha$ lines of $\mathscr{B}$ in π. If $P' \in l'$, with $l' \in \mathscr{B}$, then l' is concurrent with no line of $\mathscr{B}$ in π, a contradiction. Hence $m = 0$. By Theorem 26.8.4, since $\mathscr{S}$ is not a partial geometry, we now have $\mathscr{S} = NQ^-(3, 2)$. $\square$

Remark: The determination of all projective (0, 2)-geometries with ambient space $PG(n, 2)$ is quite complicated. However, this problem has been completely solved, and the main results are given in §26.9.

Open problems: Concerning the determination of all projective $(0, \alpha)$-geometries with ambient space $PG(n, s)$, for given α, n, and s, the following problems are still open:

(a) Do there exist projective $(0, \alpha)$-geometries with ambient space $PG(3, s)$ and $m > 2$? If yes, classify them.

(b) Determine all projective partial quadrangles (here $\alpha = 1$) with ambient space $PG(n, s)$, $n \geq 4$.

(c) Determine all projective dual partial quadrangles (here $\alpha = 1$) with ambient space $PG(n, s)$, $n \geq 3$. An infinite class of projective dual partial quadrangles, which are not generalized quadrangles, was given by example (h), at the end of §26.7.

26.9 Notes and references

§26.1. Theorem 26.1.1 is due to Tits (1974), and Theorem 26.1.2 is due to Buekenhout and Shult (1974). The example (e) for a generalized quadrangle is taken from Hall (1971). Lemma 26.1.3 and Theorem 26.1.4 are from Buekenhout and Shult (1974).

We now state two important characterizations of Grassmann varieties, in which generalized quadrangles play a central role. Let $\mathscr{S} = (\mathscr{P}, \mathscr{B})$ be a pair consisting of a non-empty finite set $\mathcal{P}$ of *points* and a set $\mathcal{B}$ of distinguished subsets of cardinality at least three of $\mathcal{P}$ called *lines*. For any point P let $P^\perp$ be the set of all points collinear with P, and for any line l let $l^\perp$ be the set of all points collinear with each point of l. Suppose that $\mathcal{S}$ is connected, that $\mathcal{P}$ contains at least two non-collinear points, and that $P^\perp \backslash \{P\}$ is connected for each point P. A *subspace* of $\mathcal{S}$ is a set $\mathcal{X}$ of pairwise collinear points such that any line intersecting $\mathcal{X}$ in more than one point is contained in $\mathcal{X}$. A subspace that is not properly contained in a larger subspace is called a *max space*. Let us now consider the following conditions:

(i) for any point P and any line l, the point P is collinear with 0, 1, or all points of l;

(ii) if P and P' are non-collinear points such that $|P^\perp \cap P'^\perp| \geq 2$, then $P^\perp \cap P'^\perp$ provided with the lines it contains is a generalized quadrangle;

(iii) if $P \in \mathcal{P}$ and $l \in \mathcal{B}$ such that $P^\perp \cap l = \varnothing$ but $P^\perp \cap l^\perp \neq \varnothing$, then $P^\perp \cap l^\perp$ is a line;

(iii)′ each line is contained in exactly two max spaces.

Theorem 26.9.1: *If $\mathcal{S}$ satisfies* (i), (ii), *and* (iii), *then either $\mathcal{S}$ is a non-degenerate Shult space of rank* 3, *or $\mathcal{S}$ is the incidence structure formed by all points and all lines of a Grassmann variety $\mathcal{G}_{r,n}$ with $n \geq 3$ and $1 \leq r \leq n-2$.* □

Theorem 26.9.2: *If $\mathcal{S}$ satisfies* (i), (ii), *and* (iii)′, *then $\mathcal{S}$ is the incidence structure formed by all points and all lines of a Grassmann variety $\mathcal{G}_{r,n}$ with $n \geq 3$ and $1 \leq r \leq n-2$.* □

In the case of the Grassmann variety $\mathcal{G}_{r,n}$, $n \geq 3$ and $1 \leq r \leq n-2$, the generalized quadrangle of (ii) is always the classical grid $\mathcal{Q}(3, q)$. Further, the corresponding versions for the infinite case of Theorems 26.9.1 and 26.9.2 have also been proved. Theorem 26.9.1 is due to Cooperstein (1977) and Cohen (1983), and Theorem 26.9.2 is due to Hanssens (1984) and Hanssens and Thas (1987).

§26.2. This is taken from Payne and Thas (1984).

Generalized quadrangles were introduced by Tits (1959). The classical generalized quadrangles, all of which are associated with classical groups, were first recognized as generalized quadrangles also by Tits.

Higman (1971, 1974) first proved Theorem 26.2.3 by a complicated matrix-theoretic method. The argument given here was used by Bose and Shrikhande (1972) to show that in case $t = s^2 > 1$ each triad has $1 + s$ centres, and Cameron (1975) apparently first observed that the above technique also provides the inequality.

The generalized quadrangle $\mathcal{S}' = (\mathcal{P}', \mathcal{B}', \mathrm{I}')$ of order (s', t') is called a *subquadrangle* of the generalized quadrangle $\mathcal{S} = (\mathcal{P}, \mathcal{B}, \mathrm{I})$ of order (s, t) if

$\mathscr{P}' \subset \mathscr{P}$, $\mathscr{B}' \subset \mathscr{B}$, and if I$'$ is the restriction of I to $(\mathscr{P}' \times \mathscr{B}') \cup (\mathscr{B}' \times \mathscr{P}')$. If $\mathscr{S}' \neq \mathscr{S}$, then we say that $\mathscr{S}'$ is a *proper* subquadrangle of $\mathscr{S}$. A point of $\mathscr{P}$ which is incident with no line of $\mathscr{B}'$ is called *external* to $\mathscr{S}'$. The following theorem is due to Payne (1973a); see also Thas (1974d) and Payne and Thas (1984).

Theorem 26.9.3: *Let $\mathscr{S}'$ be a proper subquadrangle of $\mathscr{S}$, with notation as above. Then either $s = s'$ or $s \geq s't'$. If $s = s'$, then each external point is collinear with $1 + st'$ points of $\mathscr{S}'$; if $s = s't'$, then each external point is collinear with $1 + s'$ points of $\mathscr{S}'$. The dual holds similarly.* □

Now, we briefly describe generalized quadrangles with small parameters. The detailed proofs of all these results can be found in Payne and Thas (1984). Let $\mathscr{S} = (\mathscr{P}, \mathscr{B}, \mathrm{I})$ be a generalized quadrangle of order (s, t), with $2 \leq s \leq 4$ and $s \leq t$. By Theorems 26.2.2 and 26.2.3, $(s, t) \in \{(2, 2), (2, 4), (3, 3), (3, 5), (3, 6), (3, 9), (4, 4), (4, 6), (4, 8), (4, 11), (4, 12), (4, 16)\}$. An easy proof shows that up to isomorphism there is only one generalized quadrangle of order 2. The uniqueness of the generalized quadrangle of order (2, 4) was proved independently at least five times, by Seidel (1968), Shult (1972), Thas (1974e), Freudenthal (1975), and Dixmier and Zara (1976). Payne (1975) and independently Dixmier and Zara (1976) showed that a generalized quadrangle of order 3 is isomorphic to $\mathscr{W}(3)$ or to its dual $\mathscr{Q}(4, 3)$. The uniqueness of a generalized quadrangle of order (3, 5) was proved by Dixmier and Zara (1976), and the proof of the uniqueness of the generalized quadrangle of order (3, 9) is due to Dixmier and Zara (1976) and independently to Cameron (see Payne and Thas (1976)). The non-existence of a generalized quadrangle of order (3, 6) was proved by Dixmier and Zara (1976). Further, Payne (1977b,c) proved up to isomorphism that there is just one generalized quadrangle of order 4. The long proof of this last theorem is divided into a large number of steps. At one point in the published proof of Payne (1977b) the argument is incomplete, and it was Tits who provided the necessary correction to it (see Payne and Thas (1984)). Finally, nothing is known about generalized quadrangles with order (4, 11) or (4, 12), and single examples are known in the cases (4, 6), (4, 8), and (4, 16).

§26.3. Theorems 26.3.29, 26.3.30, and 26.3.31 are taken from Buekenhout and Lefèvre (1974, 1976) and Lefèvre-Percsy (1977b). Theorem 26.3.32 is due to Thas (1974c).

All finite projective generalized quadrangles were first determined by Buekenhout and Lefèvre (1974) by a proof most of which is valid in the infinite case. Independently Olanda (1973, 1977) has given a typically finite proof, and Thas and De Winne (1977) have given a different combinatorial proof under the assumption that the three-dimensional case is already settled. The infinite case was settled by Dienst (1980a,b). For projective Shult spaces of rank at least three, the infinite case was completely solved by Buekenhout and Lefèvre (1976) and Lefèvre-Percsy (1977b). Because the generalized quadrangles, and more generally the Shult spaces, in this book are finite, we have modified the presentation of Buekenhout and Lefèvre somewhat.

All finite generalized quadrangles embedded in the affine space $AG(d, q)$, $d \geq 2$, were determined by Thas (1978a). The three-dimensional case was independently settled by Bichara (1978).

Weak projective Shult spaces were considered by Lefèvre-Percsy (1981c,d,e,g). Here the lines of the Shult space are subsets of the lines of $PG(d, q)$.

§26.4. The detailed proofs of Theorems 26.4.1 to 26.4.25 can be found in Payne and Thas (1984). Theorem 26.4.2, which is probably the oldest combinatorial characterization of a class of generalized quadrangles, was discovered independently by several authors, for example by Singleton (1966), Benson (1970), and Tallini (1971c). Theorem 26.4.3 is due to Thas (1977c), Theorems 26.4.5 and 26.4.6 to Thas (1973d), Theorems 26.4.7 and 26.4.8 to Payne and Thas (1976), and Theorem 26.4.8 independently to Mazzocca (1974a), Theorems 26.4.9, 26.4.10, and 26.4.11 to Thas (1978b), and Theorem 26.4.9(i) independently to Mazzocca (1974b), Theorem 26.4.12 to Thas (1981c), Theorem 26.4.13 to Tallini (1971c), Theorem 26.4.14 to Thas (1976), Theorem 26.4.15 to Payne and Thas (1976), Theorem 26.4.16 to Thas (1977c), Theorems 26.4.17 and 26.4.18 to Thas and Payne (1976) and Thas (1977c), and Theorems 26.4.19, 26.4.20, and 26.4.21 to Thas (1977c). Theorem 26.4.22, which is a characterization in terms of matroids, is taken from Mazzocca and Olanda (1979b). A considerable shortening of the original proof was given by Payne and Thas (1984). Theorem 26.4.23 is due to Ronan (1980b). Ronan's treatment includes infinite generalized quadrangles and relies on topological methods. Payne and Thas (1984) offer an 'elementary' treatment which is more combinatorial than topological, and which corrects a slight oversight in the case $t = 2$. Theorem 26.4.24 is taken from Thas (1981c) and Theorem 26.4.25 from Payne and Thas (1984).

Theorem 26.4.26 was proved by Ealy (1977) and Theorem 26.4.28 by Thas (1985a, 1986b). Using the language of BN pairs, Fong and Seitz (1973, 1974) obtained a characterization of the finite classical generalized polygons, in particular the finite classical generalized quadrangles. Tits (1976a,b, 1983, 1990) determined all finite and infinite generalized polygons, satisfying what he called the 'Moufang condition'; the finite case of Tits' fundamental work, in particular Theorem 26.4.28, is essentially the theorem of Fong and Seitz. Finally, we mention that Payne and Thas (1984) had hoped to obtain an independent and purely geometric proof of Theorem 26.4.28, or equivalently a purely geometric proof of the Fong–Seitz theorem in the case of finite generalized quadrangles. In fact, they came very close and to complete a proof of the theorem it would have been sufficient to show that if $(\mathcal{S}^{(P)}, G)$

is a translation generalized quadrangle of order (s, s^2), $s \neq 1$, for each point P of $\mathscr{S}$, then $\mathscr{S} \cong \mathscr{Q}(5, s)$.

In order to state Theorem 26.4.13 we used the notion '*linear space*'. Here a linear space is an incidence structure $\mathscr{S}^* = (\mathscr{P}, \mathscr{B}^*, \in)$ with $\mathscr{P}$ a non-empty set, $\mathscr{B}^*$ a non-empty set of subsets of $\mathscr{P}$ where each of the subsets has cardinality at least two, and having the property that any two distinct elements (points) of $\mathscr{P}$ are contained in a unique element (line) of $\mathscr{B}^*$.

§26.5. Partial geometries were introduced by Bose (1963). Theorem 26.5.3 is taken from Cameron, Goethals, and Seidel (1978). The proper partial geometries $\mathscr{S}(\mathscr{K})$ are due to Thas (1973b, 1974b) and independently to Wallis (1973); the proper partial geometries $\mathscr{T}_2^*(\mathscr{K})$ are due to Thas (1973b, 1974b). For a survey on partial geometries we refer to De Clerck (1978, 1979), Thas (1977b), and Brouwer and van Lint (1984).

§26.6. Theorem 26.6.1 is taken from De Clerck and Thas (1978) and Theorem 26.6.2 is taken from Thas and De Clerck (1977).

All finite partial geometries embedded in the affine space $AG(d, q)$, $d \geq 2$, were determined by Thas (1978a).

§26.7. Partial quadrangles were introduced by Cameron (1975), semi-partial geometries by Debroey and Thas (1978a), and $(0, \alpha)$-geometries by De Clerck and Thas (1983). Theorems 26.7.1 to 26.7.5 are taken from Debroey and Thas (1978a). Theorem 26.7.6 is due to Debroey (1978), but in §26.7 a new and simpler proof is given. Most of the examples can be found in Brouwer and van Lint (1984), Cameron (1975), Debroey and Thas (1978a), De Clerck and Thas (1983), and Hall (1983b).

§26.8. Lemmas 26.8.1, 26.8.2 and Theorems 26.8.4, 26.8.5, 26.8.11 are taken from De Clerck and Thas (1983). Theorem 26.8.3 is due to Debroey and Thas (1978c), and Theorems 26.8.6 and 26.8.7 are unpublished results of Thas. Lemma 26.8.8 and Theorems 26.8.9, 26.8.10 are taken from Thas, Debroey, and De Clerck (1984).

Farmer and Hale (1980) proved that any projective $(0, s)$-geometry with ambient space $PG(n, s)$, with $s > 2$ and $n \geq 3$, is a $W(n, 2k, s)$ with $2k \in \{2, 4, \ldots, n + 1\}$. In the original proof of Theorem 26.8.9 this result was used. However, since the theorem of Farmer and Hale has not been proved in this volume, nor in the previous two volumes, we preferred to rely on Theorem 26.3.30 concerning projective Shult spaces.

Let $\mathscr{S} = (\mathscr{P}, \mathscr{B}, \mathrm{I})$ be a projective $(0, 2)$-geometry with ambient space $PG(n, 2)$, $n \geq 3$. If P, P' are distinct points of $\mathscr{P}$ with $PP' \notin \mathscr{B}$, then we write $P \approx P'$ if for any point $P'' \in \mathscr{P} \backslash \{P, P'\}$ we have $PP'' \in \mathscr{B}$ if and only if $P'P'' \in \mathscr{B}$. By definition, let $P \approx P$ for all $P \in \mathscr{P}$. Clearly $\approx$ is an equivalence relation. The $(0, 2)$-geometry $\mathscr{S}$ is called *reduced* if all its $\approx$-classes have size one. The following theorem was proved independently by Hall (1983b) and Thas, Debroey, and De Clerck (1984).

Theorem 26.9.4: *There exist subspaces* $PG(d, 2), PG^{(1)}(d+1, 2), PG^{(2)}(d+1, 2), \ldots, PG^{(r)}(d+1, 2)$ *of* $PG(n, 2)$ *for which* $PG^{(i)}(d+1, 2) \cap PG^{(j)}(d+1, 2) = PG(d, 2)$, *for all* $i \neq j$ *and such that* $PG^{(1)}(d+1, 2) \setminus PG(d, 2), \ldots, PG^{(r)}(d+1, 2) \setminus PG(d, 2)$ *are exactly the* $\approx$*-classes of* $\mathcal{S}$. *If* $PG(n-d-1, 2)$ *is a subspace of* $PG(n, 2)$ *which is skew to* $PG(d, 2)$, *if* $\{P_i\} = PG^{(i)}(d+1, 2) \cap PG(n-d-1, 2)$ *with* $i = 1, 2, \ldots, r$, *and if, for any line* $l \in \mathcal{B}$, l^* *is the union of the* $\approx$*-classes having a non-empty intersection with* l, *then all points* P_i *together with all lines* $l^* \cap PG(n-d-1, 2)$ *constitute a reduced projective* $(0, 2)$*-geometry with ambient space* $PG(n-d-1, 2)$. *It is also clear how to apply the converse. Hence, in determining all projective* $(0, 2)$*-geometries we may restrict ourselves to the reduced geometries.* □

Let $\mathcal{S} = (\mathcal{P}, \mathcal{B}, \mathrm{I})$ be a reduced projective $(0, 2)$-geometry with ambient space $PG(n, 2)$, $n \geq 3$. In the cotriangle theorem, Shult (1975) has shown that $\mathcal{S}$ is isomorphic to one of $NQ^+(2n'-1, 2)$ with $n' > 2$, $NQ^-(2n'-1, 2)$ with $n' \geq 2$, $NQ(2n', 2)$ with $n' \geq 2$ (recall that $NQ(2n', 2)$ is isomorphic to $\overline{W(2n'-1, 2)}$), or $\mathcal{S}_h$ (which is example (i) of §26.7). The following theorems are taken from Hall (1983b).

Theorem 26.9.5: *If* $\mathcal{S}$ *is isomorphic to* $NQ^-(2n'-1, 2)$, $n' \geq 2$, *then* $n = 2n'-1$ *and* $\mathcal{S} = NQ^-(n, 2)$. □

Theorem 26.9.6: *If* $\mathcal{S}$ *is isomorphic to* $NQ^+(2n'-1, 2)$, $n' > 3$, *then* $n = 2n'-1$ *and* $\mathcal{S} = NQ^+(n, 2)$. *If* $\mathcal{S}$ *is isomorphic to* $NQ^+(5, 2)$, *then* $n \in \{5, 6\}$ *with two different embeddings for* $n = 5$ *and exactly one for* $n = 6$. □

Theorem 26.9.7: *If* $\mathcal{S}$ *is isomorphic to* $NQ(2n', 2)$, $n' \geq 2$, *then either* $n = 2n'$ *and* $\mathcal{S} = NQ(n, 2)$, *or* $n = 2n'-1$ *and* $\mathcal{S} = \overline{W(n, 2)}$. □

Theorem 26.9.8: *The determination of all reduced projective* $(0, 2)$*-geometries isomorphic to* $\mathcal{S}_h$ *is equivalent to the determination of all binary, even, linear codes of length* h *with minimal distance at least six.* □

Finally, all finite semi-partial geometries embedded in the affine space $AG(d, q)$, with $d \in \{2, 3\}$, were determined by Debroey and Thas (1978b).

27

ARCS AND CAPS

27.1 Introduction

A $(k; r, s; n, q)$*-set* $\mathcal{K}$ is defined to be a set of k points in $PG(n, q)$ with at most r points in any s-space such that $\mathcal{K}$ is not contained in a proper subspace. This is a slight variation on the definition of §3.3, where the last condition is not present. The large question is to describe all such sets. Four questions particularly are of interest. The set $\mathcal{K}$ is *complete* if it is not contained in a $(k + 1; r, s; n, q)$-set.

I. Find the maximum value $m(r, s; n, q)$ of k.
II. Characterize the sets, the *maximum sets*, with this value of k.
III. Find the size $m'(r, s; n, q)$ of the second largest, complete $(k; r, s; n, q)$-set.
IV. Characterize the complete $(k; r, s; n, q)$-sets.

Question IV includes I, II, and III. The importance of III is that if $\mathcal{K}$ is a $(k; r, s; n, q)$-set with $k > m'(r, s; n, q)$, then $\mathcal{K}$ is contained in a maximum set. So upper bounds on $m'(r, s; n, q)$ permit inductive arguments.

In fact, we examine these questions only when $r = s + 1$ and then only in the cases $s = 1$ and $s = n - 1$.

A $(k; 2, 1; n, q)$-set is a k-set with at most two points on any line of $PG(n, q)$ and is also called a *k-cap*. The number $m(2, 1; n, q)$ is written as $m_2(n, q)$. The only precise values known are the following:

$$m_2(n, 2) = 2^n; \tag{27.1}$$

$$m_2(2, q) = \begin{cases} q + 1, & q \text{ odd} \\ q + 2, & q \text{ even}; \end{cases} \tag{27.2}$$

$$m_2(3, q) = q^2 + 1, \quad q > 2; \tag{27.3}$$

$$m_2(4, 3) = 20; \tag{27.4}$$

$$m_2(5, 3) = 56. \tag{27.5}$$

Upper bounds for $m_2(n, q)$ are determined in §§27.2–27.4.

A $(k; n, n - 1; n, q)$-set is a k-set not contained in a prime with at most n points in any prime of $PG(n, q)$ and is also called a *k-arc*; for $n = 2$, a k-cap and a k-arc are equivalent. The number $m(n, n - 1; n, q)$ is written $m(n, q)$;

by definition $m(2, q) = m_2(2, q)$. Values obtained in previous chapters are as follows:

$$m(n, q) = n + 2 \qquad \text{for } q \leq n + 1; \tag{27.6}$$

$$m(2, q) = \begin{cases} q + 1, & q \text{ odd} \\ q + 2, & q \text{ even}; \end{cases} \tag{27.7}$$

$$m(3, q) = q + 1 \qquad \text{for } q > 3. \tag{27.8}$$

All other values known and all the values determined in §§27.5–27.7 are $m(n, q) = q + 1$ or $q + 2$.

In deciding the value of $m(n, q)$, the value of $m'(2, q)$, the size of the second largest complete arc in $PG(2, q)$, is crucial. It is also useful to write, for $q \geq 5$,

$$f(q) = q - m'(2, q). \tag{27.9}$$

For $q \leq 5$, there is only one complete plane arc and $m'(2, q)$ is not defined. For other small values of q we have the following results:

q	7	8	9	11	13	16
$m'(2, q)$	6	6	8	10	12	13
$f(q)$	1	2	1	1	1	3

Also

$$m'(2, q) \leq q - \tfrac{1}{4}\sqrt{q} + \tfrac{25}{16}, \quad q \text{ odd}; \tag{27.10}$$

the slightly weaker result

$$m'(2, q) \leq q - \tfrac{1}{4}\sqrt{q} + \tfrac{7}{4}, \quad q \text{ odd}, \tag{27.11}$$

has been frequently used. Further,

$$m'(2, q) \leq q - \sqrt{q} + 1, \quad q \text{ even}, q > 2; \tag{27.12}$$

$$m'(2, q) = q - \sqrt{q} + 1, \quad q = 2^{2m}, m > 1; \tag{27.13}$$

$$m'(2, q) \leq \tfrac{44}{45}q + \tfrac{8}{9}, \qquad q \text{ prime}; \tag{27.14}$$

$$m'(2, q) \leq q - \tfrac{1}{4}\sqrt{(pq)} + \tfrac{29}{16}p + 1, \quad q = p^{2m+1}, m \geq 1, p \text{ odd}; \tag{27.15}$$

$$m'(2, q) \leq q - \sqrt{(2q)} + 2, \quad q = 2^{2m+1}, m \geq 1. \tag{27.16}$$

It is convenient to record an elementary result which is frequently applied in §27.3.

Lemma 27.1.1: *If A, B, C are three sets such that $C \supset A \cup B$, then*

$$|A \cap B| \geq |A| + |B| - |C|.$$

Proof:

$$|A \cap B| = |A| + |B| - |A \cup B| \geq |A| + |B| - |C|. \quad \square$$

27.2 Caps and codes

Let $\mathscr{K}$ be the k-cap $\{P_1, \ldots, P_k\}$, where $P_i = \mathbf{P}(a_{i0}, a_{i1}, \ldots, a_{in})$. Then the $k \times (n+1)$ matrix

$$A = [a_{ij}] \quad i = 1, \ldots, k; j = 0, \ldots, n$$

is called a *matrix of* $\mathscr{K}$. Any permutation of the rows of A or multiplication of a row of A by an element of γ_0 gives another matrix of $\mathscr{K}$. For any projectivity $\mathfrak{T}$, write $\mathscr{K}\mathfrak{T} = \{P_1\mathfrak{T}, \ldots, P_k\mathfrak{T}\}$. So the caps $\mathscr{K}_1$ and $\mathscr{K}_2$ are (projectively) equivalent if $\mathscr{K}_1\mathfrak{T} = \mathscr{K}_2$ for some projectivity $\mathfrak{T}$.

Lemma 27.2.1: *Let $\mathscr{K}_1$ and $\mathscr{K}_2$ be caps with respective matrices A_1 and A_2. Then $\mathscr{K}_1$ and $\mathscr{K}_2$ are equivalent if and only if A_2 can be obtained from A_1 via a sequence of operations of the following type*:

(C1) *multiplication of a column by a non-zero scalar*;
(C2) *interchange of two columns*;
(C3) *addition of a scalar multiple of one column to another*;
(R1) *multiplication of a row by a non-zero scalar*;
(R2) *interchange of two rows*.

Proof: The operation of $\mathfrak{T}$ on $\mathscr{K}_1$ corresponds to a sequence of operations of types (C1), (C2), and (C3). The operations (R1) and (R2) leave $\mathscr{K}_1$ fixed. $\square$

If A_2 can be obtained from A_1 as in the lemma, then A_1 and A_2 are *C-equivalent*.

A linear (N, d)-code over $GF(q)$ is usually defined to be a d-dimensional subspace of the N-dimensional vector space $V(N, q)$. In this section, it is more convenient to replace a vector by its non-zero multiples. So a *linear* (N, d)-*code* is a Π_{d-1} of $PG(N-1, q)$. It is also convenient to represent the points of an (N, d)-code by column vectors rather than row vectors. A *generator matrix* A for an (N, d)-code C is thus an $N \times d$ matrix whose columns generate C. Other generator matrices of C are obtained from A via operations of types (C1), (C2), (C3).

Two codes are *equivalent* if one can be obtained from the other by a permutation of the coordinate indices combined with multiplication of some coordinates by a non-zero scalar.

Lemma 27.2.2: *Let C_1 and C_2 be codes with generator matrices A_1 and A_2. Then C_1 and C_2 are equivalent if and only if A_1 and A_2 are C-equivalent.*

Proof: From the definition of equivalence of codes, a generator matrix of C_2 is obtained from a generator matrix of C_1 by operations (R1) and (R2). □

If $\mathscr{K}$ is a k-cap in $PG(n, q)$ with matrix A, the *code C of $\mathscr{K}$* is the $(k, n + 1)$-code with generator matrix A; such a code is a *cap-code*. It is assumed that $\mathscr{K}$ is not contained in a subspace of lower dimension than n.

Theorem 27.2.3: *Let $\mathscr{K}_1$ and $\mathscr{K}_2$ be caps with codes C_1 and C_2. Then $\mathscr{K}_1$ is equivalent to $\mathscr{K}_2$ if and only if C_1 is equivalent to C_2.*

Proof: This follows from Lemmas 27.2.1 and 27.2.2. □

An (N, d)-code is *projective* if the rows of a generator matrix are distinct points of $PG(N - 1, q)$. Any cap-code is projective.

With, as usual, $\theta(n) = |PG(n, q)|$, given a $(k, n + 1)$-code C, denote by $M(C)$ a $k \times \theta(n)$ matrix whose columns are the points of C. Given a linearly independent set $\{X_0, \ldots, X_t\}$ of $PG(k - 1, q)$, let $M(X_0, \ldots, X_t)$ denote a $k \times \theta(t)$ matrix whose columns are the points X_j.

Lemma 27.2.4: *The number of zeros in the ith row of the $k \times \theta(t)$ matrix $M(X_0, \ldots, X_t)$ is $\theta(t)$ if all the X_j have zero in the ith row and is $\theta(t - 1)$ otherwise.*

Proof: Let Π_t be the subspace spanned by $X_0, \ldots, X_t$. Intersect Π_t with $\mathbf{V}(x_i)$. Then $\Pi_t \cap \mathbf{V}(x_i) = \Pi_t$ if $\Pi_t \subset \mathbf{V}(x_i)$ and $\Pi_t \cap \mathbf{V}(x_i) = \Pi_{t-1}$ if $\Pi_t \not\subset \mathbf{V}(x_i)$. □

Given a projective $(k, n + 1)$-code C, let M_1 be the matrix obtained from $M(C)$ by omitting one row and also those columns having a non-zero entry in that row. By Lemma 27.2.4, M_1 is a $(k - 1) \times \theta(n - 1)$ matrix whose columns are the vectors of a $(k - 1, n)$-code C_1. The code C_1 is a *residual* code of C. The code C has k residuals, one for each row of C; some or all of these residuals may be equivalent codes. By identifying the vectors of C_1 with those of C from which one zero entry has been omitted, C_1 can be regarded as a (k, n)-subcode of C.

If A_1 is a generator matrix for a residual code C_1 of C, then C is equivalent to a code with generator matrix

$$A = \begin{bmatrix} 0 & 1 \\ A_1 & * \end{bmatrix}.$$

The matrix A_1 is a *residual* of A. If C_1 and A_1 are residuals of C and A, then C and A are *extensions* of C_1 and A_1.

Theorem 27.2.5: *A projective code C is a cap-code if and only if every residual code of C is projective.*

Proof: Suppose C_1 with matrix A_1 is a non-projective residual of C with matrix A an extension of A_1. Then two rows, the ith and jth, say, of A_1 are the same, up to a scalar multiple; so the first, $(i+1)$th and $(j+1)$th rows of A are collinear. So C is not a cap-code.

Conversely, suppose C is not a cap-code and let A be any generator matrix of C. Then three rows, say the first, second, and third, of A are collinear. However, using suitable column operations, A is C-equivalent to a matrix whose first row is $(0, 0, \ldots, 0, 1)$. Since column operations preserve the collinearity of rows 1, 2, and 3, the residual obtained by omitting the first row is non-projective. □

To fix ideas, let us look at a trivial example. Consider the 4-arc (=4-cap) $\mathcal{K}$ in $PG(2, 2)$ with points $\mathbf{P}(1, 0, 0)$, $\mathbf{P}(0, 1, 0)$, $\mathbf{P}(0, 0, 1)$, $\mathbf{P}(1, 1, 1)$. A matrix for $\mathcal{K}$ is

$$A = \begin{bmatrix} 1 & 0 & 0 \\ 0 & 1 & 0 \\ 0 & 0 & 1 \\ 1 & 1 & 1 \end{bmatrix}.$$

It is a generator matrix for a $(4, 3)$-code C for which a suitable $M(C)$ is

$$M(C) = \begin{bmatrix} 1 & 0 & 0 & 1 & 1 & 0 & 1 \\ 0 & 1 & 0 & 1 & 0 & 1 & 1 \\ 0 & 0 & 1 & 0 & 1 & 1 & 1 \\ 1 & 1 & 1 & 0 & 0 & 0 & 1 \end{bmatrix}.$$

The residuals with respect to the four rows of $M(C)$ are

$$\begin{bmatrix} 1 & 0 & 1 \\ 0 & 1 & 1 \\ 1 & 1 & 0 \end{bmatrix}, \quad \begin{bmatrix} 1 & 0 & 1 \\ 0 & 1 & 1 \\ 1 & 1 & 0 \end{bmatrix}, \quad \begin{bmatrix} 1 & 0 & 1 \\ 0 & 1 & 1 \\ 1 & 1 & 0 \end{bmatrix}, \quad \begin{bmatrix} 1 & 1 & 0 \\ 1 & 0 & 1 \\ 0 & 1 & 1 \end{bmatrix}.$$

Now, we consider the weight distribution of a cap-code. Let X be a vector (that is, a column) of the code C. The *weight* of X, denoted $w(X)$, is the number of non-zero entries in X. Denote the vectors of a $(k, n+1)$-code C by $X_1, X_2, \ldots, X_{\theta(n)}$ and let $w_i = w(X_i)$, $i \in \mathbf{N}_{\theta(n)}$. Order the X_i so that $w_1 \geq w_2 \geq \cdots \geq w_{\theta(n)}$. The ordered $\theta(n)$-tuple $(w_1, \ldots, w_{\theta(n)})$ is the *weight distribution* of C. Equivalent codes have the same weight distribution.

It turns out that cap-codes have large minimum weight. For a $(k, n+1)$-code C, let $C^\perp$ be the *dual code*; that is, $C^\perp$ is the $(k, k-n-1)$-code consisting of points Y in $PG(k-1, q)$ such that $XY^* = 0$ for all X in C. If

C is a code of the cap $\mathcal{K}$, then $C^{\perp}$ has minimum weight at least four; for, if any vector in $C^{\perp}$ had three or less non-zero entries, its orthogonality with C would force C to be non-projective or would induce a collinearity of the corresponding three rows of C. Conversely if $C^{\perp}$ has minimum weight at least four, then C is a cap-code. Thus $m(n, q)$ is the maximum value of N for which one error can be corrected and three detected with certainty by a linear $(N, N - n - 1)$-code.

Theorem 27.2.6: *Let $\mathcal{K}$ be a k-cap in $PG(n, q)$ with code C. Then the minimum weight of C, as well as that of any residual, is at least $k - m_2(n - 1, q)$.*

Proof: Let X be any vector in C and suppose X has t zeros. Let A be a generator matrix of C with X as first column. Since the rows of A form a cap and since any subset of a cap is also a cap, it follows that those rows having a zero as first coordinate form a t-cap in a prime of $PG(n, q)$. Hence $t \leq m_2(n - 1, q)$, and so $w(X) = k - t \geq k - m_2(n - 1, q)$. The result holds for a residual C_1, since any vector in C_1 is obtained from a vector in C by omitting one zero and so leaving the weight unchanged. □

Lemma 27.2.7: *Let C be a projective $(k, n + 1)$-code with weight distribution $(w_1, \ldots, w_{\theta(n)})$. Then*

$$\text{(i)} \sum w_i = kq^n; \tag{27.17}$$

$$\text{(ii)} \sum w_i^2 = kq^{n-1}\{k(q - 1) + 1\}. \tag{27.18}$$

Proof: (i) The number of non-zero entries in $M(C)$ is $\sum w_i$ (summing over columns) and by Lemma 27.2.4 is also kq^n (summing over rows).

(ii) Let $Z_1, Z_2, \ldots, Z_k$ be the rows of $M(C)$ and let B be the $k(k - 1)(q - 1) \times \theta(n)$ matrix with rows $Z_i + \lambda Z_j$, $(i, j) \in \mathbf{N}_k^2$, $i \neq j$, $\lambda \in \gamma_0$. Since C is projective, the rows of B are all non-zero; so by Lemma 27.2.4, each row has $\theta(n - 1)$ zeros. The ith column of B has $w_i(w_i - 1) + (k - w_i)(k - w_i - 1)(q - 1)$ zeros. Counting the zero entries of B via columns and rows thus gives

$$q \sum w_i^2 - \{1 + (2k - 1)(q - 1)\} \sum w_i + \theta(n)k(k - 1)(q - 1)$$
$$= k(k - 1)(q - 1)\theta(n - 1). \tag{27.19}$$

Substituting $\sum w_i$ from (27.17) gives the result. □

Now, we write $m_1 = m_2(n - 1, q)$. Then Theorem 27.2.6 says that, for a cap-code, $w_i \geq k - m_1$. So consider the *amended weights* u_i given by

$$u_i = w_i - (k - m_1). \tag{27.20}$$

Then $u_1 \geq u_2 \geq \cdots \geq u_{\theta(n)} \geq 0$ and $(u_1, \ldots, u_{\theta(n)})$ is the *amended weight distribution* of C. For a residual code C_1 of C, let $(v_1, \ldots, v_{\theta(n-1)})$ consist of the amended weights of the corresponding columns of $M(C)$ with $v_1 \geq v_2 \geq \cdots \geq v_{\theta(n-1)} \geq 0$; hence each v_i is some u_j. By abuse of language we will say that $(v_1, \ldots, v_{\theta(n-1)})$ is the *amended weight distribution of* C_1.

Lemma 27.2.8: *Let C be a projective $(k, n+1)$-cap-code with amended weight distribution $(u_1, u_2, \ldots, u_{\theta(n)})$. Then*

$$\text{(i) } \sum u_i = m_1\theta(n) - k\theta(n-1); \qquad (27.21)$$

$$\text{(ii) } \sum u_i^2 = k^2\theta(n-2) + k(q^{n-1} - 2m_1\theta(n-1)) + m_1^2\theta(n). \qquad (27.22)$$

For a residual code C_1 of C with amended weight distribution $(v_1, \ldots, v_{\theta(n-1)})$,

$$\text{(iii) } \sum v_i = (m_1 - 1)\theta(n-1) - (k-1)\theta(n-2); \qquad (27.23)$$

$$\text{(iv) } \sum v_i^2 = (k-1)^2\theta(n-3) + (k-1)\{q^{n-2} - 2(m_1 - 1)\theta(n-2)\} + (m_1 - 1)^2\theta(n-1). \qquad (27.24)$$

Proof: Equations (27.21)–(27.24) are just restatements of the previous lemma. □

It should be noted that (27.23) and (27.24) only hold because a residual of a cap-code is projective and are not true for general codes.

Theorem 27.2.9: *Let C be a $(k, n+1)$-cap-code with weight and amended weight distributions $(w_1, w_2, \ldots, w_{\theta(n)})$ and $(u_1, u_2, \ldots, u_{\theta(n)})$. Then*

$$\text{(i) } w_1 + w_2 \leq m_1(q-1) + k; \qquad (27.25)$$

$$\text{(ii) } u_1 + u_2 \leq m_1(q+1) - k. \qquad (27.26)$$

For a residual of C with amended distribution $(v_1, v_2, \ldots, v_{\theta(n-1)})$,

$$\text{(iii) } v_1 + v_2 \leq m_1(q+1) - q - k. \qquad (27.27)$$

Proof: (i) By Lemma 27.2.4, each row of the $k \times (q+1)$ matrix $M(X_1, X_2)$ has at least one zero. Hence, counting the zeros of $M(X_1, X_2)$ gives

$$(k - w_1) + (k - w_2) + \sum_{\lambda \in \gamma_0} \{k - w(X_1 + \lambda X_2)\} \geq k.$$

By Theorem 27.2.6, $w(X_1 + \lambda X_2) \geq k - m_1$. So

$$w_1 + w_2 \leq k + m_1(q-1).$$

(ii), (iii) These follow similarly. □

Now the bounds of Theorems 27.2.6 and 27.2.9 together with the identities of Lemma 27.2.8 give restrictions on $m_2(n, q)$.

Theorem 27.2.10: *For $n \geq 4$ and $q \neq 2$,*

$$m_2(n, q) \leq q m_2(n-1, q) - q + 1. \tag{27.28}$$

Proof: We use induction on n. Suppose $\mathcal{K}$ is a k-cap in $PG(n, q)$ with code C. A residual code C_1 is a projective $(k-1, n)$-code. So, if C_1 has weight distribution $w'_1, \ldots, w'_{\theta(n-1)}$, then Lemma 27.2.7 gives

$$\sum w'_i = (k-1)q^{n-1} \tag{27.29}$$

and Theorem 27.2.6 gives

$$\sum w'_i \geq (k - m_1)\theta(n-1). \tag{27.30}$$

Hence (27.29) and (27.30) imply that

$$k \leq m_1 q + (m_1 - q^{n-1})(q-1)/(q^{n-1} - 1). \tag{27.31}$$

Next, we deduce by induction that $m_2(n, q) \leq q^{n-1} + 1$. First, $m_2(3, q) = q^2 + 1$, Theorem 16.1.5. The induction hypothesis is that $m_1 = m_2(n-1, q) \leq q^{n-2} + 1$. Then (27.31) gives

$$k \leq q^{n-1} + 1 - (q^{n-1} - q^{n-2})/(q^{n-1} - 1)$$

and so $k \leq q^{n-1} + 1$. Hence $m_1 \leq q^{n-2} + 1$. Substituting this in the second occurrence of m_1 in (27.31) implies that

$$\begin{aligned} k &\leq m_1 q + (q^{n-2} + 1 - q^{n-1})(q-1)/(q^{n-1} - 1) \\ &\leq m_1 q - q + 1. \quad \square \end{aligned}$$

Now we improve (27.28) by showing that equality cannot hold.

Theorem 27.2.11: *For $n \geq 4$ and $q \neq 2$,*

$$m_2(n, q) \leq q m_2(n-1, q) - q. \tag{27.32}$$

Proof: Suppose $\mathcal{K}$ is a k-cap with $k = qm_1 - q + 1$. Let C_1 be any residual of the code C of $\mathcal{K}$ and let C_1 have amended weight distribution $(v_1, \ldots, v_{\theta(n-1)})$. Then (27.23) and (27.24) give

$$\sum v_i = m_1 - 1, \tag{27.33}$$

$$\sum v_i^2 = (m_1 - 1)\{q^{n-1} - (q-1)(m_1 - 1)\}. \tag{27.34}$$

Since $\sum v_i^2 \leq (\sum v_i)^2$,

$$q^{n-1} - (q-1)(m_1 - 1) \leq m_1 - 1;$$

this gives $m_1 \geq q^{n-2} + 1$. However, in Theorem 27.2.10, it was shown inductively that $m_1 \leq q^{n-2} + 1$; that is, $m_1 = q^{n-2} + 1$. By (27.32) and since $m_2(n-2, q) \leq q^{n-3} + 1$, it now follows that $m_2(n-2, q) = q^{n-3} + 1$. Proceeding in this way we obtain $m_2(s, q) = q^{s-1} + 1$ for all $3 \leq s < n$. To prove the theorem, it therefore suffices to show that $m_2(4, q) < q^3 + 1$. This is shown geometrically in Theorem 27.3.1. Alternatively, suppose $\mathscr{K}$ is a $(q^3 + 1)$-cap in $PG(4, q)$. Then (27.33) and (27.34) give

$$(\sum v_i)^2 = \sum v_i^2 = q^4.$$

So $(v_1, v_2, \ldots) = (q^2, 0, \ldots, 0)$ and $(w'_1, w'_2, \ldots) = (q^3, q^3 - q^2, \ldots, q^3 - q^2)$. Thus each of the k residuals of C contains a vector of weight q^3. Since any vector of weight q^3 is contained in $(q^3 + 1) - q^3 = 1$ residual, there are k distinct vectors of weight q^3 in C. In particular,

$$w_1 + w_2 = 2q^3.$$

However, from (27.25),

$$w_1 + w_2 \leq (q^2 + 1)(q - 1) + q^3 + 1 < 2q^3. \quad \square$$

There is a final improvement.

Theorem 27.2.12: *In $PG(n, q)$, $q > 2$,*

(i) $m_2(n, q) \leq q m_2(n-1, q) - (q+1)$ *for* $n \geq 4$; (27.35)

(ii) $m_2(n, q) \leq q^{n-4} m_2(4, q) - q^{n-4} - 2\theta(n-5) + 1$ *for* $n \geq 5$. (27.36)

Proof: (i) See §27.8.

(ii) This follows from (i) by induction. $\square$

27.3 The maximum size of a cap for q odd

In this section some upper bounds for $m_2(n, q)$ are proved. First, we give a general bound which is useful in that it holds for all $q > 2$. The first result is also implicit in Theorem 27.2.11.

Theorem 27.3.1: *For $q > 2$ and $n \geq 4$,*

$$m_2(n, q) \leq q^{n-1}. \tag{27.37}$$

Proof: Suppose there exists a k-cap $\mathscr{K}$ with $k = q^{n-1} + 1$.

First, we show that, for q even, any plane meets $\mathscr{K}$ in at most $q + 1$ points. Suppose therefore that $\pi \cap \mathscr{K}$ is a $(q+2)$-arc for some plane π, and consider the $\theta(n-3)$ solids $\Pi_3^{(i)}$ through π, $i \in \mathbf{N}_{\theta(n-3)}$. If $|\Pi_3^{(i)} \cap (\mathscr{K} \backslash \pi)| = d_i$, then

$$\sum d_i = q^{n-1} + 1 - (q+2) = q^{n-1} - q - 1. \tag{27.38}$$

Since $m_2(3, q) = q^2 + 1$,

$$d_i + q + 2 \leq q^2 + 1,$$

whence

$$d_i \leq q^2 - q - 1. \tag{27.39}$$

So

$$\begin{aligned}\sum d_i &\leq (q^2 - q - 1)\theta(n-3) \\ &= q^{n-1} - q - \theta(n-3) \\ &< q^{n-1} - q - 1\end{aligned}$$

for $n \geq 4$. This contradiction implies that $\mathscr{K}$ has no $(q+2)$-arc as a plane section.

Now, for all $q > 2$, let P_1 and P_2 be points of $\mathscr{K}$ and consider the $\theta(n-2)$ planes through the line P_1P_2. They contain at most

$$(q-1)\theta(n-2) + 2 = q^{n-1} + 1 = k$$

points. So each of these planes must meet $\mathscr{K}$ in a $(q+1)$-arc.

Consider a prime Π_{n-1} through P_1P_2. The $\theta(n-3)$ planes through P_1P_2 in Π_{n-1} all meet $\mathscr{K}\backslash\{P_1, P_2\}$ in $q-1$ points, whence

$$|\Pi_{n-1} \cap \mathscr{K}| = (q-1)\theta(n-3) + 2 = q^{n-2} + 1.$$

So $\mathscr{K}' = \Pi_{n-1} \cap \mathscr{K}$ is a $(q^{n-2}+1)$-cap.

It follows from this argument that any prime meeting $\mathscr{K}$ meets it either in a single point or in a $(q^{n-2}+1)$-cap.

Let Q be a point of $\mathscr{K}$. There are q^{n-1} bisecants of $\mathscr{K}$ through Q and so $\theta(n-1) - q^{n-1} = \theta(n-2)$ unisecants. Let l_1 and l_2 be two of them. Then every line through Q in the plane l_1l_2 is also a unisecant to $\mathscr{K}$, as otherwise the plane l_1l_2 would meet $\mathscr{K}$ in a $(q+1)$-arc with two unisecants at Q, a contradiction. So the set S of points on the unisecants through Q has the property that the line joining two points of S is in S. Thus S is a subspace and hence a prime, which we may call the *tangent prime to* $\mathscr{K}$ *at* Q.

Let P_1 and P_2 be two points of $\mathscr{K}$ with tangent primes T_1 and T_2. As T_1 and T_2 are distinct, they meet in a Π_{n-2} skew to $\mathscr{K}$. Let r be the number of primes through Π_{n-2} other than T_1 and T_2 which are tangent to $\mathscr{K}$ and s the number of other primes through Π_{n-2} meeting $\mathscr{K}$. So

$$r + s \leq q - 1. \tag{27.40}$$

Counting the points of $\mathscr{K}\backslash\{P_1, P_2\}$ gives

$$r + s(q^{n-2} + 1) = q^{n-1} - 1. \tag{27.41}$$

However, from (27.40),

$$\begin{aligned} r + s(q^{n-2} + 1) &\leq q - 1 - s + s(q^{n-2} + 1) \\ &= q - 1 + sq^{n-2} \\ &\leq q - 1 + (q - 1)q^{n-2} \\ &= (q - 1)(q^{n-2} + 1). \end{aligned} \tag{27.42}$$

But, $(q - 1)(q^{n-2} + 1) < q^{n-1} - 1$. So (27.41) and (27.42) cannot both hold. This proves the theorem. □

Now we improve Theorem 27.3.1 first for q odd and then for q even. In both cases, however, it is necessary for q to be sufficiently large.

The main result for q odd is a consequence of the following result, which appeared as the corollary of Theorem 18.4.8.

Theorem 27.3.2: *In $PG(3, q)$, q odd and $q \geq 67$, if $\mathscr{K}$ is a complete k-cap which is not an elliptic quadric, then*

$$k < q^2 - \tfrac{1}{4}q^{3/2} + 2q.$$

In fact

$$k \leq q^2 - \tfrac{1}{4}q^{3/2} + R(q)$$

where

$$R(q) = (31q + 14\sqrt{q} - 53)/16. \quad \square$$

To obtain a similar result in $PG(4, q)$, we consider a k-cap $\mathscr{K}$ and examine the sections of $\mathscr{K}$ by three solids through a plane π which has a sufficiently large intersection with $\mathscr{K}$.

Lemma 27.3.3: *In $PG(4, q)$, $q \geq 67$ and odd, let $\mathscr{K}$ be a k-cap and π a plane such that $\pi \cap \mathscr{K}$ in an s-arc with $s > q - \frac{1}{4}\sqrt{q} + \frac{7}{4}$. Then there do not exist three distinct solids $\alpha_1, \alpha_2, \alpha_3$ containing π such that $\mathscr{K}_i = \alpha_i \cap \mathscr{K}$ is a k-cap with $k_i > q^2 - \frac{1}{4}q^{3/2} + R(q)$, $i = 1, 2, 3$.*

Proof: Suppose that the lemma is false. Then, by Theorem 27.3.2, each $\mathscr{K}_i$ is contained in a (unique) elliptic quadric $\mathscr{Q}_i$. Clearly $\mathscr{Q}_1 \cap \mathscr{Q}_2 \cap \mathscr{Q}_3$ is the unique conic in π containing $\pi \cap \mathscr{K}$.

(i) *There exists a quadric $\mathscr{Q}$ meeting α_i in $\mathscr{Q}_i$, $i = 1, 2, 3$*

The set $\mathscr{M} = \mathscr{K}_1 \cup \mathscr{K}_2 \cup \mathscr{K}_3$ is an m-cap contained in $\mathscr{K}$ with

$$m = s + \sum (k_i - s) = k_1 + k_2 + k_3 - 2s.$$

As $s \le q + 1$, so

$$m > 3(q^2 - \tfrac{1}{4}q^{3/2} + R(q)) - 2(q+1)$$
$$= 3(q^2 - \tfrac{1}{4}q^{3/2}) + \tfrac{1}{16}(61q + 42\sqrt{q} - 191). \qquad (27.43)$$

There are two possibilities for $\mathscr{Q}$. Either $\mathscr{Q} = \mathscr{P}_4$, the non-singular quadric, or $\mathscr{Q} = \Pi_0\mathscr{E}_3$, the singular quadric with vertex Π_0 and base $\mathscr{E}_3$.

(a) $\mathscr{Q} = \mathscr{P}_4$

$\mathscr{P}_4$ comprises $(q^2 + 1)(q + 1)$ points on the same number of lines with $q + 1$ lines through a point. Through each point of a line l on $\mathscr{P}_4$ there pass q other lines, whence $q(q + 1)$ lines l' on $\mathscr{P}_4$ meet l. No two of these lines l' meet off l, as otherwise their plane would meet $\mathscr{P}_4$ in a cubic curve. Also $\mathscr{P}_4$ contains $q^2(q + 1)$ points off l. So through each point of $\mathscr{P}_4 \backslash l$ there is exactly one line l'. The m-cap $\mathscr{M}$ has at most two points on l and on each l', and every point of $\mathscr{M}$ lies on l or some l'. Hence

$$m \le 2 + 2q(q+1) = 2(q^2 + q + 1). \qquad (27.44)$$

From (27.43) and (27.44),

$$3(q^2 - \tfrac{1}{4}q^{3/2}) + \tfrac{1}{16}(61q + 42\sqrt{q} - 191) < m \le 2(q^2 + q + 1).$$

Hence

$$q^2 - \tfrac{3}{4}q^{3/2} + \tfrac{1}{16}(29q + 42\sqrt{q} - 223) < 0,$$

a contradiction.

(b) $\mathscr{Q} = \Pi_0\mathscr{E}_3$

Through Π_0 there are $q^2 + 1$ generators of $\mathscr{Q}$, each containing at most two points of $\mathscr{M}$. So

$$m \le 2(q^2 + 1). \qquad (27.45)$$

From (27.43) and (27.45),

$$3(q^2 - \tfrac{1}{4}q^{3/2}) + \tfrac{1}{16}(61q + 42\sqrt{q} - 191) < m \le 2(q^2 + 1).$$

Hence

$$q^2 - \tfrac{3}{4}q^{3/2} + \tfrac{1}{16}(61q + 42\sqrt{q} - 223) < 0,$$

a contradiction.

(ii) *There is a pencil Φ of quadric primals through $\mathscr{Q}_1$ and $\mathscr{Q}_2$, none of which contains $\mathscr{Q}_3$*

The members of Φ cut out on α_3 a pencil Φ' of quadric surfaces all containing the conic $\mathscr{C}$, the unique conic through $\pi \cap \mathscr{K}$. One member of Φ' is π repeated, and $\mathscr{Q}_3 \notin \Phi'$. So Φ' cuts out on $\mathscr{Q}_3$ a set of quartic curves $\mathscr{C} \cup \mathscr{C}'$ with $\mathscr{C}'$ quadric curves in planes π' of a pencil in α_3; each quadric $\mathscr{C}'$ is either a conic or a point. Denote the set of quadrics $\mathscr{C}'$ by Ψ. Then $\mathscr{C} \in \Psi$, and the planes π' have a common line l in π.

As $k_3 - (q+1) > 2(q+1) + (q-2)(q - \sqrt{q}/4)$, there are at least three planes π' other than π meeting $\mathcal{K}_3$ in a k'-arc $\mathcal{K}'$ with $k' > q - \frac{1}{4}\sqrt{q}$. Since $\mathcal{K}' \subset \mathcal{Q}_3$, each of these $\mathcal{K}'$ is contained in a conic $\mathcal{C}' = \pi' \cap \mathcal{Q}_3$. For at least one of these planes the quadric $\mathcal{V}$ of Φ meeting $\mathcal{Q}_3$ in $\mathcal{C} \cup \mathcal{C}'$ is non-singular. It will now be shown that for such a $\mathcal{K}'$ there exists a line $P'P_1P_2$, where $P' \in \mathcal{K}'$, $P_1 \in \mathcal{K}_1$, $P_2 \in \mathcal{K}_2$.

Take a point P' in $\mathcal{C}' \backslash \mathcal{C}$. Since it is simple for $\mathcal{V}$, the tangent space $T_{P'}(\mathcal{V})$ to $\mathcal{V}$ at P' meets $\mathcal{V}$ in a cone $P'\mathcal{P}_2$. So there are $q+1$ lines l' of $\mathcal{V}$ in $T_{P'}(\mathcal{V})$. As $\mathcal{C}'$ is non-singular, the space $T_{P'}(\mathcal{V})$ does not contain $\mathcal{C}$. Consequently, $T_{P'}(\mathcal{V})$ meets $\mathcal{C}$ in at most two points, whence at most two lines l' meet π. The others, in number at least $q-1$, all meet α_1 in a point P_1 of $\mathcal{Q}_1$ and α_2 in a point P_2 of $\mathcal{Q}_2$, with P_1, P_2 not in $\mathcal{C}$. Also, $P_1 \neq P_2$ since $\alpha_1 \cap \alpha_2 = \pi$ and $P_1, P_2 \notin \pi$. Further, $P_i \neq P'$, $i = 1, 2$, since every point of $\alpha_i \cap \mathcal{C}'$ lies on $\mathcal{C}$.

Let $P' \in \mathcal{K}' \backslash \mathcal{C}$ and note that $|\mathcal{K}' \backslash \mathcal{C}| > q - \frac{1}{4}\sqrt{q} - 2$. For each such P', there are at least $q-1$ points P_1. Conversely, each P_1 is derived from at most two P', namely $\mathcal{K}' \cap T_{P_1}(\mathcal{V})$, unless $T_{P_1}(\mathcal{V})$ contains π' and hence $\mathcal{K}'$. This exceptional case can only occur twice, when P_1 lies on the polar line of the plane π'. Thus each P' gives at least $q-3$ points P_1, apart from the exceptions; each P_1 comes from at most two P'. Thus, with $A = \{P_1$ in $\mathcal{Q}_1$ obtainable from some P' in $\mathcal{K}'\}$,

$$\begin{aligned} |A| &> \tfrac{1}{2}(q - \tfrac{1}{4}\sqrt{q} - 2)(q-3) \\ &= \tfrac{1}{2}q^2 - \tfrac{1}{8}q^{3/2} - \tfrac{5}{2}q + \tfrac{3}{8}\sqrt{q} + 3. \end{aligned}$$

Let $B = \mathcal{K}_1$, $C = \mathcal{Q}_1$, and $\mathcal{K}_1^* = \{P_1$ in $\mathcal{K}_1$ obtainable from some P' in $\mathcal{K}'\}$. Then $\mathcal{K}_1^* = A \cap B$. So, by Lemma 27.1.1,

$$\begin{aligned} |\mathcal{K}_1^*| &> \tfrac{1}{2}q^2 - \tfrac{1}{8}q^{3/2} - \tfrac{5}{2}q + \tfrac{3}{8}\sqrt{q} + 3 + (q^2 - \tfrac{1}{4}q^{3/2} + R(q)) - (q^2+1) \\ &= \tfrac{1}{2}q^2 - \tfrac{3}{8}q^{3/2} - \tfrac{1}{16}(9q - 20\sqrt{q} + 21). \end{aligned}$$

The line $P'P_1$ with P' in $\mathcal{K}'$ and P_1 in $\mathcal{K}_1^*$ meets α_2 in a point P_2 of $\mathcal{Q}_2 \backslash \mathcal{C}$. Such a P_2 is obtained at most twice when $|T_{P_2}(\mathcal{V}) \cap \mathcal{K}'| \leq 2$, unless $T_{P_2}(\mathcal{V}) \supset \pi'$, which can occur for at most two points P_2, where the polar line of π' meets $\mathcal{Q}_2$. Thus, with $A = \{P_2$ in $\mathcal{Q}_2$ obtainable from some $P_1P'\}$,

$$\begin{aligned} |A| &\geq 2 + \tfrac{1}{2}(|\mathcal{K}_1^*| - 2|\mathcal{K}' \backslash \mathcal{C}|) \geq 2 + \tfrac{1}{2}|\mathcal{K}_1^*| - (q+1) \\ &> \tfrac{1}{4}q^2 - \tfrac{3}{16}q^{3/2} - \tfrac{1}{32}(41q - 20\sqrt{q} - 11). \end{aligned}$$

Now, let $B = \mathcal{K}_2$, and $C = \mathcal{Q}_2$. Therefore, if $\mathcal{K}_2^* = \{P_2 \in \mathcal{K}_2 \mid P'P_1P_2$ is a line with $P' \in \mathcal{K}'$, $P_1 \in \mathcal{K}_1^*\}$, Lemma 27.1.1 gives that

$$\begin{aligned} |\mathcal{K}_2^*| &> \tfrac{1}{4}q^2 - \tfrac{3}{16}q^{3/2} - \tfrac{1}{32}(41q - 20\sqrt{q} - 11) \\ &\quad + (q^2 - \tfrac{1}{4}q^{3/2} + R(q)) - (q^2+1) \\ &= \tfrac{1}{4}q^2 - \tfrac{7}{16}q^{3/2} + \tfrac{1}{32}(21q + 48\sqrt{q} - 127) \\ &> 0. \end{aligned}$$

So there is a line meeting $\mathcal{K}'$, $\mathcal{K}_1$, $\mathcal{K}_2$ in distinct points. So $\mathcal{K}$ is not a cap, which provides the desired contradiction. □

Theorem 27.3.4: *In* $PG(n, q)$, $n \geq 4$, $q \geq 197$ *and odd,*

$$m_2(n, q) < q^{n-1} - \tfrac{1}{4}q^{n-3/2} + 2q^{n-2}.$$

In fact, for $q \geq 67$ *and odd,*

$$\begin{aligned} m_2(n, q) < q^{n-1} &- \tfrac{1}{4}q^{n-3/2} \\ &+ \tfrac{1}{16}(31q^{n-2} + 22q^{n-5/2} - 112q^{n-3} - 14q^{n-7/2} + 69q^{n-4}) \\ &- 2(q^{n-5} + \cdots + 1) + 1, \end{aligned}$$

where there is no term $-2(q^{n-5} + \cdots + 1)$ *for* $n = 4$.

Proof: Let $\mathcal{K}$ be a k-cap in $PG(n, q)$.

(i) $n = 4$

(a) There is no plane π such that $\pi \cap \mathcal{K}$ is an s-arc with

$$s > q - \tfrac{1}{4}\sqrt{q} + \tfrac{7}{4}.$$

Take a line l meeting $\mathcal{K}$ in two points. There are $q^2 + q + 1$ planes π through l each meeting $\mathcal{K}$ in an m-arc with $m \leq q - \frac{1}{4}\sqrt{q} + \frac{7}{4}$. So

$$\begin{aligned} k &\leq 2 + (q^2 + q + 1)(q - \tfrac{1}{4}\sqrt{q} - \tfrac{1}{4}) \\ &= q^3 - \tfrac{1}{4}q^{5/2} + \tfrac{1}{4}(3q^2 - q^{3/2} + 3q - \sqrt{q} - 1) \\ &< q^3 - \tfrac{1}{4}q^{5/2} + \tfrac{1}{16}(31q^2 + 22q^{3/2} - 112q - 14q^{1/2} + 85). \end{aligned}$$

(b) There is a plane π such that $\pi \cap \mathcal{K}$ is an s-arc with $s > q - \frac{1}{4}\sqrt{q} + \frac{7}{4}$. Then by Lemma 27.3.3, for $q \geq 67$, there are at most two solids through π meeting $\mathcal{K}$ in an elliptic quadric, and, for the other $q - 1$ solids α through π, $|\alpha \cap \mathcal{K}| \leq q^2 - \frac{1}{4}q^{3/2} + R(q)$. So

$$\begin{aligned} k &\leq s + 2(q^2 + 1 - s) + (q - 1)[q^2 - \tfrac{1}{4}q^{3/2} + \tfrac{1}{16}(31q + 14\sqrt{q} - 53) - s] \\ &= q^3 - \tfrac{1}{4}q^{5/2} + \tfrac{1}{16}(47q^2 + 18q^{3/2} - 84q - 14q^{1/2} + 85) - sq \\ &< q^3 - \tfrac{1}{4}q^{5/2} + \tfrac{1}{16}(47q^2 + 18q^{3/2} - 84q - 14q^{1/2} + 85) - q(q - \tfrac{1}{4}\sqrt{q} + \tfrac{7}{4}) \\ &= q^3 - \tfrac{1}{4}q^{5/2} + \tfrac{1}{16}(31q^2 + 22q^{3/2} - 112q - 14q^{1/2} + 85) \\ &< q^3 - \tfrac{1}{4}q^{5/2} + 2q^2 \quad \text{for } q \geq 197. \end{aligned}$$

(ii) $n > 4$

The induction formula of Theorem 27.2.12 gives that

$$\begin{aligned} m_2(n, q) &\le q^{n-4}m_2(4, q) - q^{n-4} - 2(q^{n-5} + \cdots + 1) + 1 \\ &< q^{n-1} - \tfrac{1}{4}q^{n-3/2} + \tfrac{1}{16}(31q^{n-2} + 22q^{n-5/2} \\ &\quad - 112q^{n-3} - 14q^{n-7/2} + 69q^{n-4}) \\ &\quad - 2(q^{n-5} + \cdots + 1) + 1 \quad \text{for } q \ge 67 \\ &< q^{n-1} - \tfrac{1}{4}q^{n-3/2} + 2q^{n-2} - q^{n-4} \\ &\quad - 2(q^{n-5} + \cdots + 1) + 1 \quad \text{for } q \ge 197 \\ &< q^{n-1} - \tfrac{1}{4}q^{n-3/2} + 2q^{n-2} \quad \text{for } q \ge 197. \quad \square \end{aligned}$$

27.4 The maximum size of a cap for q even

Before looking at an upper bound for $m_2(n, q)$ for q even and $q > 2$, it is necessary to improve the upper bound for $m_2'(3, q)$ from Theorem 18.3.2. We recall that $m_2'(3, q)$ is the size of the second largest cap in $PG(3, q)$; alternatively, if a k-cap has $k > m_2'(3, q)$, then it is contained in a $(q^2 + 1)$-cap. In Theorem 18.3.2 it was shown that, for q even with $q > 2$,

$$m_2'(3, q) \le q^2 - \tfrac{1}{2}\sqrt{q} + 1.$$

For any k-cap $\mathcal{K}$ in $PG(n, q)$, as above, a 1-secant line is called a *tangent* or a *unisecant*, a 2-secant line is a *bisecant*, and a 0-secant line is an *external line*. Also, let t be the number of tangents through a point P of $\mathcal{K}$; for a point Q not in $\mathcal{K}$, let $\sigma_1(Q)$ be the number of tangents through Q and let $\sigma_2(Q)$ be the number of bisecants through Q.

Lemma 27.4.1: *For a k-cap $\mathcal{K}$ in $PG(n, q)$,*

(i) $t + k = \theta(n-1) + 1$;
(ii) $\sigma_1(Q) + 2\sigma_2(Q) = k$. $\square$

Lemma 27.4.2: *In $PG(n, q)$ with q even, if Q is a point not on the k-cap $\mathcal{K}$ such that $\sigma_2(Q) \ge 1$, then $\sigma_1(Q) \le t$.*

Proof: See Lemma 18.3.1, where the proof is given for $PG(3, q)$, but extends immediately to $PG(n, q)$. $\square$

Corollary: *If $\mathcal{K}$ is complete, then $\sigma_1(Q) \le t$ for all points Q off $\mathcal{K}$.* $\square$

Theorem 27.4.3: *For q even, $q > 2$,*

$$m_2'(3, q) \leq q^2 - \tfrac{1}{2}q - \tfrac{1}{2}\sqrt{q} + 2.$$

Proof: Suppose that there exists a complete k-cap $\mathscr{K}$ in $PG(3, q)$ with

$$q^2 - \tfrac{1}{2}q - \tfrac{1}{2}\sqrt{q} + 2 < k < q^2 + 1. \tag{27.46}$$

We reach a contradiction in several stages. First, (27.46) and Lemma 27.4.1(i) imply that

$$t < \tfrac{3}{2}q + \tfrac{1}{2}\sqrt{q}. \tag{27.47}$$

I. *Through a tangent l there exists a plane α_0 such that $|\alpha_0 \cap \mathscr{K}| \leq q - \sqrt{q} + 1$ except possibly when $q = 8$ and $k = 61$*

Suppose that $|\pi_i \cap \mathscr{K}| > q - \sqrt{q} + 1$ for the $q + 1$ planes $\pi_0, \ldots, \pi_q$ through l. Let $l \cap \mathscr{K} = \{P\}$. By Theorem 10.3.3, Corollary 1, $\pi_i \cap \mathscr{K}$ can be completed to a $(q + 2)$-arc, which meets l in a point N_i other than P. As there are $q + 1$ points N_i and only q points on l for them to occupy, two of the N_i coincide; say $N_0 = N_1$. Then the tangents to $\mathscr{K}$ include the joins of N_0 to the points of $\pi_0 \cap \mathscr{K}$ and $\pi_1 \cap \mathscr{K}$. Hence

$$\sigma_1(N_0) > 1 + 2(q - \sqrt{q}). \tag{27.48}$$

By the corollary to Lemma 27.4.2, $\sigma_1(N_0) \leq t$. So, by (27.47) and (27.48),

$$2(q - \sqrt{q}) + 1 < \tfrac{3}{2}q + \tfrac{1}{2}\sqrt{q};$$

that is,

$$q - 5\sqrt{q} + 2 < 0,$$

which is impossible for $q > 16$.

For $q = 16$, (27.48) improves to $\sigma_1(N_0) \geq 27$ and (27.47) is $t < 26$; hence there is again a contradiction. For $q = 8$, (27.48) improves to $\sigma_1(N_0) \geq 13$ and (27.47) improves to $t \leq 13$. So $\sigma_1(N_0) = t = 13$; this gives the possibility of a 61-cap in $PG(3, 8)$, allowed in the statement of this lemma. For $q = 4$, (27.48) becomes $\sigma_1(N_0) \geq 7$ and (27.47) becomes $t \leq 6$; again giving a contradiction.

II. *There exists a plane α_1 through the tangent l such that $|\alpha_1 \cap \mathscr{K}| = q + 1$*

Suppose, on the contrary, that $|\pi_i \cap \mathscr{K}| < q + 1$ for all planes π_i through l.

First assume that $\mathscr{K}$ is not a 61-cap in $PG(3, 8)$. Let x equal the number of planes π_i meeting $\mathscr{K}$ in a q-arc. Then a count of the points of $\mathscr{K}$ in the planes π_i gives

$$\begin{aligned} k &\leq x(q - 1) + (q - x)(q - 2) + (q - \sqrt{q} + 1) \qquad (27.49) \\ &= x + q^2 - q - \sqrt{q} + 1. \end{aligned}$$

Combining (27.49) with (27.46) gives

$$x > \tfrac{1}{2}q + \tfrac{1}{2}\sqrt{q} + 1.$$

Let $\mathscr{R}$ be such a q-arc and let $\bar{\mathscr{R}}$ be the $(q + 2)$-arc containing $\mathscr{R}$. If N is the point of $l \cap \bar{\mathscr{R}}$ other than P, then, by Lemma 27.4.1(ii), there exists a tangent through N, apart from l, to any of these q-arcs. Hence

$$\begin{aligned}\sigma_1(N) &\geq q + x - 1\\ &> \tfrac{3}{2}q + \tfrac{1}{2}\sqrt{q} > t,\end{aligned}$$

a contradiction.

Now, let $\mathscr{K}$ be a 61-cap in $PG(3, 8)$. If, for some plane π_i, we have $|\pi_i \cap \mathscr{K}| \leq q - \sqrt{q} + 1$, then the argument above applies. So, assume that $|\pi_i \cap \mathscr{K}| \geq 7$, $i = 0, \ldots, 8$. Then, counting the points of $\mathscr{K}$ in the planes π_i, we obtain

$$61 = x \cdot 7 + (9 - x)6 + 1,$$

with x as above. Hence $x = 6$. With N_0 as in the proof of I, we have $\sigma_1(N_0) = 13$. By Lemma 27.4.1(ii), each of the six 8-arcs has at least two tangents through N. Hence $\sigma_1(N_0)$ is at least $13 + 6$; a contradiction.

III. *The tangents to $\mathscr{K}$ through the nucleus M of the $(q + 1)$-arc $\alpha_1 \cap \mathscr{K}$ all lie in the plane α_1*

Assume that there is a tangent l_1 to $\mathscr{K}$ through M but not in α_1. By II, there exists a plane π through l_1 such that $|\pi \cap \mathscr{K}| = q + 1$. So $\pi \cap \alpha_1 = l_2$ is a tangent of $\pi \cap \mathscr{K}$ through M. Since M lies on two distinct tangents l_1 and l_2 of the $(q + 1)$-arc $\pi \cap \mathscr{K}$, it is the nucleus of $\pi \cap \mathscr{K}$. So $\sigma_1(M) \geq 2q + 1$. Hence $2q + 1 < \tfrac{3}{2}q + \tfrac{1}{2}\sqrt{q}$, a contradiction. So l_1 does not exist.

IV. *When k is even, the theorem is true*

Since k is even, $\sigma_1(N)$ is even for any point N not in $\mathscr{K}$, by Lemma 27.4.1(ii). In particular, with M as in III, $\sigma_1(M)$ is even. But, by III, $\sigma_1(M) = q + 1$, which is odd; a contradiction.

V. *When k is odd, either $|\pi \cap \mathscr{K}| = q + 2$ or $|\pi \cap \mathscr{K}|$ is odd for any plane π*

Suppose $|\pi \cap \mathscr{K}| < q + 2$ so that π contains a tangent, which we may take as l. From II, the line l lies in the plane α_1 and the nucleus M lies on l. If $|\pi \cap \mathscr{K}|$ is even, then $\pi \neq \alpha_1$ and the number of tangents through M to $\pi \cap \mathscr{K}$ is even. So there exists a tangent to $\mathscr{K}$ through M which is not in α_1, contradicting III.

VI. *With k odd, for any tangent l the number y of planes π_i through l for which* $|\pi_i \cap \mathscr{K}| = q + 1$ *satisfies* $y > \frac{1}{4}q - \frac{1}{4}\sqrt{q} + \frac{3}{2}$

With $l \cap \mathscr{K} = \{P\}$, then $|\pi_i \cap \mathscr{K}| = q + 1$ if and only if there is exactly one tangent through P in π_i. Since $|\pi_i \cap \mathscr{K}|$ is odd, the number of tangents in π_i through P is one or at least three, by Lemma 27.4.1(i). So a count on these tangents gives

$$2(q + 1 - y) + 1 \leq t < \tfrac{3}{2}q + \tfrac{1}{2}\sqrt{q},$$

whence the result.

VII. *For any plane* α_0 *with* $0 < |\alpha_0 \cap \mathscr{K}| \leq q - \sqrt{q} + 1$, *we also have* $|\alpha_0 \cap \mathscr{K}| > \frac{1}{2}q - \frac{1}{2}\sqrt{q} + 2$

Through the tangent l of $\alpha_0 \cap \mathscr{K}$, where $l \cap \mathscr{K} = \{P\}$, there is the plane α_1, by II, with M the nucleus of $\alpha_1 \cap \mathscr{K}$. Suppose first that $|\alpha_0 \cap \mathscr{K}| = 1$. Since $k < q^2 + 1$, through P there is a tangent l' not contained in α_0. By II, there is a plane π through l' with $|\pi \cap \mathscr{K}| = q + 1$. Then, at P, the section $\pi \cap \mathscr{K}$ has at least two tangents, l' and $\pi \cap \alpha_0$; a contradiction. So $|\alpha_0 \cap \mathscr{K}| \geq 3$. Since all the tangents through M are in α_1, there is a bisecant b through M and a point of $\alpha_0 \cap \mathscr{K}$ other than P. Counting the points of $\mathscr{K}$ on the planes through b gives

$$k \leq q(q - 1) + |\alpha_0 \cap \mathscr{K}|; \qquad (27.50)$$

(27.46) and (27.50) imply that

$$|\alpha_0 \cap \mathscr{K}| > \tfrac{1}{2}q - \tfrac{1}{2}\sqrt{q} + 2.$$

For $q = 4$, this already provides a contradiction to I.

VIII. $\mathscr{K}$ *has at least* $q\sqrt{q} + 1$ *tangents in the plane* α_0

A plane m-arc has $F(m) = m(q + 2 - m)$ tangents. By VII,

$$\tfrac{1}{2}q - \tfrac{1}{2}\sqrt{q} + 2 < |\alpha_0 \cap \mathscr{K}| \leq q - \sqrt{q} + 1.$$

Now,

$$F(\tfrac{1}{2}q - \tfrac{1}{2}\sqrt{q} + 2) = \tfrac{1}{4}q^2 + \tfrac{3}{4}q + \sqrt{q},$$

$$F(q - \sqrt{q} + 1) = q\sqrt{q} + 1.$$

For $q \geq 4$,

$$\tfrac{1}{4}q^2 + \tfrac{3}{4}q + \sqrt{q} \geq q\sqrt{q} + 1.$$

So, in the interval $[\frac{1}{2}q - \frac{1}{2}\sqrt{q} + 2, q - \sqrt{q} + 1]$, it follows that $F(m) \geq q\sqrt{q} + 1$; that is, the arc $\alpha_0 \cap \mathscr{K}$ has at least $q\sqrt{q} + 1$ tangents in the plane α_0.

IX. *The theorem is true for k odd*

With α_0 as in VII, consider all the planes through the tangents of $\alpha_0 \cap \mathscr{K}$ which meet $\mathscr{K}$ in a $(q+1)$-arc. If two of the corresponding nuclei coincide at N, then $\sigma_1(N) \geq 2q+1$, which again contradicts Lemma 27.4.2 and (27.47). So all the nuclei are distinct. By VI, the number of $(q+1)$-arcs through a tangent is greater than $\frac{1}{4}q - \frac{1}{4}\sqrt{q} + \frac{3}{2}$ and, by VIII, there are at least $q\sqrt{q}+1$ tangents in α_0. Hence, the number of nuclei in α_0 is greater than

$$(\tfrac{1}{4}q - \tfrac{1}{4}\sqrt{q} + \tfrac{3}{2})(q\sqrt{q}+1).$$

So

$$(\tfrac{1}{4}q - \tfrac{1}{4}\sqrt{q}+\tfrac{3}{2})(q\sqrt{q}+1) + (\tfrac{1}{2}q - \tfrac{1}{2}\sqrt{q}+2) < |\alpha_0| = q^2+q+1; \qquad (27.51)$$

hence

$$q^2\sqrt{q} - 5q^2 + 6q\sqrt{q} - q - 3\sqrt{q} + 10 < 0,$$

which is a contradiction for $q \neq 8$. For $q = 8$, instead of (27.51) we obtain that $3.24 + 5 \leq 73$, again a contradiction.

Finally, suppose that there is no plane α_0 with $0 < |\alpha_0 \cap \mathscr{K}| \leq q - \sqrt{q} + 1$. By I, it follows that $q = 8$ and $k = 61$. So, from V, every plane through a tangent meets $\mathscr{K}$ in a 7-arc or a 9-arc. Let z be the number of 9-arcs in planes through a tangent l. Then, counting the points of $\mathscr{K}\backslash l$ in the planes through l gives

$$8z + 6(9 - z) = 60,$$

whence $z = 3$. As there are nine tangents to a 9-arc in its plane and 13 tangents to $\mathscr{K}$ through any of its points, the number of 9-arcs on $\mathscr{K}$ is 61.13.3/9. As this is not an integer, we have a contradiction. □

Theorem 27.4.4: (i) $m_2'(3,4) = 14$;

(ii) *a complete* 14-*cap in* $PG(3,4)$ *consists of the points on the generators of a cone* $P\beta$, *where* P *is the vertex and* β *is a* $PG(2,2)$, *outside the* $PG(3,2)$ *containing* P *and* β.

Proof: See §27.8. □

Now we consider the maximum number of points on a cap in $PG(4,q)$.

Theorem 27.4.5: *For q even,* $q > 2$,

$$m_2(4,q) \leq q^3 - \tfrac{1}{2}q^2 - \tfrac{1}{2}q\sqrt{q} + \tfrac{5}{2}q + \tfrac{1}{2}\sqrt{q} + c,$$

where $c = 0$ *for* $q \geq 8$, *and* $c = -3$ *for* $q = 4$.

Proof: In the proof, we ignore the case $q = 4$. For this value, the method is exactly the same, but the numerical calculations differ due to the fact that the exact value for $m_2'(3, 4)$ is not given by the same expression as the bound for $m_2'(3, q)$ when $q \geq 8$.

Suppose there exists a complete k-cap $\mathscr{K}$ in $PG(4, q)$ with

$$k > q^3 - \tfrac{1}{2}q^2 - \tfrac{1}{2}q\sqrt{q} + \tfrac{5}{2}q + \tfrac{1}{2}\sqrt{q}. \tag{27.52}$$

Then, with t the number of tangents through a point of $\mathscr{K}$, we have

$$t < \tfrac{3}{2}q^2 + \tfrac{1}{2}q\sqrt{q} - \tfrac{3}{2}q - \tfrac{1}{2}\sqrt{q} + 2 \tag{27.53}$$

by Lemma 27.4.1(i). We obtain a contradiction in several stages.

I. *$\mathscr{K}$ contains no plane q-arc*

Suppose that π is a plane such that $\pi \cap \mathscr{K}$ is a q-arc $\mathscr{L}$. Consider two subcases.

(a) Suppose there exist three solids $\delta_1, \delta_2, \delta_3$ containing the plane π such that, for $i = 1, 2, 3$,

$$|\delta_i \cap \mathscr{K}| > q^2 - \tfrac{1}{2}q - \tfrac{1}{2}\sqrt{q} + 2.$$

Then, by Theorem 27.4.3, $\delta_i \cap \mathscr{K}$ can be completed to an ovoid $\mathscr{O}_i$. So $\mathscr{O}_i \cap \pi$ is a $(q + 1)$-arc $\mathscr{L} \cup \{N_i\}$. However, since $\mathscr{L}$ can be contained in no more than two $(q + 1)$-arcs, at least two of the N_i coincide; say $N_1 = N_2$. The joins of N_1 to the points of $\delta_1 \cap \mathscr{K}$ and $\delta_2 \cap \mathscr{K}$ are all tangents to $\mathscr{K}$. Hence

$$\sigma_1(N_1) > 2(q^2 - \tfrac{3}{2}q - \tfrac{1}{2}\sqrt{q} + 2) + q. \tag{27.54}$$

Since $\mathscr{K}$ is complete, $\sigma_1(N_1) \leq t$ by Lemma 27.4.2. So, (27.53) and (27.54) imply that

$$q^2 - q\sqrt{q} - q - \sqrt{q} + 4 < 0,$$

a contradiction.

(b) Suppose there are at most two solids δ_1 and δ_2 through π such that, for $i = 1, 2$,

$$|\delta_i \cap \mathscr{K}| > q^2 - \tfrac{1}{2}q - \tfrac{1}{2}\sqrt{q} + 2.$$

Counting the points of $\mathscr{K}$ on the solids through π gives

$$\begin{aligned} k &\leq (q - 1)(q^2 - \tfrac{3}{2}q - \tfrac{1}{2}\sqrt{q} + 2) + 2(q^2 + 1 - q) + q \\ &= q^3 - \tfrac{1}{2}q^2 - \tfrac{1}{2}q\sqrt{q} + \tfrac{5}{2}q + \tfrac{1}{2}\sqrt{q}, \end{aligned}$$

in contradiction to (27.52).

II. *There exists no solid δ such that $q^2 + 1 > |\delta \cap \mathscr{K}| > q^2 - \tfrac{1}{2}q - \tfrac{1}{2}\sqrt{q} + 2$*

Suppose δ exists. Let $\delta \cap \mathscr{K} = \mathscr{K}'$. Then $\mathscr{K}'$ can be completed to an ovoid $\mathscr{O}$, by Theorem 27.4.3. Let $N \in \mathscr{O} \backslash \mathscr{K}'$ and let $N' \in \mathscr{K}'$. Consider the $q + 1$

planes of δ through NN'. Since each of these planes meets $\mathcal{O}$ in a $(q+1)$-arc, each plane meets $\mathcal{K}'$ in at most a q-arc. By I, there is no q-arc on $\mathcal{K}$; so each plane meets $\mathcal{K}'$ in at most a $(q-1)$-arc. Therefore, a count on the points of $\mathcal{K}'$ gives

$$q^2 - \tfrac{1}{2}q - \tfrac{1}{2}\sqrt{q} + 2 < |\mathcal{K}'| \leq (q+1)(q-2) + 1,$$

whence

$$q - \sqrt{q} + 6 < 0,$$

a contradiction.

III. *For a point N not in $\mathcal{K}$, there do not exist planes π_1 and π_2 such that $\pi_1 \cap \pi_2 = \{N\}$ and such that $\pi_i \cap \mathcal{K}$ is a $(q+1)$-arc with nucleus N for $i = 1, 2$*

Suppose π_1 and π_2 exist. Let δ be a solid containing π_1. Then $\delta \cap \mathcal{K}$ contains at least $q+2$ tangents through N, $q+1$ of which are in π_1 and one of which is in π_2; so $|\delta \cap \mathcal{K}| < q^2 + 1$. Suppose now that

$$|\delta \cap \mathcal{K}| \leq q^2 - \tfrac{1}{2}q - \tfrac{1}{2}\sqrt{q} + 2$$

for any such solid δ. Then a count on the points of $\mathcal{K}$ in the solids through π_1 gives

$$k \leq (q+1)(q^2 - \tfrac{3}{2}q - \tfrac{1}{2}\sqrt{q} + 1) + (q+1);$$

that is,

$$k \leq q^3 - \tfrac{1}{2}q^2 - \tfrac{1}{2}q\sqrt{q} + \tfrac{1}{2}q - \tfrac{1}{2}\sqrt{q} + 2. \tag{27.55}$$

But (27.52) and (27.55) imply that

$$2q + \sqrt{q} - 2 < 0,$$

a contradiction. Thus there exists a solid δ such that

$$q^2 + 1 > |\delta \cap \mathcal{K}| > q^2 - \tfrac{1}{2}q - \tfrac{1}{2}\sqrt{q} + 2.$$

But this contradicts II. So π_1 and π_2 do not exist.

IV. *The tangents through any point Q off $\mathcal{K}$ lie in a solid*

Let δ be a solid not containing Q and let $\mathcal{V}$ be the set of intersections of tangents to $\mathcal{K}$ through Q with δ. We show that each point of $\mathcal{V}$ is on at least two lines of $\mathcal{V}$.

Let $R \in \mathcal{V}$ and let r be the corresponding tangent. Suppose, for at most one plane π through r, that $|\pi \cap \mathcal{K}| = q + 1$. Then, since there is no q-arc on $\mathcal{K}$, counting the points of $\mathcal{K}$ on the planes through r gives

$$k \leq (q^2 + q)(q - 2) + q + 1,$$

a contradiction to (27.52).

Now, let π_1 and π_2 be planes through r meeting $\mathscr{K}$ in $(q+1)$-arcs. If Q is the nucleus of both $\pi_1 \cap \mathscr{K}$ and $\pi_2 \cap \mathscr{K}$, then there are two lines of $\mathscr{V}$ through R, namely $\pi_1 \cap \delta$ and $\pi_2 \cap \delta$. Therefore, suppose that Q is not the nucleus of $\pi_1 \cap \mathscr{K}$. If, for at most one solid δ' through π_1, we have $|\delta' \cap \mathscr{K}| = q^2 + 1$, then, by II,

$$k \leq q^2 + 1 + q(q^2 - \tfrac{1}{2}q - \tfrac{1}{2}\sqrt{q} + 2 - q - 1),$$

whence

$$k \leq q^3 - \tfrac{1}{2}q^2 - \tfrac{1}{2}q\sqrt{q} + q + 1. \tag{27.56}$$

But (27.56) contradicts (27.52). Thus there are two solids δ_1 and δ_2 through π_1 for which $\delta_i \cap \mathscr{K} = \mathscr{O}_i$ is an ovoid. Then Q is the nucleus of a $(q+1)$-arc $\mathscr{M}_i$ on $\mathscr{O}_i$ for $i = 1, 2$. The tangents of $\mathscr{M}_i$ meet δ in a line l_i through R, and the lines l_1 and l_2 are distinct since Q is not the nucleus of $\pi_1 \cap \mathscr{K}$. Thus R always lies on at least two lines of $\mathscr{V}$.

If there existed two skew lines in $\mathscr{V}$, then there would exist two planes π_1 and π_2 through Q with $\pi_1 \cap \pi_2 = \{Q\}$ and Q the nucleus of both $\pi_1 \cap \mathscr{K}$ and $\pi_2 \cap \mathscr{K}$, in contradiction to III. Thus the lines of $\mathscr{V}$ either all have a common point or all lie in a plane. Since each point of $\mathscr{V}$ is on at least two lines of $\mathscr{V}$, all lines of $\mathscr{V}$ lie in a plane. Hence $\mathscr{V}$ is a subset of a plane and the tangents to $\mathscr{K}$ through Q lie in a solid.

V. *The final contradiction is obtained by counting the tangents of* $\mathscr{K}$

Consider the function

$$G(x) = x(q^3 + q^2 + q + 2 - x);$$

it attains its maximum value for $x = \tfrac{1}{2}(q^3 + q^2 + q + 2)$. Since, by Theorem 27.3.1 and (27.52),

$$q^3 \geq k > q^3 - \tfrac{1}{2}q^2 - \tfrac{1}{2}q\sqrt{q} + \tfrac{5}{2}q + \tfrac{1}{2}\sqrt{q} > \tfrac{1}{2}(q^3 + q^2 + q + 2),$$

we have that

$$kt = k(q^3 + q^2 + q + 2 - k) \geq G(q^3) = q^3(q^2 + q + 2). \tag{27.57}$$

By IV, all tangents through a point Q off $\mathscr{K}$ lie in a solid, which contains at most $q^2 + 1$ points of $\mathscr{K}$. However, an ovoid has exactly $q + 1$ tangents through an external point. So, through Q, there are at most q^2 tangents of $\mathscr{K}$. A count of the pairs (R, r) where R is a point off $\mathscr{K}$ and r a tangent to $\mathscr{K}$ through R gives

$$(q^4 + q^3 + q^2 + q + 1 - k)q^2 \geq ktq. \tag{27.58}$$

From (27.52), (27.57), and (27.58) we have

$$(q^4 + q^3 + q^2 + q + 1 - q^3 + \tfrac{1}{2}q^2 + \tfrac{1}{2}q\sqrt{q} - \tfrac{5}{2}q - \tfrac{1}{2}\sqrt{q})q$$
$$> kt \geq q^3(q^2 + q + 2).$$

Hence

$$q^4 + \tfrac{3}{2}q^2 + \tfrac{1}{2}q\sqrt{q} - \tfrac{3}{2}q - \tfrac{1}{2}\sqrt{q} + 1 > q^4 + q^3 + 2q^2,$$

and

$$q^3 + \tfrac{1}{2}q^2 - \tfrac{1}{2}q\sqrt{q} + \tfrac{3}{2}q + \tfrac{1}{2}\sqrt{q} - 1 < 0,$$

the final contradiction. □

Finally, an upper bound for the size of a k-cap in $PG(n, q)$ is obtained.

Theorem 27.4.6: *For q even, $q > 2$, $n \geq 5$,*

$$m_2(n, q) \leq q^{n-1} - \tfrac{1}{2}q^{n-2} - \tfrac{1}{2}q^{n-5/2} + \tfrac{5}{2}q^{n-3} + \tfrac{1}{2}q^{n-7/2} + (c-1)q^{n-4} - 2(q^{n-5} + \cdots + q + 1) + 1,$$

whence $c = 0$ for $q \geq 8$, and $c = -3$ for $q = 4$.

Proof: This follows directly from Theorems 27.4.5 and 27.2.12(ii). □

Remark: Theorems (27.4.5) and (27.4.6) may be expressed more weakly but more compactly in the following form. For q even, $q > 2$, and $n \geq 4$,

$$m_2(n, q) \leq q^{n-1} - \tfrac{1}{2}q^{n-2} + dq^{n-7/2},$$

where $d = 0$ for $q \geq 32$ and $d = 4$ for $q \leq 16$.

27.5 General properties of k-arcs and normal rational curves

As in §21.1, a *rational curve* $\mathscr{C}_d$ of *order* d in $PG(n, q)$ is the set of points

$$\{P(t_0, t_1) = \mathbf{P}(g_0(t_0, t_1), \ldots, g_n(t_0, t_1)) \mid t_0, t_1 \in \gamma\} \qquad (27.59)$$

where each g_i is a binary form of degree d and a highest common factor of the g_i is 1. The curve $\mathscr{C}_d$ may also be written

$$\{P(t) = \mathbf{P}(f_0(t), f_1(t), \ldots, f_n(t)) \mid t \in \gamma^+\} \qquad (27.60)$$

where $f_i(t) = g_i(1, t)$. As the g_i have no non-trivial common factor, at least one f_i has degree d. Also $\mathscr{C}_d$ is *normal* if it is not the projection of a rational $\mathscr{C}'_d$ in $PG(n+1, q)$, where $\mathscr{C}'_d$ is not contained in a prime.

Theorem 27.5.1: *Let $\mathscr{C}_d$ be a normal rational curve in $PG(n, q)$ not contained in a prime. Then*

(i) *$q \geq n$;*
(ii) *$d = n$;*
(iii) *$\mathscr{C}_n$ is projectively equivalent to*

$$\{P(t) = \mathbf{P}(t^n, t^{n-1}, \ldots, t, 1) \mid t \in \gamma^+\}; \qquad (27.61)$$

(iv) *$\mathscr{C}_n$ consists of $q+1$ points no $n+1$ in a prime;*

(v) *if $q \geq n+2$, there is a unique $\mathscr{C}_n$ through any $n+3$ points of $PG(n, q)$ no $n+1$ of which lie in a prime;*

(vi) *there is a subgroup H of $PGL(n+1, q)$ isomorphic to $PGL(2, q)$ that acts 3-transitively on $\mathscr{C}_n$.*

Proof: (i)–(v) See Theorem 21.1.1.

(vi) With $\mathscr{C}_n$ as in (27.61), the transformation τ given by $t \to (at+b)/(ct+d)$, with $ad - bc \neq 0$, induces the transformation

$$\begin{aligned}&(t^n, t^{n-1}, \ldots, t, 1)\\ &\quad \to ((at+b)^n, (at+b)^{n-1}(ct+d), \ldots, (at+b)(ct+d)^{n-1}, (ct+d)^n)\\ &\quad = (t^n, t^{n-1}, \ldots, t, 1)T\end{aligned}$$

for a suitable non-singular matrix T. Hence $H = \{T \mid \tau \in PGL(2, q)\}$. □

Now, we consider further properties of $\mathscr{C}_n$. With $\mathscr{C}_n$ as in (27.59), let $g_j^{(i)} = \partial g_j / \partial t_i$. If, for a given i in $\{0, 1\}$ not all $g_j^{(i)}(t_0, t_1)$ are zero, then the point with $g_j^{(i)}(t_0, t_1)$ as $(j+1)$th coordinate is denoted by $P^{(i)}(t_0, t_1)$. If such is the case for both i and if $P(t_0, t_1) \neq P^{(i)}(t_0, t_1)$ also for both i, then

$$P(t_0, t_1)P^{(0)}(t_0, t_1) = P(t_0, t_1)P^{(1)}(t_0, t_1)$$

since

$$t_0 g_j^{(0)}(t_0, t_1) + t_1 g_j^{(1)}(t_0, t_1) = n g_j(t_0, t_1).$$

For at least one i in $\{0, 1\}$ the point $P^{(i)}(t_0, t_1)$ exists and is distinct from $P(t_0, t_1)$; for such an i, the line $P(t_0, t_1)P^{(i)}(t_0, t_1)$ is the *tangent of $\mathscr{C}_n$ at P* and is denoted by l_P.

Lemma 27.5.2: *Let $q \geq n \geq 3$, let $\mathscr{C}_n$ be a normal rational curve, and let P be a point of $\mathscr{C}_n$.*

(i) *The image of the projection map $\mathfrak{S}$ of $\mathscr{C}_n \backslash \{P\}$ from P onto a prime Π not containing P together with $P' = l_P \cap \Pi$ is a normal rational curve in Π, and is denoted by $\mathscr{C}_n \mathfrak{S}$.*

(ii) *No two tangents to $\mathscr{C}_n$ intersect.*

(iii) *If l_P lies in a prime Π', then $|\Pi' \cap \mathscr{C}_n| \leq n-1$.*

(iv) *$l_P \cap \mathscr{C}_n = \{P\}$.*

(v) *If $Q, R \in \mathscr{C}_n \backslash \{P\}$ then QR does not meet l_P.*

Proof: (i) We may take $\mathscr{C}_n$ in canonical form

$$\{P(t) = \mathbf{P}(t^n, t^{n-1}, \ldots, t, 1) \mid t \in \gamma^+\}.$$

By Theorem 27.5.1(vi) we may choose $P = \mathbf{U}_0$. Let $\mathfrak{S}'$ be the projection of

Π from P onto the prime $\mathbf{u}_0$. Then, for t in γ,

$$P\mathfrak{S}\mathfrak{S}' = \mathbf{P}(0, t^{n-1}, t^{n-2}, \ldots, t, 1).$$

Also, the tangent $l_P = \mathbf{U}_0\mathbf{U}_1$, which meets $\mathbf{u}_0$ in $\mathbf{U}_1$. Hence $\{P\mathfrak{S}\mathfrak{S}' \mid t \in \gamma\} \cup \{\mathbf{U}_1\}$ is the normal rational curve

$$\{\mathbf{P}(0, t^{n-1}, \ldots, t, 1) \mid t \in \gamma^+\}.$$

Thus $\mathscr{C}_n\mathfrak{S}$ is a normal rational curve of degree $n-1$ in Π.

(ii) Let $P = P(t)$ and $Q = P(s)$, $s, t \in \gamma$, $s \neq t$. To show that $l_P \cap l_Q = \emptyset$, consider the matrix

$$\begin{bmatrix} t^n & t^{n-1} & \cdots & t^3 & t^2 & t & 1 \\ nt^{n-1} & (n-1)t^{n-2} & \cdots & 3t^2 & 2t & 1 & 0 \\ s^n & s^{n-1} & \cdots & s^3 & s^2 & s & 1 \\ ns^{n-1} & (n-1)s^{n-2} & \cdots & 3s^2 & 2s & 1 & 0 \end{bmatrix}.$$

It has rank 4, since the submatrix formed by the last four columns has determinant $(t-s)^4$. So, when $P, Q \neq \mathbf{U}_0$, the lines l_P and l_Q do not meet. By the transitivity of the group, this is also true when $P = \mathbf{U}_0$.

(iii) Let $\Pi' = \mathbf{V}(a_0x_0 + a_1x_1 + \cdots + a_nx_n)$ and take $P = \mathbf{U}_0$. From (i), $l_P = \mathbf{U}_0\mathbf{U}_1$; so l_P lies in Π' if and only if $a_0 = a_1 = 0$. Hence, apart from P, a point $P(t)$ of $\mathscr{C}_n$ lies in Π' if and only if

$$a_2t^{n-2} + \cdots + a_n = 0.$$

This has at most $n-2$ solutions.

(iv) $\mathbf{U}_0\mathbf{U}_1 \cap \mathscr{C}_n = \{\mathbf{U}_0\}$.

(v) Take $P = \mathbf{U}_0$, $Q = \mathbf{U}_n$, and $R = \mathbf{U}$. □

Theorem 27.5.3: *If $q \geq n+2$, then*

(i) *the group $G(\mathscr{C}_n)$ of projectivities in $PG(n, q)$ fixing $\mathscr{C}_n$ is isomorphic to $PGL(2, q)$, given by the transformations*

$$t \to (at+b)/(ct+d)$$

with $ad - bc \neq 0$, acting on (27.61);

(ii) *the number of normal rational curves in $PG(n, q)$ is*

$$\begin{aligned} \nu_n &= q^{(n-1)(n+2)/2}[3, n+1]_- \\ &= \prod_{i=0}^{n} (q^{n+1} - q^i)/\{q(q^2-1)(q-1)\}. \end{aligned}$$

Proof: From Theorem 27.5.1(vi), there is a subgroup H of $G(\mathscr{C}_n)$ with $H \cong PGL(2, q)$. It must be shown that $H = G(\mathscr{C}_n)$.

Now, suppose that an element $\mathfrak{A}$ of $G(\mathscr{C}_n)$ is given by the matrix $A = [a_{ij}]$, $0 \le i, j \le n$, and $\mathscr{C}_n$ is taken in the form (27.61). Since

$$(t^n, t^{n-1}, \ldots, t, 1)A = \left(\sum_{i=0}^{n} a_{i0} t^{n-i}, \sum_{i=0}^{n} a_{i1} t^{n-i}, \ldots \right)$$
$$= (s^n, s^{n-1}, \ldots, s, 1),$$

there exists a permutation ρ_n of γ^+ such that $t\rho_n = s$, whence

$$t\rho_n = \sum_{i=0}^{n} a_{i0} t^{n-i} \Big/ \sum_{i=0}^{n} a_{i1} t^{n-i}.$$

It will now be shown by induction on n that there exist a, b, c, d in γ with $ad - bc \neq 0$ such that

$$t\rho_n = (at + b)/(ct + d).$$

For $n = 1$,

$$t\rho_n = (a_{00}t + a_{10})/(a_{01}t + a_{11});$$

so the result is proved. Assume that the result is true for

$$\mathscr{C}_{n-1} = \{\mathbf{P}(t^{n-1}, t^{n-2}, \ldots, t, 1, 0) \mid t \in \gamma^+\}, \quad n \ge 2.$$

By the transitivity of H, we may take $\mathbf{U}_n \mathfrak{A} = \mathbf{U}_n$; hence $a_{ni} = 0, 0 \le i \le n-1$, and $a_{n0} = a_{n1} = 0$, in particular. So

$$t\rho_n = \sum_{i=0}^{n-1} a_{i0} t^{n-i} \Big/ \sum_{i=0}^{n-1} a_{i1} t^{n-i}$$
$$= \sum_{i=0}^{n-1} a_{i0} t^{n-1-i} \Big/ \sum_{i=0}^{n-1} a_{i1} t^{n-1-i}.$$

Since the tangent at $\mathbf{U}_n$ is fixed, also $a_{n-1,i} = 0$, for $0 \le i \le n-2$.

Let $\mathfrak{S}$ be the projection map from $\mathbf{U}_n$ onto $\mathbf{u}_n$. Then, as in Lemma 27.5.2(i), let $\mathscr{C}_{n-1} = \mathscr{C}_n \mathfrak{S}$. Let $A' = [a'_{ij}]$, $0 \le i, j \le n-1$, with $a'_{ij} = a_{ij}$. Also let $\mathfrak{A}'$ be the projectivity on $\mathbf{u}_n$ corresponding to the matrix A'. Then, for $t \in \gamma^+ \backslash \{0\}$,

$$\mathbf{P}(t^n, \ldots, t, 1)\mathfrak{S}\mathfrak{A}' = \mathbf{P}(t^{n-1}, \ldots, t, 1, 0)\mathfrak{A}' = \mathbf{P}(t^n, \ldots, t, 1)\mathfrak{A}\mathfrak{S} \in \mathscr{C}_{n-1}.$$

Also $\mathbf{U}_{n-1}\mathfrak{A}' = \mathbf{U}_{n-1} \in \mathscr{C}_{n-1}$. Since $\mathscr{C}_{n-1}\mathfrak{A}'$ is also a normal rational curve of order $n-1$, so $\mathscr{C}_{n-1}\mathfrak{A}' = \mathscr{C}_{n-1}$ by Theorem 27.5.1(v). So $\mathfrak{A}'$ is a projectivity of $\mathbf{u}_n$ fixing $\mathscr{C}_{n-1}$. However,

$$t\rho_{n-1} = \sum_{i=0}^{n-1} a'_{i0} t^{n-i-1} \Big/ \sum_{i=0}^{n-1} a'_{i1} t^{n-1-i}$$
$$= \sum_{i=0}^{n-1} a_{i0} t^{n-1-i} \Big/ \sum_{i=0}^{n-1} a_{i1} t^{n-1-i}$$
$$= t\rho_n.$$

By the induction hypothesis, ρ_n has the required form.

(ii) Since $\mathscr{C}_n$ is projectively unique and since $G(\mathscr{C}_n) = PGL(2, q)$, so

$$\nu_n = |PGL(n+1, q)|/|PGL(2, q)|. \quad \square$$

Now we consider the existence of the k-arcs in $PG(n, q)$.

Theorem 27.5.4: *A k-arc in $PG(n, q)$, $k \geq n + 4$, exists if and only if a k-arc exists in $PG(k - n - 2, q)$.*

Proof: Choose $n + 1$ points of a k-arc $\mathscr{K}$ as the simplex of reference. Consider the $(k - n - 1) \times (n + 1)$ matrix M whose rows are the vectors of the other $k - n - 1$ points of $\mathscr{K}$. Since no $n + 1$ points of $\mathscr{K}$ lie in a prime, taking $n - s + 1$ points, where $0 \leq s \leq \min(k - n - 1, n + 1)$, of the simplex of reference and s other points shows that all $s \times s$ minors of M are non-zero. So now take the rows of the transpose M^* as vectors of points in $PG(k - n - 2, q)$ and add the simplex of reference in this space. This gives a k-arc $\mathscr{K}'$ in $PG(k - n - 2, q)$. The process is reversible. $\square$

As in Chapter 24, let $\mathscr{G}_{r,n}$ denote the Grassmannian of r-spaces in $PG(n, q)$. Let $\mathscr{A}_{k,n}$ be the set of all k-arcs in $PG(n, q)$. We now consider a relation between $\mathscr{G}_{n,k-1}$ and $\mathscr{A}_{k,n}$.

Let $\mathscr{K} = \{P_1, \ldots, P_k\}$ be a k-arc in $PG(n, q)$ with $k \geq n + 3$, let $G(\mathscr{K})$ be the group of projectivities fixing $\mathscr{K}$, and let $g(\mathscr{K}) = |G(\mathscr{K})|$. Let $P_i = \mathbf{P}(X_i)$. Then to $\mathscr{K}$ there correspond $(q - 1)^k k!$ matrices each with k rows and $n + 1$ columns:

$$\begin{bmatrix} \rho_1 X_{i1} \\ \rho_2 X_{i2} \\ \vdots \\ \rho_k X_{ik} \end{bmatrix}. \tag{27.62}$$

Here $\{i_1, \ldots, i_k\} = \{1, \ldots, k\}$ and $\rho_i \in \gamma_0$. Every subdeterminant of order $n + 1$ is non-zero, since $\mathscr{K}$ is a k-arc. This matrix can be denoted

$$M = M_{R,\sigma}, \tag{27.63}$$

where $R = (\rho_1, \ldots, \rho_k)$ and $\sigma = (i_1, \ldots, i_k)$. Now, take the columns of $M_{R,\sigma}$ as the vectors of $n + 1$ points in $PG(k - 1, q)$. These points define a $PG(n, q)$ which we will denote $\Pi_n(M)$. From the $(q - 1)^k k!$ matrices M we obtain $(q - 1)^k k!$ subspaces $\Pi_n(M)$ of $PG(k - 1, q)$; these are not necessarily distinct. Suppose that for two matrices M and M' we have $\Pi_n(M) = \Pi_n(M')$. Then there exists a unique non-singular $(n + 1) \times (n + 1)$ matrix A such that

$$MA = M'. \tag{27.64}$$

However (27.64) also defines a unique projectivity of $PG(n, q)$ fixing $\mathscr{K}$.

Conversely, if B is the matrix of a projectivity fixing $\mathscr{K}$, then $\Pi_n(\rho MB) = \Pi_n(M)$ for all $\rho \neq 0$; that is, $(q-1)g(\mathscr{K})$ matrices M give the same $\Pi_n(M)$. So, to $\mathscr{K}$ there correspond

$$\chi(\mathscr{K}) = (q-1)^{k-1}k!/g(\mathscr{K}) \tag{27.65}$$

distinct subspaces Π_n of $PG(k-1, q)$.

The Grassmannian $\mathscr{G}_{n,k-1}$ is embedded in $PG(N, q)$, with $N = \mathbf{c}(k, n+1) - 1$, and contains

$$[k-n, k]_-/[1, n+1]_-$$

points, §24.2. From above, it follows that to the k-arc $\mathscr{K}$ correspond $\chi(\mathscr{K})$ points of $\mathscr{G}_{n,k-1}$ lying in no face of the simplex of reference of $PG(N, q)$.

Now we consider how many k-arcs correspond to one of these $\chi(\mathscr{K})$ points Q of $\mathscr{G}_{n,k-1}$. To Q corresponds one Π_n' of $PG(k-1, q)$. The number of ordered $(n+1)$-tuples $(Q_1, \ldots, Q_{n+1})$ of linearly independent points of Π_n' is

$$\varphi = \prod_{i=0}^{n} (q^{n+1} - q^i)/(q-1)^{n+1}.$$

So to Π_n' there correspond

$$(q-1)^{n+1}\varphi = \prod_{i=0}^{n} (q^{n+1} - q^i)$$

matrices M.

Suppose now that the two $k \times (n+1)$ matrices Y and Z have, as their columns, vectors of $n+1$ linearly independent points of Π_n' and give the same k-arc $\mathscr{K}'$ of $PG(n, q)$. Then

$$YA = Z$$

for a unique non-singular $(n+1) \times (n+1)$ matrix A, which consequently defines a projectivity of $PG(n, q)$ fixing $\mathscr{K}'$. Conversely, a projectivity of $PG(n, q)$ fixing $\mathscr{K}'$ gives $q-1$ matrices corresponding to Π_n'. So $\mathscr{K}'$ comes from $(q-1)g(\mathscr{K}')$ matrices corresponding to Π_n'.

Since the k-arcs $\mathscr{K}$ and $\mathscr{K}'$ come from two ordered $(n+1)$-tuples of linearly independent points of Π_n', they are projectively equivalent; hence $g(\mathscr{K}) = g(\mathscr{K}')$. So, to Q of $\mathscr{G}_{n,k-1}$ there correspond

$$\prod_{i=0}^{n} (q^{n+1} - q^i)/[(q-1)g(\mathscr{K})]$$

k-arcs of $PG(n, q)$ and these are precisely the ones projectively equivalent to $\mathscr{K}$.

Let $\mathscr{V}_{n,k-1}$ denote the set of points of $\mathscr{G}_{n,k-1}$ on no face of the simplex of reference and let $\mathscr{V}_{n,k-1}(\mathscr{K})$ denote the set of $\chi(\mathscr{K})$ points corresponding to the k-arc $\mathscr{K}$. Hence $\mathscr{V}_{n,k-1}$ is partitioned by the sets $\mathscr{V}_{n,k-1}(\mathscr{K})$. In $\mathscr{A}_{k,n}$, which is the set of k-arcs of $PG(n, q)$, let $\mathscr{A}_{k,n}(\mathscr{K})$ denote the set of k-arcs corresponding to a point of $\mathscr{V}_{n,k-1}(\mathscr{K})$. Then the sets $\mathscr{A}_{k,n}(\mathscr{K})$ partition $\mathscr{A}_{k,n}$. If we now map $\mathscr{A}_{k,n}(\mathscr{K})$ onto $\mathscr{V}_{n,k-1}(\mathscr{K})$, then a bijection of the quotient set corresponding to the partition of $\mathscr{A}_{k,n}$ and the quotient set corresponding to the partition of $\mathscr{V}_{n,k-1}$ is obtained. The discussion gives the following results.

Theorem 27.5.5:

(i) $$|\mathscr{A}_{k,n}(\mathscr{K})| = \prod_{i=0}^{n} (q^{n+1} - q^i)/[(q-1)g(\mathscr{K})]. \tag{27.66}$$

(ii) $$|\mathscr{V}_{n,k-1}(\mathscr{K})|/|\mathscr{A}_{k,n}(\mathscr{K})| = (q-1)^k k! \Big/ \prod_{i=0}^{n} (q^{n+1} - q^i). \tag{27.67}$$

(iii) $$|\mathscr{V}_{n,k-1}|/|\mathscr{A}_{k,n}| = (q-1)^k k! \Big/ \prod_{i=0}^{n} (q^{n+1} - q^i). \quad \square \tag{27.68}$$

Theorem 27.5.6: *For* $q \geq \max(n+2, k-n)$, $n \geq 2$, $k \geq n+4$,

$$|\mathscr{A}_{k,k-2-n}|/|\mathscr{A}_{k,n}| = v_{k-2-n}/v_n$$
$$= \prod_{i=0}^{k-2-n} (q^{k-n-1} - q^i) \Big/ \prod_{i=0}^{n} (q^{n+1} - q^i). \tag{27.69}$$

Proof: To a point of $\mathscr{V}_{n,k-1}$ there corresponds a Π_n of $PG(k-1, q)$ skew to every $(k-2-n)$-dimensional edge of the simplex of reference of $PG(k-1, q)$, and conversely. By the principle of duality, the number of such Π_n is the same as the number of Π_{k-2-n} of $PG(k-1, q)$ skew to every n-dimensional edge of the simplex of reference. Hence

$$|\mathscr{V}_{n,k-1}| = |\mathscr{V}_{k-2-n,k-1}|. \tag{27.70}$$

The result now follows from (27.68), (27.70), and Theorem 27.5.1(vii). $\square$

There are numerous corollaries that can be drawn from Theorems 27.5.4 and 27.5.6. For the moment we only deduce results that follow from properties of $PG(2, q)$ and $PG(3, q)$.

Corollary 1: (i) $m(q-3, q) = q+1$ *for* $q \geq 5$;
(ii) $m(q-2, q) = q+1$ *for q odd with* $q \geq 5$;
(iii) $m(q-2, q) = q+2$ *for q even with* $q \geq 4$.

Proof: (i) For $q \geq 5$, $|\mathscr{A}_{q+2,3}| = 0 \Rightarrow |\mathscr{A}_{q+2,q-3}| = 0$.
(ii) For $q \geq 5$, $|\mathscr{A}_{q+2,2}| = 0 \Rightarrow |\mathscr{A}_{q+2,q-2}| = 0$.

(iii) For $q \geq 4$, $|\mathscr{A}_{q+2,2}| > 0 \Rightarrow |\mathscr{A}_{q+2,q-2}| > 0$;
for $q \geq 4$, $|\mathscr{A}_{q+3,3}| = 0 \Rightarrow |\mathscr{A}_{q+3,q-2}| = 0$. □

Corollary 2: *For $q \geq n + 3$, $n \geq 2$, if every $(q + 1)$-arc of $PG(n, q)$ is a normal rational curve, then every $(q + 1)$-arc of $PG(q - n - 1, q)$ is a normal rational curve.*

Proof: For $q \geq n + 3$, $n \geq 2$,

$$|\mathscr{A}_{q+1,q-n-1}| = \nu_{q-n-1} \Leftrightarrow |\mathscr{A}_{q+1,n}| = \nu_n. \quad \square$$

27.6 The maximum size of an arc and the characterization of such arcs

In §§21.2 and 21.3 it was shown that $m(3, q) = q + 1$ for $q > 3$; also, a $(q + 1)$-arc is a twisted cubic for q odd, while for q even it is of the form $\{\mathbf{P}(t^{2^m+1}, t^{2^m}, t, 1) | t \in GF(2^h) \cup \{\infty\}\}$ for some m coprime to h. Now we consider how this result generalizes to higher dimensions.

Theorem 27.6.1: *Let $\mathscr{K}$ be a k-arc in $PG(n, q)$ with $k \geq n + 3 \geq 6$. If there exist points P_0 and P_1 in $\mathscr{K}$ such that the projections $\mathscr{K}_0$ of $\mathscr{K}\backslash\{P_0\}$ from P_0 and $\mathscr{K}_1$ of $\mathscr{K}\backslash\{P_1\}$ from P_1 onto a prime Π_{n-1} are both contained in normal rational curves in Π_{n-1}, then $\mathscr{K}$ is contained in a unique normal rational curve of $PG(n, q)$.*

Proof: Let $\mathscr{L} = \{P_0, \ldots, P_{n+2}\}$ be an $(n + 3)$-arc in $\mathscr{K}$. For $i = 0$ and 1, let $\mathscr{L}_i$ and $\mathscr{K}_i$ be the projections from P_i of $\mathscr{L}\backslash\{P_i\}$ and $\mathscr{K}\backslash\{P_i\}$ onto Π_{n-1}. By Theorem 27.5.1(v) there exist unique normal rational curves $\mathscr{C}$ in $PG(n, q)$ and $\mathscr{C}^{(0)}$, $\mathscr{C}^{(1)}$ in Π_{n-1} such that $\mathscr{L} \subset \mathscr{C}$, $\mathscr{L}_0 \subset \mathscr{C}^{(0)}$, $\mathscr{L}_1 \subset \mathscr{C}^{(1)}$. Since $\mathscr{K}_0$ and $\mathscr{K}_1$ are assumed to be in normal rational curves, so $\mathscr{K}_0 \subset \mathscr{C}^{(0)}$ and $\mathscr{K}_1 \subset \mathscr{C}^{(1)}$. As $\mathscr{K}$ is contained in $P_0\mathscr{K}_0$ and in $P_1\mathscr{K}_1$, so

$$\mathscr{K}\backslash\{P_0, P_1\} \subset (P_0\mathscr{C}^{(0)} \cap P_1\mathscr{C}^{(1)})\backslash P_0P_1; \tag{27.71}$$

the right-hand side of (27.71) will be shown to lie in $\mathscr{C}$.

Let $\mathscr{C} = \{Q_j | j \in \bar{\mathbf{N}}_q\}$ with $Q_j = P_j$ for j in $\bar{\mathbf{N}}_{n+2}$. Also, let $\mathscr{D}^{(i)}$ be the projection of $\mathscr{C}\backslash\{P_i\}$ onto Π_{n-1} from P_i for $i = 0$ and 1. Since $\mathscr{D}^{(i)} \cup \{l_{P_i} \cap \Pi_{n-1}\}$ is a normal rational curve in Π_{n-1} containing $\mathscr{L}_i$ by Lemma 27.5.2(i), so it coincides with $\mathscr{C}^{(i)}$ by Theorem 27.5.1(v). Thus a line on the cone $P_i\mathscr{C}^{(i)}$ other than P_0P_1 is either the tangent l_{P_i} or a line P_iQ_j, where $j \neq 0, 1$. Let l_0 be a line on $P_0\mathscr{C}^{(0)}$ and l_1 a line on $P_1\mathscr{C}^{(1)}$ such that neither line is P_0P_1 but with l_0 and l_1 intersecting. If $l_0 = l_{P_0}$, then $l_1 \neq l_{P_1}$ by Lemma 27.5.2(ii); thus $l_1 = P_1Q_j$ for some $j \neq 0, 1$. Since the plane $\pi = l_0l_1$ contains P_0, P_1, Q_j, there exists a prime containing n points of $\mathscr{C}_n$ as well as

l_0, contradicting Lemma 27.5.2(iii). Thus $l_0 = P_0Q_u$ and $l_1 = P_1Q_v$ for $u, v \neq 0, 1$. Since π contains at most three points of $\mathscr{C}$, so $Q_u = Q_v$. Hence $l_0 \cap l_1$ is a point of $\mathscr{C}$. Thus $\mathscr{K}\backslash\{P_0, P_1\} \subset \mathscr{C}\backslash\{P_0, P_1\}$, whence $\mathscr{K} \subset \mathscr{C}$. □

Theorem 27.6.2: (i) *Let $\mathscr{K}$ be a $(q+2)$-arc in $PG(n, q)$ with $q + 1 \geq n + 3 \geq 6$. If P_0 and P_1 are points of $\mathscr{K}$ and Π_{n-1} is a prime containing neither P_0 nor P_1 then the projections $\mathscr{K}_0$ of $\mathscr{K}\backslash\{P_0\}$ and $\mathscr{K}_1$ of $\mathscr{K}\backslash\{P_1\}$ from P_0 and P_1 onto Π_{n-1} cannot both be normal rational curves.*

(ii) *If every $(q+1)$-arc in $PG(n-1, q)$, with $q + 1 \geq n + 3 \geq 6$, is a normal rational curve, then $m(n, q) = q + 1$.*

Proof: (i) Suppose $\mathscr{K}_0$ and $\mathscr{K}_1$ are normal rational curves in Π_{n-1}. For $P \in \mathscr{K}\backslash\{P_0, P_1\}$, let $\mathscr{K}'$ be the $(q+1)$-arc $\mathscr{K}\backslash\{P\}$. Then, by Theorem 27.6.1, $\mathscr{K}'$ is a normal rational curve in $PG(n, q)$. Let l_{P_i} be the tangent of $\mathscr{K}'$ at P_i for $i = 0$ and 1, and let $\mathscr{K}_i'$ be the projection of $\mathscr{K}'\backslash\{P_i\}$ from P_i onto Π_{n-1}. Then $\mathscr{K}' \cup \{l_{P_i} \cap \Pi_{n-1}\}$ is a normal rational curve in Π_{n-1} by Lemma 27.5.2(i). So $\mathscr{K}_i = \mathscr{K}_i' \cup \{l_{P_i} \cap \Pi_{n-1}\}$ since both curves have q points in common and $q \geq n - 1 + 3$. Thus $l_{P_i} = P_iP$ contradicting Lemma 27.5.2(ii).

(ii) If there is a $(q+2)$-arc $\mathscr{K}$ in $PG(n, q)$, then $\mathscr{K}_0$ and $\mathscr{K}_1$ are $(q+1)$-arcs and so normal rational curves, contradicting (i). □

Theorem 27.6.3: *In $PG(4, q)$, $q \geq 5$,*

$$m(4, q) = q + 1.$$

Proof: (i) For q odd, since every $(q+1)$-arc in $PG(3, q)$ is a twisted cubic, the result follows by Theorem 27.6.2(ii).

(ii) For q even, suppose there exists a $(q+2)$-arc $\mathscr{K} = \{P, Q, R_1, R_2, \ldots, R_q\}$. Take a solid Π_3 in $PG(4, q)$ containing neither P nor Q. Let $\mathscr{K}_1$ and $\mathscr{K}_2$ be the projections of $\mathscr{K}\backslash\{P\}$ and $\mathscr{K}\backslash\{Q\}$ onto Π_3 from P and Q respectively. In Theorem 21.3.10, it was shown that, at any point L of a $(q+1)$-arc $\mathscr{L}$, there are precisely two lines, called special unisecants, such that every plane through such a unisecant meets $\mathscr{L}$ in at most one point other than L; further, the special unisecants to $\mathscr{L}$ are the generators of a hyperbolic quadric $\mathscr{H}_3$. Let $\mathscr{H}_3^{(1)}$ and $\mathscr{H}_3^{(2)}$ be the corresponding quadrics containing $\mathscr{K}_1$ and $\mathscr{K}_2$. Also let $PQ \cap \Pi_3 = S$, let $PR_i \cap \Pi_3 = P_i$, and let $QR_i \cap \Pi_3 = Q_i$, $i \in \mathbf{N}_q$. Then $\mathscr{K}_1 = \{S, P_1, \ldots, P_q\}$ and $\mathscr{K}_2 = \{S, Q_1, \ldots, Q_q\}$. Since the plane PQR_j meets Π_3 in a line, so S, P_j, Q_j are collinear, for j in $\mathbf{N}_q$, on the line l_j. Also let l and m be the special unisecants at S to $\mathscr{K}_1$. Then, each of the q planes ll_j meets $\mathscr{K}_1$ in precisely two points S and P_j, and also meets $\mathscr{K}_2$ in precisely two points S and Q_j. So l, and similarly m, is a special unisecant to $\mathscr{K}_2$ at S. Thus the two quadric cones $P\mathscr{H}_3^{(1)}$ and $Q\mathscr{H}_3^{(2)}$ contain the planes $l(PQ)$ and $m(PQ)$. They therefore intersect residually in a quadric surface $\mathscr{W}_3$ which either (a) lies in a solid Π'_3, or (b) lies in no solid. However, $\mathscr{W}_3$ contains $R_1, \ldots, R_q$. In case (a), Π'_3 contains at most four points of $\mathscr{K}$ and so $q \leq 4$. In case (b), $\mathscr{W}_3$ is a pair of planes with just one common point, which can contain at most six points of $\mathscr{K}$, whence $q \leq 6$. Thus both (a) and (b) are impossible. □

Theorem 27.6.4: *In $PG(n, q)$, q odd, $n \geq 3$,*

(i) *if $\mathscr{K}$ is a k-arc with $k > q - \frac{1}{4}\sqrt{q} + n - \frac{1}{4}$, then $\mathscr{K}$ lies on a unique normal rational curve;*

(ii) *if $q > (4n - 5)^2$, every $(q + 1)$-arc is a normal rational curve;*

(iii) *if $q > (4n - 9)^2$,*

$$m(n, q) = q + 1.$$

Proof: (i) This follows by induction from Theorems 27.6.1 and 10.4.4.

(ii) $q + 1 > q - \frac{1}{4}\sqrt{q} + n - \frac{1}{4} \Leftrightarrow q > (4n - 5)^2$.

(iii) This follows from Theorem 27.6.2(ii) and part (ii). □

The next result shows that in part (ii) of the above theorem some restriction on q is necessary.

Theorem 27.6.5: *In $PG(4, q)$, q odd,*

(i) *for $q \leq 7$, a $(q + 1)$-arc is a normal rational curve;*

(ii) *for $q = 9$, there exist precisely two projectively distinct 10-arcs, the normal rational curve and one other.*

Proof: (i) For $q = 3$ and 5, the result is trivial. For $q = 7$, we can apply Theorem 27.5.6, Corollary 2 with $n = 2$.

(ii) With $GF(9)_0 = \{\sigma^i | i \in \bar{\mathbf{N}}_7, \sigma^2 = \sigma + 1\}$, every 10-arc in $PG(4, 9)$ other than a normal rational curve is projectively equivalent to

$$\mathscr{K} = \{\mathbf{P}(1, t, t^2 + \sigma t^6, t^3, t^4) | t \in GF(9)\} \cup \{\mathbf{P}(0, 0, 0, 0, 1)\};$$

see §27.8. The 10-arc $\mathscr{K}$ projects to the unique complete 8-arc in $PG(2, 9)$, §14.7. □

The situation is surprisingly different for $(q + 1)$-arcs in $PG(4, q)$ with q even, as we proceed to demonstrate.

Let $\mathscr{K} = \{P_0, P_1, \ldots, P_q\}$ be a $(q + 1)$-arc in $PG(4, q)$, $q = 2^h$, $h \geq 3$. At each point P_i of $\mathscr{K}$ there is an induced incidence structure $\mathscr{S}(P_i)$ isomorphic to $PG(3, q)$, whose points, lines, and planes are the lines, planes, and solids of $PG(4, q)$ through P_i; the incidence is that induced by $PG(4, q)$. As usual, a subspace of dimension r is denoted Π_r; however, we shall also denote by $\Pi_s^{(i)}$ an s-dimensional subspace of $\mathscr{S}(P_i)$. Thus a Π_r through P_i is also a $\Pi_{r-1}^{(i)}$. This notation will only be used for the remainder of this section.

From the definition of $\mathscr{K}$, any solid through P_i contains at most three other points of $\mathscr{K}$. Thus the set of q lines P_iP_j, $j \in \bar{\mathbf{N}}_q \backslash \{i\}$, is a q-arc $\mathscr{K}_i$ of $\mathscr{S}(P_i)$. By Theorem 27.7.10 for $q > 16$, this arc $\mathscr{K}_i$ can be completed to a $(q+1)$-arc $\mathscr{K}'_i$ of $\mathscr{S}(P_i)$ by adding a unique Π_1 through P_i; for $q \leq 16$, see §27.8. Let this line be denoted l_i and called the *tangent line to* $\mathscr{K}$ *at* P_i. From Theorem 21.3.10, the points of $\mathscr{K}'_i$ lie on a hyperbolic quadric, denoted $\mathscr{Q}_i$. Let $\mathscr{S}_i$ denote the quadric cone of $PG(4, q)$ whose points lie on the $\Pi_0^{(i)}$ of $\mathscr{Q}_i$; that is, $\mathscr{S}_i = P_i\mathscr{H}_3$, where $\mathscr{H}_3$ is a solid section of $\mathscr{Q}_i$ regarded as a set of Π_0.

Lemma 27.6.6: (i) *The tangent lines* l_i *of* $\mathscr{K}$ *are pairwise skew.*

(ii) *For* $i \neq j$, *the solid* l_il_j *meets* $\mathscr{K}$ *in* $\{P_i, P_j\}$.

(iii) *There is a unique plane* α_{ij} *through* P_i *and* P_j *which is both a* $\Pi_1^{(i)}$ *of* $\mathscr{Q}_i$ *and a* $\Pi_1^{(j)}$ *of* $\mathscr{Q}_j$. *Further,* P_il_j *is a* $\Pi_1^{(i)}$ *of* $\mathscr{Q}_i$ *and* P_jl_i *is a* $\Pi_1^{(j)}$ *of* $\mathscr{Q}_j$.

Proof: (i) By construction, $l_i \cap \mathscr{K} = \{P_i\}$. Suppose $l_i \cap l_j$ is a Π_0; then l_il_j is a Π_2 as well as a $\Pi_1^{(i)}$ and a $\Pi_1^{(j)}$. By Theorem 21.3.10, through P_iP_j there are exactly two special unisecants $\Pi_1^{(i)}$ of $\mathscr{K}'_i$, and these are generators of $\mathscr{Q}_i$. Any Π_3 containing a special unisecant $\Pi_1^{(i)}$ of $\mathscr{K}'_i$ through P_iP_j meets $\mathscr{K}'_i$ in at most one $\Pi_0^{(i)}$ other than P_iP_j. Hence, for $l_i \not\subset \Pi_3$, any such Π_3 meets $\mathscr{K}$ in P_i, P_j and at most one further point; for $l_i \subset \Pi_3$, it meets $\mathscr{K}$ in P_i, P_j. Since $l_i \subset \Pi_3$ if and only if $l_j \subset \Pi_3$, these two $\Pi_1^{(i)}$ of $\mathscr{Q}_i$ are also $\Pi_1^{(j)}$ of $\mathscr{Q}_j$ through P_iP_j. Thus the two cones $\mathscr{S}_i$ and $\mathscr{S}_j$ intersect in these Π_2 and hence residually in a quadric surface $\mathscr{Q}$. Since $\mathscr{K}\backslash\{P_i, P_j\}$ is in $\mathscr{Q}$ and $q + 1 \geq 9$, the surface $\mathscr{Q}$ does not contain a plane. So $\mathscr{Q}$ lies in a solid and also contains the $q - 1 \geq 7$ points of $\mathscr{K}\backslash\{P_i, P_j\}$, a contradiction.

(ii), (iii) Project $\mathscr{K}\backslash\{P_i, P_j\}$ from P_iP_j onto a plane Π_2 skew to P_iP_j; the projection of $\mathscr{K}\backslash\{P_i, P_j\}$ is a $(q-1)$-arc $\mathscr{K}'$ of Π_2. Then both $P_jl_i \cap \Pi_2 = \{Q\}$ and $P_il_j \cap \Pi_2 = \{Q'\}$ extend $\mathscr{K}'$ to a q-arc. From §10.3, $\mathscr{K}' \cup \{Q, Q'\}$ is a $(q+1)$-arc. Hence $l_il_j \cap \mathscr{K} = \{P_i, P_j\}$, P_il_j is a $\Pi_1^{(i)}$ of $\mathscr{Q}_i$, and P_jl_i is a $\Pi_1^{(j)}$ of $\mathscr{Q}_j$. Let Q'' be the unique point of Π_2 which extends $\mathscr{K}' \cup \{Q, Q'\}$ to a $(q+2)$-arc. Then the plane $Q''P_iP_j = \alpha_{ij}$ is both a $\Pi_1^{(i)}$ of $\mathscr{Q}_i$ and a $\Pi_1^{(j)}$ of $\mathscr{Q}_j$. □

Lemma 27.6.7: *For a given i, the planes* P_il_j, *for j in* $\bar{\mathbf{N}}_q\backslash\{i\}$, *are q of the* $\Pi_1^{(i)}$ *of a regulus* $\mathscr{R}_i$ *of* $\mathscr{Q}_i$.

Proof: Let l_j and l_k be distinct tangent lines of $\mathscr{K}$, with $j, k \neq i$. By Lemma 27.6.6(ii), $P_i \notin l_jl_k$, and hence P_il_j and P_il_k are skew $\Pi_1^{(i)}$; they are generators of $\mathscr{Q}_i$, by Lemma 27.6.6(iii), and so belong to a regulus $\mathscr{R}_i$. □

Lemma 27.6.8: *Let* g_i *and* g'_i *be the two* $\Pi_1^{(i)}$ *of* $\mathscr{Q}_i$ *through* l_i *and let* $g_i \in \mathscr{R}_i$. *Then* $g'_i \cap l_j$ *is a* Π_0, *for* $j \neq i$.

Proof: Let g_i' belong to the regulus $\mathscr{R}_i'$ complementary to $\mathscr{R}_i$. By the previous lemma, $P_i l_j \in \mathscr{R}_i$ for $j \neq i$. Since lines of complementary reguli meet, so $g_i' \cap P_i l_j$ is a $\Pi_0^{(i)}$. Thus $g_i' \cap l_j$ is a Π_0. □

Corollary: *Through each l_i there are two $\Pi_1^{(i)}$ of $\mathscr{Q}_i$; they are g_i, which is skew to all l_j for $j \neq i$, and g_i', which meets all l_j for $j \neq i$.* □

Lemma 27.6.9: *The $q + 1$ generators g_i' of $\mathscr{Q}_i$ contain a unique Π_1, denoted by l, which is disjoint from $\mathscr{K}$ and meets the l_i in distinct points.*

Proof: For $i \neq j$, let $g_i' \cap l_j = \{Q\}$ and $g_j' \cap l_i = \{R\}$. Since l_i and l_j are skew, so Q and R are distinct points. Since $l_i \subset g_i'$ and $l_j \subset g_j'$, so $g_i' \cap g_j' = QR$. Thus $g_0', \ldots, g_q'$ are $q + 1$ planes Π_2, meeting in pairs in a Π_1; so either they pass through a common Π_1 or they lie in a common Π_3. The latter is impossible since $\mathscr{K}$ lies in the space they generate. So they meet in a Π_1, denoted by l. Now, l is disjoint from $\mathscr{K}$, since otherwise every g_i' would contain $l \cap \mathscr{K}$. Since $l_i \subset g_i'$ and $l_i \not\subset g_j'$, $j \neq i$, so $|l_i \cap l| = 1$. Since the l_i are skew, each point of l lies on a unique tangent line l_i of $\mathscr{K}$. □

Theorem 27.6.10: *In $PG(4, q)$, $q = 2^h$, every $(q + 1)$-arc $\mathscr{K}$ is a normal rational curve.*

Proof: For $h = 1$ and 2, the result is trivial. So, let $h \geq 3$ and use the above notation. For $\mathscr{K} = \{P_0, \ldots, P_q\}$, it is possible to choose coordinates so that $P_0 = \mathbf{U}_0$, $P_1 = \mathbf{U}_4$, $A = \alpha_{01} \cap g_0 \cap g_1 = \mathbf{U}_2$, $B = l \cap l_1 = \mathbf{U}_3$, $C = l \cap l_0 = \mathbf{U}_1$, and $\mathbf{U}$ is any point of $\mathscr{K} \backslash \{P_0, P_1\}$. Note that $ll_0 = g_0'$, $ll_1 = g_1'$, $Al_0 = g_0$, $Al_1 = g_1$, and $P_0P_1A = \alpha_{01}$.

Let $\beta_1 = g_1 g_1' = \mathbf{u}_0$ and consider the $(q + 1)$-arc $\beta_1 \cap \mathscr{K}_0'$. This $(q + 1)$-arc (in a Π_3) contains the points $C = \mathbf{U}_1$, $P_1 = \mathbf{U}_4$, and $P_0\mathbf{U} \cap \beta_1 = \mathbf{P}(0, 1, 1, 1, 1)$. The special unisecants of $\beta_1 \cap \mathscr{K}_0'$ at C and P_1 are the intersections of β_1 with the $\Pi_1^{(0)}$ of $\mathscr{Q}_0$ containing l_0 and P_0P_1; these unisecants are therefore $CA = g_0 \cap \beta_1$, $CB = l = g_0' \cap \beta_1$, $P_1B = P_0 l_1 \cap \beta_1$, and $P_1A = \alpha_{01} \cap \beta_1$. Thus these unisecants intersect at $A = \mathbf{U}_2$ and $B = \mathbf{U}_3$.

It now follows from Theorem 21.3.15 and its proof that

$$\mathscr{K}_0 = \beta_1 \cap \mathscr{K}_0' = \{\mathbf{P}(0, 1, \mu, \mu^\sigma, \mu^{\sigma+1}) \mid \mu \in \gamma^+\}$$

where σ is a generator of the automorphism group of $\gamma = GF(q)$; hence $x^\sigma = x^{2^n}$ for some n coprime to h. Since, by definition, $\mathscr{K}_0$ is $l_0 \cap \beta_1$ together with a projection of $\mathscr{K} \backslash \{\mathbf{U}_0\}$ from $\mathbf{U}_0$, we obtain that

$$\mathscr{K} = \{\mathbf{P}(1, f(\mu), \mu f(\mu), \mu^\sigma f(\mu), \mu^{\sigma+1} f(\mu)) \mid \mu \in \gamma^+\}$$

for some function $f: \gamma \to \gamma$ with $f(0) = 0$, $f(1) = 1$.

Next, consider the $(q+1)$-arc $\beta_0 \cap \mathscr{K}'_1$, where $\beta_0 = g_0 g'_0 = \mathbf{u}_4$. Similarly to the above we obtain

$$\mathscr{K}_1 = \beta_0 \cap \mathscr{K}'_1 = \{\mathbf{P}(1, \lambda, \lambda^\tau, \lambda^{\tau+1}, 0) \mid \lambda \in \gamma^+\},$$

where τ is an automorphism of γ such that $x^\tau = x^{2^m}$ with m coprime to h. Since $\mathscr{K}_1$ is $l_1 \cap \beta_0$ together with a projection of $\mathscr{K} \backslash \{\mathbf{U}_4\}$ from $\mathbf{U}_4$, so

$$\mathscr{K} = \{\mathbf{P}(1, \lambda, \lambda^\tau, \lambda^{\tau+1}, f'(\lambda)) \mid \lambda \in \gamma\} \cup \{\mathbf{U}_4\},$$

where f' is a function on γ with $f'(0) = 0$, $f'(1) = 1$. The two forms for $\mathscr{K}$ are the same if, for all λ and μ in γ,

$$f(\mu) = \lambda, \quad \mu f(\mu) = \lambda^\tau, \quad \mu^\sigma f(\mu) = \lambda^{\tau+1}, \quad \mu^{\sigma+1} f(\mu) = f'(\lambda).$$

Hence

$$\mu = \lambda^{\tau-1}, \quad \mu^{\sigma-1} = \lambda, \quad \text{and} \quad \lambda = (\lambda^{\tau-1})^{\sigma-1}.$$

From the definitions of τ and σ,

$$\lambda^\tau = \lambda^{2^m}, \qquad \lambda^\sigma = \lambda^{2^n};$$

we may take $1 \le n \le m < h$. So, mod$(2^h - 1)$,

$$(2^m - 1)(2^n - 1) \equiv 1$$

$$2^{m+n} - 2^m - 2^n \equiv 0$$

$$2^m - 2^{m-n} - 1 \equiv 0.$$

Since $0 \le 2^m - 2^{m-n} - 1 < 2^h - 1$, so

$$2^m - 2^{m-n} - 1 = 0.$$

Therefore $m = n = 1$. Thus

$$\mathscr{K} = \{\mathbf{P}(1, \lambda, \lambda^2, \lambda^3, \lambda^4) \mid \lambda \in \gamma^+\},$$

where $\lambda = \infty$ gives the point $\mathbf{U}_4$. □

Theorem 27.6.11: *For $q \ge 5$, $n \ge 5$,*

$$m(n, q) \le \begin{cases} q + n - 3, & q \text{ odd} \\ q + n - 4, & q \text{ even.} \end{cases}$$

Proof: For q odd, the result follows from Theorem 27.6.2 and induction, using the fact that a $(q+1)$-arc in $PG(3, q)$ is a normal rational curve. For q even, a similar argument applies, but now the fact that a $(q+1)$-arc in $PG(4, q)$ is a normal rational curve must be used. □

27.7 Arcs and primals

In this section a connection is obtained between arcs and primals (hypersurfaces). The main aim is to obtain an upper bound for $m(n, q)$ with q even. To do this, we require a more sophisticated notion of algebraic variety than in the previous chapters.

With $\gamma = GF(q)$, let $H, H_1, \ldots, H_r$ be forms in $\Omega = \gamma[x_0, \ldots, x_n]$; in fact, $H_1, \ldots, H_r$ will always be linear. The *variety* $\mathscr{A}$ in $PG(n, q)$ defined by H and $H_1, \ldots, H_r$ is denoted

$$\mathscr{A} = \mathbf{A}(H, H_1, \ldots, H_r)$$

and consists of the pair $(\mathbf{V}(\mathscr{A}), \mathbf{I}(\mathscr{A}))$ where $\mathbf{V}(\mathscr{A}) = \mathbf{V}(H, H_1, \ldots, H_r)$ is the set of zeros of $H, H_1, \ldots, H_r$ in $PG(n, q)$ and $\mathbf{I}(\mathscr{A}) = \mathbf{I}(H, H_1, \ldots, H_r)$ is the ideal generated by $H, H_1, \ldots, H_r$ in Ω; that is,

$$\mathbf{V}(\mathscr{A}) = \{\mathbf{P}(X) \in PG(n, q) \mid H(X) = H_1(X) = \cdots = H_r(X) = 0\},$$

$$\mathbf{I}(\mathscr{A}) = \{F \in \Omega \mid F = GH + G_1H_1 + \cdots + G_rH_r \text{ for some } G, G_1, \ldots, G_r \text{ in } \Omega\}.$$

The number of points in $\mathbf{V}(\mathscr{A})$ will be denoted simply as $|\mathscr{A}|$.

If $\mathscr{A}$ and $\mathscr{B}$ are varieties in $PG(n, q)$, then $\mathscr{A}$ is *algebraically contained in* $\mathscr{B}$, denoted $\mathscr{A} \subset \mathscr{B}$, if $\mathbf{I}(\mathscr{A}) \supset \mathbf{I}(\mathscr{B})$. The varieties $\mathscr{A}$ and $\mathscr{B}$ are (*algebraically*) *equal* if $\mathbf{I}(\mathscr{A}) = \mathbf{I}(\mathscr{B})$. A variety $\mathscr{A} = \mathbf{A}(H)$ with $\deg H = d$ is a *primal* (or *hypersurface*) of *degree d*. If $\mathscr{A} = \mathbf{A}(H, H_1, \ldots, H_r)$ is a variety with $H, H_i \neq 0$, all i, and $H_{r+1}, \ldots, H_u$ are other linear forms in $\Omega \backslash \{0\}$, then

$$\mathscr{A} \cap \pi_1 \cap \cdots \cap \pi_u = \mathscr{A} \cap \pi_{r+1} \cap \cdots \cap \pi_u$$

denotes the variety $\mathbf{A}(H, H_1, \ldots, H_u)$, where π_j is the prime $\mathbf{V}(H_j)$, $j = 1, \ldots, u$. We note that $H_1, \ldots, H_r$ were initially defined to be linear. For H_j linear, we may use $\mathbf{V}(H_j)$ and $\mathbf{A}(H_j)$ indistinguishably.

Theorem 27.7.1: *In* $\Sigma = PG(n, q)$, $n \geq 3$, *let* $\mathscr{K} = \{\pi_1, \pi_2, \ldots, \pi_k\}$ *be a set of primes any three of which are linearly independent and such that to* π_i *is associated a primal* Φ_i *of* Σ *of degree d with the following properties*:

(a) $\Phi_i \cap \pi_i \cap \pi_j = \Phi_j \cap \pi_i \cap \pi_j$ *for all distinct* i, j;
(b) $|\Phi_i \cap \pi_i \cap \pi_j| < \theta(n-2)$ *for all distinct* i, j;
(c) $|\Phi_i \cap \pi_i \cap \pi_j \cap \pi_u| < \theta(n-3)$ *for all distinct* i, j, u.

Then there exists a primal Φ *in* Σ *of degree d such that, for all i*,

$$\Phi \cap \pi_i = \Phi_i \cap \pi_i. \tag{27.72}$$

Proof: We use induction on k. For $k = 1$, there is nothing to prove. Assume that $k \geq 2$, and that the statement holds for $k - 1$. Let Φ' be a primal of degree d such that, for $1 \leq i \leq k-1$,

$$\Phi' \cap \pi_i = \Phi_i \cap \pi_i. \tag{27.73}$$

Then, for $1 \leq i \leq k-1$,

$$\Phi' \cap \pi_k \cap \pi_i = \Phi_i \cap \pi_i \cap \pi_k = \Phi_k \cap \pi_k \cap \pi_i. \tag{27.74}$$

Let $\Phi' = \mathbf{A}(D')$, $\Phi_k = \mathbf{A}(D_k)$, and $\pi_j = \mathbf{A}(H_j)$, $j = 1, \ldots, k$. By (27.74), for $1 \leq i \leq k-1$,

$$\mathbf{I}(D', H_k, H_i) = \mathbf{I}(D_k, H_k, H_i).$$

So

$$D' = u_i D_k + r_i H_k + s_i H_i$$

where $u_i, r_i, s_i \in \gamma[x_0, \ldots, x_n]$. By comparing the terms of degree d, we obtain that

$$D' + t_i D_k \in \mathbf{I}(H_k, H_i) \tag{27.75}$$

with t_i in γ; here t_i is non-zero, since otherwise $|\Phi' \cap \pi_i \cap \pi_k| \geq \theta(n-2)$, whence $|\Phi_i \cap \pi_i \cap \pi_k| \geq \theta(n-2)$, a contradiction.

We show that $t_i = t_j$. From (27.75),

$$(t_i - t_j)D_k = (D' + t_i D_k) - (D' + t_j D_k) \in \mathbf{I}(H_k, H_i, H_j). \tag{27.76}$$

Since $D_k \notin \mathbf{I}(H_k, H_i, H_j)$ by (c), so (27.76) implies that $t_i - t_j = 0$. We write $t_i = \lambda$ and note that $\lambda \neq 0$. Next, choose coordinates so that $\pi_k = \mathbf{A}(x_0)$ and $\pi_i = \mathbf{A}(a_{i0}x_0 + \cdots + a_{in}x_n)$, $1 \leq i \leq k-1$. Write, for $1 \leq i \leq k-1$,

$$D' + \lambda D_k = GH_k + G_i(H_i - a_{i0}H_k),$$

where G_i is chosen in such a way that it contains no terms in x_0. Thus

$$D' + \lambda D_k - GH_k = G_i(H_i - a_{i0}H_k) = G_j(H_j - a_{j0}H_k).$$

Hence

$$G_j(H_j - a_{j0}H_k) = F \prod_{i=1}^{k-1} (H_i - a_{i0}H_k), \tag{27.77}$$

since π_k, π_i, π_j are linearly independent for distinct k, i, j.

Finally, we show that $\Phi = \mathbf{A}(D)$ with

$$D = D' - F \prod_{i=1}^{k-1} H_i$$

has the required properties. The only thing to check is that $\Phi \cap \pi_k = \Phi_k \cap \pi_k$, that is $\mathbf{I}(D, H_k) = \mathbf{I}(D_k, H_k)$. This can be shown as follows:

$$\begin{aligned}\mathbf{I}(D, H_k) &= \mathbf{I}\left(D' - F\prod_{i=1}^{k-1} H_i, H_k\right)\\ &= \mathbf{I}\left(D' - F\prod_{i=1}^{k-1}(H_i - a_{i0}H_k), H_k\right)\\ &= \mathbf{I}(GH_k - \lambda D_k, H_k)\\ &= \mathbf{I}(D_k, H_k). \quad \square\end{aligned}$$

Remark: For $k \geq 3$, hypothesis (b) follows from the others. For, if $|\Phi_i \cap \pi_i \cap \pi_j| = \theta(n-2)$, then $|\Phi_i \cap \pi_i \cap \pi_j \cap \pi_u| = \theta(n-3)$, contradicting (c).

Theorem 27.7.2: *In $\Sigma = PG(n, q)$, $n \geq 3$, let $\mathscr{K} = \{\pi_1, \pi_2, \ldots, \pi_k\}$, $k \geq n$, be a set of primes any n of which are linearly independent such that for each plane $\pi_{i_1} \cap \pi_{i_2} \cap \cdots \cap \pi_{i_{n-2}}$ there is an associated primal $\mathscr{C}_{\{i_1,\ldots,i_{n-2}\}}$ of degree d. Suppose*

(a) $\mathscr{C}_{\{i_1,\ldots,i_{n-2}\}} \cap \pi_{i_1} \cap \cdots \cap \pi_{i_{n-1}} = \mathscr{C}_{\{j_1,\ldots,j_{n-2}\}} \cap \pi_{j_1} \cap \cdots \cap \pi_{j_{n-1}}$

for all subsets $\{i_1, i_2, \ldots, i_{n-1}\} = \{j_1, \ldots, j_{n-1}\}$ of order $n-1$ of $\{1, 2, \ldots, k\}$;

(b) $$|\mathscr{C}_{\{i_1,\ldots,i_{n-2}\}} \cap \pi_{i_1} \cap \cdots \cap \pi_{i_n}| = 0$$

for any subset $\{i_1, \ldots, i_n\}$ of size n of $\{1, \ldots, k\}$.

Then there exist primals $\Phi, \Phi_1, \ldots, \Phi_k$ in Σ of degree d such that

(i) $\Phi_{i_1} \cap \pi_{i_1} \cap \cdots \cap \pi_{i_{n-2}} = \mathscr{C}_{\{i_1,\ldots,i_{n-2}\}} \cap \pi_{i_1} \cap \cdots \cap \pi_{i_{n-2}}$

for all distinct $i_1, \ldots, i_{n-2}$;

(ii) $$\Phi \cap \pi_i = \Phi_i \cap \pi_i \text{ for } 1 \leq i \leq k.$$

Proof: For $n = 3$, the statement holds by the previous theorem and the subsequent remark. Consider the 3-space $\pi_{i_1} \cap \cdots \cap \pi_{i_{n-3}}$. Again by the previous theorem and remark, in this 3-space and so in Σ, there exists a primal $\Phi_{\{i_1,\ldots,i_{n-3}\}}$ of degree d with

$$\Phi_{\{i_1,\ldots,i_{n-3}\}} \cap \pi_{i_1} \cap \cdots \cap \pi_{i_{n-2}} = \mathscr{C}_{\{i_1,\ldots,i_{n-2}\}} \cap \pi_{i_1} \cap \cdots \cap \pi_{i_{n-2}}$$

for any i_{n-2} in $\mathbf{N}_k \backslash \{i_1, \ldots, i_{n-3}\}$. By the previous theorem and remark, in each 4-space $\pi_{i_1} \cap \cdots \cap \pi_{i_{n-4}}$ and so in Σ, there exists a primal $\Phi_{\{i_1,\ldots,i_{n-4}\}}$ of degree d with

$$\Phi_{\{i_1,\ldots,i_{n-4}\}} \cap \pi_{i_1} \cap \cdots \cap \pi_{i_{n-3}} = \Phi_{\{i_1,\ldots,i_{n-3}\}} \cap \pi_{i_1} \cap \cdots \cap \pi_{i_{n-3}}$$

for any i_{n-3} in $\mathbf{N}_k \backslash \{i_1, \ldots, i_{n-4}\}$. Hence, for distinct i_{n-3}, i_{n-2} in $\mathbf{N}_k \backslash \{i_1, \ldots, i_{n-4}\}$, we have

$$\begin{aligned}\Phi_{\{i_1,\ldots,i_{n-4}\}} \cap \pi_{i_1} \cap \cdots \cap \pi_{i_{n-2}} &= \Phi_{\{i_1,\ldots,i_{n-3}\}} \cap \pi_{i_1} \cap \cdots \cap \pi_{i_{n-2}} \\ &= \mathscr{C}_{\{i_1,\ldots,i_{n-2}\}} \cap \pi_{i_1} \cap \cdots \cap \pi_{i_{n-2}}.\end{aligned}$$

Continuing in this way we obtain

$$\Phi_{i_1} \cap \pi_{i_1} \cap \pi_{i_2} = \Phi_{\{i_1,i_2\}} \cap \pi_{i_1} \cap \pi_{i_2}$$

for primals $\Phi_{i_1}, \Phi_{\{i_1,i_2\}}$ of degree d and any i_2 in $\mathbf{N}_k \backslash \{i_1\}$. Hence for distinct $i_2, \ldots, i_{n-2}$ in $\mathbf{N}_k \backslash \{i_1\}$,

$$\begin{aligned}\Phi_{i_1} \cap \pi_{i_1} \cap \cdots \cap \pi_{i_{n-2}} &= \Phi_{\{i_1,i_2\}} \cap \pi_{i_1} \cap \cdots \cap \pi_{i_{n-2}} \\ &= \mathscr{C}_{\{i_1,\ldots,i_{n-2}\}} \cap \pi_{i_1} \cap \cdots \cap \pi_{i_{n-2}}.\end{aligned}$$

Finally, we obtain a primal Φ of degree d in Σ such that, for any i_1 in $\mathbf{N}_k$ and for any $i_2, \ldots, i_{n-2}$ in $\mathbf{N}_k \backslash \{i_1\}$,

$$\Phi \cap \pi_{i_1} = \Phi_{i_1} \cap \pi_{i_1},$$

$$\begin{aligned}\Phi \cap \pi_{i_1} \cap \cdots \cap \pi_{i_{n-2}} &= \Phi_{i_1} \cap \pi_{i_1} \cap \cdots \cap \pi_{i_{n-2}} \\ &= \mathscr{C}_{\{i_1,\ldots,i_{n-2}\}} \cap \pi_{i_1} \cap \cdots \cap \pi_{i_{n-2}}. \quad \square\end{aligned}$$

The essential construction

In $PG(n, q)$, $n \geq 3$, $q = 2^h$, let $\mathscr{K} = \{\pi_1, \pi_2, \ldots, \pi_k\}$, $k \geq n+1$, be an arc of primes; that is, every $n+1$ primes in $\mathscr{K}$ are linearly independent or, equivalently, no $n+1$ primes in $\mathscr{K}$ have a point in common. For distinct $i_1, \ldots, i_{n-1}$, there are exactly $t = q + n - k$ points on the line $\pi_{i_1} \cap \cdots \cap \pi_{i_{n-1}}$ contained in no other prime of $\mathscr{K}$. With $S = \{i_1, \ldots, i_{n-1}\}$, denote this set of t points by

$$Z_S = \{Z_S^{(1)}, \ldots, Z_S^{(t)}\}.$$

In the plane $\Pi_2 = \pi_{i_1} \cap \cdots \cap \pi_{i_{n-2}}$, the other primes of $\mathscr{K}$ cut out a $(k-n+2)$-arc of lines. As $q + 2 - (k - n + 2) = t$, so by Theorem 10.3.1, the points in Z_S lie on an algebraic curve $\mathscr{C}_{\{i_1,\ldots,i_{n-2}\}}$ of degree t in Π_2 with

$$\mathscr{C}_{\{i_1,\ldots,i_{n-2}\}} \cap \pi_{i_{n-1}} = Z_S.$$

Also

$$\mathscr{C}_{\{i_1,\ldots,i_{n-2}\}} \cap \pi_{i_{n-1}} = \mathscr{C}_{\{j_1,\ldots,j_{n-2}\}} \cap \pi_{j_{n-1}}$$

for all equal subsets $\{i_1, \ldots, i_{n-1}\}$ and $\{j_1, \ldots, j_{n-1}\}$ of size $n-1$ in $\mathbf{N}_k$. Further,

$$|\mathscr{C}_{\{i_1,\ldots,i_{n-2}\}} \cap \pi_{i_{n-1}} \cap \pi_{i_n}| = 0$$

for any subset $\{i_1, \ldots, i_n\}$ of size n in $\mathbf{N}_k$.

By Theorem 27.7.2, the curves $\mathscr{C}_{\{i_1,\ldots,i_{n-2}\}}$ for a fixed i are algebraically contained in a primal Φ_{i_1} in π_{i_1} of degree t with

$$\Phi_{i_1} \cap \pi_{i_2} \cap \cdots \cap \pi_{i_{n-2}} = \mathscr{C}_{\{i_1,\ldots,i_{n-2}\}},$$

while the varieties $\Phi_1, \ldots, \Phi_k$ are algebraically contained in a primal $\Phi = \Phi(\mathscr{K})$ of $PG(n, q)$ of degree t with $\Phi \cap \pi_i = \Phi_i$ for $1 \le i \le k$. The primal $\Phi = \Phi(\mathscr{K})$ is the *primal associated to* $\mathscr{K}$.

Theorem 27.7.3: *In $PG(n, q)$, $n \ge 3$ and $q = 2^h$, let $\mathscr{K} = \{\pi_1, \ldots, \pi_k\}$, $k \ge n + 1$, be a k-arc of primes. For distinct $i_1, i_2, \ldots, i_{n-1}$ in $\mathbf{N}_k$, let $Z_{\{i_1,\ldots,i_{n-1}\}}$ denote the set of $t = q + n - k$ points on the line $\pi_{i_1} \cap \cdots \cap \pi_{i_{n-1}}$ that lie on no other prime of $\mathscr{K}$. Then*

(i) *there exists a curve $\mathscr{C}_{\{i_1,\ldots,i_{n-2}\}}$ of degree t in the plane $\pi_{i_1} \cap \cdots \cap \pi_{i_{n-2}}$ such that*

$$\mathscr{C}_{\{i_1,\ldots,i_{n-2}\}} \cap \pi_{i_{n-1}} = Z_{\{i_1,\ldots,i_{n-1}\}};$$

(ii) *for fixed i_1, the curves $\mathscr{C}_{\{i_1,\ldots,i_{n-2}\}}$ are algebraically contained in a primal Φ_{i_1} of π_{i_1} of degree t with $\Phi_{i_1} \cap \pi_{i_2} \cap \cdots \cap \pi_{i_{n-2}} = \mathscr{C}_{\{i_1,\ldots,i_{n-2}\}}$ and each variety Φ_{i_1} is algebraically contained in a primal $\Phi = \Phi(\mathscr{K})$ of $PG(n, q)$ of degree t with $\Phi(\mathscr{K}) \cap \pi_i = \Phi_i$;*

(iii) *if $k > \frac{1}{2}q + n - 1$, the primal $\Phi(\mathscr{K})$ is unique;*

(iv) *with $k > \frac{1}{2}q + n - 1$, if $\mathscr{L} = \{\pi_1, \ldots, \pi_k, \ldots, \pi_u\}$ is an arc of primes containing $\mathscr{K}$, the primal $\Phi(\mathscr{K})$ has components $\Phi(\mathscr{L})$, $\pi_{k+1}, \ldots, \pi_u$;*

(v) *if $k > \frac{1}{2}q + n - 1$, there is a bijection between primes of $PG(n, q)$ extending $\mathscr{K}$ to a $(k + 1)$-arc and linear components over $GF(q)$ of $\Phi(\mathscr{K})$.*

Proof: (i), (ii) These were proved in the previous theorem and the subsequent remarks.

(iii) Since $k - n + 2 > t = q - k + n$, it follows from Theorem 10.3.1 that the curve $\mathscr{C}_{\{i_1,\ldots,i_{n-2}\}}$ is unique. Suppose Φ and Φ' are distinct primals of degree t for which

$$\Phi \cap \pi_{i_1} \cap \cdots \cap \pi_{i_{n-2}} = \Phi' \cap \pi_{i_1} \cap \cdots \cap \pi_{i_{n-2}} = \mathscr{C}_{\{i_1,\ldots,i_{n-2}\}}.$$

Let $\Phi = \mathbf{A}(D)$ and $\Phi' = \mathbf{A}(D')$.

First, let $n = 3$ and fix an index i_1. As in Theorem 27.7.1, there exists λ in γ_0 for which $D + \lambda D'$ vanishes at all points of the $k - 1$ lines $\pi_{i_1} \cap \pi_{i_2}$ with $i_2 \ne i_1$. The surfaces Φ and Φ' both have degree $t = q + 3 - k$. Since $k > \frac{1}{2}q + 2$, we have $k - 1 > t = q + 3 - k$. So $D + \lambda D'$ vanishes at all points of π_{i_1}. Since the surface $\Phi'' = \mathbf{A}(D + \lambda D')$ contains all points of the lines $\pi_{i_1} \cap \pi_{i_2}$, it follows that Φ'' has the k planes $\pi_1, \ldots, \pi_k$ as components. However, $k > t = \deg \Phi''$, a contradiction. So $\Phi' = \Phi$.

Next, let $n > 3$ and proceed by induction on n. Since $k - 1 > \frac{1}{2}q + (n - 1) - 1$, assume that the varieties $\Phi_{i_1} = \Phi'_{i_1}$ are unique for i_1 in $\mathbf{N}_k$. Fix an index i_1. Again, as in Theorem 27.7.1, there exists λ in γ_0 such that $D + \lambda D'$

vanishes at all points of the $(n-2)$-spaces $\pi_{i_1} \cap \pi_{i_2}$, of which there are $k-1$. Now, both Φ and Φ' have degree $t = q + n - k$. Since $k > \frac{1}{2}q + n - 1$, we have $k - 1 > t$. Hence $D + \lambda D'$ vanishes at all points of π_{i_1}. Since $\Phi'' = \mathbf{A}(D + \lambda D')$ contains all points of $\pi_{i_1} \cap \pi_{i_2}$, the primal Φ'' has the k primes as components. As $k > t = \deg \Phi''$, we have a contradiction and $\Phi = \Phi'$.

(iv) $\Phi(\mathscr{K}) \cap \pi_{i_1} \cap \cdots \cap \pi_{i_{n-1}}$ consists of the set $\Phi(\mathscr{L}) \cap \pi_{i_1} \cap \cdots \cap \pi_{i_{n-1}}$ together with the points

$$\pi_{i_1} \cap \cdots \cap \pi_{i_{n-1}} \cap \pi_{k+1}, \ldots, \pi_{i_1} \cap \cdots \cap \pi_{i_{n-1}} \cap \pi_u.$$

Since $\Phi(\mathscr{K})$ is unique, the required factorization is obtained.

(v) Suppose $\mathscr{K} \cup \{\pi\}$ is a $(k+1)$-arc of primes. By (iv), π is a linear component of $\Phi(\mathscr{K})$. Conversely, let σ be a linear component over $GF(q)$ of $\mathscr{K}$. Let $\pi_{i_1}, \ldots, \pi_{i_n}$ be any n primes in $\mathscr{K}$. Then these n primes and σ have no point in common, since such a point would lie on $\Phi(\mathscr{K}) \cap \pi_{i_1} \cap \cdots \cap \pi_{i_n}$, contradicting the defining property of $\Phi(\mathscr{K})$. So no $n+1$ primes in $\mathscr{K} \cup \{\sigma\}$ have a point in common, whence $\mathscr{K} \cup \{\sigma\}$ is a $(k+1)$-arc of primes. □

Theorem 27.7.4: *Let $\mathscr{K} = \{\pi_1, \ldots, \pi_k\}$ be a k-arc of primes in $PG(n, q)$, $n \geq 3$ and $q = 2^s$. If $k > \frac{1}{2}q + n - 1$, then $\mathscr{K}$ is contained in a unique complete arc.*

Proof: Let $\mathscr{K}'$ and $\mathscr{K}''$ be distinct complete arcs of primes containing $\mathscr{K}$, and assume that $\pi \in \mathscr{K}' \backslash \mathscr{K}''$. By Theorem 27.7.3(v), π is a component of $\Phi(\mathscr{K})$. Since $\pi \notin \mathscr{K}''$, by part (iv), the prime π is a component of $\Phi(\mathscr{K}'')$. Again, by part (v) the arc $\mathscr{K}''$ can be extended to an arc $\mathscr{K}'' \cup \{\pi\}$ where $\pi \notin \mathscr{K}''$. This contradicts the completeness of $\mathscr{K}''$. □

Now, these results are applied in $PG(3, q)$, $q = 2^h$. First, we restate the necessary results for $n = 3$.

Theorem 27.7.5: *Let $\mathscr{K} = \{\pi_1, \ldots, \pi_k\}$ be a k-arc of planes in $\Sigma = PG(3, q)$, $q = 2^h$. For any two distinct planes π_i and π_j, let Z_{ij} be the set of points of $\pi_i \cap \pi_j$ in exactly two planes of $\mathscr{K}$. Then*

(i) *there exists an algebraic curve $\mathscr{C}_i$ in π_i containing all sets Z_{ij} and with $\mathscr{C}_i \cap \pi_j = Z_{ij}$;*

(ii) *there exists an algebraic surface $\Phi = \Phi(\mathscr{K})$ of degree $t = q + 3 - k$ algebraically containing the curves $\mathscr{C}_i$ and with $\Phi \cap \pi_i = \mathscr{C}_i$.*

Suppose further that $k > \frac{1}{2}q + 2$. Then

(iii) *the surface Φ is unique;*

(iv) *if $\mathscr{L} = \mathscr{K} \cup \{\pi_{k+1}, \ldots, \pi_u\}$ is a u-arc of planes, $u > k$, the surface $\Phi(\mathscr{K})$ factors into $\Phi(\mathscr{L}), \pi_{k+1}, \ldots, \pi_u$;*

(v) *there is a bijection between planes of Σ extending $\mathscr{K}$ to a $(k+1)$-arc and linear components over $GF(q)$ of $\Phi(\mathscr{K})$;*

(vi) *$\mathscr{K}$ is contained in a unique complete arc of Σ.* □

Lemma 27.7.6: *Let $k > q - \sqrt{q} + 2$, $q = 2^h$. With notation as in Theorem 27.7.5,*

(i) *the curve $\mathscr{C}_i$ of degree $t = q + 3 - k$ factors into t lines forming an arc of lines in π_i;*

(ii) *these t lines $l_{i1}, \ldots, l_{it}$, called S-lines, together with the $k - 1$ lines $\pi_i \cap \pi_j$, $j \in \mathbf{N}_k \backslash \{i\}$, form a $(q + 2)$-arc of lines in π_i;*

(iii) *each S-line lies in a unique plane of $\mathscr{K}$;*

(iv) *each point P on an S-line lies on exactly one other S-line.*

Proof: (i) This follows from §10.3.

(ii), (iii) These follow from Theorem 27.7.5.

(iv) Each point $Z_{ij}^{(u)}$ in Z_{ij} lies on one S-line l_{ia} in π_i and on one S-line l_{jb} in π_j. If P is on l_{ia} and is not of type $Z_{ij}^{(u)}$, then by (ii) P lies on exactly one other S-line, which will be of type l_{jb} with $b \neq a$. □

Theorem 27.7.7: *Let $\mathscr{K} = \{\pi_1, \pi_2, \ldots, \pi_k\}$ be an arc of planes in $\Sigma = PG(3, q)$, $q = 2^h$ and $k > \frac{1}{2}q + 2$. Assume that $\mathscr{K}$ is contained in a $(q + 1)$-arc $\mathscr{L}$ of planes in Σ. Then*

(i) *$\Phi(\mathscr{K})$ factors into $t - 2$ linear components over $GF(q)$ and one quadratic component $\mathscr{H}$, where $t = q + 3 - k$ and $\mathscr{H}$ is a hyperbolic quadric $\mathscr{H}_3$;*

(ii) *$\Phi(\mathscr{L}) = \mathscr{H}$;*

(iii) *each plane of $\mathscr{L}$ is a tangent plane to $\mathscr{H}$;*

(iv) *for any plane π in $\mathscr{L}$, the planes of $\mathscr{L} \backslash \mathscr{H}$ together with $\mathscr{H}$ cut out a $(q + 2)$-arc of lines in π.*

Proof: (i), (ii) Let $\mathscr{L} = \{\pi_1, \pi_2, \ldots, \pi_{q+1}\}$. By the definition of an arc, $q > 2$. From Theorem 27.7.5(iv), $\Phi(\mathscr{K})$ factors into $\Phi(\mathscr{L}), \pi_{k+1}, \ldots, \pi_{q+1}$ with $\Phi(\mathscr{L}) = \mathscr{H}$ of degree 2. Since $q > 2$, Theorem 21.3.8 says that $\mathscr{L}$ is complete. From Theorem 27.7.5(v), the quadric $\mathscr{H}$ is irreducible.

Let $\pi_{k+1} = \pi'_1, \ldots, \pi_{q+1} = \pi'_{t-2}$. Each plane π'_j in $\mathscr{L} \backslash \mathscr{K}$ intersects each plane π_i of $\mathscr{K}$ in a line. Also, if π'_j, π'_u, π'_v are distinct, then $\pi'_j \cap \pi'_u \cap \pi'_v \cap \pi_i = \varnothing$ since these four planes belong to an arc. Therefore the $t - 2$ planes $\pi'_1, \ldots, \pi'_{t-2}$ cut out a $(t - 2)$-arc of lines in π_i. Let $\mathscr{C}_i$ be the curve of degree t in π_i corresponding to the arc $\mathscr{K}$. We have that $\mathscr{C}_i$ has $t - 2$ linear components $\pi'_1 \cap \pi_i, \ldots, \pi'_{t-2} \cap \pi_i$ and one quadratic component $\mathscr{H}^{(i)} = \mathscr{H} \cap \pi_i$. For each $j \neq i$, with π_j in $\mathscr{K}$, each of the lines $\pi'_u \cap \pi_i$ contains exactly one of the t points $Z_{ij}^{(u)}$. If $a \neq b$ then $\pi'_a \cap \pi_i \cap Z_{ij} \neq \pi'_b \cap \pi_i \cap Z_{ij}$ since otherwise π'_a, π'_b, π_i, and π_j have a point in common, contradicting that these four planes are part of an arc of planes of Σ. Therefore $\mathscr{H}^{(i)}$ contains exactly two points of Z_{ij}, say $Z_{ij}^{(1)}$ and $Z_{ij}^{(2)}$, $i \neq j$. Then $|\mathscr{H}^{(i)}| \geq 2(k - 1) > q + 2$. Since $|\mathscr{H}^{(i)}| > q + 2$, so $\mathscr{H} \cap \pi_i$ cannot be a conic. It follows that $\mathscr{H} \cap \pi_i$ factors into a pair of distinct lines x_i, y_i with $x_i \cap Z_{ij} = Z_{ij}^{(1)}$, $y_i \cap Z_{ij} = Z_{ij}^{(2)}$, and $Z_{ij}^{(1)} \neq Z_{ij}^{(2)}$. Hence each plane of $\mathscr{K}$ contains exactly two lines of $\mathscr{H}$. Further, each line of $\mathscr{H}$ is on at most one plane of $\mathscr{K}$. So $\mathscr{H}$ contains at least $2k > q + 4$ lines. It follows that $\mathscr{H}$ cannot be a cone and is in fact a hyperbolic quadric of Σ.

(iii), (iv) Let π be any plane of $\mathscr{L}$. The planes of $\mathscr{L}\backslash\{\pi\}$ cut out a q-arc of lines in π since $\mathscr{L}$ is an arc of planes in Σ. Hence $\Phi(\mathscr{L}) \cap \pi$ is a curve $\mathscr{C}$ of degree 2. Also since $q + 1 > q - \sqrt{q} + 2$, it follows from §10.3 that $\mathscr{C}$ factors into two lines l, m, which are therefore lines of $\Phi(\mathscr{L})$. Since $\Phi(\mathscr{L})$ is a hyperbolic quadric, (iii) follows. By Lemma 27.7.6(ii), part (iv) also follows. □

Assume for the remainder of the section that $q > 2$.

Lemma 27.7.8: *Let $\mathscr{K}$ be a k-arc of planes in $PG(3, q)$, $q = 2^h$. If $(q - 1)/t^2 + 2/t > t$ with $t = q + 3 - k$, then $k > q - \sqrt{q} + 2$.*

Proof: Assume that $k \le q - \sqrt{q} + 2$; then $\sqrt{q} \le t - 1$. Therefore $(q - 1)/(q + 1 + 2\sqrt{q}) \ge (q - 1)t^2$ and so $1 > (q - 1)/t^2$. This implies that

$$t - 2/t \ge \sqrt{q} + 1 - 2/t \ge \sqrt{q} + 1 - 2/(\sqrt{q} + 1) > 1 > (q - 1)/t^2.$$

So $t - 2/t > (q - 1)/t^2$, a contradiction. □

Theorem 27.7.9: *Let $\mathscr{K} = \{\pi_1, \ldots, \pi_k\}$ be a complete k-arc of planes in $\Sigma = PG(3, q)$, $q = 2^h$. Assume that $(q - 1)/t^2 + 2/t > t$. Then each $\mathscr{C}_i$ factors into t S-lines and each S-line belongs to a hyperbolic quadric algebraically contained in $\Phi(\mathscr{K})$.*

Proof: From Lemmas 27.7.6(i) and 27.7.8, the curve $\mathscr{C}_i$ in π_i factors into t S-lines. Let l_{iu} be a fixed S-line in π_i. By Lemma 27.7.6(iv) there are exactly $q + 1$ S-lines distinct from l_{iu} having a point in common with l_{iu}; denote these by $l_1, l_2, \ldots, l_{q+1}$. Let i, j, g be distinct. The S-lines in π_j are the lines l_{jv}, and the S-lines in π_g are denoted by l_{gw}. Let f_{vw} be the number of lines l_r having a point in common with l_{jv} and l_{gw}. If l_r is not in π_j nor in π_g, then l_r is concurrent with one line of type l_{jv} and one of type l_{gw}. If l_r is in the plane π_j, then it is concurrent with t lines l_{jv} and one line l_{gw}; if l_r is in π_g, it is concurrent with t lines l_{gw} and one line l_{jv}. Consequently

$$\sum_{v,w=1}^{t} f_{vw} = q - 1 + 2t.$$

Averaging we get $\bar{f} = (q - 1)/t^2 + 2/t$. Therefore there exist two S-lines, say $l_{jv} = l'$ and $l_{gw} = l''$, for which $f_{vw} \ge \bar{f} = (q - 1)/t^2 + 2/t$. We show that $l_{iu} = l$, l', and l'' are mutually skew.

Suppose for example that l' and l'' meet in a point P. Since $f_{vw} \ge (q - 1)/t^2 + 2/t$ and since, by Lemma 27.7.6(iv), P lies on exactly two

S-lines, the plane $l'l''$ contains at least $(q-1)/t^2 + 2/t$ S-lines. Because $l'l'' \cap \Phi(\mathscr{K})$ is an algebraic curve of degree t or $l'l''$ is a component of $\Phi = \Phi(\mathscr{K})$, we have $(q-1)/t^2 + 2/t \le t$ or else the plane $l'l''$ is a component of Φ. By assumption $(q-1)/t^2 + 2/t > t$; so the plane $l'l''$ is a component of Φ. But then by Theorem 27.7.5(v) the plane $l'l''$ extends $\mathscr{K}$, contradicting the fact that $\mathscr{K}$ is complete. It follows that the lines l, l', l'' are mutually skew.

If λ is the integer defined by $(q-1)/t^2 + 2/t + 1 > \lambda \ge (q-1)/t^2 + 2/t$, then there are at least λ S-lines of the form l_i, say $l_1, l_2, \ldots, l_\lambda$ which meet l, l', l''. Hence $l_1, l_2, \ldots, l_\lambda$ belong to a regulus $\mathscr{R}$. Let m be a line of the complementary regulus $\mathscr{R}'$. Then m has at least $\lambda > t$ points in common with $\Phi = \Phi(\mathscr{K})$. Consequently m is a line of Φ. So the hyperbolic quadric $\mathscr{H}_3$ with reguli $\mathscr{R}, \mathscr{R}'$ is a component of Φ. Consequently, each S-line l_{iu} belongs to a hyperbolic quadric which is algebraically contained in Φ. □

Lemma 27.7.10: *If* $q - \frac{1}{2}\sqrt{q} + \frac{9}{4} < k < q + 1$, *then for any plane* π *of* $PG(3, q)$ *the curve* $\Phi \cap \pi$ *is reducible over the algebraic closure* $\bar{\gamma}$ *of* $\gamma = GF(q)$.

Proof: Clearly $k > q - \sqrt{q} + 2$ and since there is an integer between $q - \frac{1}{2}\sqrt{q} + \frac{9}{4}$ and $q + 1$, we necessarily have $q \ge 32$. We may assume that no S-line is contained in π. Then any S-line has exactly one point in common with π. Since the number of S-lines is equal to $k(q + 3 - k) = kt$ and each point is on either zero or two S-lines, we have

$$|\Phi \cap \pi| \ge kt/2.$$

Assume $\Phi \cap \pi$ is absolutely irreducible, that is irreducible over $\bar{\gamma}$. By the Hasse–Weil bound of Theorem 10.2.1,

$$(q + 3 - k)k/2 \le q + 1 + (q + 2 - k)(q + 1 - k)\sqrt{q}.$$

Consequently, either $k \ge q + 1$ or $k \le q - \frac{1}{2}\sqrt{q} + \frac{9}{4} - 1/(4 + 8\sqrt{q}) < q - \frac{1}{2}\sqrt{q} + \frac{9}{4}$, a contradiction. We conclude that $\Phi \cap \pi$ is reducible over $\bar{\gamma}$. □

Lemma 27.7.11: *If* $q - \frac{1}{2}\sqrt{q} + \frac{9}{4} < k < q + 1$ *and* k *is even, then, for each plane* π *of* $PG(3, q)$, *the curve* $\Phi \cap \pi = \mathscr{C}$ *contains a line as component (over* γ*).*

Proof: We may assume that no S-line is contained in π. By Lemma 27.7.10 the curve $\mathscr{C}$ is reducible over $\bar{\gamma}$. We have $2 < t < \frac{1}{2}\sqrt{q} + \frac{3}{4}$ and $q \ge 32$.

If $\mathscr{C}'$ is an absolutely irreducible component of $\mathscr{C}$ of degree m, with $m \ge 4$, then we show that $|\mathscr{C}'| < (q + 1)m/2$.

If $\mathscr{C}'$ is not defined over γ, then $|\mathscr{C}'| \le m^2$, Lemma 10.1.1; if $\mathscr{C}'$ is defined over γ then, by the Hasse–Weil bound, $|\mathscr{C}'| \le q + 1 + (m-1)(m-2)\sqrt{q}$.

Since $q + 1 + (m - 1)(m - 2)\sqrt{q} \geq m^2$, we always have $|\mathscr{C}'| \leq q + 1 + (m - 1)(m - 2)\sqrt{q}$. Assume that

$$q + 1 + (m - 1)(m - 2)\sqrt{q} \geq (q + 1)m/2.$$

Then either

$$\tfrac{1}{4}\sqrt{q} + \tfrac{3}{2} + 1/(4\sqrt{q}) + \tfrac{1}{4}\{(\sqrt{q} - 2)^2 + 2 - 4/\sqrt{q} + 1/q\}^{1/2} \leq m,$$

or

$$\tfrac{1}{4}\sqrt{q} + \tfrac{3}{2} + 1/(4\sqrt{q}) - \tfrac{1}{4}\{(\sqrt{q} - 2)^2 + 2 - 4/\sqrt{q} + 1/q\}^2 \geq m.$$

Hence either

$$m > \tfrac{1}{4}\sqrt{q} + \tfrac{3}{2} + \tfrac{1}{4}(\sqrt{q} - 2) = \tfrac{1}{2}\sqrt{q} + 1$$

or

$$m < \tfrac{1}{4}\sqrt{q} + \tfrac{3}{2} + 1/(4\sqrt{q}) - \tfrac{1}{4}(\sqrt{q} - 2) = 2 + 1/(4\sqrt{q}).$$

This contradicts $4 \leq m < t < \tfrac{1}{2}\sqrt{q} + \tfrac{3}{4}$. Hence $|\mathscr{C}'| < (q + 1)m/2$.

If $\mathscr{C}''$ is an absolutely irreducible component of $\mathscr{C}$ of odd degree m, with $m \geq 5$, then we show that $|\mathscr{C}''| < \tfrac{1}{2}(m - 3)(q + 1) + q + 1 + 2\sqrt{q}$. Note that, by the Hasse–Weil bound, $q + 1 + 2\sqrt{q}$ is the maximum number of points of an absolutely irreducible plane cubic curve over γ. As for $\mathscr{C}'$ we have $|\mathscr{C}''| \leq q + 1 + (m - 1)(m - 2)\sqrt{q}$. Since $5 \leq m \leq q + 1 - k < \tfrac{1}{2}\sqrt{q} - \tfrac{5}{4}$, we have $q \geq 256$. Assume that

$$\tfrac{1}{2}(m - 3)(q + 1) + q + 1 + 2\sqrt{q} \leq q + 1 + (m - 1)(m - 2)\sqrt{q}.$$

Then either

$$m \leq \tfrac{1}{4}\sqrt{q} + 1/(4\sqrt{q}) + \tfrac{3}{2} - \tfrac{1}{4}\{(\sqrt{q} - 6)^2 + 2 - 12/\sqrt{q} + 1/q\}^{1/2}.$$

or

$$m \geq \tfrac{1}{4}\sqrt{q} + 1/(4\sqrt{q}) + \tfrac{3}{2} + \tfrac{1}{4}\{(\sqrt{q} - 6)^2 + 2 - 12/\sqrt{q} + 1/q)\}^{1/2}.$$

Since $q \geq 256$ we have $2 - 12/(\sqrt{q}) + 1/q > 0$. Hence either

$$m < \tfrac{1}{4}\sqrt{q} + \tfrac{3}{2} + 1/(4\sqrt{q}) - \tfrac{1}{4}(\sqrt{q} - 6) = 3 + 1/(4\sqrt{q}),$$

or

$$m > \tfrac{1}{4}\sqrt{q} + \tfrac{3}{2} + \tfrac{1}{4}(\sqrt{q} - 6) = \tfrac{1}{2}\sqrt{q}.$$

This contradicts $5 \leq m < \tfrac{1}{2}\sqrt{q} - \tfrac{5}{4}$, and so $|\mathscr{C}''| < \tfrac{1}{2}(m - 3)(q + 1) + q + 1 + 2\sqrt{q}$.

Assume that $\mathscr{C}$ contains no linear component over γ, but contains β (≥ 0) linear components over $\bar{\gamma}$. Since $\mathscr{C}$ has odd degree, the number of components (over $\bar{\gamma}$) of $\mathscr{C}$ of odd degree is odd. By the preceding sections and using $2(q + 1 + 2\sqrt{q}) < 3(q + 1)$, we now have

(i) for β even,

$$\begin{aligned} |\mathscr{C}| &\leq \tfrac{1}{2}(t - \beta - 3)(q + 1) + \beta + (q + 1 + 2\sqrt{q}) \\ &\leq \tfrac{1}{2}(t - 3)(q + 1) + q + 1 + 2\sqrt{q}; \end{aligned}$$

(ii) for β odd,

$$|\mathscr{C}| \leq \tfrac{1}{2}(t - \beta)(q + 1) + \beta < \tfrac{1}{2}(t - 3)(q + 1) + q + 1 + 2\sqrt{q}.$$

Hence, in each case,

$$|\mathscr{C}| \le \tfrac{1}{2}(t-3)(q+1) + q + 1 + 2\sqrt{q}.$$

Consequently

$$\tfrac{1}{2}k(q+3-k) \le |\mathscr{C}| \le \tfrac{1}{2}(q-k)(q+1) + q + 1 + 2\sqrt{q}.$$

So, either

$$k \le q + 2 - \{(\sqrt{q}-3)^2 + 2\sqrt{q} - 7\}^{1/2}$$

or

$$k \ge q + 2 + \{(\sqrt{q}-3)^2 + 2\sqrt{q} - 7\}^{1/2}.$$

Hence either

$$k < q + 2 - (\sqrt{q} - 3) = q - \sqrt{q} + 5$$

or

$$k > q + 2 + (\sqrt{q} - 3) = q + \sqrt{q} - 1.$$

This contradicts $q - \tfrac{1}{2}\sqrt{q} + \tfrac{9}{4} < k < q + 1$.

We conclude that $\mathscr{C}$ contains a line as component over γ. □

Theorem 27.7.12: *If $q - \tfrac{1}{2}\sqrt{q} + \tfrac{9}{4} < k < q + 1$ and k is even, then Φ contains a plane as component (over γ).*

Proof: Let π_i be a plane of the k-arc. In π_i there are $q + 3 - k = t < \tfrac{1}{2}\sqrt{q} + \tfrac{3}{4}$ S-lines which form an arc of lines in π_i. Since $q \ge t(t-3)/2 + 2$, this arc is incomplete by Theorem 9.1.10; so there is a line l which intersects the t lines of the arc at t different points. Hence $|l \cap \Phi| = t$ and the t points of $\Phi \cap l$ are simple for Φ. Considering the $q + 1$ planes of $PG(3, q)$ through l and using Lemma 27.7.11, we see that at least one point P of $\Phi \cap l$ is contained in at least $(q+1)/t$ lines of Φ. Hence P is contained in at least $2\sqrt{q} - 4 + (\tfrac{1}{2}\sqrt{q} + 4)/(\tfrac{1}{2}\sqrt{q} + \tfrac{3}{4})$ lines of Φ. It follows that the tangent plane π_P of Φ at P contains more than $2\sqrt{q} - 4$ lines of Φ. Since $2\sqrt{q} - 4 > t$, the plane π_P is a component of Φ. □

Theorem 27.7.13: *Any k-arc $\mathscr{K}$ of $PG(3, q)$, with q even, k even, and $q - \tfrac{1}{2}\sqrt{q} + \tfrac{9}{4} < k < q + 1$, can be extended to a $(k+1)$-arc.*

Proof: This follows immediately from Theorems 27.7.5(v) and 27.7.12. □

Lemma 27.7.14: *If $q - \tfrac{1}{2}\sqrt{q} + \tfrac{9}{4} < k < q + 1$ and k is odd, then for each plane π of $PG(3, q)$, the curve $\Phi \cap \pi = \mathscr{C}$ either contains a line as component over γ or consists of $t/2$ conics defined over γ.*

Proof: We may assume that no S-line is contained in π. If $\mathscr{C}'$ is an absolutely irreducible component of $\mathscr{C}$ of degree m, with $q + 2 - k > m > 4$ and $q > 512$, then we show that $|\mathscr{C}'| < \frac{1}{2}(m - 4)(q + 1) + q + 1 + 2\sqrt{q}$. Note that, by the Hasse–Weil bound, $q + 1 + 2\sqrt{q}$ is the maximum number of points of an absolutely irreducible plane cubic curve over γ.

Ignore for the moment the condition $q > 512$. If $\mathscr{C}'$ is not defined over γ, then $|\mathscr{C}'| \leq m^2$; if $|\mathscr{C}'|$ is defined over γ, then $|\mathscr{C}'| \leq q + 1 + (m - 1)(m - 2)\sqrt{q}$. Hence $|\mathscr{C}'| \leq q + 1 + (m - 1)(m - 2)\sqrt{q}$. Since $5 \leq m \leq q + 1 - k < \frac{1}{2}\sqrt{q} - \frac{5}{4}$ we have $q \geq 256$. Assume that

$$\tfrac{1}{2}(m - 4)(q + 1) + q + 1 + 2\sqrt{q} \leq q + 1 + (m - 1)(m - 2)\sqrt{q}. \quad (27.78)$$

Then either

$$m \leq \tfrac{1}{4}\sqrt{q} + \tfrac{3}{2} + 1/(4\sqrt{q}) - \tfrac{1}{4}\{(\sqrt{q} - 11)^2 + 2\sqrt{q} - 20/\sqrt{q} + 1/q - 83\}^{1/2}$$

or

$$m \geq \tfrac{1}{4}\sqrt{q} + \tfrac{3}{2} + 1/(4\sqrt{q}) + \tfrac{1}{4}\{(\sqrt{q} - 11)^2 + 2\sqrt{q} - 20/\sqrt{q} + 1/q - 83\}^{1/2}.$$

For $q > 1024$ we have $2\sqrt{q} - 20/\sqrt{q} + 1/q - 83 > 0$. Hence, for $q > 1024$, either

$$m < \tfrac{1}{4}\sqrt{q} + \tfrac{3}{2} + 1/(4\sqrt{q}) - \tfrac{1}{4}(\sqrt{q} - 11) = \tfrac{17}{4} + 1/(4\sqrt{q})$$

or

$$m > \tfrac{1}{4}\sqrt{q} + \tfrac{3}{2} + \tfrac{1}{4}(\sqrt{q} - 11) = \tfrac{1}{2}\sqrt{q} - \tfrac{5}{4}.$$

This contradicts $5 \leq m < \frac{1}{2}\sqrt{q} - \frac{5}{2}$. For $q = 256$ the inequality (27.78) is satisfied; for $q = 512$, (27.78) together with $5 \leq m \leq q + 1 - k < \frac{1}{2}\sqrt{q} - \frac{5}{4}$ gives $m = 10$ and $k = 503$; for $q = 1024$, (27.78) is in contradiction to $5 \leq m < \frac{1}{2}\sqrt{q} - \frac{5}{4}$.

Let $\mathscr{C}''$ be an absolutely irreducible component of degree four of $\mathscr{C}$. If $\mathscr{C}''$ is not defined over γ, then $|\mathscr{C}''| \leq 16 < 2(q + 1)$; if $\mathscr{C}''$ is defined over γ, then $|\mathscr{C}''| \leq q + 1 + 6\sqrt{q}$, and consequently $|\mathscr{C}''| \leq 2(q + 1)$, for $q \geq 32$. Hence always $|\mathscr{C}''| \leq 2(q + 1)$.

Now assume that, over $\bar{\gamma}$, the curve $\mathscr{C}$ neither contains a line nor consists entirely of conics and irreducible quartic curves. Assume also that $q \notin \{256, 512\}$. Let β be the number of absolutely irreducible components of degree three, and let α be the number of absolutely irreducible components of degree at least five. If $\alpha = 0$ then, since $q + 3 - k = t$ is even, β is even and so $\alpha + \beta$ is even. Also $\alpha + \beta > 0$. By the preceding sections,

$$|\mathscr{C}| \leq \tfrac{1}{2}(t - 3\beta - 4\alpha)(q + 1) + (\alpha + \beta)(q + 1 + 2\sqrt{q}).$$

Further note that $2(q + 1 + 2\sqrt{q}) < 3(q + 1)$. If $\alpha + \beta$ is odd, so $\alpha \neq 0$; then

$$|\mathscr{C}| \leq \tfrac{1}{2}(t - \alpha - 3)(q + 1) + q + 1 + 2\sqrt{q} \leq \tfrac{1}{2}(t - 4)(q + 1) + q + 1 + 2\sqrt{q}.$$

If $\alpha + \beta$ is even, so $\alpha + \beta \geq 2$; then

$$\begin{aligned}|\mathscr{C}| &\leq \tfrac{1}{2}(t - \alpha - 6)(q + 1) + 2(q + 1 + 2\sqrt{q}) \\ &\leq \tfrac{1}{2}(t - 6)(q + 1) + 2(q + 1 + 2\sqrt{q}).\end{aligned}$$

Thus in both cases,

$$\begin{aligned}|\mathscr{C}| &\leq \tfrac{1}{2}(t - 6)(q + 1) + 2(q + 1 + 2\sqrt{q}) \\ &= \tfrac{1}{2}(q - k - 3)(q + 1) + 2(q + 1 + 2\sqrt{q}).\end{aligned}$$

Consequently,

$$\tfrac{1}{2}k(q + 3 - k) \leq \tfrac{1}{2}(t - 6)(q + 1) + 2(q + 1 + 2\sqrt{q}).$$

So, either

$$k \leq q + 2 - \{2q - 8\sqrt{q} + 3\}^{1/2}$$

or

$$k \geq q + 2 + \{2q - 8\sqrt{q} + 3\}^{1/2}.$$

This contradicts $q - \frac{1}{2}\sqrt{q} + \frac{9}{4} < k < q + 1$. Hence, for $q \notin \{256, 512\}$, $\mathscr{C}$ contains over $\bar{\gamma}$ either a linear component or consists entirely of conics and absolutely irreducible quartic curves. Let $\mathscr{C}$ consist of ε conics and δ absolutely irreducible quartic curves, with $\delta \geq 1$. If $q = 32$, then $t = 3$ and so k is even; hence $q \geq 64$. Since

$$\begin{aligned}&\tfrac{1}{2}(t - 4\delta)(q + 1) + \delta(q + 1 + 6\sqrt{q}) \\ &\qquad = \tfrac{1}{2}(q + 3 - k)(q + 1) + \delta(-q - 1 + 6\sqrt{q}) \\ &\qquad \leq \tfrac{1}{2}(q + 3 - k)(q + 1) - q - 1 + 6\sqrt{q},\end{aligned}$$

we have

$$\tfrac{1}{2}k(q + 3 - k) \leq |\mathscr{C}| \leq \tfrac{1}{2}(q + 3 - k)(q + 1) - q - 1 + 6\sqrt{q}.$$

Consequently, either

$$k \leq q + 2 - \{2q - 12\sqrt{q} + 3\}^{1/2}$$

or

$$k \geq q + 2 + \{2q - 12\sqrt{q} + 3\}^{1/2}.$$

This contradicts

$$q - \tfrac{1}{2}\sqrt{q} + \tfrac{9}{4} < k < q + 1.$$

So over $\bar{\gamma}$, and with $q \notin \{256, 512\}$, $\mathscr{C}$ either contains a line or consists entirely of conics.

Let l be a line of $\mathscr{C}$, and suppose that l is not defined over γ. Then $|l| \leq 1$. Let π_i be a plane of $\mathscr{K}$ not passing through a point of l over γ. The line $l' = \pi \cap \pi_i$ intersects $\pi_i \cap \Phi$, and so Φ, only in points over γ. Hence the intersection of l and l' is a point over γ, a contradiction.

Now suppose that $\mathscr{C}$ consists of $t/2$ conics, ρ of which are not defined over γ, with $\rho \geq 1$. Then

$$\tfrac{1}{2}k(q+3-k) \leq |\mathscr{C}| \leq \tfrac{1}{2}(q+3-k-2\rho)(q+1)+4\rho$$
$$\leq \tfrac{1}{2}(q+1-k)(q+1)+4.$$

Hence either

$$k \leq q+2-(2q-5)^{1/2}$$

or

$$k \geq q+2+(2q-5)^{1/2},$$

a contradiction. Consequently, for $q \notin \{256, 512\}$, $\mathscr{C}$ either contains a line over γ or consists of $t/2$ conics over γ.

Let $q = 512$. Then (27.78) together with $5 \leq m \leq q+1-k < \frac{1}{2}\sqrt{q} - \frac{5}{4}$ gives $m = 10$ and $k = 503$. So assume that $m = 10$ and $t = 12$. If $\mathscr{C}$ does not contain a line, then

$$\tfrac{1}{2}kt = \tfrac{1}{2}(503.12) \leq |\mathscr{C}| \leq (q+1)+(q+1+72\sqrt{q}) = 1026+72\sqrt{512},$$

a contradiction. Now, in the same way as for $q \notin \{256, 512\}$ we see that $\mathscr{C}$ contains a linear component over the ground field γ. If $q = 512$ and if (27.78) is not satisfied for $m > 4$, then we proceed as in the case $q \notin \{256, 512\}$.

Finally, let $q = 256$. Since $2 < t < \frac{1}{2}\sqrt{q} + \frac{3}{4}$ with t even, then, from $4 < m < t-1$, it follows that $t = 8$ and $m \in \{5, 6\}$. If $t = 8$, $m = 5$, and $\mathscr{C}$ does not contain a line, then

$$\tfrac{1}{2}kt = \tfrac{1}{2}(251.8) \leq |\mathscr{C}| \leq (q+1+12\sqrt{q})+(q+1+2\sqrt{q}) = 738,$$

a contradiction; if $t = 8$, $m = 6$, and $\mathscr{C}$ does not contain a line, then

$$\tfrac{1}{2}kt = \tfrac{1}{2}(251.8) \leq |\mathscr{C}| \leq (q+1+20\sqrt{q})+(q+1) = 834,$$

also a contradiction. In the same way as for $q \notin \{256, 512\}$ we see that $\mathscr{C}$ contains a linear component over the ground field γ. If $q = 256$ and if there is no absolutely irreducible component with $m > 4$, then we proceed as in the case $q \notin \{256, 512\}$ with $\alpha = 0$ and $\alpha + \beta = \beta \geq 2$. □

Theorem 27.7.15: *If $q - \frac{1}{2}\sqrt{q} + \frac{9}{4} < k < q+1$ and k is odd, then Φ contains a plane as component (over γ) or consists of $(q+3-k)/2$ hyperbolic quadrics (over γ).*

Proof: If Φ contains a plane ζ as component, then for each plane π_i of $\mathscr{K}$ the line $\zeta \cap \pi_i$ is an S-line; so ζ contains at least k lines over γ and consequently ζ is defined over γ.

From now on we assume that Φ does not contain a linear component. By the proof of Theorem 27.7.12 there is at least one plane which does not contain a line of Φ. Let π be a plane for which $\Phi \cap \pi = \mathscr{C}$ consists of $t/2$ conics (over γ). First we show that no two of these conics coincide. Assume that at least two of the conics do coincide. Then we have

$$\tfrac{1}{2}k(q+3-k) \leq |\mathscr{C}| \leq \tfrac{1}{2}(q+1-k)(q+1).$$

So either

$$k \leq q+2-(2q+3)^{1/2}$$

or

$$k \geq q+2+(2q+3)^{1/2}.$$

This contradicts $q - \frac{1}{2}\sqrt{q} + \frac{9}{4} < k < q+1$. Hence the $t/2$ conics are distinct. The number of points common to at least two of these conics is at most

$$4(t/2)(t/2-1)/2 = \tfrac{1}{2}t(t-2) < \tfrac{1}{2}(\tfrac{1}{2}\sqrt{q}+\tfrac{3}{4})(\tfrac{1}{2}\sqrt{q}-\tfrac{5}{4}) < q/8.$$

Let $\mathscr{C}_1$ be one of the conics and let P be a point of π, with $P \notin \mathscr{C}_1$ and P distinct from the nucleus of $\mathscr{C}_1$. Then there is at least one line l of π through P which neither contains a point over γ of $\mathscr{C}_1$ nor contains a point over $\bar{\gamma}$ common to at least two conics. For this line l each point of $l \cap \Phi$ is a simple point for Φ. Over γ we have $|l \cap \Phi| \leq t-2 < \frac{1}{2}\sqrt{q} - \frac{5}{4}$. Since Φ does not contain a linear component, at each point of $l \cap \Phi$ the tangent plane of Φ contains at most t different lines of Φ. Hence each point of $l \cap \Phi$ is contained in at most t lines of Φ. Consequently the number of planes π' through l for which $\pi' \cap \Phi$ contains a line as component over γ is at most $(t-2)t < (\frac{1}{2}\sqrt{q}-\frac{5}{4})(\frac{1}{2}\sqrt{q}+\frac{3}{4}) = \frac{1}{4}q - \frac{1}{4}\sqrt{q} - \frac{15}{16} < \frac{1}{4}q - \frac{1}{4}\sqrt{q}$. It follows that, for more than $q+1-\frac{1}{4}q+\frac{1}{4}\sqrt{q} > \frac{3}{4}q + \frac{1}{4}\sqrt{q}$ planes π' through l, the curve $\Phi \cap \pi'$ consists over γ of $t/2$ conics. Hence over $\bar{\gamma}$ the two conjugate points in the set $l \cap \mathscr{C}_1$ are contained in more than $(3q+\sqrt{q})/4$ conics of Φ, all defined over γ, lying in different planes through l. Let $\mathscr{C}_1, \mathscr{C}_2, \ldots$ be these conics, and let ζ_j be the plane of $\mathscr{C}_j$.

For any S-line l_i, let t_i be the number of conics of $\{\mathscr{C}_1, \mathscr{C}_2, \ldots\} = V$ containing at least (and so exactly one) point of l_i. The number of points of $\zeta_j \cap \Phi$ not belonging to an S-line is at most $\frac{1}{2}t(q+1) - tk/2 = t(t-2)/2 < \frac{1}{2}(\frac{1}{2}\sqrt{q}+\frac{3}{4})(\frac{1}{2}\sqrt{q}-\frac{5}{4}) < \frac{1}{8}q - \frac{1}{8}\sqrt{q}$. Hence the number of points of $\mathscr{C}_j$ belonging to an S-line is more than $q+1-\frac{1}{8}q+\frac{1}{8}\sqrt{q} > \frac{7}{8}q + \frac{1}{8}\sqrt{q}$. As each point of Φ is on zero or two S-lines it now follows that

$$\sum_i t_i > \tfrac{1}{4}(3q+\sqrt{q})\cdot\tfrac{1}{8}(7q+\sqrt{q})\cdot 2 = \tfrac{1}{16}(21q^2+10q\sqrt{q}+q). \quad (27.79)$$

The number of S-lines is equal to $(q+3-k)k$. Since the function $f(x) = (q+3-x)x$ is strictly decreasing for $x \geq (q+3)/2$, we have

$$\begin{aligned}(q+3-k)k &< (\tfrac{1}{2}\sqrt{q}+\tfrac{3}{4})(q-\tfrac{1}{2}\sqrt{q}+\tfrac{9}{4}) \\ &= \tfrac{1}{2}q\sqrt{q}+\tfrac{1}{2}q+\tfrac{3}{4}\sqrt{q}+\tfrac{27}{16}.\end{aligned} \quad (27.80)$$

From (27.79) and (27.80) it now follows that

$$\begin{aligned}\bar{t} &= (\sum t_i)/\{(q+3-k)k\} \\ &> \{21q^2 + 10q\sqrt{q} + q\}/\{8q\sqrt{q} + 8q + 12\sqrt{q} + 27\} \\ &> (21\sqrt{q})/8 - 2.\end{aligned}$$

Hence there is an S-line l' which has a point in common with more than $21\sqrt{q}/8 - 2$ conics of the set V, say with $\mathscr{C}_1, \mathscr{C}_2, \ldots, \mathscr{C}_s$.

The common points of $\mathscr{C}_1, \mathscr{C}_2, \ldots, \mathscr{C}_s$ are denoted by Q and Q'. Recall that Q, Q' are conjugate points over $\bar{\gamma}$. Let R_i be the common point of l' and $\mathscr{C}_i$, for $i = 2, 3$, and let m_2, m'_2 be the tangent lines of $\mathscr{C}_2$ at the respective points Q, Q'. The (absolutely irreducible) quadric (over γ) containing $\mathscr{C}_1, R_2, R_3$ and having m_2, m'_2 as tangent lines will be denoted by $\mathscr{Q}$. Since $R_2 \in \mathscr{Q}$ and since the tangent lines m_2, m'_2 of $\mathscr{C}_2$ are tangent lines of $\mathscr{Q}$, the conic $\mathscr{C}_2$ belongs to $\mathscr{Q}$. Since l' has at least three points in common with $\mathscr{Q}$, it also belongs to $\mathscr{Q}$. The common point of l' and $\mathscr{C}_i$ will be denoted by R_i and the tangent lines of $\mathscr{C}_i$ at the respective points Q, Q' will be denoted by m_i, m'_i, with $i = 1, 2, \ldots, s$. The tangent plane of Φ at Q is $m_1 m_2$ and the tangent plane of $\mathscr{Q}$ at Q' is $m'_1 m'_2$. Hence the tangent lines $m_i = m_1 m_2 \cap \zeta_i$ and $m'_i = m'_1 m'_2 \cap \zeta_i$ of $\mathscr{C}_i$ are also tangent lines of $\mathscr{Q}$, $i = 3, 4, \ldots, s$. Since also $R_i \in \mathscr{Q}$, the conic $\mathscr{C}_i$ belongs to $\mathscr{Q}$, $i = 3, 4, \ldots, s$. Consequently the s conics $\mathscr{C}_1, \mathscr{C}_2, \ldots, \mathscr{C}_s$ belong to $\mathscr{Q}$. As $2s > 21\sqrt{q}/4 - 4 > 2t$, we have $\mathscr{Q} \subset \Phi$ by the theorem of Bézout.

By taking for $\mathscr{C}_1$ any other conic of $\Phi \cap \pi$ we then see that Φ consists of $t/2$ absolutely irreducible quadrics over γ. For any plane $\pi_i \in \mathscr{K}$ the curve $\pi_i \cap \Phi$ consists of t different S-lines and so necessarily π_i contains exactly two different lines of any of the $t/2$ quadrics. It follows that any of the quadrics contains at least $2k > 2q - \sqrt{q} + \frac{9}{2}$ lines, and hence is hyperbolic.

We conclude that Φ either contains a plane as component (over γ) or consists of $(q + 3 - k)/2$ hyperbolic quadrics (over γ). □

Theorem 27.7.16: *Any k-arc $\mathscr{K}$ of $\Sigma = PG(3, q)$, with q even, k odd, and $q - \frac{1}{2}\sqrt{q} + \frac{9}{4} < k < q + 1$, can be extended to a $(k + 1)$-arc.*

Proof: The hypotheses imply that $q \geq 64$. By Theorem 27.7.15, Φ contains a plane as component (over γ) or consists of $(q + 3 - k)/2$ hyperbolic quadrics (over γ).

If Φ contains a plane as component (over γ), then by Theorem 27.7.5, $\mathscr{K}$ can be extended to a $(k + 1)$-arc. Now assume that Φ consists of $t/2$ hyperbolic quadrics (over γ). By Theorem 27.7.5(v) the arc $\mathscr{K}$ is complete. We work with a k-arc of planes $\mathscr{K} = \{\pi_1, \ldots, \pi_k\}$.

Let Δ_1, Δ_2 be distinct hyperbolic quadrics algebraically contained in Φ. The k planes π_i are tangent planes of Δ_1 and Δ_2. Using any correlation θ of

Σ we examine the situation that a k-arc $\mathcal{K}\theta$ of points of $PG(3, q)$ lies on the intersection $\Psi_1 \cap \Psi_2 = \mathscr{C}$ of the two quadrics Ψ_1, Ψ_2, where $\Psi_i = \Delta_i\theta$, $i = 1, 2$. The extension of the curve $\mathscr{C}$ to the algebraic closure $\overline{GF(q)}$ of $GF(q)$ is denoted by $\bar{\mathscr{C}}$. We study three possible cases.

(a) *$\bar{\mathscr{C}}$ contains as a component a line or a conic but not an irreducible cubic curve*

In this case for any k-arc in $\mathscr{C}$ we get $k \leq 8$. But since $q + 1 > k > q - \frac{1}{2}\sqrt{q} + \frac{9}{4}$ and k is odd, so $k \geq 63$ and we have a contradiction.

(b) *$\bar{\mathscr{C}}$ factors into a twisted cubic curve $\mathscr{C}'$ and a line l*

In this case $|\mathscr{C}' \cap \mathcal{K}\theta| \geq 61$; so the points of $\mathscr{C}'$ in $PG(3, q)$ form a $(q + 1)$-arc $\mathcal{K}'$. Put $\mathcal{K}'' = \mathcal{K}\theta \cap \mathscr{C}'$; then $|\mathcal{K}''| \geq k - 2 > q - \frac{1}{2}\sqrt{q} + \frac{1}{4}$. Since $q \geq 64$ we have $|\mathcal{K}''| > q - \frac{1}{2}\sqrt{q} + \frac{1}{4} > (q + 4)/2$. From the dual of Theorem 27.7.5(vi) all points of $\mathcal{K}\theta$ lie on the $(q + 1)$-arc $\mathcal{K}'$. By duality, $\mathcal{K}$ itself lies on a $(q + 1)$-arc and is not complete, a contradiction.

(c) *$\bar{\mathscr{C}}$ is an irreducible quartic*

Let π be any plane of Σ not containing P_1 where $\pi_i\theta = P_i$, $1 \leq i \leq k$, and where P_1 is non-singular for $\bar{\mathscr{C}}$ (noting that $\bar{\mathscr{C}}$ has at most one singular point). Projecting $\mathscr{C}$ from P_1 onto π gives an irreducible cubic $\mathscr{C}'$ over $GF(q)$ in π. If $P_1P_i \cap \pi = \{Q_i\}$, $i > 1$, then $\{Q_2, \ldots, Q_k\} = \mathscr{L}$ is a $(k - 1)$-arc of points of π contained in the curve $\mathscr{C}'$.

First suppose that $\mathscr{C}'$ has genus 1. Then we have from the Hasse–Weil formula, Theorem 10.2.1, that $|\mathscr{C}'| \leq q + 1 + 2\sqrt{q}$. For given $i > 1$, at least $(k - 2) - \{|\mathscr{C}'| - (k - 1)\}$ lines Q_iQ_j, $1 < j \neq i$, contain exactly two points of $\mathscr{C}'$. So at points of $\mathscr{L}$ the curve $\mathscr{C}'$ has at least $(2k - 3 - |\mathscr{C}'|)(k - 1)/2 = F(k)$ distinct tangents. From the Hasse–Weil formula we get $F(k) \geq (2k - 3 - q - 2\sqrt{q} - 1)(k - 1)/2 = G(k)$. Therefore $G(k) \leq q + 1 + 2\sqrt{q}$. Since $k > q - \frac{1}{2}\sqrt{q} + \frac{9}{4}$, we obtain $8q^2 - 28q\sqrt{q} + 10q - 64\sqrt{q} - 11 < 0$, a contradiction as $q \geq 64$.

Next suppose that $\mathscr{C}'$ has genus 0. Then, as for cubics in §11.3, we have $|\mathscr{C}'| \leq q + 2$. For given $i > 1$, at least $(k - 2) - \{|\mathscr{C}'| - (k - 1)\}$ lines Q_iQ_j, $1 < j \neq i$, contain exactly two points of $\mathscr{C}'$. Since at most one point of $\mathscr{L}$ is singular for $\mathscr{C}'$, the curve $\mathscr{C}'$ has at least $(k - 2)(2k - 4 - |\mathscr{C}'|)/2 = F(k)$ distinct tangents at points of $\mathscr{L}$ which are simple for $\mathscr{C}'$. As $|\mathscr{C}'| \leq q + 2$ we get $F(k) \geq (k - 2)(2k - 4 - q - 2)/2 = G(k)$. Since $\mathscr{C}'$ has at most $q + 1$ simple points, we have $G(k) \leq q + 1$; hence $(k - 2)(2k - q - 6) \leq 2q + 2$.

Since $k > q - \frac{1}{2}\sqrt{q} + \frac{9}{4}$, it follows that $8q^2 - 12q\sqrt{q} - 22q + 4\sqrt{q} - 19 < 0$, a contradiction.

By (a), (b), (c), Φ does not consist of $t/2$ hyperbolic quadrics; so $\mathscr{K}$ extends to a $(k + 1)$-arc. □

Theorem 27.7.17: *Let $\mathscr{K}$ be any k-arc (of points or planes) in $PG(3, q)$, q even and $q \neq 2$. If $k > q - \frac{1}{2}\sqrt{q} + \frac{9}{4}$, then $\mathscr{K}$ can be completed to a $(q + 1)$-arc $\mathscr{L}$, and $\mathscr{L}$ is uniquely determined by $\mathscr{K}$.*

Proof: Assume that $q - \frac{1}{2}\sqrt{q} + \frac{9}{4} < k < q + 1$. By Theorems 27.7.13 and 27.7.16, the k-arc $\mathscr{K}$ is not complete and so it extends to a $(k + 1)$-arc $\mathscr{K}'$. If $k + 1 = q + 1$ we are done. If $k + 1 < q + 1$ then, since $k + 1 > q - \frac{1}{2}\sqrt{q} + \frac{9}{4}$, the arc $\mathscr{K}'$ extends to a $(k + 2)$-arc $\mathscr{K}''$. Proceeding in this way we get that $\mathscr{K}$ can be extended to a $(q + 1)$-arc $\mathscr{L}$. By Theorem 27.7.5(vi), $\mathscr{L}$ is uniquely determined by $\mathscr{K}$ since $q - \frac{1}{2}\sqrt{q} + \frac{9}{4} > (q + 4)/2$. □

Before proceeding to n dimensions, it is necessary to consider the analogue of Theorem 27.7.5 in $PG(4, q)$. Then using the result of Theorem 27.6.10 that a $(q + 1)$-arc in $PG(4, q)$ is a normal rational curve, we will be able to apply Theorem 27.6.1.

First, we restate Theorems 27.7.3 and 27.7.4 for $n = 4$.

Theorem 27.7.18: *Let $\mathscr{K} = \{\pi_1, \ldots, \pi_k\}$ be a k-arc of solids in $\Sigma = PG(4, q)$, $q = 2^s$. For i, j, m distinct, let Z_{ijm} denote the set of $t = q + 4 - k$ points on the line $\pi_i \cap \pi_j \cap \pi_m$ that lie on no other solid of $\mathscr{K}$. Then*

(i) *there exists a curve $\mathscr{C}_{ij} = \mathscr{C}_{ji}$ of degree t in the plane $\pi_i \cap \pi_j$ such that $\mathscr{C}_{ij} \cap \pi_m = Z_{ijm}$;*

(ii) *for fixed i, the algebraic curves $\mathscr{C}_{ij}$ are algebraically contained in an algebraic surface Φ_i of degree t in π_i with $\Phi_i \cap \pi_j = \mathscr{C}_{ij}$;*

(iii) *the algebraic surfaces Φ_i are algebraically contained in a primal $\Phi = \Phi(\mathscr{K})$ for which $\Phi(\mathscr{K}) \cap \pi_i = \Phi_i$;*

(iv) *if $k > (q + 6)/2$, the primal $\Phi(\mathscr{K})$ is unique;*

(v) *if $\mathscr{L} = \mathscr{K} \cup \{\pi_{k+1}, \ldots, \pi_u\}$ is an arc of solids with $u > k$ and if $k > (q + 6)/2$, the primal $\Phi(\mathscr{K})$ factors into $\Phi(\mathscr{L}), \pi_{k+1}, \ldots, \pi_u$;*

(vi) *if $k > (q + 6)/2$, there is a bijection between solids of Σ extending $\mathscr{K}$ to a $(k + 1)$-arc and linear components over $GF(q)$ of Φ;*

(vii) *if $k > (q + 6)/2$, the arc $\mathscr{K}$ is contained in a unique complete arc of $PG(4, q)$.* □

Theorem 27.7.19: *Let $\mathscr{K}$ be a k-arc of solids in $\Sigma = PG(4, q)$, $q = 2^h$, with $k > q - \frac{1}{2}\sqrt{q} + \frac{13}{4}$ and $t = q + 4 - k$. Then*

(i) Φ_i *factors over* $GF(q)$ *into* $t-2$ *planes* $\alpha_{i1}, \alpha_{i2}, \ldots, \alpha_{i,t-2}$, *called S-planes, and a hyperbolic quadric* Ψ_i, *called an S-quadric*;

(ii) *the* $t-2$ *S-planes in* π_i *form an arc of planes*;

(iii) *in* $\pi_i \cap \pi_j$, $i \neq j$, *there are exactly two lines* l_{ij} *and* m_{ij} *which are lines of* Ψ_i;

(iv) *in* $\pi_i \cap \pi_j$, $i \neq j$, *the lines* l_{ij} *and* m_{ij} *together with the* $t-2$ *lines* $\pi_j \cap \alpha_{i1}, \ldots, \pi_j \cap \alpha_{i,t-2}$ *and the* $k-2$ *lines* $\pi_i \cap \pi_j \cap \pi_u$, $u \in \mathbf{N}_k \backslash \{i, j\}$, *form a* $(q+2)$*-arc of lines*;

(v) *each plane* α_{ij} *contains two lines of* Ψ_i;

(vi) *in an S-plane* α_{is} *of* π_i, *the lines* $\alpha_{is} \cap \alpha_{iu}$, $s \neq u$, *the lines* $\alpha_{is} \cap \pi_j$, $i \neq j$, *and the two lines of* Ψ_i *in* α_{is} *form a* $(q+2)$*-arc of lines*;

(vii) *the* $2(q+1)$ *generators of* Ψ_i *are the* $2(k-1)$ *lines of* $\Psi_i \cap \pi_j$, $j \neq i$, *and the* $2(t-2)$ *lines of* $\Psi_i \cap \alpha_{is}$;

(viii) *at most two members of the set* $\mathscr{V}$ *of the* $(t-2)k + k$ *planes* α_{ij} *and surfaces* Ψ_i *contain any line of* Σ;

(ix) *for any S-plane* α_{is} *of* π_i *and any point P of* α_{is}, *there are at most two S-planes containing P and meeting* α_{is} *in a line.*

Proof: By Theorem 27.6.3 we have $k \leq q+1$ for $q > 4$; for $q = 4$, we have $k \leq 6$ from §27.1. Since $k > q - \frac{1}{2}\sqrt{q} + \frac{13}{4}$, it follows that $q \geq 32$; also $k > (q+6)/2$.

(i), (ii) Since $\mathscr{K}$ is an arc, for a fixed i the $k-1$ planes $\pi_i \cap \pi_j$, $j \neq i$, form an arc $\mathscr{M}$ of planes in π_i. Since $k > q - \frac{1}{2}\sqrt{q} + \frac{13}{4}$, we have $k - 1 = |\mathscr{M}| > q - \frac{1}{2}\sqrt{q} + \frac{9}{4}$. By Theorem 27.7.17, $\mathscr{M}$ is embedded in a $(q+1)$-arc $\mathscr{L}$ of planes in π_i. The planes of $\mathscr{L} \backslash \mathscr{M}$ are the S-planes in π_i. Since $\mathscr{L}$ is a $(q+1)$-arc, then from the structure of $\Phi(\mathscr{L})$ we obtain the S-quadric in π_i as in Theorem 27.7.7(ii).

(iii) This follows from Theorem 27.7.7(iii).

(iv)–(vii) These all follow from (iii) and (iv) of Theorem 27.7.7.

(viii) Since the S-planes in a given solid π_i form an arc of planes in π_i, no three S-planes of a given solid have a common line. From (vi) we cannot have two S-planes in π_i and the S-quadric Ψ_i of π_i all containing a common line.

Let m be any line of Σ lying in an S-plane σ_1. Then σ_1 lies in a unique solid of $\mathscr{K}$, say π_i; so σ_1 is one of the S-planes α_{is} in π_i. Assume m lies in some other S-plane σ_2 not in π_i. Then σ_2 is in π_j, say, with $j \neq i$. Now, $\pi_i \cap \pi_j$ is a plane π_{ij} containing m. In π_i the plane α_{is} meets π_{ij} in exactly one line: this must be m. From (iv) there are no other S-planes or S-quadrics containing m. A similar argument handles the case that m lies in α_{is} and in Ψ_j, $j \neq i$, or lies in Ψ_i and in Ψ_j, $j \neq i$.

To summarize: if m lies in an object of $\mathscr{V}$ from π_i and in an object of $\mathscr{V}$ from π_j, $j \neq i$, then m lies in exactly two objects of $\mathscr{V}$. If the only objects of $\mathscr{V}$ containing m lie in a given arc solid, say π_i, then again m lies in at most two objects from $\mathscr{V}$.

(ix) An argument similar to that proving (viii) applies. □

Theorem 27.7.20: *Let $\mathscr{K} = \{\pi_1, \pi_2, \ldots, \pi_k\}$ be a k-arc of solids in $\Sigma = PG(4, q)$, $q = 2^h$, with $q > k > q - \frac{1}{2}\sqrt{q} + \frac{13}{4}$. Then $\mathscr{K}$ can be extended to a q-arc.*

Proof: Since $q > k > q - \frac{1}{2}\sqrt{q} + \frac{13}{4}$ we have $q \geq 128$ and $k > (q + 6)/2$. By way of contradiction assume that $\mathscr{K}$ is complete. Since $k > q - \frac{1}{2}\sqrt{q} + \frac{13}{4}$, we can use the results of the previous theorem. Any S-plane of π_j, $j > 1$, meets π_1 in a line of $\pi_1 \cap \pi_j$. This line lies either in an S-plane of π_1 or in the quadric Ψ_1. Now, there are exactly two lines of Ψ_1 in $\pi_1 \cap \pi_j$. Therefore, putting $t = q + 4 - k$, the number of S-planes not in π_1 and having a line in common with some S-plane in π_1 is at least $(k - 1)(t - 2) - (k - 1)2 = (k - 1)(t - 4)$; recall that each solid contains exactly $t - 2$ S-planes. Hence there exists an S-plane $\alpha = \alpha_{1s}$ in π_1 having a line in common with at least $(k - 1)(t - 4)/(t - 2)$ S-planes not in π_1. Thus the total number of S-planes having at least one line in common with α is at least $(k - 1)(t - 4)/(t - 2) + t - 2$. Denote the set of such planes by $V = \{\beta_1, \beta_2, \ldots\}$ with $\{\beta_1, \ldots, \beta_{t-2}\} = \{\alpha_{11}, \ldots, \alpha_{1,t-2}\}$, $\beta_1 = \alpha_{11} = \alpha$, and $\beta_i \cap \alpha = l_i$ for $\beta_i \neq \alpha$. Since $k < q$, so $t > 4$; hence $(k - 1)(t - 4)/(t - 2) \geq 1$. Let the line l_{t-1} of α lie in π_j, say, $j \neq 1$. Put $V_j = \{\alpha_{j1}, \ldots, \alpha_{j,t-2}, \Psi_j\} = \{\delta_{j1}, \ldots, \delta_{j,t-1}\}$. Let ρ_n be the number of planes of V containing a line of δ_{jn}. Since any two S-planes of V_j meet in a line and any S-plane of V_j contains a line (actually two lines) of Ψ_j by Theorem 27.7.19(v),

$$\sum \rho_n \geq (k - 1)(t - 4)/(t - 2) + (t - 2) - 1 + (t - 1).$$

Averaging gives $\bar{\rho} \geq [(k - 1)(t - 4)/(t - 2) + 2t - 4]/(t - 1)$. So there exists an element δ_{jn} for which

$$\rho_n \geq [(k - 1)(t - 4)/(t - 2) + 2t - 4]/(t - 1).$$

Note that in obtaining $\sum \rho_n$ we use the fact that each plane of V not in π_j meets exactly one element of V_j in a line.

We now consider two cases.

Case 1: $t < [(k - 1)(t - 4)/(t - 2) + 2t - 4]/(t - 1)$

Then $\delta_{jn} = \Psi_j$ or an S-plane of π_j. Assume $\delta_{jn} = \Psi_j$. The S-plane α in π_1 meets $\pi_1 \cap \pi_j$ in a unique line l_{t-1} and l_{t-1} lies on a unique S-plane η_j of π_j. Now Ψ_j meets $\pi_1 \cap \pi_j$ in exactly two lines l, m both different from l_{t-1}. The lines l, m, l_{t-1} are part of an arc of lines in $\pi_1 \cap \pi_j$ by Theorem 27.7.19(iv). So $\Psi_j \cap \alpha = \{l_{t-1} \cap l, l_{t-1} \cap m\}$. Any S-plane containing a line of Ψ_j and a line of α must pass through $l_{t-1} \cap l$ or $l_{t-1} \cap m$. One such S-plane is η_j. So by Theorem 27.7.19(ix) there are at most three S-planes altogether meeting Ψ_j and α each in lines. Therefore $3 \geq (k-1)(t-4)/(t-2) + (2t-4) > t$; so $3 > t$ contradicting the fact that $t \geq 5$ since $t = q + 4 - k$ and $k < q$ by hypothesis.

Next assume that δ_{jn} is an S-plane, say α_{j1}. If α_{j1} is the unique S-plane η_j of π_j containing l_{t-1}, then the planes α and α_{j1} lie in a Π_3. From Theorem 27.7.19(viii), l_{t-1} lies in no other S-plane. Then any other S-plane containing a line of α and α_{j1} lies in this Π_3. Therefore this Π_3 contains more than t S-planes and so is a linear component of Φ. From Theorem 27.7.18(vi), this Π_3 is a solid of Σ extending $\mathscr{K}$. This contradicts our initial assumption that $\mathscr{K}$ is complete. Therefore α_{j1} is distinct from η_j; so $\alpha_{j1} \cap \alpha$ is a point P. Any S-plane containing a line of α_{j1} and α contains P. It follows that there are more than t S-planes containing P and intersecting α in a line, contradicting Theorem 27.7.19(ix) as $t > 2$.

Case 2: $t \geq [(k-1)(t-4)/(t-2) + 2t - 4]/(t-1)$

This means that

$$t^3 - 4t^2 - (q-3)t + 4q + 4 \geq 0. \tag{27.81}$$

From the hypotheses, since $t = q + 4 - k$, we have

$$4 < t < \tfrac{1}{2}\sqrt{q} + \tfrac{3}{4}. \tag{27.82}$$

For $t = 5, 6, 7$, the inequality (27.81) implies that $q \leq 32$, a contradiction in each case. So

$$8 \leq t < \tfrac{1}{2}\sqrt{q} + \tfrac{3}{4}. \tag{27.83}$$

Rewriting (27.81), we obtain

$$(t^2 - q)t + 4q \geq 4t^2 - 3t - 4. \tag{27.84}$$

For $t \geq 2$, the right-hand side of (27.84) is positive. However, from (27.83),

$$\begin{aligned}(t^2 - q)t + 4q &< ((\tfrac{1}{2}\sqrt{q} + \tfrac{3}{4})^2 - q)t + 4q \\ &= \tfrac{3}{4}(\sqrt{q} + \tfrac{3}{4} - q)t + 4q \\ &\leq 6(\sqrt{q} + \tfrac{3}{4} - q) + 4q \\ &= -2q + 6\sqrt{q} + \tfrac{9}{2} \\ &< 0.\end{aligned}$$

This gives the desired contradiction.

It has thus been shown that $\mathscr{K}$ extends to a $(k+1)$-arc $\mathscr{K}'$. If $k + 1 < q$, then since $k + 1 > q - \frac{1}{2}\sqrt{q} + \frac{13}{4}$, the arc $\mathscr{K}'$ extends to a $(k+2)$-arc $\mathscr{K}''$. This process shows that $\mathscr{K}$ can be extended to a q-arc. □

Using the notation above we show the following result.

Theorem 27.7.21: *Any q-arc of solids in* $\Sigma = PG(4, q)$, $q = 2^h$, $q \geq 64$, *can be extended to a* $(q + 1)$*-arc.*

Proof: Let $\mathcal{K} = \{\pi_1, \ldots, \pi_q\}$ be a q-arc of solids in Σ. By way of contradiction assume that $\mathcal{K}$ is complete. The number of S-planes not in π_1 and having a line in common with some S-plane in π_1, or with Ψ_1, is exactly $(q-1)(t-2) = 2(q-1)$. So one of α_{11}, α_{12}, or Ψ_1 has a line in common with at least $2(q-1)/3$ of these S-planes. There are two possibilities:

(a) either α_{11} or α_{12} has a line in common with at least $2(q-1)/3$ of these S-planes;

(b) Ψ_1 has a line in common with at least $2(q-1)/3$ of these S-planes.

Assume (b). There are two cases.

1. *For each solid* π_i, $i > 1$, *at most one of the planes* α_{i1}, α_{i2} *contains a line of* Ψ_1

Then at least $q - 1$ of the planes α_{i1}, α_{i2}, $i > 1$, contain a line of one of α_{i1}, α_{i2}. So at least one of α_{i1}, α_{i2} contains a line of at least $(q-1)/2$ S-planes not contained in π_1.

2. *There is a solid* π_i, $i > 1$, *for which the two S-planes* α_{i1} *and* α_{i2} *contain a line of* Ψ_1

Then at least $(2(q-1)/3 - 2)/3$ S-planes not in π_1 and π_i contain a line of Ψ_1 and of one of $\alpha_{i1}, \alpha_{i2}, \Psi_i$. From Theorem 27.7.19, in $\pi_1 \cap \pi_i$ the intersection $\Phi(\mathcal{K}) \cap \pi_1 \cap \pi_i$ contains exactly four lines l_1, l_2, l_3, l_4 with no three concurrent. Also, suppose that l_1 and l_2 lie in Ψ_1 and that l_3 and l_4 lie in Ψ_2. So Ψ_1 meets Ψ_2 in four points no three of which are collinear. An analysis shows that on each of these four points there is at most one S-plane not in π_1 or π_i and having a line in common with Ψ_1 and Ψ_i. Since $(2(q-1)/3 - 2)/3 = (2q-8)/9$ is larger than four, at least $(2q-8)/9$ S-planes not in π_1 and π_i contain a line of Ψ_1 and of one of α_{i1}, α_{i2}.

Each solid of $\mathcal{K}$ contains two S-planes. Since

$$\min\{2(q-1)/3, (q-1)/2, (2q-8)/9\} + 2 = (2q+10)/9,$$

it follows from (a) and (b) that there exists an S-plane $\alpha = \alpha_{j1}$ having a line in common with at least $(2q+10)/9$ S-planes. Denote the set of these S-planes by $V = \{\beta_1, \beta_2, \ldots\}$ with $\{\beta_1, \beta_2\} = \{\alpha_{j1}, \alpha_{j2}\}$, $\beta_1 = \alpha_{j1} = \alpha$; denote $\beta_i \cap \alpha$, $\beta_i \neq \alpha$, by l_i. Note that $(2q-8)/9 > 0$. Then let l_3 lie in π_g, $g \neq j$. Put $V_g = \{\alpha_{g1}, \alpha_{g2}, \Psi_g\} = \{\delta_{g1}, \delta_{g2}, \delta_{g3}\}$. Let ρ_n be the number of planes of V

containing a line of δ_{gn}. Then, as in the previous theorem,

$$\sum \rho_n \geq (2q + 10)/9 - 1 + 3 = (2q + 28)/9.$$

Averaging gives $\bar{\rho} \geq (2q + 28)/27$. So there is an element δ_{jn} for which $\rho_n \geq (2q + 28)/27$. Note that $t = 4 < (2q + 28)/27$. Assume for example that $\delta_{gn} = \Psi_g$. We recall that α meets $\pi_j \cap \pi_g$ in just one line l_3 which, by assumption, lies in exactly one of the planes δ_{g1} or δ_{g2}. Therefore Ψ_g has no line in common with α but meets α in exactly two points lying on l_3. Then, from Theorem 27.7.19(ix), there are at most three S-planes intersecting Ψ_g and α in a line one of which is the S-plane of π_g through l_3. So $(2q + 28)/27 \leq 3$ and $q \leq 16$, contradicting $q \geq 64$. Therefore δ_{gn} is an S-plane, say α_{g1}. If α_{g1} is the (unique) S-plane of π_g containing l_3 then the planes α_{g1} and α are contained in a solid Π_3. Since, from Theorem 27.7.19(viii), l_3 lies in just two S-planes, this solid contains at least $(2q + 28)/27$ S-planes. Since $(2q + 28)/27 > 4 = t$ and $\Phi(\mathscr{K})$ has degree $t = 4$, the solid Π_3 is a linear component of $\Phi(\mathscr{K})$. From Theorem 27.7.18(vi) this solid extends $\mathscr{K}$, contradicting that $\mathscr{K}$ is complete. Therefore α_{g1} does not contain the line l_3; so $\alpha_{j1} \cap \alpha$ is a point P. Any S-plane containing a line of α_{j1} and α contains P. It follows that there are at least $(2q + 28)/27$ S-planes containing P and intersecting α in a line, contradicting Theorem 27.7.19(ix). □

Theorem 27.7.22: *Let $\mathscr{K}$ be a k-arc of points in $PG(4, q)$, $q = 2^h$, $q \neq 2$. If $k > q - \frac{1}{2}\sqrt{q} + \frac{13}{4}$, then $\mathscr{K}$ can be completed uniquely to a $(q + 1)$-arc, that is a normal rational curve.*

Proof: Since $k > q - \frac{1}{2}\sqrt{q} + \frac{13}{4}$, clearly $q \neq 4$. By Theorem 27.6.3, $k \leq q + 1$ for $q > 4$; so $k > q - \frac{1}{2}\sqrt{q} + \frac{13}{4}$ implies that $q \geq 32$ and $k \geq 33$. If $q > q - \frac{1}{2}\sqrt{q} + \frac{13}{4}$, then $q \geq 64$. From the previous two theorems, $\mathscr{K}$ lies in a $(q + 1)$-arc, which is complete by Theorem 27.6.3 and which is a normal rational curve by Theorem 26.6.10. □

This gives the climactic result of this section.

Theorem 27.7.23: *In $PG(n, q)$, $q = 2^h$, $q \neq 2$, $n \geq 4$,*

(i) *if $\mathscr{K}$ is a k-arc with $k > q - \frac{1}{2}\sqrt{q} + n - \frac{3}{4}$, then $\mathscr{K}$ lies on a unique normal rational curve;*

(ii) *if $q > (2n - \frac{7}{2})^2$, every $(q + 1)$-arc is a normal rational curve;*

(iii) *if $q > (2n - \frac{11}{2})^2$,*

$$m(n, q) = q + 1.$$

Proof: (i) This follows by induction from Theorems 27.6.1 and 27.7.22.

(ii) $q + 1 > q - \frac{1}{2}\sqrt{q} + n - \frac{3}{4} \Leftrightarrow q > (2n - \frac{7}{2})^2$.

(iii) This follows from Theorem 27.6.2(ii) and part (ii). □

Corollary: *In* $PG(n, q)$, $q = 2^h$, $q \neq 2$, $n > q - \frac{1}{2}\sqrt{q} - \frac{11}{4}$,

(i) *if* $\mathcal{K}$ *is a k-arc with* $k \geq n + 6$, *then* $\mathcal{K}$ *lies on a unique normal rational curve*;

(ii) *if* $n \leq q - 5$, *then every* $(q + 1)$*-arc is a normal rational curve*;

(iii) *if* $n \leq q - 4$, *then*

$$m(n, q) = q + 1.$$

Proof: These follow from the theorem and Theorems 27.5.4, 27.5.6, and Corollary 2 of Theorem 27.5.6. □

27.8 Notes and references

§27.1. The bound (27.1) is due to Bose (1947); it follows from Lemma 27.4.1. It is also immediate that a 2^n-cap in $PG(n, 2)$ is the complement of a prime. The bound (27.2) is also due to Bose (1947) and is discussed in §§8.1–8.2. The bound (27.3) is due to Bose (1947) for q odd and to Qvist (1952) for q even; see §16.1 for the bound, and for the characterization when q is odd or $q = 4$, and §16.4 for another example of a $(q^2 + 1)$-cap when $q = 2^h$, h odd and $h \geq 3$. For $q = 8$, every $(q^2 + 1)$-cap is one of these two types (Fellaġara 1962). For $q = 16$, every $(q^2 + 1)$-cap is an elliptic quadric (O'Keefe and Penttila 1990). The bound (27.4) is due to Pellegrino (1970) and the bound (27.5) is due to Hill (1973); the classification in $PG(4, 3)$ is due to Hill (1983) and in $PG(5, 3)$ is due to Hill (1978a).

The bounds (27.11) and (27.12) for $m'(2, q)$ are due to Segre (1967) and proved in Chapter 10. The improvement from (27.11) to (27.10) is due to Thas (1987a). The exact value for $m'(2, q)$, q an even square, is due to Fisher, Hirschfeld, and Thas (1986) and to Boros and Szönyi (1986) independently. The bound (27.14) is due to Voloch (1990b), and by similar methods (27.15) and (27.16) are due to Voloch (1991). The results (27.14)–(27.16) depend on an improvement to the Hasse–Weil theorem (see §10.2), which gives an upper bound on the number of points on a non-singular, irreducible, projective, algebraic curve with a fixed-point-free linear series. This result, due to Stöhr and Voloch (1986), depends on q, on the genus g, on the order and dimension of the linear series, and on the Frobenius order sequence.

§27.2. This is entirely based on Hill (1978a).

§27.3. The proof of Theorem 27.3.1 is taken from Tallini (1956a). The remainder of §27.3 is taken from Hirschfeld (1983a). The essence of the argument is found in Segre (1967).

The argument used to obtain the final result, Theorem 27.3.4, depends intricately on the upper bound used for $m'(2, q)$. This is both explicit in the proof of Theorem 27.3.4 and implicit in Theorem 27.3.2, on which the result heavily depends. Throughout, the bound (27.11) is used. The nature of the argument given precludes a formula or a bound for $m_2(n, q)$ in terms of $m'(2, q)$; a change in the bound for $m'(2, q)$ means a complete reworking of the argument. This can be done separately for the bounds (27.10), (27.14), or (27.15). For example, if q is a sufficiently large prime and $n \geq 4$, then J. F. Voloch (personal communication) has shown that

$$m_2(n, q) \leq \tfrac{1983}{2025}q^{n-1} + O(q^{n-2}).$$

§27.4. This is taken from Hirschfeld and Thas (1987), which also contains the details of Theorem 27.4.4. For q not a square, (27.16) could be used to improve Theorems 27.4.5 and 27.4.6.

§27.5. The first part of the section is an amalgam of §21.2 and Segre (1955b). The proof of Theorem 27.5.3(i) is based on Kaneta and Maruta (1989a). The remainder of the section is taken from Thas (1969a), although this proof of Theorem 27.5.4 is taken from Halder and Heise (1974).

§27.6. Theorem 27.6.3 is due to Segre (1955b) for q odd and to Casse (1969) for q even. Theorem 27.6.4 is due to Thas (1968a), although the treatment here and hence the necessary Theorems 27.6.1 and 27.6.2 follow Kaneta and Maruta (1989a); part (iii) is an improvement of Thas's result from $q > (4n - 5)^2$ to $(4n - 9)^2$. Theorem 27.6.5(ii) is due to Glynn (1986). Lemmas 27.6.6 to 27.6.9 and Theorem 27.6.10 follow Casse and Glynn (1984). Included in this paper is an elementary proof of the result that a q-arc in $PG(3, q)$, q even, is contained in a $(q + 1)$-arc. See also Kaneta and Maruta (1989b). Theorem 27.6.11 implies that $m(5, q) = q + 1$ for q even and $q \geq 8$. Maruta and Kaneta (1991) have shown, for q even and $q \geq 16$, that (i) in $PG(3, q)$, a $(q - 1)$-arc is contained in a unique $(q + 1)$-arc; (ii) in $PG(4, q)$, a q-arc is contained in a unique $(q + 1)$-arc; (iii) in $PG(5, q)$, a $(q + 1)$-arc is a normal rational curve; (iv) $m(6, q) = q + 1$. These results are dependent on those in §27.7.

§27.7. This is based on three papers: Bruen, Thas, and Blokhuis (1988a), Blokhuis, Bruen, and Thas (1990), and Storme and Thas (1991b).

In Blokhuis, Bruen, and Thas (1990), Theorems 27.7.1 and 27.7.2 are also applied to the case of q odd. This gives the following result analogous to Theorem 27.7.4.

Theorem 27.8.1: *Let $\mathcal{K}$ be a k-arc in $PG(n, q)$, $n \geq 3$ and q odd. If $k > \tfrac{2}{3}(q - 1) + n$, then $\mathcal{K}$ is contained in a unique complete arc.* □

APPENDIX VI

Ovoids and spreads of finite classical polar spaces

AVI.1 Finite classical polar spaces

The finite *classical polar spaces* are:

$\mathscr{W}_n(q)$: the polar space formed by the points of $PG(n, q)$, n odd and $n \geq 3$, together with all subspaces of the self-polar $(n-1)/2$-dimensional spaces of a null polarity $\mathfrak{T}$ of $PG(n, q)$; this polar space has rank $r = (n+1)/2$;
$\mathscr{P}(2n, q)$: the polar space formed by the points of a non-singular quadric $\mathscr{P}_{2n}$ of $PG(2n, q)$, $n \geq 2$, together with the projective subspaces on $\mathscr{P}_{2n}$; here the rank is $r = n$;
$\mathscr{H}(2n+1, q)$: the polar space formed by the points of a non-singular quadric $\mathscr{H}_{2n+1}$ of $PG(2n+1, q)$, $n \geq 1$, together with the projective subspaces on $\mathscr{H}_{2n+1}$; here the rank is $r = n+1$;
$\mathscr{E}(2n+1, q)$: the polar space formed by the points of a non-singular quadric $\mathscr{E}_{2n+1}$ of $PG(2n+1, q)$, $n \geq 2$, together with the projective subspaces on $\mathscr{E}_{2n+1}$; here the rank is $r = n$;
$\mathscr{U}(n, q^2)$: the polar space formed by the points of a non-singular Hermitian variety $\mathscr{U}_n$ of $PG(n, q^2)$, $n \geq 3$, together with the projective subspaces on $\mathscr{U}_n$; for n odd the rank is $r = (n+1)/2$, for n even $r = n/2$.

If the polar space $\mathscr{S}$ has rank r, then the subspaces of dimension $r-1$ of $\mathscr{S}$ are the *generators* of $\mathscr{S}$. Let $|\mathscr{S}|$ be the number of points of $\mathscr{S}$, and let $\mathscr{G}(\mathscr{S})$ be the set of all generators of $\mathscr{S}$.

Theorem AVI.1.1: *The numbers of points of the finite classical polar spaces are as follows:*

(i) $|\mathscr{W}_n(q)| = (q^{n+1} - 1)/(q - 1)$;

(ii) $|\mathscr{P}(2n, q)| = (q^{2n} - 1)/(q - 1)$;

(iii) $|\mathscr{H}(2n+1, q)| = (q^n + 1)(q^{n+1} - 1)/(q - 1)$;

(iv) $|\mathscr{E}(2n+1, q)| = (q^n - 1)(q^{n+1} + 1)/(q - 1)$;

(v) $|\mathscr{U}(n, q^2)| = (q^{n+1} + (-1)^n)(q^n - (-1)^n)/(q^2 - 1)$.

Theorem AVI.1.2: *The numbers of generators of the finite classical polar spaces are as follows:*

(i) $|\mathcal{G}(\mathcal{W}_n(q))| = (q+1)(q^2+1)\cdots(q^{(n+1)/2}+1)$;

(ii) $|\mathcal{G}(\mathcal{P}(2n,q))| = (q+1)(q^2+1)\cdots(q^n+1)$;

(iii) $|\mathcal{G}(\mathcal{H}(2n+1,q))| = 2(q+1)(q^2+1)\cdots(q^n+1)$;

(iv) $|\mathcal{G}(\mathcal{E}(2n+1,q))| = (q^2+1)(q^3+1)\cdots(q^{n+1}+1)$;

(v) $|\mathcal{G}(\mathcal{U}(2n,q^2))| = (q^3+1)(q^5+1)\cdots(q^{2n+1}+1)$;

(vi) $|\mathcal{G}(\mathcal{U}(2n+1,q^2))| = (q+1)(q^3+1)\cdots(q^{2n+1}+1)$.

Proofs: For the results on quadrics, see Theorem 22.4.6; for Hermitian varieties, see Theorem 23.3.2; for null polarities, see §AVI.5. □

AVI.2 Ovoids and spreads of polar spaces

Let $\mathcal{S}$ be a finite classical polar space of rank $r \geq 2$. An *ovoid* O of $\mathcal{S}$ is a point set of $\mathcal{S}$, which has exactly one point in common with every generator of $\mathcal{S}$. A *spread* S of $\mathcal{S}$ is a set of generators, which constitutes a partition of the point set.

The next theorem easily follows from Theorems AVI.1.1 and AVI.1.2.

Theorem AVI.2.1: *Let O be an ovoid and let S be a spread of the finite classical polar space $\mathcal{S}$. Then*

(i) *for* $\mathcal{S} = \mathcal{W}_n(q)$, $|O| = |S| = q^{(n+1)/2}+1$;

(ii) *for* $\mathcal{S} = \mathcal{P}(2n,q)$, $|O| = |S| = q^n+1$;

(iii) *for* $\mathcal{S} = \mathcal{H}(2n+1,q)$, $|O| = |S| = q^n+1$;

(iv) *for* $\mathcal{S} = \mathcal{E}(2n+1,q)$, $|O| = |S| = q^{n+1}+1$;

(v) *for* $\mathcal{S} = \mathcal{U}(2n,q^2)$, $|O| = |S| = q^{2n+1}+1$;

(vi) *for* $\mathcal{S} = \mathcal{U}(2n+1,q^2)$, $|O| = |S| = q^{2n+1}+1$. □

AVI.3 Existence of ovoids

This is described in Table AVI.1.

Any ovaloid of $PG(3,q)$, q even, is an ovoid of some $\mathcal{W}_3(q)$, and, conversely, any ovoid of $\mathcal{W}_3(q)$, q even, is an ovaloid of $PG(3,q)$; see Hirschfeld (1985) and Thas (1972).

Open problems: Existence or non-existence of ovoids in the following cases:

(a) $\mathcal{P}(6,q)$, for q odd with $q \neq 3^h$;
$\mathcal{P}(2n,q)$, for $n > 3$ and q odd;

Table AVI.1

Polar space	Existence of ovoids	References
$\mathscr{W}_3(q)$, q even	Yes	Thas (1972)
$\mathscr{W}_3(q)$, q odd	No	Thas (1972)
$\mathscr{W}_n(q)$, $n = 2t + 1$ and $t > 1$	No	Thas (1981a)
$\mathscr{P}(4, q)$	Yes	Payne and Thas (1984)
$\mathscr{P}(6, q)$, $q = 3^h$	Yes	Thas (1980c, 1981a), Kantor (1982e)
$\mathscr{P}(2n, q)$, q even and $n > 2$	No	Thas (1981a)
$\mathscr{H}(3, q)$	Yes	
$\mathscr{H}(5, q)$	Yes	Table 15.10
$\mathscr{H}(7, q)$, q even	Yes	Thas (1980c, 1981a)
$\mathscr{H}(7, q)$, q odd with q prime or $q \equiv 0$ or 2 (mod 3)	Yes	Conway, Kleidman, and Wilson (1988), Dye (1977b), Kantor (1982a, b, e), Shult (1985), Thas (1980c, 1981a) (*)
$\mathscr{H}(2n + 1, 2)$, $n \geq 4$	No	Kantor (1982e)
$\mathscr{H}(2n + 1, 3)$, $n \geq 4$	No	Shult (1989)
$\mathscr{E}(2n + 1, q)$, $n \geq 2$	No	Thas (1981a)
$\mathscr{U}(3, q^2)$	Yes	Payne and Thas (1984), Thas (1983b)
$\mathscr{U}(2n, q^2)$, $n \geq 2$	No	Thas (1981a)

(b) $\mathscr{H}(7, q)$, for q odd with $q \equiv 1 \pmod 3$ and not a prime;
$\mathscr{H}(2n + 1, q)$, for $n > 3$ and $q > 3$;
(c) $\mathscr{U}(2n + 1, q^2)$, for $n > 1$.

AVI.4 Existence of spreads

This is described in Table AVI.2.

A spread of $\mathscr{W}_n(q)$, $n = 2t + 1$, is also a t-spread of $PG(n, q)$. For every $n = 2t + 1$, the polar space $\mathscr{W}_n(q)$ has a spread which is also a regular t-spread of $PG(n, q)$; see Thas (1977a).

Any non-singular hyperbolic quadric of $PG(2n + 1, q)$, $n \geq 1$, has two families of generators, §22.4. If π, π' are generators, then they belong to the same family if and only if the dimension of $\pi \cap \pi'$ has the parity of n. It follows that $\mathscr{H}(4n + 1, q)$ has no spread.

Open problems: Existence or non-existence of spreads in the following cases:

(a) $\mathscr{P}(6, q)$, for q odd, with $q \equiv 1 \pmod 3$ and not a prime;
$\mathscr{P}(4n + 2, q)$, for $n > 1$ and q odd;
(b) $\mathscr{H}(7, q)$, for q odd, with $q \equiv 1 \pmod 3$ and not a prime;
$\mathscr{H}(4n + 3, q)$, for $n > 1$ and q odd;

Table AVI.2

Polar space	Existence of spreads	References
$\mathcal{W}_n(q)$, $n = 2t + 1$	Yes	Thas (1977a)
$\mathcal{P}(2n, q)$, q even	Yes	Dye (1977b), Thas (1980c, 1981a)
$\mathcal{P}(6, q)$, q odd with q prime or $q \equiv 0$ or 2 (mod 3)	Yes	See (*) in Table AVI.1
$\mathcal{P}(4n, q)$, q odd	No	Thas (1972, 1991c)
$\mathcal{H}(3, q)$	Yes	
$\mathcal{H}(7, q)$, q odd with q prime or $q \equiv 0$ or 2 (mod 3)	Yes	See (*) in Table AVI.1
$\mathcal{H}(4n + 3, q)$, q even	Yes	Dye (1977b), Thas (1980c, 1981a)
$\mathcal{H}(4n + 1, q)$	No	
$\mathcal{E}(5, q)$	Yes	Payne and Thas (1984), Thas (1983b)
$\mathcal{E}(2n + 1, q)$, q even	Yes	Dye (1977b), Thas (1980c, 1981a)
$\mathcal{U}(4, 4)$	No	Brouwer (1981b)
$\mathcal{U}(2n + 1, q^2)$	No	Thas (1981a, 1991c)

(c) $\mathcal{E}(2n + 1, q)$, for $n > 2$ and q odd;
(d) $\mathcal{U}(4, q^2)$, for $q > 2$;
$\mathcal{U}(2n, q^2)$, for $n > 2$.

AVI.5 Notes and references

§AVI.1. For the size of orbits of subspaces under the symplectic group, see Wan (1965a). In particular these give $|\mathcal{G}(\mathcal{W}_n(q))|$.

For similar results on orbits under the pseudo symplectic group, see Liu and Wan (1991), Pless (1964, 1965a).

APPENDIX VII

Errata for *Finite projective spaces of three dimensions* and *Projective geometries over finite fields*

Where there is no ambiguity, only the correct version is given. Here 100^{12} means 'page 100, line 12' whereas 64_6 means 'page 64, line 6 from bottom'.

Finite projective spaces of three dimensions

32^{10}	plane Π of order q, every hexagon H inscribed in a $(q+1)$-arc $\mathscr{C}$ satisfies
64_6	$\mathbf{I}(z^{p+1} + bzy^p - cy^{p+1}, \ldots$
64_3	$= z_0^{p+1} + bz_0 y_0^p - cy_0^{p+1} \cdots$
64_2	$+ z_1^{p+1} + bz_1 y_1^p - cy_1^{p+1} \cdots$
100^{12}	If $k_i = 2$, then $\sigma_1^{(i)}(Q) = 0$
141^4	The proof is wrong here; see Theorem 23.5.4 for a correct version
164^7	Lemma 19.5.4
169_9	seven singular
195^9	$T_{12,34}$, $T_{34,56}$, $T_{56,12}$
198^8	So $P = E_{31}$
242_{12}	k non-collinear points
246^{13}	$\mathbf{c}(q+3, 2)$
253^{16}	$(q^2+q+2)/2$-span
261_3	$X = O$
284_9	Glynn, D. G.
290_{21}	B. Segre's

Projective geometries over finite fields

The main list of errata is given in Appendix V in *Finite projective spaces of three dimensions.* This is a supplementary list.

52_{6-4}	If P is simple for $\mathscr{F} = \mathbf{V}(F_1, \ldots, F_r)$, then $T_P(\mathscr{F}) = T_P(\mathscr{F}_1) \cap \cdots \cap T_P(\mathscr{F}_r)$, where $\mathscr{F}_i = \mathbf{V}(F_i)$; so $T_P(\mathscr{F})$ is a proper subspace of $PG(n, K)$.
93_6	$\rho(n, q, q^k)$
93_4	$= \rho(n, q, q^k)\, s(n, q, q^k)$
107_3	In Chapter 22
110^{13}	in later volumes
112_{12}	There is a gap here in the proof of Theorem 5.3.3

115_6	in further volumes
122^5	$\{\mathbf{V}(a_0x_0^2 + a_{01}x_0x_1 + a_1x_1^2)\|$
122_{12}	$+ \bar{b}x_0\bar{x}_1$
123_8	$\cong \mathbf{Z}_p^h$
145^6	$\Delta = \det$
153_9	on $\mathbf{u}_0$
247_7	or is indeterminate, in which case $\mathscr{F}$ is singular
270^2	$c_0x_1^3$
308^7	$-(x_0^3 - 3x_0x_1^2 + dx_1^3)$
317_1	$d^2x_0x_2^2 + dx_1x_2^2$
437^{18}	$A247$
453_{16}	planes

BIBLIOGRAPHY

This continues the Bibliographies in *Projective geometries over finite fields* and *Finite projective spaces of three dimensions.* All references in these two volumes which are mentioned in this volume are included in this Bibliography. Two or more references to the same author(s) are distinguished by the year and letters, in a fashion consistent with the previous volumes.

As in these volumes, the Bibliography is as complete as possible regarding works on the geometry of $PG(n, q)$. It also contains representative works of adjacent topics. Papers in the course of publication have generally only been included if specifically cited.

Abatangelo, L. M. (1984). Una caratterizzazione gruppale delle curve hermitiane. *Le Matematiche* 39, 101–10.

—— (1988). Collineation groups of spreads. *Ars Combin.* 25B, 247–56.

Abatangelo, V. (1986). Doubly transitive $(n + 2)$-arcs in projective planes of even order q. *J. Combin. Theory Ser. A* 42, 1–8.

—— and Larato, B. (1989). A characterization of Denniston's maximal arcs. *Geom. Dedicata* 30, 197–203.

Abdul-Elah, M. S., Al-Dhahir, M. W., and Jungnickel, D. (1987). 8_3 in $PG(2, q)$. *Arch. Math.* 49, 141–50.

Ademaj, E. (1984). On a projective plane of order 11 on which operates a group of order 63 which fixes a subplane of order 2. *Glas. Mat. Ser. III* 19 (39), 217–24.

Ahrens, R. W. and Szekeres, G. (1969). On a combinatorial generalization of 27 lines associated with a cubic surface. *J. Austral. Math. Soc.* 10, 485–92.

Anderson, B. A. (1980). Hyperovals and Howell designs. *Ars Combin.* 9, 29–38.

André, J. (1954). Uber nicht-Desarguessche Ebenen mit transitiver Translationsgruppe. *Math. Z.* 50, 156–86.

Anstee, R. P., Hall, M., and Thompson, J. G. (1980). Planes of order 10 do not have a collineation of order 5. *J. Combin. Theory Ser. A* 29, 39–58.

Artin, E. (1957). *Geometric algebra.* Interscience, New York.

Artzy, R. (1988). A geometrically motivated presentation for $AG(2, 3)$. *J. Geom.* 31, 1–5.

Assmus, E. F. (1983). Applications of algebraic coding theory to finite geometric problems. *Finite geometries*, Dekker, New York, pp. 23–32.

—— (1985). The non-existence of an oval-extendable (56, 11, 2) design. *J. Geom.* 24, 168–74.

—— (1989). The coding theory of finite geometries and designs, Lecture Notes in Comput. Sci. 357, Springer, Berlin, pp. 1–6.

—— and Key, J. D. (1986). On an infinite class of Steiner systems with $t = 3$ and $k = 6$. *J. Combin. Theory Ser. A* 42, 55–60.

—— and —— (1989). Arcs and ovals in the Hermitian and Ree unitals. *European J. Combin.* 10, 297–308.

—— and —— (1990a). Affine and projective planes. *Discrete Math.* 83, 161–87.

—— and —— (1990b). Baer subplanes, ovals and unitals. *IMA Vol. Math. Appl.* 20, Springer, New York, pp. 1–8.

—— and Mattson, H. F. (1974). Coding and combinatorics. *SIAM Rev.* 16, 349–88.

——, ——, and Guza, M. (1974). Self-orthogonal Steiner systems and projective planes. *Math. Z.* 138, 89–96.

Bader, L. (1988). Some new examples of flocks of $Q^+(3, q)$. *Geom. Dedicata* 27, 213–18.

—— (1990). Derivation of Fisher flocks. *J. Geom.* 37, 17–24.

—— and Lunardon, G. (1989). On the flocks of $Q^+(3, q)$. *Geom. Dedicata* 29, 177–83.

——, ——, and Thas, J. A. (1990). Derivation of flocks of quadratic cones. *Forum Math.* 2, 163–74.

Baeza, R. (1984). On the Arf invariant of quadratic forms and of knots. *Linear and Multilinear Algebr.* 16, 247–52.

Bagchi, B. and Sastry, N. S. N. (1985). Minimum weight words of binary codes associated with finite projective geometries. *Discrete Math.* 57, 307–10.

—— and —— (1987). Even order inversive planes, generalized quadrangles and codes. *Geom. Dedicata* 22, 137–47.

—— and —— (1989). Intersection pattern of the classical ovoids in symplectic 3-space of even order. *J. Algebra* 126, 147–60.

Bagchi, S. and Bagchi, B. (1989). Designs from pairs of finite fields. I. A cyclic unital $U(6)$ and other regular Steiner 2-designs. *J. Combin. Theory Ser. A* 52, 51–61.

Baker, C. A. and Batten, L. M. (Eds) (1985). *Finite geometries*, Dekker, New York.

Baker, R. D. and Ebert, G. L. (1983). Denniston designs and the search for elliptic semiplanes. *Congr. Numer.* 39, 101–6.

—— and —— (1985). Spreads and packings for a class of $((2^n + 1)(2^{n-1} - 1) + 1, 2^{n-1}, 1)$-designs. *J. Combin. Theory Ser. A* 40, 45–54.

—— and —— (1987). A nonlinear flock in the Minkowski plane of order 11. *Congr. Numer.* 58, 75–81.

—— and —— (1988). Nests of size $q - 1$ and another family of translation planes. *J. London Math. Soc.* 38, 341–55.

—— and —— (1990). Intersection of unitals in the Desarguesian plane. *Congr. Numer.* 70, 87–94.

—— and —— (1991). On Buekenhout–Metz unitals of odd order.

——, ——, and Weida, R. (1988). Another look at Bruen chains. *J. Combin. Theory Ser. A* 48, 77–90.

Baldisserri, N. (1985). Reti di coniche in un piano sopra un campo di caratteristica 2. *Rend. Mat.* 5, 355–65.

Bannai, E. (1985). Tight t-designs in projective spaces and Newton polygons. *Ars Combin.* 20, 43–9.

——, Hao, S., and Song, S. Y. (1990). Character tables of the association schemes of finite orthogonal groups acting on the nonisotropic points. *J. Combin. Theory Ser. A* 54, 164–200.

Barker, H. A. (1986). Sum and product tables for Galois fields. *Internat. J. Math. Ed. Sci. Tech.* 17, 473–85.

Barlotti, A. (1986). Alcune questioni nella geometria dal punto di vista di von Staudt. *Combinatorica*, Symp. Math. 28, Academic Press, London, pp. 45–51.

—— and Strambach, K. (1983). The geometry of binary systems. *Adv. in Math.* 49, 1–105.
—— and —— (1985). Remarks on projectivities. *Aequationes Math.* 28, 212–28.
——, Biliotti, M., Cossu, A., Korchmáros, G., and Tallini, G. (Eds) (1986). *Combinatorics '84*, Annals of Discrete Math. 30, North-Holland, Amsterdam.
——, Marchi, M., and Tallini, G. (Eds) (1988). *Combinatorics '86*, Annals of Discrete Math. 37, North-Holland, Amsterdam.
Bartocci, U. (1983). k-insiemi densi in piani di Galois. *Boll. Un. Mat. Ital. D* 2, 71–7.
Basile, A. and Brutti, P. (1985). Equivalenza proiettiva di fibrazioni regolari di $PG(2t + 1, q)$. *Atti Accad. Sci. Lett. Arti Palermo* 2, 41–8.
Batten, L. M. (1983). Maximal affine subplanes of finite projective planes. *Ars Combin.* 16, 15–26.
—— (1984). Embedding the complement of a minimal blocking set in a projective plane. *Discrete Math.* 52, 1–5.
—— (1985). Sets of type $(1, k, n + 1)$ in projective planes of order n. *Finite geometries*, Dekker, New York, pp. 5–8.
—— (1986). *Combinatorics of finite geometries.* Cambridge University Press, Cambridge.
Benson, C. T. (1970). On the structure of generalized quadrangles. *J. Algebra* 15, 443–54.
—— and Losey, N. E. (1971). On a graph of Hoffman and Singleton. *J. Combin. Theory Ser. B* 11, 67–79.
Berardi, L. (1984a). On the cardinality of blocking sets in $PG(2, q)$. *J. Geom.* 22, 5–14.
—— (1984b). Alcune condizioni necessarie per l'esistenza di t-fibrazioni parziali di tipo $(0, m)_{2t+1}$ in $PG(n, q)$. *Rend. Mat.* 4, 59–69.
—— (1985). Some necessary conditions for the existence of t-fibrations of type $(0, m)_{2t+1}$ in $PG(n, q)$. *Rend. Mat.* 7, 59–69.
—— (1987). Irreducible blocking sets in a symmetric design. *Boll. Un. Mat. Ital. A* 1, 139–47.
—— (1988a). Blocking sets in the large Mathieu designs, I: the case $S(2, 6, 22)$. *Ann. Discrete Math.* 37, 31–42.
—— (1988b). Blocking sets in the large Mathieu designs, II: the case $S(4, 7, 23)$. *J. Inform. Optim. Sci.* 9, 263–78.
—— (1989). Constructing 3-designs from spreads and lines. *Discrete Math.* 74, 331–2.
—— and Beutelspacher, A. (1987a). Partial partitions of finite projective spaces. *Mitt. Math. Sem. Giessen* No. 180, 59–67.
—— and —— (1987b). Blocking sets of the known biplanes. *Rend. Mat. Appl.* 7, 63–76.
—— and —— (1988). Constructing the biplanes of order four from six points. *Boll. Un. Mat. Ital. A* 2, 285–9.
—— and Bichara, A. (1990). Una caratterizzazione grafica della varietà di Corrado Segre in $PG(5, q)$. *Rend. Mat.* 10, 265–78.
—— and Eugeni, F. (1988a). Blocking sets in the projective plane of order four. *Ann. Discrete Math.* 37, 43–50.
—— and —— (1988b). Blocking sets e teoria dei giochi: origini e problematiche. *Atti Sem. Mat. Fis. Univ. Modena* 34, 165–96.
—— and —— (1990). Blocking sets in the large Mathieu designs, III: the case $S(5, 8, 24)$. *Ars Combin.* 29, 33–41.

——, Beutelspacher, A., and Eugeni, F. (1983). On $(s, t; h)$-blocking sets in finite projective and affine spaces. *Atti Sem. Mat. Fis. Univ. Modena* 32, 130–57.

——, Eugeni, F., and Ferri, O. (1985). Sui blocking sets nei sistemi di Steiner. *Boll. Un. Mat. Ital. D* 3, 141–64.

Bernardi, M. P. and Torre, A. (1984). Alcuni questioni di esistenze e continuità per (m, n)-fibrazioni e semicollineazioni. *Boll. Un. Mat. Ital. B* 3, 611–22.

Bernasconi, C. (1983). Geometric designs. *Ann. Univ. Ferrara Sez. VII* 29, 129–35.

Bertini, E. (1923). *Introduzione alla geometria proiettiva degli iperspazi.* Casa Editrice Giuseppe Principato, Messina.

Beth, T. and Jungnickel, D. (1982). Variations on seven points: An introduction to the scope and methods of coding theory and finite geometries. *Aequationes Math.* 25, 177–93.

——, ——, and Lenz, H. (1984). *Design theory.* Bibliographisches Institut, Mannheim (Cambridge University Press, Cambridge).

Beutelspacher, A. (1982). *Einführung in die endliche Geometrie. I: Blockpläne.* Bibliographisches Institut, Mannheim.

—— (1983). *Einführung in die endliche Geometrie. II: Projective Räume.* Bibliographisches Institut, Mannheim.

—— (1985). Embedding finite planar spaces in projective spaces. *Finite geometries*, Dekker, New York, pp. 9–17.

—— (1986a). On complete q-arcs in projective planes of order q. *Boll. Un. Mat. Ital. A* 5, 449–54.

—— (1986b). $21 - 6 = 15$: a connection between two distinguished geometries. *Amer. Math. Monthly* 93, 29–41.

—— (1986c). Embedding linear spaces with two line degrees in finite projective planes. *J. Geom.* 26, 43–61.

—— (1987). A defense of the honour of an unjustly neglected little geometry or a combinatorial approach to the projective plane of order five. *J. Geom.* 30, 182–95.

—— (1988). Enciphered geometry. Some applications of geometry to cryptography. *Ann. Discrete Math.* 37, 59–68.

—— (1990). Partial parallelisms in finite projective spaces. *Geom. Dedicata* 36, 273–8.

—— and Eugeni, F. (1984). A lower bound for partial t-spreads of level $t - 1$ in $PG(2t + 1, q)$. *Atti Sem. Mat. Fis. Univ. Modena* 33, 221–8.

—— and —— (1985a). On the type of partial t-spreads in finite projective spaces. *Discrete Math.* 54, 241–57.

—— and —— (1985b). On blocking sets in affine and projective spaces of large order. *Rend. Mat.* 6, 587–95.

—— and —— (1985c). Sui blocking sets di date indice con particolare riguardo all'indice tre. *Boll. Un. Mat. Ital. A* 4, 441–50.

—— and —— (1986). On n-fold blocking sets. *Ann. Discrete Math.* 30, 31–8.

—— and Kersten, A. (1985). Finite semiaffine linear spaces. *Arch. Math.* 44, 557–68.

—— and Mazzocca, F. (1987). Blocking-sets in infinite projective and affine spaces. *J. Geom.* 28, 112–16.

—— and Metsch, K. (1986). Embedding finite linear spaces in projective planes. *Ann. Discrete Math.* 30, 39–56.

—— and Seeger, D. (1987). Characterizing the line set of a Baer subspace. *J. Combin. Theory Ser. A* 45, 152–6.

—— and Ueberberg, J. (1991). On the intersections of Baer subspaces. *Arch. Math.* 56, 203–8.

—— and Vedder, K. (1989). Geometric structures as threshold schemes. *Inst. Math. Appl. Conf. Ser. New Ser.* 20, Oxford University Press, New York, pp. 255–68.

——, Jungnickel, D., and Vanstone, S. A. (1989). On the chromatic index of a finite projective space. *Geom. Dedicata* 32, 313–18.

——, Tallini, G., and Zanella, C. (1991). Examples of essentially s-fold secure authentication systems with large s. *Rend. Mat.* 10, 321–6.

Bichara, A. (1978). Caratterizzazione dei sistemi rigati immersi in $A_{3,q}$. *Riv. Mat. Univ. Parma* 4, 277–90.

—— (1986). Veronese quadruples. *Ann. Discrete Math.* 30, 57–67.

—— (1989). An elementary proof of the nonexistence of a projective plane of order six. *Mitt. Math. Sem. Giessen* No. 192, 89–93.

—— and Mazzocca, F. (1982a). On a characterization of Grassmann space representing the lines in an affine space. *Simon Stevin* 56, 129–41.

—— and —— (1982b). On the independence of the axioms defining the affine and projective Grassmann spaces. *Ann. Discrete Math.* 14, 123–8.

—— and —— (1983). On a characterization of the Grassmann spaces associated with an affine space. *Ann. Discrete Math.* 18, 95–112.

—— and Migliori, G. (1983). Alcune osservazioni sui $(q-1)$-archi completi di $S_{2,q}$. *Boll. Un. Mat. Ital. D* 2, 21–7.

—— and Somma, C. (1987). On the flag space associated with an affine or projective space. *Rend. Mat.* 7, 323–34.

—— and Tallini, G. (1982). On a characterization of the Grassmann manifold representing the planes in a projective space. *Ann. Discrete Math.* 14, 129–50.

—— and —— (1983). On a characterization of Grassmann space representing the h-dimensional subspaces in a projective space. *Ann. Discrete Math.* 18, 113–32.

Bierbrauer, J. (1985a). Necessary conditions for additive loops of finite projective planes. *Geom. Dedicata* 19, 207–16.

—— (1985b). Blocking sets of 16 points in projective planes of order 10. II. *Quart. J. Math. Oxford Ser.* 36, 383–91.

—— (1985c). Blocking sets of 16 points in projective planes of order 10. III. *Rend. Sem. Mat. Univ. Padova* 74, 163–74.

—— (1991a). Weighted arcs, the finite Radon transform and a Ramsey problem. *Graphs Combin.*

—— (1991b). On projective planes of orders 28 or 32 whose binary code contains vectors of small weight. *Mitt. Math. Sem. Giessen.*

Biliotti, M. and Korchmáros, G. (1986a). Collineation groups strongly irreducible on an oval. *Ann. Discrete Math.* 30, 85–97.

—— and —— (1986b). Collineation groups which are primitive on an oval of a projective plane of odd order. *J. London Math. Soc.* 33, 525–34.

—— and —— (1987). Hyperovals with a transitive collineation group. *Geom. Dedicata* 24, 269–81.

—— and —— (1989a). Collineation groups preserving a unital of a projective plane of odd order. *J. Algebra* 122, 130–49.

—— and —— (1989b). Collineation groups preserving a unital of a projective plane of even order. *Geom. Dedicata* 31, 333–44.

—— and —— (1990a). Collineation groups preserving an oval in a projective plane of odd order. *J. Austral. Math. Soc. Ser A* 48, 156–70.

—— and —— (1990b). Some finite translation planes arising from A_6-invariant ovoids of the Klein quadric. *J. Geom.* 37, 29–47.

Bilo, J. (1980). A geometrical problem depending on the equation $ax - xa = 2$ in skew fields. *Quart. J. Pure Appl. Math.* 54, 27–62.

Biondi, P. (1987). A characterisation of the Grassmann space representing the 2-dimensional subspaces of projective space. *Boll. Un. Mat. Ital.* 7, 713–27.

—— (1988). On finite Grassmann spaces. *Ann. Discrete Math.* 37, 69–73.

—— (1989). On incidence structures defined by projective spaces. *Simon Stevin* 63, 141–55.

—— and Melone, N. (1985). Incidence structures of affine and projective types. *J. Geom.* 25, 178–91.

—— and —— (1986). On sets of Plücker class two in $PG(3, q)$. *Ann. Discrete Math.* 30, 99–103.

Black, S. C. and List, R. J. (1990). On certain abelian groups associated with finite projective geometries. *Geom. Dedicata* 33, 13–19.

Blake, I. F., Fuji-Hara, R., Mullin, R. C., and Vanstone, S. A. (1984). Computing logarithms in finite fields of characteristic two. *SIAM J. Algebraic Discrete Methods* 5, 276–85.

——, Mullin, R. C., and Vanstone, S. A. (1985). Computing logarithms in $GF(2^2)$. *Advances in cryptology*, Lecture Notes in Comput. Sci. 196, Springer, New York, pp. 73–82.

Blokhuis, A. (1984). On subsets of $GF(q^2)$ with square differences. *Ned. Akad. Wetensch. Proc. Ser. A* 87, 369–72.

—— (1991). Characterization of seminuclear sets in a finite projective plane. *J. Geom.* 40, 15–19.

—— and Brouwer, A. E. (1986). Blocking sets in desarguesian projective planes. *Bull. London Math. Soc.* 18, 132–4.

—— and Bruen, A. A. (1989). The minimal number of lines intersected by a set of $q + 2$ points, blocking sets, and intersecting circles. *J. Combin. Theory Ser. A* 50, 308–15.

—— and Mazzocca, F. (1991). On maximal sets of nuclei in $PG(2, q)$ and quasi-odd sets in $AG(2, q)$. *Advances in finite geometries and designs*, Oxford University Press, Oxford, pp. 27–34.

—— and Wilbrink, H. A. (1987). A characterization of exterior lines of certain sets of points in $PG(2, q)$. *Geom. Dedicata* 23, 253–4.

——, Brouwer, A. E., Delandtsheer, A., and Doyen, J. (1987). Orbits on points and lines in finite linear and quasilinear spaces. *J. Combin. Theory Ser. A* 44, 159–63.

——, ——, and Wilbrink, H. (1989). Heden's bound on maximal partial spreads. *Discrete Math.* 74, 335–9.

——, ——, and —— (1991). Hermitian unitals are codewords. *Discrete Math.*

——, Bruen, A. A., and Thas, J. A. (1990). Arcs in $PG(n, q)$, MDS-codes and three fundamental problems of B. Segre—some extensions. *Geom. Dedicata* 35, 1–11.

——, Seress, À., and Wilbrink, H. A. (1991a). Characterization of complete exterior sets of conics. *Combinatorica.*

——, ——, and —— (1991b). On sets of points in $PG(2, q)$ without tangents. *Mitt. Math. Sem. Giessen.*

Bolker, E. D. (1987). The finite Radon transform. *Contemp. Math.* 63, 27–50.

Bollinger, R. C. and Burchard, C. L. (1990). Lucas's theorem and some related results for extended Pascal triangles. *Amer. Math. Monthly* 97, 198–204.

Bombieri, E. (1974). Counting points on curves over finite fields. *Seminaire Bourbaki* 430, Lecture Notes in Math. 383, Springer, Berlin, pp. 234–41.
Bonisoli, A. (1988). On the sharply 1-transitive subsets of $PG(2, p^m)$. *J. Geom.* 31, 32–41.
—— and Korchmáros, G. (1991). Flocks of hyperbolic quadrics and linear groups containing homologies.
Boros, E. (1988). $PG(2, p^s)$, $p > 2$, has property $B(p + 2)$. *Ars Combin.* 25, 111–13.
—— and Szönyi, T. (1986). On the sharpness of a theorem of B. Segre. *Combinatorica* 6, 261–8.
——, Füredi, Z., and Kahn, J. (1989). Maximal intersecting families and affine regular polygons in $PG(2, q)$. *J. Combin. Theory. Ser. A* 52, 1–9.
——, Szönyi, T., and Wettl, F. (1987). Sperner extensions of affine spaces. *Geom. Dedicata* 22, 163–72.
Boros, I., Moreno, C. J., and Porta, H. (1975). Elliptic curves over finite fields. II. *Math. Comp.* 29, 951–64.
Bose, R. C. (1947). Mathematical theory of the symmetrical factorial design. *Sankhyā* 8, 107–66.
—— (1963). Strongly regular graphs, partial geometries and partially balanced designs. *Pacific J. Math.* 13, 389–419.
—— and Chakravarti, I. M. (1966). Hermitian varieties in a finite projective space $PG(N, q^2)$. *Canad. J. Math.* 18, 1161–82.
—— and Shrikhande, S. S. (1972). Geometric and pseudo-geometric graphs $(q^2 + 1, q + 1, 1)$. *J. Geom.* 2, 75–94.
Brouwer, A. E. (1981b). Private communication.
—— (1985). Some new two-weight codes and strongly regular graphs. *Discrete Appl. Math.* 10, 111–14.
—— (1986). An inequality in binary vector spaces. *Discrete Math.* 59, 315–17.
—— (1991). A nondegenerate generalized quadrangle with lines of size four is finite. *Advances in finite geometries and designs*, Oxford University Press, Oxford, pp. 47–9.
—— and Cohen, A. M. (1985). Computation of some parameters of Lie geometries. *Ann. Discrete Math.* 26, 1–48.
—— and van Lint, J. H. (1984). Strongly regular graphs and partial geometries. *Enumeration and design*, Academic Press, Toronto, pp. 85–122.
—— and Wilbrink, H. A. (1984). A symmetric design with parameters 2-(49, 16, 5). *J. Combin. Theory Ser. A* 37, 193–4.
—— and —— (1990). Ovoids and fans in the generalized quadrangle $Q(4, 2)$. *Geom. Dedicata* 36, 121–4.
——, Cohen, A. M., and Neumaier, A. (1989). *Distance regular graphs.* Springer, Berlin.
Brown, J. M. N. (1988). Partitioning the complement of a simplex in $PG(e, q^{d+1})$ into copies of $PG(d, q)$. *J. Geom.* 33, 11–16.
Bruck, J. and Blaum, M. (1989). Neural networks, error-correcting codes and polynomials over the binary n-cube. *IEEE Trans. Inform. Theory* 35, 976–87.
Bruck, R. H. (1960). Quadratic extensions of cyclic planes. *Proc. Sympos. Appl. Math.* 10, 15–44.
—— and Bose, R. C. (1966). Linear representations of projective planes in projective spaces. *J. Algebra* 4, 117–72.

Bruen, A. A. (1972c). Permutation functions on a finite field. *Canad. Math. Bull.* 15, 595–7.

—— (1986). Arcs and multiple blocking sets. *Combinatorica*, Symp. Math. 28, Academic Press, London, pp. 15–29.

—— (1989). Kummer configurations and designs embedded in planes. *J. Combin. Theory Ser. A* 52, 154–7.

—— (1990). Nuclei of sets of $q + 1$ points in $PG(2, q)$ and blocking sets of Redei type. *J. Combin. Theory Ser. A* 55, 130–2.

—— and Hirschfeld, J. W. P. (1986). Intersections in projective space I: Combinatorics. *Math. Z.* 193, 215–25.

—— and —— (1988). Intersections in projective space II: Pencils of quadric surfaces. *European J. Combin.* 9, 255–71.

—— and Ott, U. (1990). On the p-rank of incidence matrices and a question of E. S. Lander. *Finite geometries and combinatorial designs*, American Mathematical Society, Providence, RI, pp. 39–45.

—— and Rothschild, B. L. (1982). Lower bounds on blocking sets. *Pacific J. Math.* 118, 303–11.

—— and Silverman, R. (1987). Arcs and blocking sets, II. *European J. Combin.* 8, 351–6.

—— and —— (1988). On extendable planes, M.D.S. codes and hyperovals in $PG(2, q)$, $q = 2^t$. *Geom. Dedicata* 28, 31–43.

—— and Thas, J. A. (1976). Partial spreads, packings and Hermitian manifolds in $PG(3, q)$. *Math. Z.* 151, 207–14.

——, ——, and Blokhuis, A. (1988a). On M.D.S. codes, arcs in $PG(n, q)$ with q even, and a solution of three fundamental problems of B. Segre. *Invent. Math.* 92, 441–59.

——, ——, and —— (1988b). MDS codes and arcs in projective space I; II. *C.R. Math. Rep. Acad. Sci. Canada* 10, 225–30; 233–5.

Buekenhout, F. (1969a). Ensembles quadratiques des espaces projectifs. *Math. Z.* 110, 306–18.

—— (1979b). Diagrams for geometries and groups. *J. Combin. Theory Ser. A* 27, 121–51.

—— (1987). On affine quadratic sets. *Atti Sem. Mat. Fis. Univ. Modena* 35, 71–6.

—— (1990). On the foundations of polar geometry II. *Geom. Dedicata* 33, 21–6.

—— and Buset, D. (1988). On the foundations of incidence geometry. *Geom. Dedicata* 25, 269–96.

—— and Lefèvre, C. (1974). Generalized quadrangles in projective spaces. *Arch. Math.* 25, 540–52.

—— and —— (1976). Semi-quadratic sets in projective spaces. *J. Geom.* 7, 17–42.

—— and Shult, E. E. (1974). On the foundations of polar geometry. *Geom. Dedicata* 3, 155–70.

——, Delandtsheer, A., and Doyen, J. (1988). Finite linear spaces with flag-transitive groups. *J. Combin. Theory Ser. A* 49, 268–93.

——, ——, ——, Kleidman, P. B., Liebeck, M. W., and Saxl, J. (1990). Linear spaces with flag-transitive automorphism groups. *Geom. Dedicata* 36, 89–94.

Buratti, M. (1988). Bruck–Ryser abstract theorem and symmetric designs. *Geom. Dedicata* 27, 241–50.

Burau, W. (1961). *Mehrdimensionale projektive und höhere Geometrie.* VEB Deutscher Verlag der Wissenschaften, Berlin.

Calderbank, A. R. and Goethals, J.-M. (1984). Three-weight codes and association schemes. *Philips J. Res.* 39, 143–52.
—— and —— (1985). On a pair of dual subschemes of the Hamming scheme $H_n(q)$. *European J. Combin.* 6, 133–47.
—— and Kantor, W. M. (1986). The geometry of two-weight codes. *Bull. London Math. Soc.* 18, 97–122.
—— and Morton, P. (1990). Quasi-symmetric 3-designs and elliptic curves. *SIAM J. Discrete Math.* 3, 178–96.
—— and Wales, D. B. (1984). The Haemers partial geometry and the Steiner system $S(5, 8, 24)$. *Discrete Math.* 51, 125–36.
Cameron, P. J. (1975). Partial quadrangles. *Quart. J. Math. Oxford Ser.* 26, 61–74.
—— (1985). Four lectures on projective geometry. *Finite geometries*, Dekker, New York, pp. 27–63.
—— and Mazzocca, F. (1986). Bijections which preserve blocking sets. *Geom. Dedicata* 21, 219–29.
—— and van Lint, J. H. (1980). *Graphs, codes and designs.* Cambridge University Press, Cambridge.
——, Goethals, J.-M., and Seidel, J. J. (1978). Strongly regular graphs having strongly regular subconstituents. *J. Algebra* 55, 257–80.
——, Hughes, D. R., and Pasini, A. (1990). Extended generalized quadrangles. *Geom. Dedicata* 35, 193–228.
——, Mazzocca, F., and Meshulam, R. (1988). Dual blocking sets in projective and affine planes. *Geom. Dedicata* 27, 203–7.
——, Thas, J. A., and Payne, S. E. (1976). Polarities of generalized hexagons and perfect codes. *Geom. Dedicata* 5, 525–8.
Capodaglio (di Cocco), R. (1982). Elations in a projective pappian plane. *Boll. Un. Mat. Ital. D* 1, 107–16.
—— (1986). On thick $(q + 2)$-sets. *Ann. Discrete Math.* 30, 115–24.
Casse, L. R. A. (1969). A solution to Beniamino Segre's 'Problem $I_{r,q}$' for q even. *Atti Accad. Naz. Lincei Rend.* 46, 13–20.
—— and Glynn, D. G. (1984). On the uniqueness of $(q + 1)_4$-arcs of $PG(4, q)$, $q = 2^h$, $h \geq 3$. *Discrete Math.* 48, 173–86.
—— and O'Keefe, C. M. (1990a). Projective spread sets. *J. Geom.* 37, 55–76.
—— and —— (1990b). Indicator sets for t-spreads of $PG((s + 1)(t + 1) - 1, q)$. *Boll. Un. Mat. Ital. B* 4, 13–33.
——, ——, and Wild, P. R. (1990). Maximal (k, r)-sets of $PG(3n - 1, q)$ with regular projections. *Simon Stevin* 64, 89–96.
——, Thas, J. A., and Wild, P. R. (1985). $(q^n + 1)$-sets of $PG(3n - 1, q)$, generalized quadrangles and Laguerre planes. *Simon Stevin* 59, 21–42.
Cassels, J. W. S. (1966). Diophantine equations with special reference to elliptic curves. *J. London Math. Soc.* 41, 193–285.
Ceccherini, P. V. (1984). A q-analogous of the characterization of hypercubes as graphs. *J. Geom.* 22, 57–74.
Chakravarti, I. M. (1990). Families of codes with few distinct weights from singular and nonsingular Hermitian varieties and quadrics in projective geometries and Hadamard difference sets and designs associated with two-weight codes. *IMA Vol. Math. Appl.* 20, Springer, New York, pp. 35–50.
Chan, A. H. and Games, R. A. (1986). On the linear span of binary sequences obtained

from finite geometries. *Advances in cryptology*, Lecture Notes in Comput. Sci. 263, Springer, New York, pp. 183–96.

Cherowitzo, W. (1985a). Harmonic ovals of even order. *Finite geometries*, Dekker, New York, pp. 65–81.

—— (1985b). Translation *B*-ovals. *Congr. Numer.* 47, 161–72.

—— (1986). Hyperovals in Desarguesian planes of even order. *Ann. Discrete Math.* 37, 87–94.

—— (1990). Ovals in Figueroa planes. *J. Geom.* 37, 84–6.

——, Kiel, D. I., and Killgrove, R. B. (1986). Ovals and other configurations in the known planes of order nine. *Congr. Numer.* 55, 167–79.

Chudnovsky, D. V. (1987). Algebraic complexity and algebraic curves over finite fields. *Proc. Nat. Acad. Sci. U.S.A.* 84, 1739–43.

—— and Chudnovsky, G. V. (1988). Algebraic complexities and algebraic curves over finite fields. *J. Complexity* 4, 285–316.

Cohen, A. (1981). Value sets of functions over finite fields. *Acta Arith.* 39, 339–59.

Cohen, A. M. (1981). A new partial geometry with parameters $(s, t, \alpha) = (7, 8, 4)$. *J. Geom.* 16, 181–6.

—— (1983). On a theorem of Cooperstein. *European J. Combin.* 4, 107–26.

—— and Wilbrink, H. A. (1980). The stabilizer of Dye's spread on a hyperbolic quadric in $PG(4n-1, 2)$ within the orthogonal group. *Atti Accad. Naz. Lincei Rend.* 69, 22–5.

Cohen, S. D. (1989). Windmill polynomials over fields of characteristic two. *Monatsh. Math.* 107, 291–301.

—— (1990). The factorable core of polynomials over finite fields. *J. Austral. Math. Soc. Ser. A* 49, 309–18.

Conway, J. H. and Sloane, N. J. A. (1988). *Sphere packings, lattices and groups.* Springer, New York.

——, Kleidman, P. B., and Wilson, R. A. (1988). New families of ovoids in O_8^+. *Geom. Dedicata* 26, 157–70.

Cooperstein, B. N. (1977). A characterization of some Lie incidence structures. *Geom. Dedicata* 6, 205–58.

—— (1990a). A note on the Weyl group of type E_7. *European J. Combin.* 11, 415–19.

—— (1990b). A sporadic ovoid in $\Omega^+(8, 5)$ and some non-Desarguesian translation planes of order 25. *J. Combin. Theory Ser. A* 54, 135–40.

Coppersmith, D. (1986). Discrete logarithms in $GF(p)$. *Algorithmica* 1, 1–15.

Coxeter, H. S. M. (1988). Regular and semi-regular polytopes. III. *Math. Z.* 200, 3–45.

Csima, J. and Furedi, Z. (1986). Colouring finite incidence structures. *Graphs Combin.* 2, 339–47.

D'Agostini, E. (1981). On caps with weighted points in $PG(t, q)$. *Discrete Math.* 34, 103–10.

Dai, Z. D. and Feng, X. N. (1965). Studies in finite geometries and the construction of incomplete block designs. IV. Some 'Anzahl' theorems in orthogonal geometry over finite fields of characteristic $\neq 2$. *Chinese Math.* 7, 265–79.

Dalla Volta, F. (1988). Regular sets for the affine and projective groups over the field of two elements. *J. Geom.* 33, 17–26.

—— (1989). Regular orbits for projective orthogonal groups over finite fields of odd characteristic. *Geom. Dedicata* 32, 229–45.

Datta, B. T. (1979). On Tutte's conjecture for tangential 2-blocks. *Graph theory and related topics*, Academic Press, New York, pp. 121–31.

Dawson, E. (1984). The binary code of the (191, 20, 2) biplane. *Ars Combin.* 17, 209–23.

Debroey, I. (1978). Semi-partiële meetkunden. *Ph.D. thesis*, State University of Ghent.

—— (1979a). Semi partial geometries satisfying the diagonal axiom. *J. Geom.* 13, 171–90.

—— (1979b). Semi partial geometries. *Bull. Soc. Math. Belg. Sér. B* 31, 183–90.

—— and Thas, J. A. (1978a). On semipartial geometries. *J. Combin. Theory Ser. A* 25, 242–50.

—— and —— (1978b). Semipartial geometries in $AG(2, q)$ and $AG(3, q)$. *Simon Stevin* 51, 195–209.

—— and —— (1978c). Semipartial geometries in $PG(2, q)$ and $PG(3, q)$. *Atti Accad. Naz. Lincei Rend.* 64, 147–51.

De Clerck, F. (1978). Een kombinatorische studie van de eindige partiële meetkunden. *Ph.D. thesis*, State University of Ghent.

—— (1979). Partial geometries—a combinatorial survey. *Bull. Soc. Math. Belg. Sér. B* 31, 135–45.

—— (1987). A characterisation of the partial geometry $T_2^*(K)$. *European J. Combin.* 8, 121–7.

—— and Mazzocca, F. (1988). The classification of polarities in reducible projective spaces. *European J. Combin.* 9, 245–9.

—— and Thas, J. A. (1978). Partial geometries in finite projective spaces. *Arch. Math.* 30, 537–40.

—— and —— (1983). The embedding of $(0, \alpha)$-geometries in $PG(n, q)$. Part I. *Ann. Discrete Math.* 18, 229–40.

—— and —— (1985). Exterior sets with respect to the hyperbolic quadric in $PG(2n-1, q)$. *Finite geometries*, Dekker, New York, pp. 83–91.

——, Gevaert, H., and Thas, J. A. (1988a). Translation partial geometries. *Ann. Discrete Math.* 37, 117–36.

——, ——, and —— (1988b). Flocks of a quadratic cone in $PG(3, q)$, $q \leq 8$. *Geom. Dedicata* 26, 215–30.

——, ——, and —— (1991). Partial geometries and copolar spaces. *Combinatorics '88*, Mediterranean Press, Cosenza.

de Finis, M. (1985). A characterisation of the generalised quadrangle $T_2^*(O)$. *Rend. Mat.* 5, 137–47.

—— (1986). On k-sets in $PG(3, q)$ of type (m, n) with respect to planes. *Ars Combin.* 21, 119–36.

Delandtsheer, A. (1986a). Basis-homogeneous geometric lattices. *J. London Math. Soc.* 34, 385–93.

—— (1986b). A geometric consequence of the classification of finite doubly transitive groups. *Geom. Dedicata* 21, 145–56.

—— (1986c). Transitivity on sets of independent points in geometric lattices. *Discrete Math.* 61, 103–5.

—— (1986d). Flag-transitive finite simple groups. *Arch. Math.* 47, 395–400.

—— and Doyen, J. (1990). A classification of line-transitive maximal (v, k)-arcs in finite projective planes. *Arch. Math.* 55, 187–92.

Dembowski, P. (1963). Inversive planes of even order. *Bull. Amer. Math. Soc.* 69, 850–4.

—— (1964). Möbiusebenen gerader Ordnung. *Math. Ann.* 157, 179–205.

—— (1968). *Finite geometries.* Springer, Berlin.

Dempwolff, U. (1984). A note on the Figueroa planes. *Arch. Math.* 43, 285–8.

—— (1985). On the automorphism group of planes of Figueroa type. *Rend. Sem. Mat. Univ. Padova* 74, 59–62.

Denniston, R. H. F. (1973a). Uniqueness of the inversive plane of order 5. *Manuscripta Math.* 8, 11–19.

—— (1973b). Uniqueness of the inversive plane of order 7. *Manuscripta Math.* 8, 21–3.

—— (1979). Some non-Desarguesian translation ovals. *Ars Combin.* 7, 221–2.

de Resmini, M. J. (1983). A characterization of the secants of an ovaloid in $PG(3, q)$, q even, $q > 2$. *Ars Combin.* 16, 33–49.

—— (1985a). A characterization of the set of lines either tangent to or lying on a nonsingular quadric in $PG(4, q)$, q odd. *Finite geometries*, Dekker, New York, pp. 271–88.

—— (1985b). On k-sets of class $[0, q/2 - 1, q/2, q/2 + 1, q]$ in a plane of even order q. *European J. Combin.* 6, 303–15.

—— (1985c). On 2-blocking sets in projective planes. *Ars Combin.* 20, 59–69.

—— (1987a). On 3-blocking sets in projective planes. *Ann. Discrete Math.* 34, 145–52.

—— (1987b). A 35-set of type (2, 5) in $PG(2, 9)$. *J. Combin. Theory Ser. A* 45, 303–5.

—— (1988). On admissible sets with two intersection numbers in a projective plane. *Ann. Discrete Math.* 37, 137–46.

—— and Migliori, G. (1986). A 78-set of type (2, 6) in $PG(2, 16)$. *Ars Combin.* 22, 73–5.

—— and Puccio, L. (1987). Subplanes of the Hughes plane of order 25. *Arch. Math.* 49, 151–65.

Derr, J. B. (1980). Stabilizers of isotropic subspaces in classical groups. *Arch. Math.* 34, 100–7.

De Soete, M. (1987a). Characterizations of $P(Q(4, q), L)$. *J. Geom.* 29, 50–60.

—— (1987b). A characterization of the generalised quadrangle $Q(4, q)$, q odd. *J. Geom.* 28, 57–79.

—— and Thas, J. A. (1984). A characterization theorem for the generalized quadrangle $T_2^*(O)$ of order $(s, s + 2)$. *Ars Combin.* 17, 225–42.

—— and —— (1986a). R-regularity and characterizations of the generalized quadrangle $P(W(s), (\infty))$. *Ann. Discrete Math.* 30, 171–84.

—— and —— (1986b). Characterizations of the generalized quadrangles $T_2^*(O)$ and $T_2(O)$. *Ars Combin.* 22, 171–86.

—— and —— (1987). A characterization of the generalized quadrangle $Q(4, q)$, q odd. *J. Geom.* 28, 57–79.

—— and —— (1988). A coordinatization of generalized quadrangles of order $(s, s + 2)$. *J. Combin. Theory Ser. A* 48, 1–11.

Deuring, M. (1941). Die Typen der Multiplikatorenringe elliptischer Funktionenkörper. *Abh. Math. Sem. Univ. Hamburg* 14, 197–272.

Deza, M. and Frankl, P. (1986). On squashed designs. *Discrete Comput. Geom.* 1, 379–90.

——, ——, and Hirschfeld, J. W. P. (1985). Sections of varieties over finite fields as large intersection families. *Proc. London Math. Soc.* 50, 405–25.

Dickson, L. E. (1901). *Linear groups with an exposition of the Galois field theory.* Teubner, Leipzig (Dover, 1958).

Dienst, K. J. (1980a). Verallgemeinerte Vierecke in projectiven Räumen. *Arch. Math.* 35, 177–86.

—— (1980b). Verallgemeinerte Vierecke in pappusschen projectiven Räumen. *Geom. Dedicata* 9, 199–206.

Dieudonné, J. A. (1971). *La géométrie des groupes classiques*. Third edition, Springer, Berlin.

Dillon, J. F. (1986). Graphs, codes and designs in quadrics. *Congr. Numer.* 55, 15–22.

Di Paola, J. W. (1985). The shape of minimum blocking sets in small planes. *Ars Combin.* 20, 15–26.

—— (1987). Designs from a minimum blocking set of $PG(2, 8)$. *Congr. Numer.* 58, 303–11.

Dixmier, J. (1987). On the projective invariants of quartic plane curves. *Adv. in Math.* 64, 279–304.

Dixmier, S. and Zara, F. (1976). Etude d'un quadrangle généralisé autour de deux de ses points non-liés. Preprint.

Dorwart, H. L. (1985). Configurations: a case study in mathematical beauty. *Math. Intelligencer* 7, 39–48.

Drake, D. A. (1985). A covering theorem for Baer subplanes in a cyclic projective plane. *Geom. Dedicata* 18, 181–90.

—— (1990). A bound for blocking sets in finite projective planes. *Finite geometries and combinatorial designs*, American Mathematical Society, Providence, RI, pp. 93–7.

—— and Ho, C. Y. (1988). Projective extensions of Kirkman systems as substructures of projective planes. *J. Combin. Theory Ser. A* 48, 197–208.

Driencourt, Y. (1987a). Un exemple de codes géométriques: les codes elliptiques. *Traitement Signal* 4, 147–53.

—— (1987b). Elliptic codes over fields of characteristic 2. *J. Pure Appl. Algebra* 45, 15–39.

Dur, A. (1987). The automorphism groups of Reed–Solomon codes. *J. Combin. Theory Ser. A* 44, 69–82.

—— (1988). On linear MDS codes of length $q + 1$ over $GF(q)$ for even q. *J. Combin. Theory Ser. A* 49, 172–4.

Du Val, P. (1979). Beniamino Segre. *Bull. London Math. Soc.* 11, 215–35.

Dye, R. H. (1970a). On the transitivity of the orthogonal and symplectic groups in projective space. *Proc. Cambridge Philos. Soc.* 68, 33–43.

—— (1977b). Partitions and their stabilizers for line complexes and quadrics. *Ann. Mat. Pura Appl.* 114, 173–94.

—— (1985). Maximal subgroups of the finite orthogonal and unitary groups stabilizing anisotropic subspaces. *Math. Z.* 189, 111–29.

—— (1986a). Scherk's theorem on orthogonalities revisited. *Geom. Dedicata* 20, 349–56.

—— (1986b). Maximal subgroups of $PSp_{6n}(q)$ stabilizing spreads of totally isotropic planes. *J. Algebra* 99, 191–209.

—— (1986c). Maximal subgroups of finite orthogonal groups stabilizing spreads of lines. *J. London Math. Soc.* 33, 279–93.

—— (1988a). Twelve hexagons associated with the 10-point conic and the isomorphism $PSL_2(9) \cong A_6$. *J. London Math. Soc.* 37, 437–46.

—— (1988b). A quick geometrical proof that $G_2(K)$ is maximal in $P\Omega_7(K)$. *Geom. Dedicata* 26, 361–4.

—— (1991). Hexagons, conics, A_5 and $PSL_2(K)$. *J. London Math. Soc.*

Ealy, C. E. (1977). Generalized quadrangles and odd transpositions. *Ph.D. thesis*, University of Chicago.

Ebert, G. L. (1978). Amusing configurations of disjoint circles. *Ars Combin.* 6, 197–207.

—— (1985a). Partitioning projective geometries into caps. *Canad. J. Math.* 37, 1163–75.

—— (1985b). Spreads obtained from ovoidal fibrations. *Finite geometries*, Dekker, New York, pp. 117–26.

—— (1985c). The completion problem for partial packings. *Geom. Dedicata* 18, 261–7.

—— (1988). Nests, covers, and translation planes. *Ars Combin.* 25, 213–33.

—— (1989). Spreads admitting regular elliptic covers. *European J. Combin.* 10, 319–30.

Edge, W. L. (1985). $PGL(2, 11)$ and $PSL(2, 11)$. *J. Algebra* 97, 492–504.

Egawa, Y. (1985). Association schemes of quadratic forms. *J. Combin. Theory Ser. A* 38, 1–4.

Eugeni, F. (1984). Sulla esistenze di t-fibrazioni in $PG(r, q)$ di fissato tipo. *Boll. Un. Mat. Ital.* 3-D, 19–43.

—— and Innamorati, S. (1986). On sets of fixed parity in Steiner systems. *Ann. Discrete Math.* 37, 157–68.

—— and —— (1987). Arcs and blocking sets in symmetric designs. *Ars Combin.* 24 A, 29–45.

—— and Mayer, E. (1986). Blocking sets of index two. *Ann. Discrete Math.* 37, 169–76.

Faina, G. (1985). The subgroup generated by regular involutions of a doubly transitive B-oval. *Rend. Sem. Mat. Univ. Padova* 74, 139–45.

—— (1986). Pascalian configurations in projective planes. *Ann. Discrete Math.* 30, 203–15.

—— (1989). Complete caps having less than $(q^2 + 1)/2$ points in common with an elliptic quadric of $PG(3, q)$, q odd. *Rend. Mat. Appl.* 8, 277–81.

—— and Korchmáros, G. (1989). On sharply 3-transitive permutation sets. *Atti Sem. Mat. Fis. Univ. Modena* 37, 95–103.

Farinola, A. and Leuci, M. (1983). Su un problema di Bumcrot. *Note Mat.* 3, 55–82.

Farmer, K. B. and Hale, M. P. (1980). Dual affine geometries and alternative bilinear forms. *Linear Algebra Appl.* 30, 183–99.

Feit, W. (1990). Finite projective planes and a question about primes. *Proc. Amer. Math. Soc.* 108, 561–4.

—— and Higman, G. (1964). The nonexistence of certain generalized polygons. *J. Algebra* 1, 114–31.

Fellagara, G. (1962). Gli ovaloidi di uno spazio tridimensionale di Galois di ordine 8. *Atti Accad. Naz. Lincei Rend.* 32, 170–6.

Feng, G. L. (1982). A discriminant for a polynomial of degree 3 over $GF(2^m)$ has three distinct roots in $GF(2^m)$ (Chinese). *J. Math.* (*Wuhan*) 2, 307–12.

Feng, X. N. and Dai, Z. D. (1965). Studies in finite geometries and the construction of incomplete block designs. V. Some 'Anzahl' theorems in orthogonal geometry over finite fields of characteristic 2. *Chinese Math.* 7, 392–410.

Fernandes, O. (1985). Harmonic points and the intersections of ovals. *Geom. Dedicata* 19, 271–6.

Ferri, O. (1976). Su di una caratterizzazione grafica della superficie di Veronese di un $S_{5,q}$. *Atti Accad. Naz. Lincei Rend.* 61, 603–10.

—— (1980a). Le calotte a due caratteri rispetto ai piani in uno spazio di Galois $S_{3,q}$. *Riv. Mat. Univ. Parma* 6, 55–63.
—— (1985). Una caratterizzazione di q^2-calotte in $AG(3, q)$. *Rend. Mat.* 5, 285–9.
—— (1986). Su una caratterizzazione dei paraboloidi iperbolici di $AG(3, q)$. *Rend. Mat.* 6, 239–43.
—— (1987). Proprietà grafiche caratterizzando gli iperboloidi iperbolici di $AG(3, q)$. *Rend. Mat.* 7, 131–7.
Fisher, J. C. (1988). Conics, order, and k-arcs in $AG(2, q)$ with q odd. *J. Geom.* 32, 21–40.
——, Hirschfeld, J. W. P., and Thas, J. A. (1986). Complete arcs in planes of square order. *Ann. Discrete Math.* 30, 243–50.
Fisher, P. H., Penttila, T., Praeger, C. E., and Royle, C. F. (1989). Inversive planes of odd order. *European J. Combin.* 10, 331–6.
Földes, S. (1990). Decidability of configuration theorems in projective planes and other incidence structures. *Ars Combin.* 29, 277–81.
Fong, P. and Seitz, G. (1973). Groups with a BN-pair of rank 2, I. *Invent. Math.* 21, 1–57.
—— and —— (1974). Groups with a BN-pair of rank 2, II. *Invent. Math.* 24, 191–239.
Fowler, J. C. (1985). A note on complete (k, μ)-arcs. *Congr. Numer.* 48, 101–3.
—— (1987). On intersecting maximal arcs. *Ars Combin.* 24, 179–83.
Frankl, P. (1986). Finite projective spaces and intersecting hypergraphs. *Combinatorica* 6, 335–54.
—— and Graham, R. L. (1985). Intersection theorems for vector spaces. *European J. Combin.* 6, 183–7.
Freudenthal, H. (1975). Une étude de quelques quadrangles généralisés. *Ann. Mat. Pura Appl.* 102, 109–33.
Fuji-Hara, R. (1986). Mutually 2-orthogonal resolutions of finite projective space. *Ars Combin.* 21, 179–87.
—— (1988). Hyperplane skew resolutions and their applications. *J. Combin. Theory Ser. A* 47, 134–44.
——, Jimbo, M., and Vanstone, S. (1986). Some results on the line partitioning problem in $PG(2k, q)$. *Utilitas Math.* 30, 235–42.
Füredi, Z. (1988). Matchings and covers in hypergraphs. *Graphs Combin.* 4, 115–206.
—— (1989). A projective plane is an outstanding 2-cover. *Discrete Math.* 74, 321–4.
—— (1990). Covering pairs by $q^2 + q + 1$ sets. *J. Combin. Theory Ser. A* 54, 248–71.
—— and Rosenberg, I. G. (1988). Multicoloured lines in a finite geometry. *Discrete Math.* 71, 149–63.
Games, R. A. (1986a). The geometry of quadrics and correlations of sequences. *IEEE Trans. Inform. Theory* 32, 423–6.
—— (1986b). The geometry of m-sequences: three valued cross correlations and quadrics in finite projective geometry. *SIAM J. Algebraic Discrete Methods* 7, 43–52.
Garcia, A. and Viana, P. (1986). Weierstrass points on certain non-classical curves. *Arch. Math.* 46, 315–22.
—— and Voloch, J. F. (1987). Wronskians and linear independence over finite fields. *Manuscripta Math.* 59, 457–69.
—— and —— (1988). Fermat curves over finite fields. *J. Number Theory* 30, 345–56.
Gevaert, H. and Johnson, N. L. (1988). On maximal partial spreads in $PG(3, q)$ of cardinalities $q^2 - q + 1$, $q^2 - q + 2$. *Ars Combin.* 26, 191–6.

——, ——, and Thas, J. A. (1988). Spreads covered by reguli. *Simon Stevin* 62, 51–62.
Ghinelli-Smit, D. (1987). A new result on difference sets with -1 as multiplier. *Geom. Dedicata* 23, 309–17.
——, Melone, N., and Ott, U. (1989). On abelian cubic arcs. *Geom. Dedicata* 32, 31–52.
Gluck, D. (1990a). Affine planes and permutation polynomials. *IMA Vol. Math. Appl.* 21, Springer, New York, pp. 97–100.
—— (1990b). A note on permutation polynomials and finite geometries. *Discrete Math.* 80, 97–100.
Glynn, D. G. (1983). On the characterization of certain sets of points in finite projective geometry of dimension three. *Bull. London Math. Soc.* 15, 31–4.
—— (1984a). The Hering classification for inversive planes of even order. *Simon Stevin* 58, 319–53.
—— (1984b). A representation of $PSp(4, q)$, q even. *Simon Stevin* 58, 87–90.
—— (1985). An introduction to half-planes. *Ars Combin.* 19, 309–42.
—— (1986). The non-classical 10-arc of $PG(4, 9)$. *Discrete Math.* 59, 43–51.
—— (1987a). On finite division algebras. *J. Combin. Theory Ser. A* 44, 253–66.
—— (1987b). Rings of geometries I. *J. Combin. Theory Ser. A* 44, 34–48.
—— (1988a). On a set of lines of $PG(3, q)$ corresponding to a maximal cap contained in the Klein quadric of $PG(5, q)$. *Geom. Dedicata* 26, 273–80.
—— (1988b). Rings of geometries II. *J. Combin. Theory Ser. A* 49, 26–66.
—— (1989). A condition for the existence of ovals in $PG(2, q)$, q even. *Geom. Dedicata* 32, 247–52.
Godeaux, L. (1948). *Géométrie algébrique* (two volumes). Sciences et Lettres, Liège.
Gomez-Calderon, G. and Madden, D. J. (1988). Polynomials with small value set over finite fields. *J. Number Theory* 28, 167–88.
Goppa, V. D. (1984). Codes and information. *Russian Math. Surveys* 39, 87–141.
—— (1988). *Geometry and codes* (translated by N. G. Shartse). Kluwer, Dordrecht.
Gordon, C. and Killgrove, R. (1990). Representative arcs in field planes of prime order. *Congr. Numer.* 71, 73–85.
Graham, R. L. and Rothschild, B. (1971). Rota's geometric analogue to Ramsey's theorem. *Combinatorics*, American Mathematical Society, Providence, RI, pp. 101–4.
——, Leeb, K., and Rothschild, B. L. (1972). Ramsey's theorem for a class of categories. *Adv. in Math.* 8, 417–33.
Greene, J. (1987). Lagrange inversion over finite fields. *Pacific J. Math.* 130, 313–25.
Griffith, G. J. (1981). The 'Coxeter curves' $x^{2/p} + y^{2/p} + z^{2/p} = 0$ for prime values of p. *J. Geom.* 17, 24–34.
Grundhöfer, T. (1986). A synthetic construction of the Figueroa planes. *J. Geom.* 26, 191–201.
Grüning, K. (1987). A class of unitals of order q which can be embedded in two different planes of order q^2. *J. Geom.* 29, 61–77.
Haemers, W. and Roos, C. (1981). An inequality for generalized hexagons. *Geom. Dedicata* 10, 219–22.
—— and van Lint, J. H. (1982). A partial geometry $pg(9, 8, 4)$. *Ann. Discrete Math.* 15, 205–12.
Halder, H.-R. and Heise, W. (1974). On the existence of finite chain-m-structures and k-arcs in finite projective space. *Geom. Dedicata* 3, 483–6.

Hall, J. I. (1982). Classifying copolar spaces and graphs. *Quart. J. Math. Oxford Ser.* 33, 421–49.

—— (1983b). Linear representations of cotriangular spaces. *Linear Algebra Appl.* 49, 257–73.

—— (1988). The hyperbolic lines of finite symplectic spaces. *J. Combin. Theory Ser. A* 47, 284–98.

Hall, M. (1971). Affine generalized quadrilaterals. *Studies in pure mathematics*, Academic Press, London, pp. 113–16.

—— (1972). Incidence axioms for affine geometry. *J. Algebra* 21, 535–47.

Hamada, N. (1987a). Characterization of min.hypers in a finite projective geometry and its applications to error-correcting codes. *Bull. Osaka Women's Univ.* 24, 1–24.

—— (1987b). Characterization of $\{12, 2; 2, 4\}$-min.hypers in a finite projective geometry $PG(2, 4)$. *Bull. Osaka Women's Univ.* 24, 25–31.

—— (1989a). Characterization of $\{(q+1)+2, 1; t, q\}$-min.hypers and $\{(2(q+1)+2, 2; 2, q\}$-min.hypers in a finite projective geometry. *Graphs Combin.* 5, 63–81.

—— (1989b). Characterization of $\{v_{\mu+1}+2v_\mu, v_\mu+2v_{\mu-1}; t, qb\}$-min.hypers and its applications to error-correcting codes. *Graphs Combin.* 5, 137–47.

—— and Deza, M. (1988a). Characterization of $\{2(q+1)+2, 2; t, q\}$-min.hypers in $PG(t, q)$ $(t \geq 3,\ q \geq 5)$ and its applications to error-correcting codes. *Discrete Math.* 71, 219–31.

—— and —— (1988b). A characterization of some $(n, k, d; q)$-codes meeting the Griesmer bound for given integers $k \geq 3$, $q \geq 5$ and $d = q^{k-1} - q^\alpha - q^\beta - q^\gamma$ $(0 \leq \alpha \leq \beta < \gamma < k-1$ or $0 \leq \alpha < \beta \leq \gamma < k-1)$. *Bull. Inst. Math. Acad. Sinica* 16, 321–38.

—— and —— (1989a). A characterization of $\{v_{\mu+1}+\varepsilon, v_\mu; t, q\}$-min.hypers and its applications to error-correcting codes and factorial designs. *J. Statist. Plann. Inference* 22, 323–36.

—— and —— (1989b). A survey of recent works with respect to a characterization of an $(n, k, d; q)$-code meeting the Griesmer bound using a min.hyper in a finite projective geometry. *Discrete Math.* 77, 75–87.

—— and Helleseth, T. (1990). A characterization of some minihypers in a finite projective geometry $PG(t, 4)$. *European J. Combin.* 11, 541–8.

Hanssens, G. (1984). Punt-rechte meetkunden van sferische gebouwen. *Ph.D. thesis*, State University of Ghent.

—— (1985). On a problem of partial linear spaces arising from the classification theory of point–line geometries. *Ars Combin.* 19, 17–24.

—— (1986). A characterization of buildings of spherical type. *European J. Combin.* 7, 333–47.

—— and Thas, J. A. (1987). Pseudopolar spaces of polar rank three. *Geom. Dedicata* 22, 117–35.

—— and Van Maldeghem, H. (1988). Coordinatization of generalized quadrangles. *Ann. Discrete Math.* 37, 195–207.

Heden, O. (1986). Maximal partial spreads and two-weight codes. *Discrete Math.* 62, 277–93.

—— (1990). No maximal partial spread of size 10 in $PG(3, 5)$. *Ars Combin.* 29, 297–8.

Hefez, A. (1989). Non-reflexive curves. *Compositio Math.* 69, 3–35.

—— and Voloch, J. F. (1990). Frobenius nonclassical curves. *Arch. Math.* 54, 263–73.

Heider, F.-P. and Kolvenbach, P. (1984). The construction of $SL(2, 3)$-polynomials. *J. Number Theory* 19, 392–411.

Heise, W. (1974). Minkowski-Ebenen gerader Ordnung. *J. Geom.* 5, 83.
—— (1976). Optimal codes, *n*-arcs and Laguerre geometry. *Acta Inform.* 6, 403–6.
Herzer, A. (1982). Die Schmieghyperebenen an die Veronese-Mannigfaltigkeit bei beliebiger Charakteristik. *J. Geom.* 18, 140–54.
—— (1984). On a projective representation of chain geometries. *J. Geom.* 22, 83–99.
—— and Lunardon, G. (1984). Una caratterizzazione delle fibrazioni di Galois. *Atti Accad. Naz. Lincei Rend.* 77, 151–4.
Higman, D. G. (1971). Partial geometries, generalized quadrangles and strongly regular graphs. *Geometria combinatoria e sue applicazioni*, University of Perugia, Perugia, pp. 263–93.
—— (1974). Invariant relations, coherent configurations and generalized polygons. *Combinatorics, part 3*, Math. Centre Tracts 57, Amsterdam, pp. 27–43.
—— (1978). (Appendix by D. E. Taylor). *Classical groups.* Department of Mathematics, Technological University, Eindhoven.
Hill, R. (1973). On the largest size of cap in $S_{5,3}$. *Atti Accad. Naz. Lincei Rend.* 54, 378–84.
—— (1978a). Caps and codes. *Discrete Math.* 22, 111–37.
—— (1983). On Pellegrino's 20-caps in $S_{4,3}$. *Ann. Discrete Math.* 18, 433–47.
—— (1986). *A first course in coding theory.* Oxford University Press, Oxford.
Hiramine, Y. (1989). A conjecture on affine planes of prime order. *J. Combin. Theory Ser. A* 52, 44–50.
Hirschfeld, J. W. P. (1979). *Projective geometries over finite fields.* Oxford University Press, Oxford.
—— (1983a). Caps in elliptic quadrics. *Ann. Discrete Math.* 18, 449–66.
—— (1984). Linear codes and algebraic curves. *Geometrical combinatorics*, Research Notes in Mathematics 114, Pitman, London, pp. 35–53.
—— (1985). *Finite projective spaces of three dimensions.* Oxford University Press, Oxford.
—— (1986). Quadrics over finite fields. *Combinatorica*, Symp. Math. 28, Academic Press, London, pp. 53–87.
—— (1990a). Codes and curves. *Finite buildings, related geometries and applications*, Oxford University Press, Oxford, pp. 129–44.
—— (1990b). Hermitian varieties. *Mitt. Math. Sem. Giessen.*
—— (1991). Projective spaces of square size. *Simon Stevin.*
—— and Szönyi, T. (1991a). Sets in a finite plane with few intersection numbers and a distinguished point. *Discrete Math.*
—— and —— (1991b). A problem on squares in a finite field and its applications to geometry. *Advances in finite geometries and designs*, Oxford University Press, Oxford, pp. 169–76.
—— and Thas, J. A. (1980a). Sets of type $(1, n, q+1)$ in $PG(d, q)$. *Proc. London Math. Soc.* 41, 254–78.
—— and —— (1980b). The characterization of projections of quadrics over finite fields of even order. *J. London Math. Soc.* 22, 226–38.
—— and —— (1985). The generalized hexagon $H(q)$ and the associated generalized quadrangle $K(q)$. *Simon Stevin* 59, 407–35.
—— and —— (1987). Linear independence in finite spaces. *Geom. Dedicata* 23, 15–31.
—— and —— (1990). Sets with more than one representation as an algebraic curve of degree three. *Finite geometries and combinatorial designs*, American Mathematical Society, Providence, RI, pp. 99–110.

—— and Voloch, J. F. (1988). The characterization of elliptic curves over finite fields. *J. Austral. Math. Soc. Ser. A* 45, 275–86.

——, Hughes, D. R., and Thas, J. A. (Eds) (1991). *Advances in finite geometries and designs.* Oxford University Press, Oxford.

——, Storme, L., Thas, J. A., and Voloch, J. F. (1991). A characterization of Hermitian curves. *J. Geom.* 41, 72–7.

Ho, C. Y. (1984). Hamming spaces and maximal self dual codes over $GF(q)$, q = odd. *Bol. Soc. Brasil. Mat.* 15, 7–24.

—— (1986a). Characterization of projective planes of small prime orders. *J. Combin. Theory Ser. A* 41, 189–200.

—— (1986b). Involutory collineations of finite planes. *Math. Z.* 193, 235–40.

—— (1989). On multiplier groups of finite cyclic planes. *J. Algebra* 122, 250–9.

—— (1990). On the order of a finite projective plane and its collineation group. *Finite geometries and combinatorial designs*, American Mathematical Society, Providence, RI, pp. 299–301.

Hodge, W. V. D. and Pedoe, D. (1947). *Methods of algebraic geometry, volume I.* Cambridge University Press, Cambridge.

—— and —— (1953). *Methods of algebraic geometry, volume II.* Cambridge University Press, Cambridge.

—— and —— (1954). *Methods of algebraic geometry, volume III.* Cambridge University Press, Cambridge.

Homma, M. (1989). A souped-up version of Pardini's theorem and its application to funny curves. *Compositio Math.* 71, 295–302.

Huang, J. F., Shiva, S. S., and Seguin, G. (1984). On certain projective geometry codes. *IEEE Trans. Inform. Theory* 30, 385–8.

Huber, M. (1985). A characterization of Baer cones in finite projective spaces. *Geom. Dedicata* 18, 197–211.

—— (1987). Baer cones in finite projective spaces. *J. Geom.* 28, 128–44.

Hughes, D. R. and Piper, F. C. (1985). *Design theory.* Cambridge University Press, Cambridge.

Ivanov, A. A. and Shpectorov, S. V. (1989). Characterization of the association schemes of Hermitian forms over $GF(2^2)$. *Geom. Dedicata* 30, 23–33.

Jagannathan, T. V. S. (1985). The ternary rings of Desarguesian and Pappian planes—a simple proof using perspectivities. *Geom. Dedicata* 18, 191–5.

Janko, Z. (1985). On planar collineations of order 13 acting on projective planes of order 16. *Rad. Mat.* 1, 163–72.

Janwa, H. (1990). Some optimal codes from algebraic geometry and their covering radii. *European J. Combin.* 11, 249–66.

Jepsen, C. H. (1987). On colourings of finite projective planes. *Discrete Math.* 63, 95–6.

Jeurissen, R. H. (1985). The Petersen graph as a box of Pandora. *Nieuw Arch. Wisk.* 3, 219–33.

Jha, V. (1985). On spreads in $PG(2, 2^s)$ that admit projective groups of order 2^s. *Proc. Edinburgh Math. Soc.* 29, 355–60.

—— and Johnson, N. L. (1986). On regular r-packings. *Note Mat.* 6, 121–37.

—— and —— (1989). Nests of reguli and flocks of quadratic cones. *Simon Stevin* 63, 311–38.

Johnson, B. (1986). Cryptography, prime numbers and the Riemann hypothesis. *Normat* 34, 2–16.

Johnson, N. L. (1987). Semifield flocks of quadratic cones. *Simon Stevin* 61, 3–4.

—— (1990). Flocks and partial flocks of quadric sets. *Finite geometries and combinatorial designs*, American Mathematical Society, Providence, RI, pp. 111–16.

Jordan, C. (1870). *Traité des substitutions et des équations algébriques*. Gauthier-Villars, Paris.

Jungnickel, D. (1984). Maximal partial spreads and translation nets of small deficiency. *J. Algebra* 90, 119–32.

—— (1987). Divisible semiplanes, arcs, and relative difference sets. *Canad. J. Math.* 39, 1001–24.

—— (1989a). Some self-blocking block designs. *Discrete Math.* 77, 123–35.

—— (1989b). Partial spreads over $\mathbb{Z}_q$. *Linear Algebra Appl.* 14, 95–102.

—— (1989c). An elementary proof of Wilbrink's theorem. *Arch. Math.* 52, 615–17.

—— (1989d). Design theory: an update. *Ars Combin.* 28, 129–99.

—— and Vedder, K. (1984). Generalized homologies. *Mitt. Math. Sem. Giessen* No. 166, 103–25.

—— and —— (1987). Simple quasidoubles of projective planes. *Aequationes Math.* 43, 96–100.

Justesen, J., Larsen, K. L., Jansen, H. E., Havemose, A., and Hoholt, T. (1988). Construction and decoding of a class of algebraic geometry codes. *IEEE Trans. Inform. Theory* 35, 811–21.

Kahn, J. (1984). A geometric approach to forbidden minors for $GF(3)$. *J. Combin. Theory Ser. A* 37, 1–12.

Kaneta, H. and Maruta, T. (1989a). An elementary proof and extension of Thas' theorem on k-arcs. *Math. Proc. Cambridge Philos. Soc.* 105, 459–62.

—— and —— (1989b). An algebraic geometrical proof of the extendability of q-arcs in $PG(3, q)$ with q even. *Simon Stevin* 64, 363–6.

—— and —— (1991). The discriminant of a cubic curve. *Advances in finite geometries and designs*, Oxford University Press, Oxford, pp. 237–49.

Kantor, W. M. (1982a). Spreads, translation planes and Kerdock sets. I. *SIAM J. Algebraic Discrete Methods* 3, 151–65.

—— (1982b). Spreads, translation planes and Kerdock sets. II. *SIAM J. Algebraic Discrete Methods* 3, 308–18.

—— (1982e). Ovoids and translation planes. *Canad. J. Math.* 34, 1195–1203.

—— (1985a). Generalized quadrangles and translation planes. *Algebras, Groups and Geometries* 2, 313–22.

—— (1985b). Homogeneous designs and geometric lattices. *J. Combin. Theory Ser. A* 38, 66–74.

—— (1985c). Flag-transitive planes. *Finite geometries*, Dekker, New York, pp. 179–81.

—— (1986). Some generalized quadrangles with parameters q^2, q. *Math. Z.* 192, 45–50.

—— (1987). Some locally finite flag-transitive buildings. *European J. Combin.* 8, 429–36.

Karzel, H. (1987). Uber einen Fundamentalsatz der synthetischen algebraischen Geometrie von W. Burau und H. Timmermann. *J. Geom.* 28, 86–101.

Katsmann, G. L., Tsfasman, M. A., and Vladut, S. G. (1984). Modular curves and codes with a polynomial construction. *IEEE Trans. Inform. Theory* 30, 353–5.

Keedwell, A. D. (1988). Simple constructions for elliptic cubic curves with specified small numbers of points. *European J. Combin.* 9, 463–81.

—— (1989). A theorem concerning the embedding of graphic arcs in algebraic plane curves. *Ann. New York Acad. Sci.* 555, 241–7.

Kestenband, B. C. (1982). Finite projective geometries that are incidence structures of caps. *Linear Algebra Appl.* 48, 303–13.

—— (1986). Balanced incomplete block designs on $q + 1$ elements. *J. Statist. Plann. Inference* 13, 45–50.

—— (1988). Partitioning projective planes into arcs. *Math. Proc. Cambridge Philos. Soc.* 104, 435–40.

—— (1989a). Partitions of Desarguesian and Hughes planes into complete and incomplete arcs. *J. Geom.* 36, 91–8.

—— (1989b). A family of complete arcs in finite projective planes. *Colloq. Math.* 57, 59–67.

—— (1990). Correlations whose squares are projectivities. *Geom. Dedicata* 33, 289–315.

Key, J. D. (1987). Incidence structures with regular sets. *Congr. Numer.* 60, 211–20.

—— (1988). Regular sets in geometries. *Ann. Discrete Math.* 37, 217–23.

—— and MacKenzie, K. (1991). An upper bound for the p-rank of a translation plane. *J. Combin. Theory Ser. A* 65, 297–302.

—— and Siemons, J. (1987). Regular sets and geometric groups. *Resultate Math.* 11, 97–116.

——, ——, and Wagner, A. (1986). Regular sets on the projective line. *J. Geom.* 27, 188–94.

Killgrove, R., Sternfeld, R., and Gordon, C. (1990). Arcs in projective planes and MOLS. *Congr. Numer.* 72, 61–70.

Kitto, C. (1989). A bound for blocking sets of Redei type in finite projective planes. *Arch. Math.* 52, 203–8.

Kiyek, K. (1985). Einfache Kurvensingularitäten in beliebiger Charakteristik. *Arch. Math.* 45, 565–73.

Kleidman, P. B. (1988). The 2-transitive ovoids. *J. Algebra* 117, 117–35.

Koblitz, N. (1980). The p-adic approach to solutions of equations over finite fields. *Amer. Math. Monthly* 87, 45–60.

—— (1984). *Introduction to elliptic curves and modular forms.* Springer, New York.

—— (1989). Three practical applications of algebraic curves over finite fields. *Sichuan Daxue Xuebao* 26, 28–35.

Konvalina, J. (1985). A note on models of finite geometries arising from difference sets. *European J. Combin.* 6, 189–91.

Korchmáros, G. (1986a). Inherited arcs in affine planes. *J. Combin. Theory Ser. A* 42, 140–3.

—— (1986b). Recenti risultati di geometria combinatoria. *Combinatorica*, Symp. Math. 28, Academic Press, London, pp. 113–25.

—— (1991). Collineation groups of $(q + t, t)$-arcs of type $(0, 2, t)$. *Advances in finite geometries and designs*, Oxford University Press, Oxford, pp. 257–64.

—— and Mazzocca, F. (1990). On $(q + t)$-arcs of type $(0, 2, t)$ in a Desarguesian plane of order q. *Math. Proc. Cambridge Philos. Soc.* 108, 445–59.

Koyima, K. (1978). Algebraic curves with non-classical types of gap sequences for genus three and four. *Hiroshima Math. J.* 8, 371–400.

Kramer, E. (1984). A new result about projective planes of order 20. *Rad. Jugoslav. Akad. Znan. Umjet.* 413, 31–7.

—— (1985). On projective planes of order 20. *Rad. Mat.* 1, 173–7.

Kramer, E. S. (1984). The t-designs using $M_{21} \simeq PSL_3(4)$ on 21 points. *Ars Combin.* 17, 79–90.

Kugurakov, V. S. (1986). Remark on the solution of quadratic equations in a finite field of characteristic 2. *Probabilistic Methods and Cybernetics* 21, 107–8.

Lachaud, G. (1986). Les codes géométriques de Goppa. *Séminaire Bourbaki* 1984/85, *Astérisque* No. 133–4, 189–207.

—— (1987). Sommes d'Eisenstein et nombre de points de certaines courbes algébriques sur les corps finis. *C.R. Acad. Sci. Paris Ser. I Math.* 305, 729–32.

—— (1989). Exponential sums and the Carlitz–Uchiyama bound. Lecture Notes in Comput. Sci. 388, Springer, New York, pp. 63–75.

—— and Wolfmann, J. (1987). Sommes de Kloosterman, courbes elliptiques et codes cycliques en caractéristique 2. *C.R. Acad. Sci. Paris Ser. I Math.* 305, 881–3.

Lam, C. W. H., Crossfield, S., and Thiel, L. H. (1985). Estimates of a computer search for a projective plane of order 10. *Congr. Numer.* 48, 253–63.

——, Thiel, L. H., and Swiercz, S. (1986). The non existence of code words of weight 16 in a projective plane of order 10. *J. Combin. Theory Ser. A* 42, 207–14.

——, ——, and —— (1988). A computer search for a projective plane of order 10. *London Math. Soc. Lecture Note Series* 131, Cambridge University Press, Cambridge, pp. 155–65.

——, ——, and —— (1989). The nonexistence of finite projective planes of order 10. *Canad. J. Math.* 41, 1117–23.

Larato, B. (1983). Una caratterizzazione degli unitals parabolici di Buekenhout–Metz. *Matematiche* (Catania) 38, 95–8.

Laskar, R. C. and Sherk, F. A. (1985). Generating sets in finite projective planes. *Finite geometries*, Dekker, New York, pp. 183–97.

Lefèvre(-Percsy), C. (1975). Tallini sets in projective spaces. *Atti Accad. Naz. Lincei Rend.* 59, 392–400.

—— (1977b). Sur les semi-quadriques en tant qu'espaces de Shult projectifs. *Acad. Roy. Belg. Bull. Cl. Sci.* 63, 160–4.

—— (1980). An extension of a theorem of G. Tallini. *J. Combin. Theory Ser. A* 29, 297–305.

—— (1981c). Espaces polaires dégénérés des espaces projectifs. *Simon Stevin* 55, 237–46.

—— (1981d). Projectivités conservant un espace polaire faiblement plongé. *Acad. Roy. Belg. Bull. Cl. Sci.* 67, 45–50.

—— (1981e). Quadrilatères généralisés faiblement plongés dans $PG(3, q)$. *European J. Combin.* 2, 249–55.

—— (1981g). Espaces polaires faiblement plongés dans un espace projectif. *J. Geom.* 16, 126–37.

—— (1985). Characterization of finite Hermitian semi-quadrics. *J. London Math. Soc.* 31, 150–62.

Lenstra, H. W. and Schoof, R. J. (1987). Primitive normal bases for finite fields. *Math. Comp.* 48, 217–31.

Lenz, H. (1989). Variations on the projective plane of order four. *Mitt. Math. Sem. Giessen* No. 192, 79–84.

Limbos, M. (1981). A characterization of the embeddings of $PG(m, q)$ into $PG(n, q^r)$. *J. Geom.* 16, 50–5.

Litsyn, S. N. and Tsfasman, M. A. (1986). A note on lower bounds. *IEEE Trans. Inform. Theory* 32, 705–6.

Liu, T. and Wan, Z. (1991). Pseudo symplectic geometries over finite fields of characteristic 2. *Advances in finite geometries and designs*, Oxford University Press, Oxford, pp. 265–88.

Lopez, A. V. and Lopez, J. V. (1985). Classification of finite groups according to the number of conjugacy classes. *Israel J. Math.* 51, 305–38.

Lord, E. A. (1988). Geometry of the Mathieu groups and Golay codes. *Proc. Indian Acad. Sci. Math. Sci.* 98, 153–77.

Lo Re, P. M. and Olanda, D. (1981). Grassmann spaces. *J. Geom.* 17, 50–60.

—— and —— (1986). On $\{0, 2\}$-semiaffine planes. *Simon Stevin* 60, 157–82.

—— and —— (1988). On $\{1, 2, s\}$-semiaffine planes. *Simon Stevin* 62, 29–49.

Lorimer, P. (1974). A projective plane of order 16. *J. Combin. Theory Ser. A* 16, 334–47.

Luisi, G. and Rella, L. (1984). Weighted line spaces. *Rend. Mat.* 4, 257–64.

Lunardon, G. (1984a). Fibrazioni planari e sottovarietà algebriche della varietà di Grassmann. *Geom. Dedicata* 16, 291–313.

—— (1984b). Insiemi indicatori proiettivi e fibrazioni planari di uno spazio proiettive finite. *Boll. Un. Mat. Ital. B* 3, 717–35.

—— (1984c). On regular parallelisms in $PG(3, q)$. *Discrete Math.* 561, 229–35.

—— (1986). Varietà di Segre e ovoidi dello spazio polare $Q^+(7, q)$. *Geom. Dedicata* 20, 121–31.

McEliece, J. R. (1987). *Finite fields for computer scientists and engineers.* Kluwer, Boston.

MacWilliams, F. J. K. and Sloane, N. J. A. (1977). *The theory of error-correcting codes.* North-Holland, Amsterdam.

Magliveras, S. S. and Leavitt, D. W. (1984). Simple 6-(33, 8, 36) designs from $P\Gamma L_2(32)$. *Computational group theory*, Academic Press, London, pp. 337–52.

Manickam, N. (1987). W-complement d-spreads, Singleton systems. *European J. Combin.* 8, 437–9.

Manin, Y. I. and Tsfasman, M. A. (1986). Rational varieties: algebra, geometry and arithmetic. *Russian Math. Surveys* 41, 51–116.

Maruta, T. (1990). On Singleton arrays and Cauchy matrices. *Discrete Math.* 81, 33–6.

—— (1991). A geometric approach to semi-cyclic codes. *Advances in finite geometries and designs*, Oxford University Press, Oxford, pp. 311–18.

—— and Kaneta, H. (1991). On the uniqueness of $(q + 1)$-arcs of $PG(5, q)$, $q = 2^h$, $h \geq 4$. *Math. Proc. Cambridge Philos. Soc.* 110, 91–4.

Masol, V. I. (1989). Some applications of algorithms for constructing subspaces over a finite field (Russian). *Ukrain. Mat. Zh.* 1146–8.

Matulic-Bedenic, I. (1984). Projective planes of order 11 with a collineation group of order 5. *Rad. Jugoslav. Akad. Znan. Umjet.* 413, 39–43.

—— (1985). The classification of projective planes of order 11 which possess an involution. *Rad. Mat.* 1, 149–57.

Maurras, J. F. (1983). Sur une propriété extrème des plans projectifs finis. *Ann. Discrete Math.* 17, 459–63.

Mazzocca, F. (1974a). Immergibilità in $S_{4,q}$ di certi sistemi rigati di seconda specie. *Atti Accad. Naz. Lincei Rend.* 56, 189–96.

—— (1974b). Caratterizzazione dei sistemi rigati isomorfi ad una quadrica ellittica dello $S_{5,q}$, con q dispari. *Atti Accad. Naz. Lincei Rend.* 57, 360–8.

—— (1991). Blocking sets with respect to special families of lines and nuclei of Θ_n-sets in finite n-dimensional projective and affine spaces.

—— and Melone, N. (1984). Caps and Veronese varieties in projective Galois spaces. *Discrete Math.* 48, 243–52.

—— and Olanda, D. (1979b). Sistemi rigati in spazi combinatori. *Rend. Mat.* 12, 221–9.

—— and —— (1983). A graphic characterization of the lines of an affine space. *Ann. Discrete Math.* 18, 625–34.

—— and Tallini, G. (1985). On the non existence of blocking-sets in $PG(n, q)$ and $AG(n, q)$, for all large enough n. *Simon Stevin* 59, 43–50.

Melone, N. (1983). Veronese spaces. *J. Geom.* 20, 169–80.

—— (1985). The linear line geometry in $PG(3, q)$ from a synthetic point of view. *Simon Stevin* 59, 305–19.

—— and Olanda, D. (1981). Spazi pseudoprodotto e varietà di C. Segre. *Rend. Mat.* 1, 381–97.

—— and —— (1984). A characteristic property of the Grassmann manifold representing the lines of a projective space. *European J. Combin.* 5, 323–30.

Mencerrey, Y. (1988). Miquelian inversive planes which admit an orthogonality both in the sense of Dembowski and Hughes and of Benz. *Geom. Dedicata* 27, 199–202.

Mendelsohn, N. S. (1985). A search for a non-Desarguesian plane of prime order. *Finite geometries*, Dekker, New York, pp. 199–208.

—— and Padmanabhan, R. (1989). Self-inscribed polygons with vertices on nonsingular cubic curves. *Linear Algebra Appl.* 114, 603–11.

——, ——, and Wolk, B. (1987a). Designs embeddable in a plane cubic curve. *Note Mat.* 7, 113–48.

——, ——, and —— (1987b). Planar projective configurations I. *Note Mat.* 7, 91–112.

——, ——, and —— (1987c). Some remarks on n-clusters on cubic curves. *Ann. Discrete Math.* 34, 371–8.

——, ——, and —— (1988). Straight edge constructions on planar cubic curves. *C.R. Math. Rep. Acad. Sci. Canada* 10, 77–82.

Menghini, M. (1987). Endliche Geometrien: Didaktische Aspekte. *Math. Semesterber.* 34, 44–60.

Menichetti, G. (1986). Roots of affine polynomials. *Ann. Discrete Math.* 30, 303–9.

Metsch, K. (1988). Embedding locally projective planar spaces into projective spaces. *Ann. Discrete Math.* 37, 293–5.

—— (1989). Embedding finite planar spaces into 3-dimensional projective spaces. *J. Combin. Theory Ser. A* 51, 161–8.

Michon, J.-F. (1989). Codes and curves. Lecture Notes in Comput. Sci. 357, Springer, Berlin, pp. 22–30.

Migliori, G. (1987). Insiemi di tipo $(0, 2, q/2)$ in un piano proiettivo e sistemi di terne di Steiner. *Rend. Mat.* 7, 77–82.

Mizuno, H. and Ando, K. (1989). Algebraic-geometric codes on X_s: $x^s y + y^s z + z^s x = 0$ (Japanese). *Bull. Univ. Electro-Comm.* 2, 297–304.

Mollin, R. A. (1987). On permutation polynomials over finite fields. *Internat. J. Math. Sci.* 10, 535–43.

Montaron, B. (1985). On incidence matrices of finite projective planes. *Discrete Math.* 56, 227–37.

Moorhouse, G. E. (1989). $PSL(2, q)$ as a collineation group of projective planes of small order. *Geom. Dedicata* 31, 63–88.

Moran, G. (1984). Chords in a circle and linear algebra over $GF(2)$. *J. Combin. Theory Ser. A* 37, 239–47.

Moreno, C. J. (1991). *Algebraic curves over finite fields*. Cambridge University Press, Cambridge.

Mullen, G. L. and Niederreiter, H. (1985). The structure of a group of permutation polynomials. *J. Austral. Math. Soc. Ser. A* 38, 164–70.

Namba, M. (1984). *Geometry of projective algebraic curves*. Dekker, New York.

Neumaier, A. (1984). Some sporadic geometries related to $PG(3, 2)$. *Arch. Math.* 42, 89–96.

Nöbauer, R. (1986). Uber die minimale Fixpunktanzahl von Dickson-Permutationen auf Galois-feldern. *Monatsh. Math.* 101, 193–210.

Novillo Sardi, J. (1985). The (16, 6, 2) biplane with 60 ovals and the weight distribution of a code. *European J. Combin.* 6, 193–7.

Numata, M. (1985). On the graphical characterization of the projective space over a finite field. *J. Combin. Theory Ser. B* 38, 143–55.

O'Brien, G. L. (1980). Pairwise independent random variables. *Ann. Probability* 8, 170–5.

Odlyzko, A. M. (1985). Discrete logarithms in finite fields and their cryptographic significance. *Advances in cryptology*, Lecture Notes in Comput. Sci. 209, Springer, New York, pp. 224–314.

O'Keefe, C. M. (1988). Regularity of 1-spreads of $PG(2s + 1, q)$. *Ars Combin.* 26, 113–24.

—— (1990). Ovals in Desarguesian planes. *Australas. J. Combin.* 1, 149–59.

—— and Penttila, T. (1990). Ovoids of $PG(3, 16)$ are elliptic quadrics. *J. Geom.* 38, 95–106.

—— and —— (1991a). Hyperovals in $PG(2, 16)$. *European J. Combin.* 12, 51–9.

—— and —— (1991b). Polynomials for hyperovals of Desarguesian planes. *J. Austral. Math. Soc. Ser. A*.

—— and Venezia, A. (1987). Blocking sets in $AG(r, 5)$. *Simon Stevin* 61, 179–88.

——, Penttila, T., and Praeger, C. E. (1991). Stabilisers of hyperovals in $PG(2, 32)$. *Advances in finite geometries and designs*, Oxford University Press, Oxford, pp. 337–57.

Olanda, D. (1973). Sistemi rigati immersi in uno spazio proiettivo. *Relazione* 26, Istituto di Matematica, Università di Napoli.

—— (1977). Sistemi rigati immersi in uno spazio proiettivo. *Atti Accad. Naz. Lincei Rend.* 62, 489–99.

—— (1988). Seminversive planes. *Ann. Discrete Math.* 37, 311–13.

Opencomb, W. E. (1984). On the intricacy of combinatorial construction problems. *Discrete Math.* 50, 71–97.

Orlik, P. and Solomon, L. (1985). Arrangements in unitary and orthogonal geometry over finite fields. *J. Combin. Theory Ser. A* 38, 217–29.

Ott, U. (1981). Eine Bemerkung über Polaritäten eines verallgemeinerten Hexagons. *Geom. Dedicata* 11, 341–5.

Padmanabhan, R. (1982). Logic of equality in geometry. *Ann. Discrete Math.* 15, 319–31.

—— (1985). Configuration theorems on cubic quasigroups. *Finite geometries*, Dekker, New York, pp. 209–21.

Pankratova, L. Y. (1980). Projective planes of order 9 containing subplanes of order 3. *Combin. Analysis* 5, 90–4.

Pardini, R. (1986). Some remarks on plane curves over fields of finite characteristic. *Compositio Math.* 60, 3–17.

Pasini, A. (1988). Geometric and algebraic methods in the classification of geometries belonging to Lie diagrams. *Ann. Discrete Math.* 37, 315–35.

Payne, S. E. (1973a). A restriction on the parameters of a subquadrangle. *Bull. Amer. Math. Soc.* 79, 747–8.

—— (1973b). Finite generalized quadrangles: a survey. *Projective planes*, Washington State University Press, Pullman, WA, pp. 219–61.

—— (1974). Generalized quadrangles of even order. *J. Algebra* 31, 367–91.

—— (1975). All generalized quadrangles of order 3 are known. *J. Combin. Theory* 18, 203–6.

—— (1977b). Generalized quadrangles of order 4, I. *J. Combin. Theory Ser. A* 22, 267–79.

—— (1977c). Generalized quadrangles of order 4, II. *J. Combin. Theory Ser. A* 22, 280–8.

—— (1978). An inequality for generalized quadrangles. *Proc. Amer. Math. Soc.* 71, 147–52.

—— (1985a). Hyperovals and generalized quadrangles. *Finite geometries*, Dekker, New York, pp. 251–70.

—— (1985b). A new infinite family of generalized quadrangles. *Congr. Numer.* 49, 115–28.

—— (1985c). A garden of generalized quadrangles. *Algebras, Groups and Geometries* 3, 323–54.

—— (1986). Hyperovals yield many *GQ*. *Simon Stevin* 60, 211–25.

—— (1987). Tight pointsets in finite generalized quadrangles. *Congr. Numer.* 60, 243–60.

—— (1988a). Spreads, flocks and generalized quadrangles. *J. Geom.* 33, 113–28.

—— (1988b). The Thas–Fisher generalized quadrangles. *Ann. Discrete Math.* 37, 357–66.

—— (1989). An essay on skew generalized quadrangles. *Geom. Dedicata* 32, 93–118.

—— and Thas, J. A. (1975). Generalized quadrangles with symmetry, Part I. *Simon Stevin* 49, 3–32.

—— and —— (1976). Generalized quadrangles with symmetry, Part II. *Simon Stevin* 49, 81–103.

—— and —— (1984). *Finite generalized quadrangles.* Pitman, London.

—— and —— (1991a). Generalized quadrangles, *BLT*-sets, and Fisher flocks. *Congr. Numer.*

—— and —— (1991b). Conical flocks, partial flocks, derivation and generalized quadrangles. *Geom. Dedicata* 38, 229–43.

Pei, D. Y. (1986). Normal basis of finite field *GF*(2). *IEEE Trans. Inform. Theory* 32, 285–7.

Pellegrino, G. (1970). Sul massimo ordine delle calotte in $S_{4,3}$. *Matematiche* (Catania) 25, 1–9.

Pellikaan, R., Shen, B. Z., and van Wee, G. J. M. (1991). Which linear codes are algebraic–geometric? *IEEE Trans. Inform. Theory* 37, 583–602.

Percsy, N. (1974). A characterization of classical Minkowski planes over a perfect field of characteristic two. *J. Geom.* 5, 191–204.

—— (1989). On the Buekenhout–Shult theorem on polar spaces. *Bull. Soc. Math. Belg.* 41, 283–94.

Perret, M. (1989). Families of codes exceeding the Varshamov–Gilbert bound. Lecture Notes in Comput. Sci. 357, Springer, Berlin, pp. 28–36.

Peterson, W. W. and Weldon, E. J. (1972). *Error-correcting codes.* MIT Press, Cambridge, MA.

Petrov, E. E. (1988). Harmonics on a finite projective space (Russian). *Mat. Zametki* 43, 31–7.

Phelps, K. T. (1985). Every finite group is the automorphism group of some linear code. *Congr. Numer.* 49, 139–41.

Pickert, G. (1985). Der Satz von Pascal, Ovale und Kubale. *Math. Semesterber.* 32, 61–83.

—— (1988). Differenzmengen und Ovale. *Discrete Math.* 73, 165–79.

Pless, V. (1964). On Witt's theorem for nonalternating symmetric bilinear forms over a field of characteristic 2. *Proc. Amer. Math. Soc.* 15, 979–83.

—— (1965a). The number of isotropic subspaces in a finite geometry. *Atti Accad. Naz. Lincei Rend.* 39, 418–21.

—— (1990). Cyclic codes and cyclic configurations. *Finite geometries and combinatorial designs*, American Mathematical Society, Providence, RI, pp. 171–7.

Pomilio, A. I. (1984). A characterization of the points of an affine space $AG(3, q)$ with q odd and prime, with respect to a nondegenerate quadric. *Rend. Mat.* 4, 313–25.

Pott, A. (1988). A note on non-abelian planar difference sets. *European J. Combin.* 9, 169–70.

Procesi Ciampi, R. and Rota, R. (1988). On k-sets of type $(0, m, n)$ in $S_{r,q}$ with three exterior hyperplanes. *Ann. Discrete Math.* 37, 377–83.

Qvist, B. (1952). Some remarks concerning curves of the second degree in a finite plane. *Ann. Acad. Sci. Fenn. Ser. A* No. 134.

Raguso, G. (1989). Sugli archi cubichi di un piano di Galois di ordine dispari. *Rend. Mat.* 9, 183–7.

—— and Rella, L. (1983a). Sui $(k, n; f)$-archi di tipo $(1, n)$ di un piano proiettivo finito. *Note Mat.* 3, 307–20.

—— and —— (1983b). Sulle $(k, n; f)$-calotte di tipo $(1, n)$. *Note Mat.* 3, 267–80.

—— and —— (1984). Graphic arcs of order 5, 6 embeddable in algebraic plane curves of the same order. *Mitt. Math. Sem. Giessen* No. 166, 167–76.

—— and —— (1985a). On the graphic arcs embeddable in algebraic plane curves. *Mitt. Math. Sem. Giessen* No. 169, 45–53.

—— and —— (1985b). Sugli archi grafici di ordine 3 di un piano di Galois di caratteristica 2. *Boll. Un. Mat. Ital. D* 4, 161–6.

—— and —— (1990). Nota su certe $\{k, 3\}$-calotte di $S_{3,q}$. *Rend. Mat.* 10, 205–8.

Rahilly, A. (1989). Embeddings of maximal arc type in finite projective planes. *J. Combin. Math. Combin. Comput.* 6, 131–41.

Rajola, S. (1988). A blocking set in $PG(3, q)$, $q \geq 5$. *Ann. Discrete Math.* 37, 391–4.

—— (1990). Sulle fibrazioni massimali di una quadrica non singolare di $PG(4, q)$. *Rend. Mat.* 10, 377–82.

Ray-Chaudhuri, D. K. and Sprague, A. P. (1976). Characterization of projective incidence structures. *Geom. Dedicata* 5, 361–76.

Rees, S. (1985). C_3 geometries arising from the Klein quadric. *Geom. Dedicata* 18, 67–85.

Reifart, A. (1984). The classification of the translation planes of order 16. *Geom. Dedicata* 17, 1–9.

Roghelia, N. N. and Sane, S. S. (1984). Classification of (16, 6, 2)-designs by ovals. *Discrete Math.* 51, 167–77.

Ronan, M. A. (1980a). A note on the ${}^3D_4(q)$ generalized hexagons. *J. Combin. Theory Ser. A* 29, 249–50.

—— (1980b). Semiregular graph automorphisms and generalized quadrangles. *J. Combin. Theory Ser. A* 29, 319–28.

—— (1980c). A geometric characterization of Moufang hexagons. *Invent. Math.* 57, 227–62.

—— (1981). A combinatorial characterization of the dual Moufang hexagons. *Geom. Dedicata* 11, 61–7.

—— (1987). Embeddings and hyperplanes of discrete geometries. *European J. Combin.* 8, 179–85.

Ronyai, L. and Szönyi, T. (1989). Planar functions over finite fields. *Combinatorica* 9, 315–20.

Roos, C. (1980). An alternative proof of the Feit–Higman theorem on generalized polygons. *Delft Progr. Rep. Ser. F: Math. Eng.* 5, 67–77.

Rosa, A. (1987). Repeated blocks in indecomposable twofold triple systems. *Discrete Math.* 65, 261–76.

Roth, R. M. and Lempel, A. (1988). Composition of Reed–Solomon codes and geometric designs. *IEEE Trans. Inform. Theory* 34, 810–16.

—— and —— (1989). A construction of non-Reed–Solomon type MDS codes. *IEEE Trans. Inform. Theory* 35, 655–7.

Rotman, J. J. (1988). Collineation groups of projective planes of order *n*. *J. Algebra* 115, 313–31.

Rück, H.-G. (1987). A note on elliptic curves over finite fields. *Math. Comp.* 49, 301–4.

Ryan, C. (1987). An application of Grassmannian varieties to coding theory. *Congr. Numer.* 57, 257–71.

Salzburg, P. M. (1985). Random number generation: a combinatorial approach. *Discrete Math.* 54, 313–20.

Sane, S. S. (1985). Affine subplanes of order three in the projective plane of order four. *Mitt. Math. Sem. Giessen* No. 169, 73–8.

Scharlau, W. (1989). Selbstduale Goppa-codes. *Math. Nachr.* 143, 119–22.

Schmidt, F. K. (1939a). Die Wronskische Determinante in beliebigen differenzierbaren Funktionenkörpern. *Math. Z.* 45, 62–74.

—— (1939b). Zur arithmetischen Theorie der algebraischen Funktionen: II. Allgemeine Theorie der Weierstrasspunkte. *Math. Z.* 45, 75–96.

Schoof, R. (1987). Nonsingular plane cubic curves over finite fields. *J. Combin. Theory Ser. A*, 46, 183–211.

Schrijver, A. and van Lint, J. H. (1981). Construction of strongly regular graphs, two-weight codes and partial geometries by finite fields. *Combinatorica* 1, 63–73.

Schröder, E. M. (1988). A characterization of ovoidal quadrics by plane sections. *J. Geom.* 31, 151–8.

Schulz, R. H. (1974). Zur Geometrie der $PSU(3, q^2)$: eine Klasse von Steinerschen Systemen. *Geom. Dedicata* 3, 11–19.

—— (1977). Zur Geometrie der $PSU(3, q^2)$. *Beiträge zur geometrischen Algebra*, Birkhäuser, Basel, pp. 293–8.

Segre, B. (1955b). Curve razionali normali e *k*-archi negli spazi finiti. *Ann. Mat. Pura Appl.* 39, 357–79.

—— (1959a). Le geometrie di Galois. *Ann. Mat. Pura Appl.* 48, 1–97.

—— (1960a). *Lectures on modern geometry* (with an appendix by L. Lombardo-Radice), Cremonese, Rome.

—— (1964b). Teoria di Galois, fibrazioni proiettive e geometrie non desarguesiane. *Ann. Mat. Pura Appl.* 64, 1–76.

—— (1965a). Forme e geometrie hermitiane, con particolare riguardo al caso finito. *Ann. Mat. Pura Appl.* 70, 1–201.

—— (1967). Introduction to Galois geometries (edited by J. W. P. Hirschfeld). *Atti Accad. Naz. Lincei Mem.* 8, 133–236.

Segre, C. (1891). Sulle varietà che rappresentano le coppie di punti di due piani o spazi. *Rend. Circ. Mat. Palermo* 5, 192–204.

Seidel, J. J. (1968). Strongly regular graphs with $(-1, 1, 0)$ adjacency matrix having eigenvalue 3. *Linear Algebra Appl.* 1, 281–98.

Semple, J. G. and Roth, L. (1949). *Introduction to algebraic geometry.* Oxford University Press, Oxford.

Seroussi, G. and Roth, R. M. (1986). On MDS extensions of generalized Reed–Solomon codes. *IEEE Trans. Inform. Theory* 32, 349–54.

Shaw, R. (1989). Clifford algebras and finite geometries. *Nuovo Cimento B* 103, 655–7.

—— and Jarvis, T. M. (1990). Finite geometries and Clifford algebras. II. *J. Math. Phys.* 31, 1315–24.

Shen, H. (1985). An enumeration theorem for unitary geometries over a finite field and the construction of PBIB designs. *Acta Mat. Sinica* 28, 747–55.

—— (1987). The use of finite geometries to construct block designs. *J. Shanghai Jiaotong Univ.* 2, 29–37.

—— (1988). Finite geometries and the construction of BIB designs. *Chinese Ann. Math. Ser. A* 9, 546–54.

Sherk, F. A. (1986). The geometry of $GF(q^3)$. *Canad. J. Math.* 38, 672–96.

—— (1990). Cubic surfaces in $AG(3, q)$ and projective planes of order q^3. *Geom. Dedicata* 34, 1–11.

Sherman, B. F. (1990). Ovoids in finite uniform linear spaces. *J. Geom.* 37, 159–70.

Shult, E. E. (1972). Characterizations of certain classes of graphs. *J. Combin. Theory Ser. B* 13, 142–67.

—— (1975). Groups, polar spaces and related structures. *Combinatorics, part 3*, Math. Centre Tracts 57, Amsterdam, pp. 130–61.

—— (1983). Characterizations of the Lie incidence geometries. *Surveys in combinatorics*, London Math. Soc. Lecture Note Series 82, Cambridge University Press, Cambridge, pp. 157–86.

—— (1985). A sporadic ovoid in $\Omega^+(8, 7)$, *Algebras, Groups and Geometries* 2, 495–513.

—— (1989). Nonexistence of ovoids in $\Omega^+(10, 3)$. *J. Combin. Theory Ser. A* 51, 250–7.

Silverman, J. A. (1986). *The arithmetic of elliptic curves.* Springer, New York.

Simmons, G. J. (1990). Sharply focused sets of lines on a conic in $PG(2, q)$. *Congr. Num.* 73, 181–204.

Singerman, D. (1986). Klein's Riemann surface of genus 3 and regular imbeddings of finite projective planes. *Bull. London Math. Soc.* 18, 364–70.

Singleton, R. R. (1966). Minimal regular graphs of maximal even girth. *J. Combin. Theory* 1, 306–32.

Skorobogatov, A. N. and Vladut, S. G. (1990). On the decoding of algebraic–geometric codes. *IEEE Trans. Inform. Theory* 36, 1050–60.

Smeichel, E. (1989). On the cycle structure of finite projective planes. *Ann. New York Acad. Sci.* 555, 368–74.

Spanicciati, R. (1984). On affine plane q-sets. *Rend. Mat.* 4, 525–30.

—— (1985). Nonesistenza di affini q^d-insiemi nonbanali di uno spazio affine $A_{d+1,q}$ con $d + 1 \geq 3$. *Rend. Mat.* 5, 243–9.

Sprague, A. P. (1978). Characterization of projective graphs. *J. Combin. Theory Ser. B* 24, 294–300.

—— (1981a). Incidence structures whose planes are nets. *European J. Combin.* 2, 193–204.

—— (1981b). Pasch's axiom and projective spaces. *Discrete Math.* 33, 79–87.

—— (1984a). A characterization of special Laguerre planes and extended dual affine planes. *Geom. Dedicata* 16, 149–56.

—— (1984b). Extended dual affine planes. *Geom. Dedicata* 16, 107–21.

—— (1985). Rank 3 incidence structures admitting dual-linear, linear diagram. *J. Combin. Theory Ser. A* 38, 254–9.

Stichtenoth, H. (1988a). A note on Hermitian codes over $GF(q^2)$. *IEEE Trans. Inform. Theory* 34, 1345–8.

—— (1988b). Self-dual Goppa codes. *J. Pure Appl. Algebra* 55, 199–211.

—— (1989). Which extended Goppa codes are cyclic? *J. Combin. Theory Ser. A* 51, 205–20.

—— (1990a). On automorphisms of geometric Goppa codes. *J. Algebra* 130, 113–21.

—— (1990b). On the dimension of subfield codes. *IEEE Trans. Inform. Theory* 36, 90–3.

Stinson, D. R. and Vanstone, S. A. (1986). Orthogonal packings in $PG(5, 2)$. *Aequationes Math.* 31, 159–68.

Stöhr, K. O. and Voloch, J. F. (1986). Weierstrass points and curves over finite fields. *Proc. London Math. Soc.* 52, 1–19.

Storme, L. and Thas, J. A. (1991a). Generalized Reed–Solomon codes and normal rational curves: an improvement of results by Seroussi and Roth. *Advances in finite geometries and designs*, Oxford University Press, Oxford, pp. 361–89.

—— and —— (1991b). MDS codes and arcs in $PG(n, q)$ with q even: An improvement of the bounds of Bruen, Thas and Blokhuis. *J. Combin. Theory Ser. A*.

Street, A. P. (1985). A survey of irreducible balanced incomplete block designs. *Ars Combin.* 19, 215–35.

Sturmfels, B. and White, N. (1990). All 11_3 and 12_3-configurations are rational. *Aequationes Math.* 39, 254–60.

Sun, B. M. and Zhu, X. L. (1984). PBIB designs using a class of hyperbolic planes in finite orthogonal geometries. *Dongbei Shida Xuebao* No. 2, 11–15.

Sun, Q. and Xiao, R. (1989a). A class of good elliptic curves used in cipher systems (Chinese). *Kexue Tongbao* 34, 237.

—— and —— (1989b). Two kinds of elliptic curves over F_q used to set up cryptosystems (Chinese). *Sichuan Daxue Xuebao* 26, 39–43.

Sved, M. (1985). On the Baer structure of finite projective spaces of square order. *Ars Combin.* 19, 215–35.

Szönyi, T. (1985). Small complete arcs in Galois planes. *Geom. Dedicata* 18, 161–72.

—— (1987a). Note on the order of magnitude of k for complete k-arcs in $PG(2, q)$. *Discrete Math.* 66, 279–82.

—— (1987b). Arcs in cubic curves and 3-independent subsets of abelian groups. *Colloq. Math. Soc. János Bolyai* 52, 499–508.

—— (1988). Complete arcs in non-Desarguesian planes. *Ars Combin.* 25, 169–78.

—— (1989a). Arcs and k-sets with large nucleus-set in Hall planes. *J. Geom.* 34, 187–94.

—— (1989b). Complete arcs in non-Desarguesian planes. *Confer. Sem. Mat. Univ. Bari* No. 233.

—— (1991). Note on the existence of large minimal blocking sets in $PG(2, q)$, $q \equiv 1 \pmod 4$. *Combinatorica.*

—— and Wettl, F. (1988a). On complexes in a finite abelian group. I. *Proc. Japan Acad. Ser. A Math. Sci.* 64, 245–8.

—— and —— (1988b). On complexes in a finite abelian group. II. *Proc. Japan Acad. Ser. A Math. Sci.* 64, 286–7.

Tallini, G. (1956a). Sulle k-calotte di uno spazio lineare finito. *Ann. Mat. Pura Appl.* 42, 119–64.

—— (1957a). Caratterizzazione grafica delle quadriche ellittiche negli spazi finiti. *Rend. Mat. e Appl.* 16, 328–51.

—— (1958a). Una proprietà grafica caratteristica delle superficie di Veronese negli spazi finiti (Note I; II). *Atti Accad. Naz. Lincei Rend.* 24, 19–23; 135–8.

—— (1971c). Strutture di incidenza dotate di polarità. *Rend. Sem. Mat. Fis. Milano* 41, 3–42.

—— (1981b). On a characterization of the Grassmann manifold representing the lines in a projective space. *Finite geometries and designs*, London Math. Soc. Lecture Note Series 49, Cambridge University Press, Cambridge, pp. 354–8.

—— (1982c). k-insiemi e blocking sets in $PG(r, q)$ e $AG(r, q)$. *Sem. Geom. Combin.*, Quad. 1, Università de l'Aquila.

—— (1983). B. Segre. *Ann. Discrete Math.* 18, 5–12.

—— (1984). Spazi finiti di rette e k-insiemi di $PG(r, q)$. *Confer. Sem. Mat. Univ. Bari* No. 192.

—— (1985). Fibrazioni mediante rette in $PG(r, q)$. *Matematiche* 37, 8–27.

—— (1986a). Ovoids and caps in planar spaces. *Ann. Discrete Math.* 30, 347–54.

—— (1986b). Campi di Galois non standard. *Conf. Sem. Mat. Univ. Bari* No. 209.

—— (1986c). Partial line spaces and algebraic varieties. *Combinatorica*, Symp. Math. 28, Academic Press, London, pp. 203–17.

—— (1987a). Sugli insiemi di rette di tipo pari di uno spazio di Galois $PG(3, q)$. *Rend. Mat.* 7, 1–16.

—— (1987b). Some new results on sets of type (m, n) in projective planes. *J. Geom.* 29, 191–9.

—— (1987c). Spazi parziali di rette e codici correttori. *Riv. Mat. Pura Appl.* 1, 43–69.

—— (1987d). Linear codes associated with geometric structures. *Resultate Math.* 12, 411–12.

—— (1988). On blocking sets in finite projective and affine spaces. *Ann. Discrete Math.* 37, 433–50.

—— (1989a). Spazi geometrici e codici lineari associati. *Rend. Sem. Mat. Fis. Milano* 57, 321–36.

—— (1989b). Fibrazioni per rette in una quadrica nonsingolare $Q_{4,q}$ di $PG(4, q)$. *Atti Accad. Peloritana Pericolanti Cl. Sci. Fis. Mat. Natur.* 66, 127–46.

Tallini Scafati, M. (1967a). Caratterizzazione grafica delle forme hermitiane di un $S_{r,q}$. *Rend. Mat. Appl.* 26, 273–303.

—— (1981). The theory of k-sets in a Galois space. *Atti Soc. Peloritana Sci. Fis. Mat. Natur.* 27, 57–65.

—— (1985). The k-sets of $PG(r, q)$ from the character point of view. *Finite geometries*, Dekker, New York, pp. 321–6.

—— (1987). Recent results on (m, n)-type k-sets in an affine plane α_q. *J. Geom.* 29, 94–100.

Taussky, O. (1987). Nonsingular cubic curves as determinantal loci. *J. Math. Phys. Sci.* 21, 665–78.

Tecklenberg, H. (1987). A proof of the theorem of Pappus in finite Desarguesian affine planes. *J. Geom.* 30, 172–81.

Teirlinck, L. (1975). On linear spaces in which every plane is either projective or affine. *Geom. Dedicata* 4, 39–44.

—— (1987). Non-trivial t-designs without repeated blocks exist for all t. *Discrete Math.* 65, 301–11.

Thas, J. A. (1968a). Normal rational curves and k-arcs in Galois spaces. *Rend. Mat.* 1, 331–4.

—— (1969a). Connection between the Grassmannian $G_{k-1;n}$ and the set of the k-arcs of the Galois space $S_{n,q}$. *Rend. Mat.* 2, 121–34.

—— (1972). Ovoidal translation planes. *Arch. Math.* 23, 110–12.

—— (1973b). Construction of partial geometries. *Simon Stevin* 46, 95–8.

—— (1973d). On 4-gonal configurations. *Geom. Dedicata* 2, 317–26.

—— (1974b). Construction of maximal arcs and partial geometries. *Geom. Dedicata* 3, 61–4.

—— (1974c). On semi ovals and semi ovoids. *Geom. Dedicata* 3, 229–31.

—— (1974d). A remark concerning the restriction on the parameters of a 4-gonal subconfiguration. *Simon Stevin* 48, 65–8.

—— (1974e). On 4-gonal configurations with parameters $r = q^2 + 1$ and $k = q + 1$, Part I. *Geom. Dedicata* 3, 365–75.

—— (1974f). Translation 4-gonal configurations. *Atti Accad. Naz. Lincei Rend.* 56, 303–14.

—— (1975c). On 4-gonal configurations with parameters $r = q^2 + 1$ and $k = q + 1$, Part II. *Geom. Dedicata* 4, 51–9.

—— (1976). On generalized quadrangles with parameters $s = q^2$ and $t = q^3$. *Geom. Dedicata* 5, 485–96.

—— (1977a). Two infinite classes of perfect codes in metrically regular graphs. *J. Combin. Theory Ser. B* 23, 236–8.

—— (1977b). Combinatorics of partial geometries and generalized quadrangles. *Higher combinatorics*, Reidel, Dordrecht, pp. 183–99.

—— (1977c). Combinatorial characterizations of the classical generalized quadrangles. *Geom. Dedicata* 6, 339–51.

—— (1978a). Partial geometries in finite affine spaces. *Math. Z.* 158, 1–13.

—— (1978b). Combinatorial characterizations of generalized quadrangles with parameters $s = q$ and $t = q^2$. *Geom. Dedicata* 7, 223–32.

—— (1980c). Polar spaces, generalized hexagons and perfect codes. *J. Combin. Theory Ser. A* 29, 87–93.

—— (1980d). A remark on the theorem of Yanushka–Ronan characterizing the generalized hexagon $H(q)$ arising from the group $G_2(q)$. *J. Combin. Theory Ser. A* 29, 361–2.

—— (1981a). Ovoids and spreads of finite classical polar spaces. *Geom. Dedicata* 10, 135–44.

—— (1981c). New combinatorial characterizations of generalized quadrangles. *European J. Combin.* 2, 299–303.
—— (1983b). Semi-partial geometries and spreads of classical polar spaces. *J. Combin. Theory Ser. A* 35, 58–66.
—— (1984a). 3-regularity in generalized quadrangles of order (s, s^2). *Geom. Dedicata* 17, 33–6.
—— (1984b). The theorems of Dembowski and Heise–Percsy from the point of view of generalized quadrangles. *Sem. Geom. Combin.*, Quad. 6, Università de l'Aquila.
—— (1985a). Characterizations of generalized quadrangles by generalized homologies. *J. Combin. Theory Ser. A* 40, 331–41.
—— (1985b). Circle geometries and generalized quadrangles. *Finite geometries*, Dekker, New York, pp. 327–52.
—— (1986a). Extensions of finite generalized quadrangles. *Combinatorica*, Symp. Math. 28, Academic Press, London, pp. 127–43.
—— (1986b). The classification of all (x, y)-transitive generalized quadrangles. *J. Combin. Theory Ser. A* 42, 154–7.
—— (1987a). Complete arcs and algebraic curves in $PG(2, q)$. *J. Algebra* 106, 451–64.
—— (1987b). Generalized quadrangles and flocks of cones. *European J. Combin.* 8, 441–52.
—— (1989a). A note on spreads and partial spreads of Hermitian varieties. *Simon Stevin* 63, 101–5.
—— (1989b). Interesting pointsets in generalized quadrangles and partial geometries. *Linear Algebra Appl.* 114/115, 103–31.
—— (1990a). Flocks, maximal exterior sets and inversive planes. *Finite geometries and combinatorial designs*, American Mathematical Society, Providence, RI, pp. 187–218.
—— (1990b). Solution of a classical problem on finite inversive planes. *Finite buildings, related geometries and applications*, Oxford University Press, Oxford, pp. 145–59.
—— (1991a). Maximal exterior sets of hyperbolic quadrics: the complete classification. *J. Combin. Theory Ser. A* 56, 303–8.
—— (1991b). Recent results on flocks, maximal exterior sets and inversive planes. *Combinatorics* '88, Mediterranean Press, Cosenza.
—— (1991c). Old and new results on spreads and ovoids of finite classical polar spaces. *Combinatorics* '90.
—— and De Clerck, F. (1977). Partial geometries satisfying the axiom of Pasch. *Simon Stevin* 51, 123–37.
—— and De Winne, P. (1977). Generalized quadrangles in finite projective spaces. *J. Geom.* 10, 126–37.
—— and Payne, S. E. (1976). Classical finite generalized quadrangles: a combinatorial study. *Ars Combin.* 2, 57–110.
—— and Van Maldeghem, H. (1991). Generalized Desargues configurations in generalized quadrangles. *Bull. Soc. Math. Belg.* 42, 713–22.
——, Debroey, I., and De Clerck, F. (1984). The embedding of $(0, \alpha)$-geometries in $PG(n, q)$: Part II. *Discrete Math.* 51, 283–92.
——, Payne, S. E., and Gevaert, H. (1988). A family of ovals with few collineations. *European J. Combin.* 9, 353–62.
——, ——, and Van Maldeghem, H. (1991). Half Moufang implies Moufang for generalized quadrangles. *Invent. Math.* 105, 153–6.

Thomas, S. (1987). Designs over finite fields. *Geom. Dedicata* 24, 237–42.

Tiersma, H. J. (1987). Remarks on codes from Hermitian curves. *IEEE Trans. Inform. Theory* 33, 605–9.

Tits, J. (1959). Sur la trialité et certains groupes qui s'en déduisent. *Inst. Hautes Etudes Sci. Publ. Math.* 2, 13–60.

—— (1974). *Buildings of spherical type and finite BN-pairs*, Lecture Notes in Math. 386, Springer, Berlin.

—— (1976a). Classification of buildings of spherical type and Moufang polygons: a survey. *Teorie combinatorie, volume I*, Accademia Nazionale dei Lincei, Rome, pp. 229–46.

—— (1976b). Quadrangles de Moufang, I. Preprint.

—— (1983). Moufang octagons and the Ree groups of type ${}^2F_4$. *Amer. J. Math.* 105, 539–94.

—— (1990). Private communication.

Todorov, D. T. (1985). Three mutually orthogonal Latin squares of order 14. *Ars Combin.* 20, 45–7.

Tonchev, V. D. (1984). The isomorphism of the Cohen, Haemers–van Lint and the De Clerck–Dye–Thas partial geometries. *Discrete Math.* 49, 213–17.

—— (1988). *Combinatorial configurations.* Wiley, New York.

Tsfasman, M. A., Vladut, S. G., and Zink, T. (1982). Modular curves, Shimura curves and Goppa codes, better than Varshamov–Gilbert bound. *Math. Nachr.* 109, 21–8.

Tzanakis, N. and Wolfskill, J. (1987). The diophantine equation $x^2 = 4q^{a/2} + 4q + 1$ with an application to coding theory. *J. Number Theory* 26, 96–116.

Ueberberg, J. (1990). A characterization of the line set of an odd-dimensional Baer subspace. *J. Geom.* 37, 171–80.

Ughi, E. (1986). Sul numero delle soluzioni di certi sistemi algebrici di congruenze in un campo di Galois. *Note Mat.* 6, 139–54.

—— (1987a). Construzione di $\{k, n\}$-blocking sets con n relativamente piccolo in $PG(2, q)$ e $PG(3, q)$. *Atti Sem. Mat. Fis. Univ. Modena* 34, 223–31.

—— (1987b). Saturated configurations of points on projective Galois spaces. *European J. Combin.* 8, 325–34.

—— (1988). On (k, n)-blocking sets which can be obtained as a union of conics. *Geom. Dedicata* 26, 241–6.

—— and Vipera, M. C. (1988). Sul minimo numero di radici 'libere' di un polinomio semplice. *Atti Sem. Mat. Fis. Univ. Modena* 36, 395–404.

Vainsencher, I. (1989). On Stöhr-Voloch's proof of Weil's theorem. *Manuscripta Math.* 64, 121–6.

van de Craats, J. and Simonis, J. (1986). Affinely regular polygons. *Nieuw Arch. Wisk.* 4, 225–40.

Vanden Cruyce, P. (1985). Geometries related to $PSL(2, 19)$. *European J. Combin.* 6, 163–73.

van Lint, J. H. (1982). *Introduction to coding theory.* Springer, New York.

—— (1990). Algebraic geometric codes. *IMA Vol. Math. Appl.* 20, Springer, New York, pp. 137–62.

—— and Springer, T. A. (1987). Generalised Reed–Solomon codes from algebraic geometry. *IEEE Trans. Inform. Theory* 33, 305–9.

—— and van der Geer, G. (1988). *Introduction to coding theory and algebraic geometry.* Birkhäuser, Basel.

Van Maldeghem, H., Payne, S. E., and Thas, J. A. (1991). Desarguesian finite generalized quadrangles are classical or dual classical. *Designs, Codes and Cryptography*.

van Oorschot, P. C. and Vanstone, S. A. (1989). A geometric approach to root finding in $GF(q^m)$. *IEEE Trans. Inform. Theory* 35, 444–53.

Varshamov, R. R. (1984). A method for constructing irreducible polynomials over finite fields. *Akad. Nauk. Armyan. SSR Dokl.* 79, 245–51.

Vecchi, I. (1984). Some results on coverings of Galois spaces with ovoids and related BIB designs. *J. Statist. Plann. Inference* 10, 219–25.

Veldkamp, F. D. (1959). Polar geometry I–IV. *Indag. Math.* 21, 512–51.

—— (1960). Polar geometry V. *Indag. Math.* 22, 207–12.

Vladut, S. G. (1990). On the decoding of algebraic–geometric codes over F_q for $q \geq 16$. *IEEE Trans. Inform. Theory* 36, 1461–3.

—— and Drinfeld, V. G. (1983). Number of points on an algebraic curve. *Functional Analysis* 17, 53–4.

—— and Manin, Y. I. (1985). Linear codes and modular curves. *J. Soviet Math.* 30, 2611–43. (French translation by Deza, M. and Le Brigand, D. (1984). *Publications Mathématiques de l'Université Pierre et Marie Curie* No. 72.)

Voigt, B. (1985). On the evolution of finite affine and projective spaces. *IX Symposium on operations research, part I*, Athenaum, Konigstein, pp. 313–27.

Voloch, J. F. (1987). On the completeness of certain plane arcs. *European J. Combin.* 8, 453–6.

—— (1988). A note on elliptic curves over finite fields. *Bull. Soc. Math. France* 116, 455–8.

—— (1989). On the number of values taken by a polynomial over a finite field. *Acta Arith.* 52, 197–201.

—— (1990a). On the completeness of certain plane arcs II. *European J. Combin.* 11, 491–6.

—— (1990b). Arcs in projective planes over prime fields. *J. Geom.* 38, 198– 200.

—— (1991). Complete arcs in Galois planes of non-square order. *Advances in finite geometries and designs*, Oxford University Press, Oxford, pp. 401–6.

Voss, C. and Stichtenoth, H. (1990). Asymptotically good families of subfield codes of geometric Goppa codes. *Geom. Dedicata* 33, 111–16.

Wallis, W. D. (1973). Configurations arising from maximal arcs. *J. Combin. Theory* 15, 115–19.

—— (1990). Finite planes and clique partitions. *Finite geometries and combinatorial designs*, American Mathematical Society, Providence, RI, pp. 279–85.

Wan, D. (1985). On a problem of Niederreiter and Robinson about finite fields. *J. Austral. Math. Soc.* 41, 336–8.

—— (1988). Zeros of diagonal equations over finite fields. *Proc. Amer. Math. Soc.* 103, 1049–52.

—— (1990). Permutation polynomials and resolution of singularities over finite fields. *Proc. Amer. Math. Soc.* 110, 303–9.

Wan, Z. (1965a). Studies in finite geometries and the construction of incomplete block designs. I. Some 'Anzahl' theorems in symplectic geometry over finite fields. *Chinese Math.* 7, 55–62.

—— (1989). A new derivation of an Anzahl theorem of V. Pless. *Sichuan Daxue Xuebao* 26, 104–6.

—— and Yang, B. (1965b). Studies in finite geometries and the construction of incomplete block designs. III. Some 'Anzahl' theorems in unitary geometry over finite fields and their applications. *Chinese Math.* 7, 252–64.

Ward, R. K. (1984). Error correction and detection, a geometric approach. *Comput. J.* 27, 203–5.

Waterhouse, W. C. (1987). How often do determinants over finite fields vanish? *Discrete Math.* 65, 103–4.

Weida, R. A. (1989). An extension of Bruen chains. *Geom. Dedicata* 30, 11–21.

Wells, A. L. (1983). Universal projective embeddings of the Grassmannian, half spinor, and dual orthogonal geometries. *Quart. J. Math. Oxford Ser.* 34, 375–86.

Wertheimer, M. A. (1990). Oval designs in quadrics. *Finite geometries and combinatorial designs*, American Mathematical Society, Providence, RI, pp. 287–97.

Wettl, F. (1985). On the nuclei of a pointset of a finite projective plane. *J. Geom.* 30, 157–63.

—— (1991). Internal nuclei of k-sets in finite projective spaces of three dimensions. *Advances in finite geometries and designs*, Oxford University Press, Oxford, pp. 407–19.

Whiteley, W. (1985). The projective geometry of rigid frameworks. *Finite geometries*, Dekker, New York, pp. 353–70.

Whittle, G. (1987). Modularity in tangential k-blocks. *J. Combin. Theory Ser. B* 42, 24–35.

—— (1988). Quotients of tangential k-blocks. *Proc. Amer. Math. Soc.* 102, 1088–98.

—— (1989). Dowling group geometries and the critical problem. *J. Combin. Theory Ser. B* 47, 80–92.

Wilbrink, H. A. (1985). A note on planar difference sets. *J. Combin. Theory Ser. A* 38, 94–5.

—— and Brouwer, A. E. (1984). A characterization of two classes of semi partial geometries by their parameters. *Simon Stevin* 58, 273–88.

Wild, P. (1984). Higher-dimensional ovals and dual translation planes. *Ars Combin.* 17, 105–12.

Wilson, B. J. (1986). $(k, n; f)$-arcs and caps in finite projective spaces. *Ann. Discrete Math.* 30, 355–62.

Wirtz, M. (1988). On the parameters of Goppa codes. *IEEE Trans. Inform. Theory* 34, 1341–3.

Witczynski, K. (1988). A geometrical proof of the generalized Pappus theorem. *Demonstratio Math.* 21, 1105–14.

Wolfmann, J. (1986). Recent results on coding and algebraic geometry. Lecture Notes in Comput. Sci. 229, Springer, Berlin, pp. 167–84.

—— (1987a). A group algebra construction of binary even self dual codes. *Discrete Math.* 65, 81–9.

—— (1987b). Nombre de points rationnels de courbes algébriques sur des corps finis associées a des codes cycliques. *C.R. Acad. Sci. Paris Ser. I. Math.* 305, 345–8.

—— (1989a). The weights of orthogonals of certain cyclic codes or extended Goppa codes. Lecture Notes in Comput. Sci. 357, Springer, Berlin, pp. 476–80.

—— (1989b). New bounds on cyclic codes from algebraic curves. Lecture Notes in Comput. Sci. 388, Springer, New York, pp. 47–62.

—— (1989c). The number of points on certain algebraic curves over finite fields. *Comm. Algebra* 17, 2055–60.

Yang, B. F. and Wei, W. D. (1989a). Finite unitary geometry and PBIB designs. I. *J. Combin. Math. Combin. Comput.* 6, 51–61.

—— and —— (1989b). Orthogonal geometries over finite fields with characteristic $\neq 2$ and block designs. *Sichuan Daxue Xuebao* 26, 107–11.

—— and —— (1990). Finite orthogonal geometries with characteristic $\neq 2$ and PBIB designs. *Linear Algebra Appl.* 134, 1–23.

Yanushka, A. (1981). On order in generalized polygons. *Geom. Dedicata* 10, 451–8.

Yusty Pita, J. (1982). On the nine-point line (Spanish). *Gaceta Mat.* 34, 92–8.

Zink, T. (1985). Degeneration of Shimura surfaces and a problem in coding theory. *Fundamentals of computation theory*, Lecture Notes in Comput. Sci. 199, Springer, New York, pp. 503–11.

Zirilli, F. (1966). Sulla struttura del gruppo delle proiettività del piano. *Ist. Veneto Sci. Lett. Arti Atti* 124, 415–42.

INDEX OF NOTATION

Where appropriate, the page number gives the first occurrence of the notation. First general and then chapter specific notation is given.

Operations on sets and groups

$\lvert X \rvert$	number of elements in the set X
$X \backslash Y$	the set of elements of X not in Y
$X \subset Y$	X is a subset of Y
$G \cong H$	the groups G and H are isomorphic
$G < H$	G is a subgroup of H
$G \times H$	the direct product of the groups G and H

Miscellaneous sets and numbers

$\mathbf{N}$	the natural numbers $\{1, 2, 3, \ldots\}$
$\mathbf{N}_r$	$\{1, 2, \ldots, r\}$
$\bar{\mathbf{N}}_r$	$\{0, 1, \ldots, r\}$
$\theta(n)$	$(q^{n+1} - 1)/(q - 1)$
$\mathbf{c}(n, r)$	$n(n-1)\cdots(n-r+1)/r!$
$((r))$	$2!3!\cdots r!$
$[r, s]_-$	$\prod_{i=r}^{i=s} (q^i - 1)$ for $s \geq r$
$[r, s]_-$	1 for $s < r$
$[r, s]_+$	$\prod_{i=r}^{i=s} (q^i + 1)$ for $s \geq r$
$[r, s]_+$	1 for $s < r$
$\overline{[r, s]}$	$\prod_{i=r}^{i=s} ((\sqrt{q})^i - (-1)^i)$ for $s \geq r$
$\overline{[r, s]}$	1 for $s < r$
$(m_1, \ldots, m_r)$	greatest common divisor of $m_1, \ldots, m_r$

Fields and rings

K	an arbitrary field	
γ, $GF(q)$	the Galois field of $q = p^h$ elements	
γ_0	$\gamma \backslash \{0\}$	
$\bar{K}$	the algebraic closure of K	
$\bar{\gamma}$	the algebraic closure of γ	
$\mathbf{C}$	the complex numbers	
$\mathbf{R}$	the real numbers	
$C(t)$, $T(t)$	$t + t^2 + t^4 + \cdots + t^{2^{h-1}}$, $q = 2^h$	5, 93
$C(t)$	$(1 - t^{(q-1)/2})/2$, $q = p^h$, $p > 2$	5

$\mathscr{C}_0$	$\{t \in \gamma_0 \mid C(t) = 0\}$, q odd
$\mathscr{C}_1$	$\{t \in \gamma_0 \mid C(t) = 1\}$, q odd
$\mathscr{C}_0$	$\{t \in \gamma \mid C(t) = 0\}$, q even
$\mathscr{C}_1$	$\{t \in \gamma \mid C(t) = 1\}$, q even
$\bar{x}$	$x^{\sqrt{q}}$ when q is a square
$K[x_0, \ldots, x_n]$	ring of polynomials in $x_0, \ldots, x_n$ over K
A^*	transpose of the matrix A

Spaces, subspaces, and their numbers

$V(n+1, K)$	$(n+1)$-dimensional vector space over K
$V(n+1, q)$	$V(n+1, K)$ when $K = \gamma$
$PG(n, K)$	n-dimensional projective space over K
$PG(n, q)$	$PG(n, K)$ when $K = \gamma$
$AG(n, K)$	n-dimensional affine space over K
$AG(n, q)$	$AG(n, K)$ when $K = \gamma$
Π_r	an r-dimensional subspace of $PG(n, K)$, specific or generic
$PG^{(r)}(n, q)$	set of Π_r in $PG(n, q)$
$\mathbf{P}(X)$	point of $PG(n, K)$ with vector $X = (x_0, \ldots, x_n)$
$\boldsymbol{\pi}(U)$	prime of $PG(n, K)$ with equation $u_0x_0 + u_1x_1 + \cdots + u_nx_n = 0$, $U = (u_0, u_1, \ldots, u_n)$
$\mathbf{U}_i$	$\mathbf{P}(0, \ldots, 0, 1, 0, \ldots, 0)$ with 1 in $(i+1)$th place
$\mathbf{U}$	$\mathbf{P}(1, 1, \ldots, 1)$
$\mathbf{u}_i$	$\boldsymbol{\pi}(0, \ldots, 0, 1, 0, \ldots, 0)$ with 1 in $(i+1)$th place
$\mathbf{u}$	$\boldsymbol{\pi}(1, 1, \ldots, 1)$
$\theta(n, q)$, $\theta(n)$	$\lvert PG(n, q)\rvert$
$\phi(r; n, q)$	$\lvert PG^{(r)}(n, q)\rvert$
$\chi(s, r; n, q)$	$\lvert\{\Pi_r \text{ in } PG(n, q) \text{ containing a fixed } \Pi_s\}\rvert$
$\psi_{12}(t, s, r; n, q)$	$\lvert\{\Pi_r \text{ in } PG(n, q) \text{ meeting a fixed } \Pi_s \text{ in a fixed } \Pi_t\}\rvert$

Groups

$\mathbf{Z}$	the additive group of integers
$\mathbf{Z}_n$	$\mathbf{Z}/n\mathbf{Z}$
$PGL(n+1, q)$	group of projectivities of $PG(n, q)$
$P\Gamma L(n+1, q)$	group of collineations of $PG(n, q)$
$G(\mathscr{F})$	subgroup of $PGL(n+1, q)$ fixing the set $\mathscr{F}$

Varieties

$\mathbf{V}(F)$	$\{\mathbf{P}(X) \in PG(n, q) \mid F(X) = 0\}$ } F and F_i forms in
$\mathbf{V}(F_1, \ldots, F_r)$, $\mathbf{V}_{n,K}(F_1, \ldots, F_r)$	$\mathbf{V}(F_1) \cap \mathbf{V}(F_2) \cap \cdots \cap \mathbf{V}(F_r)$ } $K[x_0, \ldots, x_n]$

$\mathscr{F}_1 \sim \mathscr{F}_2$	the varieties $\mathscr{F}_1$ and $\mathscr{F}_2$ are projectively equivalent	
$\Pi_r\mathscr{F}$	cone with vertex Π_r and base $\mathscr{F}$	3

Chapter 22

$\mathscr{W}_n$	a quadric in $PG(n, q)$	3
$\mathscr{Q}_n$	a non-singular quadric in $PG(n, q)$	3
$\mathscr{H}_n$	$\mathscr{Q}_n$ hyperbolic	3
$\mathscr{E}_n$	$\mathscr{Q}_n$ elliptic	3
$\mathscr{P}_n$	$\mathscr{Q}_n$ parabolic	3
$\Pi_k\mathscr{Q}_s$	cone with vertex Π_k and base $\mathscr{Q}_s$ in a Π_s skew to Π_k	3
$T_P(\mathscr{Q}_n)$	tangent space to $\mathscr{Q}_n$ at the point P	10
$T_{\Pi_r}(\mathscr{Q}_n)$	tangent space to $\mathscr{Q}_n$ along the subspace $\Pi_r \subset \mathscr{Q}_n$	12
g, $g(\mathscr{Q}_n)$	projective index of $\mathscr{Q}_n$	15
g, $g(\mathscr{W}_n)$	projective index of $\mathscr{W}_n$	15
w, $w(\mathscr{Q}_n)$	character of $\mathscr{Q}_n$	16
w, $w(\mathscr{W}_n)$	character of $\mathscr{W}_n$	16
$\mathscr{G}$, $\mathscr{G}(\mathscr{Q}_n)$	set of generators of $\mathscr{Q}_n$	17
$\kappa(n; w)$	$\lvert\mathscr{G}\rvert$	17
$\rho(d, n; w)$	$\lvert\{\Pi_g$ in $\mathscr{G}$ containing a fixed $\Pi_d\}\rvert$	17
$\lambda(d, n; w)$	$\lvert\{\Pi_g$ in $\mathscr{G}$ meeting a fixed generator in some $\Pi_d\}\rvert$	17
$\mu(c)$, $\mu(c, n; w)$	$\lvert\{\Pi_g$ in $\mathscr{G}$ meeting a fixed generator in a fixed $\Pi_{g-c}\}\rvert$	17
$N(\Pi_m, \mathscr{Q}_n)$	number of Π_m on $\mathscr{Q}_n$	22
$N(m; n, w)$	number of Π_m on $\mathscr{Q}_n$ of character w	22
$N(\Pi_m, \mathscr{W}_n)$	number of Π_m on $\mathscr{W}_n$	22
$G(\mathscr{Q}_n)$	subgroup of $PGL(n + 1, q)$ fixing $\mathscr{Q}_n$	24
$PGO(n + 1, q)$	$G(\mathscr{Q}_n)$ when $\mathscr{Q}_n = \mathscr{P}_n$	24
$PGO_+(n + 1, q)$	$G(\mathscr{Q}_n)$ when $\mathscr{Q}_n = \mathscr{H}_n$	24
$PGO_-(n + 1, q)$	$G(\mathscr{Q}_n)$ when $\mathscr{Q}_n = \mathscr{E}_n$	24
$\mathscr{N}(\mathscr{Q}_n)$	set of $\mathscr{Q}_n$ in $PG(n, q)$ under the action of $PGL(n + 1, q)$	24
$\mathscr{S}(m, t, v; n, w)$	set of Π_m in $PG(n, q)$ such that $\Pi_m \cap \mathscr{Q}_n$ is of type $\Pi_{m-t-1}\mathscr{Q}_t$, where $\mathscr{Q}_n$ and $\mathscr{Q}_t$ have respective characters w and v	27
$N(m, t, v; n, w)$,	$\lvert\mathscr{S}(m, t, v; n, w)\rvert$	27
$N(\Pi_{m-t-1}\mathscr{Q}_t, \mathscr{Q}_n)$		27
T	$n + t - 2m$	28

$N_+(m, t, 1; n, 1)$	number of external Π_m meeting $\mathscr{P}_n$ in a section of type $\Pi_{m-t-1}\mathscr{P}_t$	43
$N_-(m, t, 1; n, 1)$	number of internal $\mathscr{P}_m$ meeting $\mathscr{P}_n$ in a section of type $\Pi_{m-t-1}\mathscr{P}_t$	43
$N_0(m, t, 1; n, 1)$	number of nuclear Π_m meeting $\mathscr{P}_n$ in a section of type $\Pi_{m-t-1}\mathscr{P}_t$	43
$N_1(m, t, 1; n, 1)$	number of non-nuclear Π_m meeting $\mathscr{P}_n$ in a section of type $\Pi_{m-t-1}\mathscr{P}_t$	43
T_P, $T_P(\mathscr{K})$	tangent space to $\mathscr{K}$ at P	45
Π_P	union of lines of $\mathscr{K}$ through P and a point of the subgenerator Π, P not in Π	48
S_{PQ}	$T_P \cap T_Q$	52
τ	triality	55

Chapter 23

$\mathscr{U}_n$, $\mathscr{U}_{n,q}$	non-singular Hermitian variety in $PG(n, q)$	59
$\Pi_{n-r-1}\mathscr{U}_r$	cone with vertex Π_{n-r-1} and base $\mathscr{U}_r$ in Π_r skew to Π_{n-r-1}	59
T_P, $T_P(\mathscr{U}_n)$	tangent prime to $\mathscr{U}_n$ at P	62
T_{Π_r}	tangent space to $\mathscr{U}_n$ along Π_r	62
$\mathfrak{U}$	polarity of $\mathscr{U}_n$	61
μ_n	$\lvert\mathscr{U}_n\rvert$	63
$\mu_n^{(t)}$	$\lvert\Pi_t\mathscr{U}_{n-t-1}\rvert$	63
$\mathscr{N}(\mathscr{U}_n)$	set of $\mathscr{U}_n$ in $PG(n, q)$	64
$G(\mathscr{U}_n)$	subgroup of $PGL(n+1, q)$ fixing $\mathscr{U}_n$	64
$PGU(n+1, q)$	$G(\mathscr{U}_n)$	64
g, $g(\mathscr{U}_n)$	projective index of $\mathscr{U}_n$	65
$N(\Pi_r, \mathscr{U}_n)$	number of Π_r on $\mathscr{U}_n$	66
$\nu_{r,n}$	$N(\Pi_r, \mathscr{U}_n)$	66
$N(\Pi_r\mathscr{U}_{m-r-1}, \mathscr{U}_n)$	number of spaces Π_m meeting $\mathscr{U}_n$ in a section of type $\Pi_r\mathscr{U}_{m-r-1}$	70
T	$n - 2m + s$, $r + s = m - 1$	67, 70
$\rho_{s,r}$	$N(\mathscr{U}_s, \mathscr{U}_r)$	70
$\mathscr{R}_n$	projection of $\mathscr{Q}_{n+1}$ onto a prime	86
$\mathscr{R}_n^+$	$\mathscr{R}_n$ when $\mathscr{Q}_{n+1} = \mathscr{H}_{n+1}$	87
$\mathscr{R}_n^-$	$\mathscr{R}_n$ when $\mathscr{Q}_{n+1} = \mathscr{E}_{n+1}$	87
$k_{r,n,q}$	k-set of type $(1, r, q+1)$ in $PG(n, q)$ with $1 \le r \le q$ and at least one line meeting the k-set in r points	71
T_P	tangent space at P	79
$\mathscr{J}$	residual of $\mathscr{K}$	85
$\mathscr{S}_P$	support of P	86

Chapter 24

$\mathbf{\Pi}_r(L)$	the subspace Π_r with coordinate vector L	101
T_x	$[x_j^i]$, $0 \le i \le r$, $0 \le j \le n$, i indexing rows and j columns	100
$(i_0 i_1 \cdots i_r)_x$, $(i_0 i_1 \cdots i_r)$	subdeterminant of T_x of order $r + 1$	100
T_u	$[u_j^i]$, $0 \le i \le n - r - 1$, $0 \le j \le n$, i indexing rows and j columns	101
$\mathscr{G}_{r,n}$, $\mathscr{G}_{r,n,K}$	Grassmannian or Grassmann variety of Π_r in $PG(n, K)$	107
$\mathscr{G}_{r,n}$, $\mathscr{G}_{r,n,q}$	$\mathscr{G}_{r,n,K}$ where $K = \gamma$	107
$\mathscr{T}_r(\Pi_g, \Pi_h)$	$\{\Pi_r \mid \Pi_g \subset \Pi_r \subset \Pi_h;\ \Pi_g, \Pi_h \text{ fixed}\}$	115
(Π_{r-1}, Π_{r+1})	$\mathscr{T}_r(\Pi_{r-1}, \Pi_{r+1})$, a pencil	110
$\bar{\mathscr{G}}_{r,n}$, $\bar{\mathscr{G}}_{r,n,K}$	$\mathscr{G}_{r,n,\bar{K}}$	108
δ	fundamental polarity associated to $\mathscr{G}_{r,2r+1}$	109
η	correlation $\rho x_i = u_i$, $i = 0, \ldots, 2r + 1$, of $PG(2r + 1, K)$	110
ζ	projectivity of $PG(N, K)$ induced by η	110
$\mathfrak{G}$	Grassmann mapping from $PG^{(r)}(n, K)$ to $\mathscr{G}_{r,n,K}$	107
$\mathscr{S}_{\mathrm{L}}$	set of Latin spaces, $\{\mathscr{T}_r(\Pi_{r-1}, \Pi_n)\mathfrak{G}\}$	114
$\mathscr{S}_{\mathrm{G}}$	set of Greek spaces, $\{\mathscr{T}_r(\Pi_{-1}, \Pi_{r+1})\mathfrak{G}\}$	114
$\mathscr{G}_{r,n}^{(s)}$	$\{\Pi_s \mid \Pi_s \subset \mathscr{G}_{r,n}\}$	115
$G(\mathscr{G}_{r,n})$	subgroup of $PGL(N + 1, K)$ fixing $\mathscr{G}_{r,n}$	119
$G(\mathscr{G}_{r,n,q})$	$G(\mathscr{G}_{r,n})$ when $K = \gamma$	121
θ	maps each element of $PGL(n + 1, K)$ onto the corresponding element of $G(\mathscr{G}_{r,n})$	119
$\mathscr{P}$	set of points of Grassmann space	123
$\mathscr{B}$	set of lines of Grassmann space	123
$\mathscr{S}$, $\mathscr{T}$	sets of maximal subspaces of Grassmann space	123
$\mathscr{C}$	set of projective planes belonging to elements of $\mathscr{T}$	125
$\mathscr{R}(P, \pi)$	set of elements of $\mathscr{S}$ meeting π at lines through P, π in $\mathscr{C}$ and P in π	126
$\mathscr{R}$	set with elements the sets $\mathscr{R}(P, \pi)$	126
$\mathscr{S}_P$	set of elements of $\mathscr{S}$ containing P	127
$\mathscr{S}(\pi, \pi')$	set of elements in $\mathscr{S}$ on the line $\pi\pi'$ of $(\mathscr{S}, \mathscr{R})$ or meeting both π and π' at points of $\mathscr{P} \backslash \{P\}$, $P \in \pi \cap \pi'$	130
$\tilde{\mathscr{S}}$, $\tilde{\mathscr{T}}$	sets of maximal subspaces of $(\mathscr{S}, \mathscr{R})$	130, 131
$\mathrm{wt}(P)$	weight of P	142

Chapter 25

$\mathscr{V}, \mathscr{V}_n, \mathscr{V}_n^{2^n}$	Veronesean of quadrics of $PG(n, K)$	145
$\mathscr{V}_2^4$	Veronesean of plane quadrics	147
$\mathscr{M}_4^3$	union of conic planes of $\mathscr{V}_2^4$	150
$\pi(P)$	tangent plane of $\mathscr{V}_2^4$ at P	152
ζ	map from $PG(n, K)$ to $\mathscr{V}_n$	146
$G(\mathscr{V}_n)$	subgroup of $PGL(N+1, q)$ fixing $\mathscr{V}_n$	149
$\mathscr{F}$	set of q^2+q+1 planes in $PG(5, q)$, q odd, satisfying (i), (ii), (iii)	156
	set of k planes of $PG(m, q)$, q odd, satisfying (i)″, (ii)′, (iii)	163
$T(\pi)$	contact point of tangent plane π	156
$\mathscr{V}$	$\{T(\pi)\}$	156
$\mathscr{F}^*$	set of planes $\bar{\pi}$	159
$\mathscr{L}$	set of q^2+q+1 planes in $PG(5, q)$, q odd, satisfying (i)′, (ii), (iii)′	163
	set of k subspaces Π_{m-3} of $PG(m, q)$, q odd, satisfying (i)‴, (ii)″, (iii)′	165
$\mathscr{K}$	set of k points of $PG(m, q)$ satisfying (i), (ii)	166
	set of k points of $PG(5, q)$ satisfying (i)′, (ii)	170
	set of k points of $PG(N, q)$ satisfying (i), (ii), (iii)	179
	set of k points of $PG(N, q)$ satisfying (i)–(iv)	185
$t(P, \mathscr{C})$	Γ-tangent to $\mathscr{C}$ at P	179
$[P, P']$	Γ-conic through P and P'	180
$\pi(P, \mathscr{C})$	plane of tangents at P of Γ-conics through P and a point of $\mathscr{C}$	180
$T(P)$	$\bigcup t(P, \mathscr{C})$, tangent space at P	180
$\mathscr{S}_{n_1;n_2;\ldots;n_k}$	Segre variety of $PG(n_1, K), PG(n_2, K), \ldots, PG(n_k, K)$	189
δ	bijection to Segre variety	190
$G(\mathscr{S}_{n_1;n_2})$	subgroup of $PGL((n_1+1)(n_2+1), q)$ fixing $\mathscr{S}_{n_1;n_2}$	193
Σ_i	system of Π_{n_i} on $\mathscr{S}_{n_1;\ldots;n_k}$	190
$\tilde{\zeta}$	element of $G(\mathscr{S}_{n;n})$ defined by $x'_{sr} = x_{rs}$, with $r, s = 0, 1, \ldots, m$	194
$\mathscr{R}(\Pi_n, \Pi'_n, \Pi''_n)$	n-regulus containing Π_n, Π'_n, Π''_n	200
$\mathscr{S}(l_0, \ldots, l_n)$	n-spread of $PG(2n+1, q)$	201
$G(\mathscr{S})$	subgroup of $PGL(2n+2, q)$ fixing the spread $\mathscr{S}$	204

Chapter 26

$\mathcal{S}$, ($\mathcal{P}$, $\mathcal{B}$, I)	incidence structure, generalized quadrangle, partial geometry, semi-partial geometry, (0, α)-geometry, partial quadrangle	209
$\mathcal{S}$	polar space, Shult space	209
n	rank of polar space or Shult space	209
$P \mathrm{I} l$, $l \mathrm{I} P$	P is incident with l, l is incident with P	209
$s + 1$	number of points incident with a line	209
$t + 1$	number of lines incident with a point	209
b	number of lines	214
v	number of points	214
$\mathcal{T}_2^*(\mathcal{O})$	generalized quadrangle constructed from oval $\mathcal{O}$, q even	210
$P_1 \sim P_2$	P_1 is collinear with P_2	211
$\rho(P)$	equivalence class of P under the relation ρ	211
$\mathcal{R}$	radical	211
$\mathcal{Q}(3, q)$	generalized quadrangle on $\mathcal{H}_3$ in $PG(3, q)$	212
$\mathcal{Q}(4, q)$	generalized quadrangle on $\mathcal{P}_4$ in $PG(4, q)$	212
$\mathcal{Q}(5, q)$	generalized quadrangle on $\mathcal{E}_5$ in $PG(5, q)$	212
$\mathcal{U}(3, q^2)$	generalized quadrangle on $\mathcal{U}_3$ in $PG(3, q^2)$	213
$\mathcal{U}(4, q^2)$	generalized quadrangle on $\mathcal{U}_4$ in $PG(4, q^2)$	213
$\mathcal{W}(q)$	generalized quadrangle in $PG(3, q)$ formed from a linear complex	213
$l \sim l'$	l and l' are concurrent	213
PP'	line incident with two collinear points P, P'	213
$l \cap l'$	point incident with two concurrent lines l, l'	213
$P^\perp$	set of points collinear with P	213
$\{P, P'\}^\perp$, tr(P, P')	trace of $\{P, P'\}$	213
$\mathcal{A}^\perp$	$\bigcap_{P \in \mathcal{A}} P^\perp$	213
sp(P, P'), $\{P, P'\}^{\perp\perp}$	span of $\{P, P'\}$	214
$\langle \mathcal{V}_1, \ldots, \mathcal{V}_k \rangle$	subspace generated by $\mathcal{V}_1, \ldots, \mathcal{V}_k$	219
$\mathcal{S}(P)$	tangent set (hyperplane) of $\mathcal{S}$ at P	221
$\mathcal{S}_P$	collar of $\mathcal{S}$ for P	224
$P\zeta$	polar of P with respect to $\mathcal{S}$	224
$\bar{\mathcal{D}}$	(linear) closure of $\mathcal{D}$	228
cl(P, P')	closure of $\{P, P'\}$	235
$\pi(P, P')$	affine plane associated to an anti-regular point P and a point P' in $P^\perp \backslash \{P\}$	235
(D), $(D)'_P$, $(D)''_P$	axioms	239, 240
$\mathcal{B}^{\perp\perp}$	set of all spans $\{P, P'\}^{\perp\perp}$, $P \nsim P'$	240

$\mathcal{S}^{\perp\perp}$	$(\mathcal{P}, \mathcal{B}^{\perp\perp}, \in)$	240
$(A)_P$, (A)	axioms	240
$(\hat{A})_l$, $(\hat{A})$	duals of $(A)_P$, (A)	240
$\mathcal{B}^*$	set of all spans $\{P, P'\}^{\perp\perp}$	240
$\mathcal{S}^*$	$(\mathcal{P}, \mathcal{B}^*, \in)$	241
(H)	axiom	241
$(\mathcal{S}^{(P)}, G)$,	elation generalized quadrangle with elation group G	
$\mathcal{S}^{(P)}$	and base point P	244
$H(P, P')$	group of all homologies with centres P, P', $P \nsim P'$	245
α	non-zero number of points incident with a line l and collinear with a point P, $P \not\mathrm{I}\, l$	246
H_q^n	dual net formed by the lines of $PG(n, q)$ skew to a given Π_{n-2}	247
$\mathcal{S}(\mathcal{K})$, $\mathcal{T}_2^*(\mathcal{K})$	partial geometries associated to a maximal arc $\mathcal{K}$	247
(Pa)	axiom of Pasch	251
μ	number of points collinear with two non-collinear points	252
$[P, l]$	number of points incident with a line l and collinear with a point P, $P \not\mathrm{I}\, l$	253
v_l	number of points P with $[P, l] = \alpha$	253
b_P	number of lines l with $[P, l] = \alpha$	253
$\mathcal{S}_h$	semi-partial geometry associated to any set of size h	258
$\overline{W(n, 2k, q)}$	$(0, \alpha)$-geometry arising from null polarity	259
$\overline{W(n, q)}$	$\overline{W(n, n+1, q)}$ semi-partial geometry	259
$NQ^-(n, 2)$	semi-partial geometry from $\mathcal{E}_n$ in $PG(n, 2)$	259
$NQ^+(n, 2)$	semi-partial geometry from $\mathcal{H}_n$ in $PG(n, 2)$	259
$NQ(n, 2)$	semi-partial geometry from $\mathcal{P}_n$ in $PG(n, 2)$	259
$NQ^+(3, 2^h)$	$(0, 2^{h-1})$-geometry from $\mathcal{H}_3$ in $PG(3, 2^h)$	259
$\mathcal{S}(\pi)$	partial geometry induced in plane π	260
$s(\pi) + 1$	number of points on line of $\mathcal{S}(\pi)$	260
$t(\pi) + 1$	number of lines of $\mathcal{S}(\pi)$ through one of its points	260
$m(\pi)$, m	number of isolated points in π	260
$\mathcal{K}(\pi)$	maximal arc formed by points of π not in $\mathcal{S}(\pi)$	260
ρ_a, ρ_b, ρ_c	number of common points of $\mathcal{S}$ and π	261
$\approx$	equivalence relation in $(0, 2)$-geometry	283

Chapter 27

$m_2(n, q)$	maximum size of a k-cap in $PG(n, q)$	285
$m(n, q)$	maximum size of a k-arc in $PG(n, q)$	285
$m'(2, q)$	size of second largest complete k-arc in $PG(2, q)$	286
$M(C)$	matrix whose columns are the points of the code C	288
$M(X_1, \ldots, X_t)$	matrix whose columns are the points X_j	288
$w(X_i)$, w_i	weight of X_i	289
$(w_1, \ldots, w_{\theta(n)})$	weight distribution	289
$C^\perp$	dual of code C	289
t	number of tangents to a k-cap $\mathscr{K}$ through a point of $\mathscr{K}$	299
$\sigma_1(Q)$	number of tangents to $\mathscr{K}$ through $Q \notin \mathscr{K}$	299
$\sigma_2(Q)$	number of bisecants to $\mathscr{K}$ through $Q \notin \mathscr{K}$	299
$\mathscr{C}_n$	normal rational curve in $PG(n, q)$	307
l_P	tangent to $\mathscr{C}_n$ at P	308
$\mathfrak{S}$	projection map of $\mathscr{C}_n \backslash \{P\}$ from P onto a prime π, $P \in \mathscr{C}_n$, $P \notin \pi$	308
$\mathscr{C}_n\mathfrak{S}$	image of projection map together with $l_P \cap \pi$	308
$G(\mathscr{C}_n)$	subgroup of $PGL(n + 1, q)$ fixing $\mathscr{C}_n$	309
v_n	number of $\mathscr{C}_n$ in $PG(n, q)$	309
$\Pi_n(M)$	$PG(n, q)$ defined by the columns of the matrix M	311
$\mathscr{V}_{n,k-1}$	subset of $\mathscr{G}_{n,k-1}$ on no face of simplex of reference	312
$\mathscr{V}_{n,k-1}(\mathscr{K})$	subset of $\mathscr{V}_{n,k-1}$ corresponding to $\mathscr{K}$	312
$G(\mathscr{K})$	subgroup of $PGL(n + 1, q)$ fixing $\mathscr{K}$	311
$g(\mathscr{K})$	$\lvert G(\mathscr{K})\rvert$	311
$\chi(\mathscr{K})$	$(q-1)^{k-1}k!/g(\mathscr{K})$ with $k = \lvert\mathscr{K}\rvert$	312
$\mathscr{A}_{k,n}$	set of k-arcs in $PG(n, q)$	311
$\mathscr{A}_{k,n}(\mathscr{K})$	set of k-arcs corresponding to a point of $\mathscr{V}_{n,k-1}(\mathscr{K})$	313
$\mathscr{A}$	$(\mathbf{V}(\mathscr{A}), \mathbf{I}(\mathscr{A})) = \mathbf{A}(H, H_1, \ldots, H_r)$ with $H_1, \ldots, H_r$ linear forms	320
$\mathbf{V}(\mathscr{A})$	$\mathbf{V}(H, H_1, \ldots, H_r)$	320
$\mathbf{I}(\mathscr{A})$	$\mathbf{I}(H, H_1, \ldots, H_r)$, ideal of $\Gamma = \gamma[x_0, \ldots, x_n]$ generated by $H, H_1, \ldots, H_r$	320
$\lvert\mathscr{A}\rvert$	number of points in $\mathscr{A}$	320
$A \subset B$	$\mathbf{I}(A) \supset \mathbf{I}(B)$	320
$\mathscr{A} \cap \pi_1 \cap \cdots \cap \pi_u$, $\mathscr{A} \cap \pi_{r+1} \cap \cdots \cap \pi_u$	$\mathbf{A}(H, H_1, \ldots, H_u)$ with $\mathscr{A} = \mathbf{A}(H, H_1, \ldots, H_r)$, H_j a linear form and $\pi_j = \mathbf{V}(H_j)$	320

Φ, $\Phi(\mathscr{K})$	primal associated to the k-arc $\mathscr{K}$ of primes	324
l_{ij}	S-line	326
α_{ij}	S-plane	338
Ψ_i	S-quadric	338

Appendix VI

$\mathscr{W}_n(q)$	polar space arising from a null polarity of $PG(n, q)$, n odd	345
$\mathscr{P}(2n, q)$	polar space arising from $\mathscr{P}_{2n}$ in $PG(2n, q)$	345
$\mathscr{H}(2n+1, q)$	polar space arising from $\mathscr{H}_{2n+1}$ in $PG(2n+1, q)$	345
$\mathscr{E}(2n+1, q)$	polar space arising from $\mathscr{E}_{2n+1}$ in $PG(2n+1, q)$	345
$\mathscr{U}(n, q^2)$	polar space arising from $\mathscr{U}_n$ in $PG(n, q^2)$	345
r	rank of polar space	345
$\lvert\mathscr{S}\rvert$	number of points of polar space $\mathscr{S}$	345
$\mathscr{G}(\mathscr{S})$	set of generators of $\mathscr{S}$	345
S	spread of polar space	346
O	ovoid of polar space	346

Other symbols

$\mathscr{A}\mathscr{B}\mathscr{C}\mathscr{D}\mathscr{E}\mathscr{F}\mathscr{G}\mathscr{H}\mathscr{I}\mathscr{J}\mathscr{K}\mathscr{L}\mathscr{M}\mathscr{N}\mathscr{O}\mathscr{P}\mathscr{Q}\mathscr{R}\mathscr{S}\mathscr{T}\mathscr{U}\mathscr{V}\mathscr{W}\mathscr{X}\mathscr{Y}\mathscr{Z}$	script alphabet
$\mathfrak{A}\mathfrak{B}\mathfrak{C}\mathfrak{D}\mathfrak{E}\mathfrak{F}\mathfrak{G}\mathfrak{H}\mathfrak{I}\mathfrak{J}\mathfrak{K}\mathfrak{L}\mathfrak{M}\mathfrak{N}\mathfrak{O}\mathfrak{P}\mathfrak{Q}\mathfrak{R}\mathfrak{S}\mathfrak{T}\mathfrak{U}\mathfrak{V}\mathfrak{W}\mathfrak{X}\mathfrak{Y}\mathfrak{Z}$	German alphabet
$\square$	end of proof of theorem, corollary, or lemma

AUTHOR INDEX

This index contains all authors cited either in the Preface or in the Notes and References section at the end of the chapters, providing the author's work is given there or in the Bibliography. Other authors cited are included in the General Index.

GENERAL INDEX